T0302359

SPICE and LTspice for Power Electronics and Electric Power

Power electronics can be a difficult course for students to understand and for professional professors to teach, simplifying the process for both. *SPICE* and *LTspice for Power Electronics and Electric Power, Fourth Edition* illustrates methods of integrating industry-standard LTspice software for design verification and as a theoretical laboratory bench.

Based on the author Muhammad H. Rashid's considerable experience merging design content and SPICE into a power electronics course, this vastly improved and updated edition focuses on helping readers integrate the LTspice simulator with a minimum amount of time and effort. Giving users a better understanding of the operation of a power electronic circuit, the author explores the transient behavior of current and voltage waveforms for every circuit element at every stage. The book also includes examples of common types of power converters as well as circuits with linear and nonlinear inductors.

New in this edition:

- Helpful LTspice software and program files available for download
- Changes to run on OrCAD SPICE, or LTspice IV or higher
- Students' learning outcomes (SLOs) listed at the start of each chapter
- Abstracts of chapters
- List the input side and output side performance parameters of the converters
- The characteristics of power semiconductors—diodes, BJTs, MOSFETs, and IGBTs
- Generating PWM and sinusoidal PWM gating signals
- Evaluating the power efficiency of converters
- Monte Carlo analysis of converters
- Worst-case analysis of converters
- Nonlinear transformer model
- Evaluate user-defined electrical quantities (.MEASURE)

This book demonstrates techniques for executing power conversion and ensuring the quality of output waveform rather than the accurate modeling of power semiconductor devices. This approach benefits students, enabling them to compare classroom results obtained with simple switch models of devices.

Power Electronics and Applications Series
Series Editor: Muhammad H. Rashid
University of West Florida, Pensacola, USA

Complex Behavior of Switching Power Converters
Chi Kong Tse

DSP-Based Electromechanical Motion Control
Hamid A. Toliyat and Steven Campbell

Advanced DC/DC Converters
Fang Lin Luo and Hong Ye

Renewable Energy Systems: Design and Analysis with Induction Generators
M. Godoy Simoes and Felix A. Farret

Uninterruptible Power Supplies and Active Filters
Ali Emadi, Abdolhosein Nasiri, and Stoyan B. Bekiarov

Modern Electric, Hybrid Electric, and Fuel Cell Vehicles: Fundamentals, Theory, and Design
Mehrdad Eshani, Yimin Gao, Sebastien E. Gay, and Ali Emadi

Electric Energy: An Introduction
Mohamed EI-Sharkawi

Electrical Machine Analysis Using Finite Elements
Nicola Bianchi

For more information about this series, please visit: https://www.routledge.com/Power-Electronics-and-Applications-Series/book-series/CRCPOWELEAPP

SPICE and LTspice for Power Electronics and Electric Power

Fourth edition

Muhammad H. Rashid

CRC Press is an imprint of the
Taylor & Francis Group, an **informa** business

Fourth edition published 2025
by CRC Press
2385 NW Executive Center Drive, Suite 320, Boca Raton FL 33431

and by CRC Press
4 Park Square, Milton Park, Abingdon, Oxon, OX14 4RN

CRC Press is an imprint of Taylor & Francis Group, LLC

First edition published by Prentice Hall 1990

Third edition published by CRC Press 2012

ISBN: 978-1-032-25661-0 (hbk)
ISBN: 978-1-032-25662-7 (pbk)
ISBN: 978-1-003-28445-1 (ebk)

DOI: 10.1201/9781003284451

Typeset in Times
by KnowledgeWorks Global Ltd.

Access the Support Material: www.routledge.com/9781032256610

To my parents, my wife Fatema,

my children: Faeza, Farzana, and Hasan, grandchildren:
Hannah, Laith, Laila, Nora, Amal, Isac, Inara, and Nyle

Contents

Preface

Power electronics is normally offered as a technical elective. It is an application-oriented and interdisciplinary course that requires a background in mathematics, electrical circuits, control systems, analog and digital electronics, microprocessors, electric power, and electrical machines.

The understanding of the operation of a power electronics circuit requires a clear knowledge of the transient behavior of current and voltage waveforms for each circuit element at every instant of time.

These features make power electronics a difficult course for students to understand and for professors to teach. A laboratory helps in understanding power electronics and its control interfacing circuits. Development of a power electronics laboratory is expensive compared to other courses in electronics or electric power (EE) curriculum. However, power electronics is playing a key role in industrial power control.

The Engineering Accreditation Commission of the Accreditation Board for Engineering and Technology (EAC/ABET) requirements specify the computer integration and design content in the EE curriculum. To be competitive, a power electronics course should integrate design content and an extensive use of computer-aided analysis.

The LTspice software, which is available free to students and professionals, is ideal for classroom use and for assignments requiring computer-aided simulation and analysis. Without any additional resources and lecture time, LTspice can also be integrated into power electronics.

The graphical postprocessor in LTspice is very useful in plotting the results of the simulation. Especially with the capability of arithmetic operation, it can be used to plot impedance, power, and so on. Once the students gain experience in simulating on LTspice, they really appreciate the advantages of the graphical postprocessor.

The graphical processor is like a theoretical oscilloscope with special features to perform arithmetic operations. It can be used as a laboratory bench to view the waveforms of currents, voltages, power, power factor, and so on, with Fourier analysis giving the total harmonic distortion (THD) of any waveform.

The capability of a graphical postprocessor, along with other data representation features such as Table, Value, Function, Polynomial, Laplace, Param, and Step, makes LTspice a versatile simulation tool for EE courses. Students can design power electronics circuits, use the LTspice simulator to verify the design and make necessary design modifications. In the absence of a dedicated power electronics laboratory, the laboratory assignments could be design problems to be simulated and verified by LTspice.

This book is based on the author's experience in integrating design content and LTspice on a power electronics course of three credit hours. The students were assigned design problems and asked to use LTspice to verify their designs by plotting the output waveforms and to confirm the ratings of devices and components by plotting the instantaneous voltage, current, and power. The objective of this book is to integrate the LTspice simulator with a power electronics course at the junior level or senior level with a minimum amount of time and effort. This book assumes no prior knowledge about the LTspice simulator and introduces the applications of various LTspice commands through numerous examples of power electronics circuits.

This book can be divided into nine parts: (1) introduction to SPICE simulation—Chapters 1–3; (2) source and element modeling—Chapters 4 and 5; (3) SPICE commands—Chapter 6; (4) rectifiers—Chapters 7 and 11; (5) DC–DC converters—Chapter 8; (6) inverters—Chapters 9 and 10; (7) AC voltage controllers—Chapter 12; (8) control applications—Chapters 13 and 14; and (9) difficulties—Chapter 15. Chapters 7–12 use simple models for power semiconductor switches, leaving the complex models for special projects and assignments. Chapter 14 uses the simple circuit models of DC motors and AC inductor motors to predict their control characteristics. Two reference tables are included to aid in choosing a device, component, or command.

This book is intended to demonstrate the techniques for power conversions and the quality of the output waveforms, rather than the accurate modeling of power semiconductor devices. This approach has the advantage that the students can compare the results with those obtained in a classroom environment with simple switch models of devices.

This book can be used as a textbook on LTspice for students specializing in power electronics and power systems. It can also be a supplement to any standard textbook on power electronics and power systems. The following sequence is recommended:

1. Supplement to a basic power systems (or electrical machine) course with three hours of lectures (or equivalent lab hours) and self-study assignments from Chapters 1–6. Starting from Chapter 2, the students should work with PCs or Laptops.
2. Continue as a supplement to a power electronics course with two hours of lectures (or equivalent lab hours) and self-study assignments from Chapters 7–14.

Without any prior experience with LTspice and integrating SPICE at the power electronics level, two hours of lectures (or equivalent lab hours) are recommended for Chapters 1–6. Chapters 7–14 could be left for self-study assignments. From the author's experience in the class, it has been observed that after two lectures of 50 minutes duration, all students could solve assignments independently without any difficulty. The class could progress in a normal manner with one assignment per week on power electronics circuits simulation and analysis with SPICE.

The book has sections on suggested laboratory experiments and design problems on power electronics. The complete laboratory guidelines for each experiment are presented. Thus, the book can also be used as a laboratory manual for power electronics. The design problems can be used as assignments for a design-oriented simulation laboratory.

Although the materials in this book have been developed for engineering students, the book is also strongly recommended for EET students specializing in power electronics and power systems.

The key changes in the fourth edition of this book are as follows:

- Changes to run on OrCAD SPICE, or LTspice IV or higher
- Students; learning outcomes (SLOs) are listed at the start of each chapter
- Abstracts of chapters
- List the input side and output side performance parameters of the converters
- The characteristics of power semiconductors—diodes, BJTs, MOSFETs, and IGBTs
- Generating PWM and sinusoidal PWM gating signals
- Evaluating the power efficiency of converters
- Monte Carlo analysis of converters
- Worst-case analysis of converters
- Nonlinear transformer model
- Evaluate user-defined electrical quantities (.MEASURE)

LTspice can be downloaded from:

- https://www.analog.com/en/resources/design-tools-and-calculators/ltspice-simulator.html
- https://ltspice-iv.software.informer.com/

Any comments and suggestions regarding this book are welcome and should be sent to the author at the following address:

Dr. Muhammad H. Rashid
Professor of Electrical Engineering
Florida Polytechnic University
4700 Research Way, Lakeland, FL 33805
E-mail: mrashid@floridapoly.edu

Author

Muhammad H. Rashid is employed by the Florida Polytechnic University as a professor of Electrical Engineering. He is founding chair of Electrical and Computer Engineering for 3/12 years. Previously he worked for the University of West Florida, Pensacola. He was also employed by the University of Florida as professor and director of UF/UWF Joint Program. Rashid received B.Sc. degree in Electrical Engineering from the Bangladesh University of Engineering and Technology, and M.Sc. and Ph.D. degrees from the University of Birmingham in UK. Previously, he worked as professor of Electrical Engineering and the chair of the Engineering Department at Indiana University- Purdue University at Fort Wayne. Also, he worked as visiting assistant professor of Electrical Engineering at the University of Connecticut, associate professor of Electrical Engineering at Concordia University (Montreal, Canada), professor of Electrical Engineering at Purdue University Calumet, and visiting professor of Electrical Engineering at King Fahd university of Petroleum and Minerals (Saudi Arabia), as a design and development engineer with Brush Electrical Machines Ltd. (England, UK), a research engineer with Lucas Group Research Centre (England, UK), a lecturer and head of Control Engineering Department at the Higher Institute of Electronics (in Libya & Malta).

Dr. Rashid is actively involved in teaching, researching, and lecturing in electronics, power electronics, and professional ethics. He has published 29 books and more than 160 technical papers. He is a fellow of the Institution of Engineering & Technology (IET, UK) and a life fellow of the Institute of Electrical and Electronics Engineers (IEEE, USA). He was elected as an IEEE fellow with the citation "*Leadership in power electronics education and contributions to the analysis and design methodologies of solid-state power converters*."

Dr. Rashid is an ABET program evaluator and was an engineering evaluator for the Southern Association of Colleges and Schools (SACS, USA). He is also an ABET program evaluator for (general) engineering program. He is the series editor of *Power Electronics and Applications*, and *Nanotechnology and Applications* with the CRC Press. He serves as the editorial advisor of *Electric Power and Energy* with Elsevier Publishing. He lectures and conducts workshops on Outcome-Based Education (OBE) and its implementations including assessments.

Dr. Rashid is a distinguished lecturer for the IEEE Education Society and a regional speaker (previously Distinguished Lecture) for the IEEE Industrial Applications Society. He also authored a book on "The Process of Outcome-Based Education - Implementation, Assessment and Evaluations". 2012 UiTM Press, Malaysia.

1 Introduction

After completing this chapter, students should be able to do the following:

- Describe the general features and the types of SPICE software.
- Describe the types of analysis that can be performed on electronic and electrical circuits.
- Describe the limitations of PSpice® software.
- List the online resources on SPICE.

1.1 INTRODUCTION

Electronic circuit design requires accurate methods of evaluating circuit performance. Because of the enormous complexity of modern integrated circuits, computer-aided circuit analysis is essential and can provide information about circuit performance that is almost impossible to obtain with laboratory prototype measurements. Computer-aided analysis makes possible the following procedures:

1. Evaluation of the effects of variations in elements, such as resistors, transistors, and transformers
2. Assessment of performance improvements or degradations
3. Evaluation of the effects of noise and signal distortion without the need for expensive measuring instruments
4. Sensitivity analysis to determine the permissible bounds determined by the tolerances of all element values or parameters of active elements
5. Fourier analysis without expensive wave analyzers
6. Evaluation of the effects of nonlinear elements on circuit performance
7. Optimization of the design of electronic circuits in terms of circuit parameters

SPICE (simulation program with integrated circuit emphasis) is a general-purpose circuit program that simulates electronic circuits. It can perform analyses on various aspects of electronic circuits, such as the operating (or quiescent) points of transistors, time-domain response, small-signal frequency response, and so on. SPICE contains models for common circuit elements, active as well as passive, and it is capable of simulating most electronic circuits. It is a versatile program and is widely used in both the industry and academic institutions.

Until recently, SPICE was available only on mainframe computers. In addition to the cost of the computer system, such a machine can be inconvenient for classroom use. In 1984, MicroSim introduced the PSpice simulator, which is similar to the Berkeley version of SPICE and runs on an IBM-PC or compatible, and is available free of charge to students for classroom use. PSpice thus widens the scope for the integration of computer-aided circuit analysis into electronic circuits courses at

DOI: 10.1201/9781003284451-1

the undergraduate level. Other versions of PSpice, which run on the Macintosh II, 486-based processor, VAX, SUN, NEC, and other computers, are also available.

1.2 DESCRIPTIONS OF SPICE

PSpice is a member of the SPICE family of circuit simulators, all of which originate from the SPICE2 circuit simulator, whose development spans a period of about 30 years. During the mid-1960s, the program ECAP was developed at IBM [1]. In the late 1960s, ECAP served as the starting point for the development of the program CANCER at the University of California (UC) at Berkeley. Using CANCER as the basis, SPICE was developed at Berkeley in the early 1970s. During the mid-1970s, SPICE2, which is an improved version of SPICE, was developed at UC–Berkeley. The algorithms of SPICE2 are robust, powerful, and general in nature, and SPICE2 has become an industry-standard tool for circuit simulation. SPICE3, a variation of SPICE2, is designed especially to support computer-aided design (CAD) research programs at UC–Berkeley. As the development of SPICE2 was supported using public funds, this software is in the public domain, which means that it may be used freely by all US citizens.

SPICE2, referred to simply as SPICE, has become an industry standard. The input syntax for SPICE is a free-format style that does not require data to be entered in fixed column locations. SPICE assumes reasonable default values for unspecified circuit parameters. In addition, it performs a considerable amount of error checking to ensure that a circuit has been entered correctly.

PSpice, which uses the same algorithms as SPICE2, is equally useful for simulating all types of circuits in a wide range of applications. A circuit is described by statements stored in a file called the *circuit file*. The circuit file is read by the SPICE simulator. Each statement is self-contained and independent of every other statement and does not interact with other statements. SPICE (or PSpice) statements are easy to learn and use.

A schematic editor can be used to draw the circuit and create a Schematics file, which can then be read by PSpice for running the simulation.

1.3 TYPES OF SPICE

The commercially supported versions of SPICE2 can be classified into two types: mainframe versions and PC-based versions. Their methods of computation may differ, but their features are almost identical. However, some may include such additions as a preprocessor or shell program to manage input and provide interactive control, as well as a postprocessor to refine the normal SPICE output. A person used to one SPICE version (e.g., PSpice) should be able to work with other versions.

Mainframe versions are as follows:

HSPICE (from Meta-Software), which is for integrated circuit design with special device models

RAD-SPICE (from Meta-Software), which simulates circuits subjected to ionizing radiation

IG-SPICE (from A.B. Associates), which is designed for "interactive" circuit simulation with graphics output
I-SPICE (from NCSS Time Sharing), which is designed for "interactive" circuit simulation with graphics output
Precise (from Electronic Engineering Software)
PSpice (from MicroSim)
AccuSim (from Mentor Graphics)
Spectre (from Cadence Design)
SPICE-Plus (from Valid Logic)

The PC versions include the following:

AllSpice (from Acotech)
Is-Spice (from Intusoft)
Z-SPICE (from Z-Tech)
SPICE-Plus (from Analog Design Tools)
DSPICE (from Daisy Systems)
PSpice (from MicroSim)
OrCAD (from Cadence)
Spice (from KEMET)
B2 Spice A/D (from Beige Bag Software)
AIM-Spice (from AIM-Software)
VisualSpice (from Island Logix)
Spice3f4 (from Kiva Design)
OrCAD SPICE (from OrCAD)
MDSPICE (from Zeland Software, Inc.)
Ivex Spice (from Ivex Design)
LTspice (from Linear Technology Corporate)
LTspice (Analog Devices)

1.4 TYPES OF ANALYSIS

PSpice allows various types of analysis. Each analysis is invoked by including its command statement. For example, a statement beginning with the .DC command invokes the DC sweep. The types of analysis and their corresponding .(dot) commands are described in the following text.

DC analysis is used for circuits with time-invariant sources (e.g., steady-state DC sources). It calculates all node voltages and branch currents for a range of values, and their quiescent (DC) values are the outputs. The dot commands and their functions are as follows:

- DC sweep of an input voltage or current source, a model parameter, or temperature over a range of values (.DC)
- Determination of the linearized model parameters of nonlinear devices (.OP)
- DC operating point to obtain all node voltages

- Small-signal transfer function with small-signal gain, input resistance, and output resistance (Thevenin's equivalent; .TF)
- DC small-signal sensitivities (.SENS)

Transient analysis is used for circuits with time-variant sources (e.g., AC sources and switched DC sources). It calculates all node voltages and branch currents over a time interval, and their instantaneous values are the outputs. The dot commands and their functions are as follows:

- Circuit behavior in response to time-varying sources (.TRAN)
- DC and Fourier components of the transient analysis results (.FOUR)

AC analysis is used for small-signal analysis of circuits with sources of variable frequencies. It calculates all node voltages and branch currents over a range of frequencies, and their magnitudes and phase angles are the outputs. The dot commands and their functions are as follows:

- Circuit response over a range of source frequencies (.AC)
- Noise generation at an output node for every frequency (.NOISE)

In Schematics versions, the commands are invoked from the setup menu, as shown in Figure 1.1. The analysis setup menus for the OrCAD versions are shown in Figure 1.2.

LTspice: The types of analyses are similar to other SPICE software tools as shown in Figure 1.3(a). The Fast Fourier Transformer (FFT) is performed on a waveform from the View menu as shown in Figure 1.3(b). You select the particular plot or the trace and then FFT from the view menu as shown in Figure 1.3(b). LTspice, however, does not support sensitivity analysis directly.

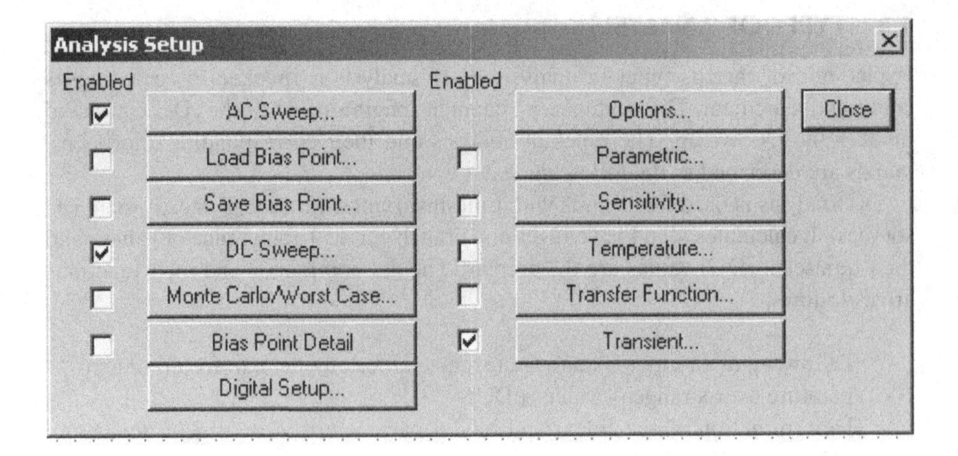

FIGURE 1.1 Analysis setup in PSpice Schematics versions.

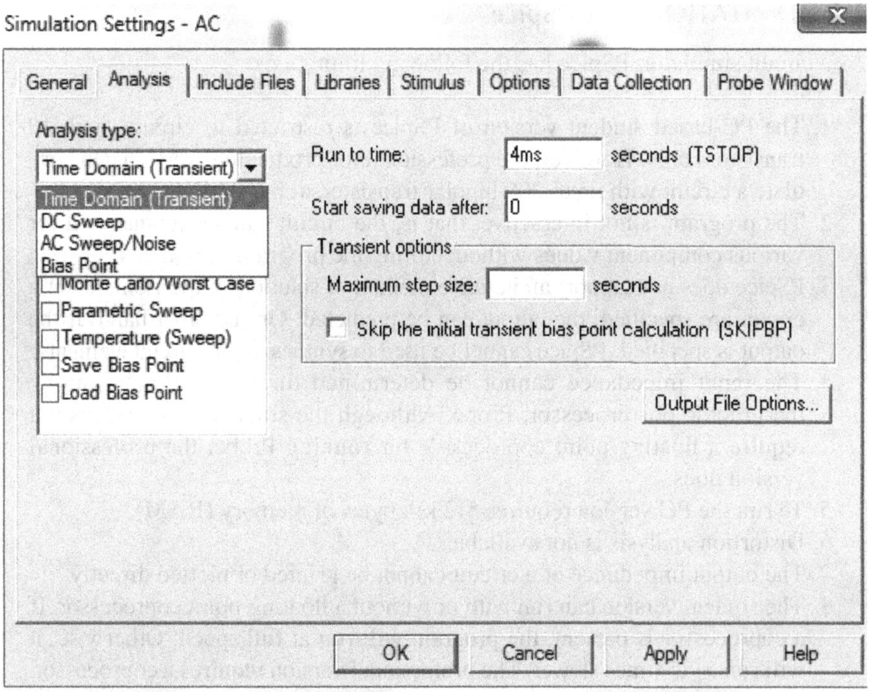

Simulation Settings - AC

General | Analysis | Include Files | Libraries | Stimulus | Options | Data Collection | Probe Window

Analysis type:

Time Domain (Transient)

Time Domain (Transient)
DC Sweep
AC Sweep/Noise
Bias Point

Monte Carlo/Worst Case
Parametric Sweep
Temperature (Sweep)
Save Bias Point
Load Bias Point

Run to time: 4ms seconds (TSTOP)

Start saving data after: 0 seconds

Transient options

Maximum step size: seconds

Skip the initial transient bias point calculation (SKIPBP)

Output File Options...

OK Cancel Apply Help

FIGURE 1.2 Analysis setup in OrCAD versions.

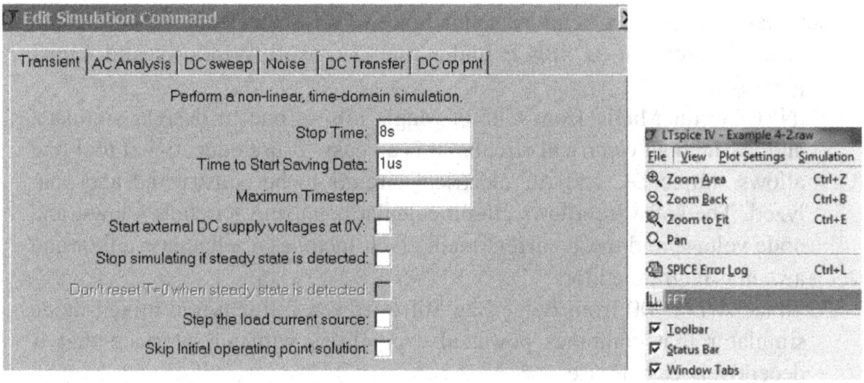

Edit Simulation Command

Transient | AC Analysis | DC sweep | Noise | DC Transfer | DC op pnt

Perform a non-linear, time-domain simulation.

Stop Time: 8s

Time to Start Saving Data: 1us

Maximum Timestep:

Start external DC supply voltages at 0V:

Stop simulating if steady state is detected:

Don't reset T=0 when steady state is detected:

Step the load current source:

Skip Initial operating point solution:

LTspice IV - Example 4-2.raw

File View Plot Settings Simulation

Zoom Area Ctrl+Z

Zoom Back Ctrl+B

Zoom to Fit Ctrl+E

Pan

SPICE Error Log Ctrl+L

FFT

Toolbar

Status Bar

Window Tabs

(a) (b)

FIGURE 1.3 LTspcie analysis setup menu. (a) Edit simulation menu. (b) FFT menu.

1.5 LIMITATIONS OF PSpice

As a circuit simulator, PSpice has the following limitations:

1. The PC-based student version of PSpice is restricted to circuits with 10 transistors only. However, the professional (or production) version can simulate a circuit with up to 200 bipolar transistors (or 150 MOSFETs).
2. The program is not interactive; that is, the circuit cannot be analyzed for various component values without editing the program statements.
3. PSpice does not support an iterative method of solution. If the elements of a circuit are specified, the output can be predicted. On the other hand, if the output is specified, PSpice cannot be used to synthesize the circuit elements.
4. The input impedance cannot be determined directly without running the graphic postprocessor, Probe. Although the student version does not require a floating-point coprocessor for running Probe, the professional version does.
5. To run the PC version requires 512 kilobytes of memory (RAM).
6. Distortion analysis is not available.
7. The output impedance of a circuit cannot be printed or plotted directly.
8. The student version can run with or without a floating-point coprocessor. If a coprocessor is present, the program will run at full speed. Otherwise, it will run 5–15 times slower. The professional version requires a coprocessor.

Note: If you are not sure about the limitations, just try and run the simulations. PSpice schematic and OrCAD will give you the error messages (syntax errors) if any.

1.6 DESCRIPTIONS OF SIMULATION SOFTWARE TOOLS

There are many simulation software tools [1] in addition to SPICE; the following are some examples:

Automatic Integrated Circuit Modeling Spice (AIM-Spice) from AIM-Software is a new version of SPICE under the Microsoft Windows and Linux platforms.

AKNM Circuit Magic from Circuit Magic allows one to design, simulate, and learn about electrical circuits. It is an easy-to-use educational tool that allows simple DC and AC electrical circuits to be constructed and analyzed. The software allows circuit calculations using Kirchoff's laws, and node voltage and mesh current methods. It includes a schematic editor and a vector diagram editor.

B2 Spice A/D 2000 from Beige Bag Software is a full-featured mixed-mode simulator that combines powerful capabilities with an interface that is deceptively easy to use.

Electronics Workbench Suite from Electronics Workbench is a professional circuit design solution with a suite of integrated tools that includes schematic capture, simulation, layout, and autorouting for printed circuit boards (PCBs) and programmable logic devices such as FPGAs and CPLDs.

SPICE simulation, analog and mixed-signal circuit design, magnetics transformer design, and test program development tools are provided by Intusoft.
VisualSpice from Island Logix is an electronic circuit design and simulation software. It is a completely integrated, modern-user-interface circuit design environment that allows one to quickly and easily capture schematic designs, perform simulations, and analyze the results.
Ivex Design provides Windows-based EDA tools, schematic capture, SPICE tools, and PCB layout.
Schematic CAD software and EDA software include circuit simulation, schematic entry, PCB layout, and Gerber Viewer.

LTspice Software: This program is a high-performance SPICE simulator, schematic capture, and waveform viewer with enhancements and models for easing the simulation of switching regulators. Although it is designed for switching regulator design program by Linear Technology, it can be used for other electrical and electronic circuits. Included in this download are LTspice IV, Macro Models for 80% of Linear Technology's switching regulators, over 200 op amp models, as well as resistors, transistors, and MOSFET models.

MacroSim Digital Simulator from Visionics is a sophisticated software tool that integrates the process of designing and simulating the operation of digital electronic circuitry. It has been developed for professional engineers, hobbyists, and tertiary and late-secondary students.
MDSPICE from Zeland Software is a mixed-frequency and time-domain SPICE simulator for predicting the time-domain response of high-speed networks, high-frequency circuits, and nonlinear devices directly using S-parameters.

Multisim Software (Previously Known as Electronic Workbench): Multisim software provides SPICE simulation, analysis, and PCB tools to help you quickly iterate through designs and improve prototype performance.

Micro-Cap from Spectrum Software is an analog or digital simulation that is compatible with SPICE and PSpice.
NOVA-686 Linear RF Circuit Simulation is a shareware, RF circuit simulation program for the RF design engineer, radio amateur, and hobbyist.
SIMetrix from Catena Software is an affordable mixed-mode circuit simulator designed for professional circuit designers.

1.7 PSpice PLATFORM

The platform depends on the SPICE version. There are four platforms for PSpice, as follows:

1. PSpice A/D or OrCAD PSpice A/D (version 9.1 or above)
2. PSpice Schematics (version 9.1 or below)
3. OrCAD Capture Lite (version 9.2 or above)
4. LTspice (version IV or above)

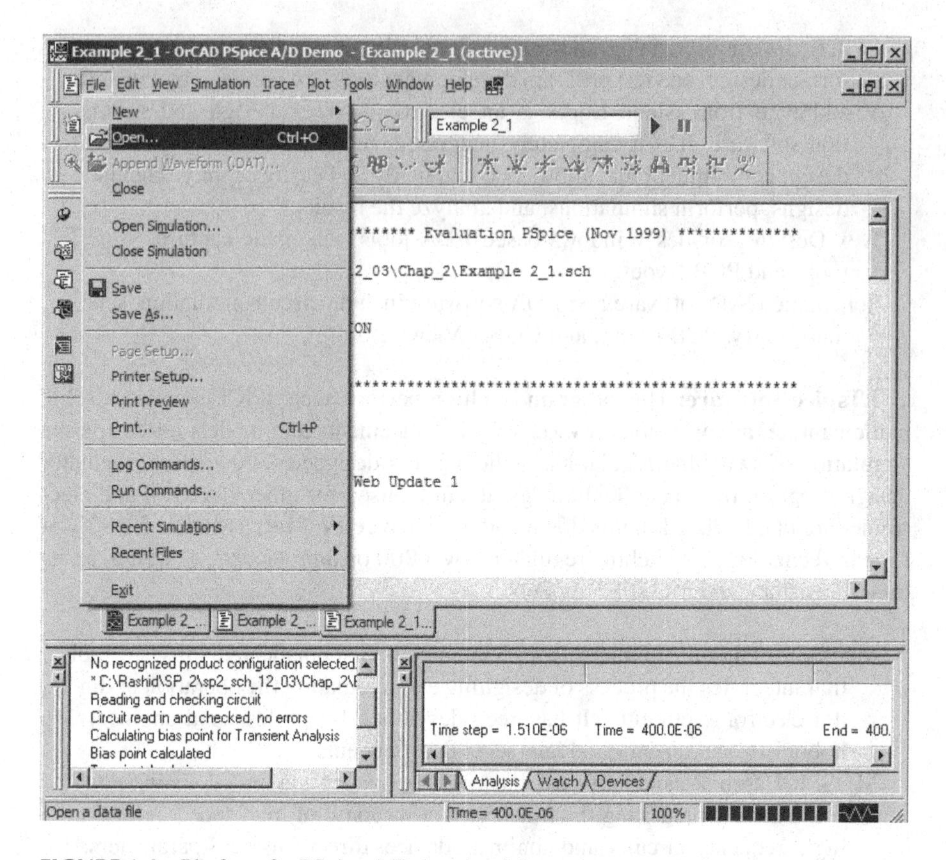

FIGURE 1.4 Platform for PSpice A/D (version 9.1).

1.7.1 PSpice A/D

The platform for PSpice A/D is shown in Figure 1.4. A circuit described by statements and analysis commands is simulated by a run command from the platform. The output results can be displayed and viewed from platform menus.

1.7.2 PSpice Schematics

The platform for PSpice Schematics is shown in Figure 1.5. The circuit that is drawn on the platform is run from the analysis menu. The simulation type and settings are specified from the analysis menu. After the simulation run is completed, PSpice automatically opens PSpice A/D for displaying and viewing the output results.

1.7.3 OrCAD Capture

The platform for OrCAD Capture, which is similar to that of PSpice Schematics and has more features, is shown in Figure 1.6. The circuit that is drawn on the platform

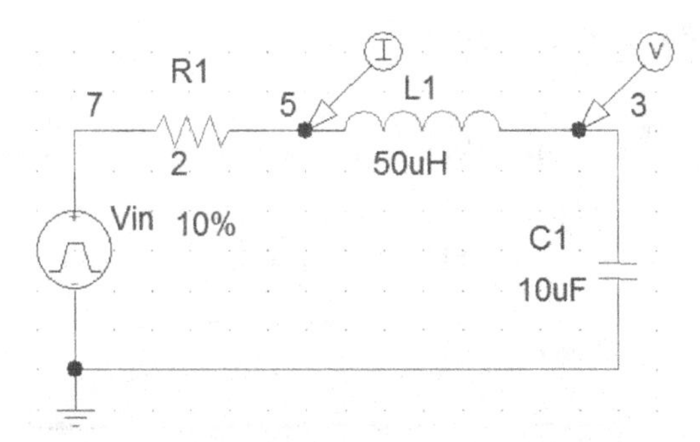

FIGURE 1.5 Platform for PSpice Schematics (version 9.1).

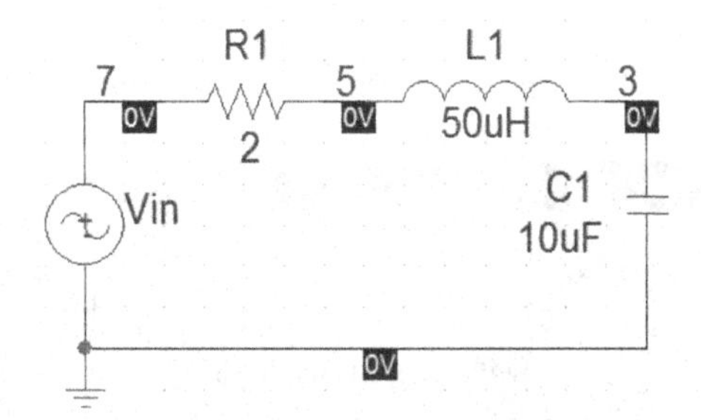

FIGURE 1.6 Platform for OrCAD Capture (version 9.2).

is run from the PSpice menu. The simulation type and settings are specified in the PSpice menu. After the simulation run is completed, Capture automatically opens PSpice A/D for displaying and viewing the output results.

Note: If you have a latter version of OrCAD, the platform would be almost identical not significantly different. You can run an OrCAD file drawn and analysis setup in version 9.2 in the latter versions but not necessarily vice versa. Thus, it is convenient for learning and teaching purposes to use an earlier version, and then you can have the option to run in later versions once.

1.8 PSpice SCHEMATICS VERSUS OrCAD CAPTURE

OrCAD Capture has some new features, and the platform is similar to that of PSpice Schematics. Schematics files (with extension .SCH) can be imported to OrCAD Capture (with extension .OPJ). However, OrCAD files cannot be run on PSpice Schematics. Therefore, it is advisable that those readers familiar with PSpice Schematics use PSpice Schematics version 9.1, which has a platform similar to that of version 8.0. However, its PSpice A/D is similar to that of OrCAD Capture. PSpice Schematics (version 9.1), as shown in Figure 1.7, can be downloaded from the Cadence Design Systems website, http://www.cadence.com.

(a)

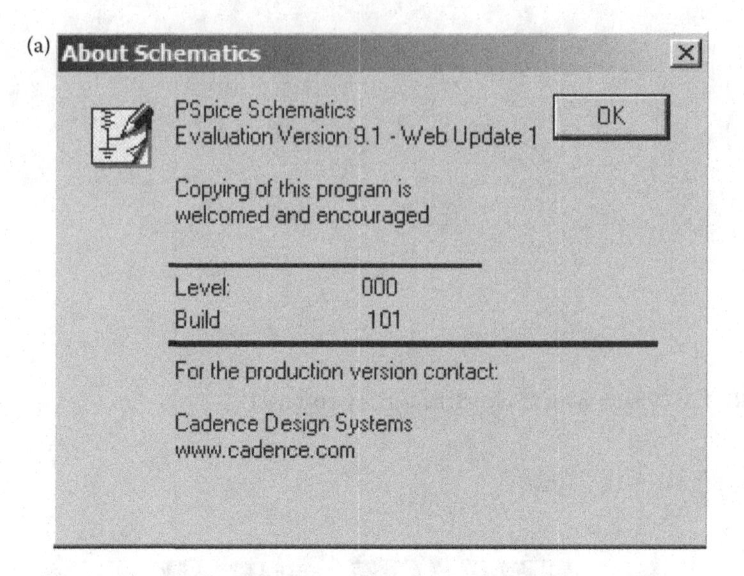

(b)

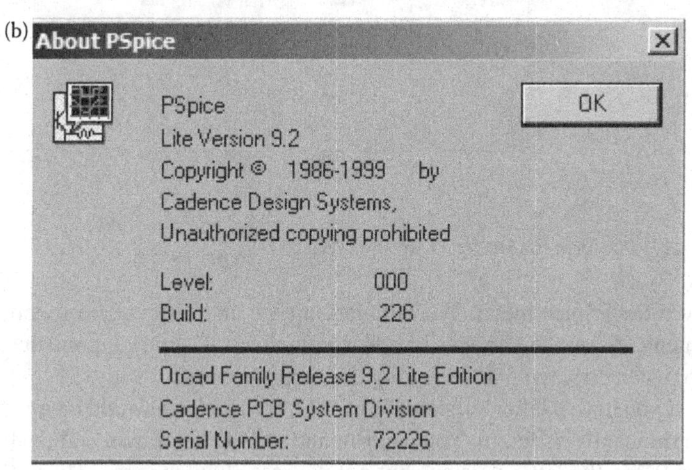

FIGURE 1.7 Information about Schematics. (a) PSpice Schematics and (b) OrCAD.

Notes:

- If you have drawn a schematic in PSpice version 9.1 and have completed the circuit analysis in PSpice 9.1, it is convenient to use the Save As feature in the file menu (in version 9.1) to modify an existing schematic to create a similar schematic.
- Although there are later versions of OrCAD, the 9.2 version has proven to work well. If you have a schematic in lower version, you can run at a higher version, not vice versa. Thus, the circuit files were drawn in lower versions 9.1 and 9.2 so that these files can be run on higher versions of PSpice or OrCAD software.

FIGURE 1.8 LTspcie New Schematic menu.

1.9 LTspice PLATFORM

The LTspice platform is shown in Figure 1.8. All of the actions such as selecting component, drawing, wiring, editing, viewing, analysis, and run can be performed from the opening platform once we select *New Schematic* from the File menu as shown in Figure 1.8.

1.10 LTspice APPLICATION EXAMPLES

LTspice comes with numerous application examples. They are applications of different fields of electronics for education purposes. These illustrate the applications of the LTspcie commands in simulating electrical and electronic circuits. The files can be found at the LTspice following the LTspcie directory.

LTC\LTspiceiV\examples\educational\

And it has three subdirectories: contrib, FRA, and PAsystem.

1.10.1 Subdirectory: LTC\LTspiceiV\examples\educational\contrib\

This subdirectory "contrib" contains the schematic files as listed in Table 1.1 [7, 18].

1.10.2 Subdirectory: LTC\LTspiceiV\examples\educational\FRA

The subdirectory "FRA" for Frequency Response Analysis contains the schematic files as listed in Table 1.2 [7, 18]. These use .meas command to study counter reaction and servo-controls in order to obtain an open loop Bode diagram from closed loop measurements.

TABLE 1.1
LTspice Schematic Files in Subdirectory, contrib

File name	Description
elip_grd.asc	Elliptic filter.
gr_del.asc	Group delay correction filter.
qztst.asc.	Crystal model for easily measurable parameters.

TABLE 1.2
LTspice Schematic Files in Subdirectory, FRA

#	File name	Description
1.	Eg1.asc	Display response open loop Bode diagram of an operational amplifier with a counter reaction from closed loop measurements.
2.	Eg2.asc	Display response open loop Bode diagram of an SMPS servo-control from closed loop measurements (with LTC3611).
3.	Eg3.asc	Display response open loop Bode diagram of an SMPS servo-control from closed loop measurements (with LTC1735).

TABLE 1.3
LTspice Schematic Files in Subdirectory, PAsystem

#	File name	Description
1.	HandsFreePreamp.asc	Microphone preamplifier.
2.	HandsFreelayout.asc	Components placement and w1ring of the microphone preamplifier.
3.	PowerAmp.asc	Low-frequency amplifier.
4.	PowerAmplayout.asc	Placement and wiring of low-frequency amplifier with bridgeable output.

1.10.3 Subdirectory: LTC\LTspiceiV\examples\educational\PAsystem\

The subdirectory "PAsystem" (for Public Address Systems) contains the schematic files as listed in Table 1.3 [1, 2]. These are examples of complete circuits. This directory also contains all symbols and libraries associated with the simulation of these circuits.

1.10.4 Subdirectory: LTC\LTspiceiV\examples\educational\

The subdirectory "Educational" contains 62 schematic files as listed in Table 1.4 [7, 18]. These are examples of complete schematic circuits that supplied by Linear Technology for education purposes and illustrate the instructions in the LTspice manual.

1.11 SPICE RESOURCES

There are many online resources. Some of them are listed in the following subsections.

1.11.1 Websites with Free Spice Models

Analog Devices
http://products.analog.com/products_html/list_gen_spice.html
Apex Microtechnology
http://eportal.apexmicrotech.com/mainsite/index.asp
Coilcraft
http://www.coilcraft.com/models.cfm

TABLE 1.4
LTspice Schematic Files in Subdirectory, Educational

#	File Name	Description
1.	160.asc	Example of a TIL circuit, the 74160, BCD counter .**PWL** sources, **transient analysis.**
2.	1563.as	Second-order filter, **AC analysis, SPICE directive. Include.**
3.	Astable.asc	Astable multivibrator with bipolar transistors **transient analysis**.
4.	Audioamp.asc	Study of a small audio amplifier **AC analysis, transient analysis** and measurement of **harmonic distortion** displayed in the file SPICE Error Log.
5.	BandGaps.asc	Study of the temperature stability of a voltage reference, **DC sweep analysis with temperature as parameter**.
6.	Butter.asc	RLC band-pass filter, **AC analysis.**
7.	Clapp.asc	Clapp oscillator, **transient analysis and SPICE directive .options.**
8.	Cohn.asc	High-frequency band-pass filter, **AC analysis and SPICE directive .param.**
9.	Colpits.asc	Colpits oscillator, first version, **transient analysis and SPICE directive .options.**
10.	Colpits2.asc	Colpits oscillator, second version, **transient analysis and SPICE directive .options.**
11.	Curvetrace.asc	Display the characteristics network of a bipolar transistor, **DC sweep analysis** and current reading **.lc.**
12.	DCopPnt.asc	Example of the search for an amplifier's DC operatmg point values, **analyse .op.**
13.	Dimmer.asc	Dimmer on mains power (US standards) **transient analysis and SPICE directives** of DIAC TRIAC subcircuits + **.step** showing the impact of different values of R.
14.	Fc.asc	Application of **capometer, transient analysis and SPICE directive .Include and .params.**
15.	GFT.asc	10 MHz amplifier, model using controlled current sources, **AC analysis and SPICE directive .param and .step.**
16.	HalfSiope.asc	Transfer function: **transient analysis and AC, use of Laplace transform**
17.	Hartly.asc	Hartly oscillator, **transient analysis.**
18.	Howland.asc	Phase shifter, **transient analysis and directive .Step.**
19.	IdeaiTransformer.asc	Ideal impulsion transformer**, transient analysis and PULSE source**
20.	LM78XX.asc	LM7805, 12 and 15 V regulator equivalent schematic, **transient analysis.**
21.	LM308.asc	Operational amplifier circuit amplifying a sine-wave voltage, **transient analysis with Startup option.**
22.	LM741.asc	Operational amplifier amplifying a sine-wave voltage, **transient analysis without Startup option.**
23.	Logamp.asc	Logarithmic circuit with operational amplifier, transient analysis and B source (arbitrary).
24.	Loopgain.asc	First method to assess an operational amplifier open loop gain, **AC analysis**
25.	Loopgain2.asc	Second method to assess an operational amplifier open loop gain, **AC analysis and use of SPICE directive .Step.**
26.	MC1648.asc	Start-up of a 35 MHz VCO, transient analysis with Startup option.
27.	MeasureBW.asc	Band-pass measurement 3 dB, **AC analysis and use of SPICE directive .Meas** (visible in the file SPICE Error Log).

(Continued)

TABLE 1.4 ***(Continued)***
LTspice Schematic Files in Subdirectory, Educational

#	File Name	Description
28.	MonteCarlo.asc	Simulation of the consequences of the variation of values of all components of a band-pass filter, **AC analysis and use of directives .Param and .Step**.
29.	NE555.asc	Schematic and simulation of the NE555 timer, **transient analysis and PULSE source.**
30.	Noise.asc	RMS measure of noise integral, *Noise analysis and SPICE directive .Meas.*
31.	NoiseFigure.asc	Measurement of noise factor, **Noise analysis**.
32.	NonlinearTransformer.asc	Real transformer approach by addition of saturable inductor, **transient analysis and SPICE directive .Step.**
33.	Notch.asc	4 circuits configured with RC band-stop filter, **AC analysis and use of SPICE directive .Param.**
34.	Opamp.asc	Filter **AC analysis and use of SPICE directive .Include.**
35.	P2.asc	Operational amplifier equivalent schematic, **transient analysis and SPICE directive .Param.**
36.	Passive.asc	PB filter, **AC analysis and use of SPICE directive .Param.**
37.	Phaseshift.asc	RC network oscillator, **transient analysis with Startup.**
38.	Phaseshift2.asc	RC network oscillator, **transient analysis with Startup and directives .Param and .options.**
39.	Phono.asc	RIAA equalization, **AC analysis**.
40.	Pierce.asc	Pierce crystal oscillator, **transient analysis and SPICE directive .Options** to set calculation method.
41.	PLL.asc	Phase locked loop by use of modulate block fulfilling VCO function simulation**, transient analysis and SPICE directive .Params.**
42.	PLL2.asc	Second example of the special function modulate, **transient analysis and directives .Params.**
43.	Relax.asc	RC oscillator around a LT1001 amplifier, **transient analysis.**
44.	Royer.asc	SMPS with LT1184F to supply a fluorescent tube from a 28 V DC source, **transient analysis.**
45.	SampleAndHold.asc:	Sample & hold (two running modes), **transient analysis.**
46.	S-param.asc	RF filter simulation, **AC analysis and SPICE directive .Net**
47.	StepAC.asc	Band-pass tested with 3 capacitor values, **AC analysis and SPICE directive .Step**
48.	Stepmodelparam.asc:	Characteristics of a transistor according to a model parameter: VAF, **DC Sweep analysis and SPICE directive .Step**
49.	Stepnoise.asc	Influence of the impedance of a differential amplifier in terms of noise, **Noise analysis and SPICE directives .Param and.Step**.
50.	Steptemp.asc	Output offset shift of an amplifier depending on temperature, **DC op analysis and SPICE directives .Step + temp**.
51.	Transformer.asc	Pulse transformer with a secondary winding, **transient analysis and PULSE source.**
52.	Transformer2.asc	Pulse transformer with 2 secondary windings, **transient analysis and PULSE source.**
53.	Transmissionlinelnverter. asc	Model of a coaxial cable with SPICE model, **transient analysis and PULSE source.**
54.	TwoTau.asc	Model of a RC filter with E source, **transient analysis and AC analysis, use of Laplace transform**.

(Continued)

TABLE 1.4 *(Continued)*
LTspice Schematic Files in Subdirectory, Educational

#	File Name	Description
55.	Universal0pamp2.asc	Influence of an op amplifier's internal compensation, **AC analysis.**
56.	Varactor.asc	Band-pass filter with varicap diode, **AC analysis and SPICE directives .Param and .Step.**
57.	Varactor2.asc	Band-pass filter with varicap diode, **AC analysis and SPICE directives .Param and .Save**
58.	Varistor.asc	Application examples of the special function varistor, **transient analysis.**
59.	Vswitch.asc	Example of voltage controlled switch, **transient analysis and creation of a .Model.**
60.	Wavein.asc	Example of use of a **.wav file**, source modulation, **transient analysis.**
61.	Waveout.asc	Example of saving a **.wav file**, resulting from a simulation, **transient analysis and SPICE directive .wave**
62.	Wien.asc	Regulated Wien bridge oscillator, **transient analysis with Startup.**

Comlinear
http://www.national.com/models
Elantec
http://www.elantec.com/pages/products.html
Epcos Electronic Parts and Components
http://www.epcos.de/web/home/html/home_d.html
Fairchild Semiconductor Models and Simulation Tools
http://www.fairchildsemi.com/models/
Infineon Technologies AG
http://www.infineon.com/
International Rectifier
http://www.irf.com/product-info/models/
Intersil Simulation Models
http://www.intersil.com/design/simulationModels.asp
Johanson Technology
http://www.johansontechnology.com/
Linear Technology
http://www.linear-tech.com/software/
Maxim
http://www.maxim-ic.com/
Microchip
https://www.microchipusa.com/
Motorola Semiconductor Products
http://www1.motorola.com/
National Semiconductor
http://www.national.com/models
Philips Semiconductors
http://www.semiconductors.philips.com/

Polyfet
http://www.polyfet.com/
Teccor
http://www.teccor.com/asp/sitemap.asp?group=downloads
Texas Instruments
http://www.ti.com/sc/docs/msp/tools/macromod.htm#comps
Würth Elektronik GmbH & Co. KG, Germany. LTspice & WE Component Library
http://www.we-online.com/ltspice
Zetex
http://www.zetex.com/

1.11.2 Websites with Spice Models

Analog & RF Models
http://www.home.earthlink.net/~wksands/
Analog Innovations
http://www.analog-innovations.com/
Duncan's Amp Pages
http://www.duncanamps.com/
EDN Magazine
http://www.e-insite.net/ednmag/
Intusoft Free SPICE Models
http://www.intusoft.com/models.htm
Linear Technology Corporation
http://www.linear.com/designtools/software/
MOSIS IC Design Models
http://www.mosis.org/
Planet EE
http://www.planetee.com/
PSpice.com
http://www.pspice.com/
SPICE Models from Symmetry
http://www.symmetry.com/
SPICE Model Index
http://homepages.which.net/~paul.hills/Circuits/Spice/ModelIndex.html

1.11.3 Spice and Circuit Simulation Information Sites

AboutSpice.com
http://www.aboutspice.com/
Artech House Publishers—Books and Software for High-Technology Professionals
http://www.artech-house.com/
EDTN Home Page
http://www.edtn.com/

E/J Bloom Associates Home Page—SMPS Books & Software
http://www.ejbloom.com/
MOSIS IC Foundry
http://www.mosis.org/
NCSU SPICE Benchmarks
http://www.cbl.ncsu.edu/pub/Benchmark_dirs/
NIST Modeling Validation Group
http://ray.eeel.nist.gov/modval.html
Norman Koren Vacuum Tube Audio Page
http://www.normankoren.com/Audio/
PSpice.com
http://www.pspice.com/
Ridley Engineering PWM Simulation
http://www.ridleyengineering.com/
SGS-Thomson
http://us.st.com/stonline/index.shtml
SPICE Simulations Dr Vincent G Bello
http://www.spicesim.com/
SPICE Simulation One Trick Pony (Japanese)
Temic (Siliconix)
http://www.temic.com/index_en.html?
University of Exeter's Online SPICE3 User's Manual
http://newton.ex.ac.uk/teaching/CDHW/Electronics2/userguide/
Virtual Library Electrical Engineering
http://webdiee.cem.itesm.mx/wwwvlee/
Yahoo Club Circuit Simulation Chat Room
http://login.yahoo.com/config/login?.intl=uk&.src=ygrp&.done=http://uk.groups.yahoo.com%2Fclubs%2Felectroniccircuitsimulation

1.11.4 Engineering Magazines with Spice Articles

EDN Home Page
http://www.e-insite.net/ednmag/
Electronic Design
http://www.elecdesign.com/
PCIM Home Page
http://www.pcim.com/
Personal Engineering & Instrumentation
http://www.pcim.com/
Planet EE
http://www.planetee.com/

SUGGESTED READING

1. M.H. Rashid, *SPICE for Circuits and Electronics Using LTSpice® Schematics, PSpice® Schematics, and OrCAD® Capture*, Fourth Edition, India: Cengage Learning, 2019.

2. M.H. Rashid, *Introduction of PSpice Using Orcad for Circuits and Electronics*, Englewood Cliffs, NJ: Prentice-Hall, 2004.
3. M.H. Rashid, *Microelectronics Laboratory using Software Tools: PSpice, Orcad, Multisim*, First Edition, India: Cengage Learning, 2016.
4. M.H. Rashid, *SPICE for Power Electronics and Electric Power*, Englewood Cliffs, NJ: Prentice-Hall, 2003.
5. M.H. Rashid, *Introduction of PSpice Using Orcad for Circuits and Electronics*, Englewood Cliffs, NJ: Prentice-Hall, 2004.
6. E. Brumgnach, *PSpice for Windows*, New York: Delmar Publishers, 1995.
7. Gilles Brocard, *The LTSpice IV Simulator: Manual, Methods and Applications*. Waldenburg, Germany: 2016 Würth Elektronik GmbH & Co, 2011.
8. Dennis Fitzpatrick, *Analog Design and Simulation Using OrCAD Capture and PSpice*, Oxford, UK; Waltham, MA: Newnes, 2012.
9. R.W. Goody, *PSpice for Windows—A Circuit Simulation Primer*, Englewood Cliffs, NJ: Prentice-Hall, 1995.
10. R.W. Goody, *PSpice for Windows Vol II: Operational Amplifiers and Digital Circuits*, Englewood Cliffs, NJ: Prentice-Hall, 1996.
11. R.W. Goody, *OrCAD PSpice for Windows Volume 1: DC and AC Circuits*, Englewood Cliffs, NJ: Prentice-Hall, 2000.
12. R.W. Goody, *OrCAD PSpice for Windows Volume II: Devices, Circuits, and Operational Amplifiers*, Englewood Cliffs, NJ: Prentice-Hall, 2000.
13. M.E. Herniter, *Schematic Capture with Cadence PSpice*, Englewood Cliffs, NJ: Prentice-Hall, 2001.
14. J. Keown, *PSpice and Circuit Analysis*, Third Edition, New York: Merrill (Macmillan Publishing Company), 1997.
15. P.G. Krol, *Inside Orcad Capture*, Clifton Park, NY: OnWord Press, 1998.
16. R. Lamey, *The Illustrated Guide to PSpice for Windows*, New York: Delmar Publishers, 1995.
17. Yim-Shu Lee, *Computer-Aided Analysis and Design of Switch-Mode Power Supplies*, New York: Marcel Dekker, 1993.
18. *LTspice IV Manual*, Milpitas, CA: Linear Technology Corporation, 2013.
19. G. Massobrio and P. Antognetti, *Semiconductor Device Modeling with SPICE*, Second Edition, New York: McGraw-Hill, 1993.
20. A.J. Okyere, *PSPICE and MATLAB for Electronics: An Integrated Approach*, Boca Raton, FL: CRC Press, 2002.
21. T.E. Price, *Analog Electronics: An Integrated PSpice Approach*, Englewood Cliffs, NJ: Prentice-Hall, 1996.
22. R. Ramshaw and D. Schuurman, *PSpice Simulation of Power Electronics Circuits*, New York: Kluwer Academic Publishers, 1997.
23. C. Schroeder, *Inside Orcad Capture for Windows*, Burlington, MA: Newnes, 1998.
24. G.W. Roberts and A.S. Sedra, *SPICE*, New York: Oxford University Press, 1997.
25. J.A. Stuller, *Basic Introduction to PSpice*, New York: John Wiley & Sons, 1995.
26. J.A. Svoboda, *PSpice for Linear Circuits*, New York: John Wiley & Sons, 2002.
27. P. Tuinenga, *SPICE: A Guide to Circuit Simulation and Analysis Using PSpice*, Third Edition, Englewood Cliffs, NJ: Prentice-Hall, 1995.

2 Circuit Descriptions

After completing this chapter, students should be able to do the following:

- Describe circuits for PSpice simulation.
- Create input schematic files that can be read by PSpice.
- Obtain output results and plots of simulations.
- Perform PSpice simulation for finding the transient plots and total harmonic distortion (THD) of voltages and currents.

Notes:

- For circuits to be simulated, refer to any textbook on Circuit Analysis.
- Students are encouraged to run the LTspcie schematic files to compare the results with those from PSpice/OrCAD.

2.1 INTRODUCTION

PSpice is a general-purpose circuit program that can be applied to simulate electronic and electrical circuits. A circuit must be specified in terms of element names, element values, nodes, variable parameters, and sources. The input to the circuit shown in Figure 2.1(a) is a pulse voltage as shown in Figure 2.1(b). The circuit is to be simulated for calculating and plotting the transient response from 0 to 400 μs with an increment of 1 μs. The Fourier series coefficients and THD are to be printed. We discuss how to (1) describe this circuit to PSpice, (2) specify the type of analysis to be performed, and (3) define the output variables required. A circuit that consists of elements is described in SPICE by the element name, the type, the values and the connecting nodes listed as follows:

Input files
Nodes
Element values
Circuit elements
Element models
Sources
Output variables
Types of analysis
PSpice output commands
Format of circuit files
Format of output files

DOI: 10.1201/9781003284451-2

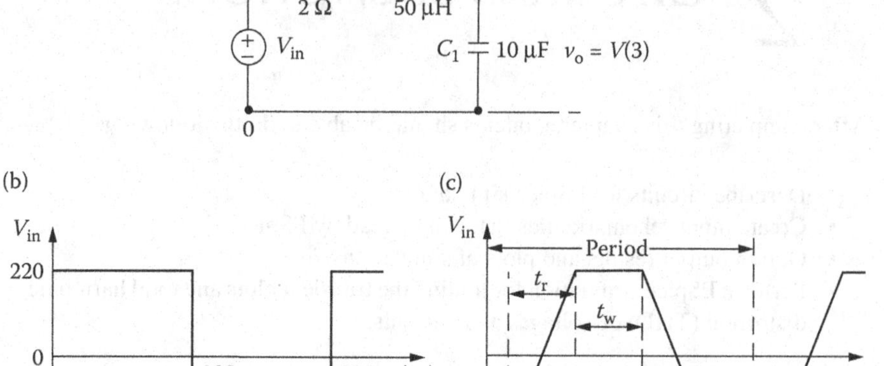

FIGURE 2.1 *RLC* circuit with pulse input. (a) Circuit. (b) Ideal pulse input. (c) Practical pulse.

2.2 INPUT FILES

The input to the SPICE simulation can be either a Schematics file or a netlist file (also known as the *circuit file*). In a circuit file, the user assigns the node numbers to the circuit as shown in Figure 2.1(a). The ideal and practical pulses are shown in Figure 2.1(b) and (c). The nodes connect the circuit elements, the semiconductor devices, and the sources. If a resistor R is connected between two nodes 7 and 5, SPICE relates the voltage v_R across R and the current i through R by

$$v_R = V(7) - V(5) = Ri$$

where the current i flows from node 7 to 5. If an inductor L is connected between two nodes 5 and 3, SPICE relates the voltage v_L across L and the current i through L by

$$v_L = V(5) - V(3) = L\frac{di}{dt}$$

where the current i flows from node 5 to 3. If a capacitor C is connected between two nodes 3 and 0, SPICE relates the voltage v_C across C, and the current i through C by

$$v_C = V(3) - V(0) = \frac{1}{C}\int i\, dt + v_C(t = 0)$$

where the current i flows from node 3 to 0, $v_C(t = 0)$ is the initial condition (or voltage) at $t = 0$, and $V(0) = 0$ is the ground potential.

TABLE 2.1
Names and Functions of Programs within PSpice

Program Name	Program Function
Schematics	Schematic circuit entry, symbol editor, and design management
PSpice A/D	Mixed analog and digital circuit simulation
Probe	Presentation and postprocessing of simulation results
Stimulus editor	Graphic creation of input signals

From these descriptions of the circuit elements in the form of a netlist, SPICE develops a set of matrices and solves for all voltages and currents for specified input sources. The circuit file can be simulated by PSpice A/D. Table 2.1 shows the names and functions of programs within PSpice. Probe is a graphical postprocessor for displaying output variables such as voltages, currents, powers, and impedances.

For a schematic input file, PSpice Schematics creates the netlist automatically. Once the descriptions of the circuit and the type of analysis are specified, PSpice Schematics can simulate the schematic file and then solve for all voltages and currents.

Note: PSpice Schematics creates automatically a netlist from the schematic input file. After you have run the simulation, you can find the netlist in the output file.

2.3 NODES

For PSpice A/D: Node numbers, which must be integers from 0 to 9999 but need not be sequential, are assigned to the circuit of Figure 2.1(a). Elements are connected between nodes. The node numbers are specified after the name of the element connected to the node. Node 0 is predefined as the ground. All nodes must be connected to at least two elements and should therefore appear at least twice. All nodes must have a DC path to the ground node. This condition, which is not met in all circuits, is normally satisfied by connecting very large resistors (see Section 15.10).

For PSpice Schematics: Node numbers are assigned by PSpice and are usually alphanumeric such as $N_0005 and $N_0003, as listed in the following table:

First Node	Second Node
$N_0005	$N_0003
0	+
–	Out
$N_0003	–
0	Out

Note: PSpice Schematics assigns nodes automatically from the schematic file. After you have run the simulation, you can find the assigned nodes in the output file.

For a schematic with N nodes, LTspice assigns numbers in ascending order 1, 2, 3, …, N as follows:

Element	First Node	Second Node	Element values
R1	N001	N002	1k
R2	N002	0	2k
Vs	N001	0	9V

We can assign node numbers to a schematic otherwise, LTspice automatically assigns nodes.

2.4 ELEMENT VALUES

The value of a circuit element is written after the nodes to which the element is connected. The values are written in standard floating-point notation with optional scale and unit suffixes. Some values without suffixes that are allowed by PSpice are as follows:

```
55. 5.0 5E+3 5.0E+3 5.E3
```

There are two types of suffixes: the scale suffix and the units suffix. The scale suffixes multiply the numbers that they follow. Scale suffixes recognized by PSpice are given below:

F	1E–15
F	1E–12
N	1E–9
U	1E–6
MIL	25.4E–6
M	1E–3
K	1E3
MEG	1E6
G	1E9
T	1E12

The units suffixes that are normally used are as follows:

V	volt
A	ampere
HZ	hertz
OHM	ohm
H	henry
F	farad
DEG	degree

The first suffix is always the scale suffix; the units suffix follows the scale suffix. In the absence of a scale suffix, the first suffix may be a units suffix provided it is not a symbol of a scale suffix such F, N, U, M, etc. The units suffixes are always ignored by PSpice. If the value of an inductor is 15 μH, it is written as 15U or 15UH. In the absence of scale and units suffixes, the units of voltage, current, frequency, inductance, capacitance, and angle are by default volts, amperes, hertz, henrys, farads, and degrees, respectively. PSpice ignores any units suffix, and the following values are equivalent:

```
25E-3 25.0E-3 25M 25MA 25MV 25MOHM 25MH
```

Note the following:

1. The scale suffixes can be all uppercase or lowercase letters.
2. M means "milli," not "mega." 2 MΩ is written as 2MEG or 2MEGOHM.

2.5 CIRCUIT ELEMENTS

For PSpice A/D: Circuit elements are identified by name. A name must start with a letter symbol corresponding to the element, but after that, it can contain either letters or numbers. Names can be up to eight characters long. Table 2.2 shows the first letter of elements and sources. For example, the name of a capacitor must start with a C.

TABLE 2.2
Symbols of Circuit Elements and Sources

First Letter	Circuit Elements and Sources
B	GaAs MES field-effect transistor
C	Capacitor
D	Diode
E	Voltage-controlled voltage source
F	Current-controlled current source
G	Voltage-controlled current source
H	Current-controlled voltage source
I	Independent current source
J	Junction field-effect transistor
K	Mutual inductors (transformer)
L	Inductor
M	MOS field-effect transistor
Q	Bipolar junction transistor
R	Resistor
S	Voltage-controlled switch[a]
T	Transmission line
V	Independent voltage source
W	Current-controlled switch[a]

[a] Not available in SPICE2 but available in SPICE3.

The format for describing passive elements is as follows:

```
<element name> <positive node> <negative node> <value>
```

where the current is assumed to flow from the positive node N+ to the negative node N−. The formats for passive elements are described in Chapter 5. The passive elements of Figure 2.1(a) are described as follows.

The statement that R_1 has a value of 2 Ω and is connected between nodes 7 and 5 is

R_1 7 5 2

The statement that L_1 has a value of 50 μH and is connected between nodes 5 and 3 is

L_1 5 3 50UH

The statement that C_1 has a value of 10 μF and is connected between nodes 3 and 0 is

C_1 3 0 10UF

For PSpice Schematics: The name of an element begins with a specific letter, as shown in Table 2.2. The part symbols can be obtained from the Place menu as shown in Figure 2.2(a). The symbols for passive elements are obtained from the analog.slb library as shown in Figure 2.2(b). The element name as shown in Figure 2.3(a) and its value can be edited or changed within the schematic as shown in Figure 2.3(b).

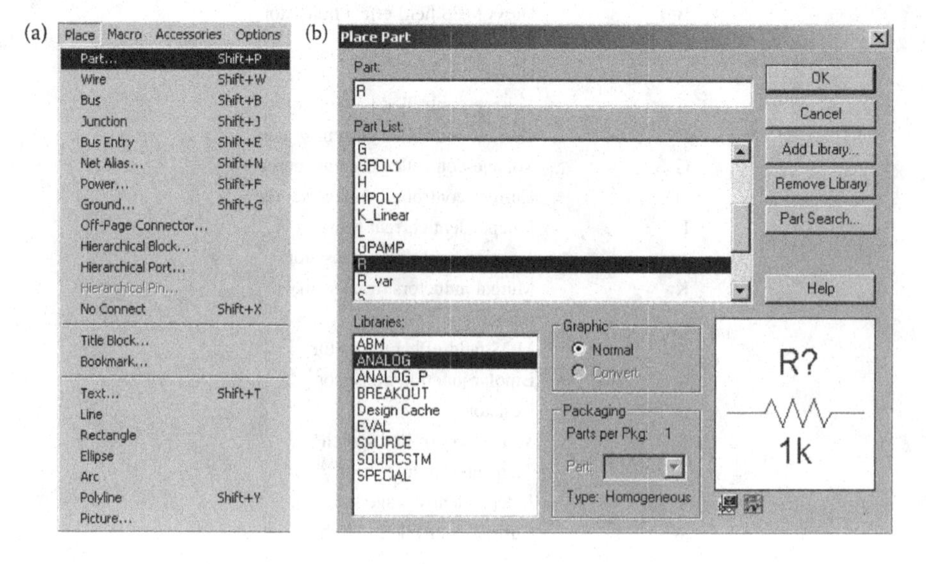

FIGURE 2.2 Place and Part menus from the analog.slb library file. (a) Place menu for getting parts. (b) Symbols from the analog.slb library file.

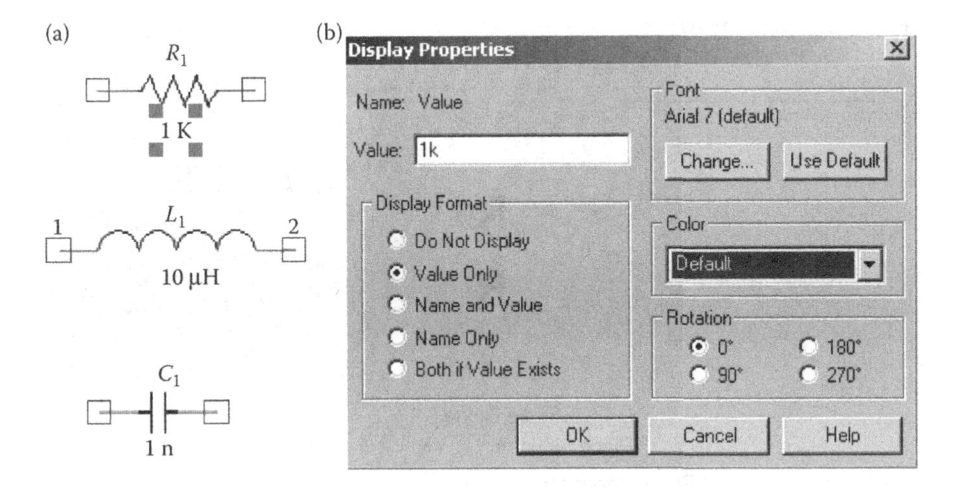

FIGURE 2.3 Symbols and names of typical elements. (a) Symbols. (b) Display properties.

The netlist is created by PSpice automatically, and a typical listing is shown in the following table:

Element Name	First Node	Second Node	Value
R_R1	$N_0002	$N_0001	2
R_Rx	0	+	1k
R_RF	–	out	5k
C_C1	$N_0001	–	10 μF
R_RL	0	out	20 k

Note: When you type the element value, there should be no space between the value and the scale suffix (as shown in the table).

2.6 LTspice SCHEMATIC

For the LTspice, the symbols for the R, L, and C elements as shown in Figure 2.4(a) can be placed directly from the main toolbar menu as shown in Figure 2.4(b). LTspice created the following statements of the nest listing for the dc circuit as shown Figure. 2.1. The listing does not include the dc voltage sources Vx and Vy.

Element	First Node	Second Node	Element values
R1	N002	N001	500
R2	N003	N002	800
R3	N002	0	1k
R4	N003	0	200
Vs	N001	0	20V
I1	0	N003	50mA

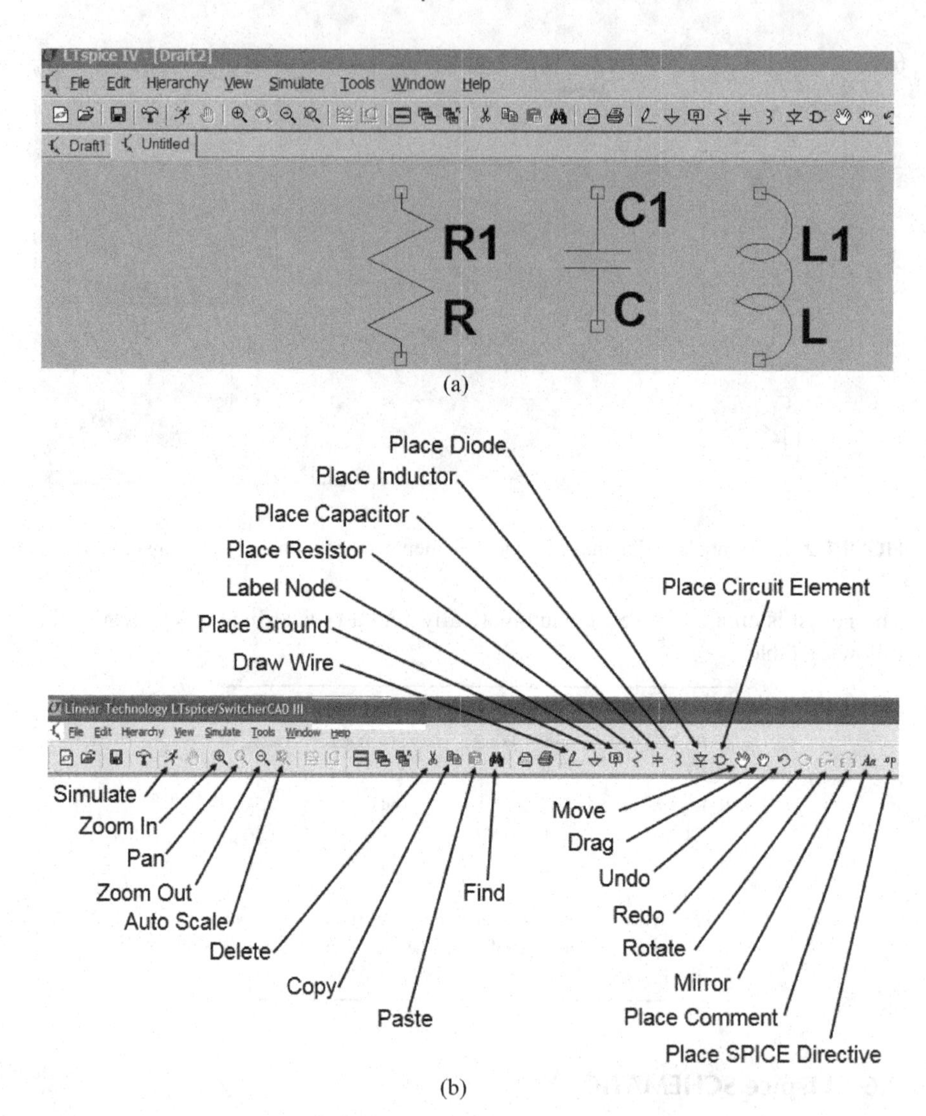

FIGURE 2.4 LTspice tool bar menu. (a) RLC schematic menu. (b) Toolbar summary.

2.7 ELEMENT MODELS

The values of some circuit elements are dependent on other parameters; for example, the inductance of an inductor depends on the initial conditions, the capacitance on voltage, and the resistance on temperature. Models may be used to assign values to the various parameters of circuit elements. The techniques for specifying models of sources, passive elements, and dot commands are described in Chapters 4–6.

We shall represent the source voltage by a pulse, which has a model of the form

```
PULSE(-VS +VS TD TR TF PW PER)
```

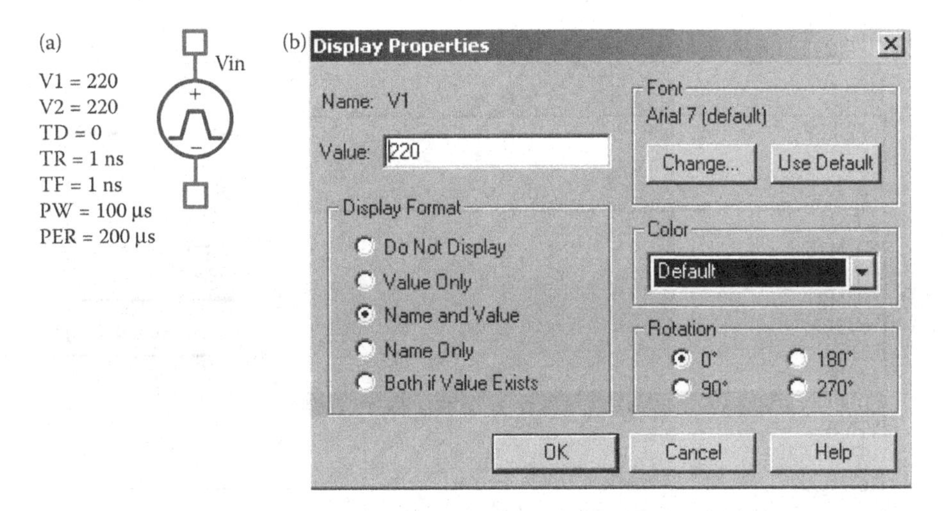

FIGURE 2.5 Model parameters of a pulse voltage. (a) Pulse parameters. (b) Display properties.

where

–VS, +VS	=	negative and positive values of the pulse, respectively
TD, TR, TF	=	delay time, rise time, and fall time, respectively
PW	=	width of the pulse
PER	=	period of the pulse

In practice, it is not possible to generate a pulse with a zero value for the TD and TF. Thus, TR and TF should have small but finite values. A practical pulse is shown in Figure 2.1(c). Let us assume that TD = 0 and TR = TF = 1 ns. The model for the input voltage of Figure 2.1(b) becomes:

```
PULSE (-220V +220V 0 1NS 1NS 100US 200US)
```

For PSpice Schematics: Model parameters are often specifications of the sources, as shown in Figure 2.5(a) for a pulse voltage. The values can be changed from the Display Properties menu as shown in Figure 2.5(b).

2.8 SOURCES

For PSpice A/D: Voltage (or current) sources can be either dependent or independent. Letter symbols are also listed in Table 2.2 for the types of sources. An independent voltage (or current) source can be DC, sinusoidal, pulse, exponential, polynomial, piecewise linear, or single-frequency frequency modulation. Models for describing source parameters are discussed in Chapter 4.

The format for a source is

```
<source name> <positive node> <negative node> <source model>
```

where the current is assumed to flow into the source from positive node N+ to negative node N–. The order of nodes N+ and N– is critical. Assuming that node 7

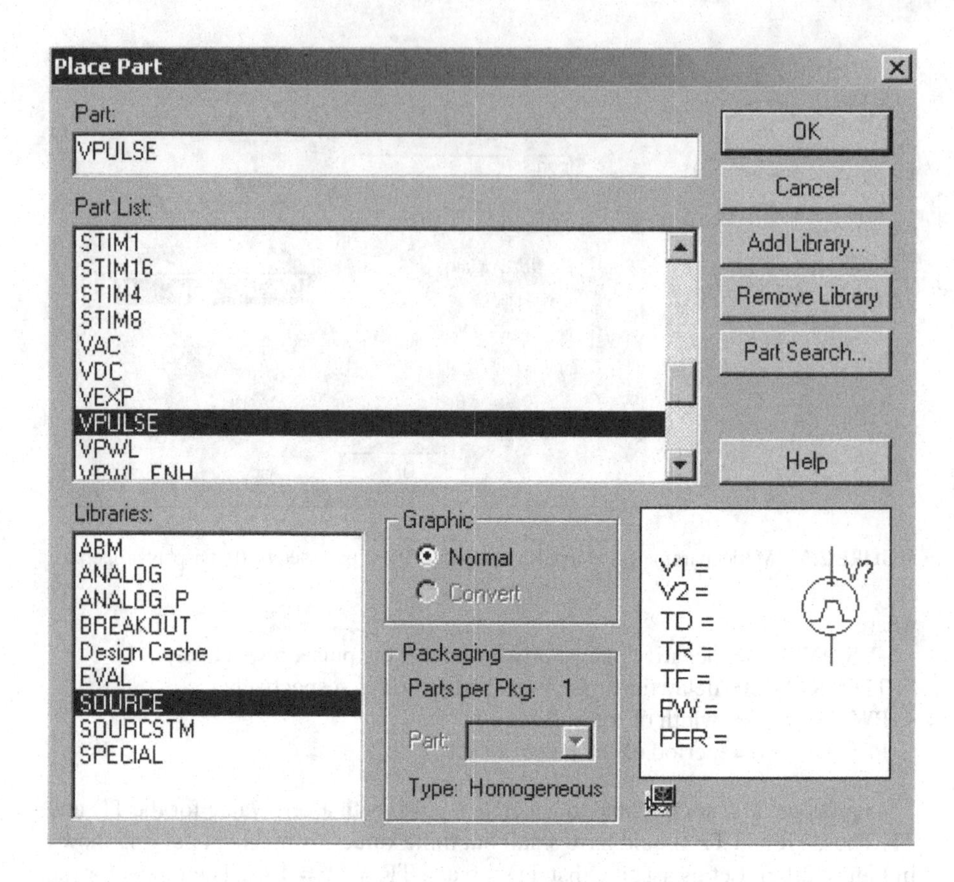

FIGURE 2.6 Source symbols for voltages and currents from the source.slb library.

has a higher potential than node 0, the statement for the input source v_{in} connected between nodes 7 and 0 is

```
VIN 7 0 PULSE (-220V +220V 0 1NS 1NS 100US 200US)
```

For PSpice Schematics: The name of a source with the specified letter is shown in Table 2.2. The source symbols for voltages and currents can be obtained from the source.slb library as shown in Figure 2.6.

2.9 OUTPUT VARIABLES

For PSpice A/D: PSpice has some unique features for printing and plotting output voltages and currents. The various types of output variables permitted by PSpice are discussed in *Chapter 3*. For example, the voltage of node 3 with respect to node 0 is specified by V(3, 0) or V(3).

For PSpice Schematics: PSpice assigns the node numbers. Thus, the user does not know the node numbers and, hence, cannot refer the output voltages to the node numbers. Instead, the output voltage is referred to one of two terminals of an element. For example, the output voltage at terminal 1 of a resistor R1 is specified by V(R1:1), at terminal 2 by V(R1:2), and between terminals 1 and 2 by V(R1:1, R1:2).

Note: Often, you do not need the node numbers. You can use voltage, voltage difference, or current markers.

2.10 TYPES OF ANALYSIS

For PSpice A/D: PSpice allows various types of analysis. Each type is invoked by including its command statement. For example, a statement beginning with a .DC command will cause a DC sweep to be carried out. The types of analysis and their corresponding dot commands are

DC analysis

DC sweep of an input voltage or current source, a model parameter, or temperature (.DC)
Linearized device model parameterization (.OP)
DC operating point (.OP)
Small-signal transfer function (Thévenin's equivalent; .TF)
Small-signal sensitivities (.SENS)

Transient analysis

Time-domain response (.TRAN)
Fourier analysis (.FOUR)

AC analysis

Small-signal frequency response (.AC)
Noise analysis (.NOISE)

It should be noted that the dot is an integral part of a command. The various dot commands are discussed in detail in Chapter 6.

The format for performing a transient response is:

```
.TRAN TSTEP TSTOP
```

where TSTEP is the time increment and TSTOP is the final (stop) time. Therefore, the statement for the transient response from 0 to 400 μs with a 1-μs increment is

```
.TRAN 1US 400US
```

PSpice performs a Fourier analysis from the results of the transient analysis. The command for the Fourier analysis of voltage V(N) is

```
.FOUR FREQ V(N)
```

where FREQ is the fundamental frequency. Thus, the duration of the transient analysis must be at least one period. Here, PERIOD = 1/FREQ. The command for the Fourier analysis of voltage V(3) with PERIOD = 200 μs and FREQ = 1/200 μs = 5 kHz is

```
.FOUR 5KHZ V(3)
```

For PSpice Schematics: The type of analysis to be performed on a circuit is chosen from the Simulation Settings menu (shown in Figure 2.7(b)) within the Edit Simulation Profile menu (shown in Figure 2.7(a)).

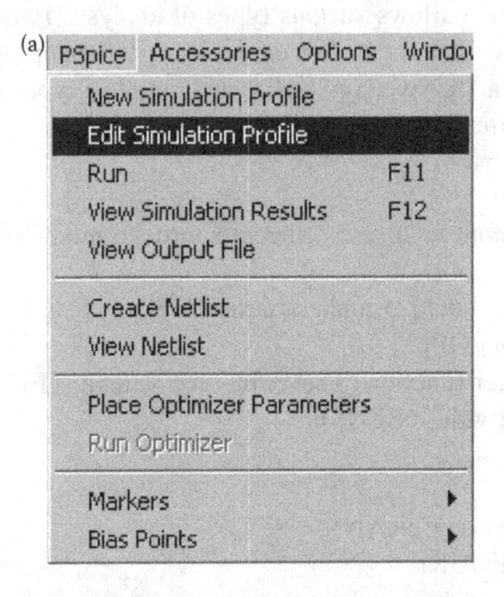

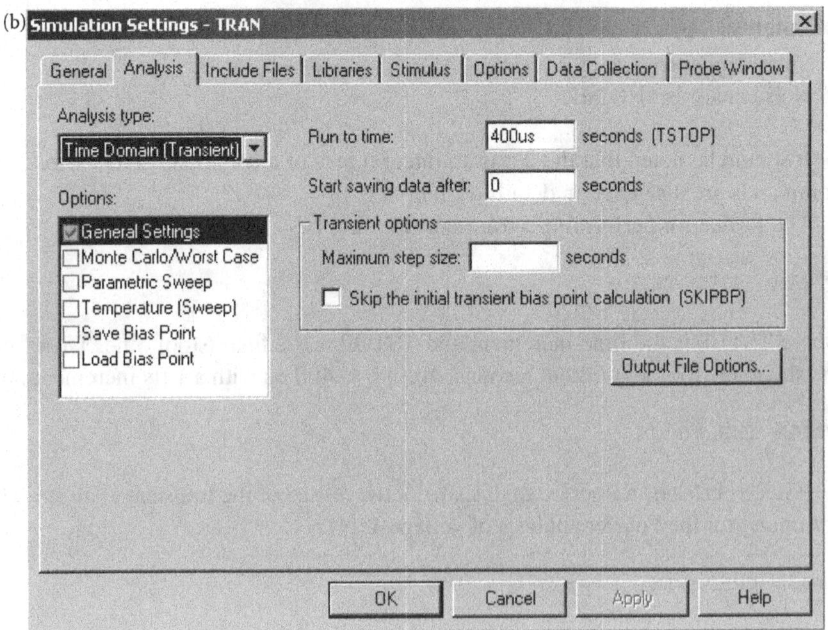

FIGURE 2.7 Analysis Settings menu. (a) Edit Simulation Profile menu. (b) Selecting the analysis type: Transient analysis.

2.11 PSpice OUTPUT COMMANDS

The most common forms of output are print tables and plots. The transient response (.TRAN), DC sweep (.DC), frequency response (.AC), and noise analysis (.NOISE) can produce output in the form of print tables and plots. The command for output in the form of tables is .PRINT, that for output plots is .PLOT, and that for graphical output is

```
.PROBE
```

The statement for the plots of V(3) and V(7) from the results of transient analysis is

```
.PLOT TRAN V(3)V(7)
```

The statement for the tables of V(3) and V(7) from the results of transient analysis is

```
.PRINT TRAN V(3)V(7)
```

For PSpice Schematics: Place the "VPRINT1" symbol as shown in Figure 2.8(a) from the SPECIAL library. Place r on the node where voltage should be printed.

The "IPRINT" should be placed in series with an element through which we want to print current. The VPRINT and IPRINT must be connected as circuit.

In order to assign the type of analysis and the output variable, double-click the mouse on the VPRINT (or IPRINT) printer. The "Property Editor" screen will open as shown in Figure 2.8(b) and assign the output variable to the printed, under the "analysis" block the string of parameters, for example, "I(R_R1), V(R_R1), I(R_R2), V(R_R2)." If the "DC or AC or TRAN" column is not present as shown in Figure 2.8(c), then add it by clicking on the "New Column ..." button.

The output of .PRINT and .PLOT commands are stored in an output file created automatically by PSpice.

Probe is the graphics postprocessor of PSpice, and the statement for this command is

```
.PROBE
```

This causes the results of a simulation to be available in the form of graphical outputs on the display and also as a hard copy. After executing the .PROBE command, Probe will display a menu on the screen to obtain graphical output. It is very easy to use Probe: With the .PROBE command, there is no need for the .PLOT command. .PLOT generates the plot on the output file, whereas .PROBE gives graphical output on the monitor screen, which can be dumped directly to a plotter or printer. The output commands are discussed in Section 6.3.

Probe is normally used for graphical outputs, instead of the print command.

For PSpice Schematics: After the simulation is run, PSpice opens with a probe menu which allows adding a trace(s) of the output variables as shown in Figure 2.9.

(a)

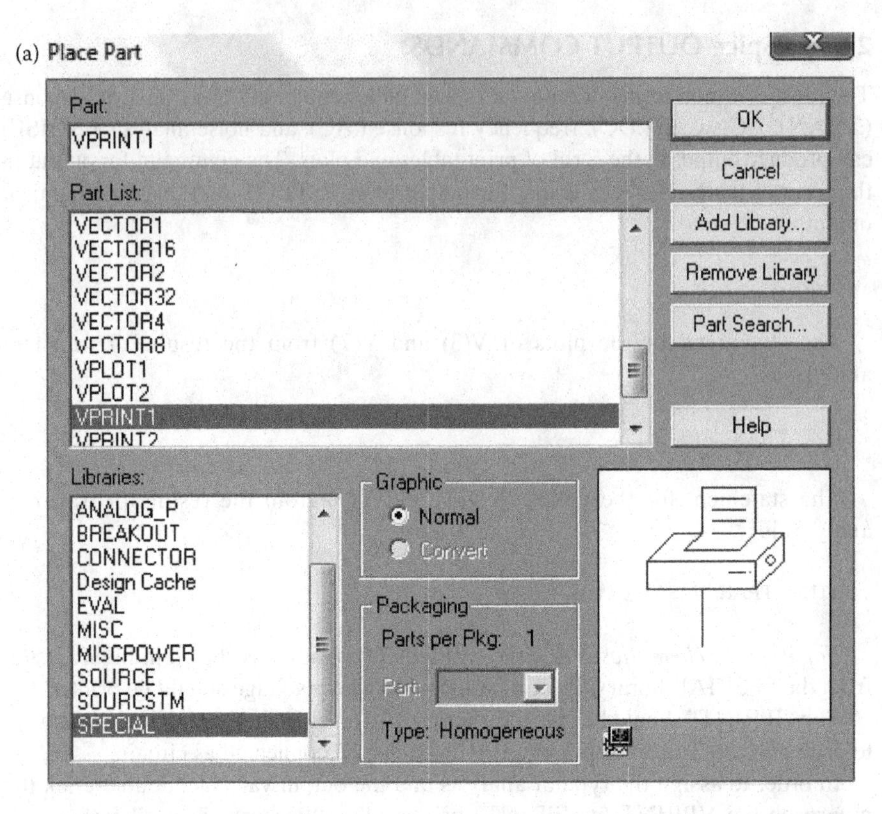

(b)

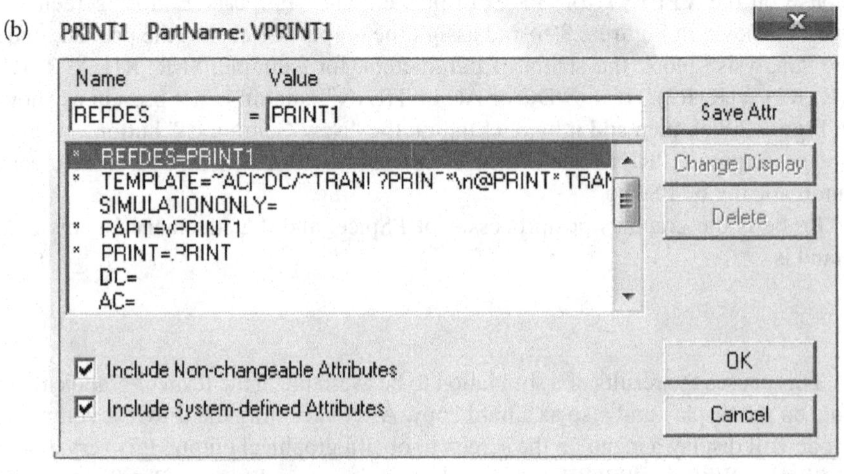

(c) New Column... | Apply | Display... | Delete Property | Filter by: Orcad-PSpice | Help

		Reference	Value	AC	DB	DC	IMAG	MAG	PHASE	PRINT	REAL	Source Part	TRAN
1	⊞ EXAMPLE B-16 : PAGE1 : PRINT1	PRINT1	VPRINT1			IRK)				.PRINT		VPRINT1.Normal	

FIGURE 2.8 Print menu settings. (a) Placing print symbol. (b) Print variables in schematic. (c) Print variables in OrCAD.

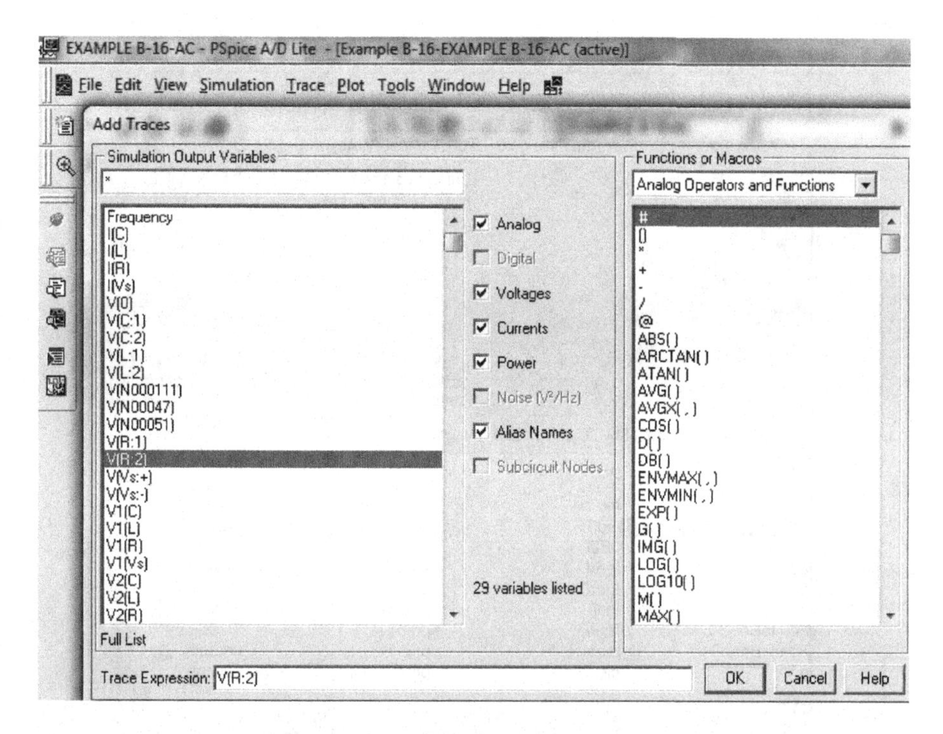

FIGURE 2.9 Adding a trace(s) of the output variables.

2.12 LTspice SCHEMATIC

LTspice has an extensive list of both customized and no-customized symbols and both analog and digital components. The current, voltage, and other symbols can be obtained from the symbol for the place circuit element menu bar symbol as shown in Figure 2.10.

- The analysis types are similar to PSpice/OrCAD. The analysis command (.OP) is set from the simulation menu as shown in Figure 2.11(a) and the submenu as shown in Figure 2.11(b). The command can be placed anywhere in the schematic.
- LTspice also assigns the node numbers. The output voltages refer to the node voltages and the output currents refer to the current through a device or a circuit element. For example, the voltage at node 2 is defined as V(n002) and the current through resistor R1 is defined as I(R1).
- The .op command also directs the node voltages and the currents through the circuit elements to the output file The LTspice opens automatically the window for graphical plots.
- LTspice creates the SPICE netlist from the schematic and the netlist can be viewed from the view menu as shown in Figure 2.12(a). The SPICE netlist for the schematic of Figure 2.1 is shown in Figure 2.12(b).

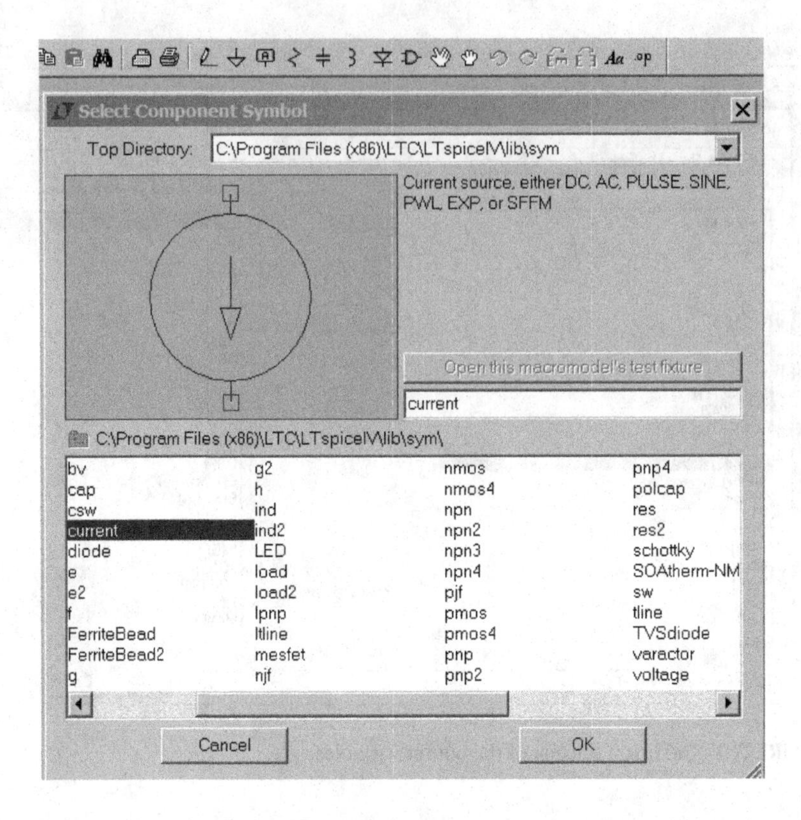

FIGURE 2.10 Place circuit element menu.

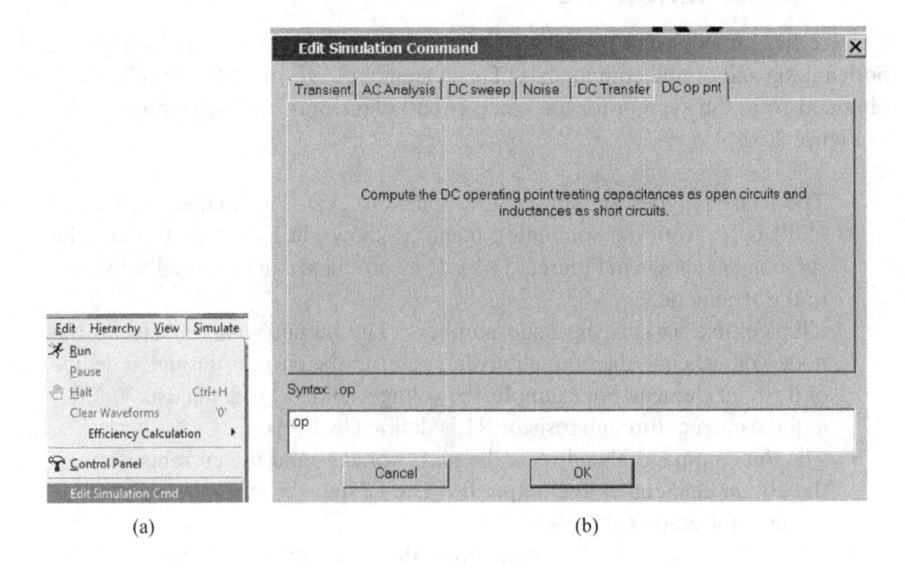

FIGURE 2.11 Simulation and Command menu. (a) Simulation menu. (b) Analysis command menu.

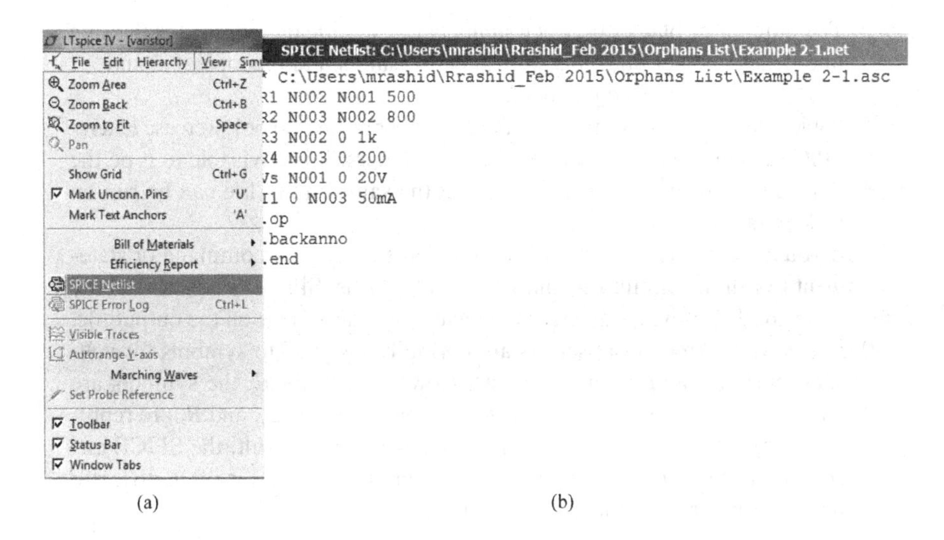

FIGURE 2.12 View menu and SPICE netlist. (a) View menu. (b) SPICE Netlist for Figure 2.1.

2.13 FORMAT OF CIRCUIT FILES

For PSpice A/D: A circuit file that is read by PSpice may be divided into five parts: (1) the title, which describes the type of circuit, or any comments, (2) the circuit description, which defines the circuit elements and the set of model parameters, (3) the analysis description, which defines the type of analysis, (4) the output description, which defines the way the output is to be presented, and (5) the end of program. Therefore, the format for a circuit file is as follows:

Title
Circuit description
Analysis description
Output description
End-of-file statement (.END)

Note the following:

1. The first line is the title line, and it may contain any type of text.
2. The last line must be the .END command.
3. The order of the intervening lines is not important and does not affect the results of the simulation.
4. If a PSpice statement is longer than one line, it can be continued in the next line. A continuation line is identified by a plus sign (+) in the first column of the next line. The continuation lines must follow one another in the proper order.
5. A comment line may be included anywhere and should be preceded by an asterisk (*).

6. The number of blanks between items is not significant (except for the title line). Tabs and commas are equivalent to blanks. For example, “ ” and “ ” and “,” and “,” are all equivalent.
7. PSpice statements or comments can be in either uppercase or lowercase letters.
8. SPICE2 statements must be in uppercase letters. It is advisable to type the PSpice statements in uppercase so that the same circuit file can be run on SPICE2 also.
9. If you are not sure of a command or statement, use that command or statement to run the circuit file and see what happens. SPICE is a user-friendly software that provides an error message or a syntax error on the output file.
10. In electrical circuits, subscripts are normally assigned to symbols for voltages, currents, and circuit elements. However, in SPICE, the symbols are represented without subscripts. For example, v_s, i_s, L_1, C_1, and R_1 are represented by VS, IS, L1, C1, and R1, respectively. As a result, the SPICE circuit description of voltages, currents, and circuit elements is often different from that in terms of the circuit symbols.

The menu for PSpice A/D is shown in Figure 2.13. A new circuit file can be created, or an existing file can be opened by using the PSpice File menu, as shown in Figure 2.14(a). The file can also be created by using any text file editor (e.g., Notepad or a word processor saved as a .text file). The circuit file can be run from the PSpice Simulation menu, as shown in Figure 2.14(c). The output file can be viewed from the PSpice View menu, as shown in Figure 2.14(b), and the graphical plot can be viewed by selecting Simulation Results from the View menu, as shown in Figure 2.14b.

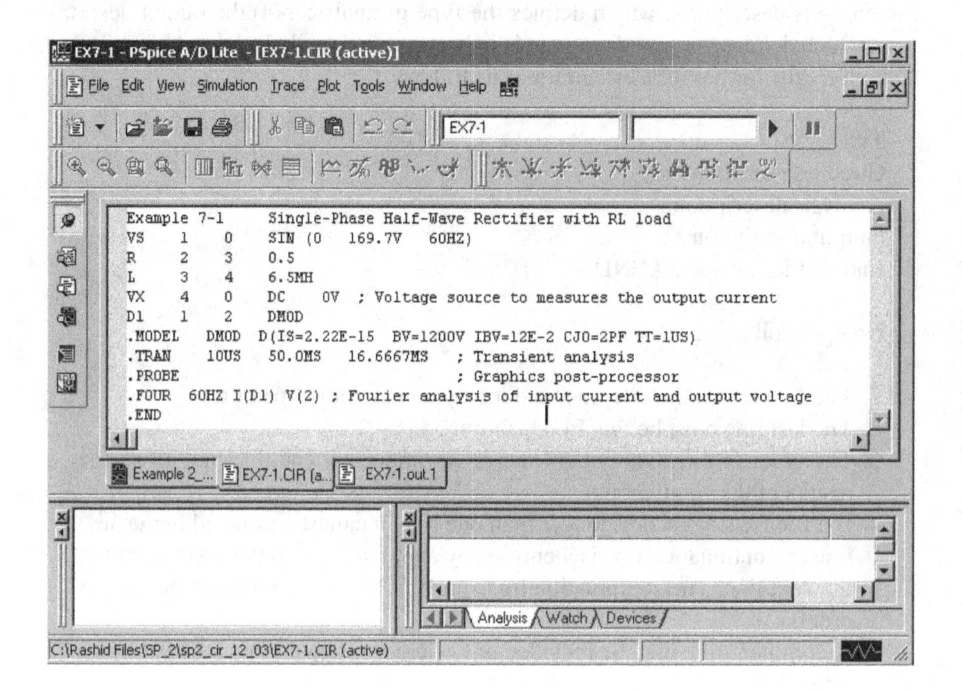

FIGURE 2.13 The PSpice A/D menu.

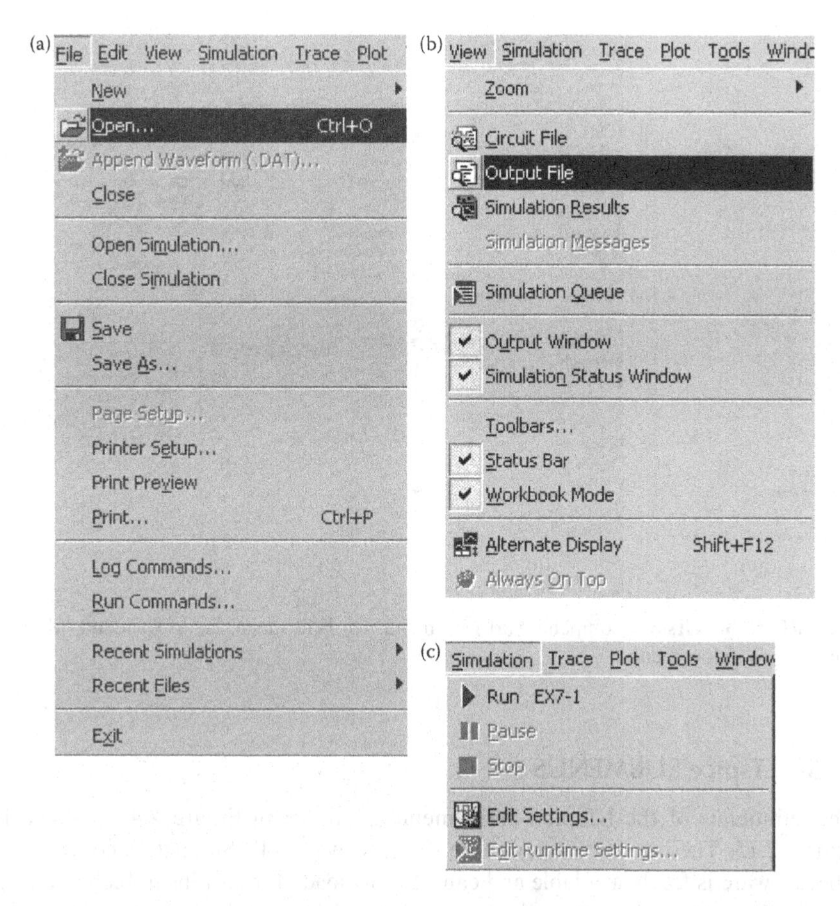

FIGURE 2.14 Submenus of the PSpice A/D. (a) File menu. (b) View menu. (c) Simulation menu.

2.14 FORMAT OF OUTPUT FILES

The results of a simulation by PSpice are stored in an output file. It is possible to control the type and amount of output by various commands. PSpice will indicate any error in the circuit file by displaying a message on the screen and will suggest looking at the output file for details. The output is of four types:

1. A description of the circuit that includes the netlist, the device list, the model parameter list, and so on.
2. Direct output from some of the analyses without the .PLOT and .PRINT commands. This includes the output from the .OP, .TF, .SENS, .NOISE, and. FOUR analyses.
3. Prints and plots resulting from .PRINT and .PLOT commands, including output from the .DC, .AC, and .TRAN analyses.
4. Run statistics, which includes various types of summary information about the entire run, including times required by various analyses and the amount of memory used.

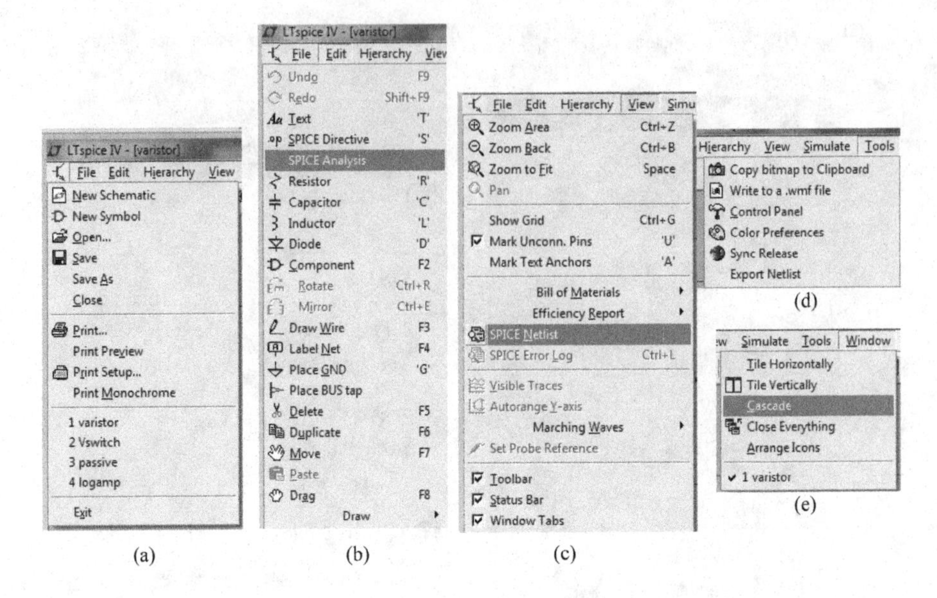

FIGURE 2.15 LTspice submenus. (a) File menu. (b) Edit menu. (c) View menu. (d) Tool menu. (e) Window menu.

2.15 LTspice SUBMENUS

The submenus of the LTspice main menu as shown in Figure 2.4 is shown in Figure 2.15. These menus include File, Edit, View, Tools, Simulate, and Window. The software is freely available and can be downloaded from Linear Technology. It has a variety of custom design simulation tools and device models to allow even novice designers to quickly and easily evaluate circuits using high-performance switching regulators, amplifiers, data converters, filters, and more.

2.16 EXAMPLES OF PSpice SIMULATIONS

We have discussed the details of the circuit in Figure 2.1(a) as a PSpice input file. We will illustrate the PSpice simulation with four examples.

2.16.1 Pulse and Step Responses of *RLC* Circuits

Example 2.1:

Transient Pulse Response of an* RLC *Circuit

The *RLC* circuit of Figure 2.1(a) is to be simulated on PSpice to calculate and plot the transient response from 0 to 400 μs with an increment of 1 μs. The capacitor voltage, V(3), and the current through R_1, I(R1), are to be plotted.

The Fourier series coefficients and THD are to be printed. The circuit file's name is EX2.1.CIR, and the outputs are to be stored in the file EX2.1.OUT. The .PROBE command will make the results available both in display form and as hard copy.

SOLUTION

The PSpice schematic is shown in Figure 2.16(a). The voltage and current markers display the output waveforms in Probe at the end of the simulation. The input voltage is specified by a pulse source as shown in Figure 2.16(b). The transient analysis is set at the Analysis Setup menu as shown in Figure 2.16(c), and its specifications are set at the Transient menu as shown in Figure 2.16(d). The circuit file contains the following statements:

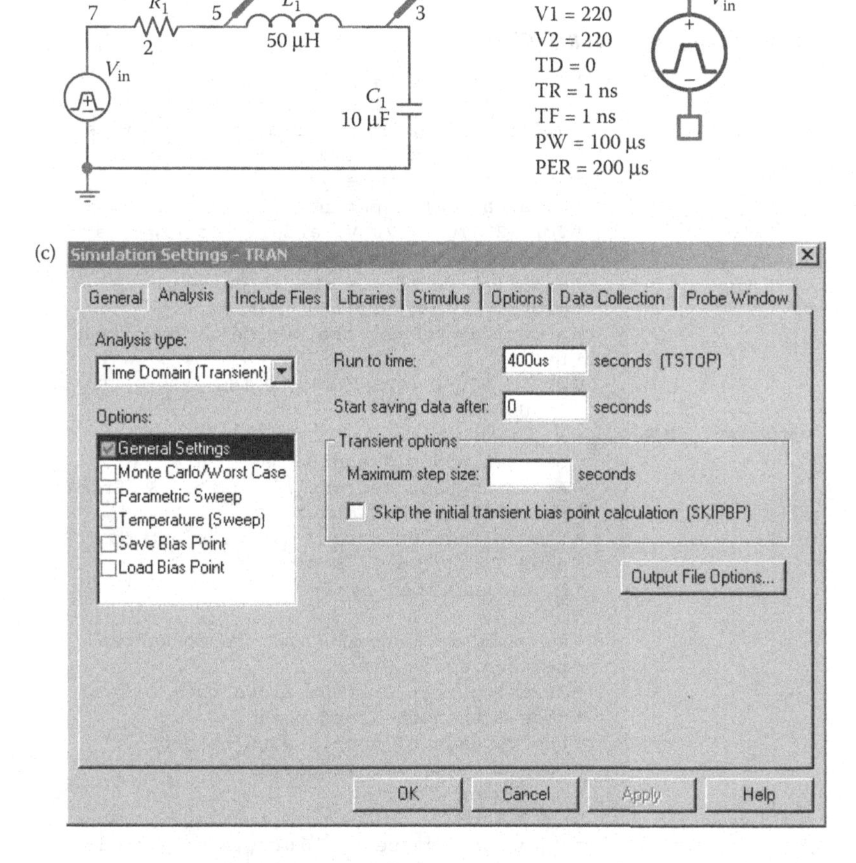

FIGURE 2.16 PSpice schematic for Example 2.1. (a) PSpice schematic. (b) Specifications of pulse voltage. (c) Analysis and runtime setup. (d) Transient output options. *(Continued)*

(d)

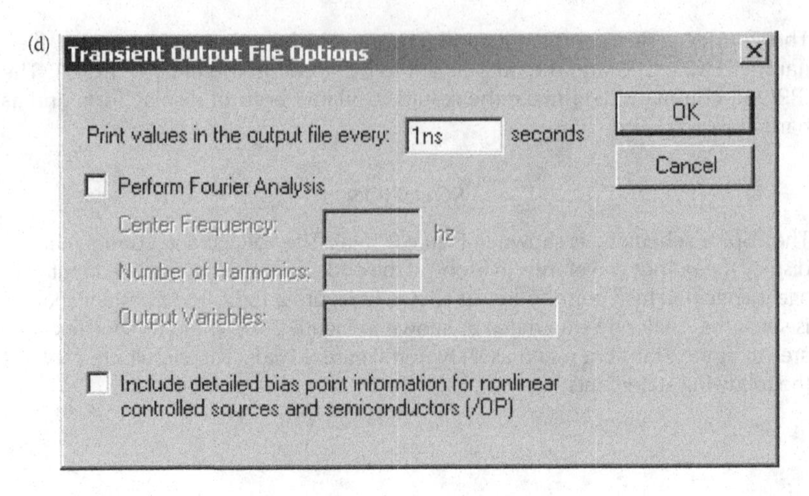

FIGURE 2.16 *(Continued)*

Pulse Response of an *RLC* Circuit

SOURCE ■

```
* The format for a pulse source is
* PULSE (-V_S +V_S TD TR TF PW PER)
* Refer to Chapter 4 for modeling sources.
* v_in is connected between nodes 7 and 0,
* assuming that node 7
* is at a higher potential than node 0.
* For -220 to +220 V, a delay of t_d = 0, a
* rise time of t_r = 1 ns,
* a fall time of t_f = 1 ns, a pulse
* width = 100 µs, and
* a period = 200 µs, the source is described
* by
VIN 7 0 PULSE (-220V 200V 0 1NS 1NS 100US
+ 200US)
```

CIRCUIT ■■

```
* R_1, with a value of 2 Ω, is connected
* between nodes 7 and 5.
* Assuming that current flows into R_1 from
* node 7 to node 5 and that
* the voltage of node 7 with respect to
* node 5, V(7, 5), is positive,
* R_1 is described by
R1 7 5 2
* L_1, with a value of 50 µH, is connected
* between nodes 5 and 3.
* Assuming that current flows into L_1 from
* node 5 to node 3 and that
* the voltage of node 5 with respect to
* node 3, V(5, 3), is positive,
* L_1 is described by
L1 5 3 50UH
* C_1, with a value of 10 µF, is connected
* between nodes 3 and 0.
* Assuming that current flows into C_1 from
```

```
                 * node 3 to node 0 and that
                 * the voltage of node 3 with respect to
                 * node 0, V(3), is positive,
                 * C1 is described by
                 C1 3 0 10UF
ANALYSIS ■■■     * Transient analysis is invoked by the
                 * .TRAN command, whose simple format is
                 .TRAN TSTEP TSTOP
                 * Refer to Chapter 6 for dot commands. For
                 * transient analysis
                 * from 0 to 400 µs with an increment of
                 * 1 µs, the statement is
                 .TRAN 1US 400US
                 * The following statement prints the
                 * results of transient analysis for
                 * V(R1), V(L1), and V(C1):
                 * .PRINT TRAN V(R1) V(L1) V(C1)
                 * The following statement plots the results
                 * of transient analysis for V(3) and
                 * I(R1):
                 .PLOT TRAN V(3)I(R1)
                 * Refer to Chapter 6 for dot commands. For
                 * Fourier analysis
                 * of V(3) at a fundamental frequency of
                 * 5 kHz, the statement is
                 .FOUR 5 KHZ V(3)
                 * Graphic output can be obtained simply by
                 * invoking the .PROBE command
                 * (refer to Chapter 6 for dot commands):
                 .PROBE
                 * The end of program is invoked by the .END
                 * command:
                 .END
```

Note: Some statements in the example run over to the next line. They should all be in the same line.

If the PSpice programs are loaded in a fixed (hard) disk and the circuit file is stored in a floppy diskette on drive A:, the general command to run the circuit file is

```
PSPICE A: <input file> A: <output file>
```

For an input file EX2.1.CIR and the output file EX2.1.OUT, the command is

```
PSPICE A:EX2.1.CIR A:EX2.1.OUT
```

If the output file's name is omitted, the results are stored by default on an output file that has the same name as the input file and is in the same drive but with an extension .OUT. It is a good practice to have .CIR and .OUT extensions on circuit files so that the circuit file and the corresponding output file can be identified. Thus, the command can simply be

```
PSPICE A:EX2.1.CIR
```

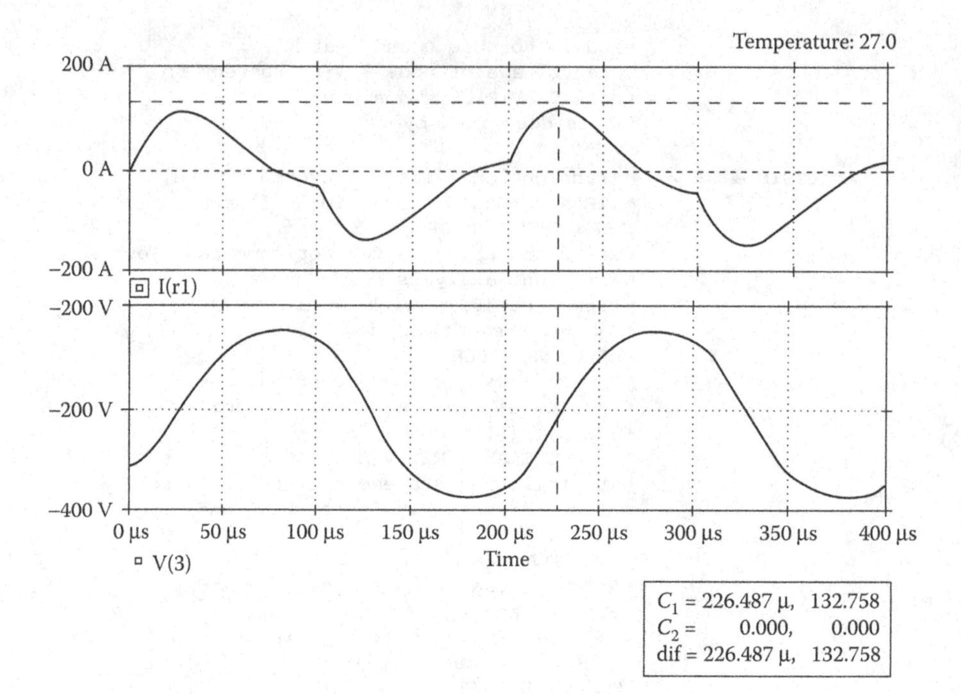

FIGURE 2.17 Pulse response for Example 2.1.

The results of the transient response that are obtained on the display by the .PROBE command are shown in Figure 2.17. We get the peak current of 132.75 A from the cursor. I(r1) is the instantaneous current through r1 and V(3) is the output voltage at node 3. The results of the .PRINT statement can be obtained by printing the contents of the output file EX2.1.OUT. From this output file, the results of the Fourier analysis are as follows:

Fourier Components of Transient Response V(3)

DC Component = −1.726830E−01

Harmonic No	Frequency (Hz)	Fourier Component	Normalized Component	Phase (deg)	Normalized Phase (deg)
1	5.000E+03	3.502E+02	1.000E+00	−5.171E+01	0.000E+00
2	1.000E+04	2.718E−01	7.762E−04	−1.795E+02	−1.278E+02
3	1.500E+04	2.245E+01	6.410E−02	−1.517E+02	−1.000E+02
4	2.000E+04	8.506E−02	2.429E−04	1.708E+02	2.225E+02
5	2.500E+04	4.411E+00	1.259E−02	−1.601E+02	−1.084E+02
6	3.000E+04	5.024E−02	1.435E−04	1.753E+02	2.270E+02
7	3.500E+04	1.683E+00	4.806E−03	−1.595E+02	−1.078E+02
8	4.000E+04	3.617E−02	1.033E−04	1.791E+02	2.309E+02
9	4.500E+04	8.936E−01	2.552E−03	−1.608E+02	−1.090E+02

Total harmonic distortion = 6.555345E+00%

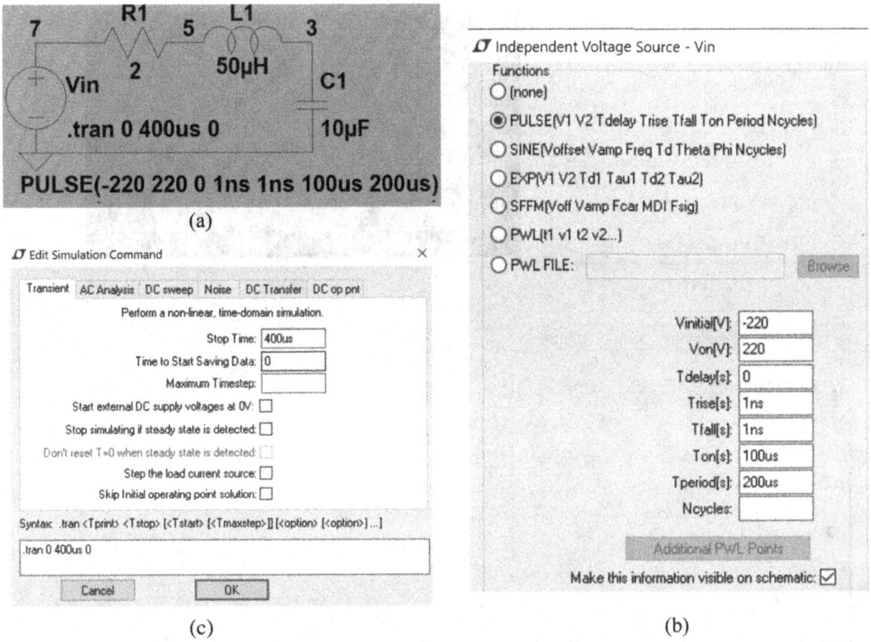

(a)

(c)

(b)

FIGURE 2.18 LTspice schematic for Example 2.1. (a) LTspice schematic. (b) Voltage source menu. (c) Simulation menu.

From the Fourier component, we get

DC component voltage = –0.172 V
Peak fundamental component at 5 kHz = 350.2 V
Fundamental phase delay = –51.71°
Total harmonic distortion = 6.55%

LTspice: LTspice schematic for Example 2.1 is shown in Figure 2.18(a). The setup parameters for the pulse source in Figure 2.18(b) and the setup parameters for the transient analysis in Figure 2.18(c).

The schematic is simulated by selecting Run from the Simulation menu as shown in Figure 2.19(a), and the voltage across node 3 and the current through R1 are shown in Figure 2.19(b). Figure 2.19(c) shows the Trace menu, Figure 2.19(d) Trace variables, Figure 2.19(e) Window menu, and Figure 2.19(f) Tools menu. We can add select the Trace variables, alternatively put the cursor on the node to select the voltage or on the circuit element to select the current through it.

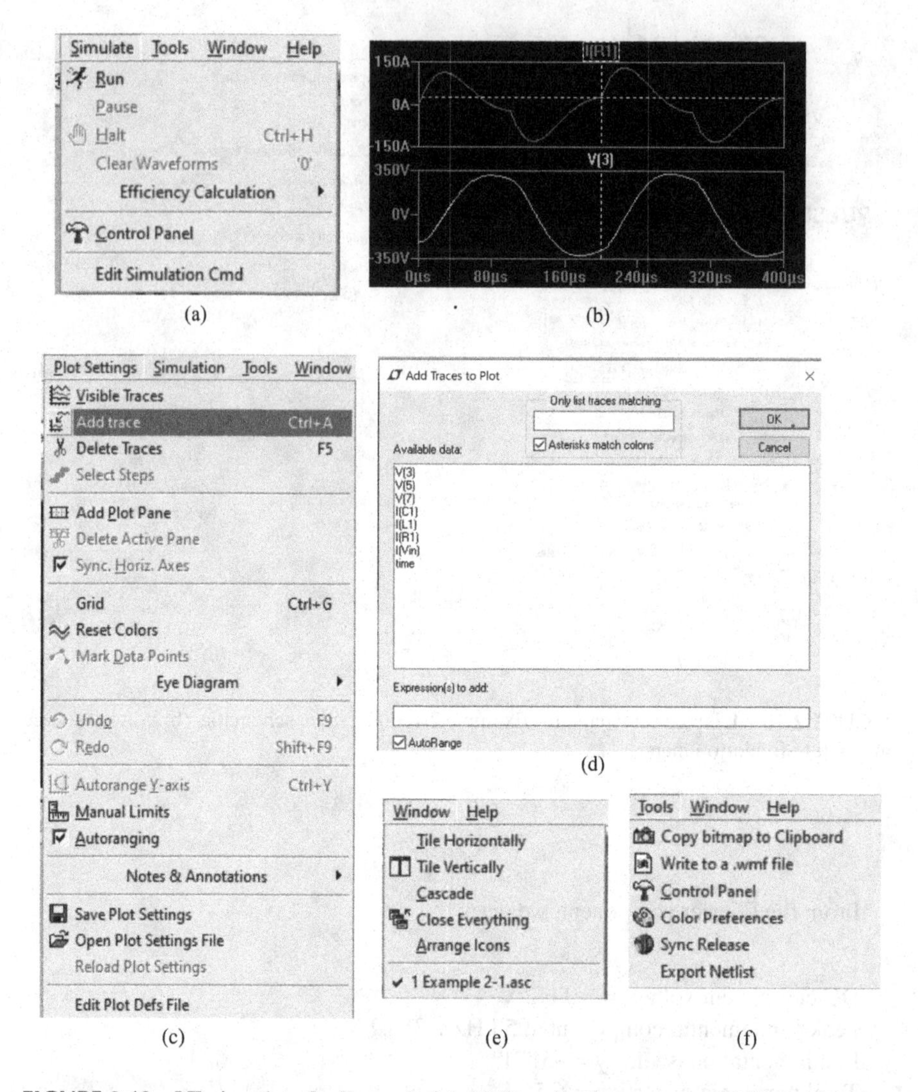

FIGURE 2.19 LTspice plots for Example 2.1. (a) Run simulation menu. (b) Plots of output voltage V(3) and current through R1. (c) Trace menu. (d) Trace variables. (e) Window menu, (f) Tools menu.

Example 2.2:

Effect of Resistors on the Transient Pulse Response of an RLC *Circuit*

Three *RLC* circuits with $R = 2$, 1, and 8 Ω are shown in Figure 2.20(a). The inputs are identical step voltages, as shown in Figure 2.20(b). Using PSpice, the transient response from 0 to 400 μs with an increment of 1 μs is to be calculated and plotted. The capacitor voltages are the outputs V(3), V(6), and V(9), which are to be plotted. The circuit is to be stored in the file EX2.1.CIR, and the outputs are to be

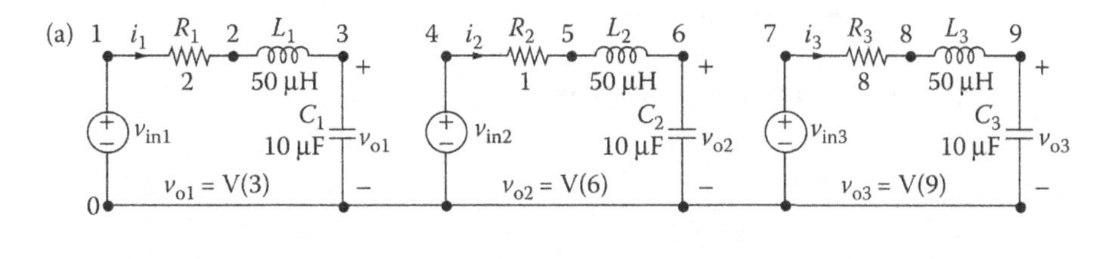

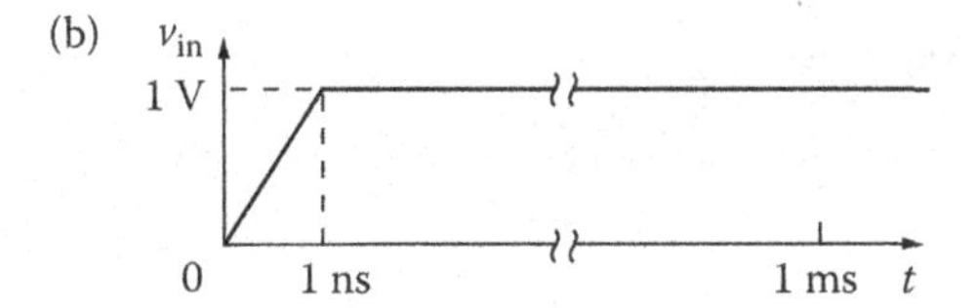

FIGURE 2.20 RLC Circuits. (a) Circuit. (b) step voltage.

stored in the file EX2.2.OUT. The results should also be available for display and as hard copy, using the .PROBE command.

SOLUTION

The PSpice schematic is shown in Figure 2.21. The voltage and current markers display the output waveforms in Probe at the end of the simulation. The input voltages are specified by three pulse sources that are similar to that in Figure 2.16(b). Transient analysis is set at the Analysis Setup menu as shown in Figure 2.16(c), and its specifications are set at the Transient menu as shown in Figure 2.16(d).

The description of the circuit file is similar to that of Example 2.1, except that the input is a step voltage rather than a pulse voltage. The circuit may be regarded as three *RLC* circuits having three separate inputs. The step signal can be represented by a piecewise linear source, and it is described in general by

```
PWL  (T1 V1 T2 V2 ... TN VN)
```

where VN is the voltage at time TN. Assuming a rise time of 1 ns, the step voltage of Figure 2.13b can be described by

```
PWL(0 0 1NS 1V IMS 1V)
```

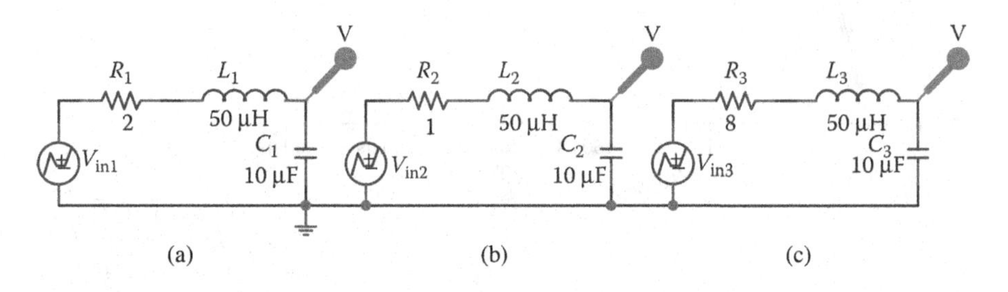

FIGURE 2.21 PSpice schematic for Example 2.2.

The listing of the circuit file is as follows:

Step Response of Series *RLC* Circuits

```
SOURCE      ■     V11 1 0 PWL (0 0 1NS 1V 1MS 1V) ;
                  Step of 1 V
                  V12 4 0 PWL (0 0 1NS 1V 1MS 1V)  ;
                  Step of 1 V
                  V13 7 0 PWL (0 0 1NS 1V 1MS 1V)  ;
                  Step of 1 V
CIRCUIT     ■■    R1 1 2 2
                  L1 2 3 50UH
                  C1 3 0 10UF
                  R2 4 5 1
                  L2 5 6 50UH
                  C2 6 0 10UF
                  R3 7 8 8
                  L3 8 9 50UH
                  C3 9 0 10UF
ANALYSIS    ■■■   .PLOT TRAN V(3) V(6) V(9)  ;
                  Plot output voltages
                  .TRAN  1US  400US      ; Transient analysis
                  .PROBE  ; Graphics post-processor
         .END
```

The results of the transient analysis that are obtained on the display by the .PROBE command are shown in Figure 2.22. The results of the .PRINT statement can be obtained by printing the contents of the output file EX2.2.OUT.

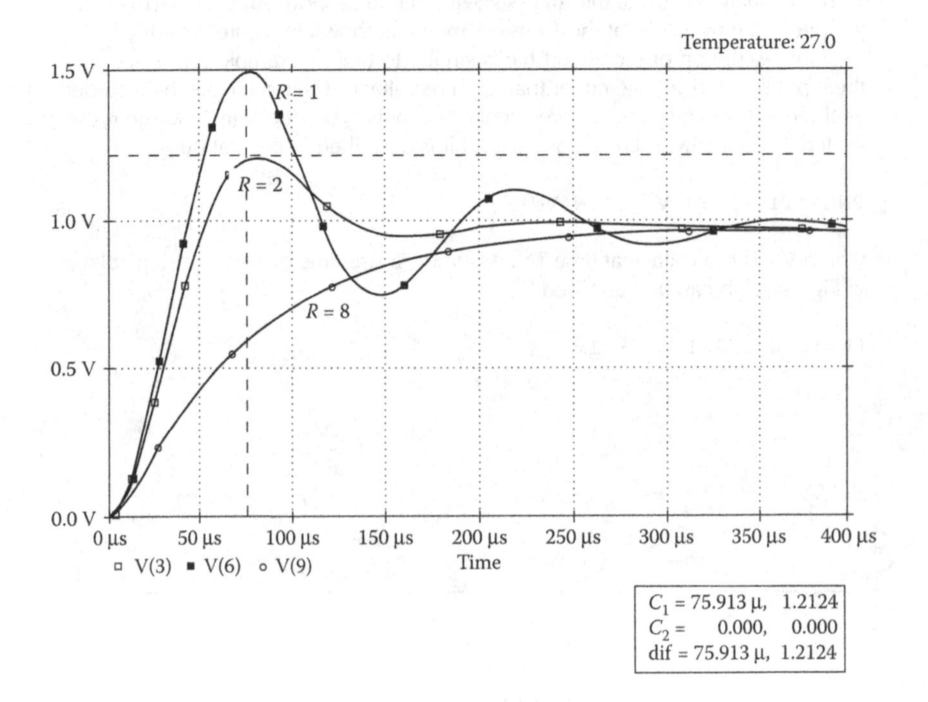

FIGURE 2.22 Step response of RLC circuits for Example 2.2.

Expected Value: In power electronic circuits, the values of *R*, *L*, and *C* are such that they form an underdamped circuit and the output voltage can be found from [5]

$$v_o(t) = 1 - e^{-\alpha t} \times \frac{\sqrt{\alpha^2 + \omega_r^2}}{\omega_r} \times \sin(\omega_r t + \phi) \tag{2.1}$$

where

$$\alpha = \frac{R}{2L}, \quad \omega_r = \left(\frac{1}{LC} - \frac{R^2}{4L^2}\right)^{1/2}, \quad \phi = \tan^{-1}\frac{\omega_r}{\alpha}$$

Maximum overshoot (OS) and the time T_p at which it occurs are given by

$$OS = e^{-\alpha\pi/\omega_r}, \quad T_p = \frac{\pi}{\omega_r} \tag{2.2}$$

Using the circuit values (for $R = 1\Omega$) gives the peak output voltage as $V_{o(peak)} =$ 1.486 V at 72.07 μs (compared to 1.492 V at 73.99 μs with PSpice).

Note: The circuit can be analyzed more effectively if we use the "PARAM" command in Chapter 6, in which case we can vary or assign the value of *R*.

LTspice: LTspice schematic for Example 2.2 is shown in Figure 2.23(a). The setup parameters for the pulse source in Figure 2.23(b) and the setup parameters for the transient analysis in Figure 2.23(c).

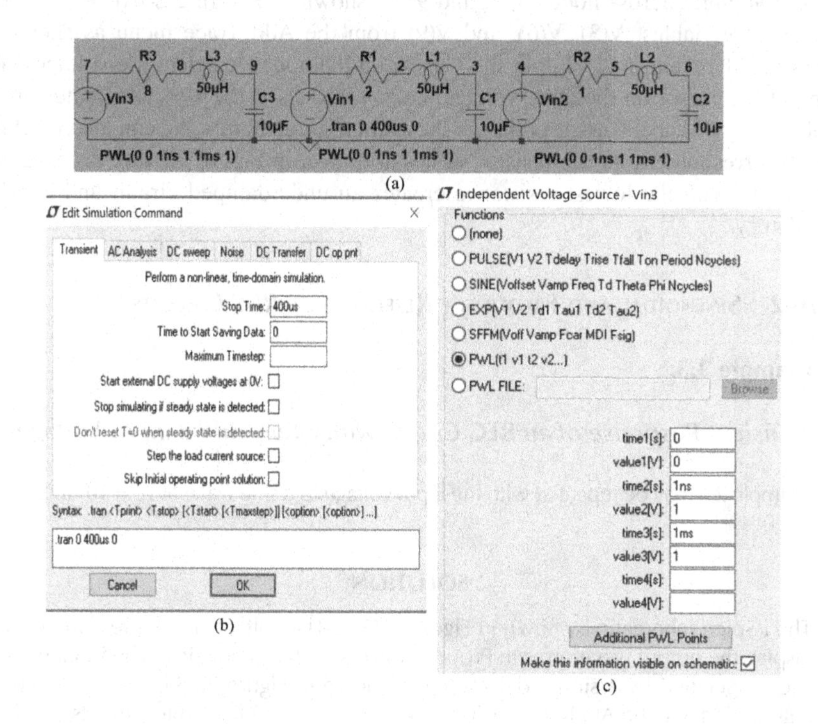

FIGURE 2.23 LTspice schematic for Example 2.21. (a) LTspice schematic. (b) Simulation menu. (c) Voltage source menu.

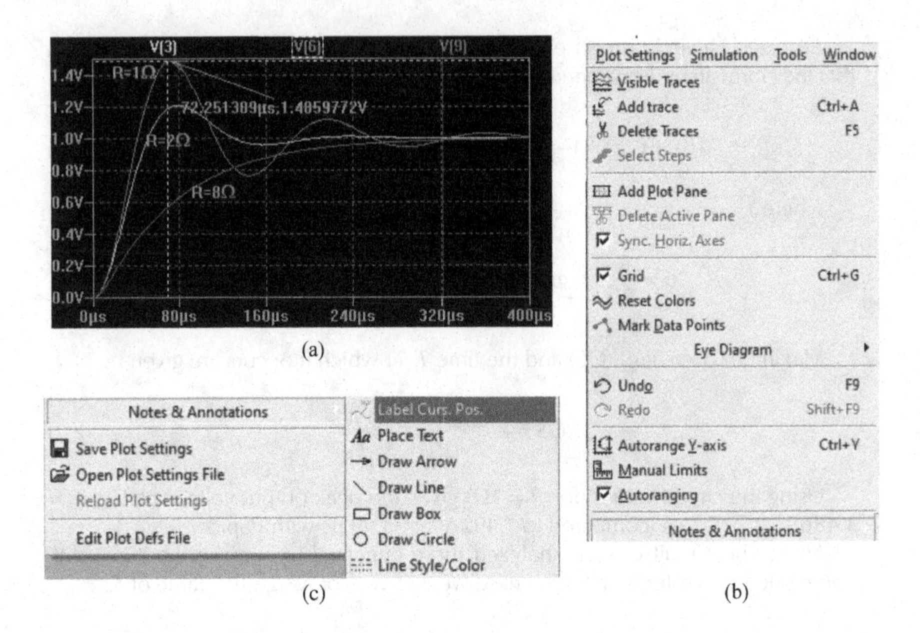

FIGURE 2.24 LTspice plots for Example 2.2. (a) Voltages across nodes 3, 6, and 9. (b) Plot setting menu. (c) Place Text menu.

The voltages across nodes 3, 6, and 9 are shown in Figure 2.24(a). We can add the Trace variables, V(3), V(6), and V(9) from the Add Trace menu as shown in Figure 2.24(b), alternatively put the Cursor on the nodes to select the voltages. We can put the cursor on the node voltage, i.e., V(3) to show the x-y coordinates of the plot and select Label Cursor points as shown in Figure 2.24(c). We can also label the traces corresponding to the resistor values as shown in Figure 2.24(a). As expected the circuit with the lowest resistance creates an underdamped circuit and is more oscillatory.

2.16.2 Sinusoidal and Frequency Responses of *RLC* Circuits

Example 2.3:

Transient Response of an RLC *Circuit with a Sinusoidal Input Voltage*

Example 2.1 may be repeated with the input voltage as a sine wave of $v_{in} = 10\sin(2\pi \times 5000t)$.

SOLUTION

The PSpice schematic is shown in Figure 2.25(a). The voltage and current markers display the output waveforms in Probe at the end of the simulation. The input voltage is specified by a sinusoidal source as shown in Figure 2.25(b). The transient analysis is set at the Analysis Setup menu as shown in Figure 2.16(c), and its specifications are set at the Transient menu as shown in Figure 2.16(d).

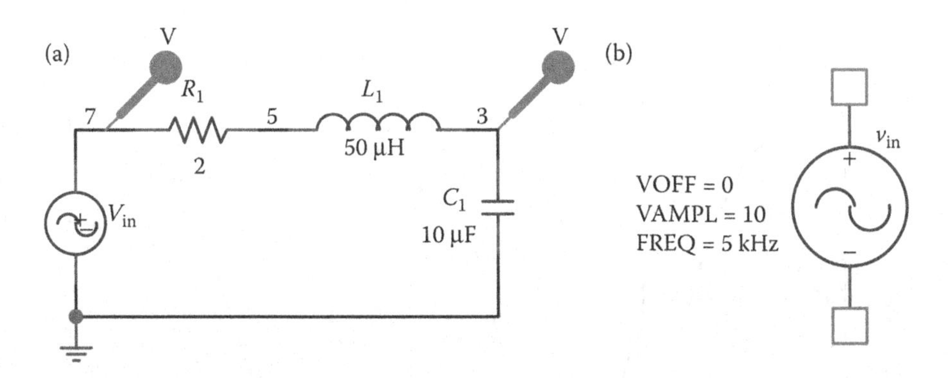

FIGURE 2.25 PSpice schematic for Example 2.3. (a) PSpice schematic. (b) Specifications of sinusoidal source.

The model for a simple sinusoidal source is

```
SIN(VO VA FREQ)
```

where

VO = offset voltage, V
VA = peak voltage, V
FREQ = frequency, Hz

For a sinusoidal voltage $v_{in} = 10 \sin(2\pi \times 5000t)$, the model is

```
SIN(0 10 V 5 KHZ)
```

The circuit file contains the following statements:

***RLC* Circuit with Sinusoidal Input Voltage**

```
SOURCE      ■      * The format for a simple sinusoidal
                   * source is
                   * SIN(VO VA FREQ)
                   * Refer to Chapter 4 for modeling sources.
                   * vin is connected between nodes 7 and 0,
                   * assuming that node 7
                   * is at a higher potential than node 0.
                   * With a peak voltage of VA = 10 V, a
                   * frequency = 5 kHz, and
                   * an offset value of Vo = 0, the source is
                   * described by
                   VIN 7 0 SIN (0 10 V 5 KHZ) ; Input voltage
CIRCUIT     ■■     R1 7 5 2
                   L1 5 3 50UH
                   C1 3 0 10UF
ANALYSIS    ■■■    .PLOT TRAN V(3) V(7)      ; Plot output
                                               voltages
                   .TRAN 1US 500US           ; Transient analysis
                   .PROBE                    ; Graphics post-
                                               processor
            .END
```

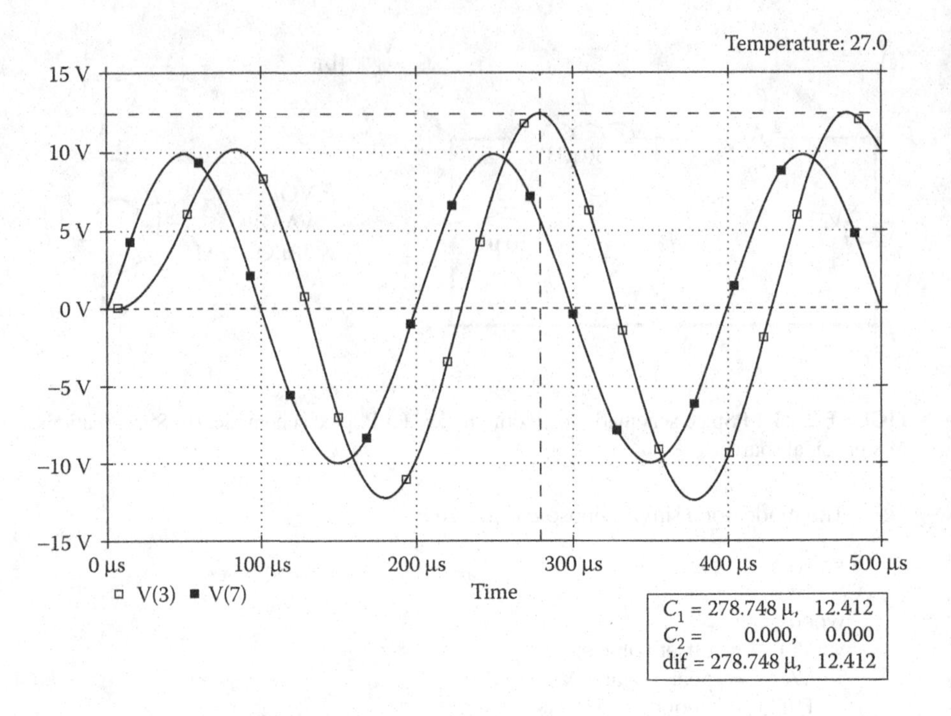

FIGURE 2.26 Transient response for Example 2.3.

The results of the transient response that are obtained on the display by the .PROBE command are shown in Figure 2.26. The results of the .PRINT statement can be obtained by printing the contents of the output file EX2.3.OUT.

Note that the output is phase shifted. The amount of the delay will depend on the values of *R, L,* and *C*.

Expected Value: The transfer function which is the ratio of the output voltage to the input voltage is given by [5]

$$G(j\omega) = \frac{V_o(j\omega)}{V_s(j\omega)} = \frac{1/j\omega C}{R + j\omega L + 1/j\omega C} = \frac{1}{(1-\omega^2 LC) + j\omega RC} \tag{2.3}$$

The peak magnitude *M* and the phase angle ϕ are given by

$$M = \frac{1}{\sqrt{(1-\omega^2 LC)^2 + (\omega RC)^2}}$$

$$\phi = \tan^{-1}\left(\frac{\omega CR}{1-\omega^2 LC}\right) \tag{2.4}$$

Using the circuit values (for $R = 2\ \Omega$) gives the peak output voltage as $M = 10 \times 1.239 = 12.39$ V (compared to 12.401 V with PSpice).

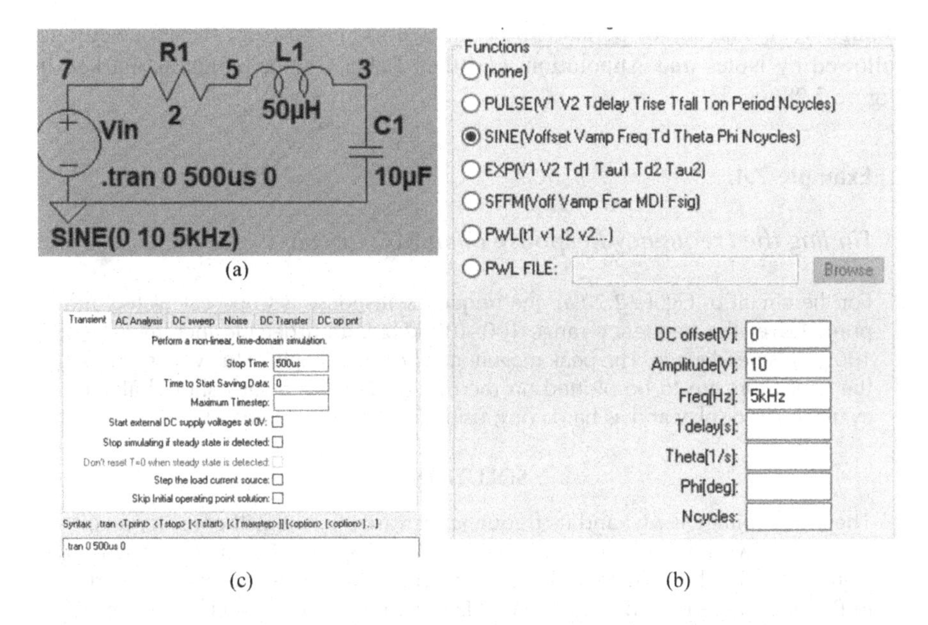

FIGURE 2.27 LTspice schematic for Example 2.3. (a) LTspice schematic. (b) Voltage source menu. (c) Simulation menu.

LTspice: LTspice schematic for Example 2.3 is shown in Figure 2.27(a). The setup parameters for the sine-wave voltage source are shown in Figure 2.27(b) and the setup parameters for the transient analysis are shown in Figure 2.27(c).

The voltages across nodes 7 and 3 are shown in Figure 2.28(a). We can add the Trace variables, V(3), and V(7) from the Add Trace menu, alternatively put the Cursor on the nodes to select the voltages. We can put the cursor on the node voltage, i.e., V(3) to show the x-y coordinates of the plot and select Label Cursor points as shown in Figure 2.28(b). As expected the output voltage V(3) lags the input

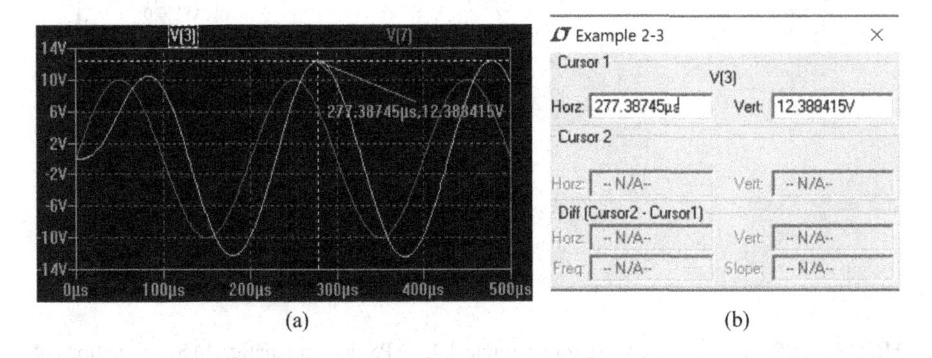

FIGURE 2.28 LTspice Plots for Example 2.3. (a) Voltages across nodes 3, and 7. (b) x-y coordinates of the plot V(3).

voltage V(7). The cursor points can be labeled as shown from the Plot Settings, followed by Notes and Annotation, and then Label Cursor points as marked in Figure 2.28(a).

Example 2.4:

Finding the Frequency Response of an RLC Circuit

For the circuit of Figure 2.20(a), the frequency response is to be calculated and printed over the frequency range 100–100 kHz with a decade increment and 100 points per decade. The peak magnitude and phase angle of the voltage across the capacitors are to be plotted on the output file. The results should also be available for display and as hard copy, using the .PROBE command.

SOLUTION

The input voltage is AC, and its frequency is variable. We will consider a voltage source with a peak magnitude of 1 V. The PSpice schematic is shown in Figure 2.29(a). The voltage and current markers display the output waveforms in Probe at the end of the simulation. The input voltage is specified by an AC source as shown in Figure 2.29(b). The AC analysis is set at the analysis setup menu as shown in Figure 2.29(c), and its specifications are set from the AC Sweep menu, also shown in Figure 2.29(c).

The circuit file is similar to that of Example 2.2, except that the statements for the type of analysis and output are different. The frequency response analysis is invoked by the .AC command, whose format is

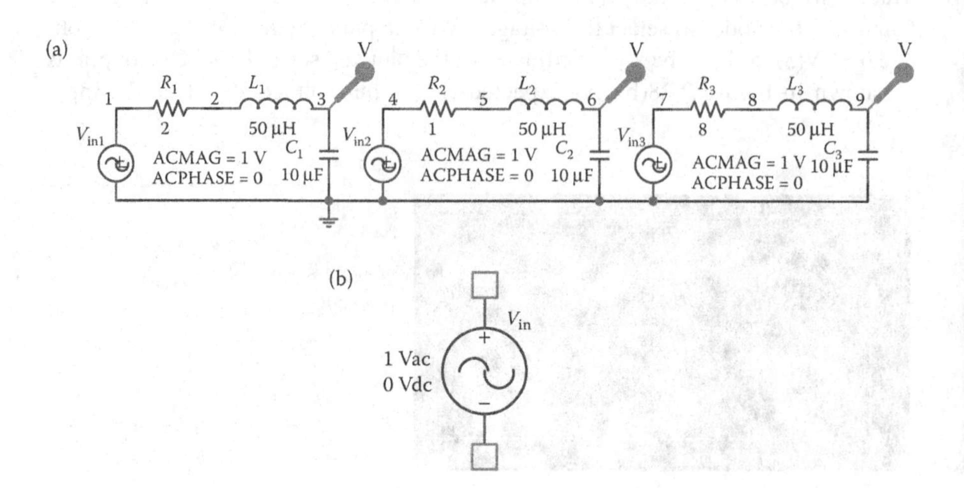

FIGURE 2.29 PSpice schematic for Example 2.4. (a) PSpice schematic. (b) Specifications of AC source. (c) AC Analysis and sweep setup. *(Continued)*

(c)

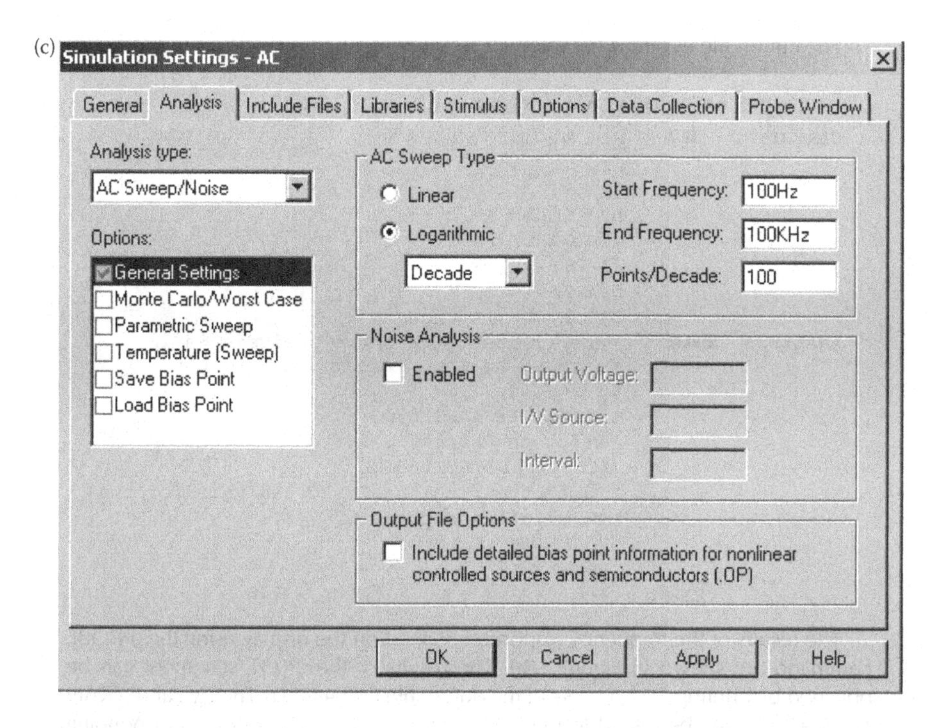

FIGURE 2.29 *(Continued)*

```
.AC DEC NP FSTART FSTOP
```

where

DEC = sweep by decade
NP = number of points per decade
FSTART = starting frequency
FSTOP = ending (or stop) frequency

For NP = 100, FSTART = 100 Hz, and FSTOP = 100 kHz, the statement is

```
.AC DEC 100 100 100 KHZ
```

The magnitude and phase of voltage V(3) are specified as VM(3) and VP(3). The statement to plot is

```
.PLOT AC VM(3) VP(3)
```

The input voltage is AC, and the frequency is variable. We can consider a voltage source with a peak magnitude of 1 V. The statement for an independent voltage source is

```
VIN 7 0 AC 1V
```

The circuit file contains the following statements:

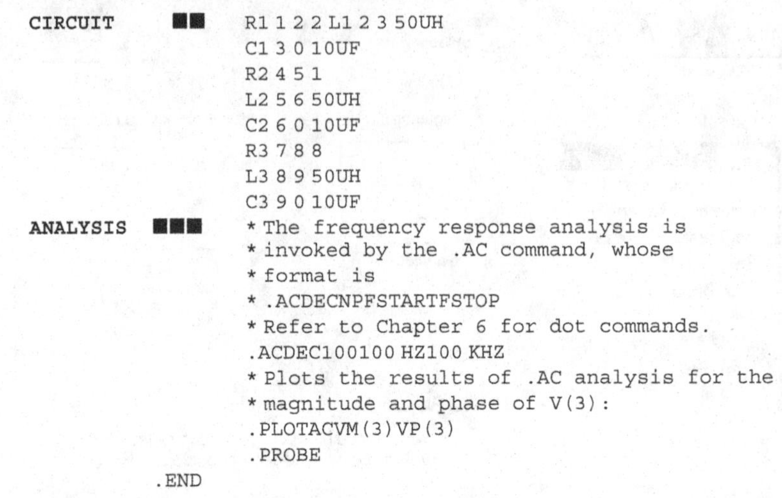

Frequency Response of *RLC* Circuits

```
CIRCUIT    ■■   R1 1 2 2 L1 2 3 50UH
                C1 3 0 10UF
                R2 4 5 1
                L2 5 6 50UH
                C2 6 0 10UF
                R3 7 8 8
                L3 8 9 50UH
                C3 9 0 10UF
ANALYSIS   ■■■  * The frequency response analysis is
                * invoked by the .AC command, whose
                * format is
                * .ACDECNPFSTARTFSTOP
                * Refer to Chapter 6 for dot commands.
                .ACDEC100100 HZ100 KHZ
                * Plots the results of .AC analysis for the
                * magnitude and phase of V(3):
                .PLOTACVM(3)VP(3)
                .PROBE
         .END
```

The results of the frequency response obtained on the display using the .PROBE command are shown in Figure 2.30. The results of the .PLOT statement can be obtained by printing the contents of the output file EX2.4.OUT. The .AC and .TRAN commands could be added to the same circuit file to perform the corresponding analyses.

Note: The peak value occurs at the resonant frequency.

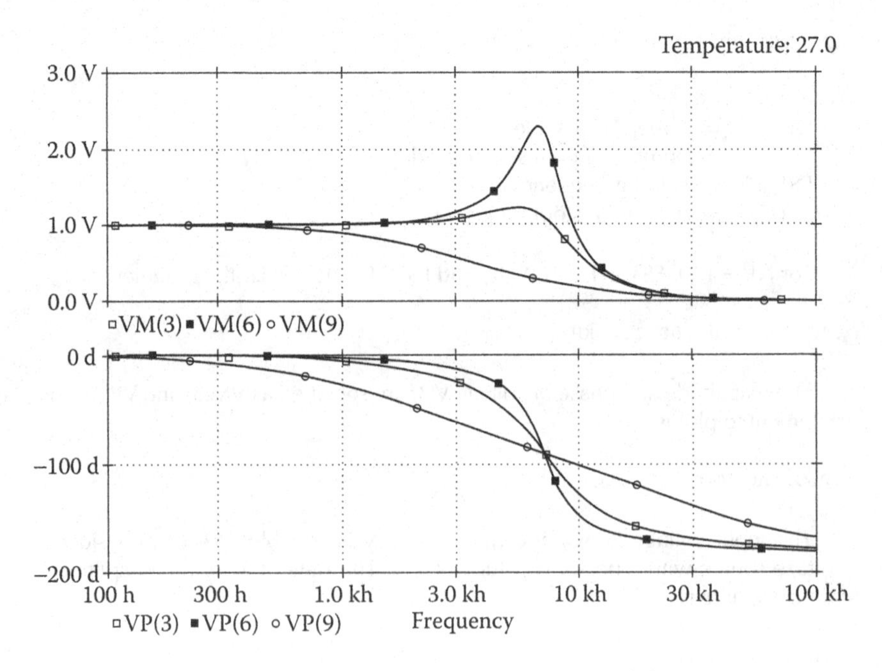

FIGURE 2.30 Frequency responses of *RLC* circuit for Example 2.4.

Expected Value: The magnitude at the resonant frequency of $f_o = (1/2\pi)(1/\sqrt{(LC)})$ can be found from Equations (2.2) and (2.3)

$$M = \frac{1}{\omega CR} = \frac{1}{R}\sqrt{\frac{L}{C}}$$
$$\phi = \frac{\pi}{2} \tag{2.5}$$

Using the circuit values (for $R = 1\ \Omega$) gives the peak output voltage as $M = 2.236$ V (compared to 2.2912 V with PSpice).

LTspice: LTspice schematic for Example 2.4 is shown in Figure 2.31(a). The setup parameters for the pulse source in Figure 2.31(b) and the setup parameters for the transient analysis in Figure 2.31(c).

The voltages across nodes 9, 6, and 3 are shown in Figure 2.32(a). We can add the Trace variables, V(3), V(6), and V(9) from the Add Trace menu, alternatively put the Cursor on the nodes to select the voltages. We can put the cursor on the node voltage, i.e., V(3) to show the x-y coordinates of the plot and select Label Cursor points as shown in Figure 2.32(b). As expected the output voltage V(3) lags the input voltage V(7). The cursor points can be labeled as shown from the Plot Settings, followed by Notes and Annotation, and then Label Cursor points as marked in Figure 2.32(a). As

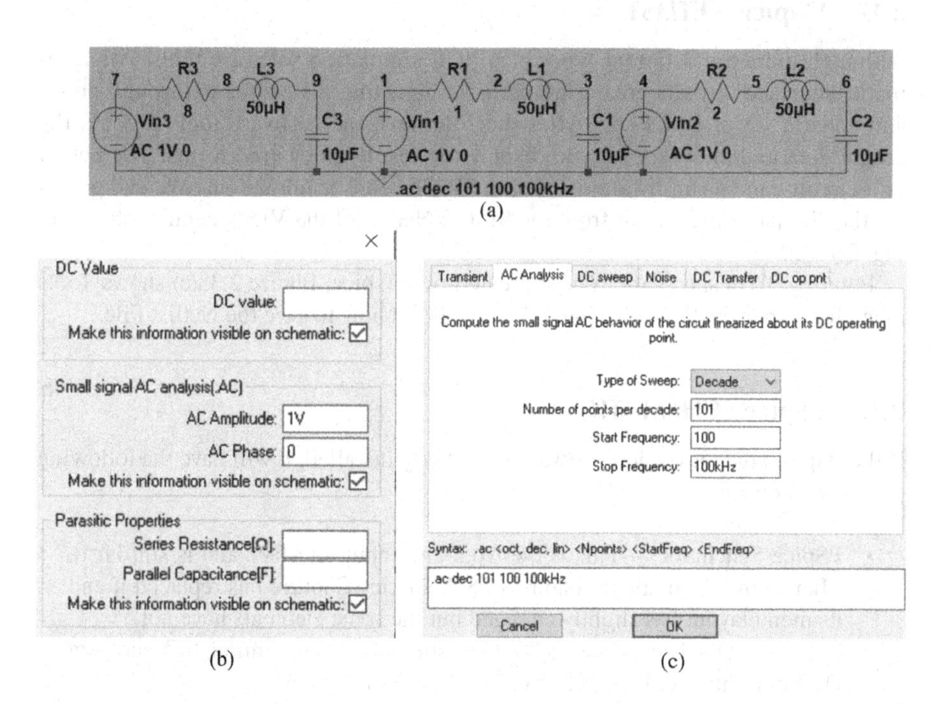

FIGURE 2.31 LTspice schematic for Example 2.4. (a) LTspice schematic. (b) AC Voltage source menu. (c) Simulation menu.

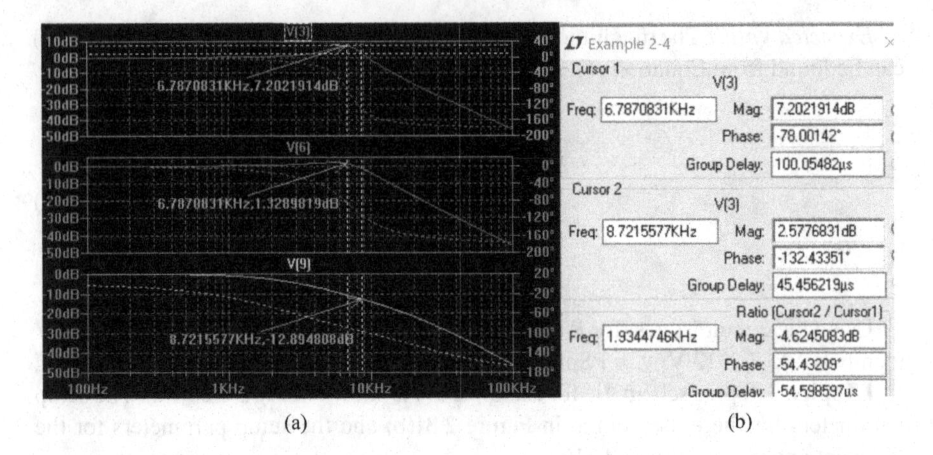

(a) (b)

FIGURE 2.32 LTspice plots for Example 2.4. (a) Voltages across nodes 3, 6, and 9. (b) x-y coordinates of V(3), V(6), and V(9).

expected, the magnitude and the phase-angle of the output voltage decrease with the increase in the series resistance.

2.17 LTspice NETLIST

Although there are different types of SPICE simulation software tools. All schematic interface circuits create the netlist connecting all elements through nodes and solve a set of equations. After the solutions, the software tools process the output data and allow the user to display the results in desired forms. The netlist of a circuit can be run by almost all the SPICE circuit simulations. We can generate the Netlist of the circuit from the SPICE Netlist of the View menu as shown in Fig. 2.14(a).

Figure 2.33(b) shows the Netlist created by LTspice, Figure 2.33(c) shows Tools menu for Export Netlist and Figure 2.33(d) shows how to save the Netlist File.

2.18 PSpice SCHEMATICS

If the PSpice student version software is properly installed, it will have the following portions modules:

- PSpice Schematics—This is the front-end input interface and is similar to MicroSim Schematics version 8.0. Schematics Capture has replaced it and its menu layout has slightly changed but the basic elements have not.
- PSpice A/D—This is a mixed-signal simulation tool similar to MicroSim Design Center, and it is relatively unchanged in OrCAD.
- Probe—This is a graphical postprocessor for viewing the simulation results, similar to MicroSim Design Center and relatively unchanged as well.

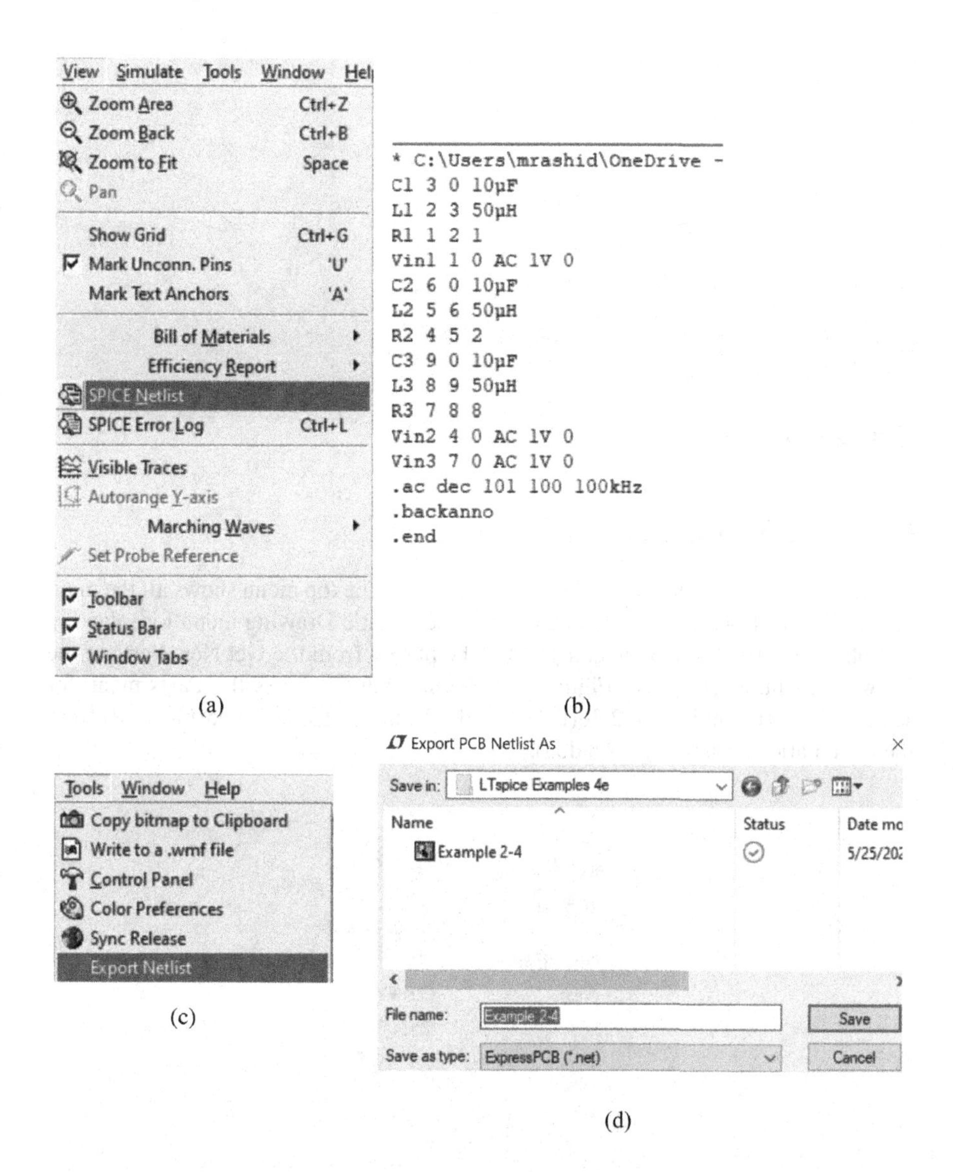

FIGURE 2.33 LTspice Netlist. (a) View menu for SPICE Netlist. (b) Netlist for Figure 2.14(a). (c) Tools menu for Export Netlist. (d) Netlist File.

The OrCAD Capture software package, which is similar to PSpice Schematics, also has three major interactive programs: Capture, PSpice A/D, and Probe. Capture is a powerful program that lets us build circuits by drawing them within a window on the monitor. PSpice A/D lets us analyze the circuit created by Capture and generate voltage and current solutions. Probe is a graphic postprocessor and lets us display plots of voltages, currents, impedance, and power.

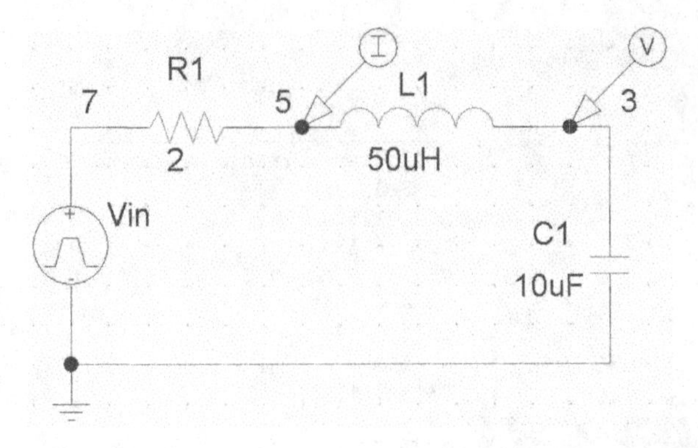

FIGURE 2.34 PSpice Schematics.

2.18.1 PSpice Schematics Layout

Figure 2.34 shows the layout of PSpice Schematics. The top menu shows all the main menus. The right-hand-side menu shows the Schematic Drawing menu for selecting and placing parts. For example, a part can be placed from the Get New Parts of the Drawing menu as shown in Figure 2.35. Figure 2.36(a) shows the Parts menu for selecting a part, and Figure 2.36(b) shows the menu for placing a part (e.g., *R*) from the Schematics library (e.g., Analog).

Draw	Navigate View Optior
Repeat	Space
Place Part	Ctrl+P
Wire	Ctrl+W
Bus	Ctrl+B
Block	
Arc	
Circle	
Box	
Polyline	
Text...	Ctrl+T
Text Box	
Insert Picture...	
Get New Part...	Ctrl+G
Rewire	Ctrl+D

FIGURE 2.35 Draw menu for "Get New Parts."

(a)

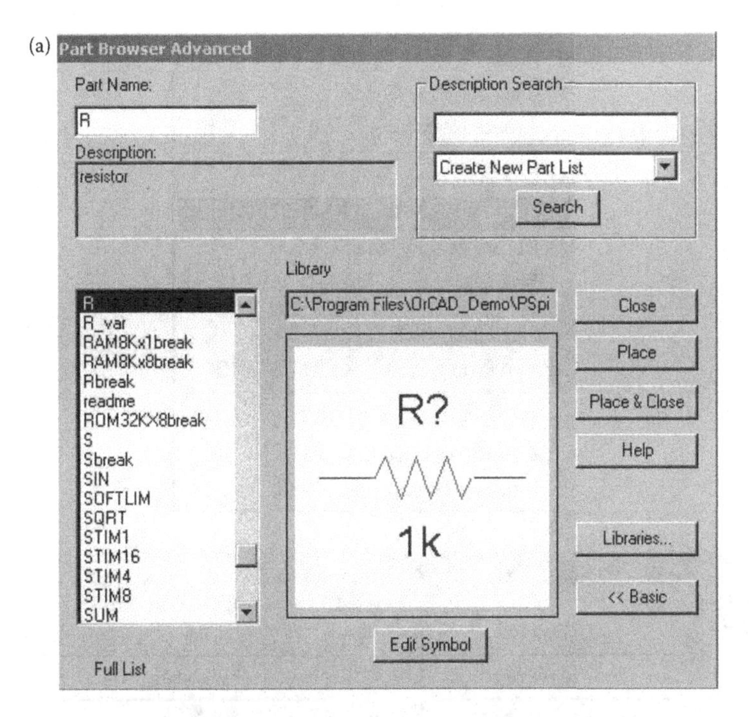

(b)

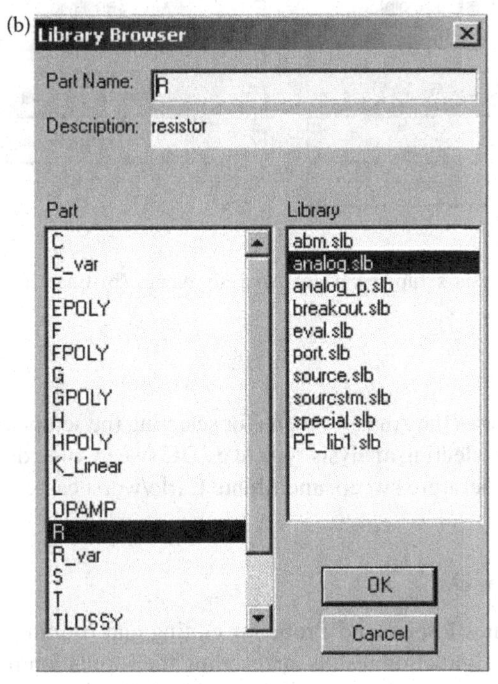

FIGURE 2.36 Parts and library menus. (a) Part menu. (b) Selecting a part from the Schematics library.

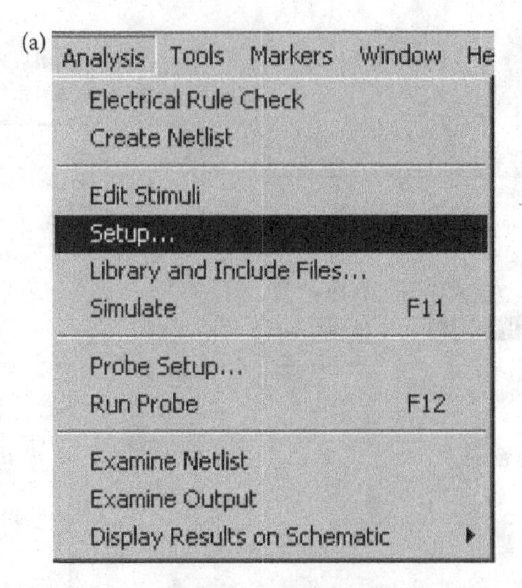

FIGURE 2.37 Analysis setup menu. (a) Analysis menu. (b) Enabling the analysis type: Transient.

Figure 2.37(a) shows the Analysis menu for selecting the setup, and Figure 2.37(b) shows the menu for selecting analysis type (e.g., DC sweep, time domain, AC sweep, and bias point), temperature sweep, and Monte Carlo/worst case.

2.18.2 PSpice A/D

PSpice A/D combines PSpice and Probe for editing and running a simulation file, viewing output and simulation results, and setting the simulation profile. Its menu is identical to that of OrCAD Capture and is shown in Figure 2.38(a). The left-hand-side menu as shown in Figure 2.38(b) allows viewing the simulation results and setting simulation profiles.

(a)

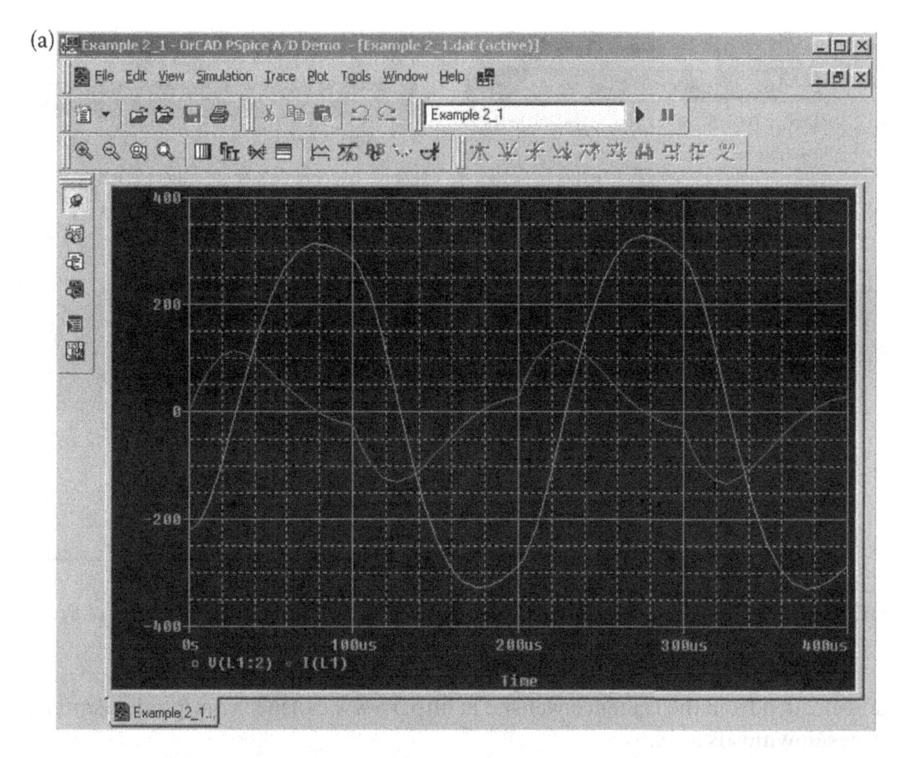

(b)

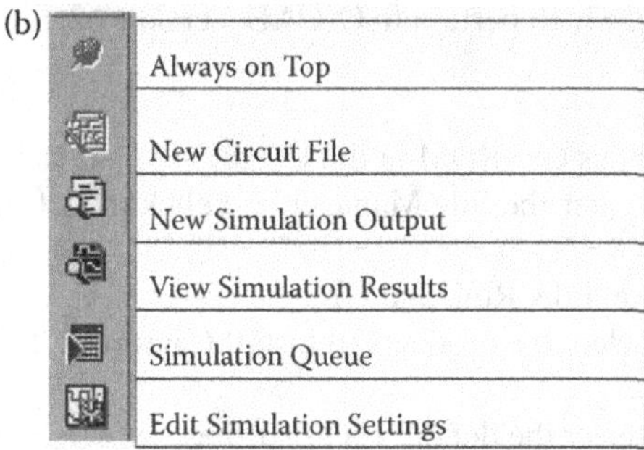

FIGURE 2.38 PSpice menu. (A) PSpice A/D menu. (b) PSpice A/D viewing and setting menus.

2.18.3 Probe

Probe menu is included in PSpice A/D. Figure 2.39(a) shows the trace menu for plotting the output variables, and Figure 2.39(b) shows the menu for setting the plot axes and the number of plots.

2.18.4 OrCAD Capture

Software Installation: The instructions for installing the OrCAD software are printed on the OrCAD installation disk or the CD-ROM. The software can be

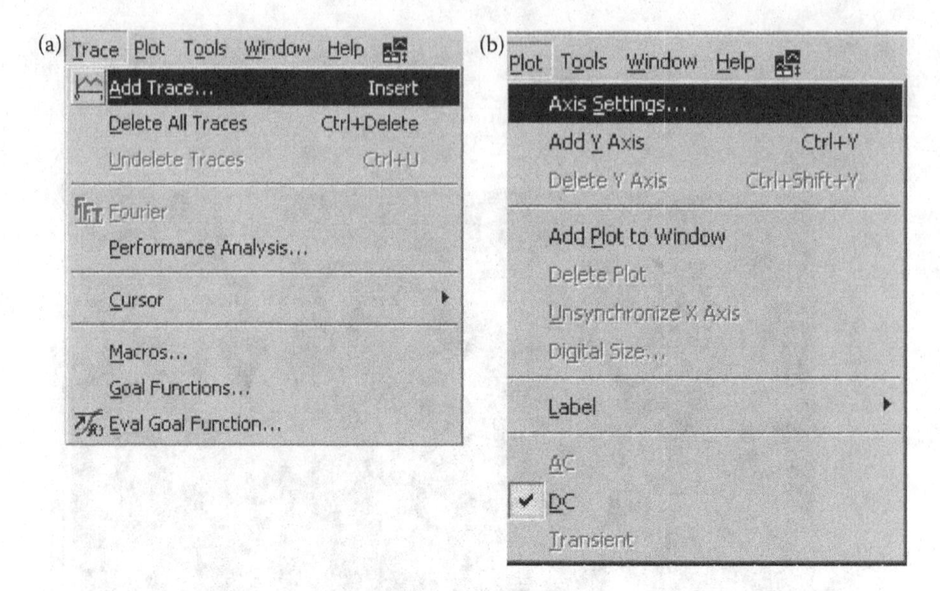

FIGURE 2.39 Probe menu of PSpice A/D. (a) Trace menu of Probe. (b) Plot menu of Probe.

downloaded from the Cadence website at http://www.cadence.com/products/orcad/Pages/downloads.aspx.

The following steps, which pertain to OrCAD Version 9.2, are applicable to other versions as well:

1. Place the schematics CD-ROM in the CD drive.
2. From Windows, enter the File Manager and click the left mouse button on the CD drive.
3. Click on setup.exe, File, Run, and OK.
4. Click on OK to select the products to install Capture and PSpice, as shown in Figure 2.40(a).
5. Click on OK to select the default C:OrcadLite.
6. Click on Yes to choose Program Folder: Orcad Family Release 9.2 Lite Edition, as shown in Figure 2.40(b).
7. Create the OrCAD Capture Lite and PSpice A/D icons from the Start menu; then go to Programs, Orcad Family, and Capture Lite menu.
8. Click the left mouse button on the Capture Lite icon once, and the window of Capture Lite will open.

Note: Although there are later versions of OrCAD, the 9.2 version has proven to be bug-free and works well.

General Layout: The general layout of Capture is shown in Figure 2.41. The top menu shows ten main choices. The right-hand-side menu shows the schematic "drawing" menu for selecting and placing parts. File, Edit, View, Place, and PSpice are most frequently used. For any help, click on the Help menu.

(a)
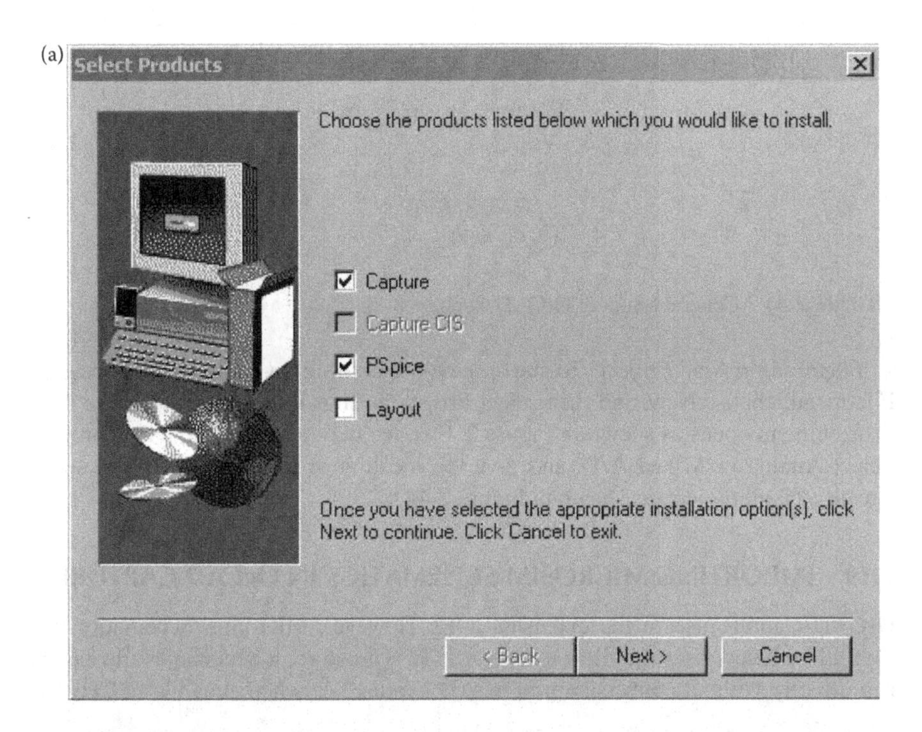

(b)
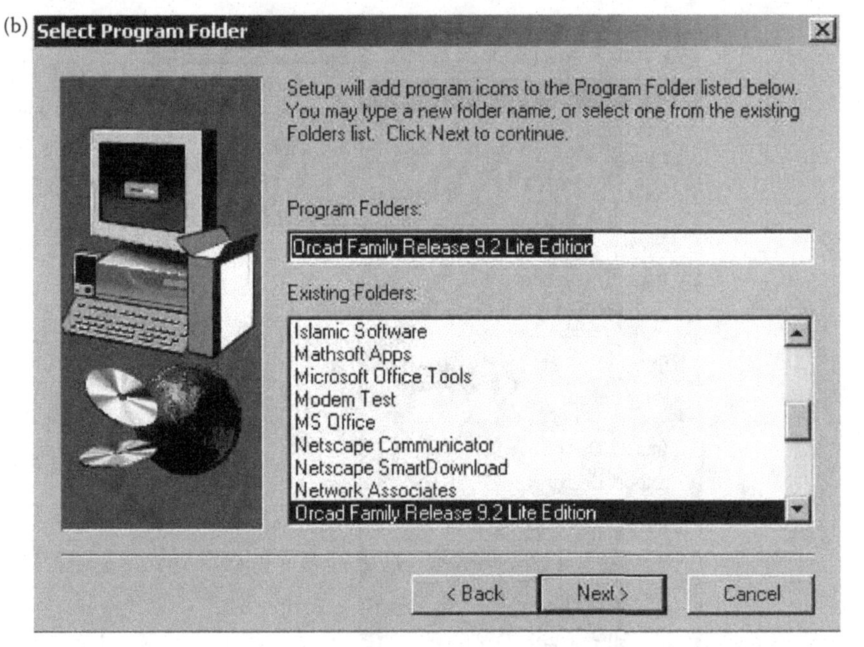

FIGURE 2.40 Installation setup for OrCAD Capture. (a) Selecting products. (b) Selecting the program folder.

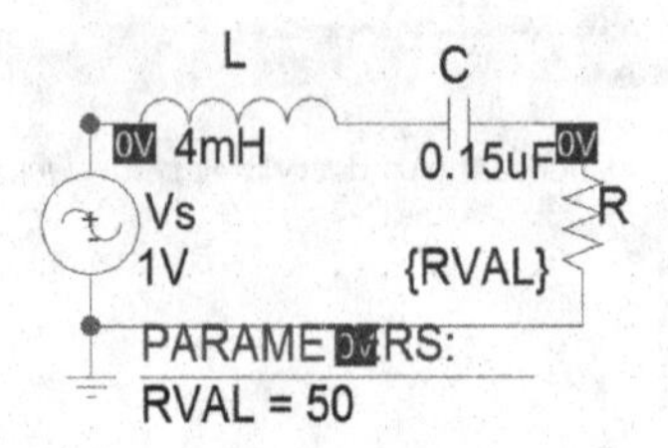

FIGURE 2.41 General layout of OrCAD Capture.

Beginning a New Project: To start a new project to draw and analyze; open the File menu, choose New, and then select Project as shown in Figure 2.42. The New Project menu opens as shown in Figure 2.43. Give the file name of the new project. Select Analog or Mixed A/D, and give the location of this new file. Next, select Create a blank project, as shown in Figure 2.44.

2.19 IMPORTING MICROSIM SCHEMATICS IN ORCAD CAPTURE

OrCAD Capture uses a file extension .OPJ. However, MicroSim Schematics version 9.1 or below uses the file extension .SCH. Therefore, a Schematics file can be run directly without conversion on OrCAD capture. A Schematics file (.SCH) can

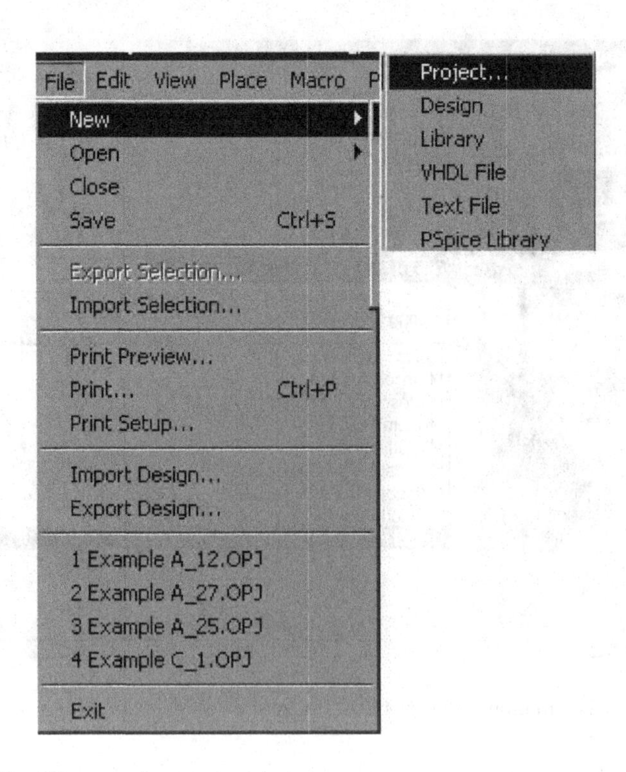

FIGURE 2.42 File menu for a new project.

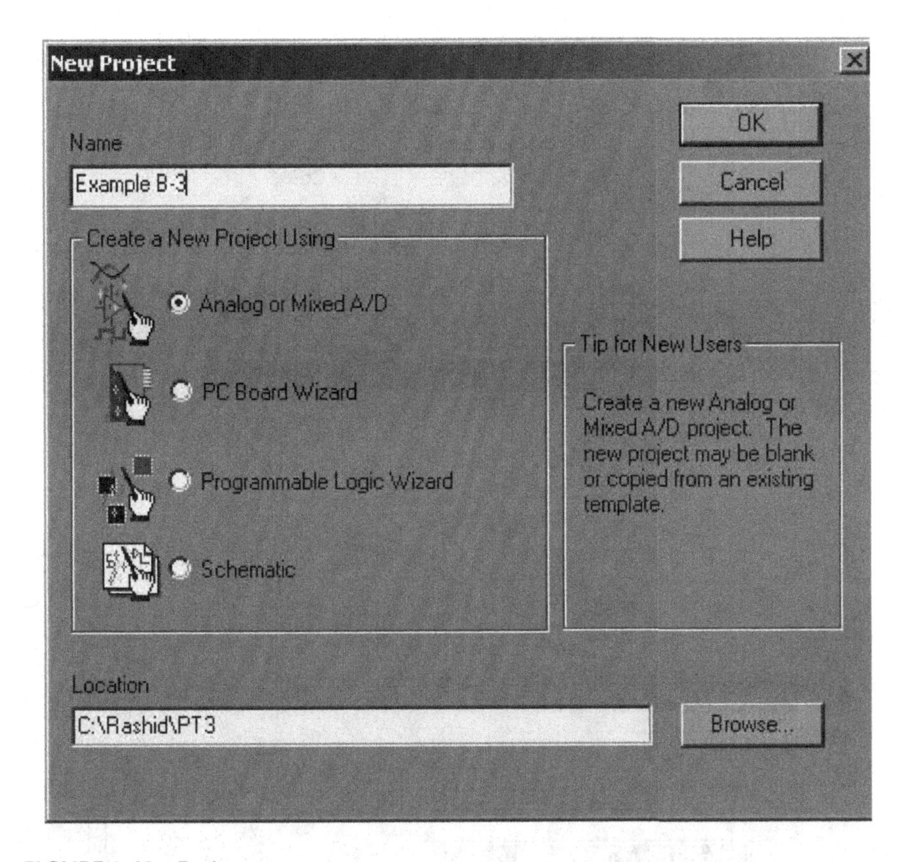

FIGURE 2.43 Project menu.

be imported to OrCAD Capture from its File menu as shown in Figure 2.45(a). It requires identifying the location of the following files:

- The name and the location of the Schematics file with .SCH extension.
- The name and the location of the Capture file with .OPJ extension, where the imported file will be saved as shown in Figure 2.45(b).

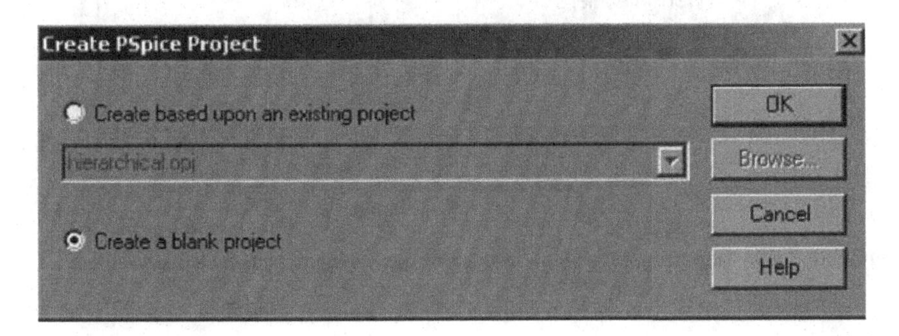

FIGURE 2.44 Create a blank project.

(a)

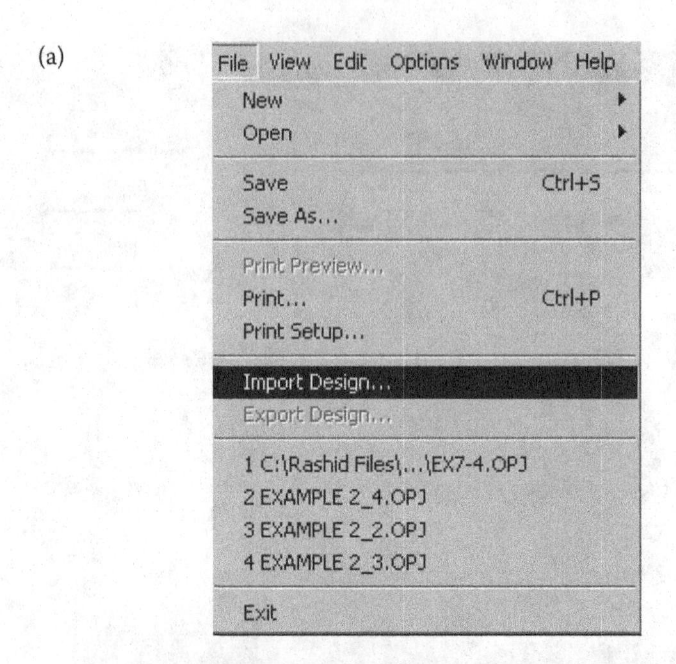

(b)

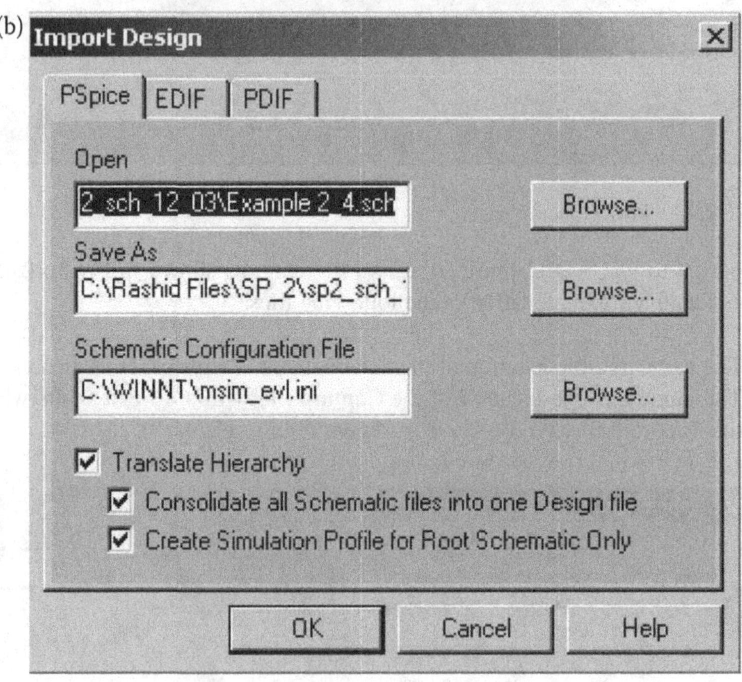

FIGURE 2.45 Importing MicroSim Schematics file. (a) File menu. (b) Design import.

- The name and the location of the schematic configuration file with .INI extension. Once the conversion is completed, the imported file with .OPJ extension can be simulated on OrCAD Capture.

PROBLEMS

2.1 The *RLC* circuit of Figure P2.1 is to be simulated to calculate and plot the transient response from 0 to 2 ms with an increment of 10 μs. The voltage across resistor *R* is the output. The input and output voltages are to be plotted on an output file. The results should also be available for display and as hard copy using the .PROBE command.

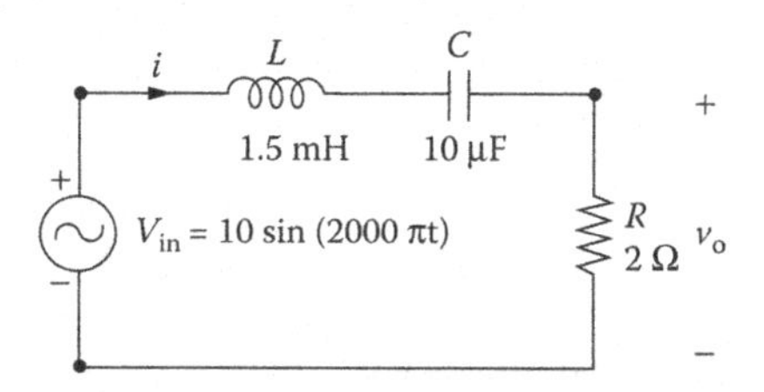

FIGURE P2.1 Series *RLC* circuit.

2.2 Repeat Problem 2.1 for the circuit of Figure P2.2, where the output is taken across capacitor *C*.

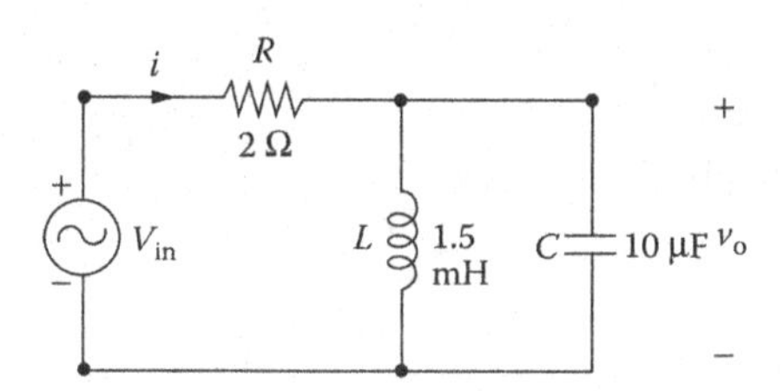

FIGURE P2.2 Series and parallel *RLC* circuit

2.3 Repeat Problem 2.1 for the circuit of Figure P2.3, where the output is the current i_s through the circuit.

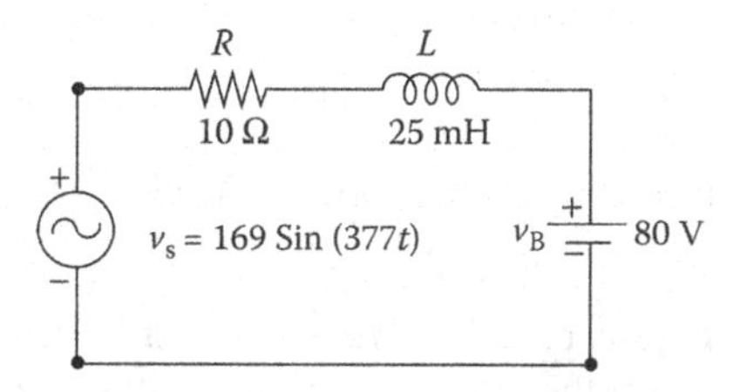

FIGURE P2.3 Series *RLC* circuit.

2.4 Repeat Problem 2.1 if the input is a step input, as shown in Figure 2.3b.

2.5 Repeat Problem 2.2 if the input is a step input, as shown in Figure 2.3b.

2.6 The *RLC* circuit of Figure P2.6a is to be simulated to calculate and plot the transient response from 0 to 2 ms with an increment of 5 μs. The input is a step current, as shown in Figure P2.6b. The voltage across resistor *R* is the output. The input and output voltages are to be plotted on an output file. The results should also be available for display and as hard copy, using the .PROBE command.

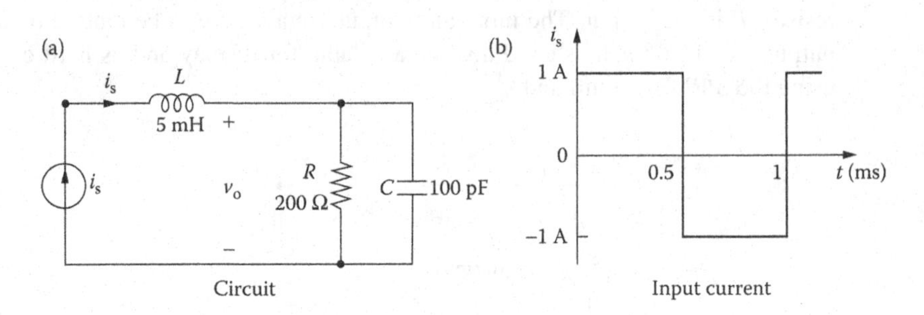

FIGURE P2.6 Series and parallel RLC circuit.

2.7 Repeat Problem 2.6 for the circuit of Figure P2.7, where the output is taken across capacitor *C*.

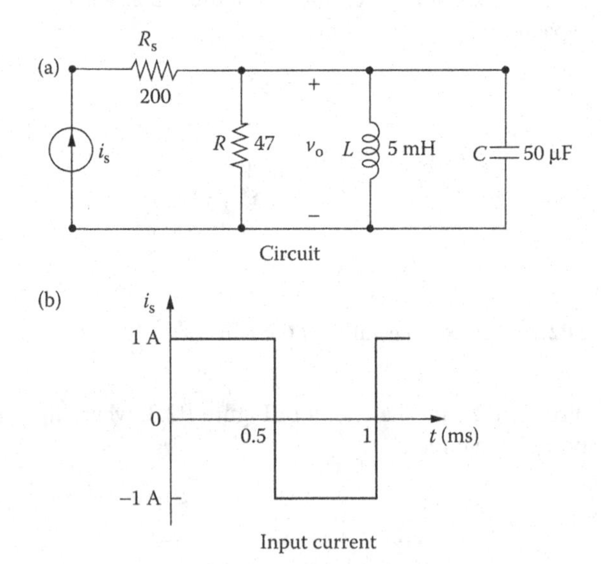

FIGURE P2.7 RLC circuit with a current source.

2.8 The circuit of Figure P2.2 is to be simulated to calculate and print the frequency response over the frequency range 10–100 kHz with a decade increment and 10 points per decade. The peak magnitude and phase angle of the voltage across the resistor are to be printed on the output file. The results should also be available for display and as hard copy, using the .PROBE command.

2.9 Repeat Problem 2.8 for the circuit of Figure P2.6.

2.10 Repeat Problem 2.8 for the circuit of Figure P2.7.

2.11 Repeat Problem 2.8 for the circuit of Figure P2.11.

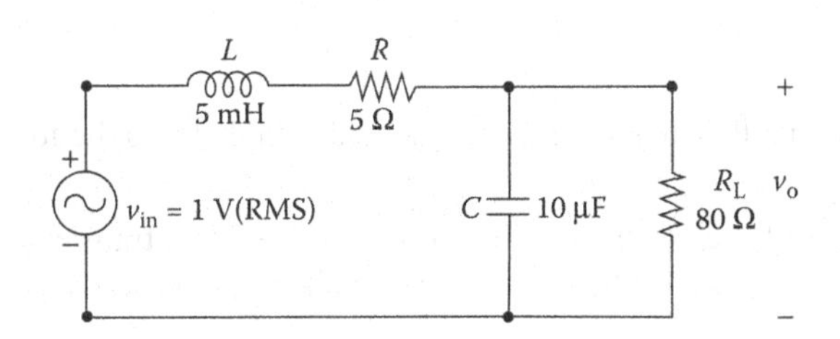

FIGURE P2.11 *RLC* circuit with a parallel load.

SUGGESTED READING

1. M.H. Rashid, *SPICE For Power Electronics and Electric Power*, Englewood Cliff, New Jersey: Prentice-Hall, Inc, 1993, 2/e.
2. M.H. Rashid, *Introduction to PSpice using OrCAD for circuits and electronics*, Upper Saddle River, NJ: Pearson/Prentice Hall, 3/e, 2004.
3. P.W. Tuinenga, *SPICE: A Guide to Circuit Simulation and Analysis Using PSPICE*, New Jersey: Prentice-Hall Inc., 1995, 3/e.
4. J.W. Nilsson and S.A. Riedel, *Introduction to PSpice manual, Electric circuits, using ORCad release 9.2*, Upper Saddle River, N.J.: Prentice Hall, 2002.
5. J.W. Nilsson and S.A. Riedel, *Introduction to PSpice manual, Electric Circuits, Using ORCad Release 9.1*, Upper Saddle River, N.J.: Prentice Hall, 2000, 4/e.
6. J.W. Nilsson and S.A. Riedel, *Introduction to PSpice Manual, Using ORCad Release 9.2., Accompany Electric Circuits*, Seventh Edition, Upper Saddle River, N.J.: Prentice Hall, 2005.
7. J.F. Morris, *Introduction to PSpice with Student Exercise Disk to Accompany Cunningham/Stuller Basic Circuit Analysis*, Boston: Houghton Mifflin, 1991.
8. S. Moslehpour, *Circuit Stimulation and Analysis: An Introduction to Computer-Aided Circuit Design Using PSpice Software*, Newington, CT: ARRL, 2013.
9. J. Keown, *OrCAD PSpice and Circuit Analysis*, Upper Saddle River, NJ: Prentice Hall, 2001, 4/e.
10. D. Fitzpatrick, *Analog Design and Simulation Using OrCAD Capture and PSpice*, Oxford, UK; Waltham, MA: Newnes, 2012.
11. K. Mitzner, *Complete PCB Design Using OrCAD Capture and PCB Editor*, Amsterdam, Boston: Newnes/Elsevier, 2009.
12. A. Vladimirescu, *The Spice Book*, New York: J. Wiley, 1994.
13. G.W. Roberts and A.S. Sedra. *Spice*, New York: Oxford University Press, 1997, 2/e
14. *PSpice Manual*, Irvine, CA: MicroSim Corporation, 1992.
15. *LTspice IV Manual*, Milpitas, CA: Linear Technology Corporation, 2013.

3 Defining Output Variables

After completing this chapter, students should be able to do the following:

- Define and specify the output variables for AC and transient analysis.
- Define and specify the magnitude and phase angles of output variables for AC analysis.
- Define and specify the output variables for noise analysis.

3.1 INTRODUCTION

PSpice has some unique features for printing or plotting output voltages or currents by **.PRINT** and **.PLOT** statements. These statements, which may have up to eight output variables, are discussed in Chapter 6. The output variables that are allowed depend on the type of analysis:

- DC sweep and transient analysis
- AC analysis
- Noise analysis

Note: The circuit elements can be described in a schematic file or a netlist file. The schematic layout might be different for each circuit simulator. However, almost all of the SPICE simulators such as PSpice, OrCAD, LTspice, Intrusoft SPICE, and Multisim can run the simulation from the netlist file.

3.2 DC SWEEP AND TRANSIENT ANALYSIS

DC sweep and transient analysis use the same type of output variables. The variables can be divided into two types: voltage output and current output. A variable can be assigned the symbol or terminal symbol of a device (or element) to identify whether the output is the voltage across the device (or element) or the current through it. Table 3.1 shows the symbols for two-terminal elements. Table 3.2 shows the symbols and terminal symbols for three- and four-terminal devices.

LTspice: By adding "X" before the element name, we can use user-defined models similar to the Breakout devices in PSpice and OrCAD. For example, resistor R1 is named as XR1 and its model parameter are specified in the model statement.

Note: Ctrl+R-Click on the element to view the component attribute editor and the initial conditions.

DOI: 10.1201/9781003284451-3

TABLE 3.1
Symbols for Two-Terminal Elements

First Letter	Element
C	Capacitor
D	Diode
E	Voltage-controlled voltage source
F	Current-controlled current source
G	Voltage-controlled current source
H	Current-controlled voltage source
I	Independent current source
L	Inductor
R	Resistor
V	Independent voltage source

TABLE 3.2
Device Symbols and Terminal Symbols for Three- and Four-Terminal Devices.

First Letter	Device	Terminals
B	GaAs MESFET	D(drain) G(gate) S(source)
J	JFET	D(drain) G(gate) S(source)
M	MOSFET	D(drain) G(gate) S(source)
Q	BJT	C(collector) B(base) E(emitter) S(substrate)
Z	IGBT	G(gate) C(collector) E(emitter)

3.2.1 Voltage Output

The output voltages for DC sweep and transient analysis can be obtained by the following statements:

V(<node>)	Voltage at <node> with respect to ground
V(N1,N2)	Voltage at node N_1 with respect to node N_2
V(<name>)	Voltage across two-terminal device, <name>

V_x(<name>) Voltage at terminal x of three-terminal device, <name>
V_{xy}(<name>) Voltage across terminals x and y of three-terminal device, <name>
V_z(<name>) Voltage at port z of transmission line, <name>

The meaning of the PSpice variables is given in the following text:

PSpice Variables	Meaning
V(0)	Voltage at the ground terminal
V(R1:1)	Voltage at terminal 1 of resistor $R1$
V(R1:2)	Voltage at terminal 2 of resistor $R1$
V1(R1)	Voltage at terminal 1 of resistor $R1$
V2(R1)	Voltage at terminal 2 of resistor $R1$
V(Vin+)	Voltage at positive terminal (+) of voltage source V_{in}
V(Vin-)	Voltage at negative terminal (–) of voltage source V_{in}
V(5)	Voltage at node 5 with respect to ground
V(4,2)	Voltage of node 4 with respect to node 2
V(R1)	Voltage of resistor R_1, where the first node (as defined in the circuit file) is positive with respect to the second node
V(L1)	Voltage of inductor L_1, where the first node (as defined in the circuit file) is positive with respect to the second node
V(C1)	Voltage of capacitor C_1, where the first node (as defined in the circuit file) is positive with respect to the second node
V(D1)	Voltage across diode D_1, where the anode positive is positive with respect to the cathode
VC(Q3)	Voltage at the collector of transistor Q_3, with respect to ground
VDS(M6)	Drain–source voltage of MOSFET M_6
VB(T1)	Voltage at port B of transmission line $T1$

Note: SPICE and some versions of PSpice do not permit measuring voltage across a resistor, an inductor, and a capacitor (e.g., V(R1), V(L1), and V(C1)). This type of statement is applicable only to outputs by **.PLOT** and **.PRINT** commands.

3.2.2 Current Output

The output currents for DC sweep and transient analysis can be obtained by the following statements:

I(<name>) Current through <name>
Ix(<name>) Current into terminal x of <name>
Iz(<name>) Current at port z of transmission line <name>

The meaning of the PSpice variables is given in the following text:

PSpice Variables	Meaning
I(R1)	Current through the resistor R_1, flowing from terminal 1 to terminal 2
I(Vin)	Current flowing from the positive terminal to the negative terminal of the source V_{in}
I(VS)	Current flowing into DC source V_s

(Continued)

PSpice Variables	Meaning
I(R5)	Current flowing into resistor R_5, where the current is assumed to flow from the first node (as defined in the circuit file) through R_5 to the second node
I(D1)	Current into diode D_1
IC(Q4)	Current into the collector of transistor Q_4
IG(J1)	Current into gate of JFET J_1
ID(M5)	Current into drain of MOSFET M_5
IA(T1)	Current at port A of transmission line T_1

Note: SPICE and some versions of PSpice do not permit measuring the current through a resistor (e.g., I(R5)). The easiest way is to add a dummy voltage source of 0 V (say, VX = 0 V) and to measure the current through that source (e.g., I(VX)).

3.2.3 Power Output

The following statement can give the instantaneous power dissipation (VI) of an element for DC and transient analysis:

W(<name>) Power absorbed by the element <name>

The meaning of the PSpice variables is given in the following text:

PSpice Variables	Meaning
W(R1)	Apparent power (VI product) of resistor R_1
W(Vin)	Apparent power (VI product) of source V_{in}

LTspice: LTspice does not support .PRTNT command for printing on the output file. For both .PRINT and .PLOT commands, LTspice opens the plot window and displays the available output variable to plot. However, .MEAS command can be used to evaluate user-defined electrical quantities. The general statement to measure a point along the abscissa (the independent variable plotted along the horizontal axis, i.e., the time axis of a .tran analysis) is given by

```
.MEAS[SURE]  [AC|DC|OP|TRAN|TF|NOISE] <name>
+  [<FIND|DERIV|PARAM> <expr>]
+  [WHEN <expr>  | AT=<expr>]]
+  [TD=<val1>]  [<RISE|FALL|CROSS>=[<count1>|LAST]]
```

For Example,

```
.MEAS TRAN res1 FIND V(out) AT=5m to print the value of V(out)
at t=5ms labeled as res1.
```

- LTspice has a built-in command to open the plot menu and display the available output variables to plot.
- LTspice has an Expression Editor and can also perform similar functions as those of .Probe. The R-click on the mouse opens the Expression Editor as shown in Figure 3.1 to type the expression.

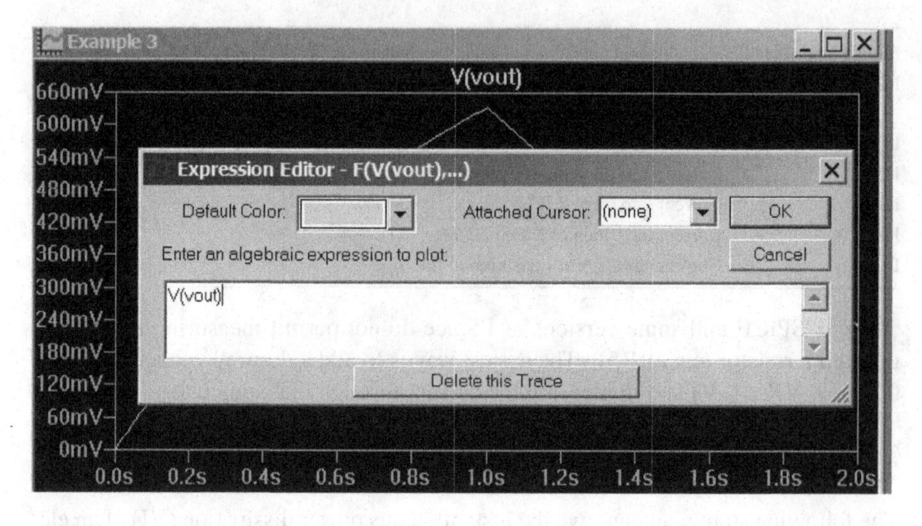

FIGURE 3.1 LTpsice Expression Editor.

Example 3.1:

Defining the Output Variables of a BJT Circuit

A DC circuit with a bipolar transistor is shown in Figure 3.2. Write the various currents and voltages in the forms that are allowed by PSpice. The DC sources of 0 V are introduced to measure currents I_1 and I_2.

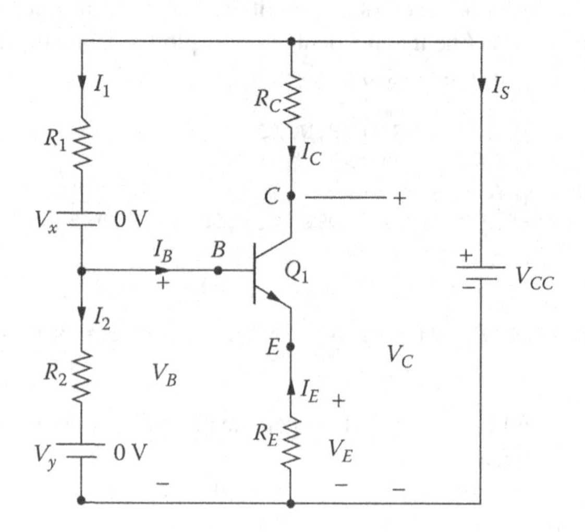

FIGURE 3.2 DC circuit with a bipolar transistor.

SOLUTION

Symbol	PSpice Variable	Meaning
I_B	IB(Q1)	Base current of transistor Q_1
I_C	IC(Q1)	Collector current of transistor Q_1
I_E	IE(Q1)	Emitter current of transistor Q_1
I_S	I(VCC)	Current through voltage source V_{cc}
I_1	I(VX)	Current through voltage source V_x
I_2	I(VY)	Current through voltage source V_y
V_B	VB(Q1)	Voltage at the base of transistor Q_1
V_C	VC(Q1)	Voltage at the collector of transistor Q_1
V_E	VE(Q1)	Voltage at the emitter of transistor Q_1
V_{CE}	VCE(Q1)	Collector–emitter voltage of transistor Q_1
V_{BE}	VBE(Q1)	Base–emitter voltage of transistor Q_1

LTspice: LTspice supports almost all of these output variables. If you are in doubt, try it, and LTspice will ignore or give a warning message in the output file. LTspice schematic for Figure 3.2 is shown in Figure 3.3. One can assign the node numbers as shown in Figure 3.3 that has the advantage of knowing which node is what. Otherwise, LTspice assigns node numbers. Whereas PSpice/OrCAD automatically assigns node numbers, and the user has no control over the numbering.

Example 3.2:

Defining the Output Variables of an RLC *Circuit*

An *RLC* circuit with a step input is shown in Figure 3.4. Write the various currents and voltages in the forms that are allowed by PSpice.

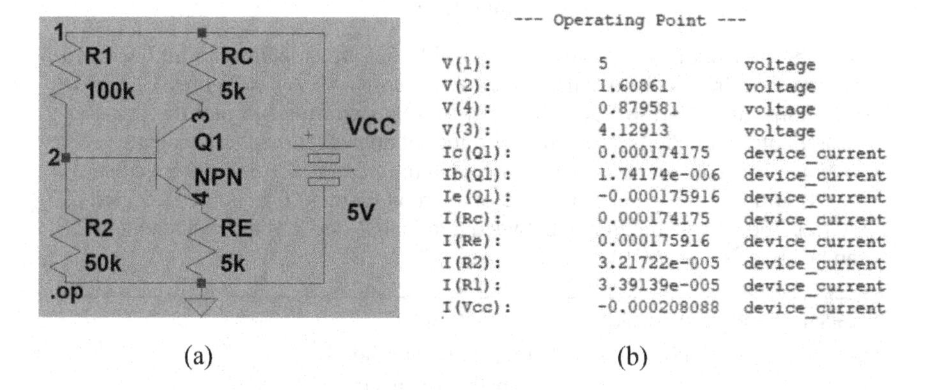

FIGURE 3.3 LTspice schematic for Figure 3.2. (a) LTspice shematic. (b) Output varibales for .op command.

FIGURE 3.4 RLC circuit with a step input.

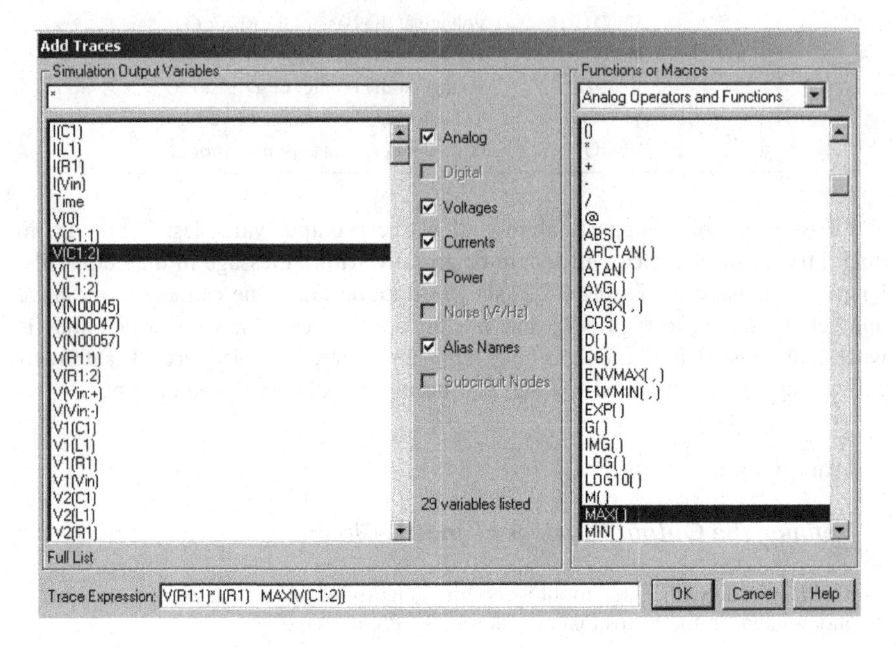

FIGURE 3.5 Simulation output variables.

SOLUTION

After the simulation is completed, all possible output variables can be displayed from the Add Traces menu of the PSpice A/D menu as shown in Figure 3.5. Analog operations, as shown in Figure 3.5, can also be performed to plot the traces of certain output functions. Note that for transient analysis, time is a variable.

Note: The voltage at terminal 1 of a circuit element C1 can be specified as V(C1:1) and the voltage at terminal 2 of a circuit element C1 can be specified as V(C1:2). The voltage difference between terminals 1 and 2 of a circuit element C1 can be specified as V(C1:1, C1:2).

Symbol	PSpice Variable	Meaning
i_R	I(R)	Current through resistor *R*
i_L	I(L)	Current through inductor *L*
i_C	I(C)	Current through capacitor *C*

(Continued)

Symbol	PSpice Variable	Meaning
i_{in}	I(VIN)	Current flowing into voltage source v_{in}
v_3	V(3)	Voltage of node 3 with respect to ground
v_{23}	V(2,3)	Voltage of node 2 with respect to node 3
v_{12}	V(1,2)	Voltage of node 1 with respect to node 2
v_R	V(R)	Voltage of resistor *R* where the first node (as defined in the circuit file) is positive with respect to the second node
v_L	V(L)	Voltage of inductor *L* where the first node (as defined in the circuit file) is positive with respect to the second node
v_C	V(C)	Voltage of inductor *L* where the first node (as defined in the circuit file) is positive with respect to the second node

Note: SPICE and some versions of PSpice do not permit measuring voltage across a resistor, an inductor, and a capacitor (e.g., V(R1), V(L1), and V(C1)). This type of statement is applicable only to outputs by **.PLOT** and **.PRINT** commands.

3.3 AC ANALYSIS

In AC analysis, the output variables are sinusoidal quantities and are represented by complex numbers. An output variable can have magnitude, magnitude in decibels, phase, group delay, real part, and imaginary part. A suffix is added to the output variables listed in Sections 3.3.1 and 3.3.2 as follows:

Suffix	Meaning
(none)	Peak magnitude
M	Peak magnitude
DB	Peak magnitude in decibels
P	Phase in radians
G	Group delay ($-\delta_{phase}/\delta_{frequency}$)
R	Real part
I	Imaginary part

3.3.1 VOLTAGE OUTPUT

The statement variables for AC analysis are similar to those for DC sweep and transient analysis, provided that the suffixes are added as illustrated:

PSpice Variable	Meaning
VP(R1:1)	Phase angle of the voltage at terminal 1 of resistor R_1
VP(R1:2)	Phase angle of the voltage at terminal 2 of resistor R_1
VM(R1:1)	Magnitude of the voltage at terminal 1 of resistor R_1
VM(R1:2)	Magnitude of the voltage at terminal 2 of resistor R_1
VP(Vin+)	Phase angle of the voltage at positive terminal (+) of voltage source V_{in}
VM(5)	Magnitude of voltage at node 5 with respect to ground

(Continued)

PSpice Variable	Meaning
VM(4,2)	Magnitude of voltage at node 4 with respect to node 2
VDB(R1)	Decibel magnitude of voltage across resistor R_1, where the first node (as defined in the circuit file) is assumed to be positive with respect to the second node
VP(D1)	Phase of anode voltage of diode D_1 with respect to cathode
VCM(Q3)	Magnitude of the collector voltage of transistor Q_3 with respect to ground
VDSP(M6)	Phase of the drain-source voltage of MOSFET M_6
VBP(T1)	Phase of voltage at port B of transmission line T_1
VR(2,3)	Real part of voltage at node 2 with respect to node 3
VI(2,3)	Imaginary part of voltage at node 2 with respect to node 3

3.3.2 Current Output

The statement variables for AC analysis are similar to those for DC sweep and transient responses. However, only the currents through the elements listed in Table 3.3 are available. For all other elements, a zero-valued voltage source must be placed in series with the device (or device terminal) of interest. Then a print or plot statement should be used to determine the current through this voltage source.

The meaning of the PSpice statement variables is given in the following text:

PSpice Variable	Meaning
IP(R1)	Phase angle of the current through the resistor R_1 flowing from terminal 1 to terminal 2
IM(R1)	Magnitude of the current through the resistor R_1 flowing from terminal 1 to terminal 2
IP(Vin+)	Phase angle of the current flowing from the positive terminal to the negative terminal of the source V_{in}
IM(R5)	Magnitude of current through resistor R_5
IR(R5)	Real part of current through resistor R_5
II(R5)	Imaginary part of current through resistor R_5
IM(VIN)	Magnitude of current through source v_{in}
IR(VIN)	Real part of current through source v_{in}
II(VIN))	Imaginary part of current through source v_{in}
IAG(T1)	Group delay of current at port A of transmission line T_1

TABLE 3.3
Current Through Elements for AC Analysis

First Letter	Element
C	Capacitor
I	Independent current source
L	Inductor
R	Resistor
T	Transmission line
V	Independent voltage source

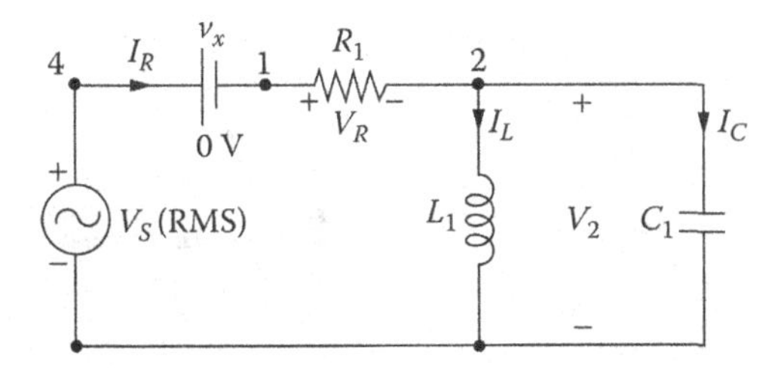

FIGURE 3.6 RLC circuit for Example 3.3.

Note: If the symbol *M* for magnitude is omitted for the AC analysis, PSpice recognizes the variable as the magnitude of a voltage or a current. That is, VM(R1:1) is identical to V(R1:1).

Example 3.3:

Defining the Output Variables of an RLC *Circuit for AC Analysis*

The frequency response is obtained for the *RLC* circuit in Figure 3.6. Write the various voltages and currents in the forms that are allowed by PSpice. The dummy voltage source of 0 V is introduced to measure current I_L.

SOLUTION

The meaning of the PSpice statement variables is given in the following table:

Symbol	PSpice Variable	Meaning
V_2	VM(2)	Peak magnitude of voltage at node 2
$\angle V_2$	VP(2)	Phase angle of voltage at node 2
V_{12}	VM(1,2)	Peak magnitude of voltage between nodes 1 and 2
$\angle V$	VP(1,2)	Phase angle of voltage between nodes 1 and 2
I_R	IM(VX)	Magnitude of current through voltage source V_x
$\angle I_R$	IP(VX)	Phase angle of current through voltage source V_x
I_L	IM(L1)	Magnitude of current through inductor L_1
$\angle L$	IP(L1)	Magnitude of current through inductor L_1
I_C	IM(C1)	Magnitude of current through capacitor C_1
$\angle I_C$	IP(C1)	Phase angle of current through capacitor C_1

3.4 OUTPUT MARKERS

In PSpice Schematics, we can select markers from the PSpice menu as shown in Figure 3.7(a) to display the voltages, currents, and power dissipation as shown in Figure 3.7(b). The advanced markers can be selected from the Markers menu to display the magnitude, phase, group delay, real part, and imaginary part as shown in Figure 3.7(c). Window Templates, as shown in Figure 3.7(d), can be used to plot certain functions such as average and derivatives as shown in Figure 3.7(e).

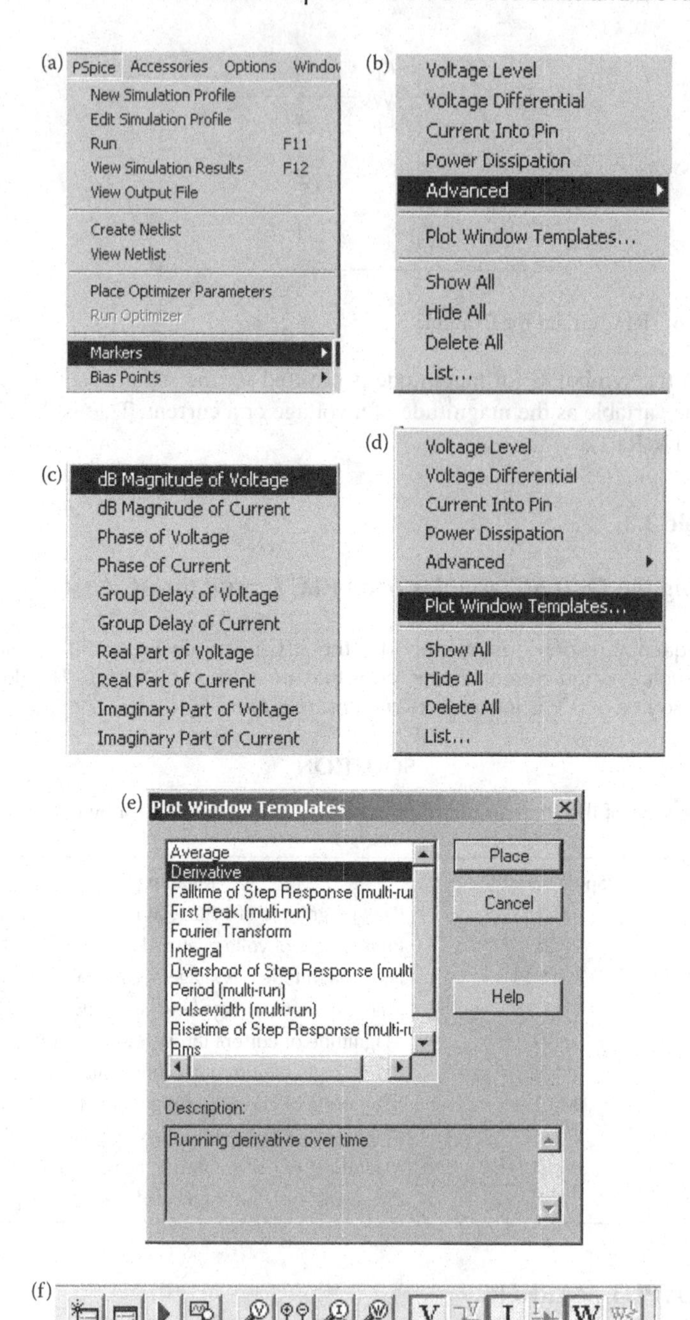

FIGURE 3.7 PSpice markers. (a) PSpice menu. (b) Markers menu. (c) Advanced menu (d) Window Templates menu. (e) Templates menu. (f) Marker and display icons.

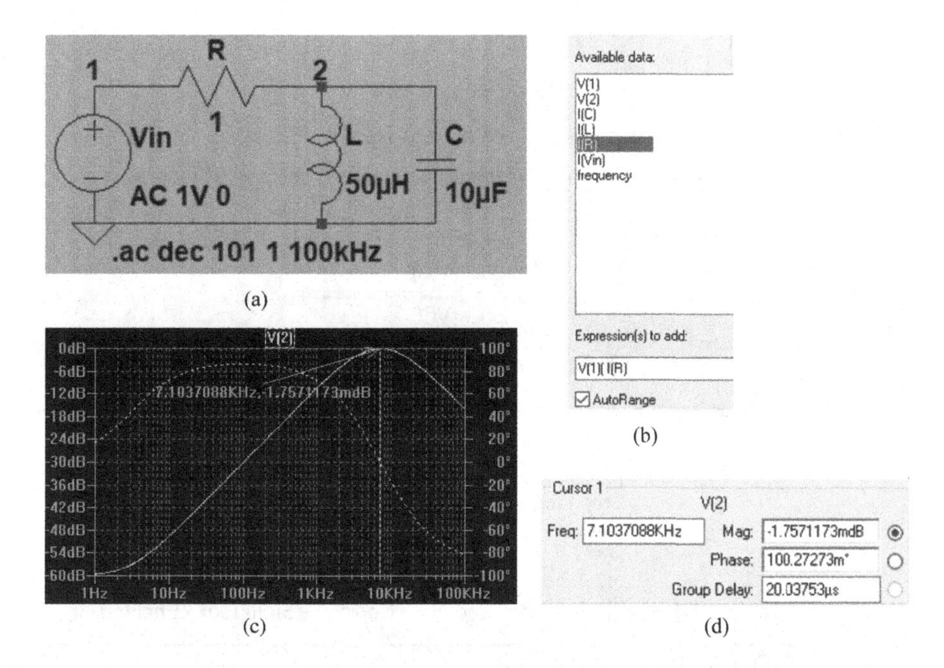

(a) (b) (c) (d)

FIGURE 3.8 LTspice AC analysis. (a) LTspice schematic. (b) Output variables. (c) Magnitude and phase angle of output voltage. (d) Magnitude and phase angle at a specified frequency.

We can also place the icons of the output markers and displays directly on the circuit terminals. PSpice will automatically translate the markers into PSpice output variables. Figure 3.7(f) shows the icons of three markers: voltage, voltage difference, and current and power dissipation, and also the icons of three displays: enable bias voltage, enable bias current, and enable bias power.

LTspice Output Variables: The LTspice schematic is shown in Figure 3.8(a). LTspice has no markers. However, after completing the simulation, it opens a menu showing the available output variables as shown in Figure 3.8(b). The magnitude and phase angle of the output voltage V(2) are shown in Figure 3.8(c) and the magnitude and phase angle at a specified frequency are shown in Figure 3.8(d).

3.5 NOISE ANALYSIS

For the noise analysis, the output variables are predefined as follows:

Output Variable	Meaning
ONOISE	Total RMS summed noise at output node
INOISE	ONOISE equivalent at the input node
DB(ONOISE)	ONOISE in decibels
DB(INOISE)	INOISE in decibels

Edit Simulation Command

Transient | AC Analysis | DC sweep | Noise | DC Transfer | DC op pnt

Perform a stochastic noise analysis of the circuit linearized about its DC operating point.

Output: V(3)
Input: VCC
Type of Sweep: Decade
Number of points per decade: 101
Start Frequency: 1Hz
Stop Frequency: 100kHz

Syntax: .noise V(<out>[,<ref>]) <src> <oct, dec, lin> <Npoints> <StartFreq> <EndFreq>

.noise V(3) VCC dec 101 1Hz 100kHz

Cancel OK

FIGURE 3.9 LTspice Noise analysis menu.

The noise output statement is

```
.PRINTNOISE INOISE ONOISE
```

Note: The noise output from only one device cannot be obtained by a **.PRINT** or **.PLOT** command. However, the print interval on the **.NOISE** statement can be used to output this information. The **.NOISE** command is discussed in Section 6.8.

LTspice noise setup menu and the noise parameters for Figure 3.3(a) are shown in Figure 3.9.

LTSpice Mouse Button Actions: The following mouse actions are commonly used in drawing schematics and plot variables (Figure 3.10):

Rotate a component: "ctrl+R" or click the rotate button ( )
Modify Component Values: R-click of the component

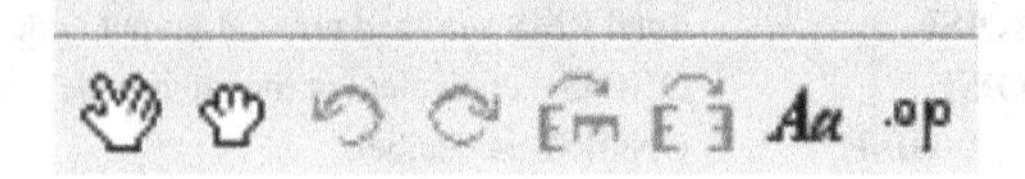

FIGURE 3.10 -LTspice mouse action buttons.

Move a component: "F7" or L-click on the component or move button ( )
Mirror a component: "Ctrl+E L-click on the component or mirror button ( )
Drag a component: Click the Drag button ( ) and select the component to move. You can drag multiple objects by dragging a box around them.
Measure voltage: Move the mouse over the node and L-Click.
Measure current: Move the mouse cursor over the component and L-Click.
Average and RMS Voltage/Current: "ctrl"+L-Click the plot variable, e.g., V(1) or I(L),
Instantaneous Power: "alt" + L-Click on the component.
Average Power: Selecting the instantaneous power of a component, then "ctrl" + L-Click on the name of the measurement in the measurement window.

SUMMARY

The PSpice variables can be summarized as follows:

V(<node>) Voltage at <node> with respect to ground
V(N1, N2) Voltage at node N_1 with respect to node N_2
V(<name>) Voltage across two-terminal device <name>
V_x(<name>) Voltage at terminal *x* of device <name>
V_{xy}(<name>) Voltage at terminal *x* with respect to terminal *y* for device <name>
V_z(<name>) Voltage at port *z* of transmission line <name>
I(<name>) Current through device <name>
I_x(<name>) Current into terminal *x* of device <name>
I_z(<name>) Current at port *z* of transmission line <name>
(none) Magnitude
M Magnitude
DB Magnitude in decibels
P Phase in radians
G Group delay($-\delta_{phase}/\delta_{frequency}$)
R Real part
I Imaginary part

4 Voltage and Current Sources

After completing this chapter, students should be able to do the following:

- Define and specify the output variables for AC and transient analysis.
- Define and specify the magnitude and phase angles of output variables for AC analysis.
- Define and specify the output variables for noise analysis.
- Model dependent and independent voltage and current sources.
- Specify dependent and independent voltage and current sources.
- Create behavioral modeling of voltage- and current-controlled sources with specifications in the form VALUE, TABLE, LAPLACE, FREQ, or as mathematical functions.

4.1 INTRODUCTION

PSpice allows generating dependent (or independent) voltage and current sources. An independent source can be time-variant. A nonlinear source can also be simulated by a polynomial. In this chapter, we explain the techniques for generating sources. The PSpice statements for various sources require the following:

- Source modeling
- Independent sources
- Dependent sources
- Behavioral device modeling

4.2 SOURCE MODELING

The independent voltage and current sources that can be modeled by PSpice are as follows:

- Pulse
- Piecewise linear
- Sinusoidal
- Exponential
- Single-frequency frequency modulation
- AC voltage

If you want to run more than one analysis type, including a transient analysis, then you need to use either a voltage source VSRC or a current source ISRC. The symbol and properties of a VSRC source are shown in Figures 4.4–4.12.

DOI: 10.1201/9781003284451-4

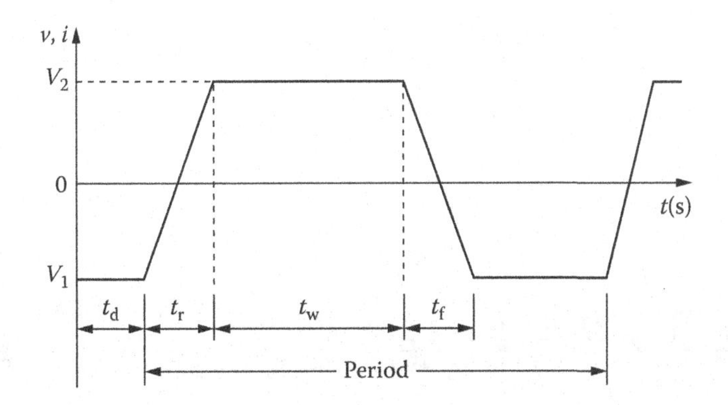

FIGURE 4.1 Pulse waveform.

4.2.1 Pulse Source

The waveform and parameters of a pulse waveform are shown in Figure 4.1 and Table 4.1. The schematic and the model parameters of a pulse source are shown in Figure 4.2(a). The model parameters that are shown in Table 4.1 can be changed from the menus as shown in Figure 4.2(b). In addition to the transient specifications, the DC (i.e., DC = 5 V) and AC (AC = 1 V) specifications can be assigned to the same source. The pulse source is used for the transient analysis of a circuit.

The symbol of a pulse source is PULSE and the general form is

```
PULSE (V1 V2 TD TR TF PW PER)
```

V1 and V2 must be specified by the user. TSTEP and TSTOP in Table 4.1 are the incrementing time and stop time, respectively, during transient (.TRAN) analysis.

4.2.1.1 Typical Statements

For $V_1 = -1$ V, $V_2 = 1$ V, $t_d = 2$ ns, $t_r = 2$ ns, $t_f = 2$ ns, pulse width = 50 ns, and period = 100 ns, the model statement is

```
PULSE (-1 1 2NS 2NS 2NS 50NS 100NS)
```

TABLE 4.1
Model Parameters of Pulse Sources

Name	Meaning	Unit	Default
V1	Initial voltage	V	None
V2	Pulsed voltage	V	None
TD	Delay time	s	0
TR	Rise time	s	TSTEP
TF	Fall time	s	TSTEP
PW	Pulse width	s	TSTOP
PER	PERIOD	s	TSTOP

(a) V1 = −220 V
V2 = 220 V
TD = 0
TR = 1 ns
TF = 1 ns
PW = 100 μs
PER = 200 μs

V1

(b) New Column... | Apply | Display... | Delete Property | Filter by: Orcad-PSpice

		Reference	Value	AC	DC	PER	PW	Source Part	TD	TF	TR	V1	V2
1	SCHEMATIC1 : PAGE1 : V1	V1	VPULSE	1	5	200us	100us	VPULSE.Normal	0	1ns	1ns	-220	220

FIGURE 4.2 PSpice schematic for a pulse source. (a) Symbol. (b) Editing model parameters.

With $V_1 = 0$, $V_2 = 1$, the model becomes

```
PULSE (0 1 2NS 2NS 2NS 50NS 100NS)
```

With $V_1 = 0$, $V_2 = -1$, the model becomes

```
PULSE (0 -1 2NS 2NS 2NS 50NS 100NS)
```

4.2.2 Piecewise Linear Source

A point in a waveform can be described by time T_i and its value V_i. Every pair of values (T_i, V_i) specifies the source value V_i at time T_i. The voltage at a time between the intermediate points is determined by PSpice by using linear interpolation.

The schematic of a piecewise linear source is shown in Figure 4.3(a), and the menu for setting the model parameters is shown in Figure 4.3(b). Up to 10 points (time, voltages, or currents) can be specified. In addition to the transient specifications, the

(a)

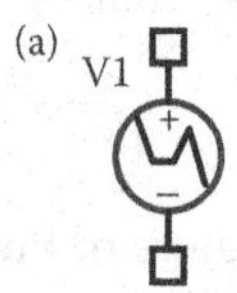

(b)

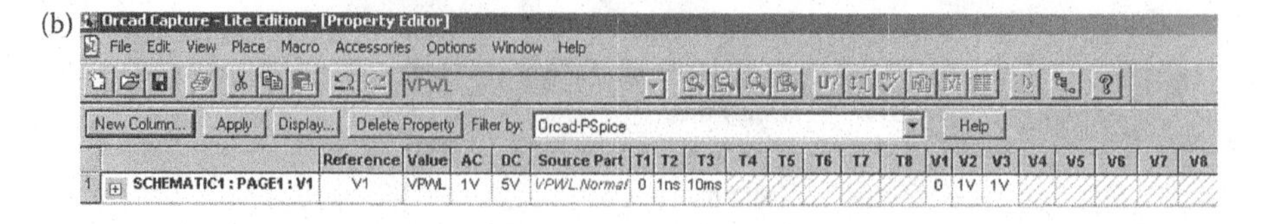

		Reference	Value	AC	DC	Source Part	T1	T2	T3	T4	T5	T6	T7	T8	V1	V2	V3	V4	V5	V6	V7	V8
1	SCHEMATIC1 : PAGE1 : V1	V1	VPWL	1V	5V	VPWL.Normal	0	1ns	10ms						0	1V	1V					

FIGURE 4.3 Piecewise linear source in PSpice Schematics. (a) Symbol. (b) Editing model parameters.

TABLE 4.2
Model Parameters of PWL Sources

Name	Meaning	Unit	Default
T_i	Time at a point	s	None
V_i	Voltage at a point	V	None

DC (i.e., DC = 5 V) and AC (AC = 1 V) specifications can be assigned to the same source.

The symbol of a piecewise linear source is PWL, and the general form is as follows (see Table 4.2):

```
PWL (T1 V1 T2 V2 ... TN VN)
```

4.2.2.1 Typical Statement

The model statement for the typical waveform of Figure 4.4 is

```
PWL (0 3 10US 3V 15US 6V 40US 6V 45US 2V 60US 0V)
```

4.2.3 Sinusoidal Source

The schematic and the model parameters of a sinusoidal source are shown in Figure 4.5(a). The model parameters that are shown in Table 4.3 can be changed from the menus as shown in Figure 4.5(b). In addition to the transient specifications, the DC (i.e., DC = 5 V) and AC (AC = 1 V) specifications can be assigned to the same source.

The symbol of a sinusoidal source is SIN and the general form is as follows (see Table 4.3):

```
SIN (VO VAMPL FREQ TD ALP THETA)
```

VO and VAMPL must be specified by the user. TSTOP in Table 4.3 is the stop time during transient (·TRAN) analysis. The waveform stays at 0 for a time of TD,

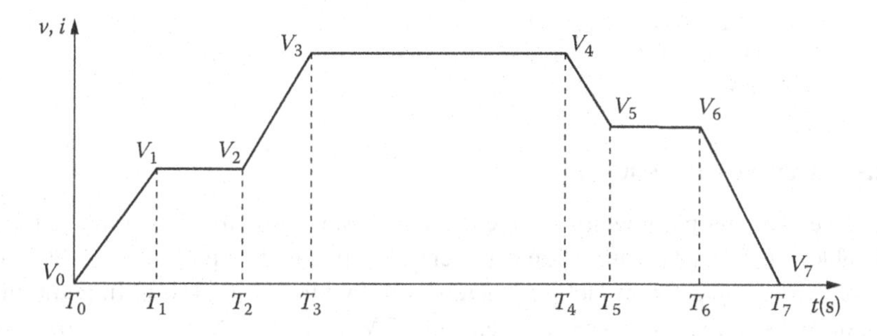

FIGURE 4.4 Piecewise linear waveform.

(a)

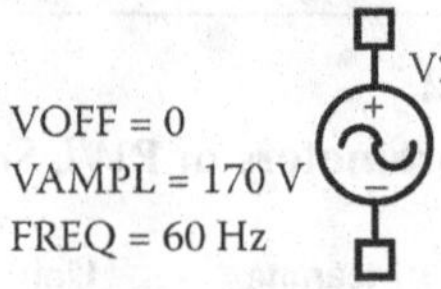

V2
VOFF = 0
VAMPL = 170 V
FREQ = 60 Hz

(b)

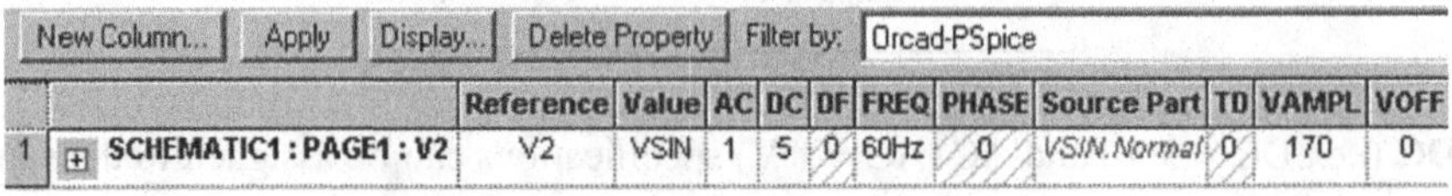

New Column... | Apply | Display... | Delete Property | Filter by: Orcad-PSpice

		Reference	Value	AC	DC	DF	FREQ	PHASE	Source Part	TD	VAMPL	VOFF
1	SCHEMATIC1 : PAGE1 : V2	V2	VSIN	1	5	0	60Hz	0	VSIN.Normal	0	170	0

FIGURE 4.5 Sinusoidal source. (a) Symbol. (b) Model parameters.

TABLE 4.3
Model Parameters of SIN sources

Name	Meaning	Unit	Default
VO	Offset voltage	V	None
VAMPL	Peak voltage	V	None
FREQ	Frequency	Hz	1/TSTOP
TD	Delay time	s	0
ALPHA	Damping factor	1/s	0
THETA	Phase delay	Degrees	0

and then the voltage becomes an exponentially damped sine wave. An exponentially damped sine wave is described by

$$V = V_0 + V_A e^{-\alpha(t-t_d)} \sin[(2\pi f(t - t_d)) - \theta]$$

and this is shown in Figure 4.6.

4.2.3.1 Typical Statements

```
SIN (0 1V 10KHZ 10US 1E5)
SIN (15V 10KHZ 01E5 30DEG)
SIN (0 2V 10KHZ 0 030DEG)
SIN (0 2V 10KHZ)
```

4.2.4 Exponential Source

The waveform and parameters of an exponential waveform are shown in Figure 4.7 and Table 4.4. The schematic of an exponential source is shown in Figure 4.8(a), and the menu for setting the model parameters is shown in Figure 4.8(b). In addition to the transient specifications, the DC (i.e., DC = 5 V) and AC (AC = 1 V) specifications can be assigned to the same source. TD1 is the rise-delay time, TC1 is the rise-time

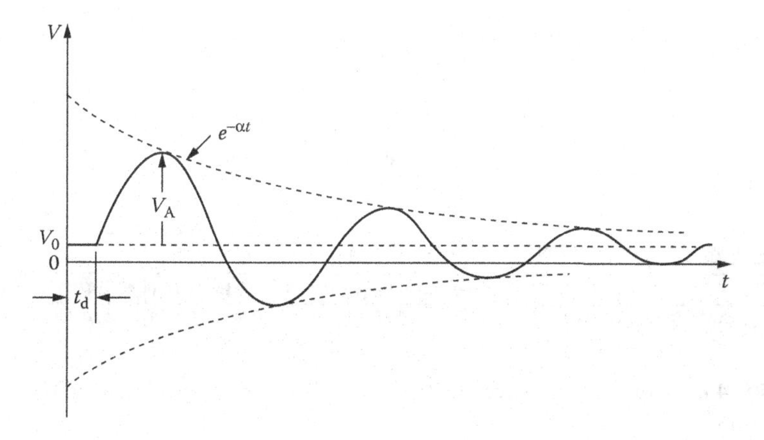

FIGURE 4.6 Damped sinusoidal waveform.

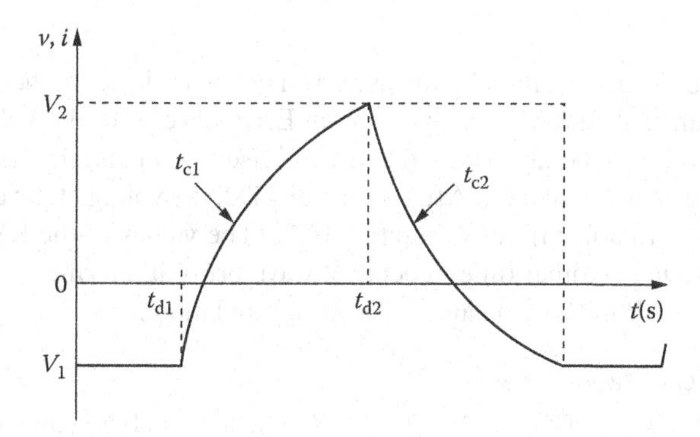

FIGURE 4.7 Exponential waveform.

TABLE 4.4
Model Parameters of EXP sources

Name	Meaning	Unit	Default
V1	Initial voltage	V	None
V2	Pulsed voltage	V	None
TD1	Rise-delay time	s	0
TC1	Rise-time constant	s	TSTEP
TD2	Fall-delay time	s	TD1 + TSTEP
TC2	Fall-time constant	s	TSTEP

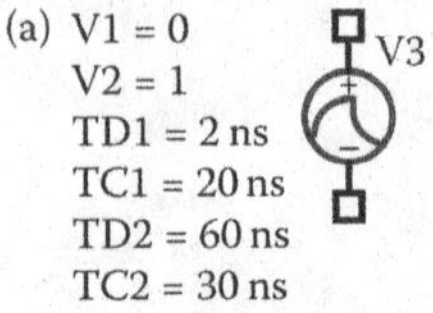

(a) V1 = 0
V2 = 1
TD1 = 2 ns
TC1 = 20 ns
TD2 = 60 ns
TC2 = 30 ns

V3

(b) New Column... Apply Display... Delete Property Filter by: Orcad-PSpice

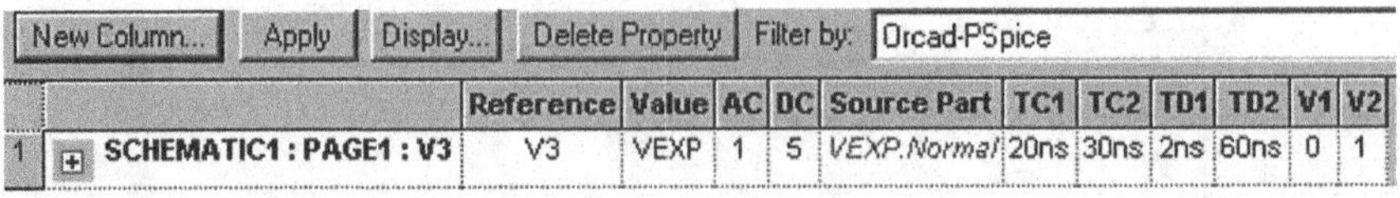

		Reference	Value	AC	DC	Source Part	TC1	TC2	TD1	TD2	V1	V2
1	SCHEMATIC1 : PAGE1 : V3	V3	VEXP	1	5	VEXP.Normal	20ns	30ns	2ns	60ns	0	1

FIGURE 4.8 Exponential source in PSpice schematic. (a) Symbol. (b) Editing model parameters.

constant, TD2 is the fall-delay time, and TD2 is the fall-time constant. The symbol of exponential sources is EXP and the general form is

```
EXP (V1 V2 TD1 TC1 TD2 TC2)
```

V1 and V2 must be specified by the user. TSTEP in Table 4.4 is the incrementing time during transient (·TRAN) analysis. In an EXP waveform, the voltage remains V1 for the first TD1 seconds. Then, the voltage rises exponentially from V1 to V2 with a rise-time constant of TC1. After a time of TD2, the voltage falls exponentially from V2 to V1 with a fall-time constant of TC2. (The values of the EXP waveform as well as the values of other time-dependent waveforms at intermediate time points are determined by PSpice by means of linear interpolation.)

4.2.4.1 Typical Statements

For V1 = 0 V, V2 = 1 V, TD1 = 2 NS, TC1 = 20 NS, TD2 = 60 NS, and TD2 = 30 NS, the model statement is

```
EXP (0 1 2NS 20NS 60NS 30NS)
```

With TRD = 0, the statement becomes

```
EXP (0 1 0 20NS 60NS 30NS)
```

With V1 = −1 V and V2 = 2 V, it is

```
EXP (-1 2 2NS 20NS 60NS 30NS)
```

4.2.5 Single-Frequency Frequency Modulation Source

The schematic of a single-frequency frequency modulation (SFFM) source is shown in Figure 4.9(a), and the menu for setting the model parameters is shown in Figure 4.9(b). In addition to the transient specifications, the DC (i.e., DC = 5 V) and AC (AC = 1 V) specifications can be assigned to the same source.

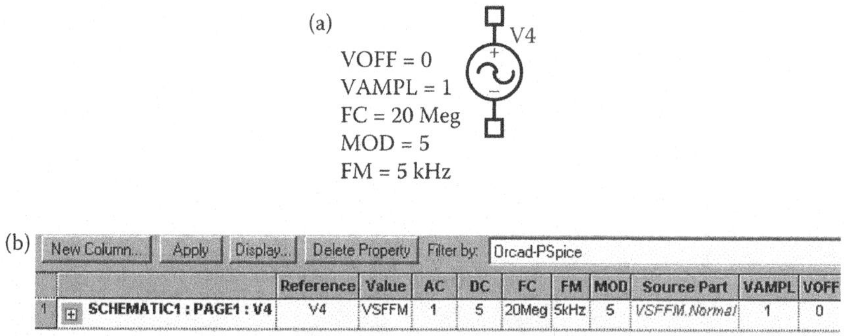

(a)

VOFF = 0
VAMPL = 1
FC = 20 Meg
MOD = 5
FM = 5 kHz

V4

(b)

New Column... | Apply | Display... | Delete Property | Filter by: Orcad-PSpice

		Reference	Value	AC	DC	FC	FM	MOD	Source Part	VAMPL	VOFF
1	SCHEMATIC1 : PAGE1 : V4	V4	VSFFM	1	5	20Meg	5kHz	5	VSFFM.Normal	1	0

FIGURE 4.9 SFFM source in PSpice Schematic. (a) Symbol. (b) Editing model parameters.

TABLE 4.5
Model Parameters of SFFM Sources

Name	Meaning	Unit	Default
VO	Offset voltage	V	None
VA	Amplitude of voltage	V	None
FC	Carrier frequency	Hz	1/TSTOP
MOD	Modulation index		0
FS	Signal frequency	Hz	1/TSTOP

The symbol of a source with single-frequency frequency modulation is SFFM, and the general form is as follows (see Table 4.5):

```
SFFM (VO VA FC MOD FS)
```

VO and VA must be specified by the user. TSTOP is the stop time during transient (·TRAN) analysis. The waveform is of the form

$$v = V_o + V_A \sin[(2\pi F_c t) + M \sin(2\pi F_s t)]$$

4.2.5.1 Typical Statements

For $V_o = 0$ V, $V_A = 1$ V, $F_C = 30$ MHz, MOD = 5, and $F_S = 5$ kHz, the model statement is

```
SFFM (0 1V 30MHZ 55KHZ)
```

With $V_o = 1$ mV and $V_A = 2$ V, the model becomes

```
SFFM (1MV 2V 30MHZ 55KHZ)
```

4.2.6 AC Sources

The AC sources are generally used for determining the frequency responses (output vs. frequency of a circuit). The schematic and the model parameters of an AC voltage source are shown in Figure 4.10(a). The menu for changing the model

(a)

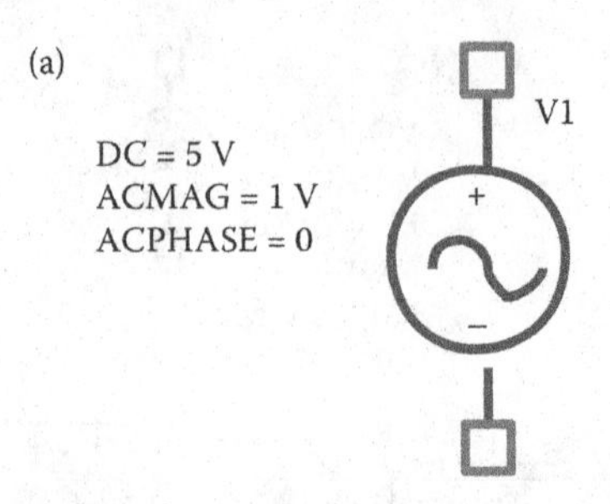

(b)

New Column... | Apply | Display... | Delete Property | Filter by: Orcad-PSpice

		Reference	Value	ACMAG	ACPHASE	DC	Source Part
1	⊞ EXAMPLE B-16 : PAGE1 : V1	V1	VAC	1V	0	5V	VAC.Normal

FIGURE 4.10 AC voltage source. (a) Symbol. (b) Editing model parameters.

parameters is shown in Figure 4.10(b). In addition to the frequency response specifications (i.e., AC = 1 V), the DC (i.e., DC = 5 V) specifications can be assigned to the same source. The symbol of an AC source is AC and the general form is

```
VAC (DC = B1 ACMAG = B2 ACPHASE = 0)
```

where ACMAG and ACPHASE are the peak value and the phase angle of an input sine wave, respectively. DC is the DC value for DC analysis and sweep.

4.3 INDEPENDENT SOURCES

The independent sources can be time-invariant and time-variant. They can be currents or voltages, as shown in Figure 4.10. The following notations are used only to explain the general format of a statement and do not appear in the PSpice statement:

(text)	Text within parentheses is a comment
[item]	Optional item
[item]*	Zero or more of optional item
<item>	Item required
<item>*	Zero or more of item required

4.3.1 Independent Voltage Source

The symbol of an independent voltage source is V, and the general form is as follows:

```
V<name>  N+ N-
+        [dc <value >]
+        [ac <(magnitude) value> <(phase) value >]
+        [(transient specifications)]
```

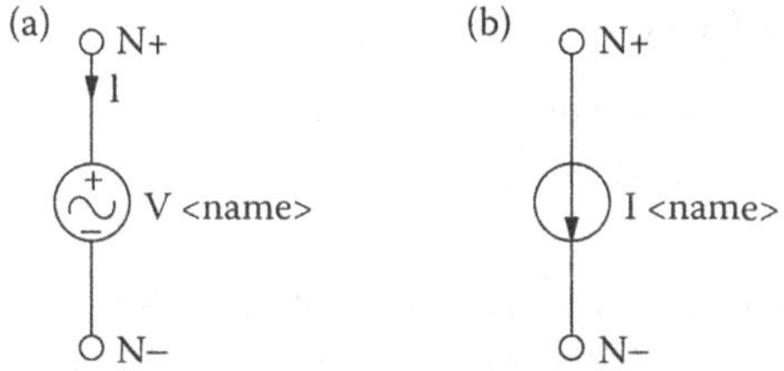

FIGURE 4.11 (a) Voltage source. (b) Current source.

Note: The first column with a + (plus) signifies the continuation of the PSpice statement. After the +sign, the statement can continue in any column.

The transient specifications must be one of the following sources:

PULSE (<parameters>)	For a pulse waveform
PWL (<parameters>)	For a piecewise linear waveform
SIN (<parameters>)	For a sinusoidal waveform
EXP (<parameters>)	For an exponential waveform
SFFM (<parameters>)	For a frequency-modulated waveform

N+ is the positive node and N− is the negative node, as shown in Figure 4.11(a). Positive current flows from node N+ through the voltage source to the negative node N−. The voltage source need not be grounded. For the **DC**, **AC**, and transient values, the default value is zero. None or all of the **DC**, **AC**, and transient values may be specified. The <(phase) value> is in degrees.

The source is set to the DC value in DC analysis. It is set to the AC value in AC analysis. If the <(phase) value> in AC analysis is omitted, the default is 0. The time-dependent source (e.g., PULSE, EXP, and SIN) is assigned for transient analysis. A voltage source may be used as an ammeter in PSpice by inserting a zero-valued voltage source into the circuit for the purpose of measuring current. Because a zero-valued source behaves as a short circuit, there will be no effect on circuit operation.

4.3.1.1 Typical Statements

```
V1      15  0   6V          ; By default, DC specification of 6 V
V2      15  0   DC      6V  ; DC specification of 6 V
VAC     5   6   AC      1V  ; AC specification of 1 V with 0°delay
VACP    5   6   AC      1V  45DEG; AC specification of 1 V with 45° delay
VPULSE  10  0   PULSE   (0 1 2NS 2NS 2NS 50NS 100NS); Transient pulse
VIN     25  22  DC 2 AC 1 30 SIN (0 2V 10KHZ)
```

Note: VIN assumes 2 V for DC analysis, 1 V with a delay angle of 30° for AC analysis, and a sine wave of 2 V at 10 kHz for transient analysis. This allows source specifications for different analyses in the same statement.

4.3.2 Independent Current Source

The symbol of an independent current source is I, and the general form is as follows:

```
I<name>  N+ N-
+        [dc <value>]
+        [ac <(magnitude) value> <(phase) value>]
+        [(transient specifications)]
```

The (transient specifications) must be one of the following sources:

PULSE (<parameters>)	For a pulse waveform
PWL (<parameters>)	For a piecewise linear waveform
SIN (<parameters>)	For a sinusoidal waveform
EXP (<parameters>)	For an exponential waveform
SFFM (<parameters>)	For a frequency-modulated waveform

N+ is the positive node and N– is the negative node, as shown in Figure 4.11(b). Positive current flows from node N+ through the current source to the negative node N–. The current source need not be grounded. The source specifications are similar to those of an independent voltage source.

4.3.2.1 Typical Statements

```
I1      15  0   2.5MA                       ; By default, DC specification of
                                            2.5 mA
I2      15  0   DC     2.5MA                ; DC specification of 2.5 mA
IAC     5   6   AC     1A                   ; AC specification of 1 A with 0°
                                            delay
IACP    5   6   AC     1A       45DEG       ; AC specification of 1 V with 45°
                                            delay
IPULSE  10  0   PULSE  (0 1A 2NS 2NS 2NS 50NS 100NS); transient pulse
IIN     25  22  DC 2A AC 1A 30DEG SIN (0 2 A 10 KHZ)
```

Note: IIN assumes 2 A for DC analysis, 1 A with a delay angle of 30° for AC analysis, and a sine wave of 2 A at 10 kHz for transient analysis. This allows source specifications for different analyses in the same statement.

4.3.3 Schematic Independent Sources

The PSpice source library source.slb is shown in Figure 4.12(a). DC voltage and current sources are shown in Figure 4.12(b) and (c). The user can change the values of the sources.

The generic source VSRC can specify transient, AC, and DC specifications in one source as shown in Figure 4.13.

4.4 DEPENDENT SOURCES

There are five types of dependent sources:

1. Polynomial source
2. Voltage-controlled voltage source
3. Current-controlled current source

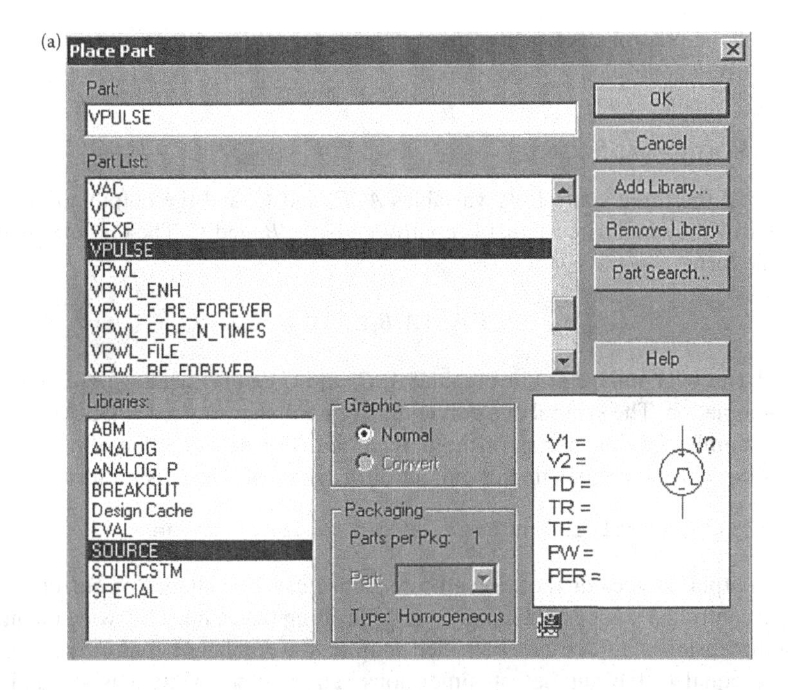

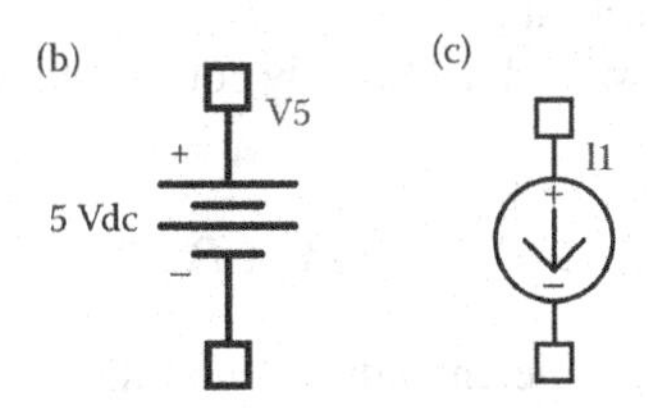

FIGURE 4.12 Independent DC sources. (a) Source menu. (b) DC voltage source. (c) DC current source.

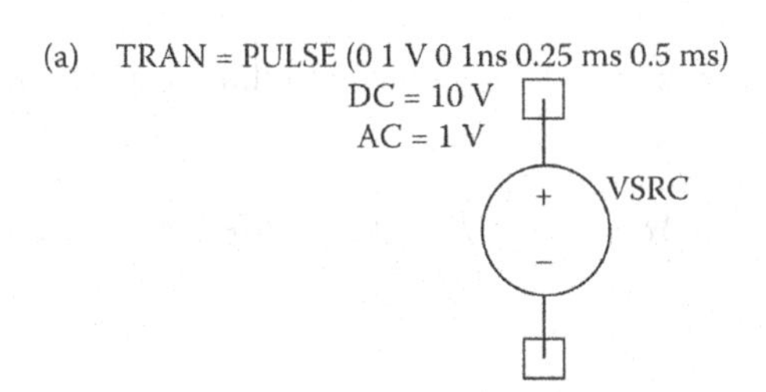

(b)

New Column... | Apply | Display... | Delete Property | Filter by: Orcad-PSpice

		Reference	Value	AC	DC	Source Part	TRAN
1	⊞ EXAMPLE B-16 : PAGE1 : Vin	Vin	VSRC	1V	10V	VSRC.Normal	PULSE 0 1V 0 1ns 1ns 0.25ms 0.5ms

FIGURE 4.13 Symbol and properties of VSRC source for types of multiple analyses. (a) VSRC schematic. (b) Editing model parameters.

4. Voltage-controlled current source
5. Current-controlled voltage source

4.4.1 Polynomial Source

Let us call the three controlling variables *A*, *B*, and *C*, and the output sources, *Y*. Figure 4.14 shows a source *Y* that is controlled by *A*, *B*, and *C*. The output source *Y* takes the form

$$Y = f(A, B, C, \ldots)$$

where *Y* can be a voltage or current, and *A*, *B*, and *C* can be a voltage or current or any combination. The symbol of a polynomial or nonlinear source is POLY(n), where n is the number of dimensions of the polynomial. The default value of *n* is 1. The dimensions depend on the number of controlling sources. The general form is

```
POLY(n) <(controlling)nodes> <(coefficient) values>
```

The output sources or the controlling sources can be voltages or currents. For voltage-controlled sources, the number of controlling nodes must be twice the number of dimensions. For current-controlled sources, the number of controlling sources must be equal to the number of dimensions. The number of dimensions and the number of coefficients are arbitrary.

For a polynomial of $n = 1$ with *A* as the only controlling variable, the source function takes the form

$$Y = P_0 + P_1 A + P_2 A^2 + P_3 A^3 + P_4 A^4 + \cdots + P_n A^n$$

where P_0, P_1, …, P_n are the coefficient values, and this is written in PSpice as

```
POLY NC1+ NC1- P0 P1 P2 P3 P4 P5 ... Pn
```

where NC1+ and NC1– are the positive and negative nodes, respectively, of controlling source *A*.

For a polynomial of $n = 2$ with *A* and *B* as the controlling sources, the source function *Y* takes the form

$$Y = P_0 + P_1 A + P_2 B + P_3 A^2 + P_4 AB + P_5 B^2 + P_6 A^3 + P_7 A^2 B + P_8 AB^2 + P_9 B^3 + \cdots$$

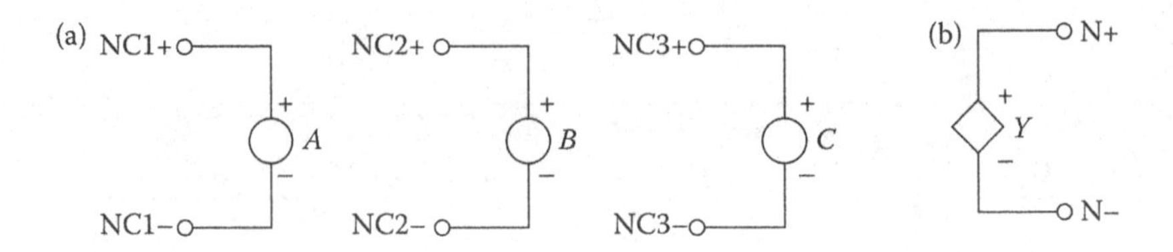

FIGURE 4.14 Polynomial source. (a) Controlling sources. (b) Output source.

and this is described in PSpice as

```
POLY(2) NC1+ NC1- NC2+ NC2- P0 P1 P2 P3 P4 P5 ... Pn
```

where NC1+, NC2+ and NC1–, NC2– are the positive and negative nodes, respectively, of the controlling sources.

For a polynomial of $n = 3$ with A, B, and C as the controlling sources, the source function Y takes the form

$$Y = P_0 + P_1A + P_2B + P_3C + P_4A^2 + P_5AB + P_6AC + P_7B^2 + P_8BC + P_9C^2 \\ + P_{10}A^3 + P_{11}A^2B + P_{12}A^2C + P_{13}AB^2 + P_{14}ABC + P_{15}AC^2 + P_{16}B^3 \\ + P_{17}B^2C + P_{18}BC^2 + P_{19}C^3 + P_{20}A^4 + \cdots$$

and this is written in PSpice as

```
POLY(3) NC1+ NC1- NC2+ NC2- NC3+ NC3- P0 P1 P2 P3 P4 P5 ... Pn
```

where NC1+, NC2+, NC3+ and NC1–, NC2–, NC3– are the positive and negative nodes, respectively, of the controlling sources.

4.4.1.1 Typical Model Statements

For Y = 2V(10), the model is

```
POLY 10 0 2.0
```

For Y = V(5) + 2[V(5)]^2 + 3[V(5)]^3 + 4[V(5)]^4, the model is

```
POLY 5 0 0.0 1.0 2.0 3.0 4.0
```

For Y = 0.5 + V(3) + 2V(5) + 3[V(3)]^2 + 4V(3)V(5), the model is

```
POLY(2) 3 0 5 0 0.5 1.0 2.0 3.0 4.0
```

For Y = V(3) + 2V(5) + 3V(10) + 4[V(3)]^2, the model is

```
POLY(3) 3 0 5 0 10 0 0.0 1.0 2.0 3.0 4.0
```

If I(VN) is the controlling current through voltage source VN, and Y = I(VN) + 2[I(VN)]^2 + 3[I(VN)]^3 + 4[I(VN)]^4, the model is

```
POLY VN 0.0 1.0 2.0 3.0 4.0
```

If I(VN) and I(VX) are the controlling currents, and Y = I(VN) + 2I(VX) + 3[I(VN)]^2 + 4I(VN)I(VX), the model is

```
POLY(2) VN VX 0.0 1.0 2.0 3.0 4.0
```

Note: If the source is of one dimension and only one coefficient is specified, as in the first example, in which Y = 2V(10), PSpice assumes that $P_0 = 0$ and the value specified is P_1. That is, $Y = 2A$.

4.4.2 Voltage-Controlled Voltage Source

The dependent sources are shown in Figure 4.15. The symbol of the voltage-controlled voltage source shown in Figure 4.15(a) is E, and it takes the linear form

```
E<name> N+ N- NC+ NC- <(voltage gain) value>
```

where N+ and N– are the positive and negative output nodes, respectively, and NC+ and NC– are the positive and negative nodes, respectively, of the controlling voltage. The nonlinear form is as follows:

```
E<name> N+ N- [POLY (polynomial specifications)]
+       [VALUE (expression)] [TABLE (expression)]
+       [LAPLACE (expression)] [FREQ (expression)]
```

The POLY is described in Section 4.4.1. The number of controlling nodes in POLY is twice the number of dimensions. A particular node may appear more than once, and the output and controlling nodes could be the same.

The VALUE, TABLE, LAPLACE, and FREQ descriptions of sources are available only with the analog behavioral modeling option of PSpice and are discussed in Section 4.5.

4.4.2.1 Typical Statements

```
EAB    1 2 4  6  1.0 ; Voltage gain of 1
EVOLT  4 7 20 22 2E5 ; Voltage gain of 2E5
```

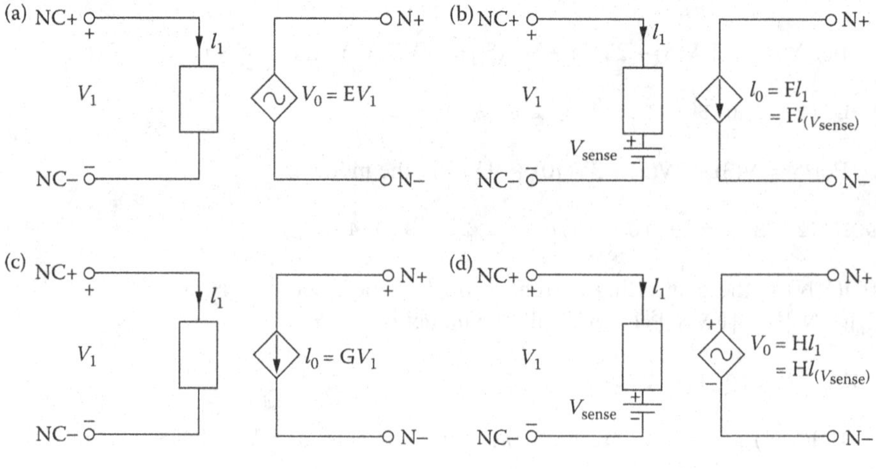

FIGURE 4.15 Dependent sources. (a) Voltage-controlled voltage source. (b) Current-controlled current source. (c) Voltage-controlled current source. (d) Current-controlled voltage source.

ENONLIN, which is connected between nodes 25 and 40, is controlled by V(3) and V(5). Its value is given by the polynomial Y = V(3) + 1.5V(5) + 1.2[V(3)]^2 + 1.7V (3)V(5), and the model becomes

```
ENONLIN 25 40POLY(2) 3 0 5 0 0.0 1.0 1.5 1.2 1.7; POLY source
```

E2, which is connected between nodes 10 and 12, is controlled by V(5), and its value is given by the polynomial Y = V(5) + 1.5[V(5)]^2 + 1.2[V(5)]^3 + 1.7[V(5)]^4, and the model becomes

```
E2 10 12 POLY 5 0 0.0 1.0 1.5 1.2 1.7; POLY source
```

4.4.3 Current-Controlled Current Source

The symbol of the current-controlled current source shown in Figure 4.15(b) is F, and it takes the linear form:

```
F <name> N+ N- VN <(current gain) value>
```

where N+ and N– are the positive and negative nodes, respectively, of the current source. VN is a voltage source through which the controlling current flows. The controlling current is assumed to flow from the positive node of VN, through the voltage source VN, to the negative node of VN. The current through the controlling voltage source, I(VN), determines the output current.

The voltage source VN that monitors the controlling current must be an independent voltage source, and it can have a finite value or zero. If the current through a resistor controls the source, a dummy voltage source of 0 V should be connected in series with the resistor to monitor the controlling current.

The nonlinear form is as follows:

```
F <name> N+ N- [POLY (polynomial specifications)]
```

The POLY source is described in Section 4.4.1. The number of controlling current sources for the POLY must be equal to the number of dimensions.

4.4.3.1 Typical Statements

```
FAB   1   2  VIN  10  ; Current gain of 10
FAMP  13  4  VCC  50  ; Current gain of 50
```

FNONLIN, which is connected between nodes 25 and 40, is controlled by the current through voltage source VN. Its value is given by the polynomial I = I (VN) + 1.5[I(VN)]^2 + 1.2[I(VN)]^3 + 1.7[I(VN)]^4, and the PSpice model becomes

```
FNONLIN  25  40  POLY  VN  0.0  1.0  1.5  1.2  1.7
```

4.4.4 Voltage-Controlled Current Source

The symbol of the voltage-controlled current source shown in Figure 4.15(c) is G, and it takes the linear form:

```
G<name> N+ N- NC+ NC- < (transconductance) value >
```

where N+ and N– are the positive and negative output nodes, respectively, and NC+ and NC– are the positive and negative nodes, respectively, of the controlling voltage.

The nonlinear form is as follows:

```
G<name>    N+ N- [POLY (polynomial specifications)]
+          [VALUE (expression)] [TABLE (expression)]
+          [LAPLACE (expression)] [FREQ (expression)]
```

The POLY is described in Section 4.4.1. The VALUE, TABLE, LAPLACE, and FREQ descriptions of sources are available only with the analog behavioral modeling option of PSpice. These are discussed in Section 4.5.

4.4.4.1 Typical Statements

```
GAB       1  2  4   6   1.0  ; Transconductance of 1
GVOLT     4  7  20  22  2E5  ; Transconductance of 2E5
```

GNONLIN, which is connected between nodes 25 and 40, is controlled by V(3) and V(5). Its value is given by the polynomial Y = V(3) + 1.5V(5) + $1.2[V(3)]^2$ + 1.7V(3)V(5), and the model becomes

```
GNONLIN 25 40 POLY(2) 3 0 5 0 0.0 1.0 1.5 1.2 1.7;POLY source
```

G2, which is connected between nodes 10 and 12, is controlled by V(5), and its value is given by the polynomial Y = V(5) + $1.5[V(5)]^2$ + $1.2[V(5)]^3$ + 1.7 $[V(5)]^4$, and the model becomes

```
G2 10 12 POLY 5 0 0.0 1.0 1.5 1.2 1.7; POLY source
```

A voltage-controlled current source can be used to simulate conductance if the controlling nodes are the same as the output nodes. This is shown in Figure 4.16(a). For example, the PSpice statement

```
GRES 4 6 4 6 0.1; transconductance of 0.1
```

is a linear conductance of 0.1 siemens (Ω^{-1} or mhos) with a resistance of 1/0.1 = 10 Ω. The PSpice statement

```
GMHO 1 2 POLY 1 2 0.0 1.5M 1.7M; POLY source
```

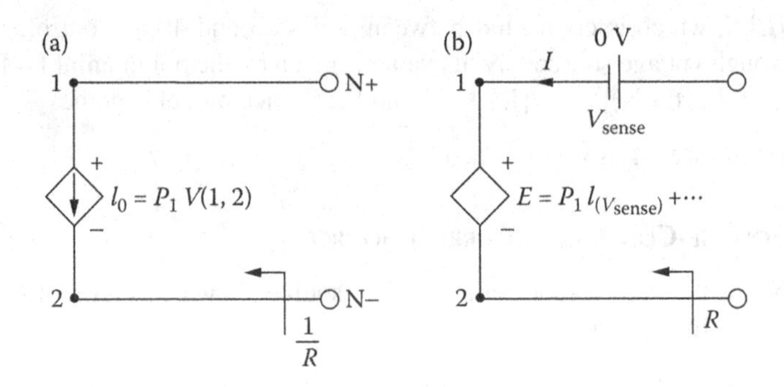

FIGURE 4.16 (a) Conductance and (b) resistance.

represents a nonlinear conductance (Ω^{-1}) of the polynomial form

$$I = 1.5 \times 1^{-3} V(1,2) + 1.7 \times 10^{-3} [V(1,2)]^2$$

4.4.5 Current-Controlled Voltage Source

The symbol of the current-controlled voltage source shown in Figure 4.15(d) is H, and it takes the linear form

```
H<name> N+ N- VN <(transresistance) value>
```

where N+ and N– are the positive and negative nodes, respectively, of the voltage source. VN is a voltage source through which the controlling current flows, and its specifications are similar to those for a current-controlled current source.

The nonlinear form is

```
H<name> N+ N- [POLY (polynomial specifications)]
```

The POLY source is described in Section 4.4.1. The number of controlling current sources for the POLY must be equal to the number of dimensions.

4.4.5.1 Typical Statements

```
HAB     1   2  VIN  10
HAMP    13  4  VCC  50
```

HNON**LIN**, which is connected between nodes 25 and 40, is controlled by I(VN). Its value is given by the polynomial V = I(VN) + 1.5$[I(VN)]^2$ + 1.2$[I(VN)]^3$ + 1.7 $[I(VN)]^4$, and the model becomes

```
HNONLIN 25 40 POLY VN 0.0 1.0 1.5 1.2 1.7; POLY source
```

A voltage-controlled current source can be applied to simulate resistance if the controlling current is the same as the current through the voltage between the output nodes. This is shown in Figure 4.15(d). For example, the PSpice statement

```
HRES 4 6 VN 0.1; Transresistance of 0.1
```

is a linear resistance of 10 Ω.

The PSpice statement

```
HOHM 1 2 POLY VN 0.0 1.5M 1.7M; POLY source
```

represents a nonlinear resistance in ohms of the polynomial form

$$H = 1.5 \times 1^{-3} I(VN) + 1.7 \times 10^{-3} [I(VN)]^2$$

4.4.6 Schematic-Dependent Sources

The PSpice analog library analog.slb is shown in Figure 4.17. The voltage-controlled voltage source (E), the current-controlled current source (F), the voltage-controlled

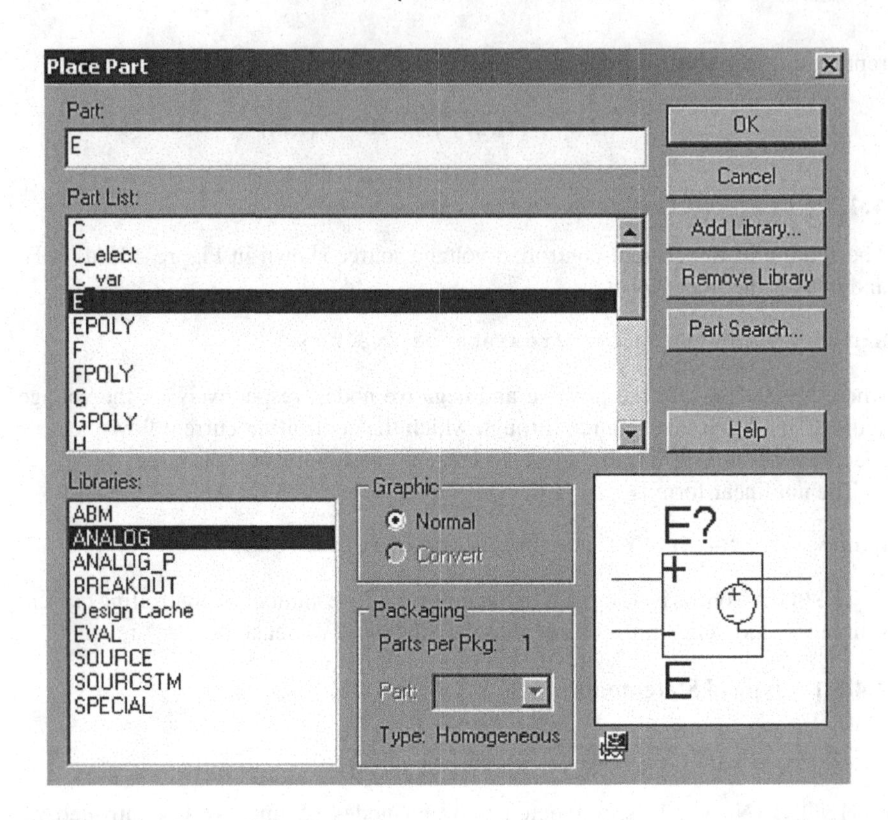

FIGURE 4.17 Analog library.

current source (G), and the current-controlled voltage source (H) of the PSpice library are shown in Figure 4.18(a)–(d).

4.5 BEHAVIORAL DEVICE MODELING

PSpice allows the characterization of devices in terms of the relation between their inputs and outputs. This relation is instantaneous. At each moment in time, there is an output for each value of the input. This representation, known as *behavioral modeling*, is available only with the analog behavioral modeling option of PSpice. This option is implemented as a set of extensions to two of the controlled sources, E and G. Behavioral modeling allows specifications in the form

VALUE
TABLE
LAPLACE
FREQ

The PSpice Schematics support many behavioral models such as DIFF, DIFFER, INTG, MULTI, SUM, SQRT, and so on. These models can be selected from ab.slb library of the library menu as shown in Figure 4.18(e).

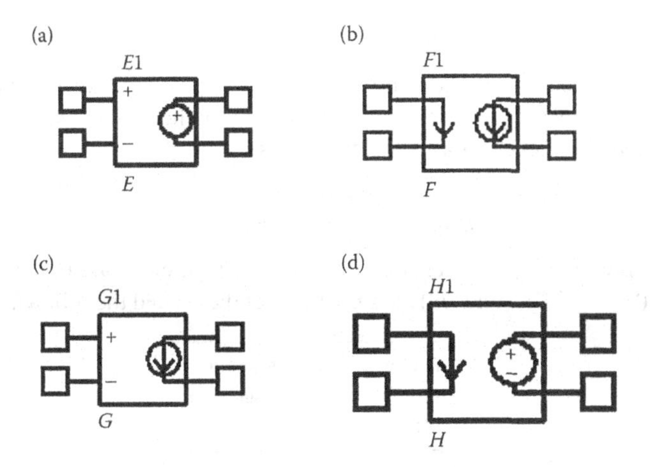

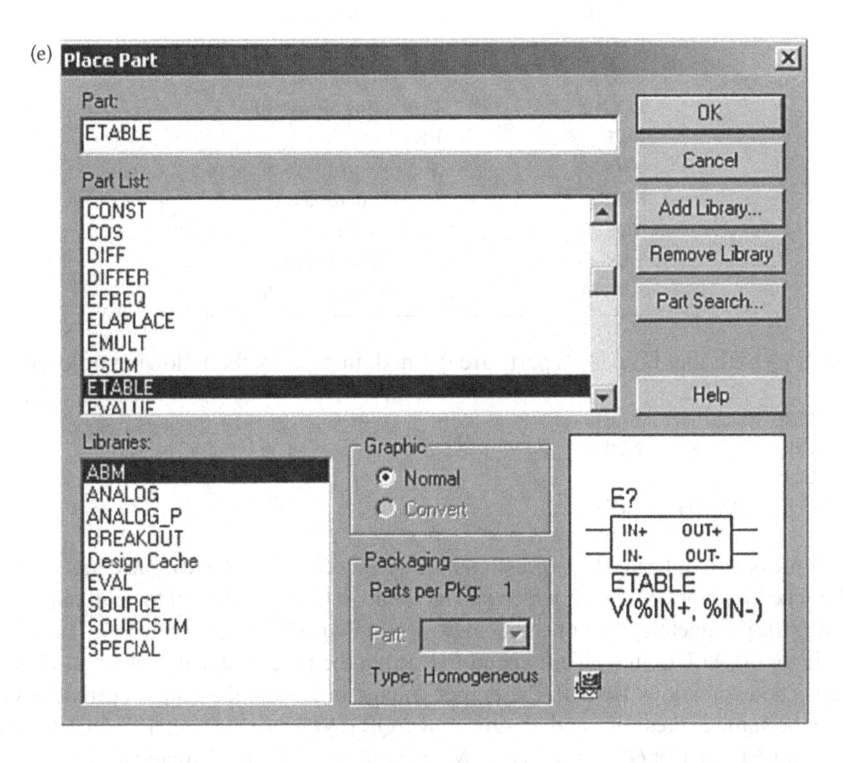

FIGURE 4.18 PSpice-dependent sources. (a) Voltage-controlled voltage source. (b) Current-controlled current source. (c) Voltage-controlled current source. (d) Current-controlled voltage source. (e) Analog behavioral models.

4.5.1 VALUE

The VALUE extension to the controlled sources of the PSpice library allows an instantaneous transfer function to be written as a mathematical expression in standard notation. The general forms are as follows:

```
E<name>        N+ N- VALUE= {<expression>}
G<name>        N+ N- VALUE= {<expression>}
```

The <expression> itself is enclosed in braces ({}). It can contain the arithmetical operators ("+," "–," "*," and "/") along with parentheses and the following functions:

Function	Meaning
ABS(x)	$\|x\|$ (absolute value)
SQRT(X)	$\sqrt{x}$

Function	Meaning
$EXP_{(x)}$	e^x
$LOG_{(x)}$	Ln(x) (log of base e)
$LOG10_{(x)}$	Log(x) (log of base 10)
$PWR_{(x,y)}$	$\|x\|^y$
$PWRS_{(x,y)}$	$+\|x\|^y$ (if $x > 0$), $-\|x\|^y$ (if $x < 0$)
$SIN_{(x)}$	sin(x) (x in radians)
$COS_{(x)}$	cos(x) (x in radians)
$TAN_{(x)}$	tan(x) (x in radians)
$ARCTAN_{(x)}$	$\tan^{-1}(x)$ (result in radians)

The EVALUE and GVALUE parts are defined, in part, by the following properties:

```
EVALUE
  EXPR  V(%IN+, %IN-)
GVALUE
  EXPR  V(%IN+, %IN-)
```

Sources are controlled by expressions which may contain voltages, currents, or both. The schematic for EVALUE, which can be found in the abm.lib, and the edit menu for setting the parameters for mathematical expressions are shown in Figure 4.19. The schematic for GVALUE and the edit menu for setting the parameters are shown in Figure 4.20. The expressions for both the output voltage in E1 and the output current in G1 have the same expressions and are given by SQRT(V(%IN+, %IN–))*I(SENCE). That is, $v_o(\text{for } E1) = i_o(\text{for } G1) = i_{in} \times \sqrt{v_{in}}$. We can use any valid mathematical expression.

4.5.1.1 Typical Statements

```
ESQROOT  2 3 VALUE = {4V*SQRT (V(5))}; Square roots
EPWR     1 2 VALUE = {V(4.3)*I(VSENSE)}; Product of v and i
ELOG     3 0 VALUE = {10V*LOG (I (VS)/10mA)}; Log of current ratio
GVCO     4 5 VALUE = {15MA*SIN (6.28*10kHz*TIME* (10V*V(7)))}
GRATIO   3 6 VALUE = {V (8, 2)/V(9)}; Voltage ratio
```

(a)

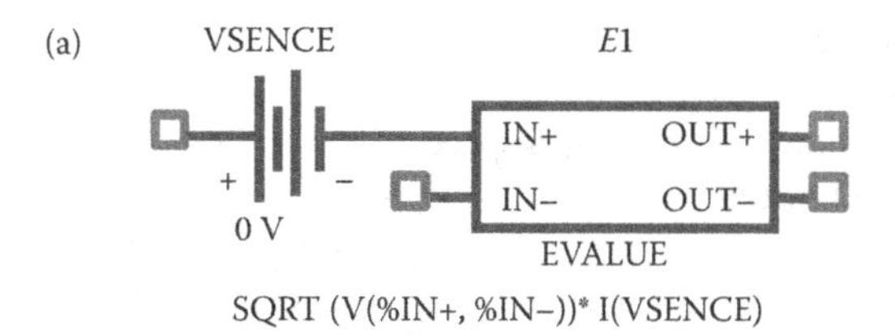

(b)

New Column... | Apply | Display... | Delete Property | Filter by: Orcad-PSpice

		Reference	Value	EXPR	Source Part
1	⊞ EXAMPLE B-16 : PAGE1 : E1	E1	EVALUE	SQRT(V(%IN+, %IN-))*I(VSENCE)	*EVALUE.Normal*

FIGURE 4.19 Schematics for EVALUE mathematical expressions. (a) EVALUE schematic. (b) Editing EVALUE model parameters.

(a)

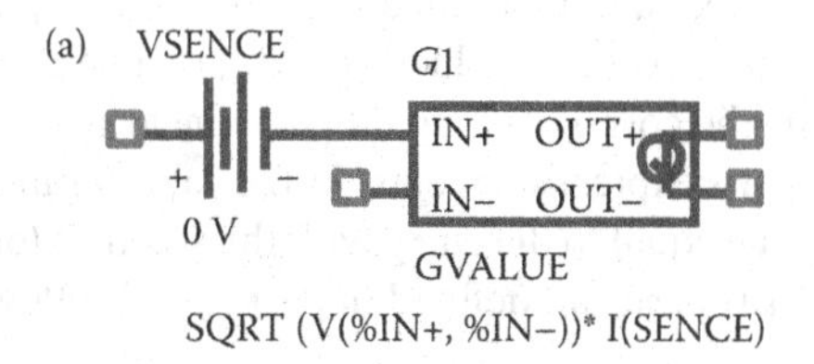

(b)

New Column... | Apply | Display... | Delete Property | Filter by: Orcad-PSpice

		Reference	Value	EXPR	Source Part
1	⊞ EXAMPLE B-16 : PAGE1 : G1	G1	GVALUE	SQRT(V(%IN+, %IN-))*I(VSENCE)	*GVALUE.Normal*

FIGURE 4.20 Schematics for GVALUE mathematical expressions. (a) GVALUE schematic. (b) Editing GVALUE model parameters.

VALUE can be used to simulate linear and nonlinear resistances (or conductances) if appropriate functions are used. A resistance is a current-controlled voltage source. For example, the statement

```
ERES 2 3 VALUE = {I (VSENSE)*5K}
```

is a linear resistance with a value of 5 kΩ. VSENSE, which is connected in series with ERES, is needed to measure the current through ERES.

A conductance is a voltage-controlled current source. For example, the statement

```
GCOND 2 3 VALUE = {V(2,3)*1M}
```

is a linear conductance with a value of 1 $m\Omega^{-1}$. The controlling nodes are the same as the output nodes.

Note the following:

1. VALUE should be followed by a space.
2. <expression> must fit on one line.

4.5.2 TABLE

The TABLE extension to the controlled sources of the PSpice library allows an instantaneous transfer function to be described by a table. This form is well suited for use with, for example, measured data. The general forms are

```
E<name>        N+      N-      TABLE {<expression>} =
+        ≪(input)value>,       <(output)value≫*
G<name>        N+      N-      TABLE {<expression>} =
+        ≪(input)value>,<(output)value≫*
```

The <expression> is evaluated, and that value is used to look up an entry in the table. The table itself consists of pairs of values. The first value in each pair is an input, and the second value is the corresponding output. Linear interpolation is done between entries. For values of <expression> outside the table's range, the device's output is a constant with a value equal to the entry with the smallest (or largest) input.

The ETABLE and GTABLE parts are defined in part by the following properties and default values:

```
ETABLE
  EXPR     V(%IN+, %IN–)
  TABLE    (-15, -15), (15,15)
GTABLE
  EXPR     V(%IN+, %IN-)
  TABLE    (-15, -15), (15,15)
```

First, EXPR is evaluated, and that value is used to look up an entry in the table. EXPR is a function of the input (current or voltage) and follows the same rules as for VALUE expressions. The table consists of pairs of values, the first of which is an input, and the second of which is the corresponding output. Linear interpolation is performed between entries. For values of EXPR outside the table's range, the device's output is a constant with a value equal to the entry with the smallest (or largest) input. This characteristic can be used to impose an upper and lower limit on the output.

The schematic for TABLE, which can be found in the abm.lib, and the edit menu for setting the parameters for expressions are shown in Figure 4.21. The input and output relationships are listed in a tabular form as shown in Figure 4.21(b). The schematic for ETABLE, which can be found in the abm.lib, and the edit menu for setting the parameters for mathematical expressions are shown in Figure 4.22. The schematic for GTABLE and the edit menu for setting the parameters are shown in Figure 4.23.

The expressions for both the output voltage in $E1$ and the output current in $G1$ have the same expressions and are given by SQRT(V(%IN+, %IN–))*I(SENCE).That is, $V_o(\text{for } E1) = i_o(\text{for } G1) = i_{in} \times \sqrt{v_{in}}$. We can use any valid mathematical expression.

(a)

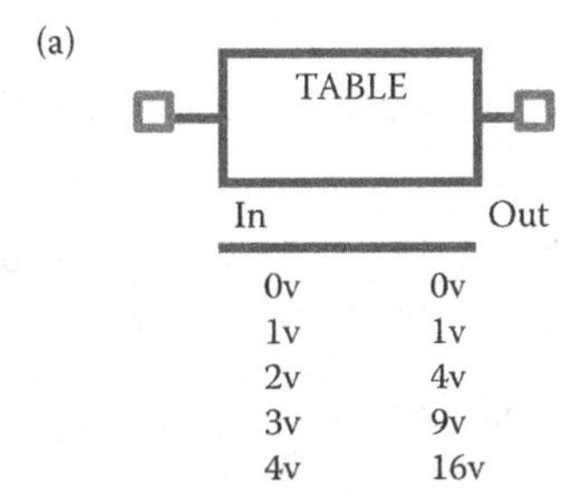

(b)

New Column... | Apply | Display... | Delete Property | Filter by: Orcad-PSpice | Help

		Reference	Value	ROW1	ROW2	ROW3	ROW4	ROW5	Source Part
1	⊞ EXAMPLE B-16 : PAGE1 : TABLE1	TABLE1	TABLE	0v 0v	1v 1v	2v 4v	3v 9v	4v 16v	TABLE.Normal

FIGURE 4.21 Schematics for TABLE mathematical expressions. (a) TABLE schematic. (b) Editing model parameters.

(a)

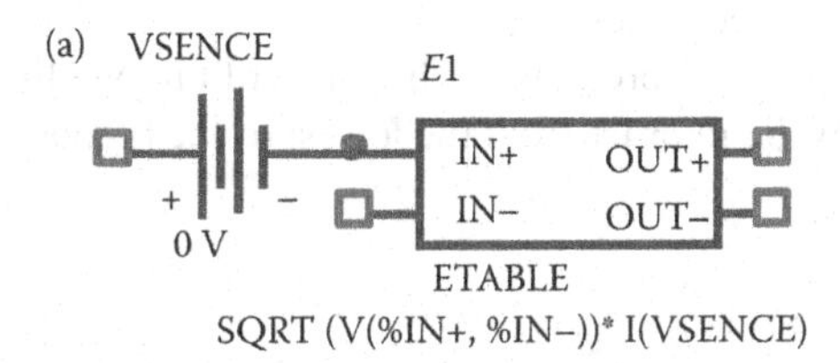

(b)

New Column... | Apply | Display... | Delete Property | Filter by: Orcad-PSpice

		Reference	Value	EXPR	Source Part	TABLE
1	⊞ EXAMPLE B-16 : PAGE1 : G1	G1	GTABLE	SQRT(V(%IN+, %IN-))*I(VSENCE)	GTABLE.Normal	(-15,-15) (15,15)

FIGURE 4.22 Schematics for ETABLE mathematical expressions. (a) ETABLE schematic. (b) Editing ETABLE model parameters.

(a)

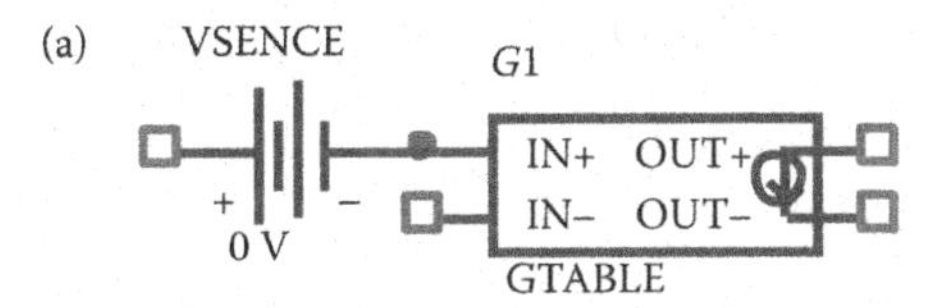

(b)

New Column... | Apply | Display... | Delete Property | Filter by: Orcad-PSpice

		Reference	Value	EXPR	Source Part	TABLE
1	⊞ EXAMPLE B-16 : PAGE1 : G1	G1	GTABLE	SQRT(V(%IN+, %IN-))*I(VSENCE)	GTABLE.Normal	(-15,-15) (15,15)

FIGURE 4.23 Schematics for GTABLE mathematical expressions. (a) GTABLE schematic. (b) Editing GTABLE model parameters.

4.5.2.1 Typical Statements

TABLE can be used to represent the voltage versus current characteristics of a diode as

```
EDIODE 5 6 TABLE{I (VSENSE)} =
+  (0.0,0.5)         (10E-3,0.870)  (20E-3,0.98)    (30E-3,1.058)
+  (40E-3,1.115)    (50E-3,1.173)  (60E-3,1.212)   (70E-3,1.250)
```

TABLE can be used to represent a constant power load $P = 400$ W with a voltage—controlled current source as

```
GCONST 2 3 TABLE {400/V(2, 3)} = (-400, -400) (400, 400)
```

GCONST tries to dissipate 400 W of power regardless of the voltage across it. But for a very small voltage, the formula 400/V(2,3) can lead to unreasonable values of current. TABLE limits the currents to between –400 and +400 A.

Note the following:

1. TABLE *must* be followed by a space.
2. The input to the table is <expression>, which must fit in one line.
3. TABLE's input *must* be in order from the lowest to the highest.

4.5.3 LAPLACE

The LAPLACE extension to the controlled sources of the PSpice library allows a transfer function to be described by a Laplace transform function. The general forms are

```
E<name>  N+  N-  LAPLACE {<expression>} = {<transform>}
G<name>  N+  N-  LAPLACE {<expression>} = {<transform>}
```

The input to the transform is the value of <expression>, which follows the same rules as in Section 4.5.1. The <transform> is an expression in the Laplace variable, *s*.

The ELAPLACE and GLAPLACE parts are defined, in part, by the following properties (default values are shown):

```
ELAPLACE
  EXPR     V(%IN+, %IN-)
  XforM    1/s
GLAPLACE
  EXPR     V(%IN+, %IN-)
  XforM    1/s
```

The LAPLACE parts use a Laplace transform description. The input to the transform is the value of EXPR, where EXPR follows the same rules as for VALUE expressions. XFORM is an expression in the Laplace variable, *s*. It follows the rules for standard expressions as described for VALUE expressions with the addition of the *s* variable.

(a)

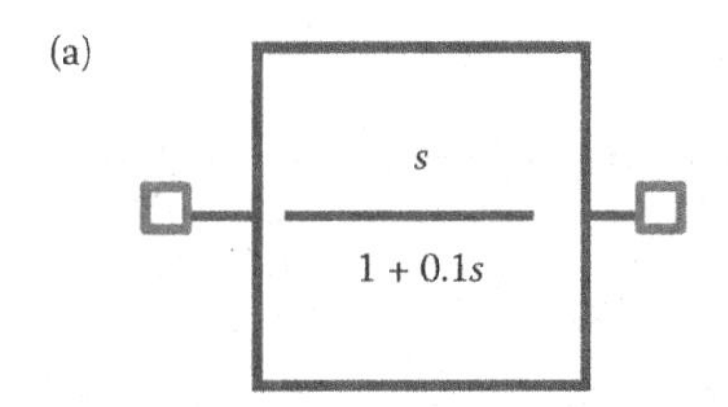

(b)

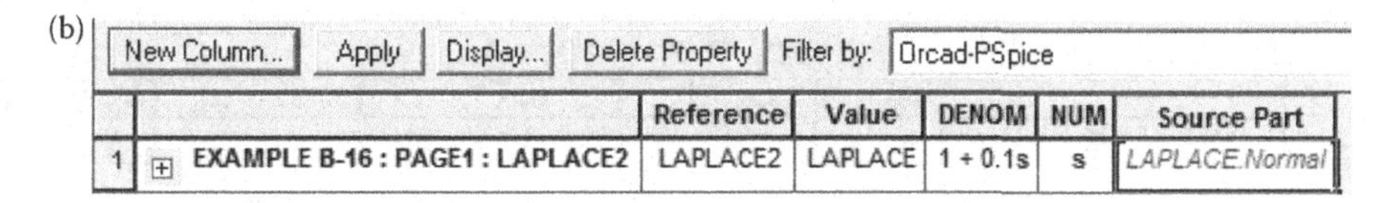

New Column... | Apply | Display... | Delete Property | Filter by: Orcad-PSpice

		Reference	Value	DENOM	NUM	Source Part
1	⊞ EXAMPLE B-16 : PAGE1 : LAPLACE2	LAPLACE2	LAPLACE	1 + 0.1s	s	LAPLACE.Normal

FIGURE 4.24 Schematics for LAPLACE expressions. (a) LAPLACE schematic. (b) Editing LAPLACE model parameters.

The schematic for LAPLACE, which can be found in the abm.lib, and the edit menu for setting the parameters for expressions are shown in Figure 4.24. The input and output relationship is expressed in Laplace's domain of "*s*" as shown in Figure 4.24(b). The schematic for ELAPLACE, which can be found in the abm.lib, and the edit menu for setting the parameters for mathematical expressions are shown in Figure 4.25. The schematic for GLAPLACE and the edit menu for setting the parameters are shown in Figure 4.26.

The expressions for both the output voltage in E1 and the output current in G1 have the same expressions and are given by SQRT(V(%IN +, %IN–))*I(SENCE).

That is, $V_0(for\ E1) = i_0(for\ G1) = i_{in\sqrt{V_{in}}} \times \frac{1}{s}$. We can use any valid mathematical expression.

(a)

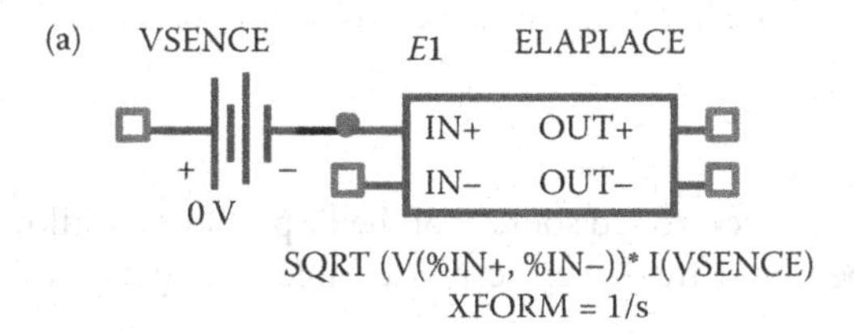

(b)

New Column... | Apply | Display... | Delete Property | Filter by: Orcad-PSpice

		Reference	Value	EXPR	Source Part	XFORM
1	⊞ EXAMPLE B-16 : PAGE1 : E1	E1	ELAPLACE	SQRT(V(%IN+, %IN-))*I(VSENCE)	ELAPLACE.Normal	1/s

FIGURE 4.25 Schematics for ELAPLACE expressions. (a) ELAPLACE schematic. (b) Editing ELAPLACE model parameters.

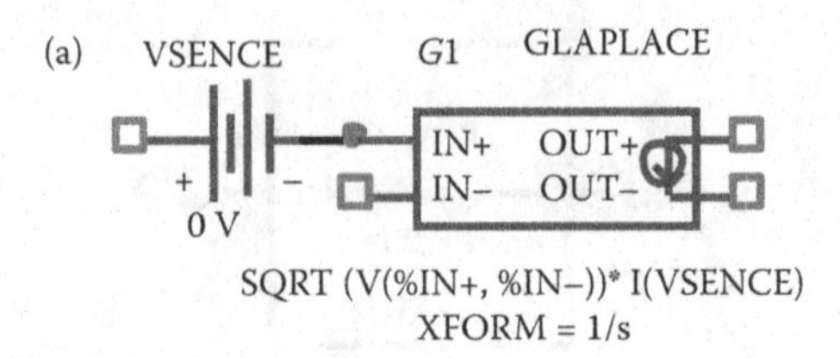

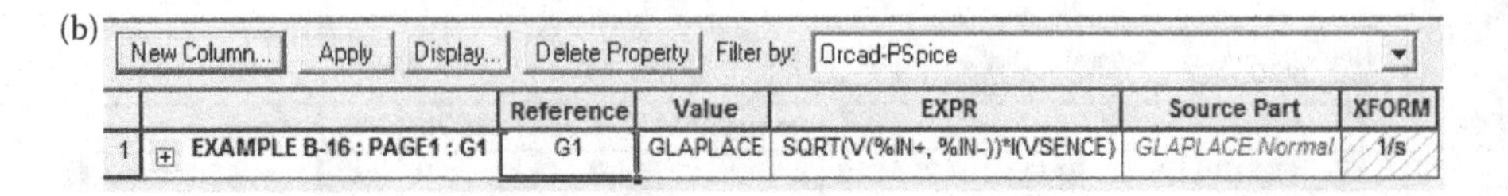
(b)

New Column... | Apply | Display... | Delete Property | Filter by: Orcad-PSpice

		Reference	Value	EXPR	Source Part	XFORM
1	⊞ EXAMPLE B-16 : PAGE1 : G1	G1	GLAPLACE	SQRT(V(%IN+, %IN-))*I(VSENCE)	GLAPLACE.Normal	1/s

FIGURE 4.26 Schematics for GLAPLACE expressions. (a) GLAPLACE schematic. (b) Editing GLAPLACE model parameters.

4.5.3.1 Typical Statements

The output voltage of a lossless integrator with a time constant of 1 ms and an input voltage V(5) can be described by

```
ERC 4 0 LAPLACE {V(5)} = {1/(1 + 0.001*sec)}
```

Frequency-dependent impedances can be simulated with a capacitor, which can be written as given below:

```
GCAP 5 4 LAPLACE {V(5,4)} = {s}
```

Note the following:

1. LAPLACE *must* be followed by a space.
2. <expression> and <transform> must each fit on one line.
3. Voltages, currents, and TIME must not appear in a Laplace transform.
4. The LAPLACE device uses much more computer memory than does the built-in capacitor (C) device and should be avoided if possible.

4.5.4 FREQ

The FREQ extension to the controlled sources of the PSpice library allows a transfer function to be described by a frequency response table. The general forms are as follows:

```
E<name>  N+  N-  FREQ {<expression>}=
+             ≪(frequency)value>,  <(magnitude in dB)value>,
              <(phase)value≫*
G<name>  N+  N-  FREQ {<expression>}=
+             ≪(frequency)value>,  <(magnitude in dB)value>,
              <(phase)value≫*
```

The input to the table is the value of <expression>, which follows the same rules as those mentioned in Section 4.5.1. The table contains the magnitude [in decibels (dB)] and the phase (in degrees) of the response for each frequency. Linear interpolation is performed between entries. The phase is interpolated linearly, and the magnitude is interpolated logarithmically with frequencies. For frequencies outside the table's range, the entry with the smallest (or largest) frequency is used.

The EFREQ and GFREQ parts are described by a table of frequency responses in either the magnitude/phase domain or the complex number domain. EFREQ and GFREQ properties are defined as follows:

Expr	Value used for table lookup; defaults to V(%IN +, %IN−) if left blank
TABLE	Series of either (input frequency, magnitude, phase) triplets or (input frequency, real part, imaginary part) triplets describing a complex value; defaults to (0,0,0) (1Meg,-10,90) if left blank
DELAY	Group delay increment; defaults to 0 if left blank
R I	Table type; if left blank, the frequency table is interpreted in the (input frequency, magnitude, phase) format; if defined with any value (such as YES), the table is interpreted in the (input frequency, real part, imaginary part) format
MAGUNITS	Units for magnitude where the value can be DB (decibels) or MAG (raw magnitude); defaults to DB if left blank
PHASEUNITS	Units for phase where the value can be DEG (degrees) or RAD (radians); defaults to DEG if left blank

If R_I, MAGUNITS, and PHASEUNITS are undefined, each table entry is interpreted as containing frequency, magnitude value in dB, and phase values in degrees. Delay defaults to 0.

The schematic for EFREQ, which can be found in the abm.lib, and the edit menu for setting the parameters for mathematical expressions are shown in Figure 4.27.

The schematic for GFREQ and the edit menu for setting the parameters are shown in Figure 4.28. The expressions for both the output voltage in E1 and the output current in G1 have the same expressions. The output for each frequency is then the input times the gain of (SQRT(V(%IN+, %IN−))*I(SENCE)) times the value of the table at that frequency.

Note: The table's frequencies must be in order from the lowest to the highest.

(a)

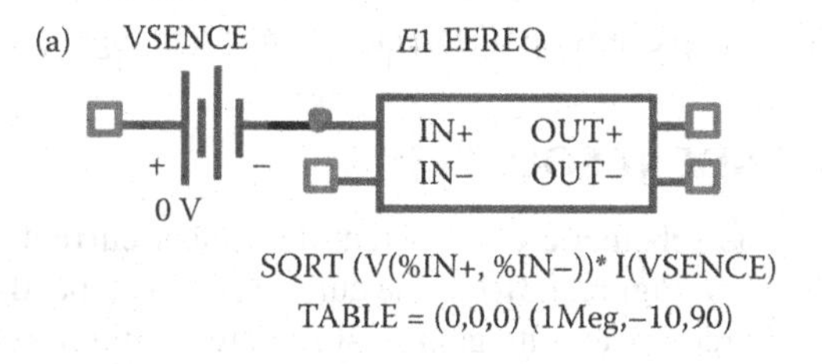

(b)

New Column... | Apply | Display... | Delete Property | Filter by: Orcad-PSpice | Help

		Reference	Value	DELAY	EXPR	MAGUNITS	PHASEUNITS	R_I	Source Part	TABLE
1	⊞ EXAMPLE B-16 : PAGE1 : E1	E1	EFREQ		SQRT(V(%IN+, %IN-))*I(VSENCE)				EFREQ.Normal	(0,0,0) (1Meg,-10,90)

FIGURE 4.27 Schematics for EFREQ mathematical expressions. (a) EFREQ schematic. (b) Editing EFREQ model parameters.

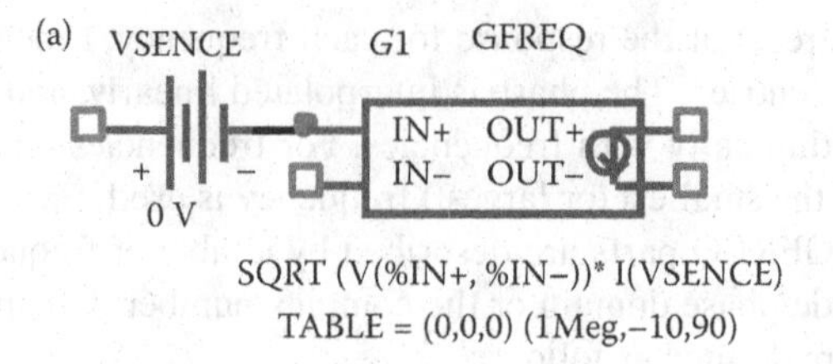

(b)

New Column... | Apply | Display... | Delete Property | Filter by: Orcad-PSpice | Help

		Reference	Value	DELAY	EXPR	MAGUNITS	PHASEUNITS	R_I	Source Part	TABLE
1	⊞ EXAMPLE B-16 : PAGE1 : G1	G1	GFREQ		SQRT(V(%IN+, %IN-))*I(VSENCE)				GFREQ.Normal	(0,0,0) (1Meg,-10,90)

FIGURE 4.28 Schematics for GFREQ mathematical expressions. (a) GFREQ schematic. (b) Editing GFREQ model parameters.

4.5.4.1 Typical Statements

The output voltage of a low-pass filter with input voltage V(2) can be expressed by

```
ELOWPASS  20  FREQ{V(2)} = (0, 0, 0)
+  (5kHz, 0, -57.6) (6kHz, 40, -69.2)
```

Note the following:

1. FREQ should be followed by a space.
2. <expression> must fit on one line.
3. FREQ frequencies must be in order from the lowest to the highest.

4.6 LTspice INDEPENDENT SOURCES

LTspice can also generate the dependent and independent voltage and current sources covered in Sections 4.2–4.5. All the sources and elements are available from the component menu icon ( ). The voltage source is shown in Figure 4.29(a) and the current source is shown in Figure 4.29(b). Both the voltage source as shown in Figure 4.29(a) and the current source as shown in Figure 4.29(b) can be assigned in different forms, e.g., DC, sinusoidal, pulse, piece-wise linear, etc. as shown in Figure 4.30(a)–(d).

4.7 LTspice DEPENDENT SOURCES

Figure 4.31(a) shows the schematic of a current-dependent current source and the attribute editor is shown in Figure 4.31(b). The current source depends on the current through the voltage source Vx and its gain is 40. There are three types of current-dependent voltage-source circuit elements. The general statement is as follows:

```
Symbol Name: F
Syntax: Fxxx n+ n- <Vnam> <gain>
Syntax: Fxxx n+ n- value={<expression>}
Syntax: Fxxx n+ n- POLY(<N>) <V1 V2 ... VN> <c0 c1 c2 c3 c4 ...>
```

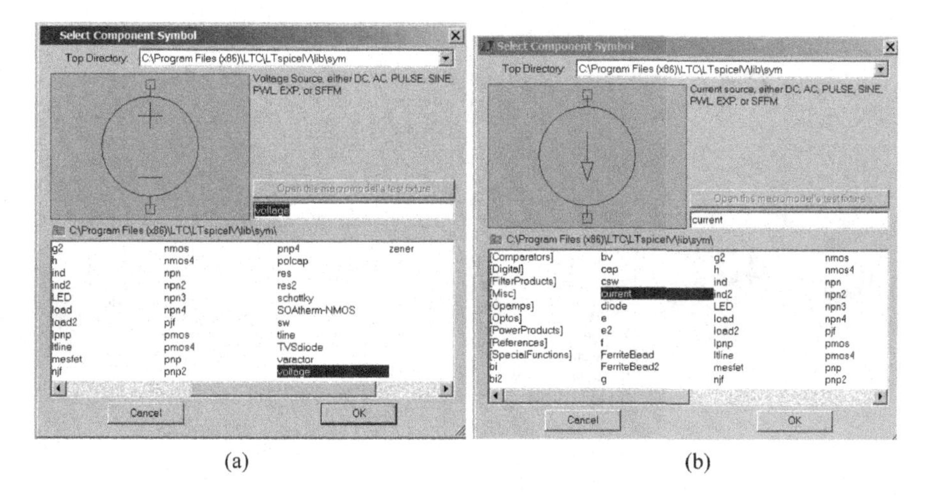

FIGURE 4.29 LTspice voltage and current sources. (a) Voltage source. (b) Current source.

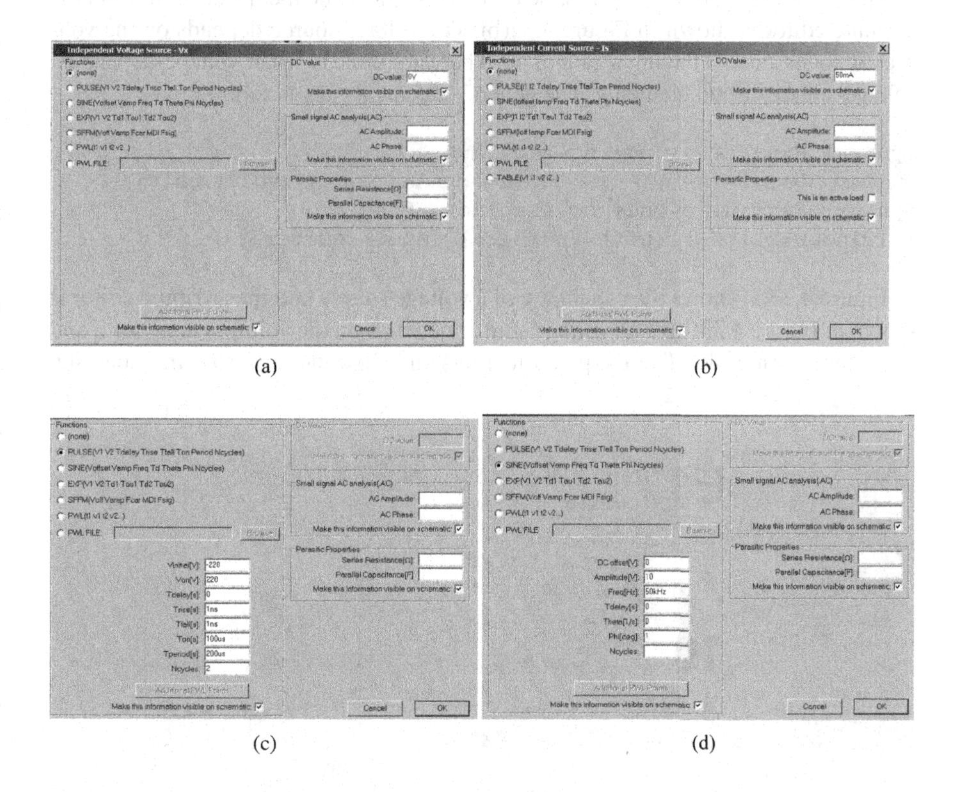

FIGURE 4.30 LTspice independent DC voltage and current sources. (a) DC Voltage. (b) DC Current source. (c) Pulse voltage. (d) Sinewave voltage.

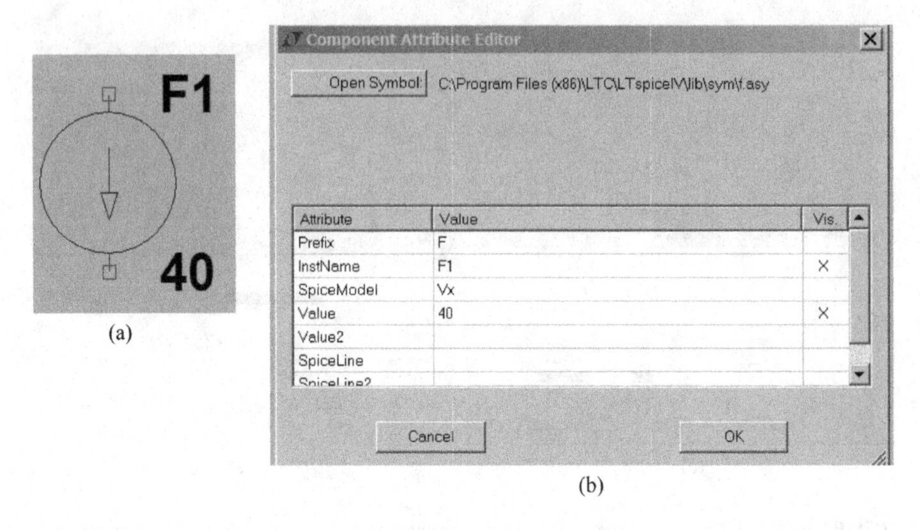

FIGURE 4.31 Current dependent current source. (a) Schematic. (b) Attribute editor.

Figure 4.32(a) shows the schematic of a voltage-dependent current source and the attribute editor is shown in Figure 4.32(b). The voltage source depends on the voltages at nodes 2 and 3 having a gain of 40. There are three types of voltage-dependent voltage-source circuit elements. The general statements are as follows:

```
Syntax: Exxx n+ n- nc+ nc- <gain>
Syntax: Exxx n+ n- nc+ nc- table=(<value pair>, <value pair>, ...)
Syntax: Exxx n+ n- nc+ nc- Laplace=<func(s)>
+ [window=<time>] [nfft=<number>] [mtol=<number>]
```

Figure 4.33(a) shows the schematic of a voltage source and the attribute editor is shown in Figure 4.33(b). The voltage source depends on the voltages at nodes 2 and 3 having a gain of 40. The LTspice functions and Operand are listed in Tables 4.6

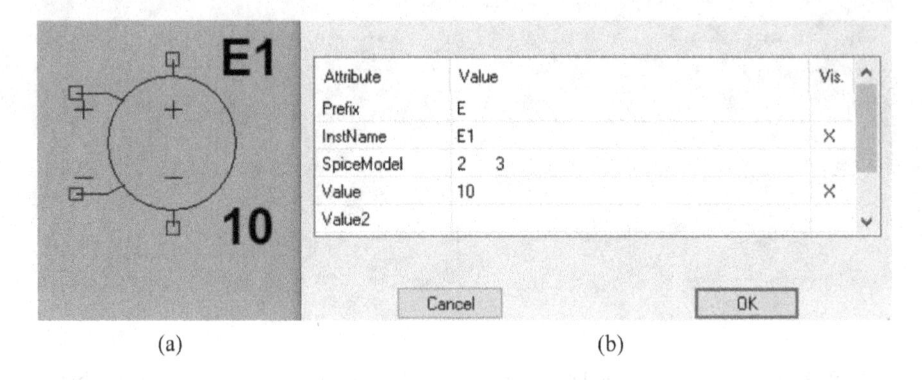

FIGURE 4.32 Voltage-dependent voltage source. (a) Schematic. (b) Attribute editor.

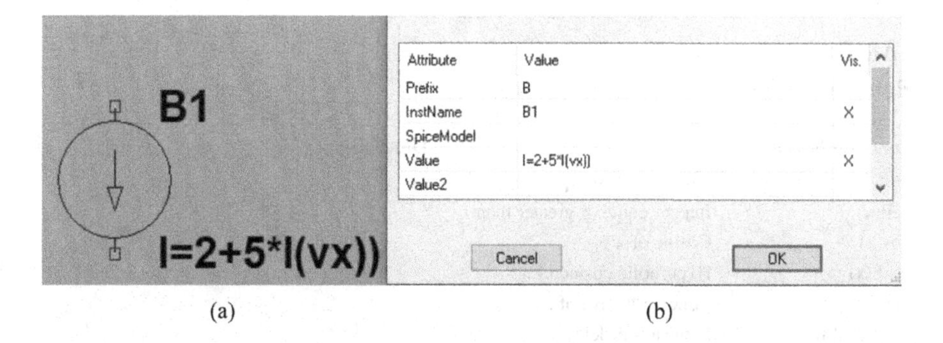

(a) (b)

FIGURE 4.33 Arbitrary behavioral voltage or current sources. (a) Schematic. (b) Attribute editor.

and 4.7, respectively. There are three types of voltage-dependent voltage-source circuit elements. The general statements are as follows:

```
Symbol names: BV, BI

Syntax: Bnnn n001 n002 V=<expression> [ic=<value>]
+ [tripdv=<value>] [tripdt=<value>]
+ [laplace=<expression> [window=<time>]
+ [nfft=<number>] [mtol=<number>]]

Bnnn n001 n002 I=<expression> [ic=<value>]
+ [tripdv=<value>] [tripdt=<value>] [Rpar=<value>]
+ [laplace=<expression> [window=<time>]
+ [nfft=<number>] [mtol=<number>]]
```

TABLE 4.6
Behavioral Functions

Function Name	Description
abs(x)	Absolute value of x
absdelay(x,t[,tmax])	x delayed by t. Optional max delay notification tmax.
acos(x)	Real part of the arc cosine of x, e.g., acos(−5) returns 3.14159, not 3.14159+2.29243i
arccos(x)	Synonym for acos()
acosh(x)	Real part of the arc hyperbolic cosine of x, e.g., acosh(.5) returns 0, not 1.0472i
asin(x)	Real part of the arc sine of x, asin(−5) is −1.57080, not −1.57080+2.29243i
arcsin(x)	Synonym for asin()
asinh(x)	Arc hyperbolic sine
atan(x)	Arc tangent of x
arctan(x)	Synonym for atan()
atan2(y,x)	Four quadrant arc tangent of y/x
atanh(x)	Arc hyperbolic tangent

(Continued)

TABLE 4.6 *(Continued)*
Behavioral Functions

Function Name	Description
buf(x)	1 if x > .5, else 0
ceil(x)	Integer equal or greater than x
cos(x)	Cosine of x
cosh(x)	Hyperbolic cosine of x
ddt(x)	Time derivative of x
delay(x,t[,tmax]	Same as absdelay()
exp(x)	e to the x
floor(x)	Integer equal to or less than x
hypot(x,y)	sqrt(x**2 + y**2)
idt(x[,ic[,a]])	Integrate x, optional initial condition ic, reset if a is true.
idtmod(x[,ic[,m[,o]]]	Integrate x, optional initial condition ic, reset on reaching modulus m, offset output by o.
if(x,y,z)	If x > .5, then y else z
int(x)	Convert x to integer
inv(x)	0. if x > .5, else 1.
limit(x,y,z)	Intermediate value of x, y, and z
ln(x)	Natural logarithm of x
log(x)	Alternate syntax for ln()
log10(x)	Base 10 logarithm
max(x,y)	The greater of x or y
min(x,y)	The smaller of x or y
pow(x,y)	Real part of x**y, e.g., pow(−1,.5)=0, not i.
pwr(x,y)	abs(x)**y
pwrs(x,y)	sgn(x)*abs(x)**y
rand(x)	Random number between 0 and 1 depending on the integer value of x.
random(x)	Similar to rand(), but smoothly transitions between values.
round(x)	Nearest integer to x
sdt(x[,ic[,assert]])	Alternate syntax for idt()
sgn(x)	Sign of x
sin(x)	Sine of x
sinh(x)	Hyperbolic sine of x
sqrt(x)	Square root of x
table(x,a,b,c,d,...)	Interpolate a value for x based on a look up table given as a set of pairs of points.
tan(x)	Tangent of x.
tanh(x)	Hyperbolic tangent of x
u(x)	Unit step, i.e., 1 if x > 0., else 0.
uramp(x)	x if x > 0., else 0.
white(x)	Random number between -.5 and .5 smoothly transitions between values even more smoothly than random().
!(x)	Alternative syntax for inv(x)
~(x)	Alternative syntax for inv(x)

TABLE 4.7
Behavioral Operand

<table>
<tr><th>Operand</th><th>Description</th></tr>
<tr><td>&</td><td>Convert the expressions to either side to Boolean, then AND.</td></tr>
<tr><td>|</td><td>Convert the expressions to either side to Boolean, then OR.</td></tr>
<tr><td>^</td><td>Convert the expressions to either side to Boolean, then XOR.</td></tr>
<tr><td>></td><td>True if expression on the left is greater than the expression on the right, otherwise false.</td></tr>
<tr><td><</td><td>True if expression on the left is less than the expression on the right, otherwise false.</td></tr>
<tr><td>>=</td><td>True if expression on the left is greater than or equal to the expression on the right, otherwise false.</td></tr>
<tr><td><=</td><td>True if expression on the left is less than or equal to the expression on the right, otherwise false.</td></tr>
<tr><td>+</td><td>Floating point addition</td></tr>
<tr><td>-</td><td>Floating point subtraction</td></tr>
<tr><td>*</td><td>Floating point multiplication</td></tr>
<tr><td>/</td><td>Floating point division</td></tr>
<tr><td>**</td><td>Raise left hand side to power of right hand side. Only the real part is returned, e.g., −1**1.5 gives zero not i.</td></tr>
<tr><td>!</td><td>Convert the following expression to Boolean and invert.</td></tr>
</table>

SUMMARY

The PSpice variables can be summarized as follows:

PULSE Pulse source

```
PULSE (V1 V2 TD TR TF PW PER)
```

PWL Piecewise linear source

```
PWL (T1 V1 T2 V2 ... TN VN)
```

SIN Sinusoidal source

```
SIN (VO VA FREQ TD ALP THETA)
```

EXP Exponential source

```
EXP (V1 V2 TRD TRC TFD TFC)
```

SFFM Single-frequency frequency modulation

```
SFFM (VO VA FC MOD FS)
```

POLY Polynomial source

```
POLY(n) <(controlling)nodes> <(coefficients)values>
```

E Voltage-controlled voltage source

```
E<name> N+ N- NC+ NC- <(voltage gain)value>
```

F Current-controlled current source

```
F <name> N + N - VN <(current gain) value >
```

G Voltage-controlled current source

```
G <name> N + N - NC + NC - <(transconductance) value>
```

H Current-controlled voltage source

```
F <name> N + N - VN <(current gain) value >
```

H Current-controlled voltage source

```
H <name> N + N - VN <(transresistance) value>
```

I Independent current source

```
I<name> N+ N- [dc <value>][ac <(magnitude)value>
+  <(phase)value>][(transient specifications)]
```

V Independent voltage source

```
V<name> N+ N- [dc <value>][ac<(magnitude)value>]
+  <(phase)value>][(transient specifications)]
```

VALUE Arithmetical function

```
E <name> N + N - VALUE = {<expression >}
G <name> N + N - VALUE = {<expression >}
```

TABLE Look-up table

```
E<name> N+ N-  TABLE{<expression>} =
   + ≪(input)value>, <(output)value≫*
G<name>  N+  N-  TABLE{<expression>} =
   + ≪(input)value>, <(output)value≫*
```

LAPLACE Laplace's transfer function

```
E<name> N+ N- LAPLACE {<expression>}={<transform>}
G<name> N+ N- LAPLACE {<expression>}={<transform>}
```

FREQ Frequency response transfer function

```
E<name> N+  N-  FREQ{<expression>} =
   + ≪(frequency)value>,  <(magnitude)
   value>,  <(phase) value≫*
G<name> N+  N-  FREQ{<expression>} =
   + ≪(frequency)value>,<(magnitude)
   value>,  <(phase) value≫*
```

PROBLEMS

Write PSpice statements for the following circuits. Assume that the first node is the positive terminal and the second node is the negative terminal.

4.1 The various voltage or current waveforms that are connected between nodes 4 and 5 are shown in Figure P4.1.

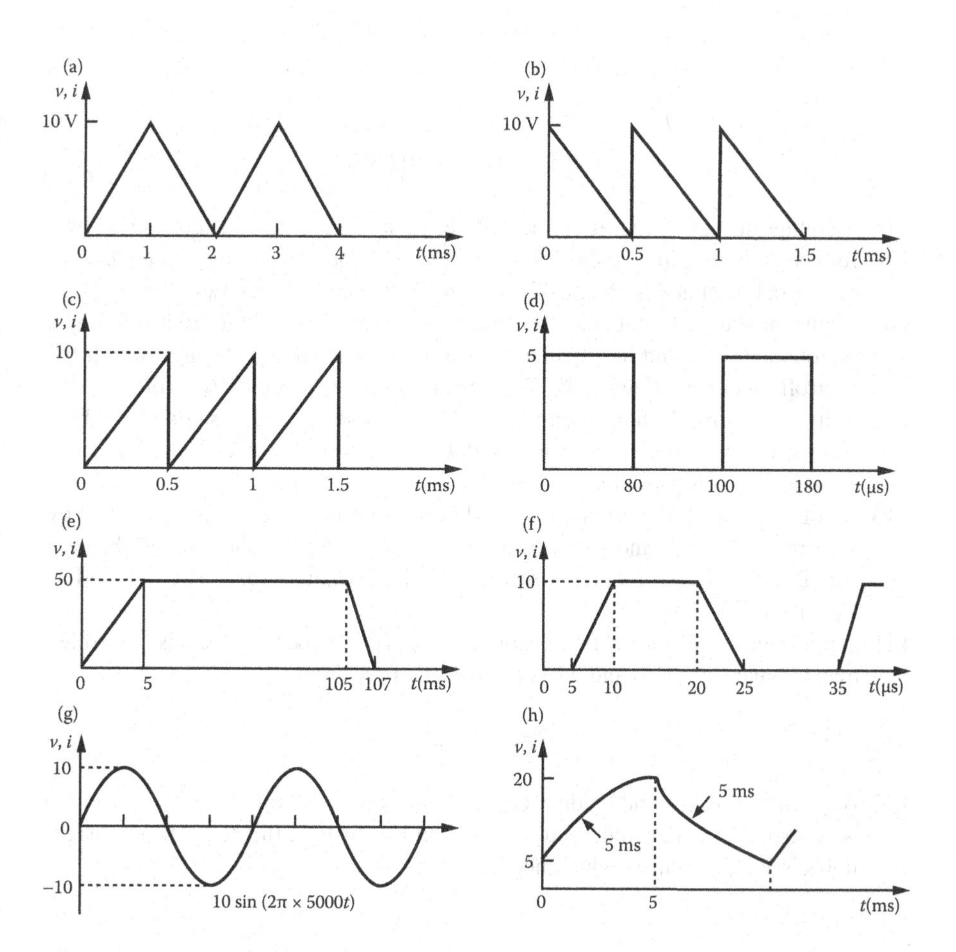

FIGURE P4.1 Problem P4.1.

4.2 A voltage source that is connected between nodes 10 and 0 has a DC voltage of 12 V for DC analysis, a peak voltage of 2 V with a 60° phase shift for AC analysis, and a sinusoidal peak voltage of 0.1 V at 1 MHz for transient analysis.

4.3 A current source that is connected between nodes 5 and 0 has a DC of 0.1 A for DC analysis, a peak current of 1 A with 60° phase shift for AC analysis, and a sinusoidal current of 0.1 A at 1 kHz for transient analysis.

4.4 A voltage source that is connected between nodes 4 and 5 is given by

$$v = 2\sin[(2\pi \times 50{,}000t) + 5\sin(2\pi \times 1000t)]$$

4.5 A polynomial voltage source Y that is connected between nodes 1 and 2 is controlled by a voltage source V_1 connected between nodes 4 and 5. The source is given by

$$Y = 0.1V_1 + 0.2V_1^2 + 0.05V_1^2$$

4.6 A polynomial current source I that is connected between nodes 1 and 2 is controlled by a voltage source V_1 connected between nodes 4 and 5. The source is given by

$$Y = 0.1V_1 + 0.2V_1^2 + 0.05V_1^2$$

4.7 A voltage source V_0 that is connected between nodes 5 and 6 is controlled by a voltage source V_1 and has a voltage gain of 25. The controlling voltage is connected between nodes 10 and 12. The source is expressed as $V_0 = 25V_1$.

4.8 A current source I_0 that is connected between nodes 5 and 6 is controlled by a current source I_1 and has a current gain of 10. The voltage through which the controlling current flows is V_c. The current source is given by $I_0 = 10I_1$.

4.9 A current source I_0 that is connected between nodes 5 and 6 is controlled by a voltage source V_1 between nodes 8 and 9. The transconductance is 0.05 Ω^{-1}. The current source is given by $I_0 = 0.051V_1$.

4.10 A voltage source V_0 that is connected between nodes 5 and 6 is controlled by a current source I_1 and has a transresistance of 150 Ω. The voltage through which the controlling current flows is V_c. The voltage source is expressed as $V_0 = 150I_1$.

4.11 A nonlinear resistance R that is connected between nodes 4 and 6 is controlled by a voltage source V_1 and has a resistance of the form

$$R = V_1 + 0.2V_1^2$$

4.12 A nonlinear transconductance G_m that is connected between nodes 4 and 6 is controlled by a current source. The voltage through the controlling current flows is V_1. The transconductance has the form

$$G_m = V_1 + 0.2V_1^2$$

4.13 The V–I characteristic of a diode is described by

$$I_D = I_S e^{v_D / n V_T}$$

where $I_S = 2.2 \times 10^{-15}$ A, $n = 1$, and $V_T = 26.8$ mV. Use VALUE to simulate the diode voltage between nodes 3 and 4 as a function of diode current.

4.14 The V–I characteristic of a diode is described by

$$I_D = I_S e^{v_D / n V_T}$$

where $I_S = 2.2 \times 10^{-15}$ A, $n = 1$, and $V_T = 26.8$ mV. Use VALUE to simulate the diode current between nodes 3 and 4 as a function of diode voltage.

4.15 The V–I characteristic of a diode is given by

i_D (A)	10	20	30	40	50	100	150	500	900
v_D (V)	0.55	0.65	0.7	0.75	0.8	0.9	1.0	1.3	1.4

Use TABLE to simulate the diode voltage between nodes 3 and 4 as a function of diode current.

4.16 The current through a varistor depends on its voltage, which is given by

i (mA)	0	0.1	0.3	1	2.5	5	15	45	100	200
v (V)	0	50	100	150	200	250	300	350	400	450

Use TABLE to simulate the varistor current between nodes 3 and 4 as a function of varistor current.

SUGGESTED READING

1. M.H. Rashid, *Introduction to PSpice Using OrCAD for Circuits and Electronics*, Third Edition, Englewood Cliffs, NJ: Prentice-Hall, 2003, Chapters 3 and 4.
2. M.H. Rashid, *SPICE for Power Electronics and Electric Power*, Englewood Cliffs, NJ: Prentice-Hall, 1993.
3. M.H. Rashid, SPICE for Circuits and Electronics Using LTspice® Schematics, PSpice® Schematics, and OrCAD® Capture, Fourth Edition, Cengage Learning, India, 2019
4. P.W. Tuinenga, *SPICE: A Guide to Circuit Simulation and Analysis Using PSpice*, Third Edition, Englewood Cliffs, NJ: Prentice-Hall, 1995.
5. MicroSim Corporation, *PSpice* Manual, Irvine, CA, 1992.
6. R. Bracewell, *The Fourier Transform and Its Applications*, Second Edition, New York: McGraw-Hill, 1986.
7. Cadence Design Systems, Inc, *PSpice User Guide*, 2000.
8. *LTspice IV Manual*, Milpitas, CA: Linear Technology Corporation, 2013.

5 Passive Elements

After completing this chapter, students should be able to do the following:

- List the model type names of elements.
- Write the general model statement of elements.
- Model passive elements.
- Model magnetic circuits and transformers.
- Model lossless transmission lines.
- Describe the parameters of switches.

5.1 INTRODUCTION

PSpice recognizes passive elements by their symbols and models. The elements can be resistors (*R*), inductors (*L*), capacitors (*C*), or switches; they can also be magnetic. The models are necessary to take into account parameter variations (e.g., the value of a resistance depends on the operating temperature). The simulation of passive elements in PSpice requires that the following be specified:

- Modeling of elements
- Operating temperature
- *RLC* elements
- Magnetic elements and transformers
- Lossless transmission lines
- Switches

5.2 MODELING OF ELEMENTS

A model that specifies a set of parameters for an element is specified in PSpice by the .MODEL command. The same model can be used for one or more elements in the same circuit. The various dot commands are explained in Chapter 6. The general form of the model statement is

```
.MODELM NAME TYPE(P1=B1 P2=B2
+       P3=B3 … … PN=BN[(tolerance specification)]*)
```

where MNAME is the name of the model and must start with a letter. Although not necessary, it is advisable to make the first letter the symbol of the element (e.g., R for resistor and L for inductor). The list of symbols for elements is shown in Table 2.2. P1, P2, … are the element parameters, and B1, B2, … are their values, respectively. TYPE is the type name of the elements as shown in Table 5.1. An element must have the correct model type name. That is, a resistor must have the type name RES, not

DOI: 10.1201/9781003284451-5

TABLE 5.1
Type Names of Elements

Breakout Device Name	Type Name	Resistor
Rbreak	RES	Resistor
Cbreak	CAP	Capacitor
Dcbreak	D	Diode
Lbreak	IND	Inductor
QbreakN	**NPN**	Bipolar junction transistor
QbreakP	PNP	Bipolar junction transistor
JbreakN	NJF	*n*-Channel junction FET
JbreakP	PJF	*p*-Channel junction FETT
MbreakN	NMOS	*n*-Channel MOSFET
MbreakP	PMOS	*p*-Channel MOSFET
Bbreak	GASFET	*n*-Channel GaAs MOSFET
Sbreak	VSWITCH	Voltage-controlled switch
Wbreak	ISWITCH	Current-controlled switch
XFRM_**LIN**EAR	None	Linear magnetic core (transformer)
XFRM_NON**LIN**EAR	CORE	Nonlinear magnetic core (transformer)
ZbreakN	NIGBT	*n*-Channel IGBT

type IND or CAP. However, there can be more than one model of the same type in a circuit with different model names.

Note: Tolerance specification is used with .MC analysis only, and it may be appended to each parameter with the following format:

```
[ DEV/<distribution name> <value in % from 0 to 9>]
[ LOT/<distribution name> <value in % from 0 to 9>]
```

where <distribution name> is one of the following:

UNIFORM Generates uniformly distributed deviations over the range of ± <value>

GAUSS Generates deviations with Gaussian distribution over the range ±4, and <value> specifies the ±1 deviation

Note: Once the schematic circuit is drawn, PSpice schematic and OrCAD create a default model of the circuit element and there is no need for defining the model statement. The user can add or modify the model parameters of the circuit elements.

In PSpice Schematics, the user can assign a model name for the breakout devices in the library breakout.slb as shown in Figure 5.1(a). The user can also specify the model parameters. PSpice can open the menu with the model name, Rbreak, as shown in Figure 5.1(b), and the model name can be changed.

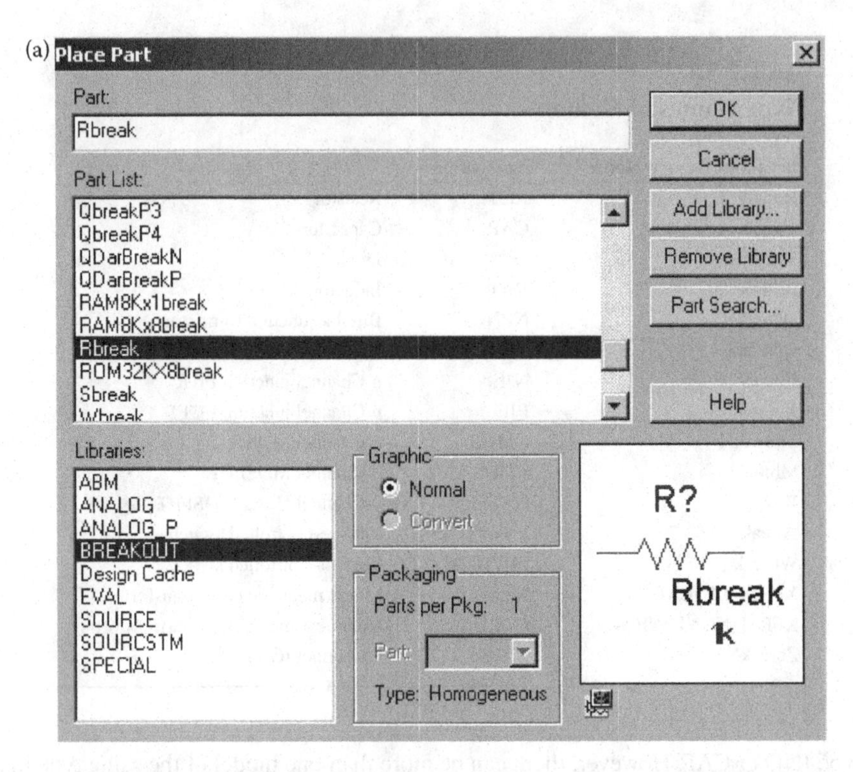

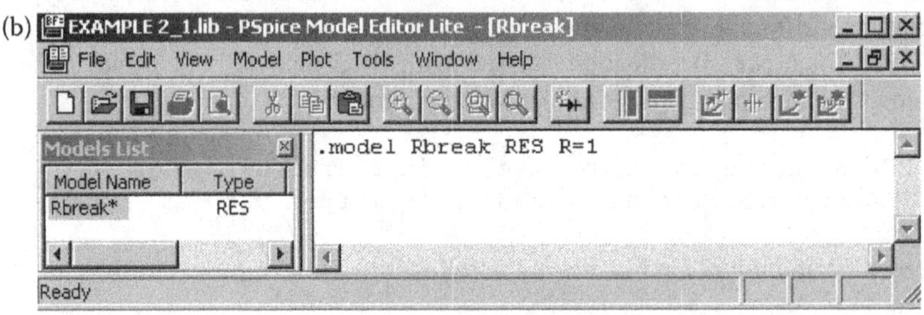

FIGURE 5.1 Breakout library for models. (a) Breakout devices. (b) PSpice model editor.

5.2.1 Some Model Statements

```
.MODEL   RLOAD      RES   (R=1 TC1 =0.02 TC2 =0.005)
.MODEL   RLOAD      RES   (R=1 DEV/GAUSS 0.5% LOT/UNIFORM 10%)
.MODEL   CPASS      CAP   (C=1 VC1 =0.01 VC2 =0.002 TC1 = 0.02 TC2 =0.005)
.MODEL   LFILTER    IND   (L=1 IL1 =0.1 IL2 =0.002 TC1 = 0.02 TC2 =0.005)
.MODEL   DNOM       D     (IS=1E-9)
.MODEL   DLOAD      D     (IS =1E-9 DEV 0.5% LOT 10%)
.MODEL   QOUT       NPN   (BF=50 IS=1E-9)
```

Note: Model type name RES is used for a resistor, CAP for capacitor, IND for inductor, D for a semiconductor diode, and **NPN** for an **NPN**-bipolar transistor. The details of the model parameters in the statements are explained under each circuit element.

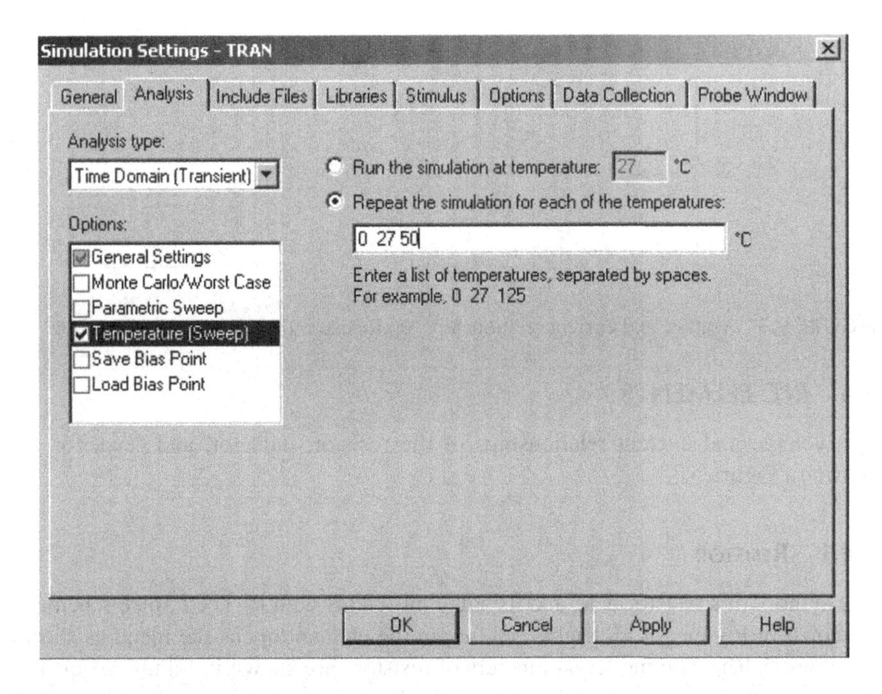

FIGURE 5.2 Setting up operating temperature.

You can create and save the .MODEL statements for different models of circuit elements and devices in a text file with the MS Notepad. Then you can include the model file in the "Include Files" menu as shown in Figure 5.2.

5.3 OPERATING TEMPERATURE

The operating temperature of an analysis can be set to any value desired by the. **TEMP** command. The general form of the statement is

```
.TEMP <(one or more temperature) values>
```

The temperatures are in degrees Celsius. If more than one temperature is specified, the analysis is performed for each temperature. The model parameters are assumed to be measured at a nominal temperature, which is by default 27°C. The default nominal temperature of 27°C can be changed by the TNOM option in the **.OPTIONS** statements, which are discussed in Section 6.5.

5.3.1 Some Temperature Statements

```
.TEMP 50
.TEMP 25 50
.TEMP 0  25 50 100
```

The operating temperature can be set from the analysis setup as shown in Figure 5.2.

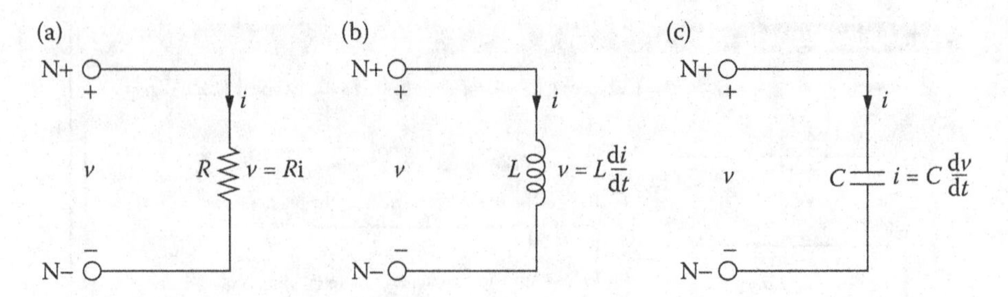

FIGURE 5.3 Voltage and current relationships. (a) Resistor. (b) Inductor. (c) Capacitor.

5.4 *RLC* ELEMENTS

The voltage and current relationships of the resistor, inductor, and capacitor are shown in Figure 5.3.

5.4.1 Resistor

The symbol of a resistor is R, and its name must start with R. The PSpice schematic is shown in Figure 5.4(a). Right-clicking on the element opens the menu as shown in Figure 5.4(b). The model parameters of resistors are shown in Figure 5.4(c). The resistor's name and its nominal value can be changed. Also, a tolerance value can be assigned to it.

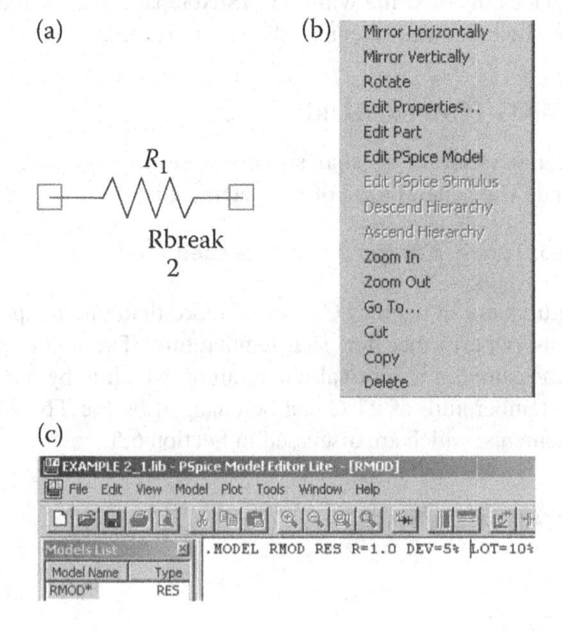

FIGURE 5.4 Resistor schematics and parameters. (a) Symbol. (b) Element menu. (c) Resistor's model parameters.

Its description takes the general form

```
R<name> N+ N- RNAME RVALUE
```

A resistor does not have a polarity, and the order of the nodes does not matter. However, by defining N+ as the positive node and N– as the negative node, the current is assumed to flow from node N+ through the resistor to node N–.

Note: Once you place a circuit element in the PSpice schematic or OrCAD, PSpice assumes that a positive current flows from the left side (say node 1) to the right side (say node 2) of the element. If you turn the element, the positive current will still flow from node 1 to node 2.

RNAME is the model name that defines the parameters of the resistor. RVALUE is the nominal value of the resistance.

Note: Some versions of PSpice or SPICE do not recognize the polarity of resistors and do not allow referring currents through the resistor [e.g., I(R1)].

The model parameters are shown in Table 5.2. If RNAME is omitted, RVALUE is the resistance in ohms, and RVALUE can be positive or negative, but must not be zero. If RNAME is included and TCE is not specified, the resistance as a function of temperature is calculated from

```
RES=RVALUE * R *[1 + TC1 * (T - TO) + TC2 * (T - TO)²]
```

If RNAME is included and TCE is specified, the resistance as a function of temperature is calculated from

```
TCE* (T - TO)
RES = RVALUE * R * 1.01
```

where T and T0 are the operating temperature and room temperature, respectively, in degrees Celsius.

5.4.1.1 Some Resistor Statements

```
R1          6       5       10k
RLOAD       13      11      ARES 2MEG
.MODEL      ARES    RES     (R=1 TC1=0.02 TC2=0.005)
RINPUT      15      14      RRES 5k
.MODEL      RRES    RES     (R=1 TCE=2.5)
```

TABLE 5.2
Model Parameters for Resistors

Name	Meaning	Unit	Default
R	Resistance multiplier		1
TC1	Linear temperature coefficient	$°C^{-1}$	0
TC2	Quadratic temperature coefficient	$°C^{-2}$	0
TCE	Exponential temperature coefficient	%/°C	0

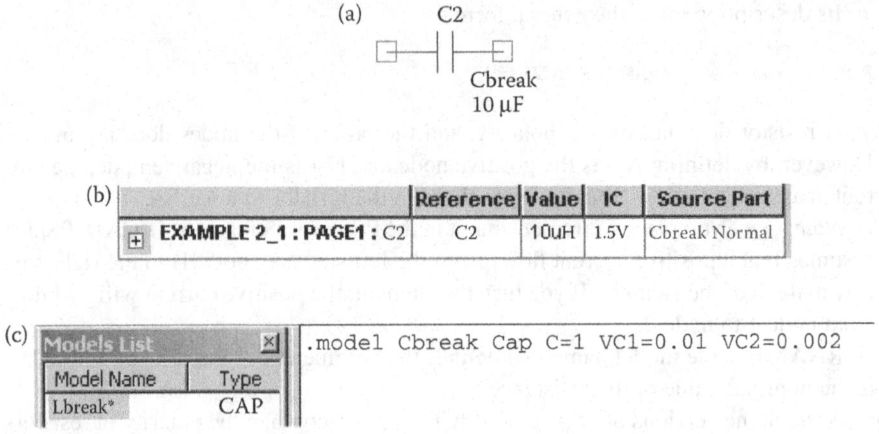

(a) C2

Cbreak
10 μF

(b)

	Reference	Value	IC	Source Part
EXAMPLE 2_1 : PAGE1 : C2	C2	10uH	1.5V	Cbreak Normal

(c)

Models List

Model Name	Type
Lbreak*	CAP

```
.model Cbreak Cap C=1 VC1=0.01 VC2=0.002
```

FIGURE 5.5 Capacitor schematics and parameters. (a) Breakout model. (b) Initial condition. (c) Capacitor's model parameters.

5.4.2 Capacitor

The symbol of a capacitor is C, and its name must start with C. The schematic of a capacitor with a breakout model is shown in Figure 5.5(a). Its value and initial condition can be changed as shown in Figure 5.5(b). The model parameters of capacitors are shown in Figure 5.5(c). The capacitor's name and its nominal value can be changed. Its description takes the general form

```
C<name> N+ N- CNAME CVALUE IC=V0
```

where N+ is the positive node and N– is the negative node. The voltage of node N+ is assumed positive with respect to node N– and the current flows from node N+ through the capacitor to node N–. CNAME is the model name and CVALUE is the nominal value of the capacitor. IC defines the initial (time-zero) voltage of the capacitor, V0.

The model parameters are shown in Table 5.3. If CNAME is omitted, CVALUE is the capacitance in farads. The CVALUE can be positive or negative but must not

TABLE 5.3
Model Parameters for Capacitors

Name	Meaning	Unit	Default
C	Capacitance multiplier		1
VC1	Linear voltage coefficient	V^{-1}	0
VC2	Quadratic voltage coefficient	V^{-2}	0
TC1	Linear temperature coefficient	$°C^{-1}$	0
TC2	Quadratic temperature coefficient	$°C^{-2}$	0

be zero. If CNAME is included, the capacitance, which depends on the voltage and temperature, is calculated from

```
CAP = CVALUE * C * (1 + VC1 * V + VC2 * V²)
                  *[1 + TC1 *(T - T0) + TC2 *
                  (T - T0)²]
```

where T is the operating temperature and T0 is the room temperature in degrees Celsius.

5.4.2.1 Some Capacitor Statements

```
C1        6    5    10UF
CLOAD     12   11   5PF    IC=2.5V
CINPUT    15   14   ACAP   10PF
C2        20   19   ACAP   20NF      IC=1.5V
.MODEL ACAP CAP (C=1 VC1=0.01 VC2=0.002 TC1=0.02 TC2=0.005)
```

Note: The initial conditions (if any) apply only if the UIC (use initial condition) opinion is specified on the .TRAN command that is described in Section 6.9.

For LTspice: A capacitor can be selected directly from the capacitor symbol (⏊) in the main menu. We can specify the capacitance, voltage, and current ratings as shown in Figure 5.6(a). It also allows to specify the internal series and parallel resistances. One can also select a standard capacitor from a list of many types and manufacturers by clicking on "Select capacitor" as shown in Figure 5.6(a). The values and model attributes, e.g., CMOD model, can be edited as shown in Figure 5.6(b).

5.4.3 Inductor

The symbol of an inductor is L, and its name must start with L. The schematic of an inductor with a breakout model is shown in Figure 5.7(a). Its value and initial condition can be changed as shown in Figure 5.7(b). The typical model parameters of inductors are shown in Figure 5.7(c). The inductor's name and its nominal value can be changed.

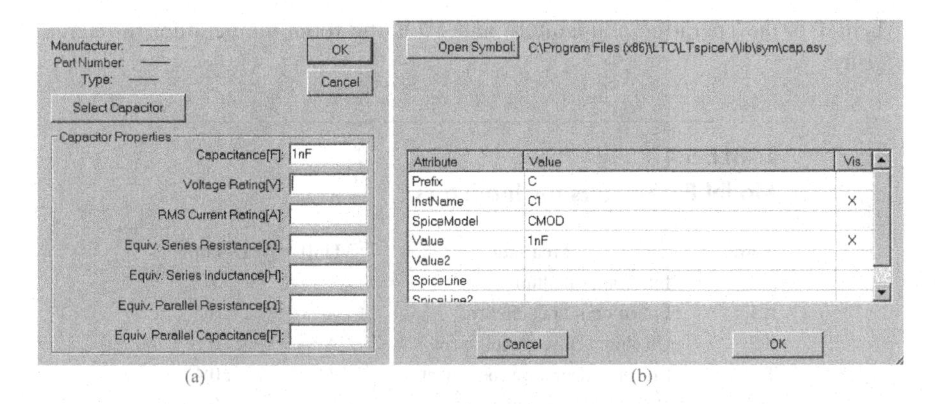

FIGURE 5.6 Capacitor menu. (a) Capacitor. (b) Capacitor attribute menu.

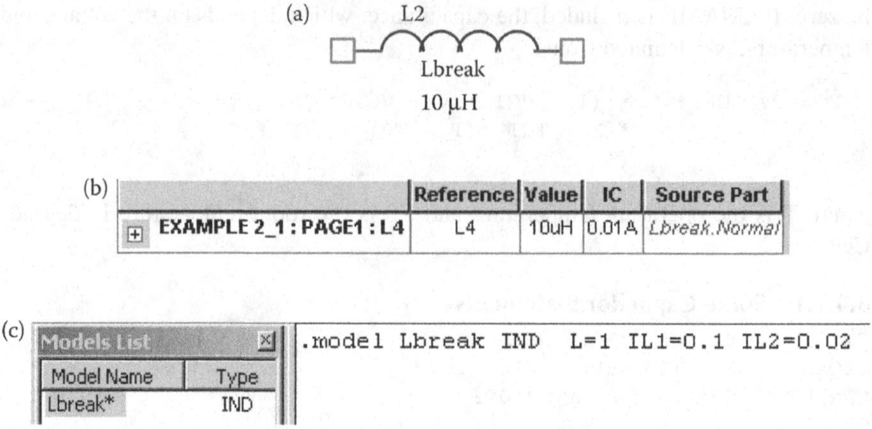

(a) L2

Lbreak

10 μH

(b)

	Reference	Value	IC	Source Part
EXAMPLE 2_1 : PAGE1 : L4	L4	10uH	0.01A	*Lbreak.Normal*

(c)

Models List

Model Name	Type
Lbreak*	IND

```
.model Lbreak IND  L=1 IL1=0.1 IL2=0.02
```

FIGURE 5.7 Inductor schematics and parameters. (a) Symbol. (b) Initial condition. (c) Inductor's model parameters.

Its description takes the general form

```
L<name> N+ N- LNAME LVALUE IC=I0
```

where N+ is the positive node and N– is the negative node. The voltage of N+ is assumed positive with respect to node N–, and the current flows from node N+ through the inductor to node N–. LNAME is the model name, and LVALUE is the nominal value of the inductor. IC defines the initial (time-zero) current I0 of the inductor.

The model parameters of an inductor are shown in Table 5.4. If LNAME is omitted, LVALUE is the inductance in henries. LVALUE can be positive or negative but must not be zero. If LNAME is included, the inductance, which depends on the current and temperature, is calculated from

```
IND = LVALUE * L * (1 + IL1 * I + IL2 * I^2)
           * [1 + TC1 * (T - T0) + TC2 *(T - T0)^2]
```

where T is the operating temperature and T0 is the room temperature in degrees Celsius.

TABLE 5.4
Model Parameters for Inductors

Name	Meaning	Unit	Default
L	Inductance multiplier		1
IL1	Linear current coefficient	A^{-1}	0
IL2	Quadratic current coefficient	A^{-2}	0
TC1	Linear temperature coefficient	$^{\circ}C^{-1}$	0
TC2	Quadratic temperature coefficient	$^{\circ}C^{-2}$	0

Note: The initial conditions (if any) apply only if the UIC (use initial condition) option is specified on the .TRAN command that is described in Section 6.9.

5.4.3.1 Some Inductor Statements

```
L1        6     5     10MH
LLOAD    12    11     5UH        IC=0.2MA
LLINE    15    14     LMOD       5MH
LCHOKE   20    19     LMOD       2UH            IC=0.5A
.MODEL LMOD IND (L=1 IL1=0.1 IL2=0.002 TC1=0.02 TC2=0.005)
```

Note: m stands for milli (10^{-3}) and MEG stands for mega (10^{6}) in PSpice statements.

Example 5.1:

Step Response of an RLC *Circuit with an Initial Inductor Current and Capacitor Voltage*

The *RLC* circuit of Figure 5.8(a) is supplied with the input voltage as shown in Figure 5.8(b). Calculate and plot the transient response from 0 to 1 ms with a time increment of 5 µs. The output voltage is taken across resistor R_2. The results should be available for display and as hard copy by using Probe. The model parameters are, for the resistor, R = 1, TC1 = 0.02, and TC2 = 0.005; and for the inductor, L = 1, IL1 = 0.1, IL2 = 0.002, TC1 = 0.02, and TC2 = 0.005. The operating temperature is 50°C.

SOLUTION

The PSpice schematic is shown in Figure 5.9(a). The voltage marker displays the output waveform in Probe at the end of the simulation. The step input voltage is specified by a PWL source as shown in Figure 5.9(b). The breakout devices are used to specify the parameters of model RMD for R_1 and R_2 as shown in Figure 5.9(c), of model LMOD for L_1 as shown in Figure 5.9(d), and of model CMOD for C_1 as shown in Figure 5.9(e). The Transient Analysis is set at the Analysis Setup as shown in Figure 2.7(a), and its specifications are set at the transient menu as shown in Figure 2.7(b). The operating temperature is set from the simulations setting menu as shown in Figure 5.2.

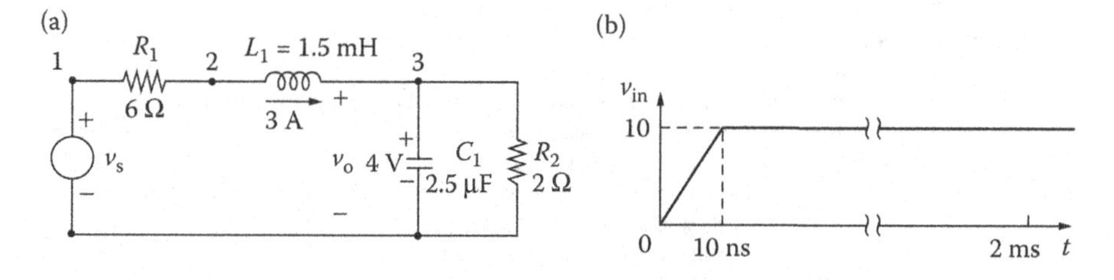

FIGURE 5.8 RLC circuit for Example 5.1. (a) Circuit. (b) Input voltage.

(a)

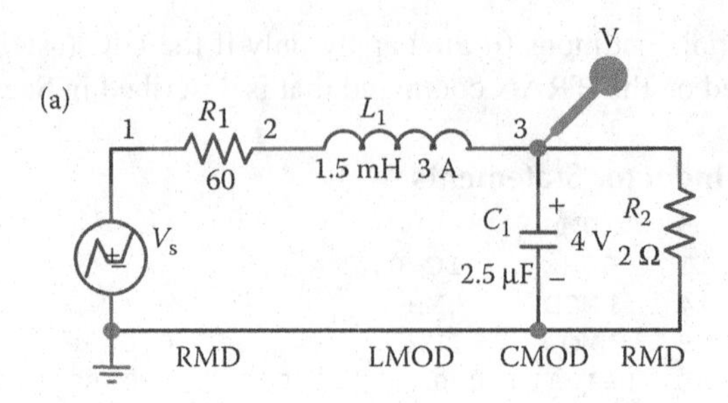

(b)

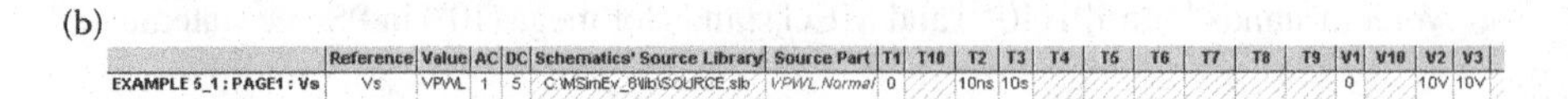

	Reference	Value	AC	DC	Schematics' Source Library	Source Part	T1	T10	T2	T3	T4	T5	T6	T7	T8	T9	V1	V10	V2	V3
EXAMPLE 5_1 : PAGE1 : Vs	Vs	VPWL	1	5	C:\MSimEv_8\lib\SOURCE.slb	VPWL.Normal	0		10ns	10s							0		10V	10V

(c)

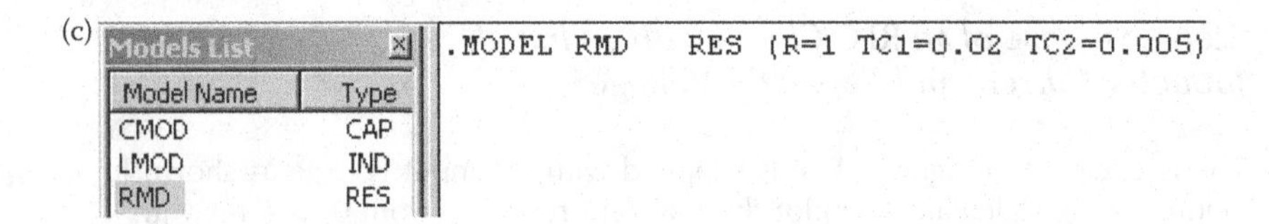

(d)

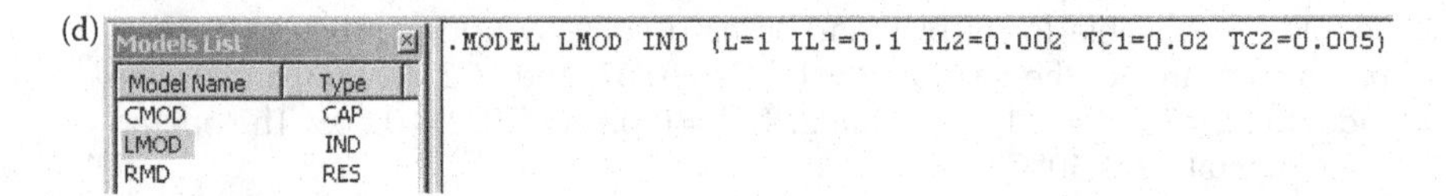

(e)

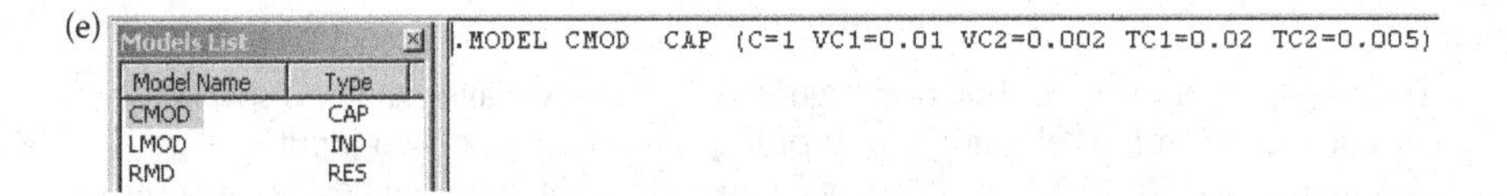

FIGURE 5.9 PSpice schematic for Example 5.1. (a) PSpice schematic. (b) PWL parameters. (c) RMD parameters. (d) LMOD parameters. (e) CMOD parameters.

The circuit file contains the following statements:

RLC Circuit

```
SOURCE    ■  * Input step voltage represented as a PWL
             * waveform: VS 1 0 PWL (0 0 10NS 10V 2MS 10V)
CIRCUIT   ■■ * R1 has a value of 60 O with model RMOD:
             R1 1 2 RMOD 60
             * Inductor of 1.5 mH with an initial current of
             * 3 A and model name LMOD:
             L1 2 3 LMOD 1.5MH IC=3A
             * Capacitor of 2.5 μF with an initial voltage of
             * 4 V and model name CMOD:
             C1 3 0 CMOD 2.5UF IC=4V
             R2 3 0 RMOD 2
             * Model statements for resistor, inductor, and
```

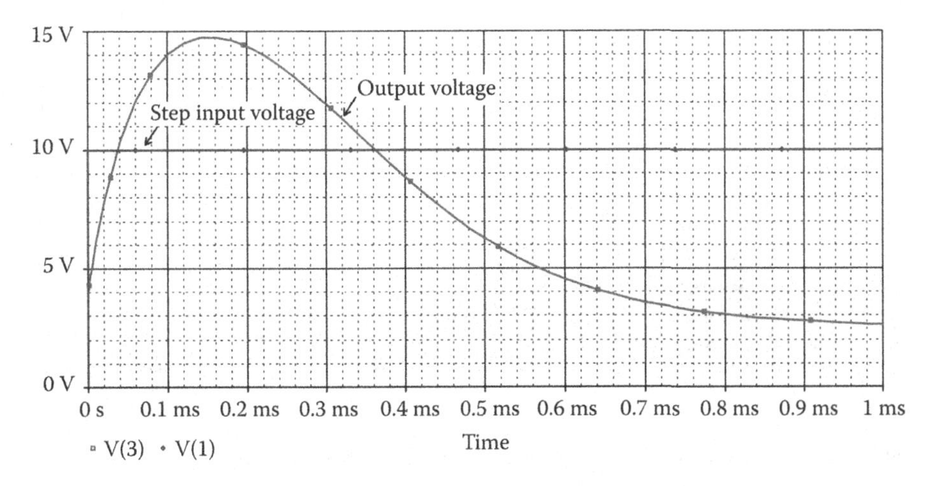

FIGURE 5.10 Transient response for Example 5.1.

```
           * capacitor:
           .MODEL RMOD RES (R=1 TC1=0.02 TC2=0.005)
           .MODEL CMOD CAP (C=1 VC1=0.01 VC2=0.002 TC1=0.02
           + TC2=0.005)
           .MODEL LMOD IND (L=1 IL1=0.1 IL2=0.002 TC1=0.02
           + TC2=0.005)
           * The operating temperature in 50°C.
           .TEMP 50
ANALYSIS ■■■ * Transient analysis from 0 to 1 ms with a 5.µs
           * time increment and using initial conditions (UIC).
           .TRAN 5US 1MS UIC
           * Plot the results of transient analysis with
           * the voltage at nodes 3 and 1.
           .PLOT TRAN V(3) V(1)
           .PROBE V(3) V(1)
           .END
```

The results of the simulation that are obtained by Probe are shown in Figure 5.10. The inductor has an initial current of 3 A, which is taken into consideration by UIC in the .TRAN command.

Note: The output has an overshoot of 47.43% above the input voltage during the transient period. With these model parameters, the resistance value will depend nonlinearly with the temperature, the capacitor value will be a nonlinear function of its voltage, and the inductor value will be a nonlinear function of its current.

LTspice: An inductor can be selected directly from the inductor symbol (3) in the main menu. We can specify the inductance and peak current ratings as shown in Figure 5.11(a). It also allows to specify the internal series and parallel resistances. We can also select a standard inductor from a list of many types and manufacturers by clicking on "Select Inductor" as shown in Figure 5.11(a). The values and model attributes, e.g., LMOD model, can also be edited as shown in Figure 5.11(b).

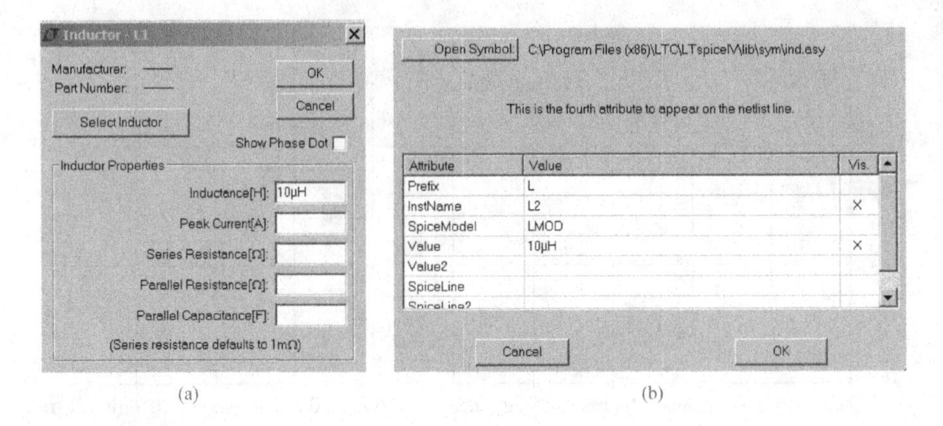

FIGURE 5.11 Inductor menu. (a) Inductor. (b) Inductor attribute menu.

5.5 MAGNETIC ELEMENTS AND TRANSFORMERS

The magnetic elements are mutual inductors (transformer). PSpice allows the simulating two types of magnetic circuits:

- Linear magnetic circuits
- Nonlinear magnetic circuits

5.5.1 Linear Magnetic Circuits

The symbol of mutual coupling is K. The general form of coupled inductors is

```
K<name> L<(first inductor) name> L<(second inductor) name>
+       <(coupling) value>
```

For linear coupled inductors, K<name> couples two or more inductors. <(coupling) value> is the coefficient of coupling, k. The value of the coefficient of coupling must be greater than 0 and less than or equal to 1: $0 < k \le 1$.

The inductors can be coupled in order either positively or negatively as shown in Figure 5.12(a) and (b), respectively. Two-coupled inductors can be connected as shown Figure 5.12(c) to simulate the function of a center-tapped transformer. In terms of the dot convention shown in Figure 5.12(a), PSpice assumes a dot on the first node of each inductor. The mutual inductance is determined from

$$M = k\sqrt{L_1 L_2}$$

In the time domain, the voltages of coupled inductors are expressed as

$$v_1 = L_1 \frac{di_1}{dt} + M \frac{di_2}{dt}$$

$$v_2 = M \frac{di_1}{dt} + L_2 \frac{di_2}{dt}$$

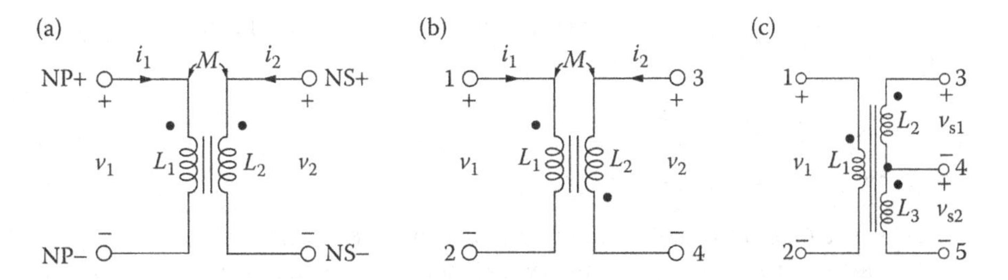

FIGURE 5.12 Coupled inductors. (a) Positively coupled. (b) Negatively coupled. (c) Single-phase transformer.

In the frequency domain, the voltages are expressed as

$$v_1 = j\omega L_1 I_1 + j\omega M I_2$$
$$v_2 = j\omega M I_1 + j\omega L_2 I_2$$

where ω is the frequency in rad/s.

Some coupled-inductor statements are

```
KTR  LA LB 0.9
KIND L1 L2 0.98
```

The PSpice schematic for coupled inductors is shown in Figure 5.13(a). The K-linear in the analog.slb library that can couple up to six inductors is shown in Figure 5.13(b) with two inductors L_1 and L_2.

The statements of the coupled inductors as shown in Figure 5.12(a) can be written as a single-phase transformer (with $k = 0.9999$):

```
* PRIMARY
L1        1  2  0.5
MH * SECONDARY
L2        3  4  0.5 MH
* MAGNETIC COUPLING
KXFRMER L1  L2  0.9999
```

(a) K *K12*

K_Linear
Coupling = 1

(b)

		Reference	Value	K13	COUPLING	L1	L2	L3	L4	L5	L6
1	⊞ EXAMPLE 5_2 : PAGE1 : K12	K12	K_Linear		0.999	L1	L2				

FIGURE 5.13 Coupled inductors with K-linear. (a) Symbol. (b) Parameters for K-linear.

If the dot in the second coil is changed as shown in Figure 5.12(b) the statements of the coupled inductors are written as

```
L1          1   2  0.5MH
L2          4   3  0.5MH
KXFRMER    L1  L2  0.9999
```

Note: The order of L_1 and L_2 does not matter in the K-statement.

A transformer with a single primary coil and center-tapped secondary as shown in Figure 5.11(c) can be written as

```
* PRIMARY
L1 1 2 0.5MH
* SECONDARY
L2 3 4 0.5MH
L3 4 5 0.5MH
* MAGNETIC COUPLING
K12 L1 L2 0.9999
K13 L1 L3 0.9999
K23 L2 L3 0.9999
```

The three statements above ($K12$, $K13$, and $K23$) can be written in PSpice as

```
KALL L1 L2 L3 0.9999
```

Note the following:

1. The name Kxx need not be related to the names of the inductors it is coupling. However, it is a good practice because it is convenient to identify the inductors involved in the coupling.
2. The polarity (or dot) is determined by the order of the nodes in the L … statements and not by the order of the inductors in the K … statement (e.g., *K*12 *L*1 *L*2 0.9999) has the same result as (*K*12 *L*2 *L*1 0.9999)].
3. If N_1 and N_2 are the number of turns of the primary and secondary sides, respectively, the secondary voltage v_2 is related to the primary voltage v_1 by

$$v_2 = \frac{N_2}{N_1} v_1$$

If N_1 and N_2 are related to the primary inductance L_1 and the secondary inductance L_2 by

$$\frac{N_2}{N_1} = \sqrt{\frac{L_2}{L_1}}$$

this relates voltage v_2 in terms of L_1 and L_2 as

$$v_2 = \sqrt{\frac{L_2}{L_1}} v_1$$

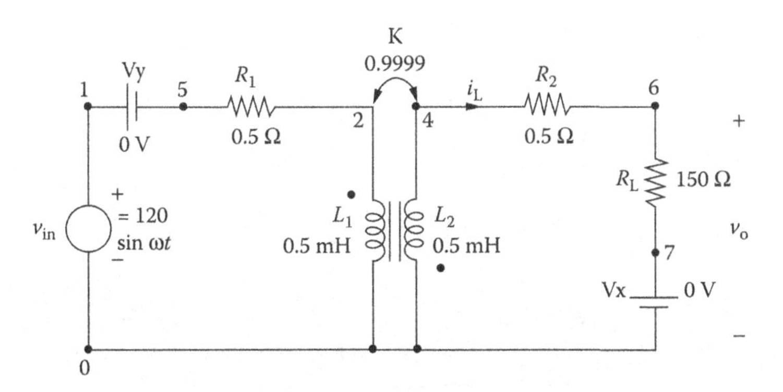

FIGURE 5.14 A circuit with two coupled inductors.

Example 5.2:

Finding the RMS Currents and the Frequency Plot of a Coupled-Inductor Circuit

A circuit with two coupled inductors is shown in Figure 5.14. The input voltage is 120 V peak. Calculate the magnitude and phase of the output current for frequencies from 60 to 120 Hz with a linear increment. The total number of points in the sweep is 2. The coefficient of coupling for the transformer is 0.999.

SOLUTION

It is important to note that the primary and the secondary have a common node. Without this common node, PSpice will give an error message because there is no DC path from the nodes of the secondary to the ground.

The PSpice schematic is shown in Figure 5.15(a). K-linear couples *L1* and *L2* with a coupling of 0.9999. The voltage marker displays the output waveforms in Probe at the end of the simulation. The PSpice plot of the input and output voltages are shown in Figure 5.15(b). The break frequency is f_b = 79.81 Hz at 84.906 V (70.7%) of the high-frequency value of 120 V.

Note: Vx and Vy are used as ammeters to measure the currents I(Vx) and I(Vy), respectively. In PSpice schematic, one could also use I(R1) and I(RL) to measure the currents.

The circuit file contains the following statements:

Coupled Linear Inductors

```
SOURCE    ■  * Input voltage is 120 V peak and 0°C phase for
             * ac analysis:
             VIN 1 0 AC 120V
CIRCUIT   ■■ R1 5 2 0.5
             * A dummy voltage source of VY = 0 is added to
             * measure the load current:
             VY 1 5 DC 0V
             * The dot convention is followed in inductors L1
             * and L2:
             L1 2 0 1mH
```

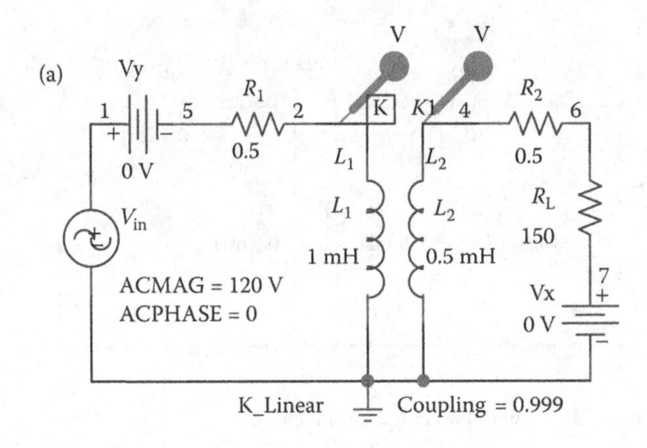

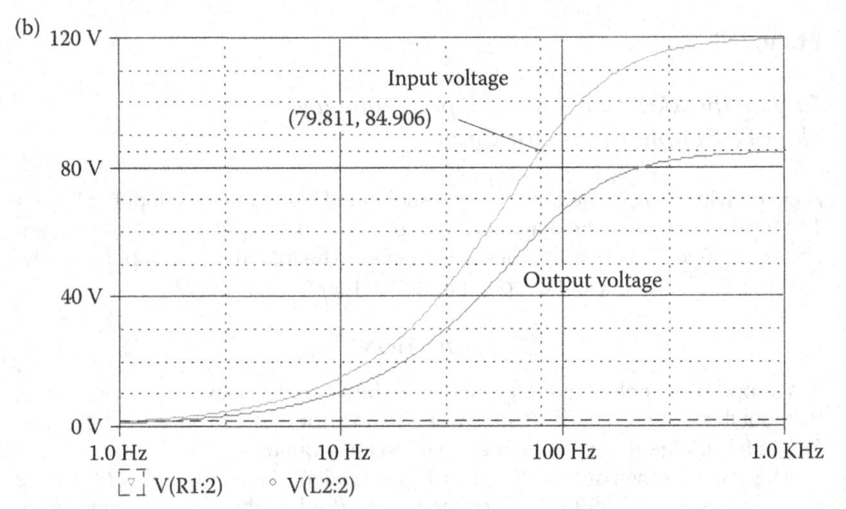

FIGURE 5.15 PSpice schematic for Example 5.2. (a) Schematic (b) Frequency responses.

```
L1 0 4 0.5mH
* Magnetic coupling coefficient is 0.999. The
* order of L1 and L2 is
* not significant.
K12 L1 L2 0.999
R2    4   6 0.5
RL    6   7 150
* A dummy voltage source of VX = 0 is added to
* measure the load current:
VX 7 0 DC 0V
```

ANALYSIS ■■■

```
* AC analysis where the frequency is varied
* linearly from 60 Hz to 120 Hz with two points:
AC LIN 260 HZ 120 HZ
* Print the magnitude and phase of output
* current. Some versions of
* PSpice and SPICE do not permit reference to
* peak currents through resistors [e.g., IM(RL),
* IP(RL)].
```

```
.PRINT AC IM(VY) IP(VY) IM(RL) IP(RL)
; Prints to the output file
.END
```

The transformer is considered to be linear and its inductances remain constant. The results of the simulation, which are stored in the output file EX5.2.OUT, are

```
FREQ       IM(VY)     IP(VY)      IM(RL)     IP(RL)      V(4)       V(2)
6.000E+01  1.915E+02  -3.699E+01  3.389E-01  -1.271E+02  5.100E+01  7.220E+01
1.200E+02  1.325E+02  -5.635E+01  4.689E-01  -1.465E+02  7.056E+01  9.989E+01
```

Note: PSpice (i.e., the net list using PSpice A/D) is better suited to printing the magnitudes and phases of voltages. Schematics is, however, better suited to drawing the schematics and plots of voltages and currents. One could also use the **VPRINT** and **IPRINT** symbols in the special.lib to print the voltage and current, respectively.

5.5.2 Nonlinear Magnetic Circuits

For a nonlinear inductor, the general form is

```
K<name> L<(inductor) name> <(coupling) value>
+ <(model) name>  [(size) value]
```

For an iron-core transformer, k is very high, greater than 0.999. The model type name for a nonlinear magnetic inductor is CORE, and the model parameters are shown in Table 5.5. [`(size) value`] scales the magnetic cross-section and defaults to 1. It represents the number of lamination layers so that only one model statement can be used for a particular lamination type of core.

If the `<(model) name>` is specified, the mutual coupling inductor becomes a nonlinear magnetic core and the inductor specifies the number of turns instead of the inductance. The list of the coupled inductors may be just one inductor.

The magnetic core's B–H characteristics are analyzed using the Jiles–Atherton model [2].

TABLE 5.5
Model Parameters for Nonlinear Magnetic

Name	Meaning	Unit	Default
AREA	Mean magnetic cross-section	cm^2	0.1
PATH	Mean magnetic path length	cm	1.0
GAP	Effective air-gap length	cm	0
PACK	Pack (stacking		1.0
MS	Magnetic saturation	A/m	1E+6
A	Shape parameter		1E+3
C	Domain wall-flexing constant		0.2
K	Domain wall-pinning constant		500

If the inductors of Figure 5.11(a) use the nonlinear core, the statements would be as follows:

```
* Inductor L1 of 100 turns:
L1 1 2 100
* Inductor L2 of 10 turns:
L2 3 4 10
* Nonlinear coupled inductors with model CMOD:
K12 L1 L2 0.9999 CMOD
* Model for the nonlinear inductors:
.MODEL CMOD CORE (AREA=2.0 PATH=62.8 GAP=0.1 PACK=0.98)
```

In PSpice Schematics, the nonlinear magnetic parameters can be described by using a K-break coupling as shown in Figure 5.16(a). The K-nonlinear model in the breakout.slb library that can couple up to six inductors is shown in Figure 5.16(b) with two inductors L_1 and L_2.

The model parameters can be adjusted to specify a B–H characteristic. The nonlinear magnetic model uses MKS (metric) units. However, the results for Probe are converted to Gauss and Oersted and may be displayed using B(Kxx) and H(Kxx). The B–H curve can be drawn by a transient run with a slowly rising current through a test inductor and then by displaying B(Kxx) against H(Kxx).

Characterizing core materials may be carried out by trial by using PSpice and Probe. The procedures for setting parameters to obtain a particular characteristic are as follows:

1. Set the domain wall-pinning constant K = 0. The curve should be centered in the B–H loop. The slope of the curve at $H = 0$ should be approximately equal to that when it crosses the H axis at $B = 0$.
2. Set the magnetic saturation MS = $B_{max}/0.01257$.
3. ALPHA sets the slope of the curve. Start with the mean-field parameter ALPHA = 0, and then vary its value to get the desired slope of the curve. It may be necessary to change MS slightly to get the desired saturation value.
4. Change K to a nonzero value to create the desired hysteresis. K affects the opening of the hysteresis loop.
5. Set C to obtain the initial permeability. Probe displays the permeability, which is $\Delta B/\Delta H$. Because Probe calculates differences, not derivatives, the curves will not be smooth. The initial value of $\Delta B/\Delta H$ is the initial permeability.

Note: A nonlinear magnetic model is not available in SPICE2.

(a) K *K12*

Kbreak

Coupling = 0.99

(b)

	Reference	Value	COUPLING	L1	L2	L3	L4	L5	L6
SCHEMATIC1 : PAGE1 : K12	K12	Kbreak	0.99	100	10				

FIGURE 5.16 Nonlinear coupled inductors. (a) Symbols. (b) Parameters for K-nonlinear.

Example 5.3:

Plotting the Nonlinear B–H *Characteristic of a Magnetic Circuit*

The coupled inductors of Figure 5.12(a) are nonlinear. The parameters of the inductors are L_1 = 200 turns, L_2 = 100 turns, and k = 0.9999. Plot the B–H characteristic of the core from the results of transient analysis if the input current is varied very slowly from 0 to −15 A, −15 to +15 A, and +15 to −15 A. A load resistance of R_L = 1 kΩ is connected to the secondary of the transformer. The model parameters of the core are AREA = 2.0, PATH = 62.73, GAP = 0.1, MS = 1.6E+6, ALPHA = 1E−3, A = 1E+3, C = 0.5, and K = 1500.

SOLUTION

The PSpice schematic is shown in Figure 5.17(a). K-break couples L_1 and L_2 with a coupling of 0.9999 and its model parameters are specified in K_CMOD. The PSpice plots of the instantaneous input current and flux density are shown in Figure 5.17(b).

The circuit file for the coupled inductors in Figure 5.17(a) would be

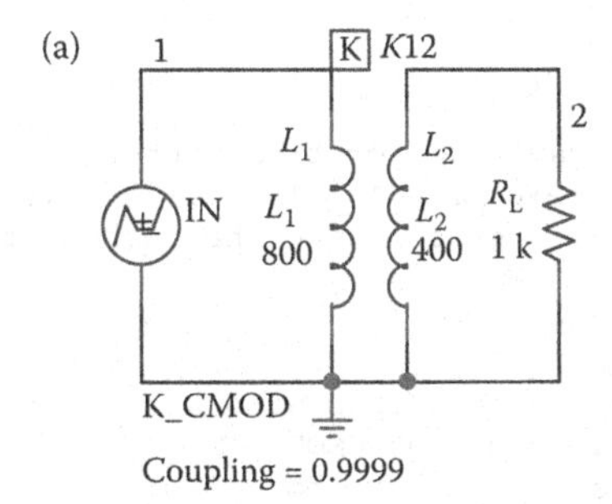

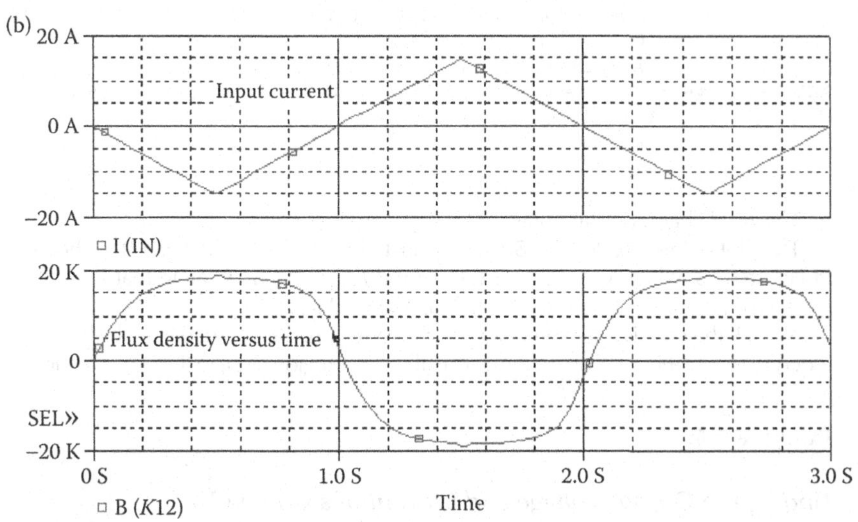

FIGURE 5.17 PSpice schematic for Example 5.3. (a) Schematic. (b) Instantaneous input current and flux density.

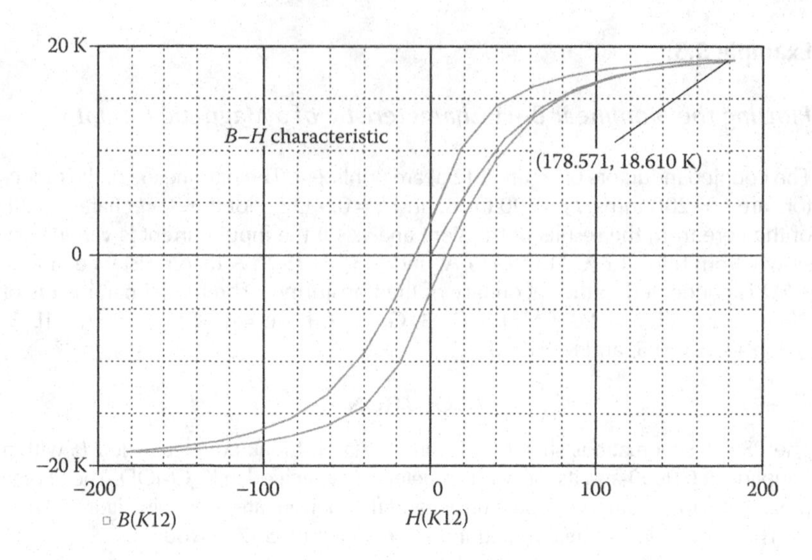

FIGURE 5.18 Typical *B–H* characteristic.

Typical *B–H* Characteristic

```
SOURCE   ■    * PWL waveform for transient analysis:
              IN 1 0 PWL (0 0 1 - 15 2 15 3 - 15)
CIRCUIT  ■■   * Inductors represent the number of turns:
              L1 1 0 800
              L2 2 0 400
              RL 2 0 1000                  ; Load resistance
              * Coupled inductors with k = 0.999 and model CMOD:
              K12 L1 L2 0.9999 CMOD
              * Model parameters for CMOD:
              .MODEL CMOD CORE (AREA=2.0 PATH=62.73
              + GAP=0.1 MS=1.6E+6)
              * A = 1E+3 C = 0.5 K = 1500)
ANALYSIS ■■■  * Transient analysis from 0 to 3 s in steps of 0.05 s:
              .TRAN 0.05S 3S
              .PROBE
              .END
```

The *B–H* characteristic obtained by Probe is shown in Figure 5.18.

The plot of the flux density, B(*K*12), against the magnetic field, H(*K*12), is shown in Figure 5.18, which gives B_{max} = 178.57 and H_{max} = 18.61 *k*. Note that by using the Probe menu, the *x*-axis has been changed to H(*K*12).

Note: If the core is operated in the saturation region, the output will be highly distorted. To avoid this, the core is normally in the linear region close to the zero-crossing.

Example 5.4:

Finding the Output Voltage and Current of a Coupled Inductor with a Nonlinear Magnetic Core

The characteristic of a nonlinear core can be included in the coupled inductor that exhibits a non-linear *B–H* characteristic. This is shown in Figure 5.19. The

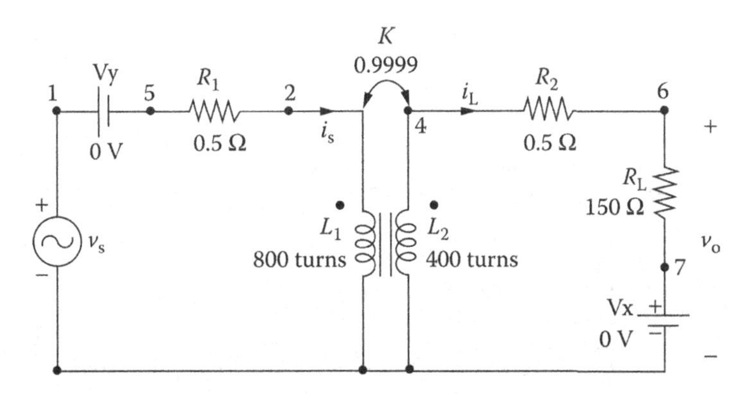

FIGURE 5.19 Circuit with nonlinear coupled inductors.

parameters of the inductors are L_1 = 800 turns, L_2 = 400 turns, and k = 0.9999. The input voltage is $v_s = 170 \sin(2\pi \times 60t)$. Plot the instantaneous values of the secondary voltage and current from 0 to 35 ms with a 10-μs increment. The results should be available for display and as hard copy by using Probe. The model parameters of the core are AREA = 2.0, PATH = 62.73, GAP = 0.1, MS = 1.6E+6, ALPHA = 1E-3, A = 1E+3, C = 0.5, and K = 1500.

SOLUTION

The primary and the secondary have a common node. Without this common node, PSpice will give an error message because there is no DC path from the nodes of the secondary to the ground. The PSpice schematic is shown in Figure 5.20(a). The K-break couples L_1 and L_2 with a coupling of 0.9999 and its model parameters are specified in K_CMOD as shown in Figure 5.20(b). The voltage and current markers display the output waveforms in Probe at the end of the simulation.

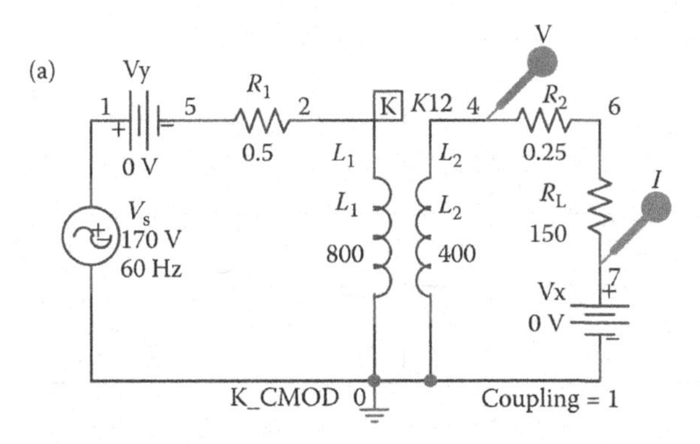

(b)

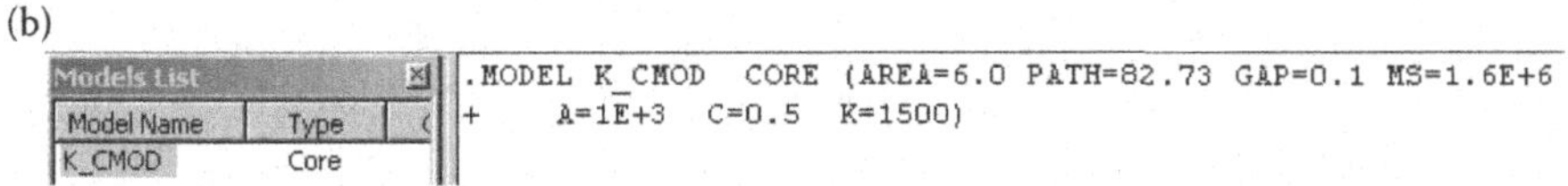

FIGURE 5.20 PSpice schematic for Example 5.4. (a) Schematic. (b) Model parameters for K_CMOD.

The circuit file contains the following statements:

Nonlinear Coupled Inductors

```
SOURCE    ■    * Input sinusoidal voltage of 170 V peak and
               * 0° phase: VS 1 0 SIN (0 170V 60HZ)
CIRCUIT   ■■   * A dummy voltage source of VY = 0 is added to
               * measure the load current:
               VY  1  5  DC 0V
               R1  5  2  0.5
               * Inductors represent the number of turns:
               L1 2 0 800
               L2 4 0 400
               * Coupled inductors with k = 0.9999 and model
               * CMOD:
               K12 L1 L2 0.9999 CMOD
               * Model parameters for CMOD:
               .MODEL CMOD CORE (AREA=2.0 PATH=62.73
               + GAP=0.1 MS=1.6E+6+A=1E+3 C=0.5 K=1500)
               R2  4 6 0.5
               RL  6 7 150
               * A dummy voltage source of VX = 0 is added to
               * measure the load current:
               VX 7 0 DC 0V
ANALYSIS  ■■■  *Transient analysis from 0 to 35 ms with a
               * 100-µs increment:
               .TRAN 10US 35MS
               * Print the output voltage and current.
               .PRINT TRAN V(4) I(VX)   ; Prints in the output
                                          file
               .PROBE                   ; Graphics
                                          post-processor
               .END
```

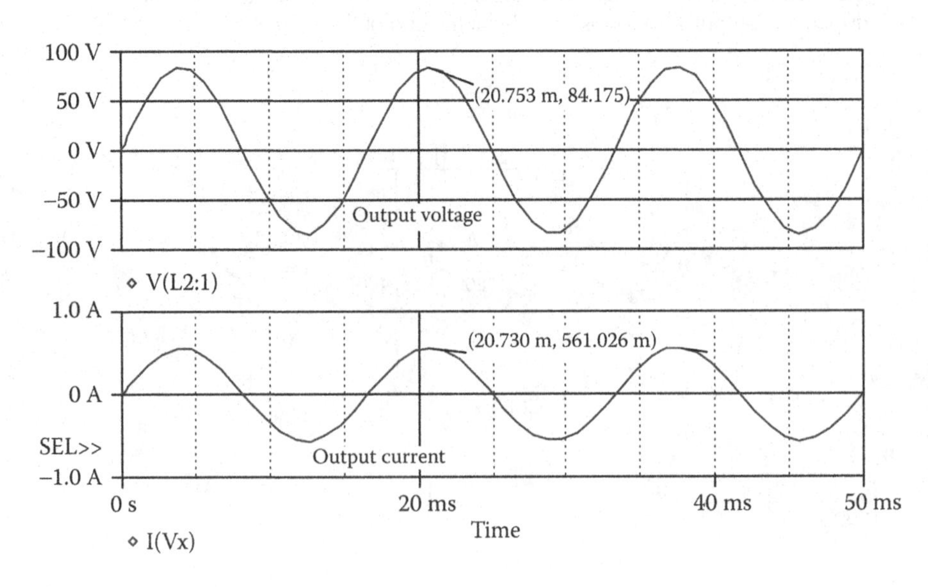

FIGURE 5.21 Transient response for Example 5.4.

The results of the simulation that are obtained by Probe are shown in Figure 5.21. Because of the nonlinear *B–H* characteristic, the input and output currents will become nonlinear and contain harmonics, depending on the number of turns and the parameters of the magnetic core.

Note: There is an amount of distortion in the waveforms. The Fourier analysis of the output voltage and the output current will give the total harmonic diction (THD) of the waveform.

5.6 LTspice MAGNETIC ELEMENTS

An inductor is made of coils wound on a magnetic core. A typical characteristic of a magnetic core is shown in Figure 5.22(a). Hc and Br are the intersections of the major hysteresis loop with the H- and B-axes. Bs is the B-axis intersection of the asymptotic line, Bsat(H) = Bs + $\mu_0 \cdot$H, approached as H goes to infinity. The inductance value is dependent on the operating point of the nonlinear characteristic. In order to minimize the complexity of a circuit analysis and the computation time, linear inductances are often used unless more accurate results are warranted.

5.6.1 FLUX DEPENDENT INDUCTANCE MODEL

The inductance is specified with an expression for the flux as shown in Figure 5.22(b). The inductor's current is referred to by the keyword "x" in the expression of the flux ϕ as given by

$$\phi = 10^{-3} \times \tanh(5x)$$

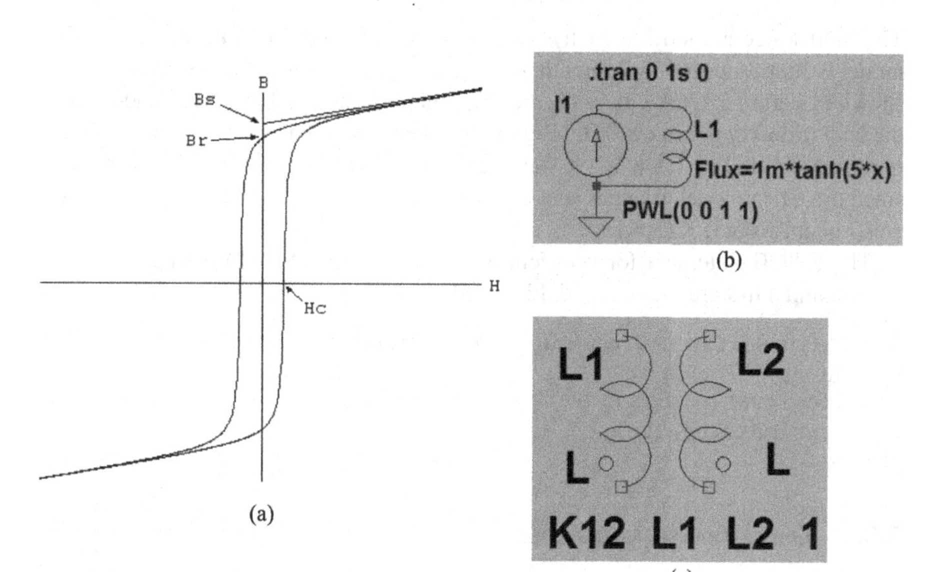

FIGURE 5.22 Typical B-H characteristic and Inductors. (a) Typical B-H characteristic. (b) Flux depended inductor. (c) Coupled inductors.

The SPICE statement for the transient analysis of Figure 5.22(b) with a unity ramp current given is listed as

```
L1 N001 0 Flux=1m*tanh(5*x)
I1 0 N001 PWL(0 0 1 1)
.tran 1
```

Since I1 supplies a unity di/dt so that the inductance can be read off as the voltage on node N001. That is, $v = L(di/dt) = L$.

5.6.2 Linear Coupled Inductors

Two or more coupled inductors can be modeled as transformers as shown in Figure 5.22(c). Adding the coupling statement to the schematic shows the coupling dots. The dots should be in opposite directions as shown in Figure 5.22(c). The turn ratio is related to the inductors by

$$\eta = \frac{N_1}{N_2} = \sqrt{\frac{L_1}{L_2}}$$

It is recommended to start a design with a mutual coupling coefficient equal to 1. This will eliminate leakage inductance that can ring at extremely high frequencies if damping is not supplied and slow the simulation.

5.6.3 Hysteretic Core Model

The inductance is modeled by hysteresis characteristic of the magnetic core. This model is first proposed by John Chan [6]. This model defines the hysteresis loop as shown in Figure 5.22(a) with only three parameters listed in Table 5.6. In addition to the core property parameters Hc, Br, and Bs, the physical dimensions of the core are required as listed in Table 5.7. K-statements don't work with the nonlinear inductor since the Chan model assumes that the field is uniform in the air; core and the field won't be uniform if K12 = 1.

The SPICE statement for transient analysis of Figure 5.22(b) with a unit ramp current and a hysteresis core model is listed as

```
L1 N001 0 Hc=16. Bs=.44 Br=.10 A=0.0000251Lm=0.0198 Lg=0.0006858
N=1000
I1 0 N001 PWL(0 0 1 1)
.options maxstep=10u
.tran 1
```

5.6.4 Linear Transformer Model

An ideal transformer consisting of two windings can be simulated with voltage-dependent current sources as shown in Figure 5.23. The cross-coupled trans-conductance has the effect of an ideal transformer. The turn ratio is specified by the parameter N. Since there is no inductance in the circuit, the simulation would be fast.

TABLE 5.6
Core Parameters of John Chan Model

Name	Description	Units
Hc	Coercive force	Amp-turns/meter
Br	Remnant flux density	Tesla
Bs	Saturation flux density	Tesla

TABLE 5.7
Physical Core Dimensions

Name	Description	Units
Lm	Magnetic Length(excl. gap)	meter
Lg	Length of gap	meter
A	Cross-sectional area	meter**2
N	Number of turns	-

5.6.5 Nonlinear Transformer Model

The ideal transformer as shown in Figure 5.23 can be made to exhibit the nonlinear characteristic of a transformer that saturates by shorting the input of an ideal transformer with an inductor that saturates as shown in Figure 5.24. The turn ratio is made to unit, N = 1. The characteristic depends on the model parameters of the inductor L1. If we run the simulation, we can observe that the simulation steps the load resistance and the peak input current goes down when the secondary isn't open circuit since the core does not saturate when the secondary is shielding the core.

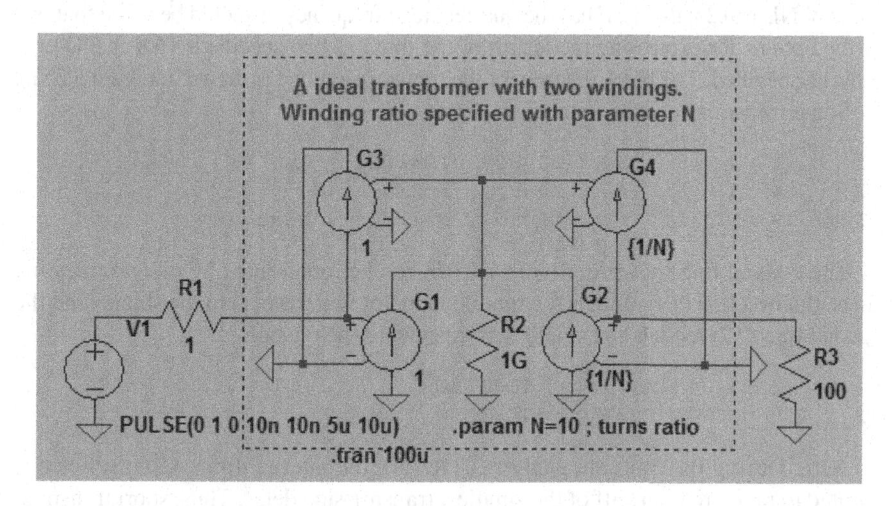

FIGURE 5.23 LTspice ideal transformer [7].

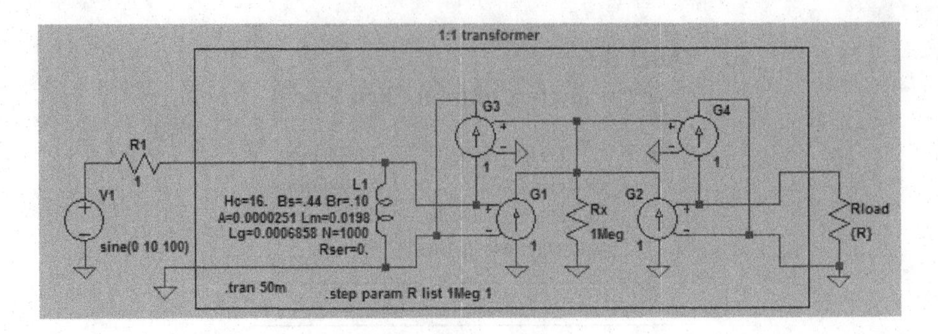

FIGURE 5.24 LTspice nonlinear transformer [10].

5.7 LOSSLESS TRANSMISSION LINES

The symbol of a lossless transmission line is T. A transmission line has two ports: input and output. The general form of a transmission line is

```
T<name> NA+ NA- NB+ NB- Z0=<value> [TD=<value>]
+          [F=<value> NL=<value>]
```

where T<name> is the name of the transmission line. NA+ and NA– are the nodes at the input port. NB+ and NB– are the nodes at the output port. NA+ and NB+ are defined as the positive nodes. NA– and NB– are defined as the negative nodes. The positive current flows from NA+ to NA– and from NB+ to NB–. Z0 is the characteristic impedance.

The length of the line can be expressed in either of the two forms: (1) the transmission delay TD may be specified or (2) the frequency F may be specified together with NL, which is the normalized electrical length of the transmission line with respect to wavelength in the line at frequency F. If the frequency F is specified but not NL, the default value of NL is 0.25; that is, F has the quarter-wave frequency. It should be noted that one of the options for expressing the length of the line must be specified. That is, TD or F must be specified. The block diagram of the transmission line is shown in Figure 5.25(a).

Some transmission statements are

```
T1      1  2   3   4    Z0=50    TD=10NS
T2      4  5   6   7    Z0=50    F=2MHZ
TTRM    9  10  11  12   Z0=50    F=2MHZ     NL=0.4
```

The coaxial line shown in Figure 5.25(b) can be represented by two propagating lines: the first line (T1) models the inner conductor with respect to the shield, and the second line (T2) models the shield with respect to the outside:

```
T1 1   2  3  4  Z0=50    TD=1.5NS
T2 2   0  4  0  Z0=150   TD=1NS
```

Note: During the transient analysis (.TRAN), the internal time step of PSpice is limited to no more than half of the smallest transmission delay. Thus, short transmission lines will cause long run times.

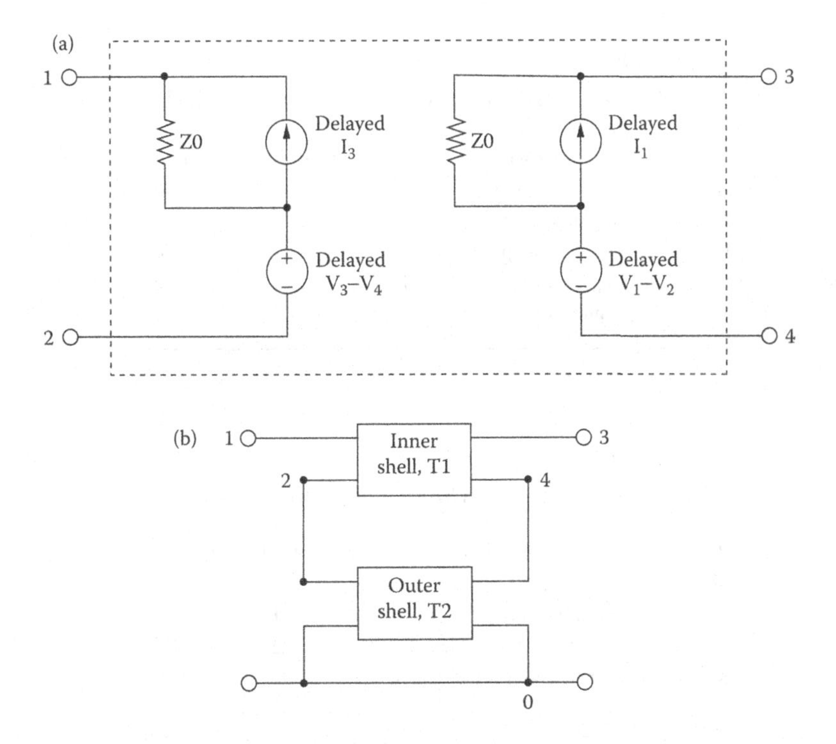

FIGURE 5.25 Transmission line. (a) Block diagram. (b) Coaxial line.

5.8 LTspice TRANSMISSION LINES

Unlike PSpice/OrCAD, LTspice has separate symbols for lossless and lossy transmission lines. The model parameters of the lossless line are similar to those of PSpice/OrCAD as shown in Figure 5.26(a). The symbol of a lossy line is shown in Figure 5.26(b). The symbol name is LTLIN and the model name is `LTRA`. The model parameters are listed in Table 5.8. The SPICE statement must start with "O" and is given by

```
Oxxx NA+ NA- NB+ NB- MNANE LTRA (model parameters)
```

where NA and NA- are the nodes at port 1. NB+ and NB-4 are the nodes at port 2. `MNANE` is the model name.

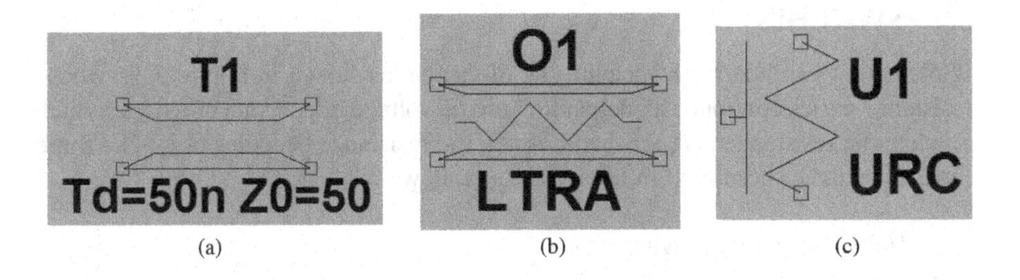

FIGURE 5.26 LTspice transmission lines. (a) `Lossles`. (b) `Lossy`. (c) `RC`.

TABLE 5.8
Model Parameters for Lossy Transmission Line LTLIN

Name	Description	Units/Type	Default
R	Resistance per unit length	Ω	0.
L	Inductance per unit length	H	0.
G	Conductivity per unit length	1/Ω	0.
C	Capacitance per unit length	F	0.
Len	Number of unit lengths	-	0.

TABLE 5.9
Model Parameters for Uniform RC-Line (URC)

Name	Description	Units	Default
K	Propagation Constant	-	2.
Fmax	Maximum Frequency of interest	Hz	1G
Rperl	Resistance per unit length	Ω	1K
Cperl	Capacitance per unit length	F	1e-15

The symbol for a Uniform RC-line is shown in Figure 5.26(c). The symbol name is URC and the model name is URC. The model parameters are listed in Table 5.9. The SPICE statement must start with "U" and is given by

```
Uxxx NA+ NA- Ncom MNAME URC (model parameters) L=<len>
[N=<lumps>]
```

where NA+ and NB– are the two element nodes the RC line connects, whereas Ncom is the node to which the capacitances are connected. MNAME is the model name and LEN is the length of the RC line in meters. Lumps, if specified, is the number of lumped segments to use in modeling the RC line.

5.9 SWITCHES

PSpice allows simulating a special type of switch, as shown in Figure 5.27, whose resistance varies continuously depending on the voltage or current. When the switch S_1 is on, the resistance is R_{ON}; when it is off, the resistance becomes R_{OFF}. The three types of switches permitted in PSpice are as follows:

- Voltage-controlled switches
- Current-controlled switches
- Time-dependent switches

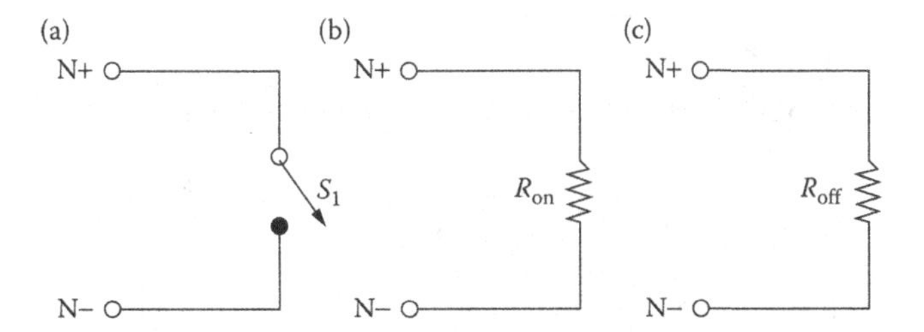

FIGURE 5.27 Switch with a variable resistance. (a) Switch. (b) On state. (c) Off state.

Note: The voltage-and current-controlled switches are not available in SPICE2. However, they are available in SPICE3.

5.9.1 Voltage-Controlled Switch

The symbol of a voltage-controlled switch is S. Its name must start with S, and it takes the general form

```
S<name> N+ N- NC+ NC- SNAME
```

where N+ and N– are the two nodes of the switch. The current is assumed to flow from N+ through the switch to node N–. NC+ and NC– are the positive and negative nodes of the controlling voltage source, as shown in Figure 5.28(a). SNAME is the model name. The resistance of the switch varies depending on the voltage across the switch. The type name for a voltage-controlled switch is VSWITCH, and the model parameters are shown in Table 5.10. The PSpice schematic is shown in Figure 5.28(b).

The voltage-controlled switch statement (for a switch which is between nodes 6 and 5, and controlled by the voltage between nodes 4 and 0) is

```
S1 6  5  4  0  SMOD
.MODEL SMOD VSWITCH (RON=0.5 ROFF=10E+6 VON=0.7
VOFF=0.0)
```

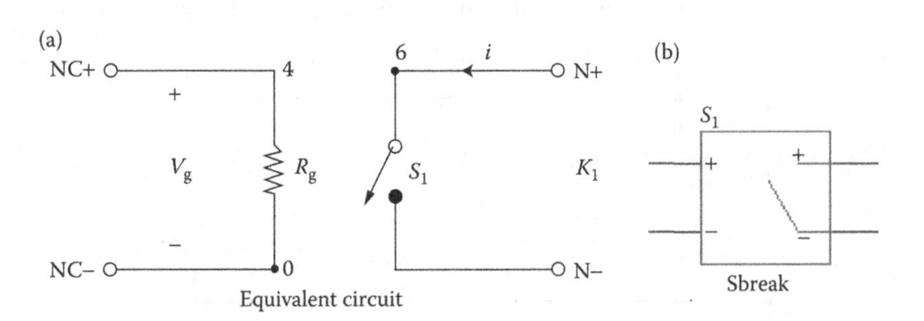

FIGURE 5.28 Voltage-controlled switch. (a) Positive and negative nodes. (b) PSpice schematic.

TABLE 5.10
Model Parameters for Voltage-Controlled Switch

Name	Meaning	Unit	Default
VON	Control voltage for on state	V	1.0
VOFF	Control voltage for off state	V	0
RON	On resistance	Ω	1.0
ROFF	Off resistance	Ω	10^6

Note the following:

1. R_{ON} and R_{OFF} must be greater than zero and less than 1/GMIN. The value of GMIN can be defined as an option as described in **.OPTIONS** command in Section 6.5. The default value of conductance, GMIN, is 1E–12 Ω^{-1}.
2. The ratio of R_{OFF} to R_{ON} should be less than 1E+12.
3. The difficulty due to the high gain of an ideal switch can be minimized by choosing the value of R_{OFF} to be as high as permissible and that of R_{ON} to be as low as possible compared with other circuit elements, within the limits of allowable accuracy.

Example 5.5:

Step Response with a Voltage-Controlled Switch

A circuit with a voltage-controlled switch is shown in Figure 5.29. The input voltage is $v_s = 200 \sin(2000\pi t)$. Plot the instantaneous voltage at node 3 and the current through the load resistor R_L for a duration of 0–1 ms with an increment of 5 µs. The model parameters of the switch are RON = 5M, ROFF = 10E+9, VON = 25M, and VOFF = 0.0. The results should be available for display by using Probe.

SOLUTION

The PSpice schematic is shown in Figure 5.30(a). The gain of the voltage-controlled voltage source E1 is set to 0.1. The voltage and current markers display the output waveform in Probe at the end of the simulation. The input voltage is

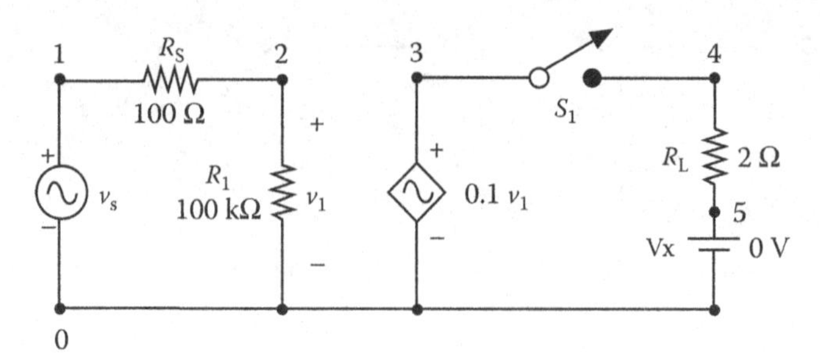

FIGURE 5.29 Circuit with a voltage-controlled switch.

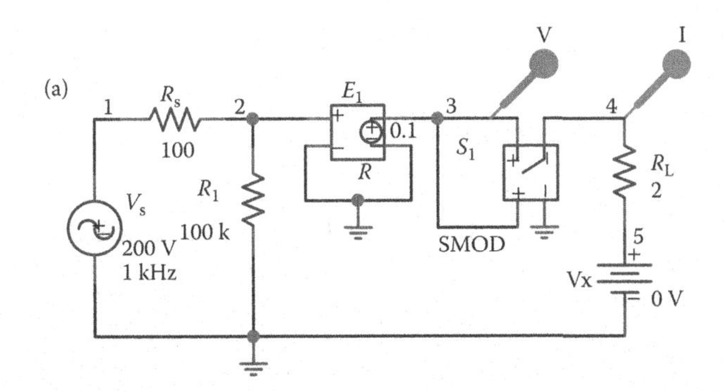

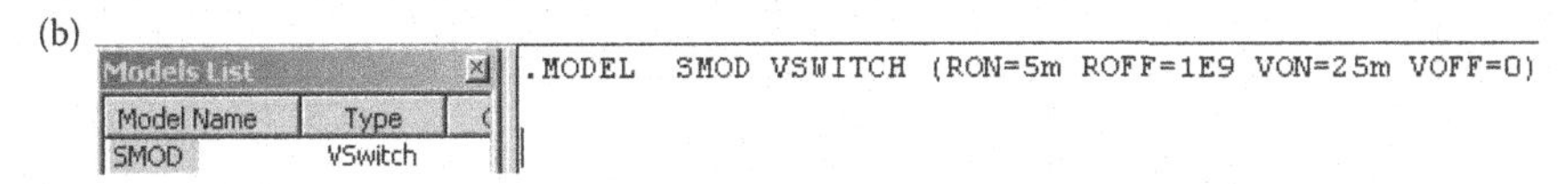

FIGURE 5.30 PSpice schematic for Example 5.5. (a) PSpice schematic. (b) Specification of VSWITCH.

specified by a sinusoidal source. The breakout voltage–controlled switch (S_1) is used to specify the switch model SMOD as shown in Figure 5.30(b) and its model parameters are RON = 5M, ROFF = 10E+9, VON = 25M, and VOFF = 0.

The voltage source VX = 0 V is inserted to monitor the output current. The listing of the circuit file is as follows:

Voltage-Controlled Switch

```
SOURCE   ■    * Sinusoidal input voltage of 200 V peak with
              * 0° phase delay:
              VS 1 0 SIN (0 200V 1KHZ)
CIRCUIT  ■■   RS 1 2 100OHM
              R1 2 0 100KOHM
              * Voltage-controlled voltage source with a
              * voltage gain of 0.1:
              E1 3 0 2 0 0.1
              RL 4 5 2OHM
              * Dummy voltage source of VX = 0 to measure the
              * load current:
              VX 5 0 DC 0V
              * Voltage-controlled switch controlled by
              * voltage across nodes 3 and 0:
              S1 3 4 3 0 SMOD
              * Switch model descriptions:
              .MODEL SMOD VSWITCH (RON=5M ROFF=10E+9
              + VON=25M VOFF=0.0)
ANALYSIS ■■■  * Transient analysis from 0 to 1 ms with a 5.
              * μs increment:
              .TRAN 5US 1MS
              * Plot the current through VS and the input
              * voltage.
              .PLOT TRAN V(3) 1(VX)      ; On the output file
              .PROBE                     ; Graphics
                                           post-processor
              .END
```

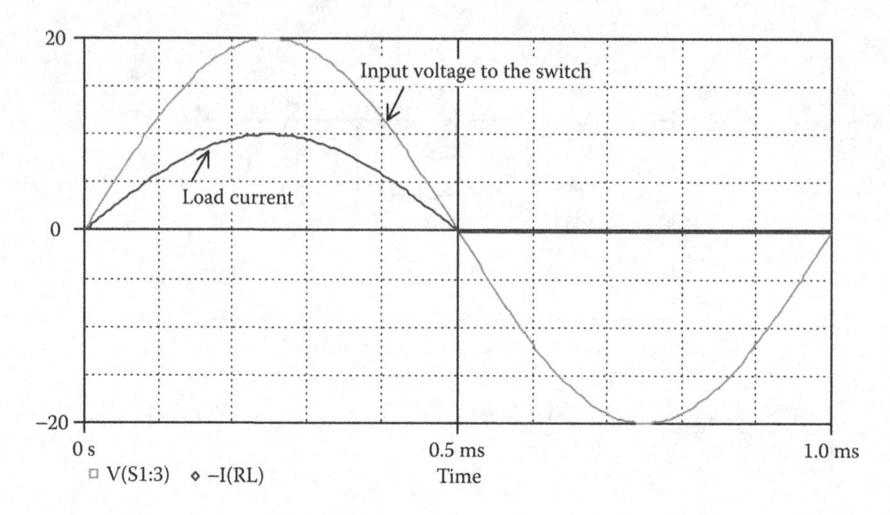

FIGURE 5.31 Transient response for Example 5.5.

The results of the simulation that are obtained by Probe are shown in Figure 5.31, which is the output of a diode rectifier. Switch S_1 behaves as a diode. That is, the switch allows current flow in the positive direction.

Current-Controlled Switch

The symbol of a current-controlled switch is W. The name of the switch must start with W, and it takes the general form

```
W<name> N+ N- VN WNAME
```

where N+ and N– are the two nodes of the switch. VN is a voltage source through which the controlling current flows as shown in Figure 5.32(a). WNAME is the model name. The resistance of the switch depends on the current through the switch. The type name for a current-controlled switch is ISWITCH and the model parameters are shown in Table 5.11. The PSpice schematic is shown in Figure 5.32(b).

The current-controlled switch statement (for a switch which is between nodes 6 and 5, and controlled by the current flowing through the voltage source V_N) is

```
W1      6  5   VN  RELAY
.MODEL RELAY  ISWITCH  (RON=0.5 ROFF=10E+6 ION=0.07
IOFF=0.0)
```

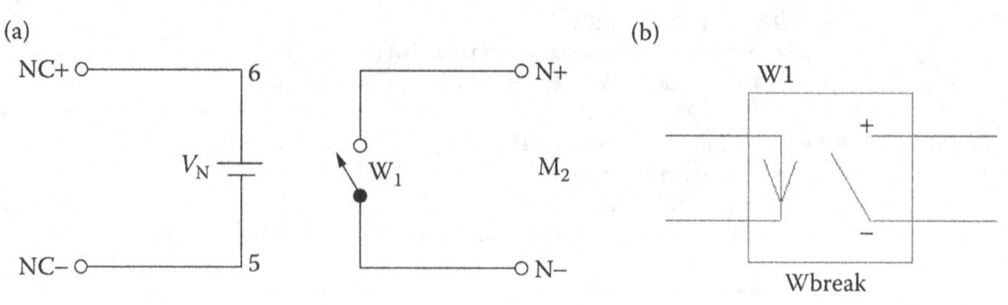

FIGURE 5.32 Current-controlled switch. (a) Equivalent circuit. (b) PSpice schematic.

TABLE 5.11
Model Parameters for Current-Controlled Switch

Name	Meaning	Unit	Default
ION	Control current for on state	A	1E-3
IOFF	Control current for off state	A	0
RON	On resistance	Ω	1.0
ROFF	Off resistance	Ω	10^6

Note: The current through voltage source VN controls the switch. The voltage source VN must be an independent source, and it can have a finite value or zero. The limitations of the parameters are similar to those for the voltage-controlled switch.

Example 5.6:

Transient Response of an LC *Circuit with a Current-Controlled Switch*

A circuit with a current-controlled switch is shown in Figure 5.33. Plot the capacitor voltage and the inductor current for a duration of 0–160 μs with an increment of 1 μs. The model parameters of the switch are RON = 1E+6, ROFF = 0.001, ION = 1MA, and IOFF = 0. The results should be available for display by using Probe.

SOLUTION

The PSpice schematic is shown in Figure 5.34(a). The voltage and current markers display the output waveforms in Probe at the end of the simulation. The initial voltage on the capacitor C1 specifies the input source. The breakout current–controlled switch (W1) is used to specify the parameters of the switch model WMOD as shown in Figure 5.34(b) and its model parameters are ION = 1MA, IOFF = 0, RON = 1E+6, and ROFF = 0.01.

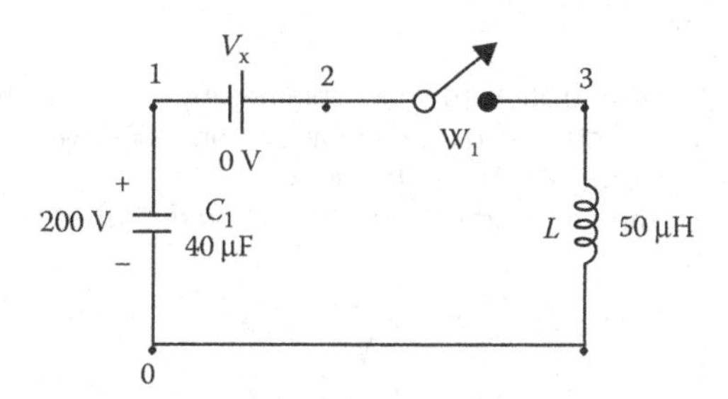

FIGURE 5.33 Circuit with a current-controlled switch.

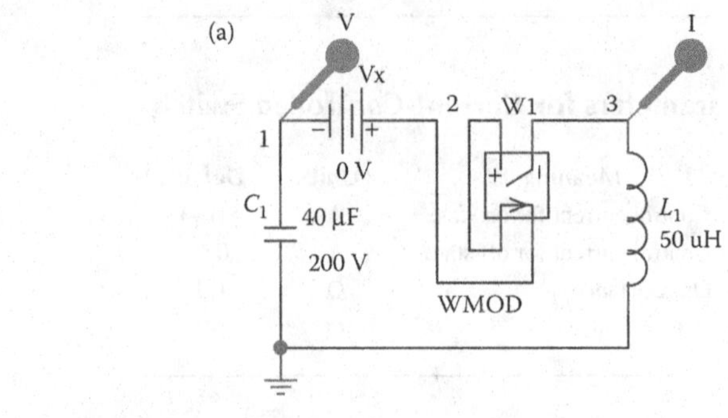

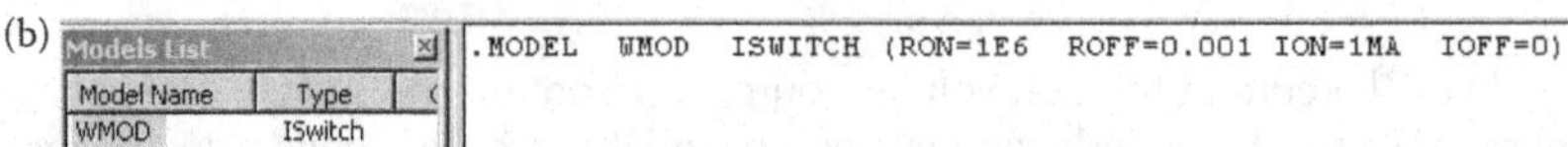

FIGURE 5.34 PSpice schematic for Example 5.6. (a) PSpice schematic. (b) Specifications of ISWITCH.

The voltage source VX = 0 V is inserted to monitor the controlling current. The listing of the circuit file is as follows:

Current-Controlled Switch

```
SOURCE  ■■   * C1 of 40 µF with an initial voltage of 200 V:
             C1 1 0 40UF IC=200
             * Dummy voltage source of VX = 0:
             VX 2 1 DC 0V
             * Current-controlled switch with model name
             * SMOD:
             W1 2 3 VX SMOD
             * Model parameters:
             .MODEL SMOD ISWITCH (RON=1E+6 ROFF=0.001
             + ION=1MA IOFF=0)
             L1 3 0 50UF
CIRCUIT ■■■  * Transient analysis with UIC (use initial
             * condition) option:
             .TRAN 1US 160US UIC
             * Plot the voltage at node 1 and the current
             * through VX.
             .PLOT TRAN V(1) I(VX)          ; On the output file
             .PROBE                          ; Graphics
                                               post-processor
             .END
```

The results of the simulation that are obtained by using Probe are shown in Figure 5.35. Switch S_1 acts as a diode and allows only positive current flow. The initial voltage on the capacitor is the driving source.

The capacitor current $i_c(t)$ and voltage $v_c(t)$ can be described by

$$i_c(t) = V_C\sqrt{\frac{L}{C}}\sin\omega t$$

$$v_c(t) = V_C\cos\omega t$$

where $\omega = 1/\sqrt{LC}$.

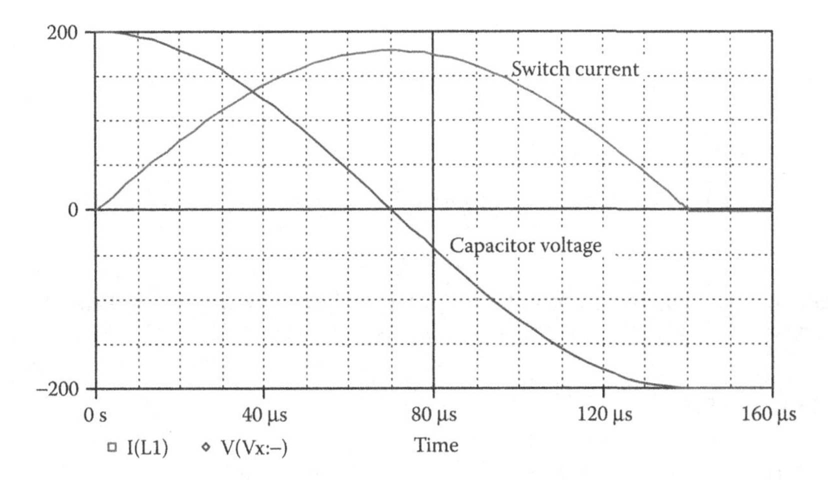

FIGURE 5.35 Transient response for Example 5.6.

The capacitor voltage is fully reversed to $-V_C$ when the capacitor current falls to zero at $\omega t = \pi$

5.9.2 Time-Dependent Switches

PSpice Schematics supports two types of time-dependent switches:

- Time-dependent close switch
- Time-dependent open switch

The closing or the opening time and the transition time of the switch are specified by the switch parameters as shown in Table 5.12.

5.9.2.1 Time-Dependent Close Switch

This switch is normally open, and setting the closing time closes it. Its schematic is shown in Figure 5.36(a). The default parameters are shown in Figure 5.36(b).

TABLE 5.12
Model Parameters for Close/Open Switch

Name	Meaning	Default
Tclose/Topen	Time at which switch begins to close/open	0
Ttran	Time required to switch states from off state to on state (must be realistic, not 0)	1 μs
Rclosed	Closed-state resistance	10 mΩ
Ropen	Open-state resistance (Ropen/Rclosed ≤ 1E+10)	1 mΩ

(a) TCLOSE = 0

1 2

U1

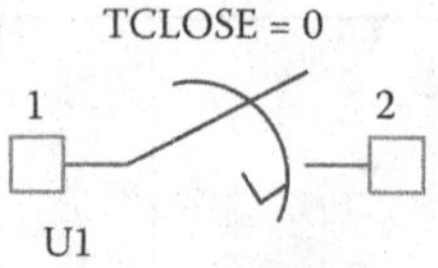

(b)

Reference	Value	K13	RCLOSED	ROPEN	Source Part	TCLOSE	TTRAN
U1	Sw_tClose		0.01	1Meg	Sw_tClose.Normal	0	1u

FIGURE 5.36 PSpice schematic for Sw_tClose switch. (a) schematic. (b) Parameters of Sw_tClose.

Example 5.7:

Plotting the Transient Response of an RC Circuit with an Sw_tClose Switch

Figure 5.37(a) shows an *RC* circuit with a step input voltage, which is generated by a PWL source. The switch Sw_tClose normally remains open. At $t = 20\ \mu s$, the switch closes, and the capacitor is charged exponentially with a time constant of R_1C_1 as shown in Figure 5.37(b).

If V_S is a step input voltage, the capacitor voltage, after the switch is closed, can be described as

$$v_C(t) = V_S\left(1 - e^{-\frac{t}{R_1C_1}}\right)$$

5.9.2.2 Time-Dependent Open Switch

This switch is normally closed, and setting the opening time opens it. Its schematic is shown in Figure 5.38(a). The default parameters are shown in Figure 5.38(b).

If V_o is the initial capacitor voltage which should equal to the input voltage V_S, the capacitor voltage, after the switch is opened, can be described as

$$v_C(t) = V_O e^{-\frac{t}{R_1C_1}}$$

Example 5.8:

Plotting the Transient Response of an RC Circuit with an Sw_tOpen Switch

Figure 5.39(a) shows an *RC* circuit with a step input voltage, which is generated by a PWL source. The switch Sw_tOpen remains normally closed. At $t = 20\ \mu s$, the switch opens and the capacitor is discharged exponentially with a time constant of R_1C_1 as shown in Figure 5.39(b).

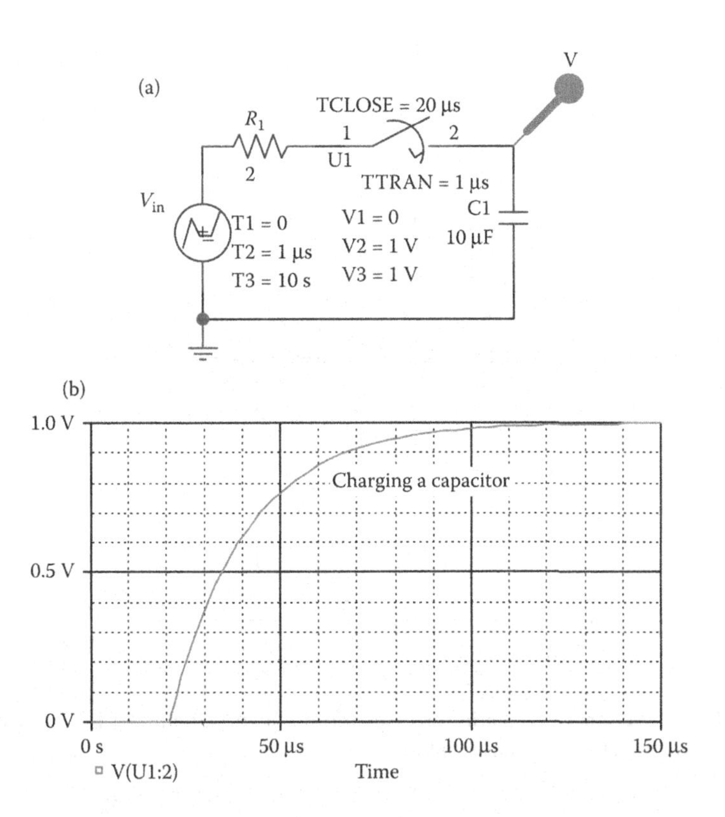

FIGURE 5.37 PSpice schematic for charging an RC circuit. (a) PSpice schematic. (b) Output voltage at switch transition (on).

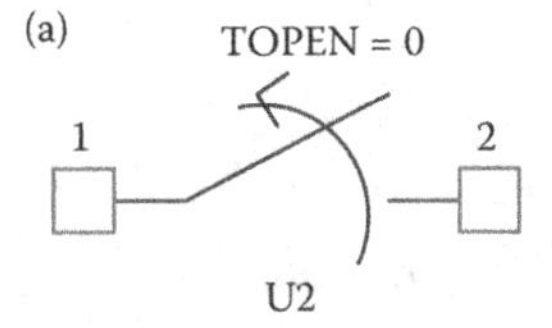

(b)

Reference	Value	K13	RCLOSED	ROPEN	Source Part	TOPEN	TTRAN
U2	Sw_tOpen		0.01	1Meg	*Sw_tOpen.Normal*	0	1u

FIGURE 5.38 PSpice schematic for discharging an RC circuit. (a) Schematic. (b) Parameters of Sw_tOpen.

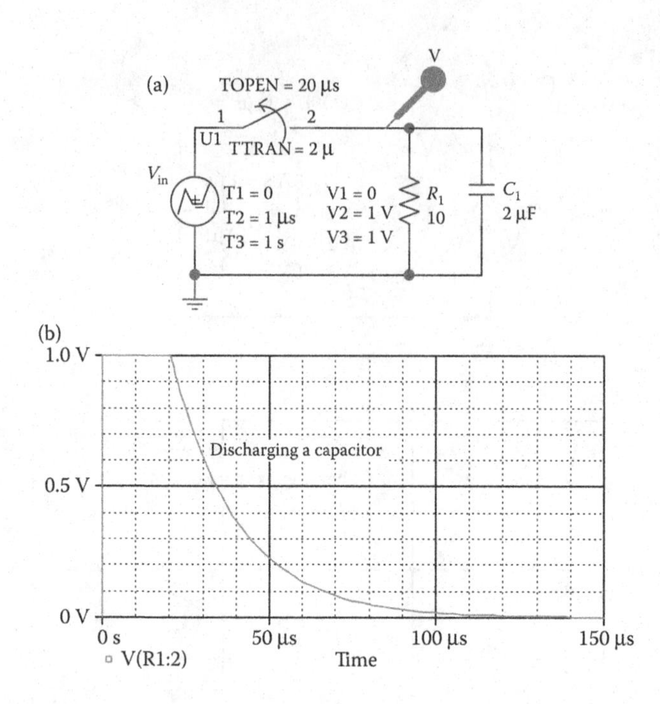

FIGURE 5.39 PSpice schematic for discharging an RC circuit. (a) PSpice schematic. (b) Output voltage at switch transition (off).

SUMMARY

The symbols of the passive elements and their statements are

C Capacitor

```
C<name> N+ N- CNAME CVALUE IC=V0
```

L Inductor

```
L<name> N+ N- LNAME LVALUE IC=10
```

K Linear mutual inductors (transformer)

```
K<name> L<(first inductor) name>
L<(second inductor)name> <value>]
```

K Nonlinear inductor

```
K<name> L<(inductor) name> <(coupling value)>
+ <(model) name> [(size) value]
```

R Resistor

```
R<name> N+ N- RNAME RVALUE
```

S Voltage-controlled switch

```
S<name> N+ N- NC+ NC- SNAME
```

T Lossless transmission lines

```
T<name> NA+ NA- NB+ NB- Z0=<value> [TD=<value>]
+ [F=<value> NL=<value>]
```

W Current-controlled switch

```
W<name> N+ N- VN WNAME
```

PROBLEMS

Write the PSpice statements for the following circuits. If applicable, the output should also be available for display and as a hard copy by using Probe.

5.1 A resistor R_1, which is connected between nodes 3 and 4, has a nominal value of $R = 10$ kΩ. The operating temperature is 55°C, and it has the form

$$R_1 = R[1 + 0.2(T - T_0) + 0.002(T - T_0)^2]$$

5.2 A resistor R_1, which is connected between nodes 3 and 4, has a nominal value of $R = 10$ kΩ. The operating temperature is 55°C, and it has the form

$$R_1 = R \times 1.01^{4.5(T - T_0)}$$

5.3 A capacitor C_1, which is connected between nodes 5 and 6, has a value of 10 pF and an initial voltage of –20 V.

5.4 A capacitor C_1, which is connected between nodes 5 and 6, has a nominal value of $C = 10$ pF. The operating temperature is $T = 55$°C. The capacitance, which is a function of its voltage and the operating temperature, is given by

$$C_1 = C(1 + 0.01V + 0.002V^2) \times [1 + 0.03(T - T_0) + 0.05(T - T_0)^2]$$

5.5 An inductor L_1, which is connected between nodes 5 and 6, has a value of 0.5 mH and carries an initial current of 0.04 mA.

5.6 An inductor L_1, which is connected between nodes 3 and 4, has a nominal value of $L = 1.5$ mH. The operating temperature is $T = 55$°C. The inductance is a function of its current and the operating temperature, and it is given by

$$L_1 = L(1 + 0.01I + 0.002I^2) \times [1 + 0.03(T - T_0) + 0.05(T - T_0)^2]$$

5.7 The two inductors, which are oppositely coupled as shown in Figure 5.10(b), are $L_1 = 1.2$ mH and $L_2 = 0.5$ mH. The coefficients of coupling are $K_{12} = K_{21} = 0.999$.

5.8 Plot the transient response of the circuit in Figure P5.8 from 0 to 5 ms with a time increment of 25 μs. The output voltage is taken across the capacitor. Use Probe for graphical output.

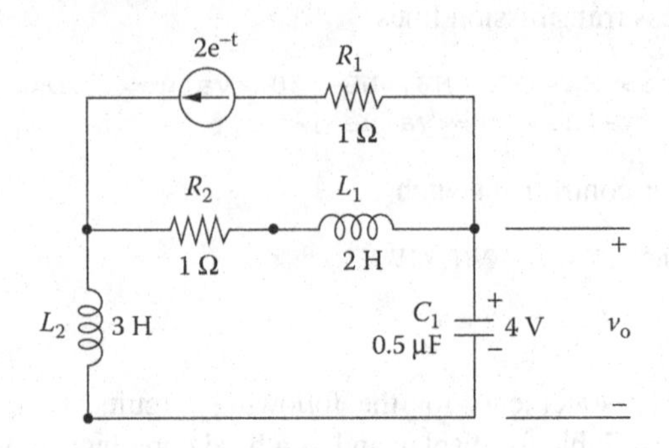

FIGURE P5.8 RLC circuit with an exponential current source.

5.9 Repeat Problem 5.8 for the circuit of Figure P5.9.

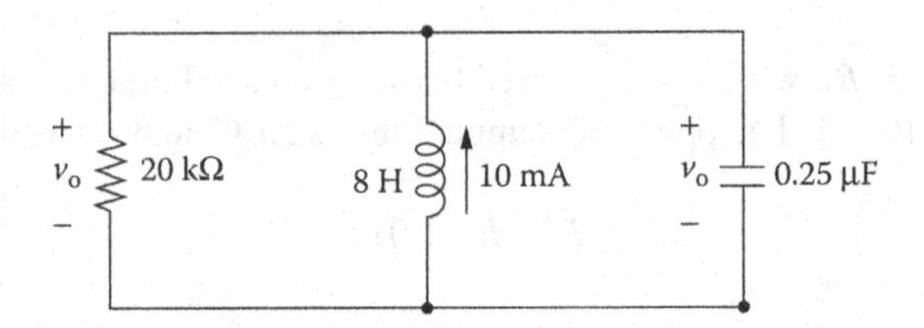

FIGURE P5.9 Parallel RLC circuit.

5.10 Plot the frequency response of the circuit in Figure P5.10 from 10 Hz to 100 kHz with a decade increment and 10 points per decade. The output voltage is taken across the capacitor. Print and plot the magnitude and phase angle of the output voltage. Assume a source voltage of 1 V peak.

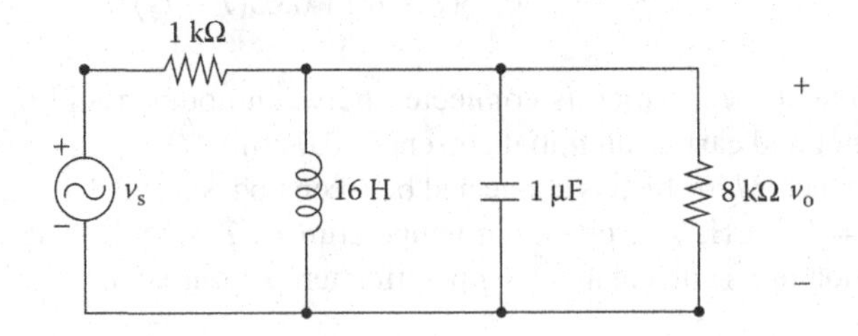

FIGURE P5.10 Series parallel circuit.

5.11 As shown in Figure P5.11, a single-phase transformer has a center-tapped primary, where $L_p = 1.5$ mH, $L_s = 1.3$ mH, and $K_{ps} = K_{sp} = 0.999$. The primary voltage is $v_p = 170 \sin(377t)$. Plot the instantaneous secondary voltage and load

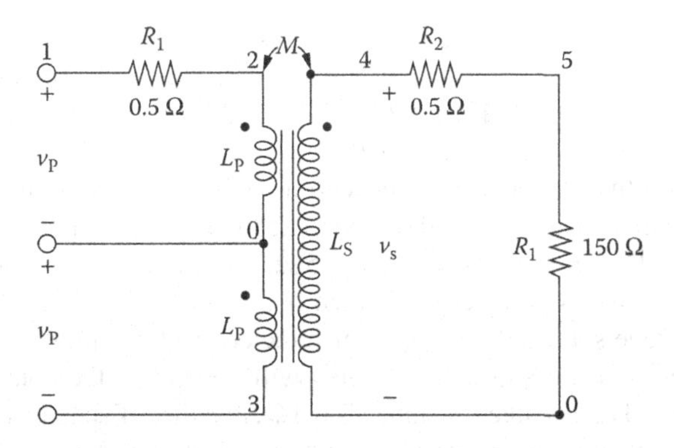

FIGURE P5.11 Center-taped transformer.

current from 0 to 35 ms with a 0.1-ms increment. The output should also be available for display and as a hard copy by using Probe.

5.12 Repeat Problem 5.11 assuming the transformer to be nonlinear with L_p = 200 turns and L_s = 100 turns. The model parameters of the core are AREA = 2.0, PATH = 62.73, GAP = 0.1, MS = 1.6E+6, ALPHA = 1E-3, A = 1E+3, C = 0.5, and K = 1500.

5.13 A three-phase transformer, which is shown in Figure P5.13(a), has $L_1 = L_2 = L_3$ = 1.2 mH and $L_4 = L_5 = L_6$ = 0.5 mH. The coupling coefficients between the primary and secondary of each phase are $K_{14} = K_{41} = K_{25} = K_{52} = K_{36} = K_{63} = 0.9999$. There is no cross-coupling with other phases. The primary phase voltage is v_p = 170 sin(377*t*). Plot the instantaneous secondary phase voltage and load phase current from 0 to 35 ms with a 0.1-ms increment. Assume balanced three-phase input voltages. Figure P5.13(b) shows the secondary side of the transformer.

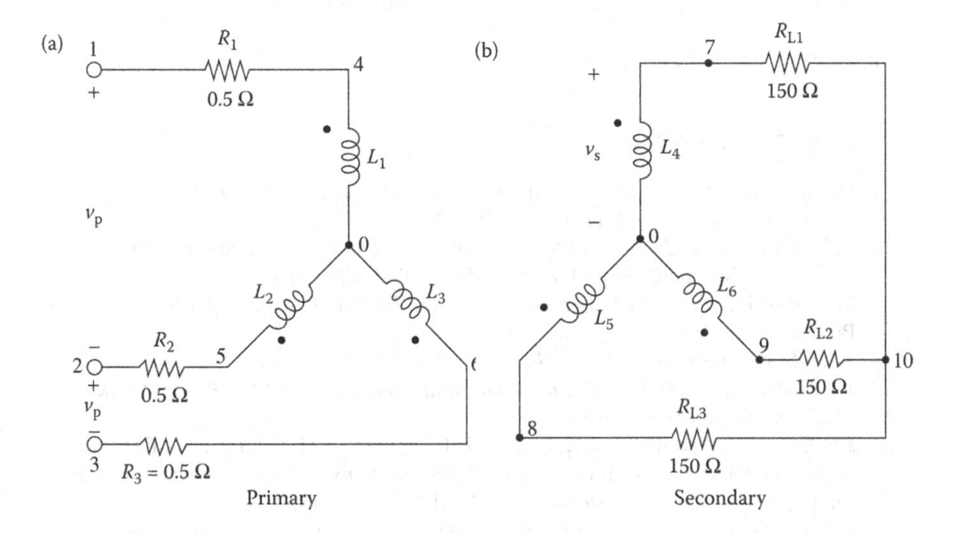

FIGURE P5.13 Three-phase transformer.

5.14 Repeat Problem 5.13 assuming the transformer to be nonlinear with $L_1 = L_2 = L_3 = 200$ turns and $L_4 = L_5 = L_6 = 100$ turns. The model parameters of the core are AREA = 2.0, PATH = 62.73, GAP = 0.1, MS = 1.6E+6, ALPHA = 1E-3, A = 1E+3, C = 0.5, and K = 1500.

5.15 A switch that is connected between nodes 5 and 4 is controlled by a voltage source between nodes 3 and 0. The switch will conduct if the controlling voltage is 0.5 V. The on-state resistance is 0.5 Ω, and the off-state resistance is 2E+6 Ω.

5.16 A switch that is connected between nodes 5 and 4 is controlled by a current. The voltage source V_1 through which the controlling current flows is connected between nodes 2 and 0. The switch will conduct if the controlling current is 0.55 mA. The on-state resistance is 0.5 Ω, and the off-state resistance is 2E+6 Ω.

5.17 for the circuit in Figure P5.17, plot the transient response of the load and source currents for five cycles of the switching period with a time increment of 10 μs. The model parameters of the voltage-controlled switches are RON = 0.025, ROFF = 1E+8, VON = 0.05, and VOFF = 0. The output should also be available for display and as a hard copy by using Probe.

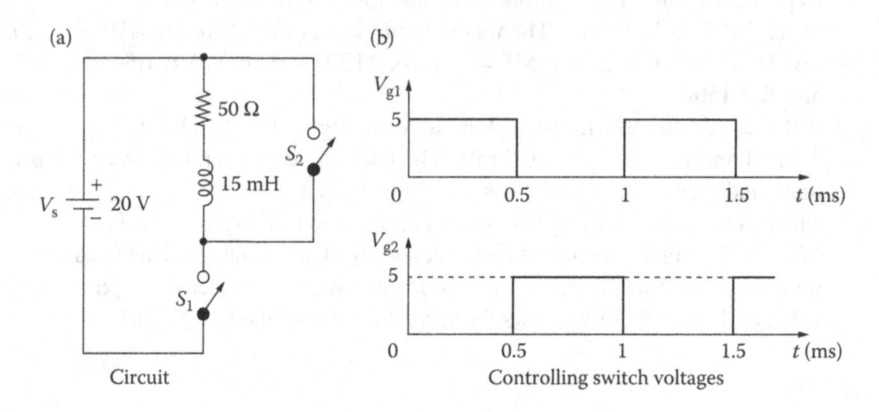

FIGURE P5.17 Switching circuit.

SUGGESTED READING

1. D.C. Jiles and D.L. Atherton, Theory of ferromagnetic hysteresis, *Journal of Magnetism and Magnetic Material*, *61*(*48*), 1986, 48–60.
2. M.H. Rashid, *Introduction to PSpice Using OrCAD for Circuits and Electronics*, Third Edition, Englewood Cliffs, NJ: Prentice-Hall, 2003, Chapter 4.
3. M.H. Rashid, *SPICE for Power Electronics and Electric Power*, Englewood Cliffs, NJ: Prentice-Hall, 1993.
4. MicroSim Corporation, *PSpice Manual*, Irvine, CA, 1992.
5. J.F. Lindsay and M.H. Rashid, *Electromechanics and Electrical Machinery*, Englewood Cliffs, NJ: Prentice-Hall, 1986.
6. J.H. Chan, A. Vladimirescu, X.-C. Gao; P. Liebmann and J. Valainis, Nonlinear transformer model for circuit simulation, *IEEE Transactions on Computer-Aided Design of Integrated Circuits and Systems*, *10*(4), April 1991, 476–482.
7. LTSpice IV: Educational Library Files, Linear Technology Corporation, 2016.

6 Dot Commands

After completing this chapter, students should be able to do the following:

- Define and call a "subcircuit."
- Define a mathematical function and assign a global node or parameter.
- Set a node at a specific voltage and include a library file.
- Use the option parameters and set their values to avoid conference problems.
- Perform parametric and step variations on the analysis.
- Assign tolerances of components and model parameters.
- Perform advanced analyses such as Fourier series of output voltages and currents, and noise, worst-case, and Monte Carlo analyses.

6.1 INTRODUCTION

PSpice has commands for performing various analyses, getting different types of output, and modeling elements. These commands begin with a dot and are known as *dot commands*. These can be selected from the PSpice Schematics menu as shown in Figure 6.1 and can be used to specify:

- Models
- Types of output
- Operating temperature and end of circuit
- Options
- DC analysis
- AC analysis
- Noise analysis
- Transient analysis
- Fourier analysis
- Monte Carlo analysis

Note: If you are not sure about a command and its effect, run a circuit file with the command and check the results. If there is a syntax error, PSpice will display a message identifying the problem.

6.2 MODELS

PSpice allows one to (1) model an element based on its parameters, (2) model a small circuit that is repeated a number of times in the main circuit, (3) use a model that is defined in another file, (4) use a user-defined function, (5) use parameters instead of number values, and (6) use parameter variations. The commands are as follows:

.MODEL	Model
.SUBCKT	Subcircuit

DOI: 10.1201/9781003284451-6

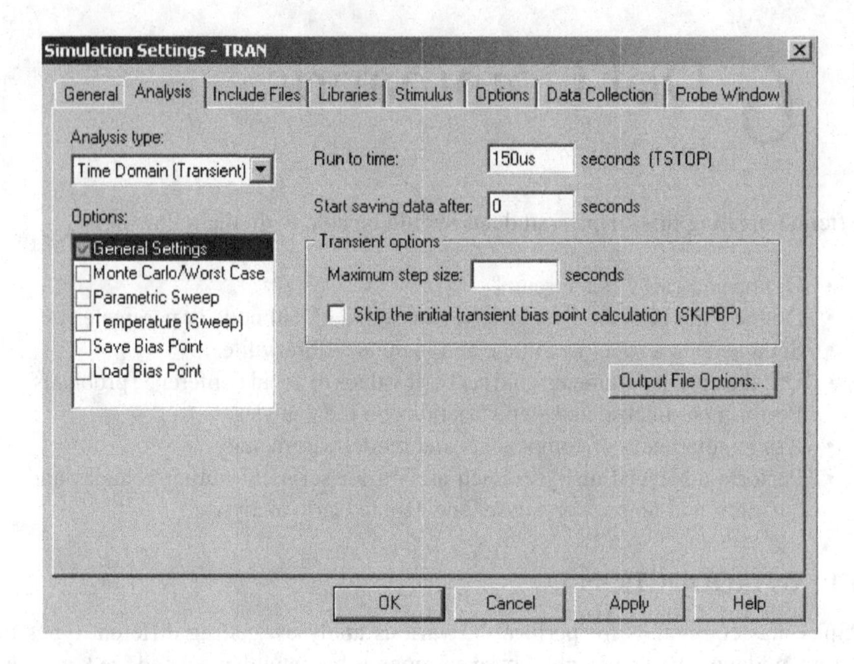

FIGURE 6.1 PSpice simulation settings menu.

.ENDS	End of subcircuit
.FUNC	Function
.GLOBAL	Global
.LIB	Library file
.INC	Include file
.PARAM	Parameter
.STEP	Parametric analysis

6.2.1 .MODEL (Model)

The .MODEL command was discussed in Section 5.2.

6.2.2 .SUBCKT (Subcircuit)

A subcircuit permits one to define a block of circuitry and then to use that block in several places. The general form for subcircuit definition (or description) is

```
.SUBCKT SUBNAME [⟨(two or more) nodes⟩]
```

The symbol of a subcircuit call is X. The general form of a call statement is

```
X ⟨name⟩ [⟨(two or more) nodes⟩] SUBNAME
```

SUBNAME is the name of the subcircuit definition, and

```
⟨(two or more) nodes⟩
```

are the nodes of the subcircuit. X⟨name⟩ causes the referenced subcircuit to be inserted into the circuit with given nodes replacing the argument nodes in the definition. The subcircuit name SUBNAME may be considered as equivalent to a subroutine name in forTRAN programing, where X⟨name⟩ is the call statement and

```
⟨(two or more) nodes⟩
```

are the variables or arguments of the subroutine.

Subcircuits may be nested, that is, subcircuit A may call other subcircuits. But the nesting cannot be circular, which means that if subcircuit A contains a call to subcircuit B, subcircuit B must not contain a call to subcircuit A. There must be the same number of nodes in the subcircuit calling statement as in its definition. The subcircuit definition should contain only element statements (statements without a dot) and may contain .MODEL statements.

6.2.3 .ENDS (End of Subcircuit)

A subcircuit must end with an .ENDS statement. The end of a subcircuit definition has the general form

```
.ENDS SUBNAME
```

SUBNAME is the name of the subcircuit, and it indicates which subcircuit description is to be terminated. If the .ENDS statement is missing, all subcircuit descriptions are terminated.

Subcircuit statements end with

```
.ENDS    OPAMP
.ENDS
```

Note: The name of the subcircuit can be omitted. However, it is advisable to identify the name of the subcircuit to be terminated, especially if there is more than one subcircuit.

Example 6.1:

Describing a Subcircuit

Write the subcircuit call and subcircuit description for the op-amp circuit of Figure 6.2.

SOLUTION

The listing of statements for the subcircuit call and description are as follows:

```
* The call statement X1 to be connected to input nodes 1 and 4
and the output nodes 7 and 9: The subcircuit name is OPAMP. Nodes
1, 4, 7, and 9 are referred to the main circuit file and do not
interact with the nodes of the subcircuit.
```

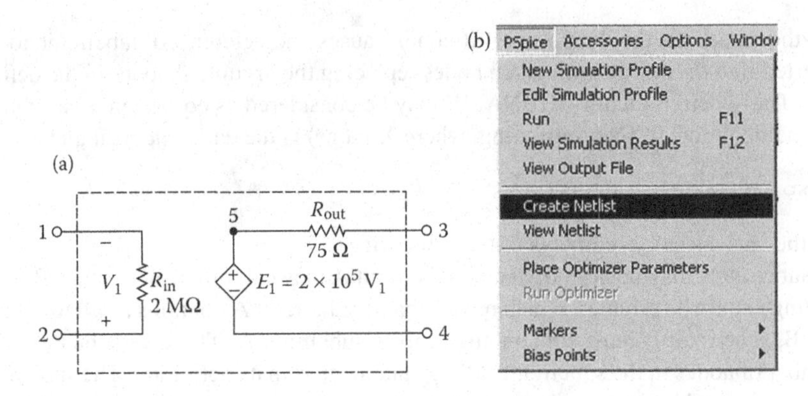

FIGURE 6.2 Op-amp subcircuit. (a) Circuit. (b) Create Netlist from PSpice menu.

```
X1        1       4       7       9       OPAMP
*         vi-     vi+     vo+     vo-     model name

* The subcircuit definition: Nodes 1, 2, 3, and 4 are referred to
the subcircuit and do not interact with the nodes of the main
circuit.

.SUBCKT        OPAMP           1       2       3       4
*          model name          vi-     vi+     vo+     vo-
RIN       1      2       2MEG
ROUT      5      3       75
E1        5      4       2          1      0.2MEG; Voltage-controlled
                                           voltage source
.ENDS     OPAMP                                ; End of subcircuit
                                               definition OPAMP
```

In PSpice Schematics, the net list for a subcircuit can be created by using the command Create Netlist from the PSpice menu.

Note: There is no interaction between the nodes in the main circuit and the subcircuit. Node numbers in the subcircuit are independent of those in the main circuit. However, the subcircuit should not have node 0 because node 0, which is considered global by PSpice, is the ground.

6.2.4 .FUNC (Function)

A function statement can be used to define functions that may be used in expressions discussed in Section 4.5.1. The functions are defined by users and are, therefore, flexible. They are also useful for overcoming the restriction that expressions be limited to a single line and for the case where there are several similar subexpressions in a single circuit file. The general form of a function statement is

```
.FUNC FNAME (arg) ⟨function⟩
```

where `FNAME (arg)` is the name of the function with argument `arg`. FNAME must not be the same as the names of the built-in functions described in Section 4.5.1, such as "sin." Up to 10 arguments may be used in a definition. The number of arguments in the use of a function must agree with the number in the definition. A function may be defined with no arguments, but the parentheses are still required. ⟨function⟩_ may refer to other functions defined previously. The .FUNC statement must precede the first use of FNAME. The users can create a file of frequently used .FUNC definitions and access them with an .INC statement (Section 6.2.7) near the beginning of the circuit file.

Some function statements are

```
FUNC E(x)                    exp (x)
.FUNC Sinh(x)                (E(x) + E(-x))/2
.FUNC MIN(C,D)               (C + D - ABS(C - D))/2
.FUNC MAX(C,D)               (C + D + ABS(C + D))/2
.FUNC IND(I(Vsense))         (A0 + A1*I(V(Sense)) +
A2*I(V(Sense)) +I(V(Sense)))
```

Note the following:

1. The definition of the ⟨function⟩_ must be restricted to one line.
2. In-line comments must not be used after the ⟨function⟩ definition.
3. The last statement illustrates a current-dependent nonlinear inductor.

6.2.5 .GLOBAL (Global)

PSpice has the capability of defining global nodes. These nodes are accessible by all subcircuits without being passed in as arguments. Global nodes may be handy for such applications as power supplies, power converters, and clock lines.

For PSpice Schematics, the symbol of a global node is shown in Figure 6.3.

The general statement is

```
.GLOBAL N
```

where N is the node number, for example,

```
.GLOBAL 4
```

makes node 4 global to the circuit file and subcircuit(s).

6.2.6 .LIB (Library File)

A library file may be referenced into the circuit file by using the statement

```
.LIB FNAME
```

where FNAME is the name of the library file to be called. A library file may contain comments, .MODEL statements, subcircuit definition, .LIB statements, and .END statements. No other statements are permitted. If FNAME is omitted, PSpice looks

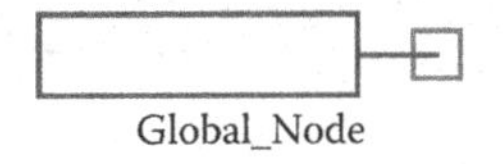

FIGURE 6.3 Global node.

for the default file, EVAL.LIB, that comes with PSpice programs. The library file FNAME may call for another library file. When a .LIB command calls for a file, it does not bring the entire text of the library file into the circuit file. It simply reads in those models or subcircuits that are called by the main circuit file. As a result, only those models or subcircuit descriptions that are needed by the main circuit file take up the main memory (RAM) space.

Some library file statements are

```
.LIB
.LIB   NOM.LIB (library file NOM.LIB is on the default drive)
.LIB   B:\LIB\NOM.LIB (library file NOM.LIB is on directory
              file LIB in drive B)
.LIB   C:\LIB\NOM.LIB (library file NOM.LIB is in directory
              file LIB on drive C)
```

In PSpice Schematics, the library files are selected from the command Library from the PSpice Simulation Settings menu as shown in Figure 6.4.

6.2.7 .INC (Include File)

The contents of another file may be included in the circuit file using the statement

```
.INC NFILE
```

where NFILE is the name of the file to be included and may be any character string that is a legal file name for computer systems. It may include a volume, directory, and version number.

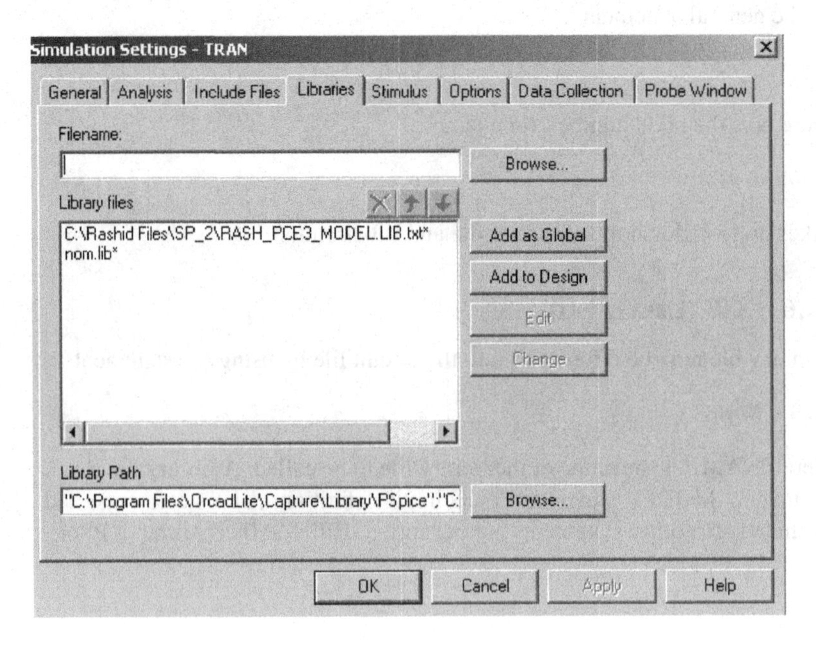

FIGURE 6.4 Libraries menu.

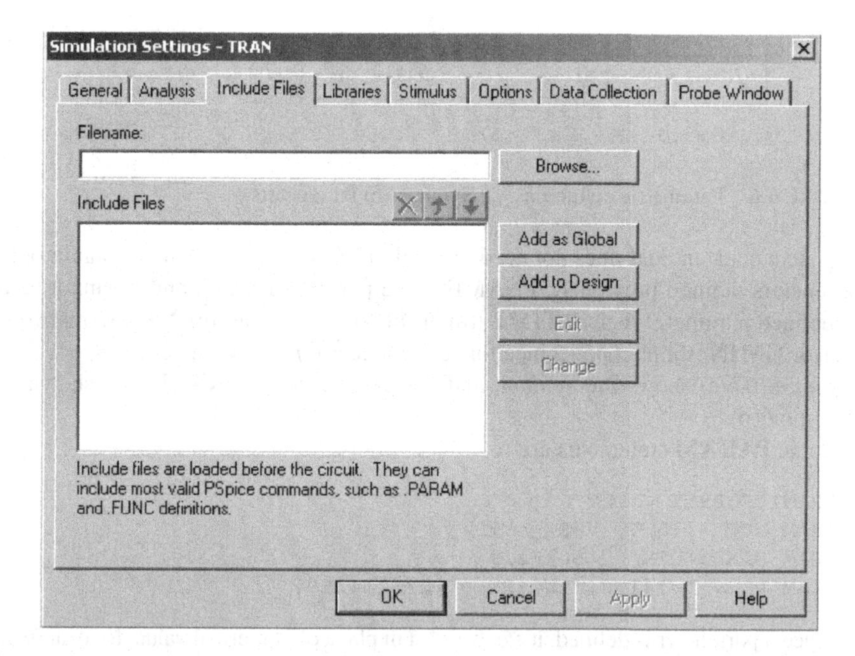

FIGURE 6.5 Include files.

Included files may contain any statements except a title line. However, a comment line may be used instead of a title line. If an .END statement is present, it only marks the end of an included file. An .INC statement may be used only up to four levels of "including." The include statement simply brings everything of the included file into the circuit file and takes up space in the main memory (RAM).

Some include file statements are

```
.INC OPAMP.CIR
.INC a:INVERTER.CIR
.INC c:\LIB\NOR.CIR
```

In PSpice Schematics, Include Files is selected from the command Include Files of the PSpice Simulation Settings menu as shown in Figure 6.5.

6.2.8 .PARAM (Parameter)

In many applications, it is convenient to use a parameter instead of a numerical value, such that the parameter can be combined into arithmetic expressions. The parameter definition is one of the forms

```
.PARAM ⟨PNAME = ⟨value⟩ or {⟨expression⟩}⟩*
```

The keyword .PARAM is followed by a list of names with values or expressions.

```
⟨value⟩
```

(a) Parameters:

LVAL = 1.5 mH
RVAL = 1 k
CVAL = 1 μF

(b)

Reference	Value	RVAL	LVAL	CVAL	Source Part
1	PARAM	1k	1.5mH	1uF	PARAM.Normal

FIGURE 6.6 Parametric definitions. (a) Symbol. (b) Parameters.

must be a constant and does not need "{" and "}". ⟨`expression`⟩ must contain only parameters defined previously. PNAME is the parameter name, and it cannot be a predefined parameter such as **TIME** (time), **TEMP** (temperature), **VT** (thermal voltage), or **GMIN** (shunt conductance for semiconductor *pn* junctions).

Figure 6.6(a) shows the symbol, and the parameters of **PARAM** are shown in Figure 6.6(b).

Some **PARAM** statements are

```
.PARAM VSUPPLY = 12V
.PARAM VCC = 15 V, VEE = -15V
.PARAM BANDWIDTH = {50kHz/5}
.PARAM PI = 3.14159, TWO-PI = {2*3.14159}
```

Once a parameter is defined, it can be used in place of numerical value; for example,

```
VCC 1 0 {SUPPLY}
VEE 0 5 {SUPPLY}
```

will change the value of **SUPPLY** in both statements. for example,

```
.FUNC IND(I(Vsense)) (A0 +A1*I(V(Sense)) +
A2*I(V(Sense)) + I(V(Sense)))
.PARAM INDUCTOR = IND (I (Vsense))
L1 1 3 {INDUCTOR}
```

will change the value of **INDUCTOR**.

Parameters defined in **.PARAM** statements are global; they can be used anywhere in the circuit, including inside subcircuits. The parameters can be made local to subcircuits by having parameters as arguments to subcircuits. for example,

```
.SUBCKT FILTER 1 2 PARAMS: CENTER=100kHz, WIDTH=10kHz
```

is a subcircuit definition for a band-pass filter with nodes 1 and 2 and with parameters **CENTER** (center frequency) and **WIDTH** (bandwidth). **PARAMs** separate the nodes list from the parameter list. When calling this subcircuit **FILTER**, the parameters can be given new values, for example,

```
X1 4 6 FILTER PARAMS: CENTER=200kHz
```

will override the default value of 100 kHz with a CENTER value of 200 kHz.

A defined parameter can be used in the following cases:

1. All model parameters
2. Device parameters, such as AREA, L, NRD, Z0, and IC values on capacitors and inductors, and TC1 and TC2 for resistors

3. All parameters of independent voltage and current sources (V and I devices)
4. Values on .NODESET statement (Section 6.6.2) and .IC statement (Section 6.9.1)

A defined parameter cannot be used for

1. Transmission line parameters NL and F
2. In-line temperature coefficients for resistors
3. E, F, G, and H device polynomial coefficient values
4. Replacing node numbers
5. Values on analysis statements (.TRAN, .AC, .DC, etc.)

6.2.9 .STEP (Parametric Analysis)

In PSpice Schematics, this command is selected from the Parametric Sweep option from the PSpice Simulation Settings menu as shown in Figure 6.7.

The effects of parameter variations can be evaluated by the **.STEP** command, whose general statement takes the general forms

```
.STEP LIN SWNAME SSTART SEND SINC
.STEP OCT SWNAME SSTART SEND NP
.STEP DEC SWNAME SSTART SEND NP
.STEP SWNAME LIST ⟨value⟩*
```

where **SWNAME** is the sweep variable name. **SSTART**, **SEND**, and **SINC** are the start value, the end value, and the increment value of the sweep variables,

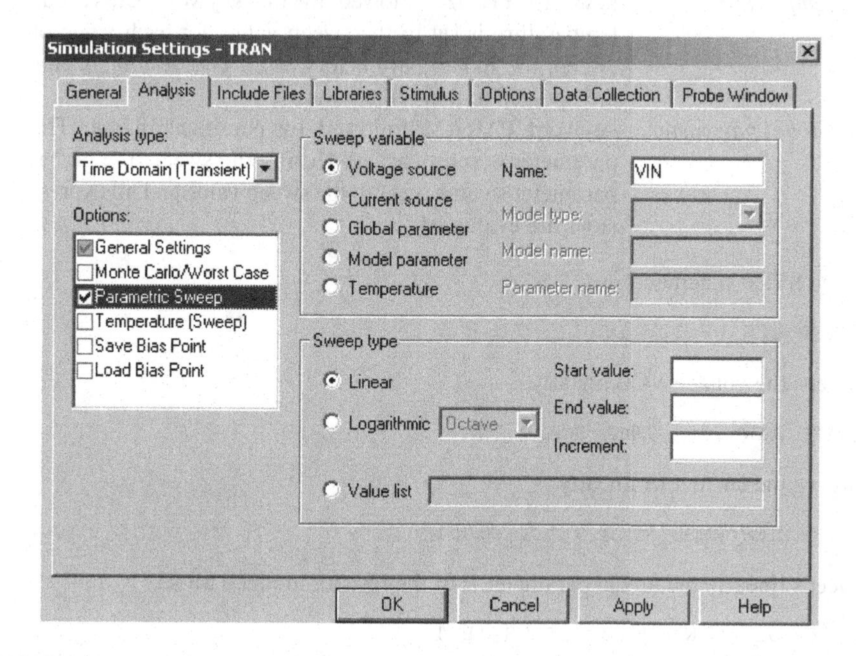

FIGURE 6.7 Parametric Sweep menu.

respectively. **NP** is the number of steps. **LIN, OCT**, or **DEC** specifies the type of sweep as follows:

LIN *Linear sweep*: **SWNAME** is swept linearly from **SSTART** to **SEND. SINC** is the step size.

OCT *Sweep by octave*: **SWNAME** is swept logarithmically by octave, and NP becomes the number of steps per octave. The next variable is generated by multiplying the present value by a constant larger than unity. OCT is used if the variable range is wide.

DEC *Sweep by decade*: **SWNAME** is swept logarithmically by decade, and NP becomes the number of steps per decade. The next variable is generated by multiplying the present value by a constant larger than unity. DEC is used if the variable range is the widest.

LIST *List of values*: There are no start and end values. The values of the sweep variables are listed after the keyword LIST.

The **SWNAME** can be one of the following types:

Source name of an independent voltage or current source. During the sweep, the source's voltage or current is set to the sweep value.

Model parameter model name type and model name followed by a model parameter name in parentheses. The parameter in the model is set to the sweep value. The model parameters L and W for a MOS device and any temperature parameters such as TC1 and TC2 for the resistor cannot be swept.

Temperature keyword **TEMP** followed by the keyword LIST. The temperature is set to the sweep value. for each value of sweep, the model parameters of all circuit components are updated to that temperature.

Global parameter keyword PARAM followed by parameter name. The parameter is set to sweep. During the sweep, the global parameter's value is set to the sweep value and all expressions are evaluated.

Some STEP statements are

```
.STEP VCE 0V 10V 1V
```

sweeps the voltage VCE linearly.

```
.STEP LIN IS - 10mA 5mA 0.1mA
```

sweeps the current IS linearly.

```
.STEP RES RMOD(R) 0.9 1.1 0.001
```

sweeps linearly the model parameter R of the resistor model RMOD.

```
.STEP DEC NPN QM(IS) 1E-18 1E-14 10
```

sweeps with a decade increment the parameter IS of the **NPN** transistor.

```
.STEP TEMP LIST 0 50 80 100 150
```

sweeps the temperature **TEMP** as listed.

```
.STEP PARAM Centerfreq 8.5kHz 10.5kHz 50Hz
```

sweeps linearly the parameter **PARAM** Centerfreq.

Note the following:

1. The sweep start value **SSTART** may be greater than or less than the sweep end value **SEND**.
2. The sweep increment **SINC** must be greater than zero.
3. The number of points **NP** must be greater than zero.
4. If the **.STEP** command is included in a circuit file, all analyses specified (**.DC**, **.AC**, **.TRAN**, etc.) are performed for each step.

6.3 TYPES OF OUTPUT

The commands that are available to obtain output from the results of simulations are

```
.PRINT Print
.PLOT Plot
.PROBE Probe output
.WIDTH Width
```

6.3.1 .PRINT (Print)

The results of **DC, AC**, transient (**.TRAN**), and noise (**.NOISE**) analyses can be obtained in the form of tables.

In PSpice Schematics, output markers rather than **PRINT** commands are normally used for plotting the transient and frequency response (AC analysis). Print commands can be used for printing data points on electronic files.

The print statement takes one of the forms

```
.PRINT DC     ⟨output variables⟩
.PRINT AC     ⟨output variables⟩
.PRINT TRAN   ⟨output variables⟩
.PRINTNOISE   ⟨output variables⟩
```

The maximum number of output variables is eight in any **.PRINT** statement. However, more than one **.PRINT** statement can be used to print all the output variables desired. The values of the output variables are printed as a table, with each column corresponding to one output Variable. The number of digits for output values can be changed with the **NUMDGT** option on the **.OPTIONS** statement as described in Section 6.4. The results of the **.PRINT** statement are stored in the output file.

Some print statements are

```
.PRINT DC V(2), V(3,5), V(R1), VCE(Q2), I(VIN), I(R1), IC(Q2)
.PRINT AC VM(2), VP(2), VM(3,5), V(R1), VG(5), VDB(5), IR(5),
II(5)
.PRINTNOISE INOISE ONOISE DB(INOISE) DB(ONOISE)
.PRINT TRAN V(5) V(4,7) (0,10V) IB(Q1) (0,50MA) IC(Q1)
(-50MA, 50MA)
```

Notes: Having two .**PRINT** statements for the same variable will not produce two tables. PSpice will ignore the first statement and produce output for the second statement. You could also use the **VPRINT** and **IPRINT** symbols in the special.lib to print the voltage and current, respectively (see Section 2.10).

6.3.2 .PLOT (Plot)

In PSpice Schematics, the **PLOT** commands are rarely used because of the availability of graphical output devices such as Probe (Section 6.3.3).

The results of **DC, AC**, transient (.**TRAN**), and noise (.**NOISE**) analyses can be obtained in the form of line printer plots. The plots are drawn by using characters and the results can be obtained in any type of printer. The plot statement takes one of the forms

```
.PLOT  DC   ⟨output variables⟩
    +  (⟨(lower limit) value⟩, ⟨(upper limit) value⟩)
.PLOT  AC   ⟨output variables⟩
    +  [⟨(lower limit) value⟩, ⟨(upper limit) value⟩]
.PLOT TRAN  ⟨output variables⟩
    +  [⟨(lower limit) value⟩, ⟨(upper limit) value⟩]
.PLOT NOISE ⟨output variables⟩
    +  [⟨(lower limit) value⟩, ⟨(upper limit) value⟩]
```

The maximum number of output variables is eight in any .**PLOT** statement. More than one .**PLOT** statement can be used to plot all the output variables desired.

The range and increment of the x-axis is fixed by the type of analysis command (e.g., .DC, .AC, .TRAN, or .**NOISE**). The range of the y-axis is set by adding (⟨(`lower limit`) `value`⟩, ⟨(`upper limit`) `value`⟩) at the end of a .**PLOT** statement. The y-axis range, (⟨(`lower limit`) `value`⟩ and ⟨(`upper limit`) `value`⟩), can be placed in the middle of a set of output variables. The output variables will follow the range specified, which comes immediately to the left.

If the y-axis range is omitted, PSpice assigns a default range determined by the range of the output variable. If the ranges of output variables vary widely, PSpice assigns the ranges corresponding to the different output variables.

Some plot statements are

```
.PLOT DC V(2), V(3,5), V(R1), VCE(Q2), I(VIN), I(R1), IC(Q2)
.PLOT AC VM(2), VP(2), VM(3,5), V(R1), VG(5), VDB(5), IR(5),
II(5)
.PLOTNOISE INOISE ONOISE DB(INOISE) DB(ONOISE)
.PLOT TRAN V(5) V(4,7) (0,10V) IB(Q1) (0, 50MA) IC(Q1)
(-50MA, 50MA)
```

Note: In the first three statements, the y axis is by default. In the last statement, the range for voltages V(5) and V(4,7) is 0 to 10 V, that for current IB(Q1) is 0 to 50 mA, and that for the current IC(Q1) is –50 to 50 mA.

6.3.3 .PROBE (Probe)

Probe is a graphics postprocessor, and it is available as an option in the professional version of PSpice. However, Probe is included in the student's version of PSpice. The results of the **DC, AC**, and transient (.TRAN) analysis cannot be used directly by Probe. First, the results have to be processed by the .PROBE command, which writes the processed data on a file, PROBE.DAT, for use by Probe. The command takes one of the forms

```
.PROBE
.PROBE ⟨one or more output variables⟩
```

In the first form, in which no output variable is specified, the .PROBE command writes all the node voltages and all the element currents to the PROBE.DAT file. The element currents are written in the forms that are permitted as output variables and are discussed in Section 3.3.2. In the second form, in which the output variables are specified, PSpice writes only the specified output variables to the PROBE.DAT file. This form is suitable for users without a fixed disk and to limit the size of the PROBE.DAT file.

6.3.3.1 Probe Statements

```
.PROBE
.PROBE V(5), V(4,3), V(C1), VM(2), I(R2), IB(Q1), VBE(Q1)
```

6.3.4 PROBE Output

It is very easy to use Probe. Once the results of the simulations are processed by the .PROBE command, the results are available for graphics output. PROBE comes with a first-level menu, as shown in Figure 6.8, from which the type of analysis can be chosen. After the first choice, the second level is the choice for the plots and coordinates of output variables as shown in Figure 6.9. After the choices, the output is displayed as shown in Figure 6.10.

With one exception, Probe does not distinguish between uppercase and lowercase variable names, that is, "V(4)" and "v(4)" are equivalent. The exception is the "m" scale suffix for numbers: "m" means "milli" (1E–3), whereas "M" means "mega" (1E+6). The suffixes "MEG" and "MIL" are not available. The units that are recognized by Probe are as follows:

V Volts
A Amperes
W Watts
s Seconds
H Hertz

```
                              Probe

               Graphics Post-Processor for PSpice
                  Version 1.13 - October 1987
     © Copyright 1985, 1986, 1987 by MicroSim Corporation
                          ---------
                       Classroom version
       Copying of this program is welcomed and encouraged

    Circuit:        EXAMPLE1 - An Illustration of all Commands
Date/Time run:    10/31/88    16:15:19                Temperature:   35.0

  0) Exit Program   1) DCSweep   2) ACSweep   3) TransientAnalysis:   1
```

FIGURE 6.8 Select analysis display for Probe.

Probe also recognizes that W = V × A, V = W/A, and A = W/V. Therefore, the addition of a trace, such as

```
VCE(Q1) *IC(Q1)
```

gives the power dissipation of transistor Q_1 and will be labeled with a "W." Arithmetic expressions of output variables are also allowed and the available

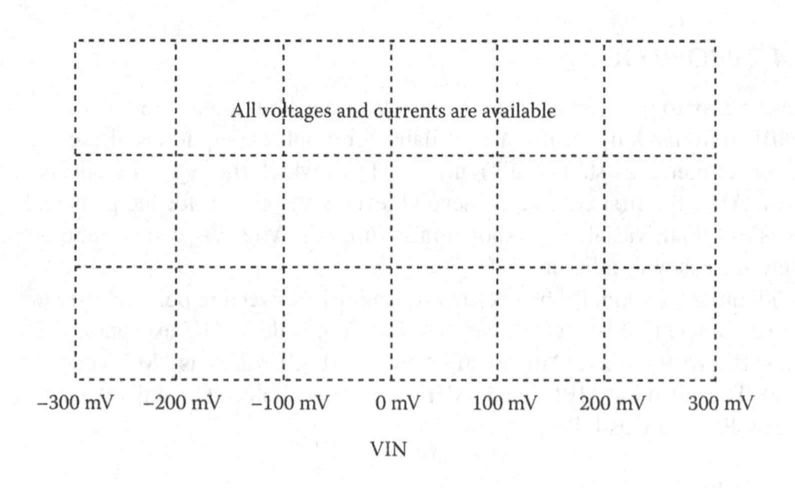

0) Exit 1) Add trace 3) X axis 4) Y axis 5) Add plot 8) Hardcopy
9) Suppress symbols : 1

FIGURE 6.9 Select plot/graphics output.

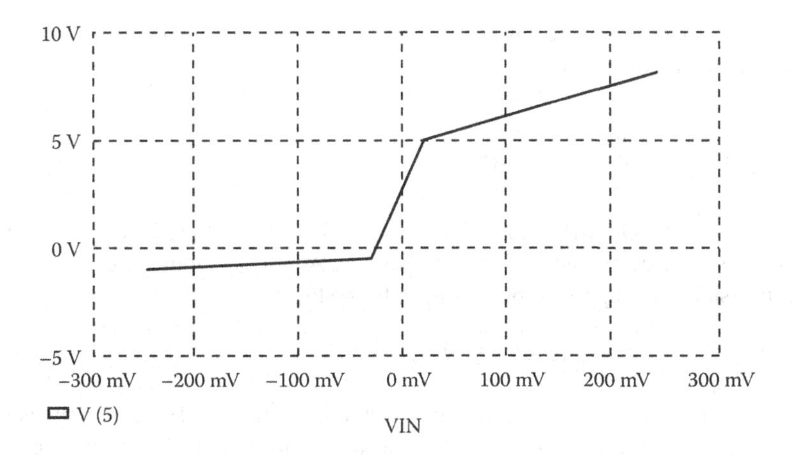

FIGURE 6.10 Output display.

operators are "+," "−," "*," and "/," along with parentheses. The available functions are as follows:

Function	Meaning
ABS(x)	$\lvert x \rvert$ (absolute value)
B(Kxy)	Flux density of coupled inductor K_{xy}
H(Kxy)	Magnetization of coupled inductor K_{xy}
SGN(x)	+1 (if $x > 0$), 0 (if $x = 0$), −1 (if $x < 0$)
EXP(x)	e^x
DB(x)	20 log($\lvert x \rvert$) (log of base 10)
LOG(x)	ln(x) (log of base e)
LOG10(x)	log(x) (log of base 10)
PWR(x, y)	$\lvert x \rvert^y$
SQRT(x)	$x^{1/2}$
SIN(x)	sin(x) (x in radians)
COS(x)	cos(x) (x in radians)
TAN(x)	tan(x) (x in radians)
ARCTAN(x)	$\tan^{-1}(x)$ (results in radians)
d(y)	Derivative of y with respect to the x-axis variable
s(y)	Integral of y over the x-axis variable
AVG(x)	Running average of x
RMS(x)	Running RMS average of x

Note: After you run the simulation, PSpice automatically opens the Probe menu. When you choose "add a trace," Probe shows all the possible output variables on the left side of the window and all the allowable functions on the right side. For example, to plot the average of a current through resistor R, you use AVG(I(R)).

For derivatives and integrals of simple variables (not expressions), the shorthand notations that are available are

dV(4) is equivalent to d(V(4))
sIC(Q3) is equivalent to s(IC(Q3))

The plot of dIC(Q2)/dIB(Q2) will give the small-signal beta value of transistor Q_2. Two or more traces can be added with only one Add Trace command, in which all the expressions are separated by " " or ","; for instance,

```
V(2) V(4), IC(M1), RMS(I(VIN))
```

will add four traces. This gives the same result as using Add Trace four times with only one trace at a time but is faster because the plot is not redrawn between the addition of each trace.

The PROBE.DAT file can contain more than one of any kind (e.g., two transient analyses with three temperatures). If PSpice is run for a transient analysis at three temperatures, the expression

```
V(1)
```

will result in Probe drawing three curves instead of the usual one curve. Entering the expression

```
V(1) @ n
```

will result in Probe drawing the curve of V(1) for the *n*th transient analysis. Entering the expression

```
V(1)@1–V(1)@2
```

will display the difference between the waveforms of the first and second temperatures, whereas the expression

```
V(1)-V(2)@2
```

will display three curves, one for each V(1).

Note the following:

1. The .PROBE command requires a math coprocessor for the professional version of PSpice but not for the student version.
2. Probe is not available on SPICE. However, the newest version of SPICE (SPICE3) has a postprocessor similar to Probe called Nutmeg.
3. It is required to specify the type of display and the type of hardcopy devices on the PROBE.DEV file as follows:
   - Display = ⟨display name⟩
   - Hard copy = ⟨port name⟩, ⟨device name⟩

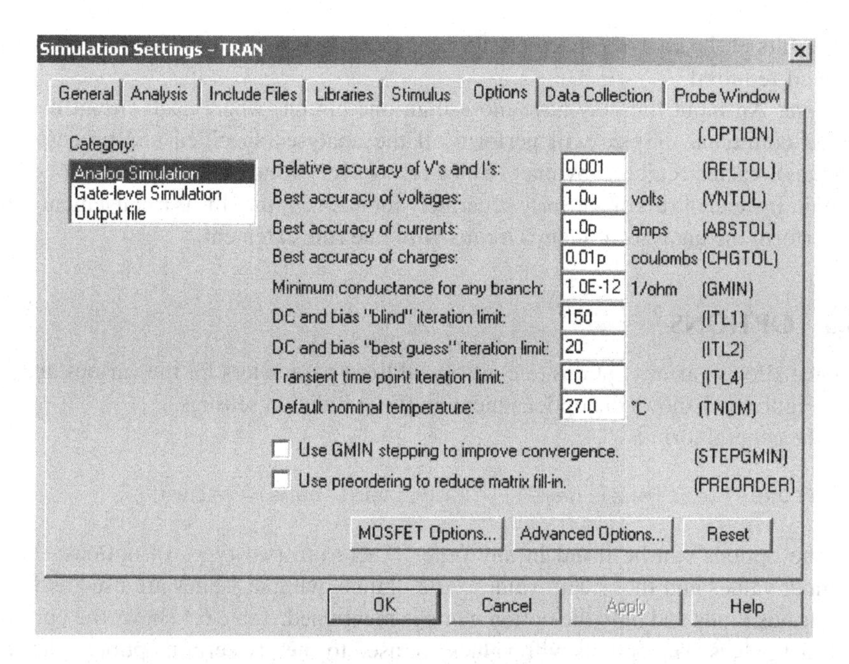

FIGURE 6.11 Options menu in the Simulation Settings.

- The details of names for display, port, and device (printer) can be found in the README.DOC file that comes with the PSpice programs or in the PSpice manual.

4. The display and hardcopy devices can be set from the display or printer setup menu.

6.3.5 .WIDTH (Width)

The width of the output in columns can be set by the .WIDTH statement, which has the general form

```
.WIDTH OUT = ⟨value⟩
```

`⟨value⟩` is in columns and must be either 80 or 132. The default value is 80.

In PSpice Schematics, the width can be set from the Output file category of the Options submenu in the Simulation Settings menu as shown in Figure 6.11.

6.4 OPERATING TEMPERATURE AND END OF CIRCUIT

The temperature command is discussed in Section 5.3. The last statement for the end of a circuit is

```
.End
```

All data and commands must come before the .END command. The .END command instructs PSpice to perform all the circuit analyses specified. After processing the results of the analysis specified, PSpice resets itself to perform further computations, if required.

Note: An input file may have more than one circuit, where each circuit has its .END command. PSpice will perform all the analyses specified and process the results of each circuit one by one. PSpice resets everything at the beginning of each circuit. Instead of running PSpice separately for each circuit, this is a convenient way to perform the analysis of many circuits with one run statement.

6.5 OPTIONS

PSpice allows various options to control and limit parameters for the various analyses. Figure 6.11 shows the options menu in the simulation settings.

The general form is

```
.OPTIONS [(options) name)] [⟨(options) name⟩=⟨value⟩]
```

The options can be listed in any order. There are two types of options: those without values and those with values. The options without values are used as flags of various kinds and only the option name is mentioned. Table 6.1 shows the options without values. The options with values are used to specify certain optional parameters. The option names and their values are specified. Table 6.2 shows the options with values. The commonly used options are **NOPAGE, NOECHO, NOMOD, TNOM, CPTIME, NUMDGT, GMIN, and LIMPTS**.

Options statements are

```
.OPTIONS NOPAGE NOECHO NOMOD DEFL=20U DEFW=15U DEFAD=50P
DEFAS=50P
.OPTIONS ACCT LIST RELTOL=.005
```

Note: In PSpice Schematics, the options can be set from the menu as shown in Figure 6.11.

TABLE 6.1
Options without Values

Option	Effects
NOPAGE	Suppresses paging and printing of a banner for each major section of output
NOECHO	Suppresses listing of the input file
NODE	Causes output of net list (node table)
MONOD	Suppresses listing of model parameters
LIST	Causes summary of all circuit elements (devices) to be output
OPTS	Causes values for all options to be output
ACCT	Summary and accounting information is output at the end of all the analysis
WIDTH	Same as .WIDTH OUT = statement

TABLE 6.2
Options with Values

Option	Effects	Unit	Default
DEFL	MOSFET channel length (L)	m	100 μm
DEFW	MOSFET channel width (W)	m	100 μm
DEFAD	MOSFET drain diffusion area (AD)	M^{-2}	0
DEFAS	MOSFET source diffusion area (AS)	M^{-2}	0
TNOM	Default temperature (also the temperature at model parameters are assumed to have been measured)	°C	27
NUMDGT	Number of digits output in print tables		4
CPTIME	Central processing unit (CPU) time allowed for a run	s	1E6
LIMPTS	Maximum points allowed for any print table or plot		201
ITL1	DC and bias-point iteration limit		40
ITL2	DC and bias-point "educated guess" iteration limit		20
ITL4	Iteration limit at any point in transient analysis		10
ITL5	Total iteration limit for all points in transient analysis (ITL5 = 0 means ITL5 = infinite)		5000
RELTOL	Relative accuracy of voltages and currents		0.001
TRTOL	Transient analysis accuracy adjustment		7.0
ABSTOL	Best accuracy of currents	A	1 pA
CHGTOL	Best accuracy of charges	C	0.01 pC
VNTOL	Best accuracy of voltages	V	1 μV
PIVREL	Relative magnitude required for pivot in matrix solution		1E-13
GMIN	Minimum conductance used for any branch	Ω^{-1}	1E-12

If the option **ACCT** is specified in the **.OPTIONS** statement, PSpice will print a *job statistics summary* displaying various statistics about the run at the end. This option is not required for most circuit simulations. The format of the summary (with an explanation of the terms appearing in it) is as follows:

Item	Meaning
NUNODS	Number of distinct circuit nodes before subcircuit expansion.
NCNODS	Number of distinct circuit nodes after subcircuit expansion. If there are no subcircuits, NCNODS = NUNODS.
NUMNOD	Total number of distinct nodes in circuit. This is NCNODS plus the internal nodes generated by parasitic resistances. If no device has parasitic resistances, NUMNOD = NCNODS.
NUMEL	Total number of devices (elements) in circuit after subcircuit expansion. This includes all statements that do not begin with "."or "X."
DIODES	Number of diodes after subcircuit expansion.
BJTS	Number of bipolar transistors after subcircuit expansion.
JFETS	Number of junction FETs after subcircuit expansion.
MFETS	Number of MOSFETs after subcircuit expansion.
GASFETS	Number of GaAs MESFETs after subcircuit expansion.

(Continued)

Item	Meaning
NUMTEM	Number of different temperatures.
ICVFLG	Number of steps of DC sweep.
JTRFLG	Number of print steps of transient analysis.
JACFLG	Number of steps of AC analysis.
INOISE	1 or 0: noise analysis was/was not done.
NOGO	1 or 0: run did/did not have an error.
NSTOP	The circuit matrix is conceptually (not physically) of dimension NSTOP × NSTOP.
NTTAR	Actual number of entries in circuit matrix at beginning of run.
NTTBR	Actual number of entries in circuit matrix at end of run.
NTTOV	Number of terms in circuit matrix that come from more than one device.
IFILL	Difference between NTTAR and NTTBR.
IOPS	Number of floating-point operations needed to do one solution of circuit matrix.
PERSPA	Percent sparsity of circuit matrix.
NUMTTP	Number of internal time steps in transient analysis.
NUMRTP	Number of times in transient analysis that a time step was too large and had to be cut back.
NUMNIT	Total number of iterations for transient analysis.
MEMUSE	Amount of circuit memory used/available in bytes. There are two memory pools.
MAXMEM	Exceeding either one will abort the run.
COPYKNT	Number of bytes that were copied in the course of doing memory management for this run.
READIN	Time spent reading and error checking the input file.
SETUP	Time spent setting up the circuit matrix pointer structure.
DCSWEEP	Time spent and iteration count for calculating DC sweep.
BIASPNT	Time spent and iteration count for calculating bias point and bias point for transient analysis.
MATSOL	Time spent solving circuit matrix (this time is also included in the time for each analysis). The iteration count is the number of times the rows or columns were swapped in the course of solving it.
ACAN	Time spent and iteration count for AC analysis.
TRANAN	Time spent and iteration count for transient analysis.
OUTPUT	Time spent preparing .**PRINT** tables and .**PLOT** plots.
LOAD	Time spent evaluating device equations (this time is also included in the time for each analysis).
OVERHEAD	Other time spent during run.
TOTAL JOB TIME	Total run time, excluding the time to load the program files PSPICE1.EXE and PSPICE2.EXE into memory.

6.6 DC ANALYSIS

In DC analysis, all the independent and dependent sources are of the DC type. The inductors and capacitors in a circuit are considered as short circuits and open circuits, respectively. This is due to the fact that at zero frequency, the impedance represented by an inductor is zero and that by a capacitor is infinite. The commands that are available for DC analyses are

```
.OP       Dc operating point
.NODESET  Node set
.SENS     Small-signal sensitivity
.TF       Small-signal transfer function
.DC       Dc sweep
```

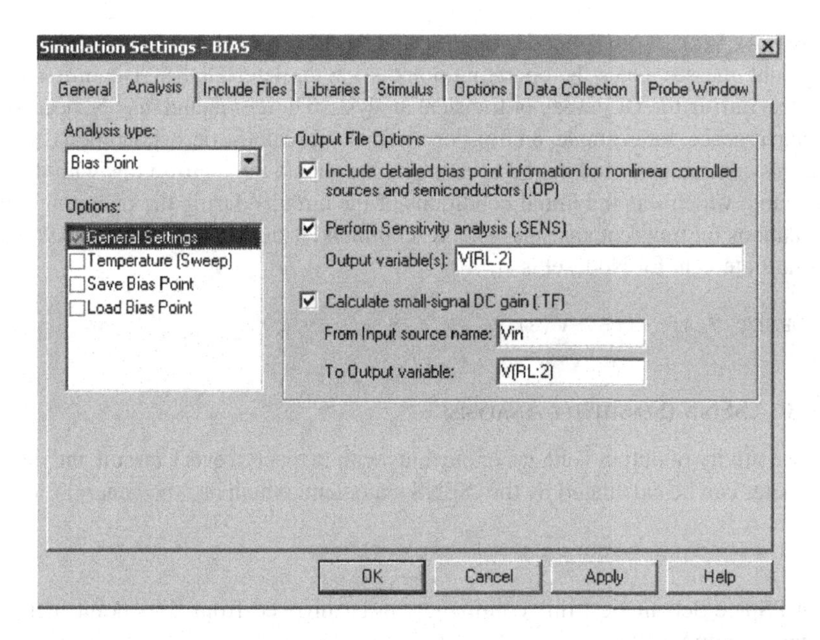

FIGURE 6.12 Setup for .OP, .TF, and .SENS commands.

In PSpice Schematics, these commands can be invoked from the Analysis type menu of the Simulation Settings as shown in Figure 6.12.

6.6.1 .OP (Operating Point)

Electronic and electrical circuits contain nonlinear devices (e.g., diodes, transistors) whose parameters depend on the *operating point*. The operating point is also known as a *bias point* or *quiescent point*. The operating point is always calculated by PSpice for determining small-signal parameters of nonlinear devices during the DC sweep and transfer function analysis. The command takes the form

```
.OP
```

In PSpice Schematics, this command can be invoked from the Bias Point menu as shown in Figure 6.12.

The .OP command controls the output of the bias point, but not the method of bias analysis or the results of the bias point. If the .OP command is omitted, PSpice prints only a list of the node voltages. If the .OP command is present, PSpice prints the currents and power dissipations of all the voltages. The small-signal parameters of all nonlinear controlled sources and all semiconductor devices are also printed in the output file.

6.6.2 .NODESET (Nodeset)

In calculating the operating bias point, some or all of the nodes of the circuit may be assigned initial guesses to help DC convergence by the statement

```
.NODESET V(1) = B1 V(2) = B2 ... V(N) = BN
```

where V(1), V(2), … are the node voltages, and B1, B2,… are their respective values of the initial guesses. Once the operating point is found, the .NODESET command has no effect during the DC sweep or transient analysis. This command may be necessary for convergence, for example, on flip-flop circuits to break the tie-in condition. In general, this command should not be necessary. One should not confuse this with the .IC command, which sets the initial conditions of the circuits during the operating point calculations for transient analysis. The .IC command is discussed in Section 6.9.1.

The statement for Nodeset is

```
.NODESET V(4) =1.5V V(6) = 0V (25)=1.5V
```

6.6.3 .SENS (Sensitivity Analysis)

The sensitivity of output voltages or currents with respect to every circuit and device parameter can be calculated by the .SENS statement, which has the general form

```
.SENS <(one or more output) variables>
```

In PSpice Schematics, this command can be invoked from Bias Point menu as shown in Figure 6.12.

The .SENS statement calculates the bias point and the linearized parameters around the bias point. In this analysis, the inductors are assumed to be short circuits and capacitors to be open circuits. If the output variable is a current, that current must be through a voltage source. The sensitivity of each output variable with respect to all the device values and model parameters is calculated. The .SENS statement prints the results automatically. Therefore, it should be noted that a .SENS statement may generate a huge amount of data if many output variables are specified.

The statement for sensitivity analysis is

```
.SENS V(5) V(2,3) I(V2) I(V5)
```

Example 6.2:

Finding the Sensitivity of the Output Voltage with Respect to Each Circuit Element

An op-amp circuit is shown in Figure 6.13. The op-amp is represented by the subcircuit of Figure 6.2. Calculate and print the sensitivity of output voltage V(4) with respect to each circuit element. The operating temperature is 40°C.

SOLUTION

The PSpice schematic is shown in Figure 6.14(a). The output is specified in the Sensitivity Analysis menu of the Analysis Setup as shown in Figure 6.14(b).

Note: The op-amp is represented as a voltage-controlled voltage source with an input resistance $R_i = 2\ M\Omega$, an open-loop voltage gain $A = 2 \times 10^6 V/V$ and an output resistance $R_0 = 75\ \Omega$.

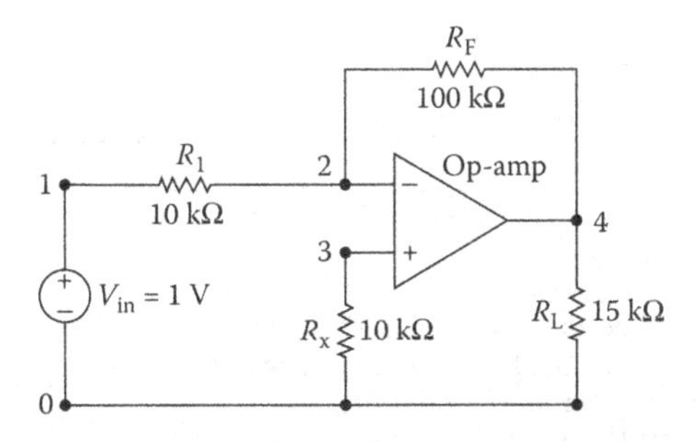

FIGURE 6.13 Op-amp circuit for Example 6.2.

(a)

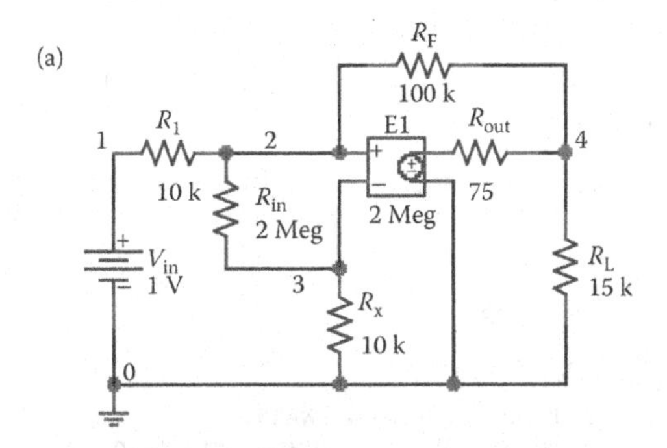

(b)

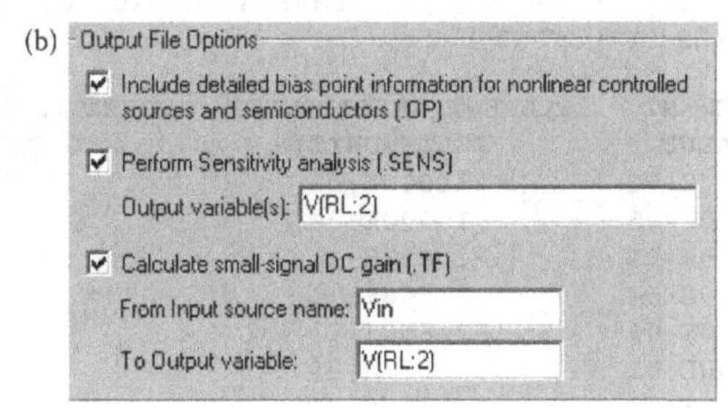

FIGURE 6.14 PSpice schematic for Example 6.2. (a) Schematic. (b) Defining the output of sensitivity analysis.

The listing of the statements for the circuit file is as follows:

DC Sensitivity Analysis

```
SOURCE    ■  * DC input voltage of 1 V:
             VIN 1 0 DC 1V
CIRCUIT   ■■ R1 1 2 10K
             RF 2 4 100K
             RL 4 0 15K
             RX 3 0 10K
             *Subcircuit call OPAMP:
             X1 2 3 4 0 OPAMP
             * Subcircuit definition:
             .SUBCKT OPAMP 1 2 3 4
             * model name
```

```
             vi- vi+ vo+ vo-
             RIN 1 2 2MEG
             ROUT 5 3 75
             E1 5 4 2 1 2MEG; Voltage-controlled voltage source
             .ENDS OPAMP; End of subcircuit definition OPAMP
ANALYSIS ■■■ * Operating temperature is 40°C:
             .TEMP 40
             * Options:
             .OPTIONS NOPAGE NOECHO
             * It calculates and prints the sensitivity
             * analysis of output
             * voltage V(4) with respect to all elements
             * in the circuit.
             .SENS V(4)
            .END
```

The results of the sensitivity analysis are shown next. The node voltages are also printed automatically.

```
****  SMALL  SIGNAL  BIAS  SOLUTION     TEMPERATURE=40.000  DEG C
NODE  VOLTAGE  NODE  VOLTAGE      NODE  VOLTAGE    NODE  VOLTAGE
(1)     1.0000  (2)  5.0 54E-06  (3)  25.14E-09  (4)    -9.9999
(X1.5)  -10.0570
VOLTAGE  SOURCE  CURRENTS
NAME         CURRENT
VIN          -1.000E-04
TOTAL POWER DISSIPATION  1.00E-04  WATTS
****    DC SENSITIVITY ANALYSIS    TEMPERATURE=40.000  DEG C
DC SENSITIVITIES OF OUTPUT V (4)

ELEMENT     ELEMENT     ELEMENT SENSITIVITY    NORMALIZED SENSITIVITY
NAME         VALUE         (VOLTS/UNIT)            (VOLTS/PERCENT)
R1          1.000E+04       1.000E-03               1.000E-01
RF          1.000E+05      -1.000E-04              -1.000E-01
RL          1.500E+04      -1.851E-11              -2.776E-09
RX          1.000E+04       2.766E-11               2.766E-09
X1.RIN      2.000E+06      -2.640E-13              -5.280E-09
X1.ROUT     7.500E+01       4.257E-09               3.193E-09
VIN         1.000E+00      -1.000E+01              -1.000E-01
```

Note: The output voltage of an inverting amplifier is more sensitive to R_i and R_F. Any increase in R_i causes the output voltage to increase whereas any increase in R_F causes the output voltage to decrease.

6.6.4 .TF (Small-Signal Transfer Function)

The small-signal transfer function capability of PSpice can be used to find the small-signal DC gain, the input resistance, and the output resistance of a circuit. If V(1) and V(4) are the input and output variables, respectively, PSpice will calculate the small-signal DC gain between nodes 1 and 4, defined by

$$A_v = \frac{\Delta V_o}{\Delta V_i} = \frac{V(4)}{V(1)}$$

as well as the input resistance between nodes 1 and 0 and the small-signal DC output resistance between nodes 4 and 0.

PSpice calculates the small-signal DC transfer function by linearizing the circuit around the operating point.

In PSpice Schematics, this command can be invoked from Bias Point menu as shown in Figure 6.12. The statement for the transfer function has one of the forms

```
.TF VOUT VIN
.TF IOUT IIN
```

where **VIN** is the input voltage and **VOUT** (or **IOUT**) is the output voltage (or output current). If the output is a current, the current must be through a voltage source. The output variable **VOUT** (or **IOUT**) has the same format and meaning as in a **.PRINT** statement. If there are inductors and capacitors in a circuit, the inductors are treated as short circuits and capacitors as open circuits.

The .TF command calculates the parameters of the Thévenin's (or Norton's) equivalent circuit of the circuit file. It prints the output automatically and does not require **.PRINT** or **.PLOT** or .PROBE statements.

Statements for transfer function analysis are

```
.TF V(10) VIN
.TF I(VN) IIN
```

Example 6.3:

Transfer Function Analysis of an Amplifier Circuit

An amplifier circuit, which is the equivalent circuit of a bipolar transistor amplifier, is shown in Figure 6.15. Calculate and print (a) the voltage gain $A_v = V(5)/v_{in}$, (b) the input resistance R_{in}, and (c) the output resistance R_o.

SOLUTION

The PSpice schematic is shown in Figure 6.16(a). The .TF command is set in the Analysis Setup. The output variable and the input source are identified in the TF set menu as shown in Figure 6.16(b).

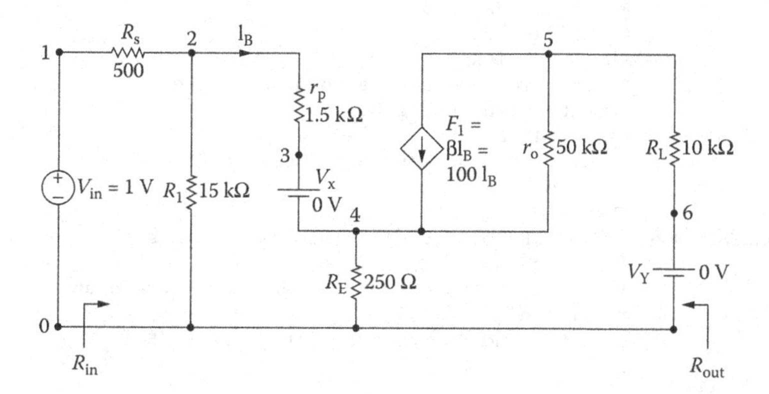

FIGURE 6.15 Amplifier circuit for Example 6.3.

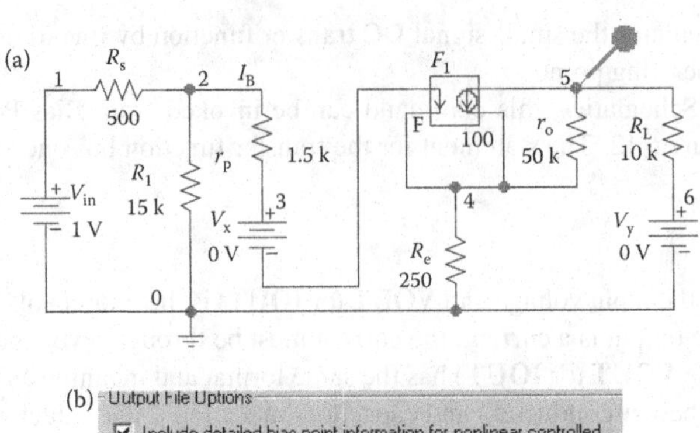

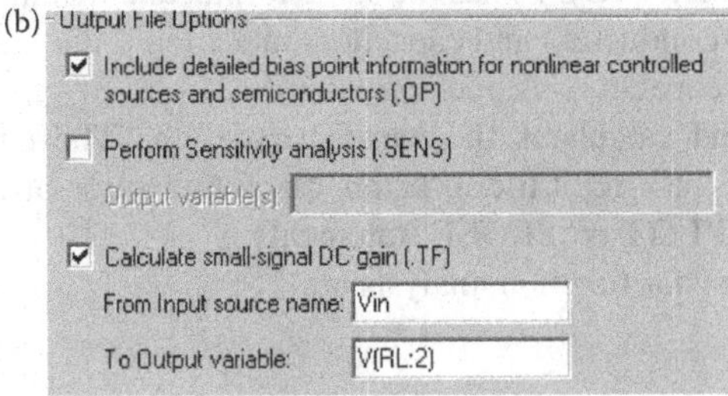

FIGURE 6.16 PSpice schematic for Example 6.3. (a) Schematic. (b) TF setup.

The listing of the circuit file is as follows:

Transfer Function Analysis

```
SOURCE    ■   * DC input voltage of 1 V:
              VIN  1   0   DC   1V
CIRCUIT   ■■  RS   1  2  500
              R1   2  0  15K
              RP   2  3  1.5K
              RE   4  0  250
              * Current-controlled current source:
              F1   5  4  VX  100
              RO   5  4  50K
              RL   5  6  10K
              * Dummy voltage source to measure the
              * controlling current:
              VX   3  4  DC  0V
              * Dummy voltage source to measure the output
              * current:
              VY   6  0  DC  0V
ANALYSIS ■■■  * The .TF command calculates and prints the
              *DC gain and
              * the input and output resistances. The input
              * voltage
              * is VIN and the output voltage is V (5).
              .TF  V(5)  VIN
       END.
```

The results of the .TF command are shown next.

```
VOLTAGE  SOURCE  CURRENTS
NAME            CURRENT
VIN             -1.053E-04
VX               4.211E-05
VY              -3.495E-03
TOTAL POWER DISSIPATION    1.05E-;04 WATTS
****       SMALL-SIGNAL CHARACTERISTICS
     V (5)/VIN = -3.495E+01
     INPUT RESISTANCE AT VIN = 9.499E+03
     OUTPUT RESISTANCE AT V (5)  = 9.839E+03
```

The input resistance can be calculated [6] as

$$\begin{aligned} R_{in} &= R_s + R_1 \| [r_p + (1+\beta_f)R_e] \\ &= 500 + 15k \| [1.5k + (1+100) \times 250] = 10.1\,k\Omega \end{aligned} \tag{6.1}$$

The voltage gain can be calculated [6] as

$$\begin{aligned} A_v &= \frac{-\beta_f R_L}{r_P + (1+\beta_f)R_E} \times \frac{R_1 \| [r_P + (1+\beta_f)R_E]}{R_{in}} \\ &= \frac{-100 \times 10k}{1.5k + (1+100) \times 250} \times \frac{15k \| [1.5k + (1+100) \times 250]}{10.1k} = -35.53\,V/V \end{aligned} \tag{6.2}$$

The output resistance can be approximately calculated as

$$R_o \approx R_L = 10\,k\Omega$$

6.6.5 .DC (DC Sweep)

DC sweep is also known as the *DC transfer characteristic*. The input variable is varied over a range of values. for each value of input variables, the DC operating point and the small-signal DC gain are computed by calling the small-signal transfer function capability of PSpice. The DC sweep (or DC transfer characteristic) is obtained by repeating the calculations of small-signal transfer function for a set of values.

An example of the DC Sweep setup in PSpice schematics is shown in Figure 6.17, and the nested sweep is set from the Secondary Sweep menu.

The statement for performing DC sweep takes one of the following general forms:

```
.DC  LIN  SWNAME  SSTART  SEND  SINC
+         [(nested sweepspecification)]
.DC  OCT  SWNAME  SSTART  SEND  NP
+         [(nested sweepspecification)]
.DC  DEC  SWNAME  SSTART  SEND  NP
+         [(nested sweepspecification)]
.DC  SWNAME  LIST  (value)*
+         [(nested sweepspecification)]
```

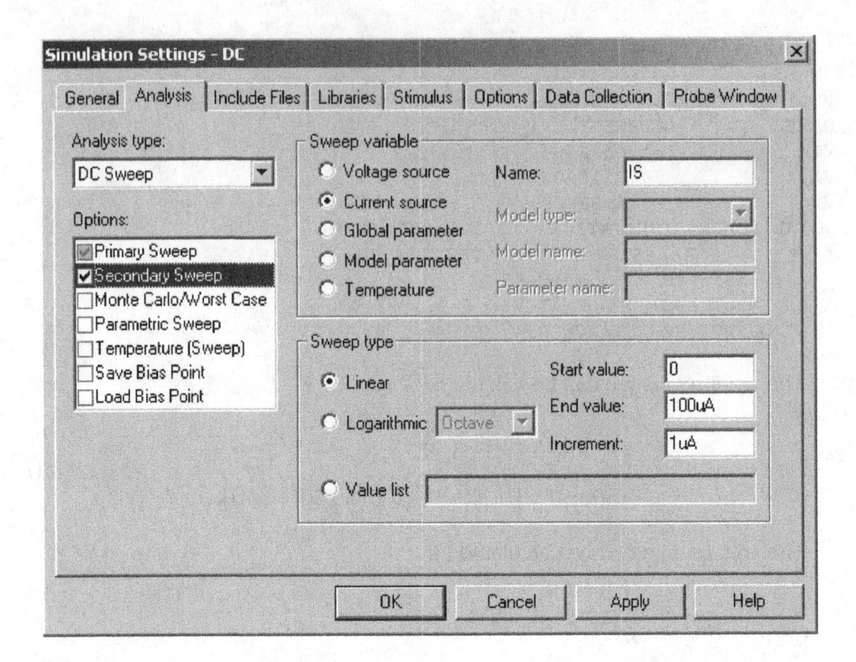

FIGURE 6.17 DC sweep in PSpice schematics.

where **SWNAME** is the sweep variable name. **SSTART, SEND**, and **SINC** are the start value, the end value, and the increment value of the sweep variables, respectively. NP is the number of steps. **LIN**, OCT, or DEC specifies the type of sweep, as follows:

LIN *Linear sweep*: **SWNAME** is swept linearly from **SSTART** to **SEND. SINC** is the step size.

OCT *Sweep by octave*: **SWNAME** is swept logarithmically by octave, and NP becomes the number of steps per octave. The next variable is generated by multiplying the present value by a constant larger than unity. OCT is used if the variable range is wide.

DEC *Sweep by decade*: **SWNAME** is swept logarithmically by decade, and NP becomes the number of steps per decade. The next variable is generated by multiplying the present value by a constant larger than unity. DEC is used if the variable range is widest.

LIST *List of values*: There are no start and end values. The values of the sweep variables are listed after the keyword LIST.

The **SWNAME** can be one of the following types:

Source Name of an independent voltage or current source. During the sweep, the source's voltage or current is set to the sweep value.

Model parameter Model name type and model name followed by a model parameter name in parentheses. The parameter in the model is set to the sweep value. The model parameters L and W for a MOS device and any temperature parameters such as TC1 and TC2 for the resistor cannot be swept.

Temperature Keyword **TEMP** followed by the keyword LIST. The temperature is set to the sweep value. for each value of sweep, the model parameters of all circuit components are updated to that temperature.

Global parameter Keyword PARAM followed by parameter name. The parameter is set to sweep. During the sweep, the global parameter's value is set to the sweep value and all expressions are evaluated.

The DC sweep can be nested, similar to a DO loop within a DO loop in forTRAN programing. The first sweep will be the inner loop and the second sweep will be the outer loop. The first sweep will be done for each value of the second sweep. ⟨nested sweep specification⟩ follows the same rules as those for the main (sweep variable).

The statements for DC sweep are

```
.DC VIN -5V 10V 0.25V
```

sweeps the voltage VIN linearly.

```
.DC LIN IIN 50MA -50MA 1MA
```

sweeps the current IIN linearly.

```
.DC VA 0 15V 0.5V IA 0 1MA 0.05MA
```

sweeps the current IA linearly within the linear sweep of VA.

```
.DC RES RMOD(R) 0.9 1.1 0.001
```

sweeps the model parameter R of the resistor model RMOD linearly.

```
.DC DEC NPN QM(IS)1E-18 1E-14 10
```

sweeps with a decade increment the parameter IS of the **NPN** transistor.

```
.DC TEMP LIST 0 50 80 100 150
```

sweeps the temperature **TEMP** as listed values.

```
.DC PARAM Vsupply -15V 15V 0.5V
```

sweeps the parameter PARAM Vsupply linearly.

PSpice does not print or plot any output by itself for DC sweep. The results of the DC sweep are obtained by **.PRINT**, **.PLOT**, or .PROBE statements. Probe allows nested sweeps to be displayed as a family of curves.

Note the following:

1. If the source has a DC value, its value is set by the sweep overriding the DC value.
2. In the third statement, the current source IA is the inner loop and the voltage source VA is the outer loop. PSpice will vary the value of current source IA from 0 to 1 mA with an increment of 0.05 mA for each value of voltage source VA, and generate an entire print table or plot for each value of voltage sweep.
3. The sweep-start value **SSTART** may be greater than or less than the sweep-end value **SEND**.
4. The sweep increment **SINC** must be greater than zero.
5. The number of points NP must be greater than zero.
6. After the DC sweep is finished, the sweep variable is set back to the value it had before the sweep started.

Example 6.4:

DC Transfer Characteristic of Varying the Load Resistor

For the amplifier circuit of Figure 6.15, calculate and plot the DC transfer characteristic, V_o versus V_{in}. The input voltage is varied from 0 to 100 mV with an increment of 2 mV. The load resistance is varied from 10 to 30 kΩ with a 10-kΩ increment.

SOLUTION

The PSpice schematic is shown in Figure 6.18(a). By enabling Bias Points from the PSpice menu, the node voltages and branch currents can also be displayed. Defining the load resistance R_L as a parameter {RVAL} as shown in Figure 6.18(b) varies its value. The sweep variable V_{in} is set from the primary sweep menu as shown in Figure 6.18(c).

Note: PARAM symbol can be drawn from the "special.lib" and it is important to note that the variable (say RVAL) be placed within {}, e.g., {RVAL}. The listing of the circuit file is as follows:

Dc Sweep

```
Source   ■   VIN 1 0 DC 100MV
CIRCUIT  ■■  * Parameter definition for VAL:
             .PARAM RVAL = 10K
             * Step variation for RVAL:
             .STEP PARAM RVAL 10K 30K 10K
             * Vary the load resistance RL:
             RL   5   6   {RVAL}
             RS   1   2   500
             R1   2   0   15K
             RP   2   3   1.5K
             RE   4   0   250
             * Current-controlled current source:
             F1   5   4   VX   100
             RO   5   4   50K
             *RL   5   6   10K
```

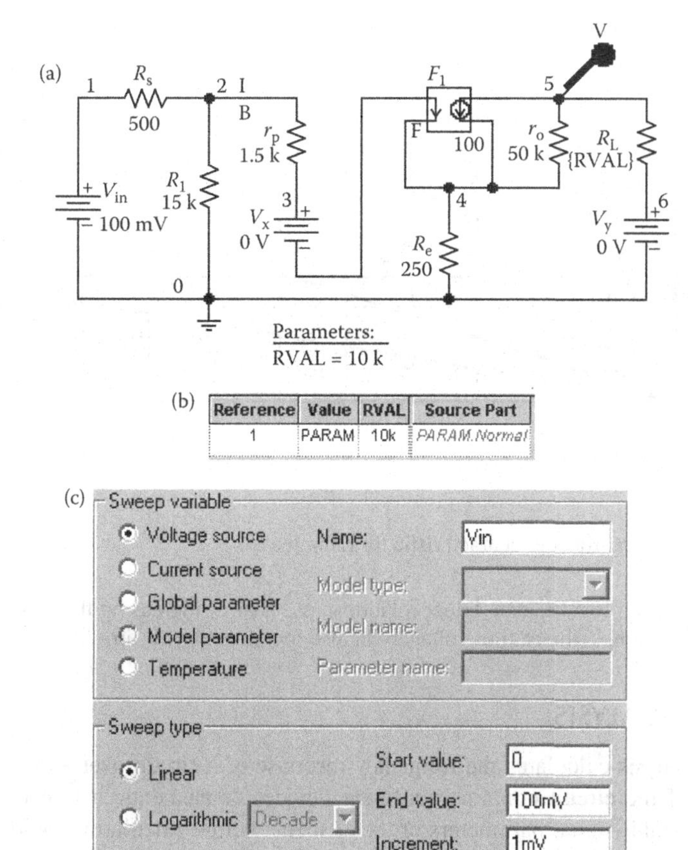

FIGURE 6.18 PSpice schematic for Example 6.4. (a) Schematic. (b) Setting PARAM values. (c) Setting sweep variable.

```
          * Dummy voltage source to measure the
          * controlling current:
          VX   3   4   DC   0V
          * Dummy voltage source to measure the output
          * current:
          VY   6   0   DC   0V
ANALYSIS ■■■ * DC sweep from 0 to 100 mV with an
          * increment of 2 mV:
          .DC   VIN   0   100MV   1 mV
          * PSpice plots the results of DC sweep.
          .PLOT DC V(5)
          .PROBE                ; Graphics post-processor
      .END
```

The transfer characteristic is shown in Figure 6.19.

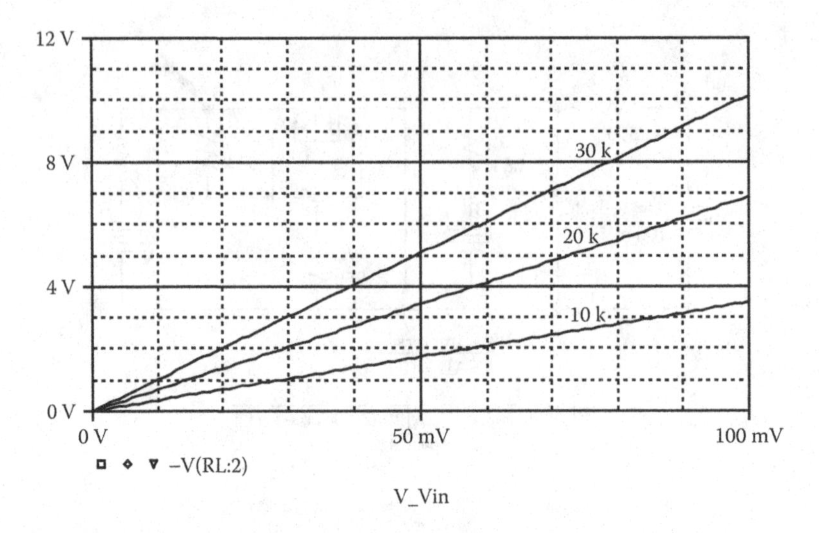

FIGURE 6.19 DC transfer characteristic for Example 6.4.

Note: The plots indicate linear relationships. That is, increasing the input voltage by two times, the output voltage will also increase by two times.

6.7 AC ANALYSIS

The AC analysis calculates the frequency response of a circuit over a range of frequencies. If the circuit contains nonlinear devices or elements, it is necessary to obtain the small-signal parameters of the elements before calculating the frequency response. Prior to the frequency response (or AC analysis), PSpice determines the small-signal parameters of the elements. The method for calculation of bias point for AC analysis is identical to that for DC analysis. The details of the bias points can be printed by an .OP command.

The AC analysis is set at the Analysis menu as shown in Figure 6.20 and its specifications are set at the AC sweep menu. The noise analysis (discussed in Section 6.13) can also be enabled.

The command for performing frequency response takes one of the general forms

```
.AC   LIN   NP   FSTART   FSTOP
.AC   OCT   NP   FSTART   FSTOP
.AC   DEC   NP   FSTART   FSTOP
```

where NP is the number of points in a frequency sweep. FSTART is the starting frequency, and FSTOP is the ending frequency. Only one of **LIN**, OCT, or DEC must be specified in the statement. **LIN**, OCT, or DEC specifies the type of sweep as follows:

LIN *Linear sweep*: The frequency is swept linearly from the starting frequency to the ending frequency, and NP becomes the total number of points in the sweep. The next frequency is generated by adding a constant to the present value. **LIN** is used if the frequency range is narrow.

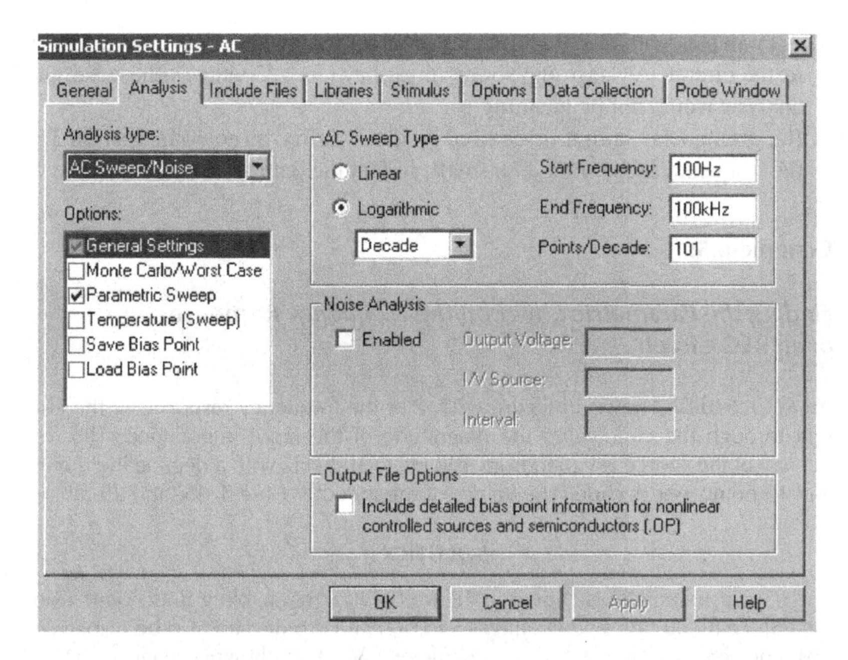

FIGURE 6.20 AC analysis setup and sweep specifications.

OCT *Sweep by octave (six times)*: The frequency is swept logarithmically by octave, and NP becomes the number of points per octave. The next frequency is generated by multiplying the present value by a constant larger than unity. OCT is used if the frequency range is wide.

DEC *Sweep by decade (10 times)*: The frequency is swept logarithmically by decade, and NP becomes the number of points per decade. DEC is used if the frequency range is widest.

PSpice does not print or plot any output by itself for AC analyses. The results of AC sweep are obtained by **.PRINT**, **.PLOT**, or .PROBE statements.

Some statements for AC analysis are

```
.AC  LIN  20  100HZ  300HZ
.AC  LIN  1  60HZ  120HZ
.AC  OCT  10  100HZ  10KHZ
.AC  DEC  100  1KHZ  1MEGHZ
```

Note the following:

1. FSTART must be less than FSTOP and must not be zero.
2. NP = 1 is permissible and the second statement calculates the frequency response at 60 Hz only.
3. Before performing the frequency response analysis, PSpice calculates, automatically, the biasing point to determine the linearized circuit parameters around the bias point.

4. All independent voltage and current sources that have AC values are inputs to the circuit. At least one source must have an AC value; otherwise, the analysis would not be meaningful.
5. If a group delay output is required by a "G" suffix, as noted in Section 3.3, the frequency steps should be small, so that the output changes smoothly.

Example 6.5:

Finding the Parametric Effect on the Frequency Response of an RLC Circuit

An *RLC* circuit is shown in Figure 6.21. Plot the frequency response of the current through the circuit and the magnitude of the input impedance. The frequency of the source is varied from 100 Hz to 100 kHz with a decade increment and 10 points per decade. The values of the inductor *L* are 5, 15, and 25 mH.

SOLUTION

The PSpice schematic is shown in Figure 6.22(a). By enabling Bias Points from the PSpice menu, the node voltages and branch currents can also be displayed. Defining the inductance L as a parameter {LVAL} as shown in Figure 6.22(b) varies its value. The AC sweep variable is set from the primary sweep menu as shown in Figure 6.22(c). LVAL is varied from the secondary sweep as shown in Figure 6.22(d).

The listing of the circuit file is as follows:

Input Impedance Characteristics

```
SOURCE   ■   VIN  1  0  AC  1V
CIRCUIT  ■■  * Parameter definition for VAL
             .PARAM  VAL = 5MH
             * Step variation for VAL:
             .STEP  PARAM  LVAL  5MH  25MH  10MH
             R  3  5  50
             * Vary the inductor:
               3  4  {LVAL}
             C  4  0  1UF
```

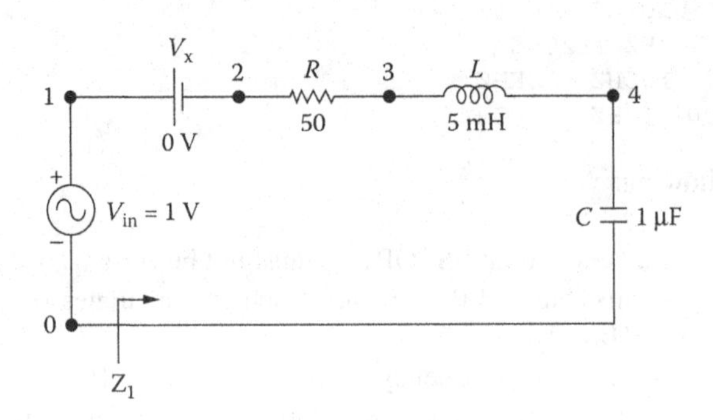

FIGURE 6.21 RLC circuit for Example 6.5.

(a)

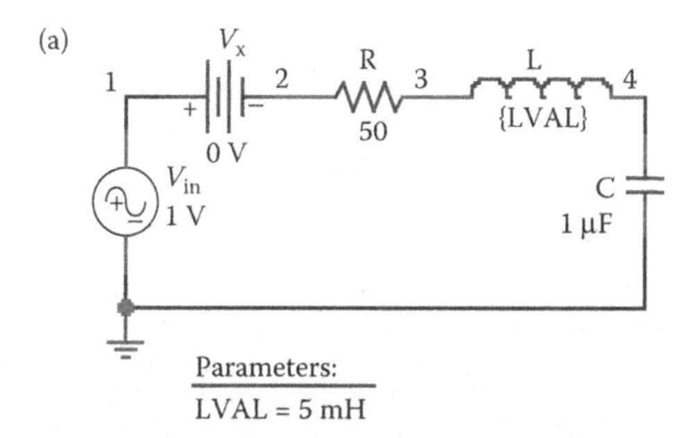

(b)

Reference	Value	LVAL	Source Part
PM1	PARAM	5mH	PARAM.Normal

(c)

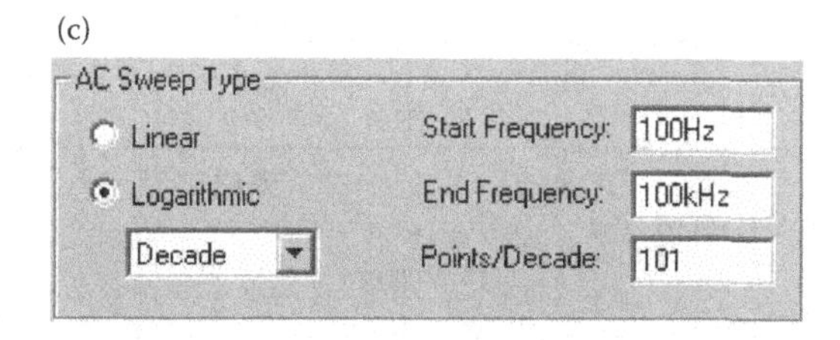

(d)

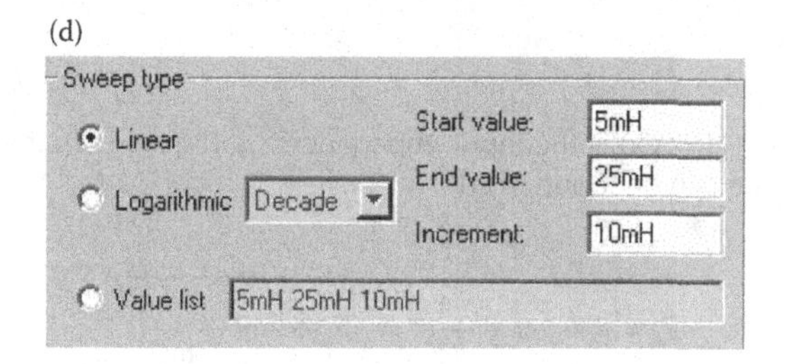

FIGURE 6.22 PSpice schematic for Example 6.5. (a) Schematic. (b) Setting PARAM values. (c) Setting sweep variable. (d) Sweep type.

```
                * Dummy voltage source to measure the
                * controlling current:
                VX  1  2  DC  0V
ANALYSIS ■■■    * AC sweep from 100 Hz to 100 kHz with 10
                * points per decade
                .AC  DEC  10  100HZ  100KHZ
                * PSpice plots the results of a DC sweep.
                .PLOT  AC  V(1)       ; Plot on the output file
                .PROBE                ; Graphics post-processor
        .END
```

The frequency response of the current through the circuit and the magnitude of the input impedance are shown in Figure 6.23. As the inductance is increased, the resonant frequency is decreased.

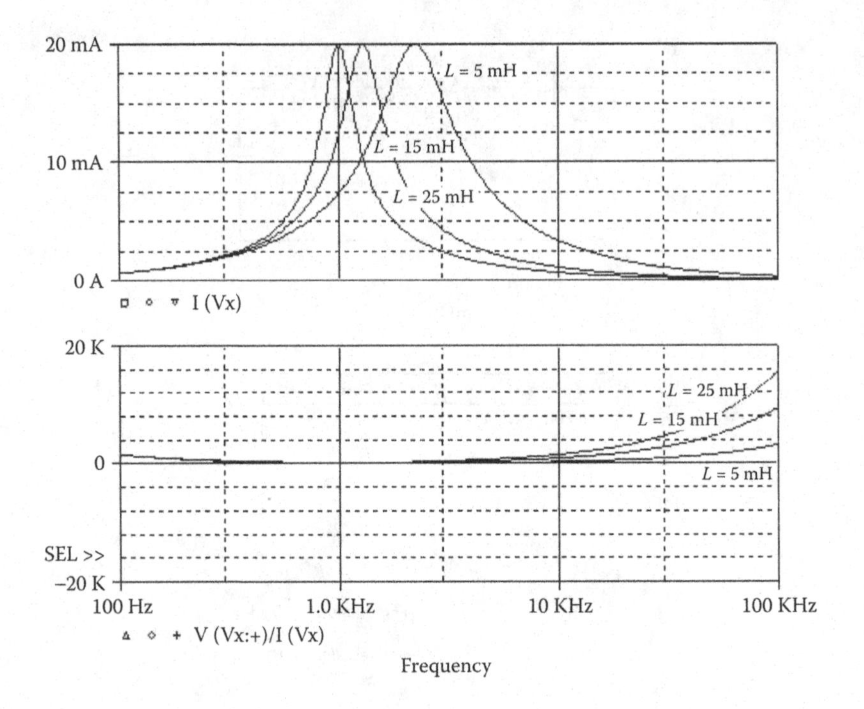

FIGURE 6.23 Frequency response for Example 6.5.

Note: V(Vx: + 1)/I(Vx) is the input impedance Z_i of the circuit. for $\omega = 2\pi f$, the input impedance can be found from

$$|Z_i| = \left|R + j\omega L + \frac{1}{j\omega C}\right| = \sqrt{R^2 + \left(\omega L - \frac{1}{\omega C}\right)^2}$$

And the input current can be described by

$$|I_i| = \frac{V_s}{\left|R + j\omega L + \frac{1}{j\omega C}\right|} = \frac{V_s}{\sqrt{R^2 + \left(\omega L - \frac{1}{\omega C}\right)^2}}$$

For LTspice: LTspice schematic is shown in Figure 6.24(a). The specifications for AC analysis are shown in Figure 6.24(b). Inductance L is defined as a parameter {RVAL} and its values are listed with a step command, starting at L = 5 mH and ending at 25 mH with an increment of 10 mH. The Bode plot of the output voltage V(4) and the input current I® are shown in Figure 6.24(c).

```
.step param LVAL 5mH 25mH 10mH
```

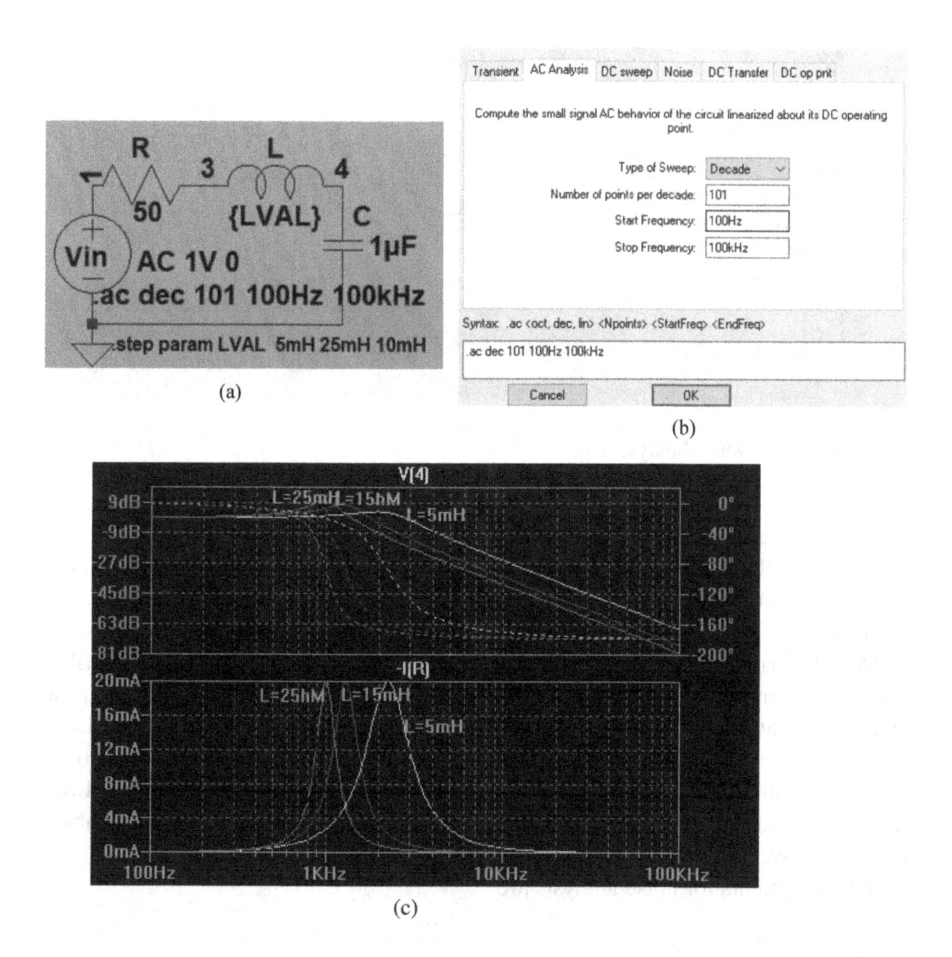

FIGURE 6.24 LTspice schematic for Example 6.5. (a) LTspice schematic. (b) AC analysis specifications. (c) Bode plot of V(4) and the input current I®.

6.8 NOISE ANALYSIS

Resistors and semiconductor devices generate various types of noise [3]. The level of noise depends on the frequency. Noise analysis is performed in conjunction with AC analysis and requires an .AC command. for each frequency of the AC analysis, the noise level of each generator in a circuit (e.g., resistors, transistor) is calculated, and their contributions to the output nodes are computed by summing the RMS noise values. The gain from the input source to the output voltage is calculated. From this gain, the equivalent input noise level at the specified source is calculated by PSpice. The statement for performing noise analysis is of the form

```
.NOISE  V(N+, N-)  SOURCE M
```

where V(N+, N–) is the output voltage across nodes N+ and N–. The output such as V(N) could be at a node N. SOURCE is the name of an independent voltage or

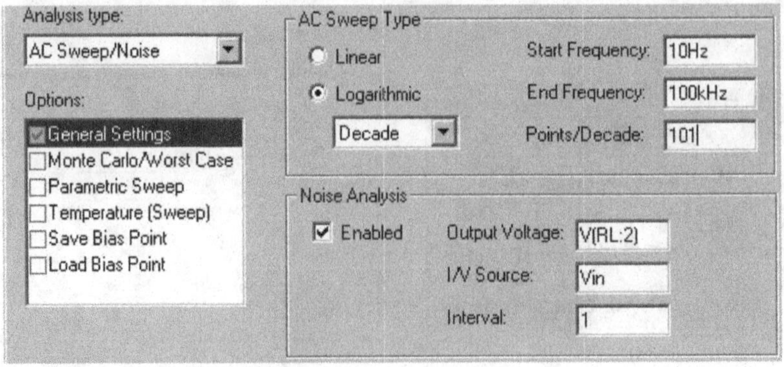

FIGURE 6.25 Noise Analysis menu.

current source at which the equivalent input noise will be generated. It should be noted that SOURCE is not a noise generator, rather, it is where the equivalent noise input must be computed. for a voltage source, the equivalent input is in $V/\sqrt{Hz}$, and for a current source, it is in $A/\sqrt{Hz}$.

M is the print interval that permits one to print a table for the individual contributions of all generators to the output nodes for every *m*th frequency. The output noise and equivalent noise of individual contributions are printed by a **.PRINT** or **.PLOT** command. If the value of M is not specified, PSpice does not print a table of individual contributions. But PSpice prints automatically a table of total contributions rather than individual contributions. In PSpice Schematics, noise analysis is enabled in the AC Sweep menu of the analysis setup as shown in Figure 6.25.

The statements for noise analysis are

```
.NOISE  V(4,5)   VIN
.NOISE  V(6)     IIN
.NOISE  V(10)    V1   10
```

Note: The .PROBE command cannot be used for noise analysis.

Example 6.6:

Finding the Equivalent Input and output Noise

For the circuit of Figure 6.13, calculate and print the equivalent input and output noise. The frequency of the source is varied from 1 Hz to 100 kHz. The frequency should be increased by a decade with 1 point per decade.

SOLUTION

The PSpice schematic is shown in Figure 6.26(a). Noise analysis, which follows AC analysis, is selected from the AC Sweep menu of the analysis setup as shown in Figure 6.26(b).

(a)

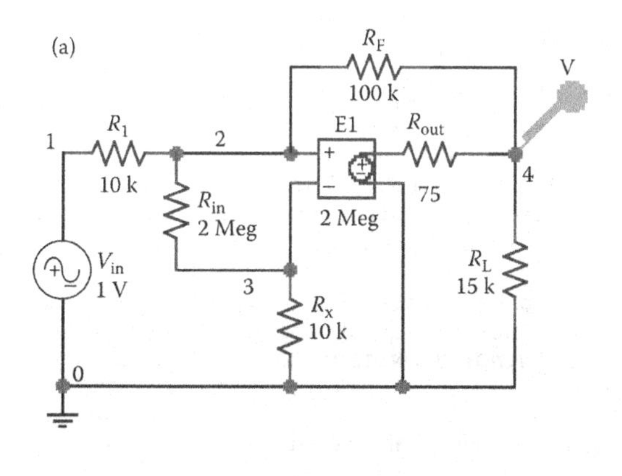

(b)

AC Sweep Type

- Linear
- Logarithmic (selected)
- Decade

Start Frequency: 1

End Frequency: 100kHz

Points/Decade: 1

Noise Analysis

- Enabled (checked)

Output Voltage: V(RL:2)

I/V Source: Vin

Interval: 1

FIGURE 6.26 PSpice schematic for Example 6.6. (a) Schematic. (b) AC and noise analysis menu.

Note: The op-amp is represented as a voltage-controlled voltage source with an input resistance $R_t = 2\ M\Omega$, an open-loop voltage gain $A = 2 \times 10^6 V/V$, and an output resistance $R_o = 75\ \Omega$.

The input source is of the AC type. The listing of the circuit file is as follows:

Noise Analysis

```
SOURCE   ■   * AC input voltage of 1 V:
             VIN  1  0  AC  1V
CIRCUIT  ■■  R1   1  2   10K
             RF   2  4   100K
             RL   4  0   15K
             RX   3  0   10K
             * Subcircuit call OPAMP:
             X1  2  3  4  0  OPAMP
             * Subcircuit definition:
             .SUBCKT    OPAMP   1       2       3       4
             *     model name  vi-  vi+  vo+  vo-
             RIN    1  2  2MEG
             ROUT   5  3  75
             E1     5  4  2  1  2MEG ; Voltage-controlled
                                       voltage source
```

```
                .ENDS OPAMP              ; End of subcircuit
                                         definition OPAMP
ANALYSIS ■■■    * AC sweep from 1 Hz to 100 kHz with a
                * decade increment and 1 point
                * per decade
                .AC  DEC 1 1HZ 100kHZ
                * Noise analysis without printing details of
                * individual contributions:
                .NOISE V(4) VIN
                * PSpice prints the details of equivalent
                * input and output noise.
                .PRINTNOISE ONOISE INOISE
            .END
```

The results of the noise analysis are shown next.

```
****        AC ANALYSIS            TEMPERATURE = 27.000  DEG C

FREQ                ONOISE (V√Hz)       INOISE (V/√Hz)
1.000E+00           1.9666Ex-07         1.966E-08
1.000E+01           1.966E-07           1.966E-08
1.000E+02           1.966E-07           1.966E-08
1.000E+03           1.966E-07           1.966E-08
1.000E+04           1.966E-07           1.966E-08
1.000E+05           1.966E-07           1.966E-08
```

Note: We could combine the .AC, **.NOISE**, and .SEN V(4) commands in the circuit file of Example 6.2 by modifying the statement as follows:

```
VIN      1  0  AC  1V  DC  1V
```

6.9 TRANSIENT ANALYSIS

A transient response determines the output in the time domain in response to an input signal in the time domain. The method for the calculation of transient analysis bias point differs from that of DC analysis bias point. The DC bias point is also known as the *regular bias point*. In the regular (DC) bias point, the initial values of the circuit nodes do not contribute to the operating point or to the linearized parameters. The capacitors and inductors are considered open- and short-circuited, respectively, whereas, in the transient bias point, the initial values of the circuit nodes are taken into account in calculating the bias point and the small-signal parameters of the nonlinear elements. The capacitors and inductors, which may have initial values, therefore, remain as parts of the circuit. Determination of the transient analysis requires statements involving

```
.IC    Initial transient conditions
.TRAN  Transient analysis
```

6.9.1 .IC (Initial Transient Conditions)

The various nodes can be assigned to initial voltages during transient analysis, and the general form for assigning initial values is

```
.IC  V(1) = B1  V(2) = B2 ...  V(N) = BN
```

where B1, B2, B3,... are the initial voltages for nodes V(1), V(2), V(3),..., respectively. These initial values are used by PSpice to calculate the transient analysis bias point and the linearized parameters of nonlinear devices for transient analysis. After the transient analysis bias point has been calculated, the transient analysis starts and the nodes are released. It should be noted that these initial conditions do not affect the regular bias-point calculation during DC analysis or DC sweep. for the .IC statement to be effective, UIC (use initial conditions) should not be specified in the .TRAN command.

The statement for initial transient conditions is

```
.IC V(1) =2.5 V(5) =1.7 VV(7) =0.5
```

Note: In PSpice Schematics and OrCAD, the .IC command and UIC option may not be available.

6.9.2 .TRAN (Transient Analysis)

Transient analysis can be performed by the .TRAN command, which has one of the general forms

```
.TRAN        TSTEP  TSTOP  [TSTART TMAX]  [UIC]
.TRAN[/OP]   TSTEP  TSTOP  [TSTART TMAX]  [UIC]
```

where TSTEP is the printing increment, TSTOP is the final time (or stop time), and TMAX is the maximum step size of internal time step. TMAX allows the user to control the internal time step. TMAX could be smaller or larger than the printing time TSTEP. The default value of TMAX is TSTOP/50.

Transient analysis always starts at time $t = 0$. However, it is possible to suppress the printing of the output for a time TSTART. TSTART is the initial time at which the transient response is printed. In fact, PSpice analyzes the circuit from t = 0 to TSTART, but it does not print or store the output variables. Although PSpice computes the results with an internal time step, the results are generated by interpolation for a printing step of TSTEP. Figure 6.27 shows the relationships of TSTART, TSTOP, and TSTEP.

In transient analysis, only the node voltages of the transient analysis bias point are printed. However, the .TRAN command can control the output for the transient response bias point. An .OP command with a .TRAN command (i.e., .TRAN/OP) will print the small-signal parameters during transient analysis.

If UIC is not specified as optional at the end of a .TRAN statement, PSpice calculates the transient analysis bias point before the beginning of transient analysis. PSpice uses the initial values specified with the .IC command. If UIC is specified as an option at the end of a .TRAN statement, PSpice does not calculate the transient analysis bias point before the beginning of transient analysis. However, PSpice uses the initial values specified with the "IC=" initial conditions for capacitors and inductors, which are discussed in Chapter 5. Therefore, if UIC is specified, the initial values of the capacitors and inductors must be supplied.

The .TRAN statements require **.PRINT** or **.PLOT** or .PROBE statements to obtain the results of the transient analysis.

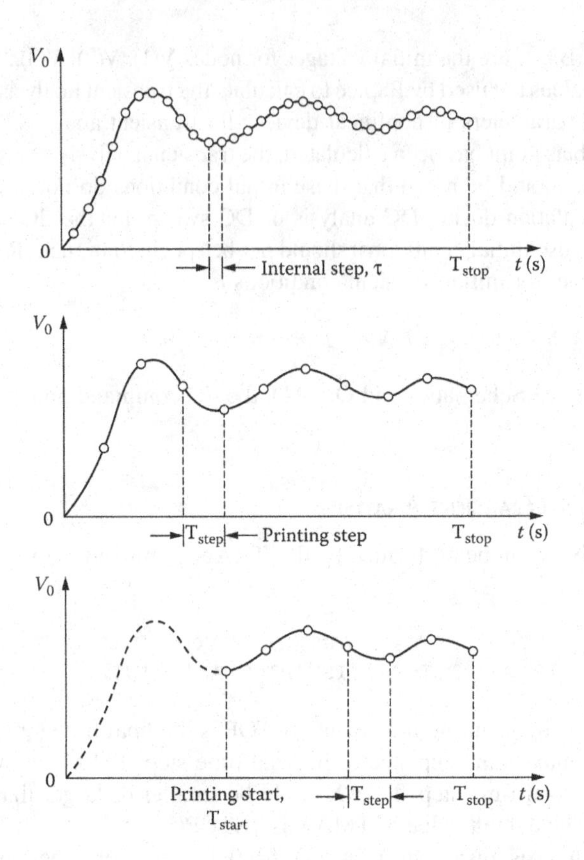

FIGURE 6.27 Response of transient analysis.

Statements for transient analysis are

```
.TRAN      5US  1MS
.TRAN      5US  1MS  200US  0.1NS
.TRAN      5US  1MS  200US  0.1NS  UIC
.TRAN      5US  1MS  200US  0.1NS  UIC
.TRAN/OP   5US  1MS  200US  0.1NS  UIC
```

In PSpice schematics, the Transient Analysis is selected from the Simulation Settings menu as shown in Figure 2.6(b). The run time is set from the General Settings menu as shown in Figure 6.28(a) and the print time is set from the Output files options as shown in Figure 6.28(b).

Note: In PSpice Schematics and OrCAD, the UIC option may not be available.

Example 6.7:

Plotting the Transient Response of an RLC *Circuit with .IC Command*

Repeat Example 5.1, if the voltage across the capacitor is set by an .IC command instead of an IC condition, and UIC is not specified.

(a)
Run to time: 1ms seconds (TSTOP)
Start saving data after: seconds
Transient options
Maximum step size: seconds
☑ Skip the initial transient bias point calculation (SKIPBP)
Output File Options...

(b)
Transient Output File Options
Print values in the output file every: 1us seconds
☐ Perform Fourier Analysis
Center Frequency: hz
Number of Harmonics:
Output Variables:

FIGURE 6.28 Transient analysis setting. (a) General settings menu. (b) Output file options.

SOLUTION

In PSpice Schematics, assigning an initial value to the capacitor and setting the initial inductor current to zero can simulate the .IC command. This is shown in Figure 6.29.

The listing of the circuit file with .IC statement and without UIC is as follows:

Transient Response of RLC Circuit

```
SOURCE   ■    * Input step voltage represented as an PWL
              * waveform:
              VS 1 0 PWL (0 0 10NS 10V 2MS 10V)
CIRCUIT  ■■   * R1 has a value of 6 with model RMOD:
              R1 1 2 RMOD 6
```

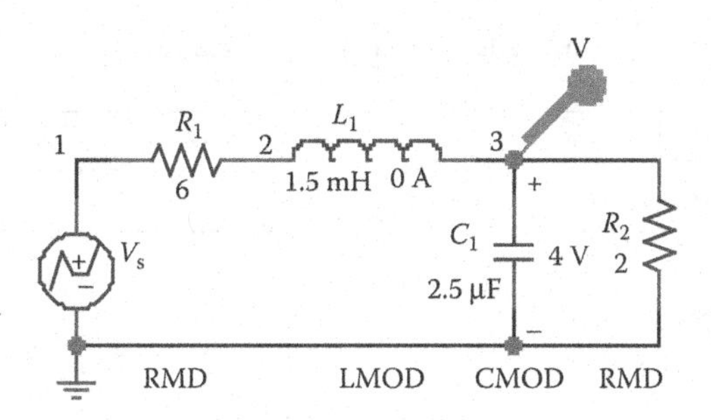

FIGURE 6.29 PSpice schematic for Example 6.7.

```
          * Inductor of 1.5 mH with an initial current
          * of 3 A and model name LMOD:
          L1 2 3 LMOD 1.5MH IC=0A
          * Capacitor of 2.5 µF with an initial
          * voltage of 4 V and model name CMOD:
          C1 3 0 CMOD 2.5UF IC=4V
          R2 3 0 RMOD 2
          * The initial voltage at node 3 is 4 V:
          .IC V(3)=4V
          * Model statements for resistor, inductor,
          * and capacitor:
          .MODEL RMOD RES (R=1 TC1=0.02 TC2=0.005)
          .MODEL CMOD CAP (C=1 VC1=0.01 VC2=0.002
          + TC1=0.02 TC2=0.005)
          .MODEL LMOD IND (L=1 IL1=0.1 IL2=0.002
          + TC1=0.02 TC2=0.005)
          * The operating temperature is 50°C:
          .TEMP 50
ANALYSIS ■■■ * Transient analysis from 0 to 1 ms with a
          * 1-µs time increment and
          * without using initial conditions (UIC):
          * that is, IC-4V has no effect
          .TRAN 1MS 1MS
          * Plot the results of transient analysis
          * with voltage at nodes 3 and 1.
          .PLOT TRAN V(3) V(1)
          .PROBE
     .END
```

The results of the simulation for the circuit in Figure 6.29 are shown in Figure 6.30. It may be noted that the response is completely different because of the assignment of an initial node voltage of 4 V on the capacitor.

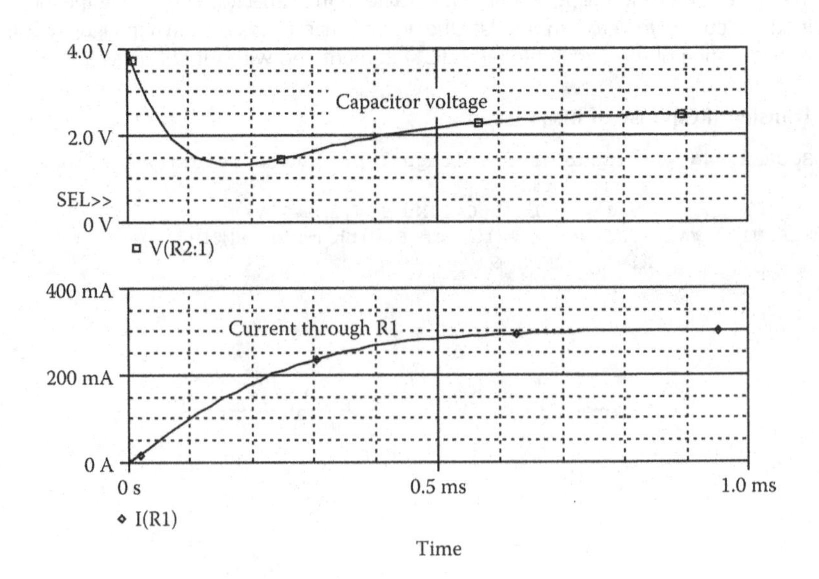

FIGURE 6.30 Transient response for Example 6.7.

6.10 FOURIER ANALYSIS

The output variables from the transient analysis are in discrete forms. These sampled data can be used to calculate the coefficients of the Fourier series. A periodic waveform can be expressed in a Fourier series as

$$v(\theta) = C_0 + \sum_{n=1}^{\infty} C_n \sin(n\theta + \phi_n)$$

where

θ = $2\pi f$
f = frequency, in hertz
C_0 = DC component
C_n = nth harmonic component

PSpice uses the results of transient analysis to perform Fourier analysis up to the ninth harmonic, or 10 coefficients. The statement takes one of the general forms

```
.FOUR  FREQ  N  V1  V2  V3 ...  VN
.FOUR  FREQ  N  I1  I2  I3 ...  IN
```

The PSpice schematics allows up to 99th harmonic. If you are not sure about the upper limit, try it and PSpice will give a syntax error.

FREQ is the fundamental frequency; V1, V2,... (or I1, I2,...) are the output voltages (or currents) for which the Fourier analysis is desired; N is the number of harmonics to be calculated. A .FOUR statement must have a .TRAN statement. The output voltages (or currents) must have the same forms as in the .TRAN statement for transient analysis.

Fourier analysis is performed over the interval TSTOP-PERIOD to TSTOP, where TSTOP is the final (or stop) time for the transient analysis and PERIOD is one period of the fundamental frequency. Therefore, the duration of the transient analysis must be at least one period, PERIOD. At the end of the analysis, PSpice determines the DC component and the amplitudes of up to the ninth harmonic (or 10 coefficients) by default, unless N is specified.

PSpice prints a table automatically for the results of Fourier analysis and does not require **.PRINT**, **.PLOT**, or .PROBE statements.

In PSpice Schematics, Fourier analysis is enabled from the Output File Options of the Transient menu as shown in Figure 6.31.

The statement for Fourier analysis is

```
.FOUR  100KHZ  V(2,3),  V(3),  I(R1),  I(VIN)
```

Example 6.8:

Finding the Fourier Coefficients

For the circuit in Example 5.5, calculate the coefficients of the Fourier series if the fundamental frequency is 1 kHz.

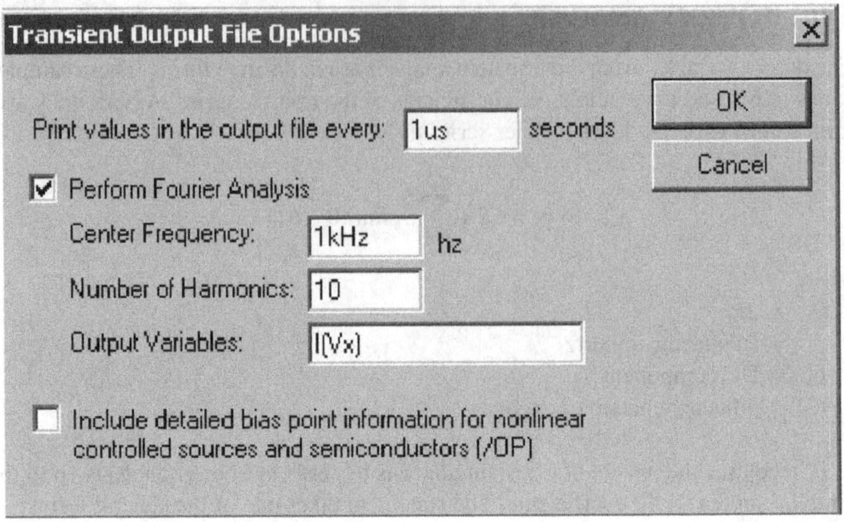

Transient Output File Options

Print values in the output file every: 1us seconds

OK

Cancel

☑ Perform Fourier Analysis

Center Frequency: 1kHz hz

Number of Harmonics: 10

Output Variables: I(Vx)

☐ Include detailed bias point information for nonlinear controlled sources and semiconductors (/OP)

FIGURE 6.31 Fourier analysis menu.

SOLUTION

The PSpice schematic is shown in Figure 6.32(a). Fourier analysis, which is performed after transient analysis, is selected from the Output File options of the Transient Analysis menu as shown in Figure 6.32(b).

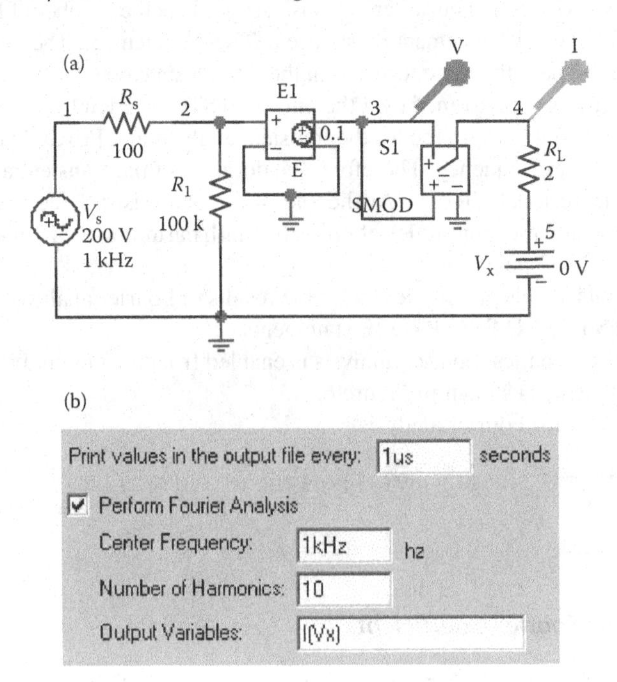

FIGURE 6.32 PSpice schematic for Example 6.8. (a) Schematic. (b) Fourier menu.

The listing of the circuit file is as follows:

Fourier Analysis

```
SOURCE    ■    * Sinusoidal input voltage of 200 V peak
               * with 0° phase delay:
               VS 1 0 SIN (0 200V 1KHZ)
CIRCUIT   ■■   RS 1 2 100OHM
               R1 2 0 100KOHM
               * Voltage-controlled voltage source with a
               * voltage gain of 0.1:
               E1 3 0 2 0 0.1
               RL 4 5 2OHM
               * Dummy voltage source of VX = 0 to measure
               * the load current:
               VX 5 0 DC 0V
               * Voltage-controlled switch controlled by
               * voltage across nodes 3 and 0:
               S1 3 4 3 0 SMOD
               * Switch model descriptions:
               .MODEL SMOD VSWITCH (RON = 5M ROFF
               = 10E+9 VON = 25M VOFF = 0.0)
ANALYSIS ■■■   * Transient analysis from 0 to 1 ms with a
               * 5-µs increment
               .TRAN 5US 1MS
               * Plot the current through VX and the input voltage.
               .PLOT TRAN V(3) I(VX); On the output file
               * Fourier analysis of load current at a
               * fundamental frequency of 1 kHz:
               .FOUR 1 KHZ I(VX)
               .PROBE        ; Graphics post-processor
         .END
```

It should be noted that we could add the statement .FOUR 1 KHZ I(VX), in the circuit file of Example 5.5. The results of the Fourier analysis are shown in the following text.

Fourier Components of Transient Response I(VX)

DC Component = 3.171401E+00

Harmonic Number	Frequency (Hz)	Fourier Component	Normalized Component	Phase (Deg)	Normalized Phase (Deg)
1	1.000E+03	4.982E+00	1.000E+00	2.615E–05	0.000E+00
2	2.000E+03	2.115E+00	4.245E–01	–9.000E+01	–9.000E+01
3	3.000E+03	9.002E–08	1.807E–08	6.216E+01	6.216E+01
4	4.000E+03	4.234E–01	8.499E–02	–9.000E+01	–9.000E+01
5	5.000E+03	9.913E–08	1.990E–08	8.994E+01	8.994E+01
6	6.000E+03	1.818E–01	3.648E–02	–9.000E+01	–9.000E+01
7	7.000E+03	7.705E–08	1.547E–08	5.890E+01	5.890E+01
8	8.000E+03	1.012E–01	2.032E–02	–9.000E+01	–9.000E+01
9	9.000E+03	9.166E–08	1.840E–08	7.348E+01	7.348E+01

Total harmonic distortion = 4.349506E+01%

(a)

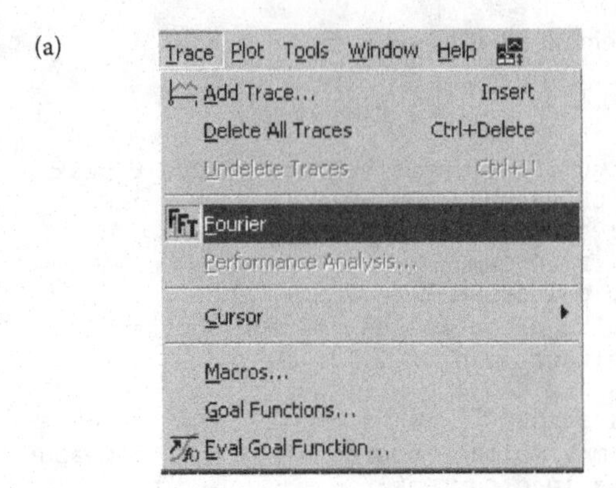

(b)

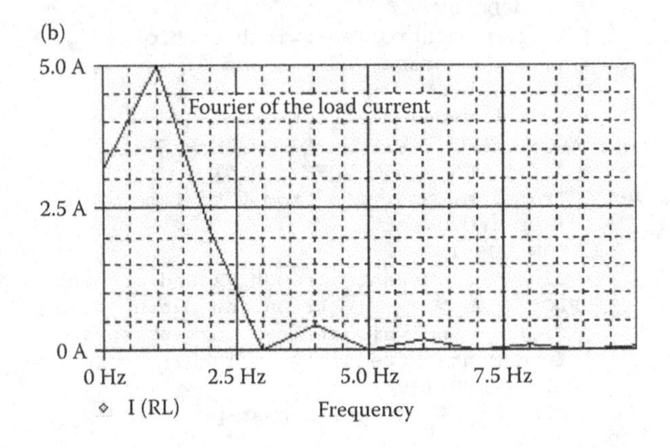

FIGURE 6.33 Fourier frequency spectrum for Example 6.8. (a) Selecting Fourier. (b) Fourier frequency spectrum.

After the Fourier analysis is completed, the Fourier spectrum shown in Figure 6.33(b) can be obtained by selecting Fourier from the Probe Trace menu as shown in Figure 6.33(a).

LTspice Fourier Analysis: The general SPICE statement for Fourier Analysis after a .TRAN Analysis is

```
.four <frequency> [Nharmonics] [Nperiods] <data trace1>
[<data trace2> ...]
```

where

Nharmonics is the number of harmonics, defaults to 9 if not specified.

Nperiods is the number of periods, an integer given after Nharmonics. If Nperiods is given as –1, the Fourier analysis is performed over the entire simulation data range.

Example: .four 1kHz 2 V(out)

6.11 MONTE CARLO ANALYSIS

PSpice allows one to perform the Monte Carlo (statistical) analysis of a circuit. The general form of the Monte Carlo analysis is

```
.MC ⟨(# runs) value⟩ ⟨(analysis)⟩ ⟨(output variable)⟩
⟨(function)⟩ + [(option)]*
```

This command performs multiple runs of the analysis selected (**DC, AC**, transient). The first run is with the nominal values of all components. Subsequent runs are with variations on model parameters as specified by DEV and LOT tolerances on each .MODEL parameter (Section 6.2).

```
⟨(# runs) value⟩
```

is the total number of runs to be performed. For printed results, the upper limit is 2000. For the output to be viewed with Probe, the limit is 100;

```
⟨(analysis)⟩
```

must be specified from one of **DC, AC**, or transient analyses.

This analysis will be repeated in subsequent passes. All analyses that the circuit contains are performed during the normal pass. Only the selected analysis is performed during subsequent passes.

```
⟨(output variable)⟩
```

is identical in format to that of a **.PRINT** output variable (in Chapter 3).

⟨(`function`)⟩ specifies the operation to be performed on the values of the ⟨(`output variable`)⟩. This value is the basis for comparisons between the nominal and subsequent runs. ⟨(`function`)⟩ must be one of the following:

YMAX	Find the greatest difference in each waveform from the nominal run.
MAX	Finds the maximum value of each waveform.
MIN	Finds the minimum value of each waveform.
RISE_EDGE (⟨value⟩)	Finds the first occurrence of the waveform crossing above the threshold ⟨value⟩. The waveform must have one or more points at or below ⟨value⟩ followed by one above. The output will be the value where the waveform increases above ⟨value⟩.
FALL_EDGE (⟨value⟩)	Finds the first occurrence of the waveform crossing below the threshold ⟨value⟩. The waveform must have one or more points at or below ⟨value⟩ followed by one below. The output will be the value where the waveform decreases below ⟨value⟩.

[(option)]* may be zero for one of the following:

LIST At the beginning of each run, print out the model parameter values actually used for each component during that run.

OUTPUT (output type) requests output from subsequent runs after the nominal (first) run. The output for any run follows the **.PRINT**, **.PLOT**, and .PROBE statements in the circuit file. If OUTPUT is omitted, only the nominal run produces output. (OUTPUT type) is one of the following:

ALL	forces all output to be generated, including the nominal run.
FIRST (value)	Generates output only during first *n* runs.
EVERY (value)	Generates output every nth run.
RUN (value)*	Does analysis and generates output only for the runs listed. Up to 25 values may be specified in the list.

RANGE (((low) value, (((high) value)) restricts the range over which ((function)) will be evaluated. An "*" can be used in place of a (value) to indicate "for all values." Examples are

YMAX RANGE (*, 0.5)	YMAX is evaluated for values of the sweep variable (time, frequency, etc.) of 0.5 or less.
MAX RANGE (-1, *)	The maximum value of the output variable is found for values of the sweep variable of –1 or more.

Some statements for Monte Carlo analysis are

```
.MC 10 TRAN   V(2)          YMAX
.MC 40 DC     IC(Q3)        YMAX LIST
.MC 20 AC     VP(3,4)       YMAX LIST OUTPUT ALL
```

In PSpice Schematics, the Monte Carlo or Worst-Case menu is selected from the Analysis Setup menu and the various options are shown in Figure 6.34.

Example 6.9:

Monte Carlo Analysis of an RLC *Circuit*

The *RLC* circuit as shown in Figure 6.29 is supplied with a step input voltage of 1 V. The circuit parameters are as follows:

$R_1 = 4\Omega \pm 20\%$ (deviation/uniform) $R_L = 500\ \Omega \pm 20\%$ (deviation/uniform)
$L_1 = 50\ \mu H \pm 15\%$ (deviation/uniform) $C_1 = 1\ \mu F \pm 10\%$ (deviation/Gauss)

The model parameters are: for the resistance, $R = 1$; for the capacitor, $C = 1$; and for the inductor, $L = 1$.

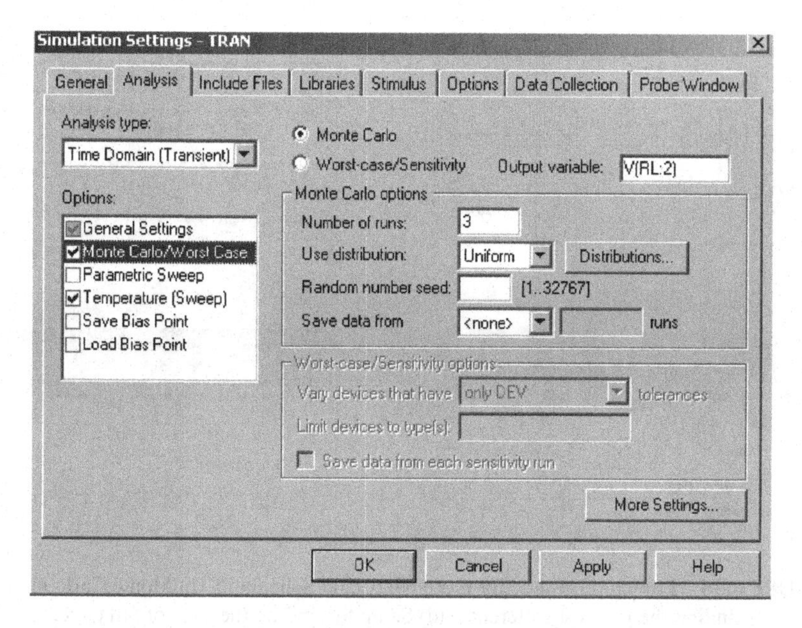

FIGURE 6.34 Monte Carlo or Worst-Case menu.

Use PSpice to perform Monte Carlo analysis of the capacitor voltage v_C from 0 to 160 μs with a time increment of 0.1 μs to find the following:

1. The greatest difference of the capacitor voltage from the nominal run to be printed for five runs.
2. The first occurrence of the capacitor voltage crossing below 1 V to be printed.

SOLUTION

The PSpice schematic is shown in Figure 6.35(a). The options for the maximum deviation (YMAX) are selected from the Output file options, as shown in Figure 6.35(c), of the Monte Carlo menu as shown in Figure 6.35(b). The options for the time of the fall edge are shown in Figure 6.35(d).

The listing of the circuit file follows:

Monte Carlo analysis

```
VIN    1 0 PWL (0 0 1NS 1V 10MS 1V) ; Step input of 1 V
    R1 1 2 Rbreak1 4                ; Resistance with model Rbreak1
    L1 2 3 Lbreak1 50UH             ; Inductance with model Lbreak1
    C1 3 0 Cbreak1 1UF              ; Capacitance with model Cbreak1
    RL 3 0 Rbreak1 500
    .MODEL Rbreak1 RES (R = 1 DEV/UNIforM = 20%)
    .MODEL Lbreak1 IND (L = 1 DEV/UNIforM = 15%)
    .MODEL Cbreak1 CAP (C = 1 DEV/GAUSS = 10%)
      .TRAN 0.11US 160US             ; Transient analysis
      *.MC 5 TRAN V(3) YMAX          ; Monte Carlo analysis
      .MC 5 TRAN V(3) FALL_EDGE(1V)  ; Monte Carlo analysis
      .PROBE                         ; Graphical waveform analyzer
  .END                               ; End of circuit file
```

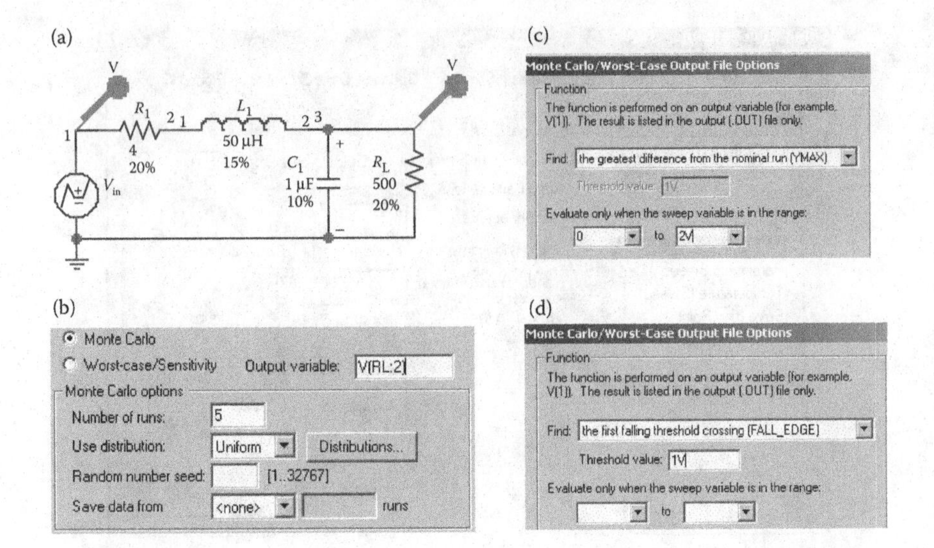

FIGURE 6.35 PSpice schematic for Example 6.9. (a) Schematic. (b) Monte Carlo menu. (c) Setup for finding the greatest difference. (d) Setup for finding the time of fall edge.

Note the following:

a. The results of the Monte Carlo analysis, which are obtained from the output file Example 6.9.out, are as follows:

```
Mean Deviation=-.0462
Sigma = .0716
Pass 3.0958        (1.34 sigma) lower at T=20.1100E-06
                   (92.054% of nominal)
Pass 4.085         (1.19 sigma) lower at T=36.1110E-06
                   (92.59% of nominal)
Pass 5.0816        (1.14 sigma) lower at T=26.5110E-06
                   (94.115% of nominal)
Pass 2.0774        (1.08 sigma) higher at T = 16.9100E-06
                   (107.89% of nominal)
```

b. To find the first occurrence of the capacitor voltage crossing below 1 V, the .MC statement is changed as follows:

```
.MC 5 TRAN V(3) FALL_EDGE(1V); Monte-Carlo Analysis
```

The results of the Monte Carlo analysis are shown below:

```
RUN     FIRST FALLING EDGE VALUE THRU 1
Pass 3    39.6120E-06
          (106.9% of nominal)
NOMINAL   37.0550E-06
Pass 5    36.9080E-06
          (99.603% of nominal)
Pass 2    34.943 0E-06
          (94.3% of nominal)
Pass 4    34.697 0E-06
          (93.635% of nominal)
```

6.12 SENSITIVITY AND WORST-CASE ANALYSIS

PSpice allows one to perform the sensitivity/worst-case analysis of a circuit for variations of model parameters. The variations are specified by DEV and LOT tolerances on model parameters. The general statement of the sensitivity/worst-case analysis is

```
.WCASE{(analysis)} {(output variable)} {(function)} {(option)}*
```

This command is similar to the .MC command, except that it does not include {(# `runs`) `value`}. With a .WCASE command, PSpice first calculates the sensitivity of all

```
{(output variable)}
```

to each model parameter. Once all the sensitivities are calculated, one final run is done with the sensitivities of the model parameters, and PSpice gives the worst-case

```
{(output variable)}.
```

Note: Both the .MC and .WCASE commands cannot be selected at the same time. Only one command is to be used at a time.

Some statements for sensitivity/worst-case analysis are

```
.WCASE TRAN V(2) YMAX
.WCASE DC V(3) YMAX LIST
.WCASE AC VP(3,4) YMAX LIST OUTPUT ALL
```

Example 6.10:

Worst-Case Analysis of an RLC *Circuit*

Run the worst-case analysis for Example 6.9.

SOLUTION

The PSpice schematics shown in Figure 6.36(a) is the same as that shown in Figure 6.35(a). The Worst-Case menu is shown in Figure 6.36(b). The option for the maximum deviation (YMAX) is selected from the Output File options, as shown in Figure 6.35(c), and the option for the time of the fall edge is shown in Figure 6.35(d).

a. To run the sensitivity/worst-case analysis greatest difference for the nominal run, the .MC statement in Example 6.9 is changed to a .WCASE statement as follows:

```
.WCASE TRAN V(3) YMAX; Sensitivity/worst-case analysis
```

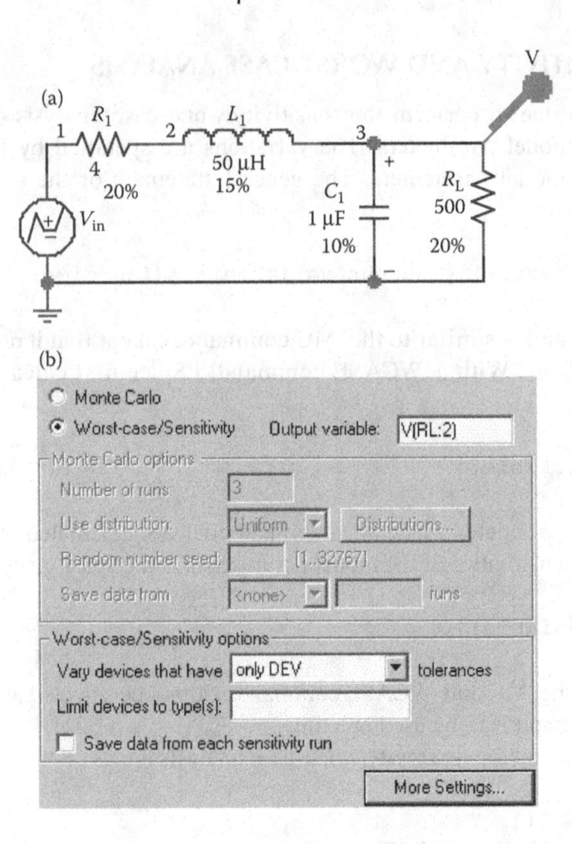

FIGURE 6.36 PSpice schematic for Example 6.10. (a) Schematic. (b) Worst-Case menu.

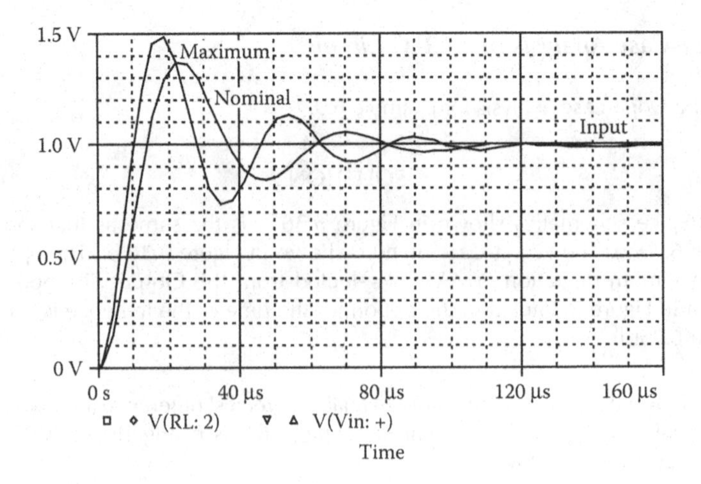

FIGURE 6.37 Plots of the greatest difference and the nominal run for Example 6.10.

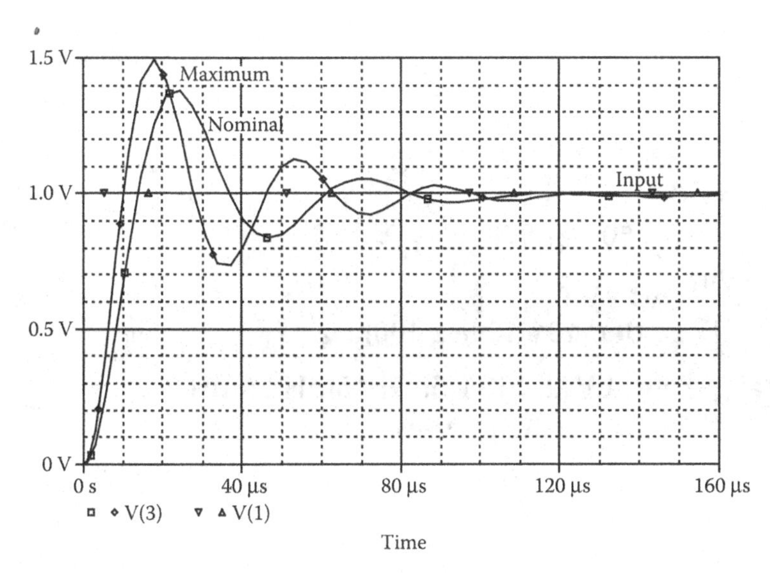

FIGURE 6.38 Plots of the first crossing below 1 V and nominal run for Example 6.10.

The plots of the greatest difference and the nominal value are shown in Figure 6.37. The results of the worst-case, obtained from the output file Example 6.10.out, are as follows:

```
**** SORTED DEVIATIONS OF V(3) TEMPERATURE=2 7.00 0 DEG C
                    WORST CASE SUMMARY
RUN              MAX DEVIATION FROM NOMINAL
ALL DEVICES        .206 higher at T=16.9100E-06
                                  (120.98% of Nominal)
```

b. To find the first occurrence of the capacitor voltage crossing below 1 V, the. WCASE statement follows:

```
.WCASE TRAN V(3) FALL_EDGE (1V); Sensitivity/worst-case
analysis
```

The plots of the greatest difference and the nominal value are shown in Figure 6.38 The results of the worst-case, obtained from the output file Example 6.10.out, are as follows:

```
**** SORTED DEVIATIONS OF V(3) TEMPERATURE = 27.000 DEG C
RUN                 FIRST FALLING EDGE VALUE THRU 1
NOMINAL             37.0550E-06
ALL DEVICES         31.8480E-06
                    (85.947% of nominal)
```

6.13 LTspice .NET COMMAND

This statement is used with a small signal (AC) analysis to compute the input and output admittance, impedance, Y-parameters, Z-parameters, H-parameters, and S-parameters of a 2-port network. It can also be used to compute the input admittance and impedance of a 1-port network. This must be used with a .AC statement,

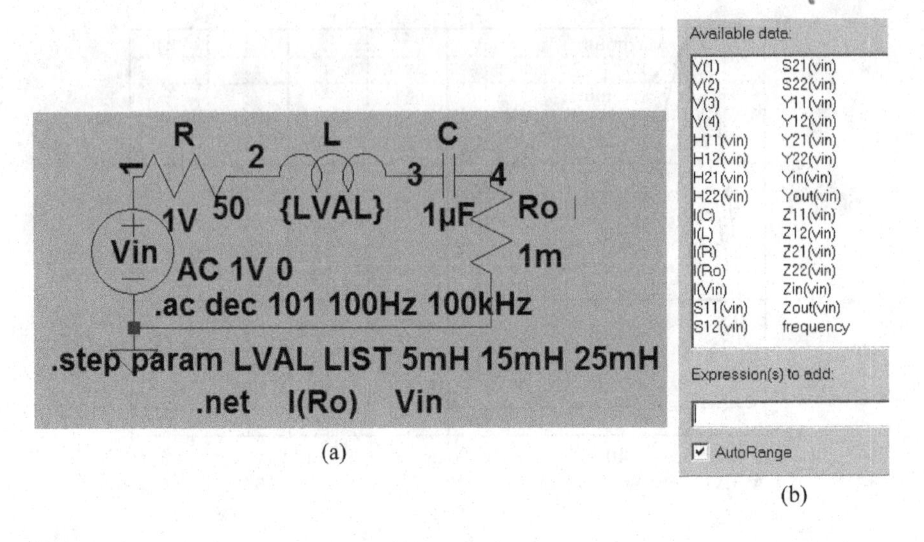

FIGURE 6.39 LTspice schematic and Impedances with .net command for Example 6.5. (a) LTspice schematic. (b) Available impedances.

which determines the frequency sweep of the network analysis. The general statement for the .net command is

```
Syntax:.net [V(out[,ref])|I(Rout)] <Vin|Iin> [Rin=<val>]
[Rout=<val>]
```

The network input is specified by either an independent voltage source, <Vin>, or an independent current source, <Iin>. The optional output port is specified either with a node, V(out), or a resistor, I(Rout). The ports will be terminated with resistances Rin and Rout. It recommends using a voltage source, V4, with Rser set as the desired source impedance and a resistor, Rout, to set the output termination with a .NET statement reading simply.

If we include the LTspice command .net statement in the circuit file for Example 6.5, we get the results as shown in Figure 6.39(a).

```
.NET I(Rout) Vin
```

Note: Ro is included in Figure 6.39(a) to measure the output current.

Where no Rin or Rout values specified are on the .net statement, the input/output devices supply default termination values. This arrangement makes the node voltages and currents of the .AC analysis corresponds to the network being terminated in the same manner as in the .NET statement. Figure 6.39(b) shows the available S, Y, and Z impedance traces to plot.

6.14 LTspice .MEAS COMMAND

The MEAS(URE) command evaluates User-Defined Electrical Quantities. The statement could refer to a point along the abscissa (the independent variable plotted along the horizontal axis, i.e., the time axis of a .tran analysis) and .MEASURE statements that refer to a range over the abscissa. It could also refer to one point on the abscissa and print a data value or expression thereof at a specific point or when a condition is met. The general statement is given by

```
.MEAS[SURE] [AC|DC|OP|TRAN|TF|NOISE] <name>
+ [<FIND|DERIV|PARAM> <expr>]
+ [WHEN <expr> | AT=<expr>]]
+ [TD=<val1>] [<RISE|FALL|CROSS>=[<count1>|LAST]]
```

The other type of .MEAS statement refers to a range over the abscissa. The general statement is given by

```
.MEAS [AC|DC|OP|TRAN|TF|NOISE] <name>
+ [<AVG|MAX|MIN|PP|RMS|INTEG> <expr>]
+ [TRIG <lhs1> [[VAL]=]<rhs1>] [TD=<val1>]
+ [<RISE|FALL|CROSS>=<count1>]
+ [TARG <lhs2> [[VAL]=]<rhs2>] [TD=<val2>]
+ [<RISE|FALL|CROSS>=<count2>]
```

The range over the abscissa is specified with the points defined by "TRIG" and "TARG.". The TRIG point defaults to the start of the simulation if omitted. Similarly, the TARG point defaults to the end of the simulation data. If all three of the TRIG, TARG, and the previous WHEN points are omitted, then the .MEAS statement operates over the entire range of data. <expr> may include the Keywords as listed in Table 6.3.

The .MEAS command to measure the maximum value of (Ro) for Example 6.5 as shown in Figure 6.40(a) is

```
.MEAS AC MAX(I(Ro))
```

.MEAS statements are usually put on the schematic as a SPICE directive or in the netlist with the rest of the simulation commands and circuit definition as shown in Figure 6.40(a). The output is put in the .log file which can be viewed with the menu command View=>SPICE Error Log as shown in Figure 6.40(c).

MEAS statements are done in post-processing after the simulation is completed. This allows us to write a script of .MEAS statements and execute them on a

TABLE 6.3
Keywords for Measure command

Keyword	Operation performed Over Interval
AVG	Compute the average of <expr>
MAX	Find the maximum value of <expr>
MIN	Find the minimum value of <expr>
PP	Find the peak-to-peak of <expr>
RMS	Compute the root mean square of <expr>
INTEG	Integrate <expr>

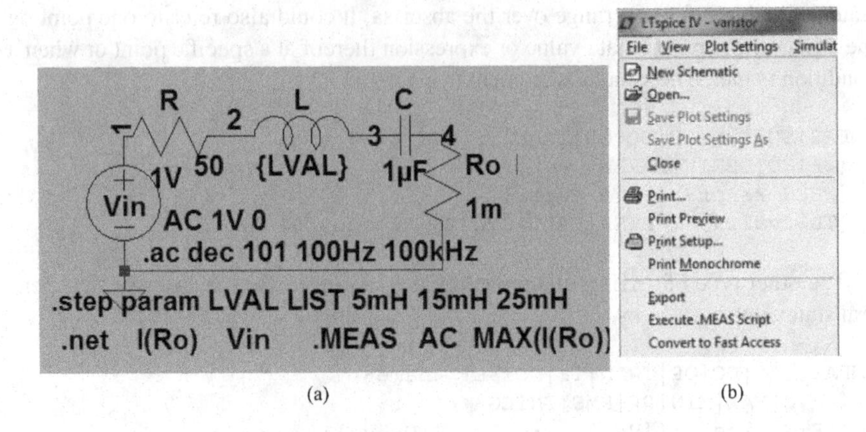

FIGURE 6.40 LTspice .MEAS command for Example 6.5. (a) LTspice schematic. (b) Execute .MEAS Script file. (c) MEAS output from View=>SPICE Error Log.

dataset. To do this, select and make the waveform window the active window and execute the menu command File=>Execute .MEAS Script from a file as shown in Figure 6.40(b).

Note: For help with any topics, search by the Help Topic in the main menu of LTspice.

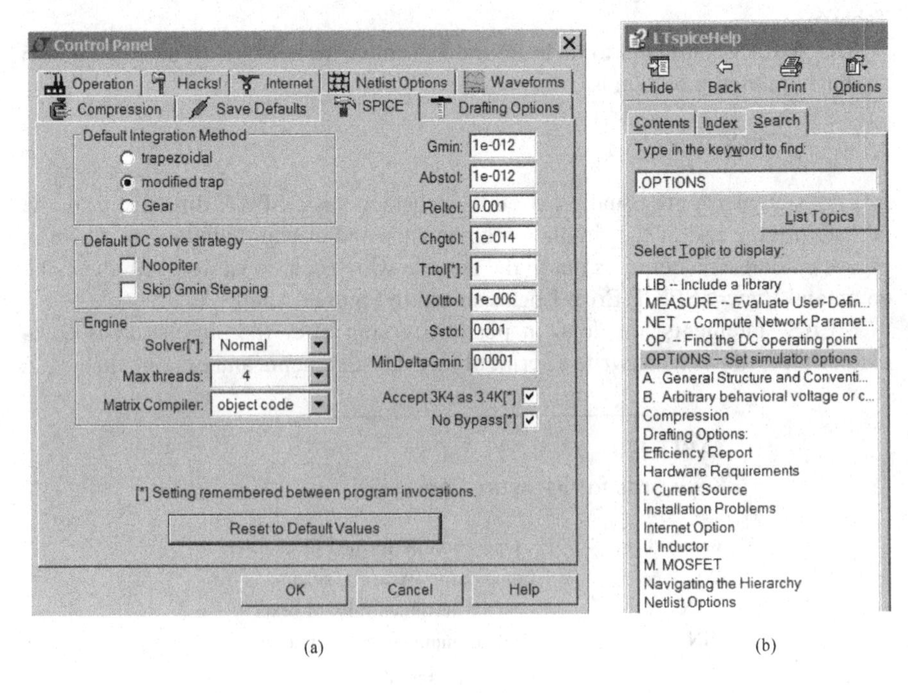

FIGURE 6.41 LTspice control panel and setting options. (a) SPICE options. (b) Setting simulator options.

6.15 LTspice OPTIONS

The SPICE options can be set from the control panel as shown in Figure 6.41(a). The options are almost identical to those of PSpice/OrCAD. The complete list of the options is found in the LTspice Help as shown in Figure 6.41(b).

SUMMARY

The PSpice dot commands can be summarized as follows:

.AC	AC analysis
.DC	DC analysis
.END	End of circuit
ENDS	End of subcircuit
.FUNC	Function
.FOUR	Fourier analysis
.GLOBAL	Global
.IC	Initial transient conditions
.INC	Include file
.LIB	Library file
.MC	Monte Carlo analysis
.MEAS	Measure
.MODEL	Model
.NET	2-port Network
.NODESET	Nodeet
.NOISE	Noise analysis
.OP	Operating point
.OPTIONS	Option
.PARAM	Parameter
.PLOT	Plot
.PRINT	Print
.PROBE	Probe
.SENS	Sensitivity analysis
.STEP	Step
.SUBCKT	Subcircuit definition
.TEMP	Temperature
.TF	Transfer function
.TRAN	Transient analysis
.WIDTH	Width

PROBLEMS

6.1 For the circuit in Figure P6.1, calculate and print the sensitivity of output voltage V_o with respect to each circuit element. The operating temperature is 50°C.

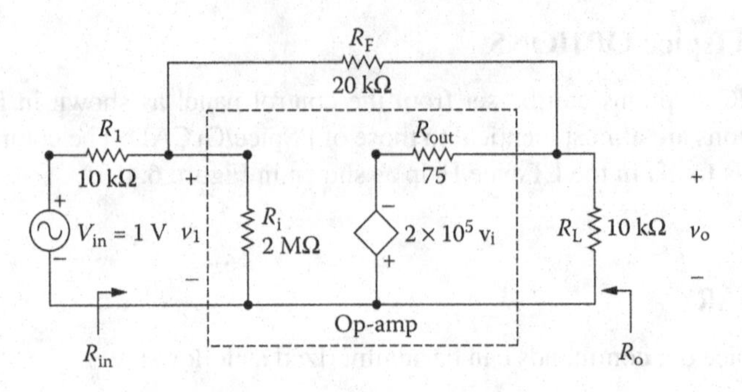

FIGURE P6.1 Amplifier circuit.

6.2 For the circuit in Figure P6.1, calculate and print (a) the voltage gain, $A_v = V_o/V_{in}$, (b) the input resistance, R_n, and (c) the output resistance, R_o.

6.3 For the circuit in Figure P6.1, calculate and plot the DC transfer characteristic, V_o versus V_{in}. The input voltage is varied from 0 to 10 V with an increment of 0.5 V.

6.4 For the circuit in Figure P6.1, calculate and print the equivalent input and output noise if the frequency of the source is varied from 10 Hz to 1 MHz. The frequency should be increased by a decade with two points per decade.

6.5 For the circuit in Figure P6.5, the frequency response is to be calculated and printed over the frequency range from 1 Hz to 100 kHz with a decade increment and 10 points per decade. The peak magnitude and phase angle of the output voltage are to be plotted on the output file. The results should also be available for display and as hard copy by using the .PROBE command.

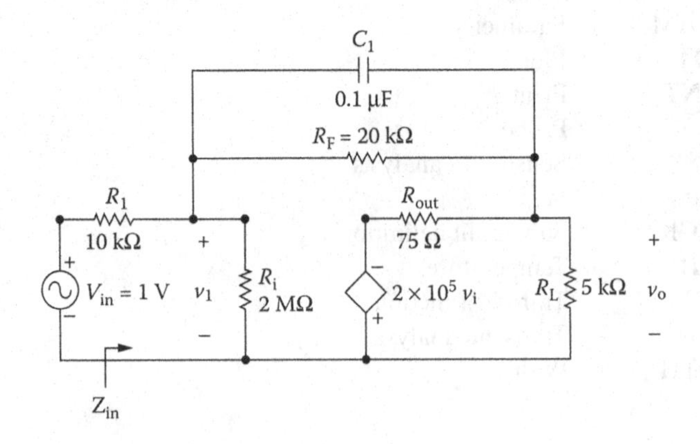

FIGURE P6.5 Integrator circuit.

6.6 Repeat Problem 6.5 for the circuit in Figure P6.6.

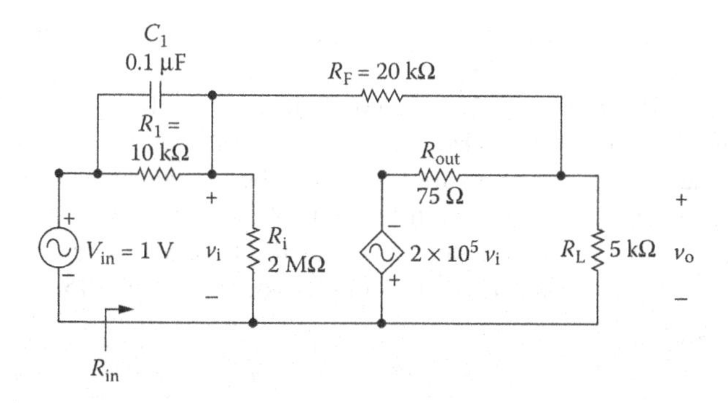

FIGURE P6.6 Low-pass filter circuit.

6.7 For the circuit in Figure P6.5, calculate and plot the transient response of the output voltage from 0 to 2 ms with a time increment of 5 μs. The input voltage is shown in Figure P6.7. The results should be available for display and as hard copy by using Probe.

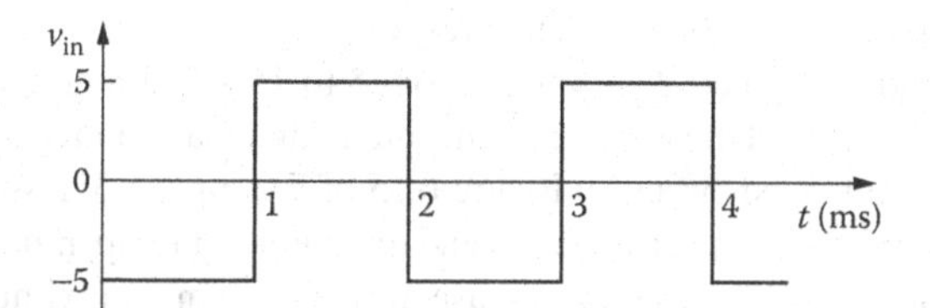

FIGURE P6.7 Pulse waveform.

6.8 Repeat Problem 6.7 for the input voltage shown in Figure P6.8.

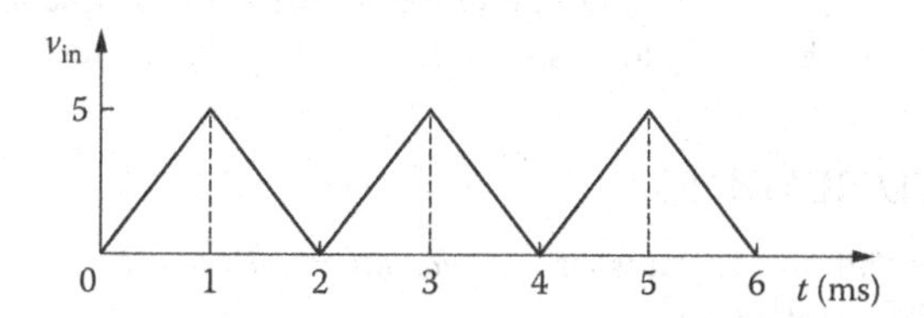

FIGURE P6.8 Triangular waveform.

6.9 For the circuit of Figure P6.6, calculate and plot the transient response of the output voltage from 0 to 2 ms with a time increment of 5 μs. The input voltage is shown in Figure P6.7. The results should be available for display and as hard copy by using Probe.

6.10 Repeat Problem 6.9 for the input voltage shown in Figure P6.8.

6.11 For Problem 6.7, calculate the coefficients of the Fourier series if the fundamental frequency is 500 Hz.

6.12 For Problem 6.8, calculate the coefficients of the Fourier series if the fundamental frequency is 500 Hz.

6.13 For Problem 6.9, calculate the coefficients of the Fourier series if the fundamental frequency is 500 Hz.

6.14 For Problem 6.10, calculate the coefficients of the Fourier series if the fundamental frequency is 500 Hz.

6.15 For the *RLC* circuit of Figure P6.15, plot the frequency response of the current i_s through the circuit and the magnitude of the input impedance. The frequency of the source is varied from 100 Hz to 100 kHz with a decade increment and 10 points per decade. The values of the inductor L are 5, 15, and 25 mH.

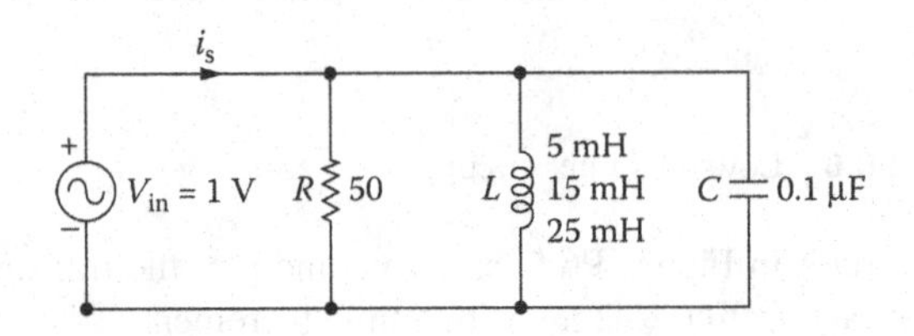

FIGURE P6.15 Triangular waveform.

6.16 Use PSpice to plot the frequency response of the output voltage in Problem 6.5 if the load resistance R_L is varied from 5 to 15 kΩ at an increment of 5 kΩ.

6.17 Use PSpice to find the worst-case minimum and maximum output voltages ($V_{o(min)}$ and $V_{o(max)}$) for the circuit in Figure 6.15 of Example 6.3. Assume uniform tolerances of ±20% for all resistances and an operating temperature of 25°C.

6.18 Use PSpice to find the worst-case minimum and maximum capacitor voltages ($V_{o(min)}$ and $V_{o(max)}$) for the circuit in Figure 6.21 of Example 6.5. Assume uniform tolerances of ±20% for all resistances, ±15% for inductors, ±10% for capacitors, and an operating temperature of 25°C.

6.19 Use PSpice to find the worst-case minimum and maximum output voltages ($V_{o(min)}$ and $V_{o(max)}$) for the circuit in Figure 6.28 of Example 6.7. Assume uniform tolerances of ±20% for all resistances and an operating temperature of 25°C.

SUGGESTED READING

1. P. Antognetti and G. Massobri, *Semiconductor Device Modeling with SPICE*, New York: McGraw-Hill, 1988.
2. M.H. Rashid, *Introduction to PSpice Using OrCAD for Circuits and Electronics*, Third Edition, Englewood Cliffs, NJ: Prentice-Hall, 2003, Chapter 6.
3. M.H. Rashid, *SPICE for Circuits and Electronics Using LTspice® Schematics, PSpice® Schematics, and OrCAD® Capture*, Fourth Edition, India: Cengage Learning, 2019.
4. M.H. Rashid, *SPICE for Power Electronics and Electric Power*, Boca Raton, FL: CRC Press, 2012 (2/e).
5. Cadence Design Systems, *PSpice Design Community*, San Jose, CA, 2001, http://www.PSpice.com.
6. *PSpice Models from Vendors*, http://www.pspice.com/models/links.asp.
7. M.H. Rashid, *Microelectronic Circuits: Analysis and Design*, Florence KY: Cengage Publishing, 2017, Chapter 8.
8. M.H. Rashid, *Power Electronics—Devices, Circuits and Applications*, NJ: Prentice-Hall, 2014, Appendix B—Magnetic Circuits.

7 Diode Rectifiers

After completing this chapter, students should be able to do the following:

- Model a diode in SPICE and specify its model parameters.
- Perform transient analysis of diode rectifiers.
- Evaluate the performance of diode rectifiers.
- Perform worst-case analysis of diode rectifiers for parametric variations of model parameters and tolerances.

7.1 INTRODUCTION

A semiconductor diode may be modeled in SPICE by a diode statement in conjunction with a model statement. The diode statement specifies the diode name, the nodes to which the diode is connected, and its model name. The model incorporates an extensive range of diode characteristics such as DC and small-signal behavior, temperature dependency, and noise generation. The model parameters take into account temperature effects, various capacitances, and physical properties of semiconductors.

A rectifier converts an alternating current (AC) voltage to a direct current (DC) voltage. It uses one or more diodes to make a unidirectional current flow from the positive terminal to the negative terminal of the output voltage. Thus, the voltage and the current on the input side are of AC types and the voltage and current on the output side are of DC types. Table 7.1 lists the performance parameters of rectifiers.

TABLE 7.1
Rectifier Performance Parameters for an Ideal No-load Rectifier

Input Side Parameters	Output Side Parameters
RMS input voltage, Vs	Average output voltage, V_{dc}
RMS input current, Is	Average output current, I_{dc}
The *transformer utilization factor*, $TUF = \frac{P_{dc}}{V_s I_s}$	The output dc power, $P_{dc} = V_{dc} I_{dc}$
The input power, $PF = \frac{P_{ac}}{V_s I_s}$	RMS output voltage, V_{rms}
	RMS output current, I_{rms}
	The output ac power, $P_{ac} = V_{rms} I_{rms} = I_{rms}^2 R_L$ for a load resistance of R_L
	The RMS ripple content of the output voltage $V_{ac} = \sqrt{V_{rms}^2 - V_{dc}^2}$
	The *ripple factor*, $RF = \frac{V_{ac}}{V_{dc}}$
	The rectification *efficiency or ratio*, $\eta = \frac{P_{dc}}{P_{ac}}$
	The *form factor*, $FF = \frac{V_{rms}}{V_{dc}}$

DOI: 10.1201/9781003284451-7

7.2 DIODE MODEL

The SPICE model for a reverse-biased diode is shown in Figure 7.1 [1–3]. The small-signal and static models that are generated by SPICE are shown in Figures 7.2 and 7.3, respectively. In the static model, the diode current, which depends on its voltage, is represented by a current source. The small-signal parameters are generated by SPICE from the operating point. SPICE generates a complex model for diodes. The model equations that are used by SPICE are described in Refs. [1,2]. In many cases, especially the level at which this book is aimed, such complex models are not necessary. Many model parameters can be ignored by the users, and SPICE assigns default values to the parameters.

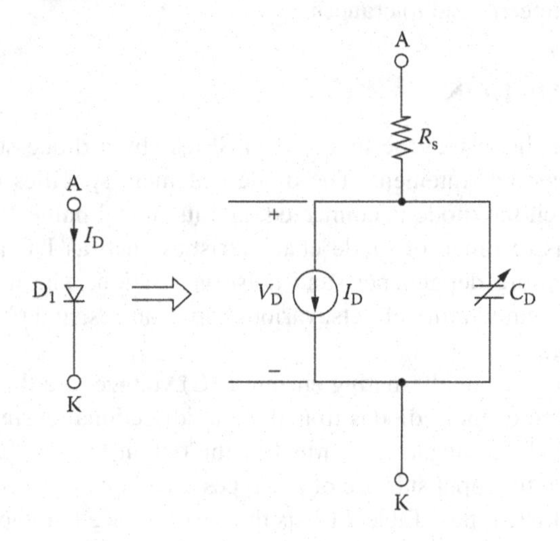

FIGURE 7.1 SPICE diode model with reverse-biased condition.

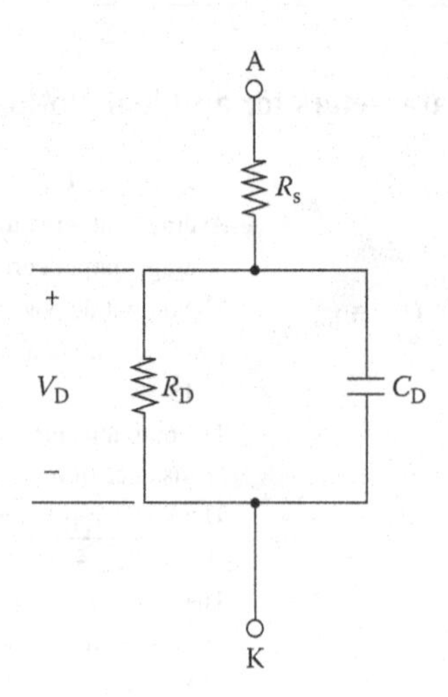

FIGURE 7.2 SPICE small-signal diode model.

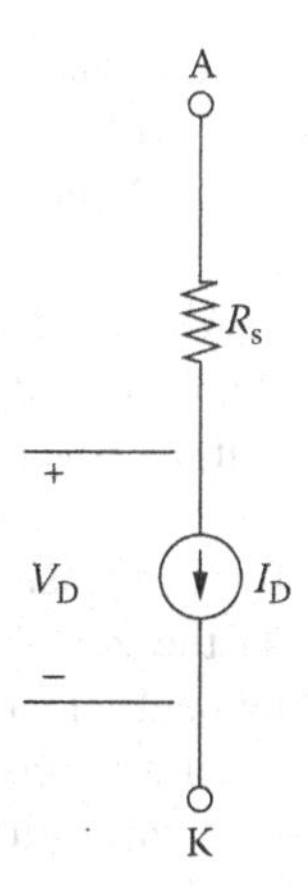

FIGURE 7.3 Static diode model with reverse-biased condition.

The model statement of a diode has the general form

```
.MODEL DNAME D(P1=B1 P2=B2 P3=B3 ... PN=BN)
```

where DNAME is the model name, and it can begin with any character, but its word size is normally limited to eight. D is the type symbol for diodes. P1, P2, … and B1, B2, … are the model parameters and their values, respectively. The model parameters are listed in Table 7.2. An area factor is used to determine the number of equivalent parallel diodes of the model specified. The model parameters that are affected by the area factor are marked with an asterisk (*) in the descriptions of the model parameters.

TABLE 7.2
Parameters of Diode Model

Name	Area	Model Parameter	Unit	Default	Typical
IS	*	Saturation current	A	1E-14	1E–14
RS	*	Parasitic resistance	Ω	0	10
N		Emission coefficient	1	1	
TT		Transit time	s	0	0.1NS
CJO	*	Zero-bias *p–n* capacitance	F	0	2PF
VJ		Junction potential	V	1	0.6
M		Junction grading coefficient		0.5	0.5
EG		Activation energy	eV	1.11	11.1
XTI		IS temperature exponent		3	3
KF		Flicker noise coefficient		0	
AF		Flicker noise exponent		1	
FC		Forward-bias depletion capacitance coefficient		0.5	
BV		Reverse breakdown voltage	V	·	50
IBV	*	Reverse breakdown current	A	1E-10	

The diode is modeled as an ohmic resistance (value = RS/area) in series with an intrinsic diode. The resistance is attached between node A and an internal anode node. `[(area) value]` scales IS, RS, CJO, and IBV, and defaults to 1. IBV and BV are both specified as positive values.

The DC characteristic of a diode is determined by the reverse saturation current IS, the emission coefficient N, and the ohmic resistance RS. Reverse breakdown is modeled by an exponential increase in the reverse diode current and is determined by the reverse breakdown voltage BV, and the current at breakdown voltage IBV. The charge storage effects are modeled by the transit time TT and a nonlinear depletion layer capacitance, which depends on the zero-bias junction capacitance CJO, the junction potential VJ, and the grading coefficient M.

The temperature of the reverse saturation current is defined by the gap activation energy (or gap energy) EG and the saturation temperature exponent XTI. The most important parameters for power electronics applications are IS, BV, IBV, TT, and CJO.

7.3 DIODE STATEMENT

The name of a diode must start with D, and it takes the general form

```
D<name> NA NK DNAME[(area) value]
```

where NA and NK are the anode node and the cathode node, respectively. The current flows from anode node NA through the diode to cathode node NK. DNAME is the model name.

Some diode statements are

```
D15 33 35 SWITCH 1.5
.MODEL SWITCH D(IS=100E-15 CJO=2PF TT=12NS BV=100 IBV=10E-3)
DCLAMP 0 8 DIN914
.MODEL DIN914 D (IS=100E-15 CJO=2PF TT=12NS BV=100 IBV=10E-3)
```

The PSpice schematic of a diode with a model name Dbreak is shown in Figure 7.4(a). The EVAL library of the PSpice student version supports a few low-power diodes,

(a)

D1

Dbreak

(b)

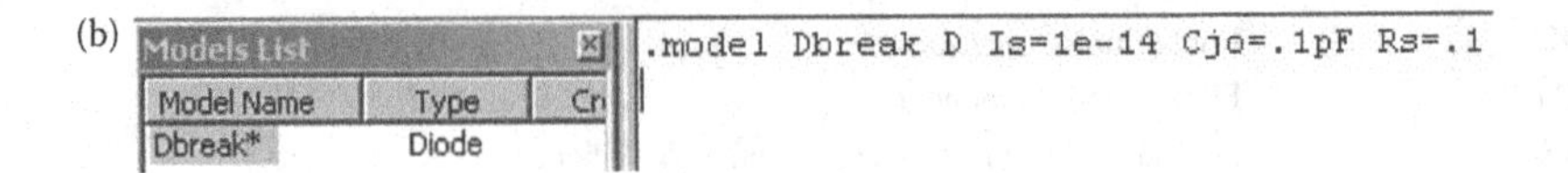

FIGURE 7.4 Diode symbol and model parameters. (a) Symbols. (b) Parameters of Dbreak model.

such as D1N4002, D1N4148, and D1N914. However, the user can change the model parameters of a Dbreak device. Click the right mouse button to open the window Edit PSpice Model as shown in Figure 7.4(b). The model name can also be changed. The breakout devices are available from the BREAKOUT Library.

7.4 DIODE CHARACTERISTICS

The typical *V–I* characteristic of a diode is shown in Figure 7.5. The characteristic can be expressed by an equation known as the *Schockley diode equation*, given by

$$I_D = I_s\left(e^{V_D/nV_T} - 1\right) \quad (7.1)$$

where

I_D = current through the diode, A
V_D = diode voltage with anode positive with respect to cathode, V
I_s = leakage (or reverse saturation) current, typically in the range of 10^{-6} to 10^{-20} A
n = empirical constant known as the *emission coefficient* (or *ideality factor*), whose value varies from 1 to 2

The emission coefficient, n, depends on the material and the physical construction of diodes. For germanium diodes, n is considered to be 1; for silicon diodes, the predicted value of n is 2, but for most silicon diodes the value of n is in the range of 1.1–1.8.

V_T in Equation 7.1 is a constant called the *thermal voltage*, and it is given by

$$V_T = \frac{kT}{q} \quad (7.2)$$

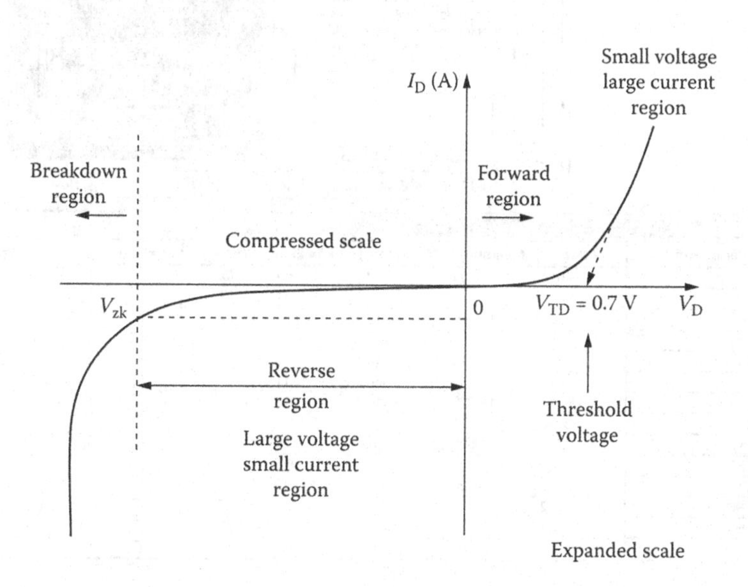

FIGURE 7.5 Diode characteristics.

where

q = electron charge: 1.6022×10^{-19} C
T = absolute temperature, kelvin (K = 273 + °C)
k = Boltzmann's constant: 1.3806×10^{-23} J/K

At a junction temperature of 25°C, Equation 7.2 gives

$$V_T = \frac{kT}{q} = \frac{1.3806 \times 10^{-23} \times (273 + 25)}{1.6022 \times 10^{-19}} \approx 25.8\text{mV}$$

At a specified temperature, the leakage current I_s remains constant for a given diode. For power diodes, the typical value of I_s is 10^{-15} A.

The plots of the diode characteristics can be viewed from the View menu of the PSpice Model Editor as shown in Figure 7.6. These include forward current versus the voltage (V_{fwd} vs. I_{fwd}), junction capacitance (V_{rev} vs. C_j), reverse leakage (V_{rev} vs. I_{rev}), reverse breakdown (V_z, I_z, Z_z), and reverse recovery (T_{rr}, I_{fwd}, I_{rev}).

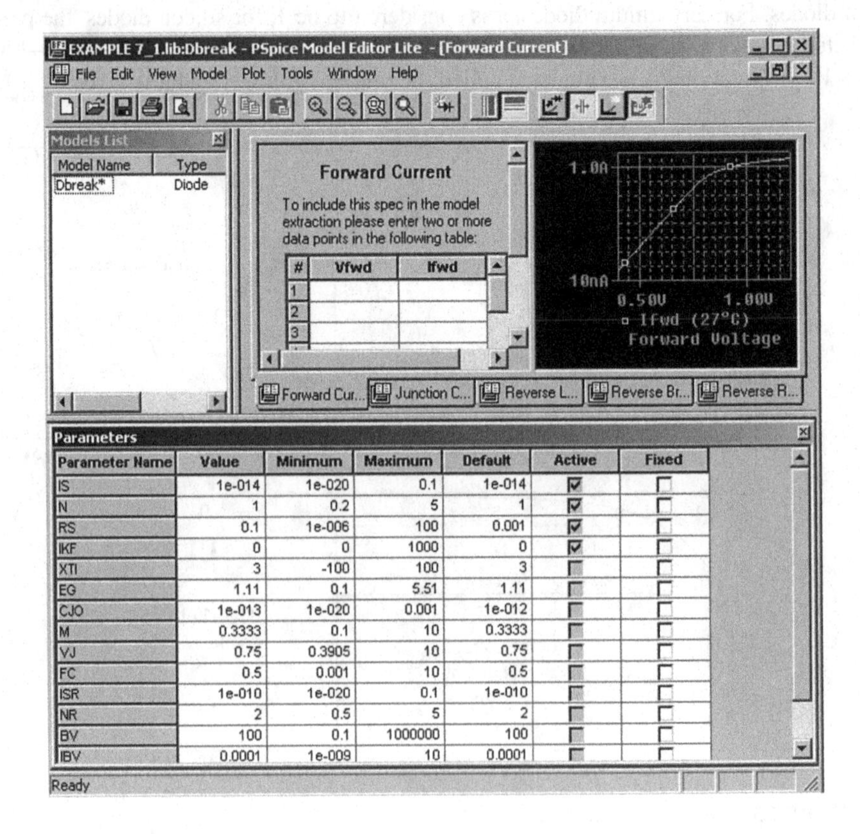

FIGURE 7.6 PSpice model editor.

7.5 DIODE PARAMETERS

The data sheet for IR diodes of type RI8 is shown in Figure 7.7. Although PSpice allows specifying many parameters, we use only the parameters that significantly affect power converter output. From the data sheet, we have

Reverse breakdown voltage, BV = 1200 V
Reverse breakdown current, IBV = 13 mA

(a)

R18C, R18S, R18CR & R18SR SERIES
1800–1200 VOLTS RANGE
185 AMP AVG STUD MOUNTED
DIFFUSED JUNCTION RECTIFIER DIODES

VOLTAGE RATINGS

VOLTAGE CODE (1)	V_{RRH}, V_R – (V) Max. rep. peak reverse and direct voltage		V_{RSH} – (V) Max. non-rep. peak reverse voltage
	T_j = 0° to 200°C	T_j = −40° to 0°C	T_j = 25° to 200°C
18A	1800	1710	1900
16A	1600	1520	1700
14B	1400	1330	1500
12B	1200	1200	1300

MAXIMUM ALLOWABLE RATINGS

PARAMETER		SERIES (2)	VALUE	UNITS	NOTES	
T_j	Junction temperature	ALL	−40 to 200	°C		
T_{stg}	Storage temperature	ALL	−40 to 200	°C		
$I_{F(AV)}$	Average current	R18C/S	185	A	180° half sine wave, T_C = 140°C	
$I_{F(AV)}$	Max. av. current (3)	ALL	200	A	180° half sine wave, T_C = 133°C, 18C/S T_C = 145°C, R18CR/SR	
$I_{F(RMS)}$	Max. RMS current (3)	ALL	314	A		
I_{FSH}	Max. peak non-rep. surge current	ALL	3820	A	50 Hz half cycle sine wave	Initial T_j = 200°C, rated V_{RPM} applied after surge.
			4000		60 Hz half cycle sine wave	
			4550		50 Hz half cycle sine wave	Initial T_j = 200°C, no voltage applied after surge.
			4750		60 Hz half cycle sine wave	
I^2t	Max. I^2t capability	ALL	73	kA^2s	t = 10 ms	Initial T_j = 200°C, rated V_{RPM} applied after surge.
			67		t = 8.3 ms	
			104		t = 10 ms	Initial T_j = 200°C, no voltage applied after surge.
			95		t = 8.3 ms	
$I^2\sqrt{t}$	Max. $I^2\sqrt{t}$ capability	ALL	1040	$kA^2\sqrt{s}$	Initial T_j = 200°C, no voltage applied after surge. I^2t for time $t_x = I^2\sqrt{t} \cdot \sqrt{t_x}$. $0.1 \le t_x \le 10$ ms.	
T	Mounting torque Min.	R18C/CR	14.1 (125)	N•m (lbf−in)	Non-lubricated threads	
	Max.		17.0 (150)			
	Min.		12.2 (108)		Lubricated threads	
	Max.		15.0 (132)			
	Min.	R18S/SR	11.3 (100)	N•m (lbf−in)	Non-lubricated threads	
	Max.		14.1 (125)			
	Min.		9.5 (85)		Lubricated threads	
	Max.		12.5 (110)			

(1) To complete the part number, refer to the Ordering Information table.

(2) R18C & R18S series have cathode-to-case polarity. R18CR & R18SR series have anode-to-case polarity.

(3) For devices assembled in Europe, max. $I_{F(AV)}$ is 175 A and max. $I_{F(RMS)}$ is 275 A.

FIGURE 7.7 (a) Data sheets, (b–d) Characteristics for IR diodes of type R18. (Courtesy of International Rectifier.)

(b) **CHARACTERISTICS**

PARAMETER		SERIES	MIN.	TYP.	MAX.	UNITS	TEST CONDITIONS
V_{FM}	Peak forward voltage	ALL	—	1.30	1.42	V	Initial $T_j = 25°C$, 50–80 Hz half sine, $I_{peak} = 628$ A.
$V_{F(TO)1}$	Low-level threshold	ALL	—	—	0.703	V	$T_j = 200°C$ Av. power = $V_{F(TO)} * I_{F(AV)} + r_F * (I_{F(RMS)})^2$ Use low level values for $I_{FM} \le \pi I_{F(AV)}$
$V_{F(TO)2}$	High-level threshold		—	—	0.738		
r_{F1}	Low-level resistance	ALL	—	—	1.100	πΩ	
r_{F2}	High-level resistance		—	—	1.080		
t_a	Reverse current rise	ALL	—	18.0	—	μs	$T_j = 175°C$, $I_{FM} = 200$ A, $di_R/dt = 1.0$ A/μs.
t_b	Reverse current fail	ALL	—	3.5	—	μs	
$I_{RM(REC)}$	Reverse current	ALL	—	18	—	A	
Q_{RR}	Recovered charge	ALL	—	194	—	μc	
I_{RM}	Peak reverse current	ALL	—	13	20	mA	$T_j = 175°C$. Ma⊠. rated V_{RRM}.
R_{thJC}	Thermal resistance, junction-to-case	R18C/S	—	—	0.250	°C/W	DC operation
		R18CR/SR	—	—	0.200		
		R18C/S	—	—	0.271	°C/W	180° sine wave
		R18CR/SR	—	—	0.221		
		R18C/S	—	—	0.275	°C/W	120° rectangular wave
		R18CR/SR	—	—	0.225		
R_{thCS}	Thermal resistance, case-to-sink	ALL	—	—	0.10	°C/W	Mtg. surface smooth, flat and greased.
wt	Weight	ALL	—	110(3.5)	—	g(oz.)	
	Case style	R18C/CR		DO-205AC (DO-30)		JEDEC	
		R18S/SR		DO-205AA (DO-8)			

FIGURE 7.7 *(Continued)*

Instantaneous voltage, $v_F = 1$ V at $i_F = 150$ A
Reverse recovery charge, $Q_{RR} = 194$ μC at $I_{FM} = 200$ A

From the data for the *V–I* forward characteristic of a diode, it is possible to determine the value of n, V_T, and I_s of the diode [4]. Assuming that $n = 1$ and $V_T = 25.8$ mV, we can apply Equation 7.1 to find the saturation current I_S:

$$I_D = I_S\left(e^{V_D/nV_T} - 1\right)$$

$$150 = I_S\left(e^{1/(25.8\times10^{-3})} - 1\right)$$

which gives $I_S = 2.2\text{E}{-15}$ A.

Let us call the diode model name DMOD. The values of TT and CJO are not available from the data sheet. Some versions of SPICE (e.g., PSpice) support device library files. The software PARTS of PSpice can generate SPICE models from the data sheet parameters of transistors and diodes. The transit time τ_T can be calculated approximately from

$$\tau_T = \frac{Q_{RR}}{I_{FM}} = \frac{194\mu\text{C}}{200} \approx 1\,\mu\text{s}$$

(c)

R18C, R18S, R18CR & R18SR SERIES
1800–1200 VOLTS RANGE

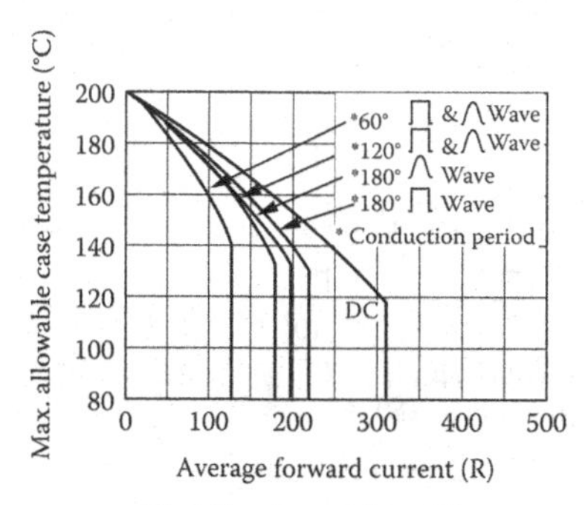

Fig. 1 Case temperature ratings—R18C and R18S series

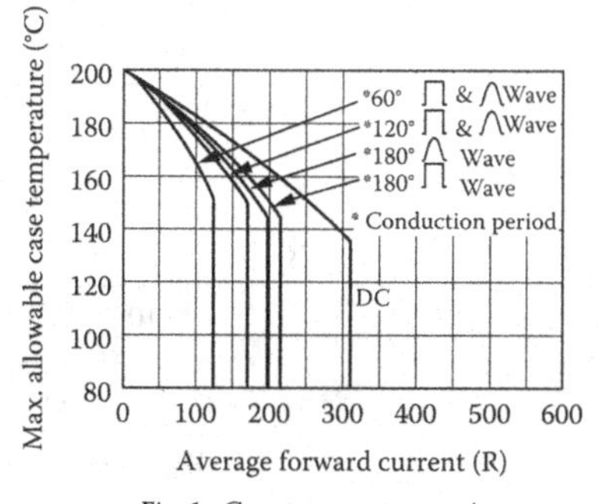

Fig. 1a Case temperature ratings—R18CR and R18SR series

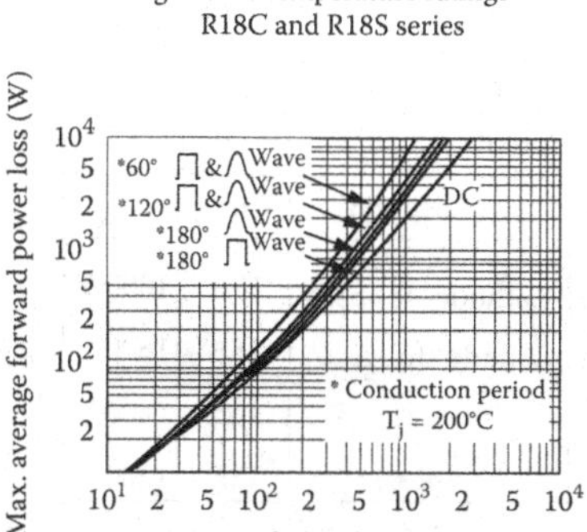

Fig. 2 Power loss characteristics—R18C, R18S, R18CR, and R18SR series

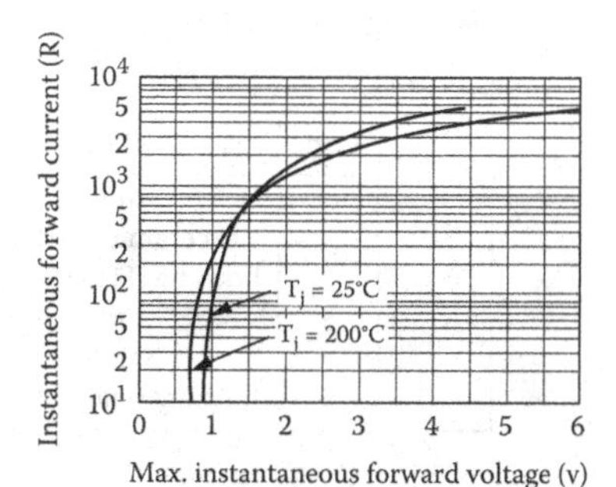

Fig. 3 Forward characteristics—R18C, R18S, R18CR, and R18SR series

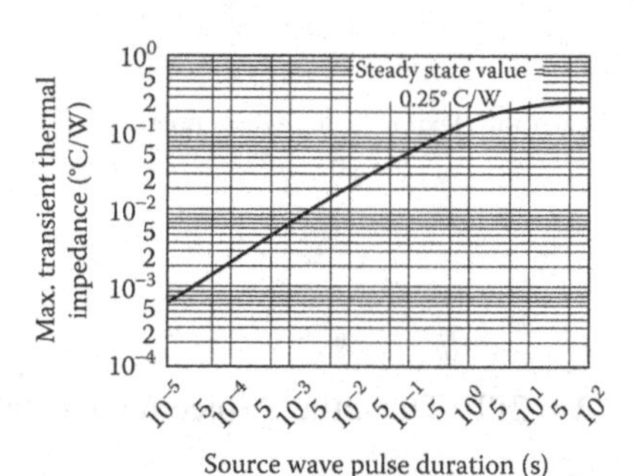

Fig. 4 Transient thermal impedance junction-to-case—R18C and R18S series

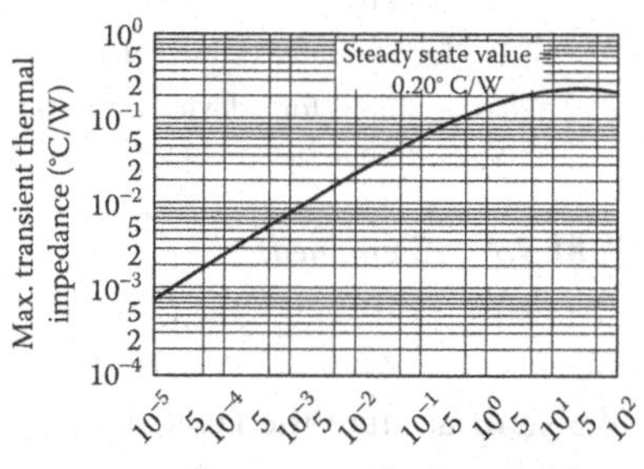

Fig. 4a Transient thermal impedance junction-to-case—R18CR and R18SR series

FIGURE 7.7 *(Continued)*

(d)

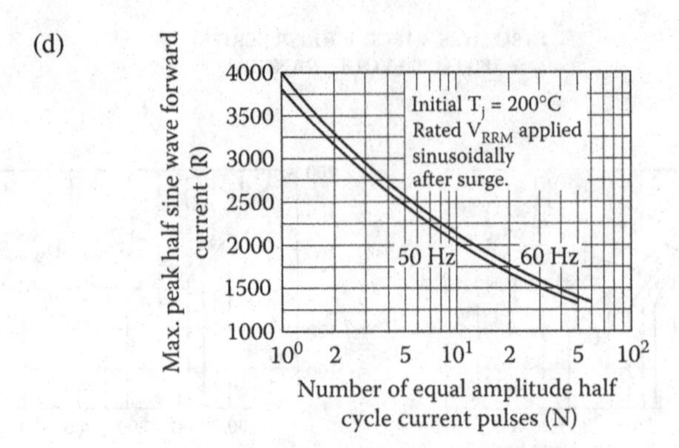

Fig. 5 Non-repetitive surge current ratings—
R18C, R18S, R18CR, and R18SR series

Ordering information

Type	Package (1)		Polarity		Voltage	
	Code	Description	Code	Description	Code	V_{RRM}
R18	C	1/2" stud, ceramic housing (Fig. 1)	—	Cathode–to–case	18A	1800 V
	S	3/8" stud, ceramic housing (Fig. 2)	R	Anode–to–case	18A	1600 V
					14B	1400 V
					12B	1200 V

(1) Other packages are also available:
- stud base with flag terminal.
- stud base with threaded top terminal.
- flat base.

For further details contact factory.

For example, for a device with 1/2" stud base and flexible lead, reverse polarity, V_{RRM} = 1600 V, order as: R18CR16A.

FIGURE 7.7 *(Continued)*

We shall assume the typical value CJO = 2PF. Thus, the PSpice model statement is

```
.MODEL DMODD (IS=2.22E-15 BV=1200 IBV=13E-3 CJO=2PF TT=1US)
```

7.5.1 Modeling Zener Diodes

To model a Zener diode, the breakdown voltage parameter BV in the model statement is set to the Zener voltage V_z. That is, BV = V_z. For a Zener diode of 220 V, the diode model statement is

```
.MODEL DMOD D(IS=2.22E-15 BV=220V IBV=12E-2 CJO=2PF TT=1US)
```

TABLE 7.3
Typical I-V Data of a Power Diode

I_D(A)	0	20	40	100	500	800	1000	1600	2000	3000	3900
V_D(V)	0	0.8	0.9	1.0	1.26	1.5	1.7	2.0	2.3	3	3.5

7.5.2 Tabular Data

The current versus voltage (*I–V*) characteristic of a diode can be found in the data sheet provided by the manufacturer, https://www.alldatasheet.com/. The data sheet for a diode is shown in Figure 7.7. This characteristic can be represented in a tabular form as shown in Table 7.3. This table can represent a current-controlled voltage source (see Section 4.5.2).

Example 7.1:

Describing the Diode Characteristics of Tabular Data

The diode circuit shown in Figure 7.8(a) has V_{DC} = 220 V and R_L = 0.5 Ω. Use PSpice to calculate the diode current and the diode voltage. Use the diode characteristics in Table 7.2.

SOLUTION

The diode is modeled as a current-controlled voltage source as shown in Figure 7.8(b) to represent the table of the *I–V* characteristic. The PSpice schematic is shown in Figure 7.9(a). The diode is modeled by an ETABLE from the abm.slb library. The current is related to the voltage in table form as shown in Figure 7.9(b).

The listing of the circuit file follows:

Diode Circuit

```
VDD 1 0 DC 220 V; DC voltage of 15 V
VX  3 2 DC 0 V  ; measures the diode current ID
RL  2 0 0.4
* The diode is represented by a table
Ediode 1 3 TABLE {I(VX)} =
```

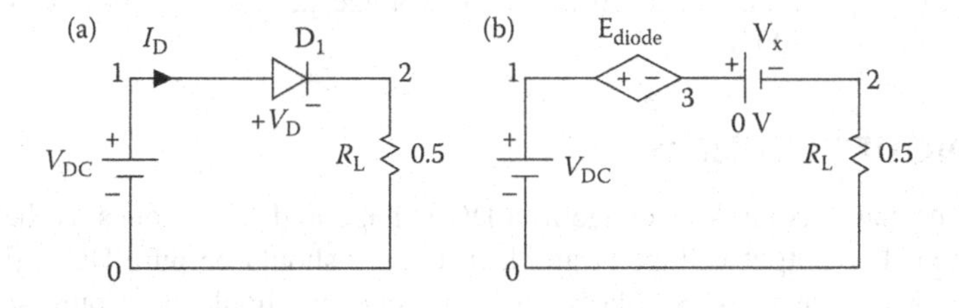

FIGURE 7.8 Diode circuit. (a) Circuit. (b) PSpice circuit.

(a)

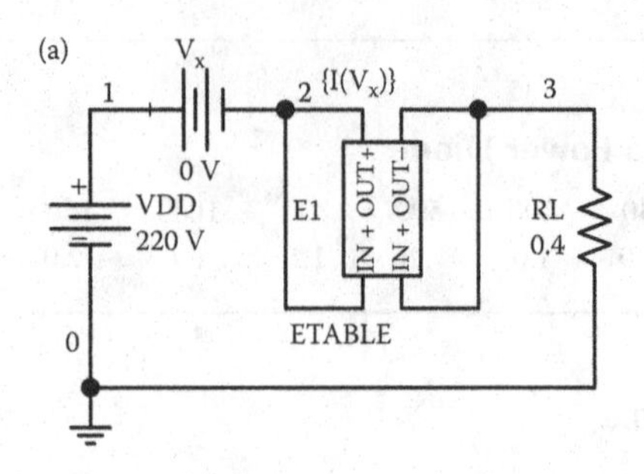

(b)

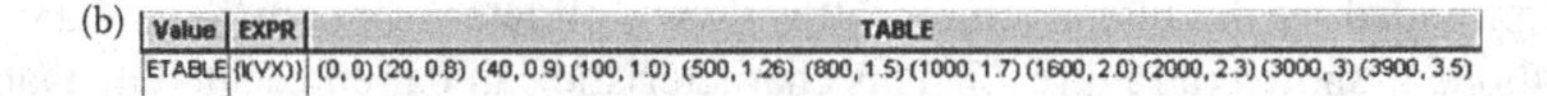

Value	EXPR	TABLE
ETABLE	{I(VX)}	(0, 0) (20, 0.8) (40, 0.9) (100, 1.0) (500, 1.26) (800, 1.5) (1000, 1.7) (1600, 2.0) (2000, 2.3) (3000, 3) (3900, 3.5)

FIGURE 7.9 PSpice schematic for Example 7.1. (a) ETABLE. (b) ETABLE parameters.

```
+ (0,0) (20,0.8) (40,0.9) (100,1.0) (500,1.26) (800,1.5)
+ (1000,1.7) (1600,2.0) (2000,2.3) (3000,3) (3900,3.5)
.OP              ; Prints the details of operating point
.END             End of circuit file
```

The information about the operating point, which is obtained from the output file EX7. 1.OUT, is as follows:

**** OPERATING POINT INFORMATION TEMPERATURE = 27.000 DEG C

NAME Ediode		
V-SOURCE	1.297E+00	(V_D = 1.297 V)
I-SOURCE	5.468E+02	(I_D = 546.8 A)

Note: The diode current is listed in terms of (i, v) data points.

LTspice: LTspice schematic for plotting the v–i characteristics listed in Table 7.2 is shown in Figure 7.10(a). The input voltage is varied from 0 to 3.5 V with an increment of 0.01. The v–i relationship of Table 7.2 is represented as a current-dependent voltage source H1 as a Tabular function as shown in Figure 7.10(b). The plot of the voltage across terminal 3, V(3) on the x-axis and the current through V_x, I(V_x) on the y-axis is shown in Figure 7.10(c). Note that the voltage source V_x acts as the current measuring device and its current produces a voltage to the current-dependent function H1. For example, at 561.8 A, the voltage across the diode as shown in Figure 7.10(c) is 1.309 V.

7.6 DIODE RECTIFIERS

A rectifier converts an AC voltage to a DC voltage and uses diodes as the switching device. The output voltage of an ideal rectifier should be pure DC and contain no harmonics or ripples. Similarly, the input current should be a pure sine wave and contain no harmonics. That is, the THD of the input current and output voltage should be zero, and the input power factor should be unity.

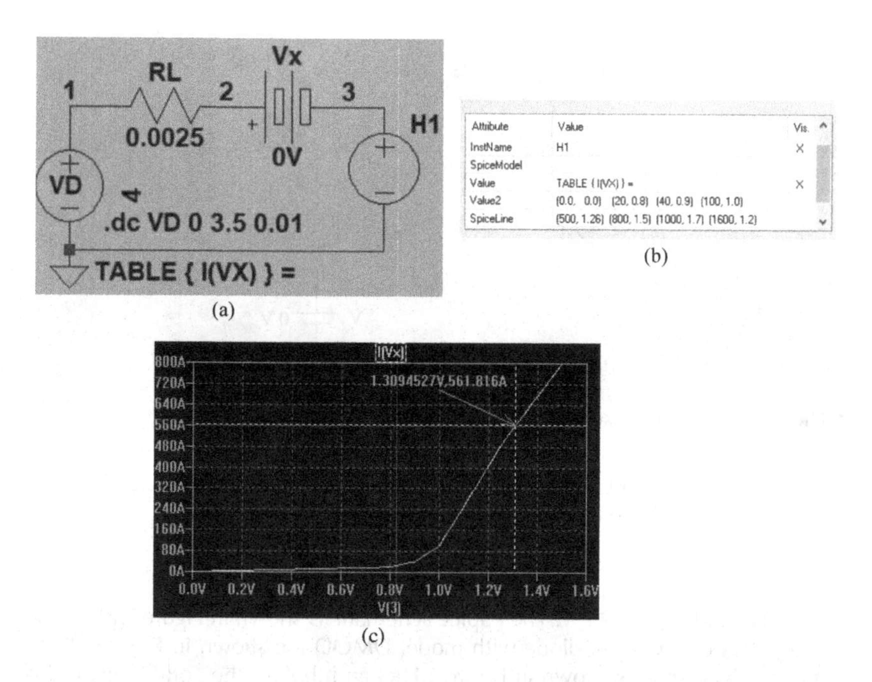

FIGURE 7.10 Plot of diode v-i characteristics. (a) LTspice schematic. (b) Tabular values of H1 function. (c) Plot of v–i characteristics.

The output voltage, the output current, and the input current of a rectifier contain harmonics. The input power factor PF_i can be determined from the THD_i of the input current as follows:

$$PF_i = \frac{I_{1(RMS)}}{I_s}\cos\phi_1 = \frac{1}{\sqrt{1+\left(\frac{\%THD}{100}\right)^2}}\cos\phi_1 \tag{7.3}$$

where

$I_{1(RMS)}$ = RMS value of the fundamental input current
I_s = RMS value of the input current
ϕ_1 = angle between the fundamental component of the input current and the fundamental component of the input voltage
$\%THD$ = percentage total harmonic distortion of the input current

7.6.1 Examples of Single-Phase Diode Rectifiers

Example 7.2:

Finding the Performance of a Half-Wave Diode Rectifier

A single-phase half-wave rectifier is shown in Figure 7.11. The input voltage is sinusoidal with a peak of 169.7 V, 60 Hz. The load inductance L is 6.5 mH, and the load resistance R is 0.5 Ω. Use PSpice (a) to plot the instantaneous output voltage

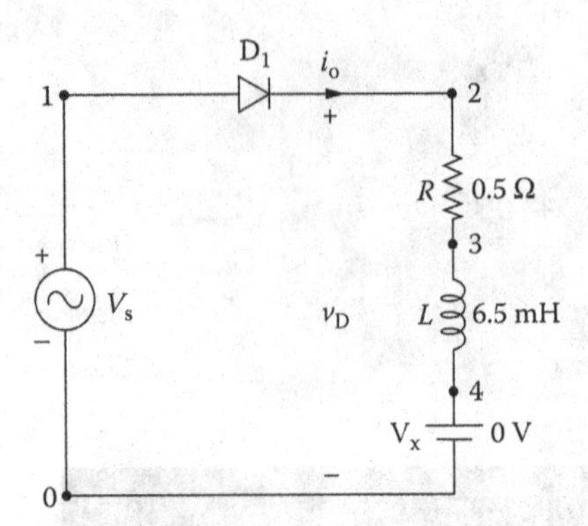

FIGURE 7.11 Single-phase half-wave rectifier for PSpice simulation.

v_o and the load current i_o, (b) to calculate the Fourier coefficient of the output voltage, and (c) to find the input power factor.

SOLUTION

$f = 60$ Hz and $V_m = 169.7$ V. The PSpice schematic is shown in Figure 7.12(a). The parameters of a Dbreak diode with model DMOD are shown in Figure 7.12(b). The transient setup is shown in Figure 7.11(c) and that for the Fourier analysis is shown in Figure 7.12(d).

Note: V_x as shown in Figure 7.11 is used as an ammeter and the current flows from the positive side to the negative side indicating a positive current. We could also print I(R:2) instead of I(V_x), but we will need to sort out the positive direction of the current.

The listing of the circuit file is as follows:

Single-Phase Half-Wave Rectifier With *RL* Load

```
SOURCE    ■     VS 1 0 SIN (0 169.7V 60HZ)
CIRCUIT   ■■    R 2 3 0.5
                L 3 4 6.5MH
                VX 4 0 DC 0V ;Voltage source to measure the
                output current
                D1 1 2 DMOD
                .MODEL DMOD D(IS=2.22E-15 BV=1200V
                IBV=13E-3 CJO=2PF TT=1US)
ANALYSIS  ■■■   .TRAN 10US 50.0MS 16.6667MS 10US ; Transient
                                                    analysis
                .FOUR 60HZ I(D1) V(2)      ; Fourier analysis of
                                             input current and
                                             output voltage
                .PROBE                       ; Graphics
                                               post-processor
                .OPTIONS ABSTOL = 1.0N RELTOL = .01 VNTOL = 1
                .0M ITL5 = 10000;
                *                                 Convergence
          .END
```

Note the following:

a. The PSpice plots of the instantaneous output voltage V(2) and load current I(V_x) are shown in Figure 7.13. The output voltage becomes negative

(a)

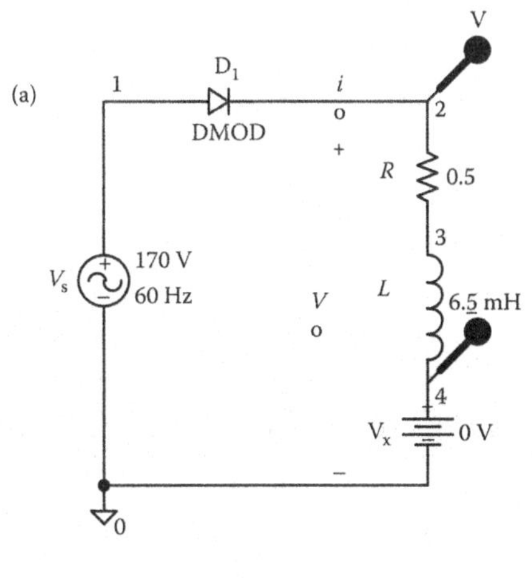

(b)

```
.MODEL DMOD D(IS=2.22E-15 BV=1200V IBV=12E-2 CJO=2PF TT=1US)
```

Diode model parameters

(c)

Run to time: 50ms seconds (TSTOP)

Start saving data after: 16.666ms seconds

Transient options

Maximum step size: seconds

☐ Skip the initial transient bias point calculation (SKIPBP)

Output File Options...

(d) Transient Output File Options

Print values in the output file every: 10us seconds

OK

Cancel

☑ Perform Fourier Analysis

Center Frequency: 60 hz

Number of Harmonics: 10

Output Variables: I(D1) V(R:2)

☐ Include detailed bias point information for nonlinear controlled sources and semiconductors (/OP)

FIGURE 7.12 PSpice schematic for Example 7.2. (a) Schematic. (b) Diode model parameters. (c) Transient setup. (d) Fourier setup.

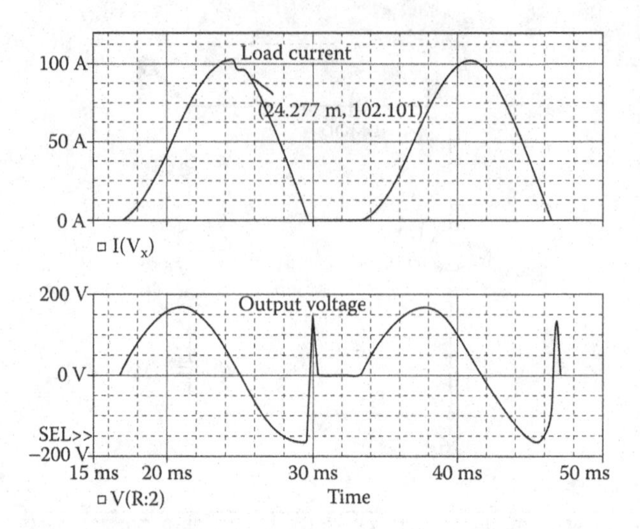

FIGURE 7.13 Plots for Example 7.2.

due to the inductive load because the current has to fall to zero before the diode can cease to conduct.

The voltage across an inductor is given by

$$v_L = L\frac{di}{dt} \tag{7.4}$$

During the time when the load current starts to fall, the voltage across the inductor L becomes negative and the diode will continue to conduct until the diode current falls to zero. However, there will be transient spikes at the output voltage due to a finite amount of transit time (TT) and diode junction capacitances (CJO). This is shown in Figure 7.13.

b. Fourier coefficients and THD will depend slightly on the internal time step TMAX discussed in Section 6.9.2. Fourier components of the output voltage and the load current can be found in the PSpice output file.

Fourier Components of Transient Response v(2)

DC Component = 2.218909E+01

Harmonic Number	Frequency (HZ)	Fourier Component	Normalized Component	Phase (Deg)	Normalized Phase (Deg)
1	6.000E+01	1.377E+02	1.000E+00	1.119E+01	0.000E+00
2	1.200E+02	3.646E+01	2.647E−01	−1.642E+02	−1.754E+02
3	1.800E+02	2.651E+01	1.925E−01	−1.076E+02	−1.188E+02
4	2.400E+02	1.649E+01	1.197E−01	−4.324E+01	−5.442E+01
5	3.000E+02	9.585E+00	6.958E−02	3.968E+01	2.849E+01
6	3.600E+02	8.108E+00	5.887E−02	1.366E+02	1.254E+02
7	4.200E+02	8.486E+00	6.161E−02	−1.431E+02	−1.542E+02
8	4.800E+02	7.607E+00	5.523E−02	−7.090E+01	−8.209E+01>
9	5.400E+02	5.897E+00	4.281E−02	6.598E+00	−4.591E+00

Total harmonic distortion=3.720415E+01%

c. To find the input power factor, we need to find the Fourier series of the input current, which is the same as the current through diode D_1.

Fourier components of transient response I (D1)

DC Component = 4.438586E+01

Harmonic Number	Frequency (HZ)	Fourier Component	Normalized Component	Phase (Deg)	Normalized Phase (Deg)
1	6.000E+01	5.507E+01	1.000E+00	−6.738E+01	0.000E+00
2	1.200E+02	7.448E+00	1.352E−01	1.107E−02	1.781E+02
3	1.800E+02	3.634E+00	6.598E−02	1.681E+02	2.355E+02
4	2.400E+02	1.606E+00	2.916E−02	−1.317E+02	−6.432E+01
5	3.000E+02	8.217E−01	1.492E−02	−5.059E+01	1.678E+01
6	3.600E+02	5.774E−01	1.048E−02	5.326E+01	1.206E+02
7	4.200E+02	4.422E−01	8.030E−03	1.264E+02	1.938E+02
8	4.800E+02	4.177E−01	7.584E−03	−1.633E+02	−9.597E+01
9	5.400E+02	2.892E−01	5.251E−03	−7.644E+01	−9.061E+00

Total harmonic distortion=1.548379E+01%

From the Fourier components of I(D1), we can find

DC input current $I_{in(DC)} = 44.39$ A
RMS fundamental input current, $I_{1(RMS)} = \frac{55.07}{\sqrt{2}} = 38.94$ A
THD of input current THD = 15.48% = 0.1548
Harmonic input current, $I_{h(RMS)} = I_{1(RMS)} \times \text{THD} = 38.94 \times 0.1548 = 6.028$ A
RMS input current $I_s = [I^2_{in(DC)} + I^2_{r(RMS)} + I^2_{h(RMS)}]^{1/2}$
$= (44.39^2 + 38.94^2 + 6.028^2)^{1/2} = 59.36$ A
Displacement angle $\phi_1 = -64.38°$
Displacement factor $DF = \cos\phi_1 = \cos(-67.38) = 0.3846$ (lagging)

Thus, the input power factor is given [1] by

$$PF = \frac{I_{1(RMS)}}{I_s}\cos\phi_1 = \frac{38.94}{59.36} \times 0.3846 = 0.2523 \text{ (lagging)}$$

The power factor can be determined directly from the *THD* as follows:

$$PF = \frac{I_{1(RMS)}}{I_s}\cos\phi_1 = \frac{1}{[1+(\%THD/100)^2]^{1/2}}\cos\phi_1$$
$$= \frac{1}{1+(0.1548^2)^{1/2}} \times 0.3846 = 0.3801(\text{lagging}) \tag{7.5}$$

This gives a higher value and cannot be applied if there is a significant amount of DC component.

Note: The load current is discontinuous. When the diode turns off, there is a voltage transient. If an antiparallel diode (also known as the freewheeling diode) is connected across the load (terminals 2 and 0), the load current will be smoother. As a result, the power factor will improve. Students are encouraged to simulate the circuit with an antiparallel diode to verify this.

The power factor PF can be determined approximately from the input and output power balances. That is,

$$I_s V_S PF = I_s^2 R_L \tag{7.6}$$

This gives the power factor PF as

$$PF = \frac{I_s R_L}{V_s} = \frac{59.36 \times 0.5}{120} = 0.247$$

This is low.

Example 7.3:

Finding the Performance of a Single-Phase Bridge Rectifier

A single-phase bridge rectifier is shown in Figure 7.14. The sinusoidal input voltage has a peak of 169.7 V, 60 Hz. The load inductance L is 6.5 mH, and the load resistance R is 0.5 Ω. Use PSpice (a) to plot the instantaneous output voltage v_o and the load current i_o and (b) to calculate the Fourier coefficients of the input current and the input power factor.

SOLUTION

V_m = 169.7 V and f = 60 Hz. The PSpice schematic is shown in Figure 7.15(a). The transient setup is shown in Figure 7.15(b) and that for the Fourier analysis of the input current is shown in Figure 7.15(c).

Note: V_x and V_y as shown in Figure 7.15 are used as ammeters to monitor the load current and the input side current, respectively.

The listing of the circuit file is as follows:

Single-Phase Bridge Rectifier with *RL* Load

```
SOURCE   ■   VS 1 0 SIN (0169.7V60HZ)
CIRCUIT  ■■  R 3 5 0.5
             L 5 6 6.5MH
             VX 6 4 DC 0V ; Voltage source to measure the
                            output current
             VY 1 2 DC 0V ; Voltage source to measure the
                            output current
             D1 2 3 DMOD
             D3 0 3 DMOD
             D2 4 0 DMOD
             D4 4 2 DMOD
             .MODEL DMOD D (IS=2.22E-15 BV=1200V IBV=13E-3
             CJO=2PF TT=1US)
ANALYSIS ■■■ TRAN 10US 50MS 33.3333MS 10US ; Transient analysis
             .FOUR 60HZ 1 (VY)            ; Fourier analysis of
             *                              input current
             .PROBE                       ; Graphic POST-
                                            post-processor
             .OPTIONS ABSTOL = 1.0 N RELTOL = .01 BNTOL = 1.0M
             ITL5 = 10000 ;
             *                              Convergence
         .END
```

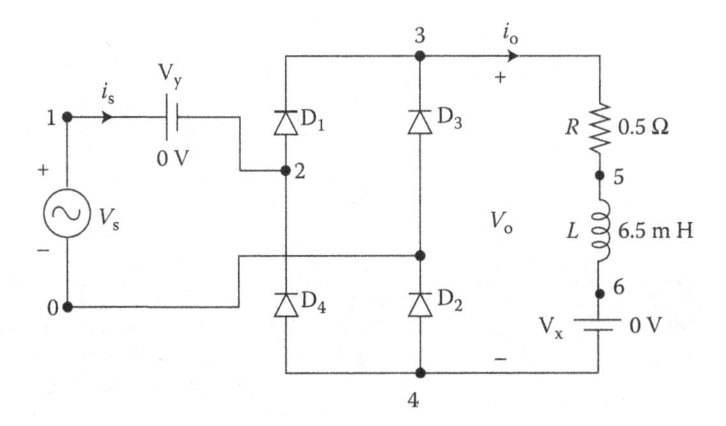

FIGURE 7.14 Single-phase bridge rectifier.

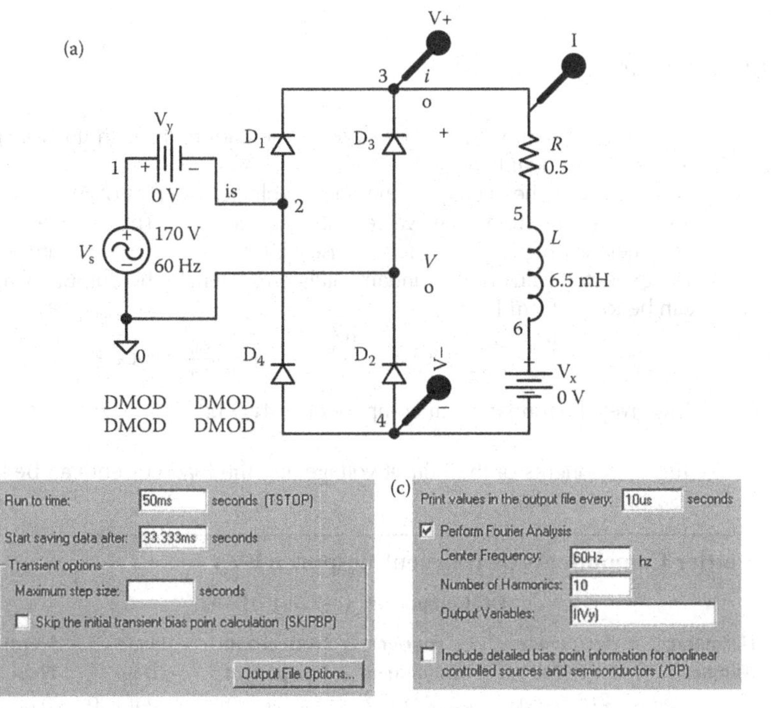

FIGURE 7.15 PSpice schematic for Example 7.3. (a) Schematic. (b) Transient setup. (c) Fourier setup.

Note the following:

a. The PSpice plots of the instantaneous output voltage V(3,4) and load current $I(V_x)$ are shown in Figure 7.16. One of the diode pairs always conducts. The load current contains ripples and has not reached steady-state conditions.

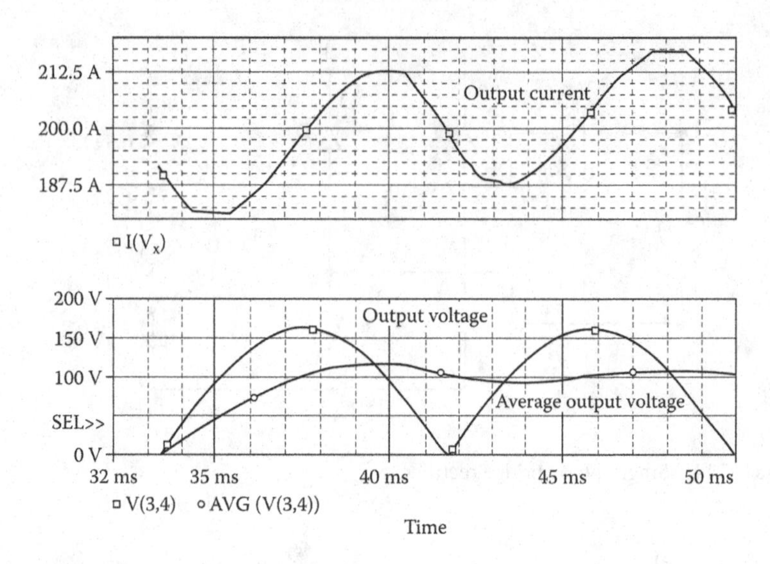

FIGURE 7.16 Plots for Example 7.3.

b. The input current, which is the same as the current through the voltage source V_y, is equal to $I(V_y)$.
c. Probe can plot the average of the output voltage. The plot of AVG(V(7,4)) is the instantaneous average value of the voltage V(7,4). The average value becomes very high at near zero because it is the time average of a function.
d. The expected value of the fundamental component of the output voltage can be found from [1]

$$v_o(t) = \frac{2V_m}{\pi} - \frac{4V_m}{3\pi}\cos 2\omega t - \frac{4V_m}{15\pi}\cos 4\omega t - \frac{4V_m}{35\pi}\cos 6\omega t - \cdots \tag{7.7}$$

This gives the fundamental component as $4V_m/3\pi$.

Fourier components of the output voltage and the load current can be found in the PSpice output file.

Fourier Components of Transient Response $I(V_y)$

DC Component = −2.56451E+00

Harmonic Number	Frequency (HZ)	Fourier Component	Normalized Component	Phase (Deg)	Normalized Phase (Deg)
1	6.000E+01	2.595E+02	1.000E+00	−3.224E+00	0.000E+00
2	1.200E+02	7.374E−01	2.842E−03	1.410E+02	1.442E+02
3	1.800E+02	8.517E+01	3.282E−01	4.468E+00	7.693E+00
4	2.400E+02	5.856E−01	2.257E−03	1.199E+02	1.232E+02
5	3.000E+02	5.118E+01	1.972E−01	3.216E+00	6.440E+00
6	3.600E+02	3.600E+02	2.130E−03	1.111E+02	1.143E+02
7	4.200E+02	3.658E+01	1.410E−01	2.868E+00	6.092E+00
8	4.800E+02	5.406E−01	2.083E−03	1.065E+02	1.097E+02
9	5.400E+02	2.846E+01	1.097E−01	2.822E+00	6.047E+00

Total harmonic distortion=4.225668E+01%

From the Fourier components of $I(V_y)$, we can find

DC input current $I_{in(DC)} = -2.56$ A, which should ideally be zero
RMS fundamental input current $I_{1(RMS)} = \frac{259.5}{\sqrt{2}} = 183.49$ A
THD of input current $THD = 42.26\% = 0.4226$
RMS harmonic current $I_{h(RMS)} = I_{1(RMS)} \times THD = 183.49 \times 0.4226 = 77.54$ A
RMS input current $I_s = (I_{in(DC)}^2 + I_{1(RMS)}^2 + I_{h(RMS)}^2)^{1/2}$
$= (2.56^2 + 183.49^2 + 77.54^2)^{1/2} = 199.22$ A
Displacement angle $\phi_1 = -3.22$
Displacement factor $DF = \cos\phi_1 = \cos(-3.22) = 0.998$ (lagging)

Thus, the input power factor is

$$PF = \frac{I_{1(RMS)}}{I_s}\cos\phi_1 = \frac{183.49}{199.22} \times 0.998 = 0.9192(\text{lagging})$$

Assuming that $I_{in(DC)} = 0$, Equation 7.3 gives the power factor as

$$PF = \frac{1}{\frac{(1+0.4226^2)1}{2}} \times 0.9981 = 0.9193(\text{lagging})$$

Note: The load current is continuous. The input power factor (0.919) is much higher compared to that (0.3802) of the half-wave rectifier.

LTspice: The LTspice schematic for Example 7.3 is shown in Figure 7.17(a), the plots of instantaneous load current, load voltage, and average load power in Figure 6.17(b) and the average and RMS load voltage in Figure 7.17(c), the average and RMS load current in Figure 7.17(d), and the average load power in Figure 7.17(e). The plots of the instantaneous input current and the input power are shown in Figure 7.17(f), the average and RMS input current in Figure 7.17(g), and the average input power in Figure 7.17(h). We can note that the input power to the rectifier is $P_{in} = 15.608$ kW and the output load power is $P_{out} = 13.558$ kW. The power difference is the caused due to losses in the power diodes. We can find the power efficiency of the rectifier $\eta = \frac{P_{out}}{P_{in}} = \frac{13558}{15608} = 86.87\%$
And the input power factor as $PF = \frac{P_{in}}{V_s I_s} = \frac{15608}{120\times150.04} = 0.869$
Notes:

- For the SPICE command menu, we need to select SPICE directive otherwise, the statement will be used as a comment, not as a directive.

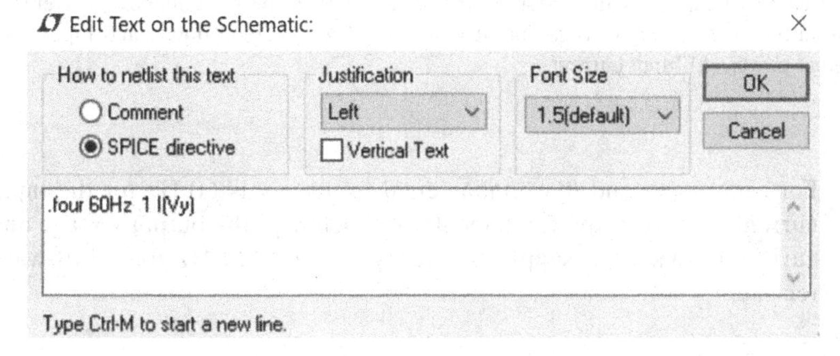

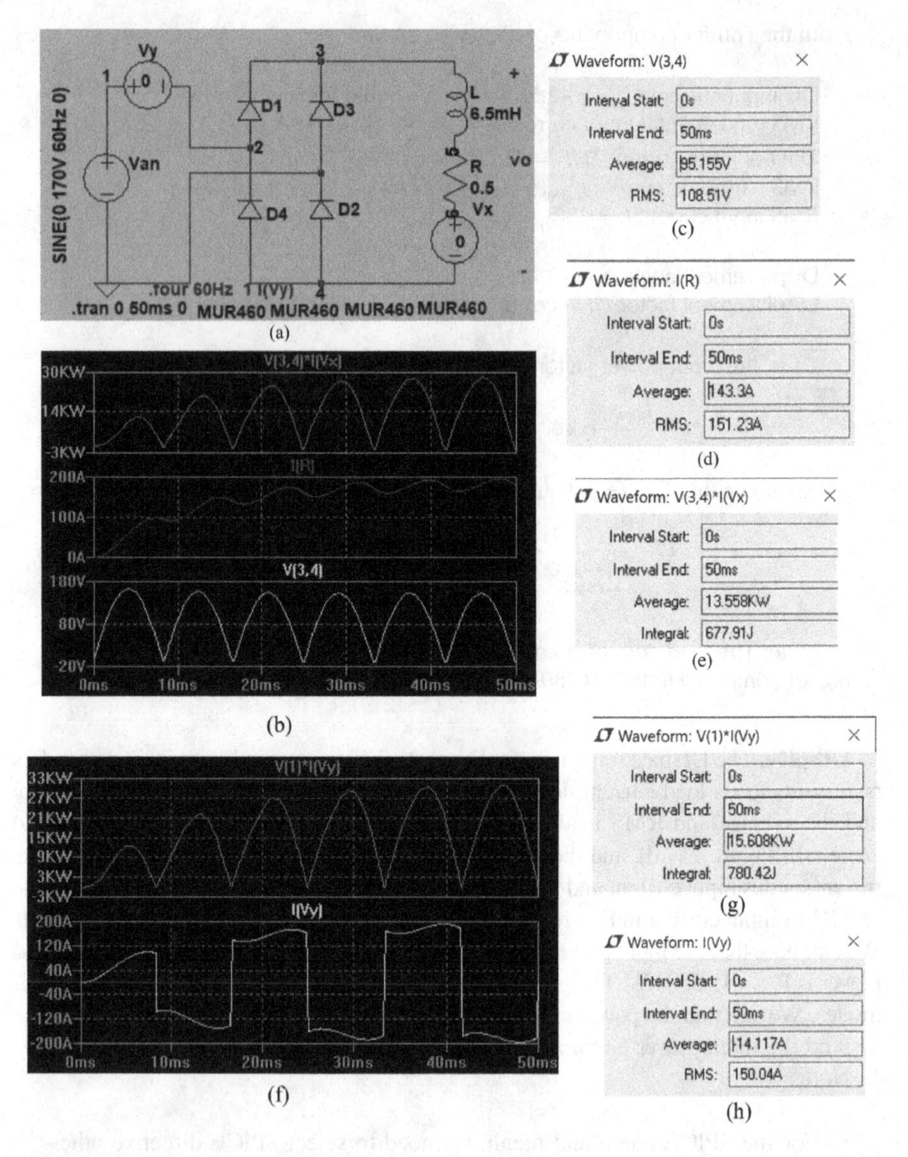

FIGURE 7.17 LTspice plots for Example 7.3. (a) LTspice schematic. (b) Plots of output power, current I®, and voltage V(3,4). (c) Average and RMS output voltage. (d) Average and RMS output current. (e) Average output power. (f) Average input power and input current. (g) Input power. (h) Input current.

- For .four command, the fundamental frequency is 60 Hz for the input current. However, the fundamental frequency of the output voltage and current is twice the supply frequency, that is, 120 Hz for a full-wave rectifier.

Example 7.4:

Finding the Performance of a Single-Phase Bridge Rectifier with an Output LC Filter

A single-phase bridge rectifier with an *LC* filter is shown in Figure 7.18. The sinusoidal input voltage has a peak of 169.7 V, 60 Hz. The load inductance *L* is 10 mH, and the load resistance *R* is 40 Ω. The filter inductance L_e is 30.83 mH, and the filter capacitance C_e is 326 μF. Use PSpice (a) to plot the instantaneous output voltage v_o and the load current i_o, (b) to calculate the Fourier coefficients of the output voltage, (c) to calculate the Fourier coefficients of the input current and input power factor, and (d) plot the instantaneous output voltage for C_e = 1, 100, and 326 μF.

SOLUTION

The PSpice schematic is shown in Figure 7.19(a). The capacitor C_e is defined as a variable CVAL and the setup for parametric sweep is shown in Figure 7.19(b).

Peak voltage V_m = 169.7 V and f = 60 Hz. The listing of the circuit file is as follows:

Single-Phase Bridge Rectifier with *RL* Load

```
SOURCE   ■    VS 1 0 SIN (0 169.7V 60Hz)
CIRCUIT  ■■   LE 3 7 30.83MH
              .PARAM CVAL = 326UF
              CE 7 4 326UF
              R 7 5 40
              L 5 6 10MH
              VX 6 4 DC 0V ; Voltage source to measure the
              output current
              VY 1 2 DC 0V ; Voltage source to measure the input
              current
              D1 2 3 DMOD
              D3 0 3 DMOD
              D2 4 0 DMOD
              D4 4 2 DMOD
              .STEP PARAM LIST 1UF 100UF 3264F
              .MODEL DMOD D(IS=2.22E-15 BV=1200V
              IBV=13E-3 TT=1US)
```

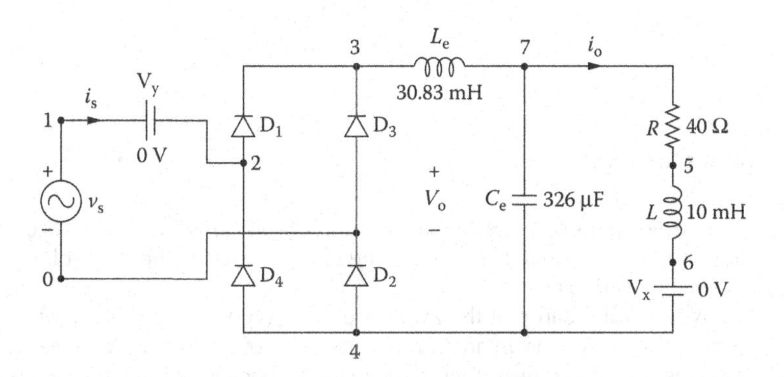

FIGURE 7.18 Single-phase bridge rectifier with load filter.

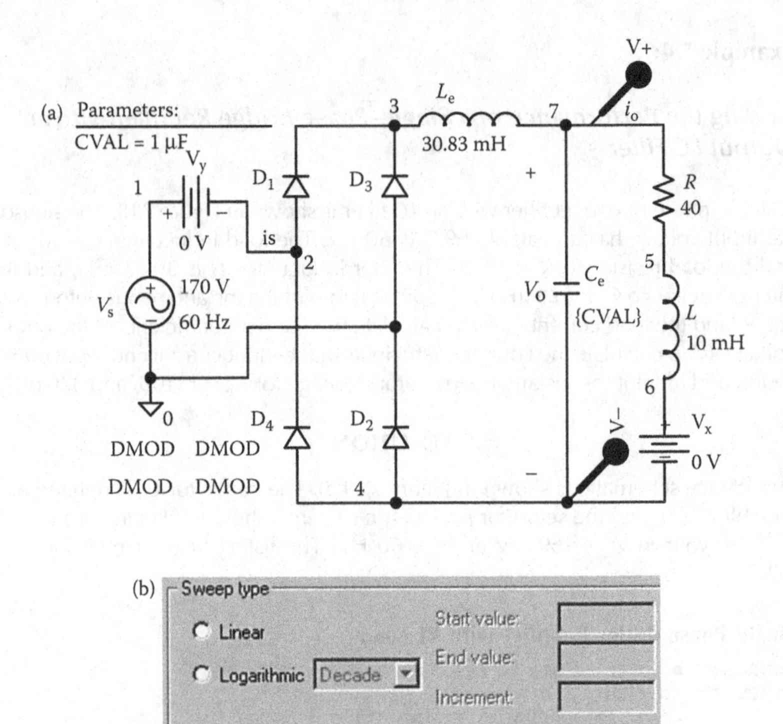

FIGURE 7.19 PSpice schematic for Example 7.4. (a) Schematic. (b) Parametric setup.

```
ANALYSIS ■■■  .TRAN 10US 50MS 33.3333MS 10US ; Transient
                                               analysis
              .FOUR 120Hz V(7,4)             ; Fourier
                                               analysis of
                                               output voltage
              *
              .PROBE
              .OPTIONS ABSTOL = 1.0 NVNTOL = .01M ITL5 = 10000 ;
                                             Convergence
              *
          .END
```

Note the following:

a. The PSpice plots of the instantaneous output voltage V(7,4) and load current I(V_x) are shown in Figure 7.20. The *LC* filter smooths the load voltage and reduces ripples.

   Note: Probe can plot the average of the output voltage. The plot of AVG(V(7,4)) is the instantaneous average value of the voltage V(7,4). The average value becomes very high at near zero because it is the time average of the instantaneous quantity.

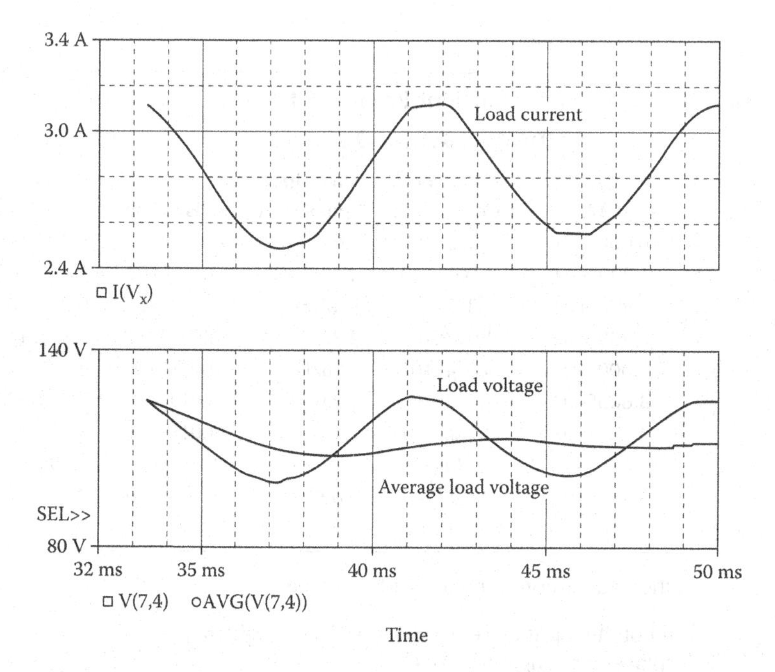

FIGURE 7.20 Plots for Example 7.4.

b. Fourier components of the output voltage and the load current can be found in the PSpice output file. The Fourier coefficients of the output voltage are as follows:

Fourier Components of Transient Response V(7,4)

DC Component = 1.143072E+02

Harmonic Number	Frequency (Hz)	Fourier Component	Normalized Component	Phase (Deg)	Normalized Phase (deg)
1	1.200E+02	1.306E+01	1.000E+00	1.034E+02	0.000E+00
2	2.400E+02	6.509E−01	4.983E−02	1.225E+02	1.907E+01
3	3.600E+02	2.315E−01	1.772E−02	9.039E+01	−1.305E+01
4	4.800E+02	1.617E−01	1.238E−02	4.774E+01	−5.570E+01
5	6.000E+02	1.316E−01	1.007E−02	2.218E+01	−8.126E+01
6	7.200E+02	1.050E-01	8.039E−03	8.698E+00	−9.474E+01
7	8.400E+02	8.482E−02	6.494E−03	2.760E+00	−1.007E+02
8	9.600E+02	7.149E−02	5.473E−03	5.647E−02	−1.034E+02
9	1.080E+03	6.137E−02	4.699E−03	−2.062E+00	−1.055E+02

Total harmonic distortion = 5.666466E+00%

c. Current, which is the same as the current through voltage source V_y, is equal to $I(V_y)$. After running PSpice to obtain the Fourier series of the input current, using the command.

```
FOUR 60HZ I(VY); Fourier analysis of input current
```

we get the following:

Fourier Components of Transient Response $I(V_Y)$

DC Component = −2.229026E+00

Harmonic Number	Frequency (HZ)	Fourier Component	Normalized Component	Phase(Deg)	Normalized Phase(Deg)
1	6.000E+01	2.555E+02	1.000E+00	−3.492E+00	0.000E+00
2	1.200E+02	1.146E+00	4.486E−03	1.192E+02	1.227E+02
3	1.800E+02	8.372E+01	3.277E−01	3.125E+00	6.617E+00t
4	2.400E+02	1.092E+00	4.273E−03	1.044E+02	1.079E+02
5	3.000E+02	5.024E+01	1.967E−01	1.036E+00	4.528E+00
6	3.600E+02	1.082E+00	4.237E−03	9.819E+01	1.017E+02
7	4.200E+02	3.586E+01	1.404E−01t	−1.011E−01	3.391E+00
8	4.800E+02	1.070E+00	4.187E−03	9.446E+01	9.795E+01
9	5.400E+02	2.788E+01	1.091E−01	−9.085E−01	2.583E+00

Total harmonic distortion=4.216000E+01%

From the Fourier components of $I(V_y)$, we get

THD of the input current THD = 42.16% = 0.4216
Displacement angle $\phi = -3.492°$
Displacement factor DF = $\cos\phi = \cos(-3.492) = 0.9981$ (lagging)

Neglecting the DC input current $I_{in(DC)} = -2.229$ A, which is small relative to the fundamental component, we can find the power factor from Equation 7.3 as

$$PF = \frac{1}{\left(1+0.4216^2\right)^{1/2}} \times 0.9981 = 0.9197\,\text{(lagging)}$$

d. The effects of the filter capacitances on the load voltage are shown in Figure 7.21. With a higher value of the filter capacitor, the steady-state peak-peak ripple on the output voltage is reduced.

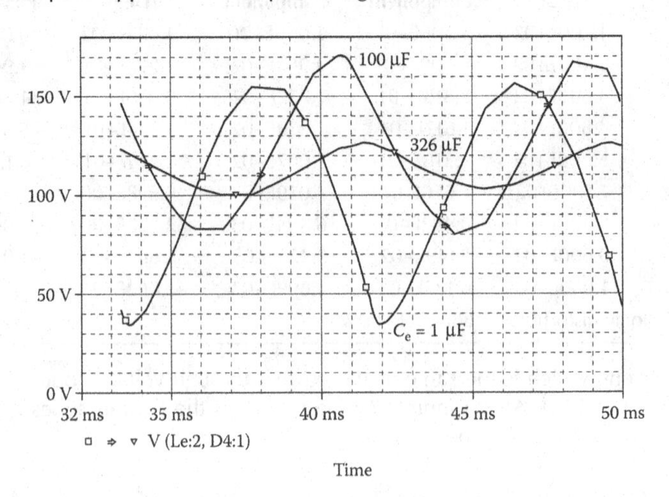

FIGURE 7.21 Effects of filter capacitances for Example 7.4.

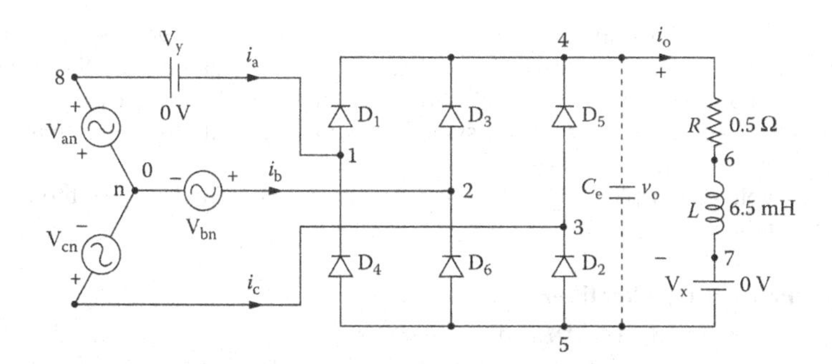

FIGURE 7.22 Three-phase bridge rectifier.

7.6.2 Examples of Three-Phase Diode Rectifiers

Example 7.5:

Finding the Performance of a Three-Phase Bridge Rectifier

A three-phase bridge rectifier is shown in Figure 7.22. The rectifier is supplied from a balanced three-phase balanced supply whose per-phase voltage has a peak of 169.7 V, 60 Hz. The load inductance L is 6.5 mH, and the load resistance R is 0.5 Ω. Use PSpice (a) to plot the instantaneous output voltage v_o and line (phase) current i_a, (b) to plot the RMS and average currents of diode D_1, (c) to plot the average output power, and (d) to calculate the Fourier coefficients of the input current and the input power factor.

SOLUTION

The PSpice schematic is shown in Figure 7.23. V_x and V_y as shown in Figure 7.23 are used as ammeters to monitor the load current and the input side current,

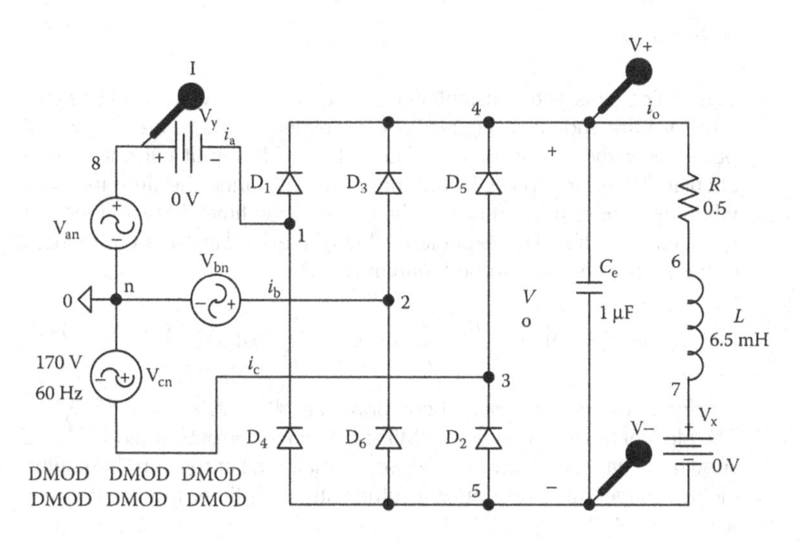

FIGURE 7.23 Three-Phase Bridge Rectifier schematic for Example 7.5.

respectively. for a balanced three-phase supply, each phase voltage should be phase shifted by 120°. A small-capacitance C_e is connected across the load to avoid any convergence problems. Without this capacitor, the voltage across or the current through the diodes would switch very rapidly, and the PSpice may find difficulties in finding a solution.

Peak voltage per phase V_m = 169.7 V, and f = 60 Hz. The listing of the circuit file is as follows:

Three-Phase Bridge Rectifier

```
SOURCE    ■    Van 8 0 SIN (0 169.7V 60 HZ)
               Vbn 2 0 SIN (0 169.7V 60 Hz 0 0 120DEG)
               Vcn 3 0 SIN (0 169.7V 60 Hz 0 0 240DEG)
CIRCUIT   ■■   CE 4 5 1UF ; Small capacitance to aid convergence
               R 4 6 0.5
               L 6 7 6.5MH
               VX 7 5 DC 0V ; Voltage source to measure the output
                              current
               VY 8 1 DC 0V ; Voltage source to measure the input
                              current
               D1 1 4 DMOD
               D3 2 4 DMOD
               D5 3 4 DMOD
               D2 5 3 DMOD
               D6 5 2 DMOD
               D4 5 1 DMOD
               .MODELDMODD (IS=2.2 2E-15 BV=1200V IBV=13E-3
               CJO=2PF TT=1US)
ANALYSIS ■■■   .TRAN 10US 33.3333MS 0 10US ; Transient analysis
               .FOUR 60Hz 1 (VY)       ; Fourier analysis of line
                                         current
               .PROBE                  ; Graphics post-processor
               .OPTIONS ABSTOL=1.0N RENTOL=1.0M VNTOL=1.0M
                                    ITL5=10000 ; Convergence
               *
          .END
```

Note the following:

a. The PSpice plots of the instantaneous output voltage V(4,5) and line current $I(V_y)$ are shown in Figure 7.24. As expected, there are six output pulses over the period of the input voltage. The input current is rectangular. There are spikes in the input current during the time intervals when the diodes turn off due to the diode transit time (TT) and junction capacitance (CJO). The expected value of the fundamental component of the output voltage can be found in Ref. [1].

$$v_o(t) = 0.9549V_m\left(1 + \frac{2}{35}\cos 6\omega t - \frac{2}{143}\cos 12\omega t + \cdots\right) \tag{7.8}$$

   This gives the fundamental component as $4V_m/3\pi$.

b. The plots of the instantaneous RMS and average currents of diode D_1 are shown in Figure 7.25. Averaging over a small time at the very beginning yields a large value. But after a sufficiently long time, it gives the true average or RMS values.

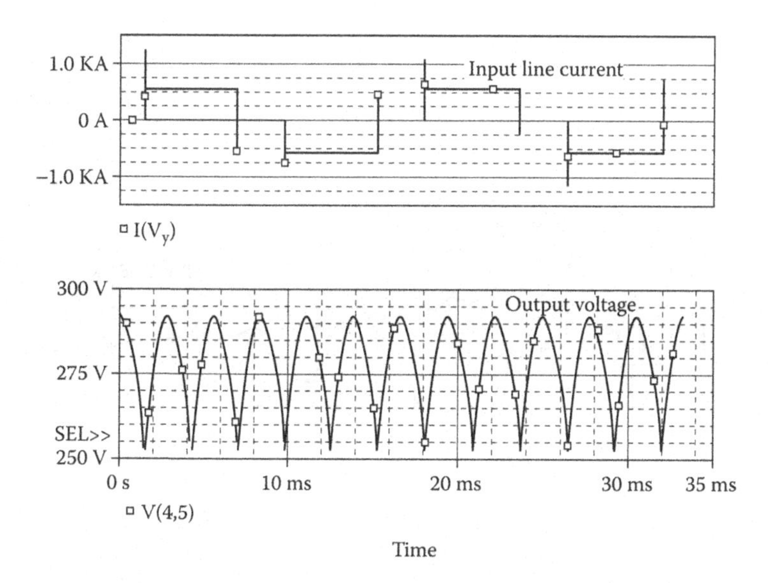

FIGURE 7.24 Plots of output voltage (4,5) and line current $I(V_y)$ for Example 7.5.

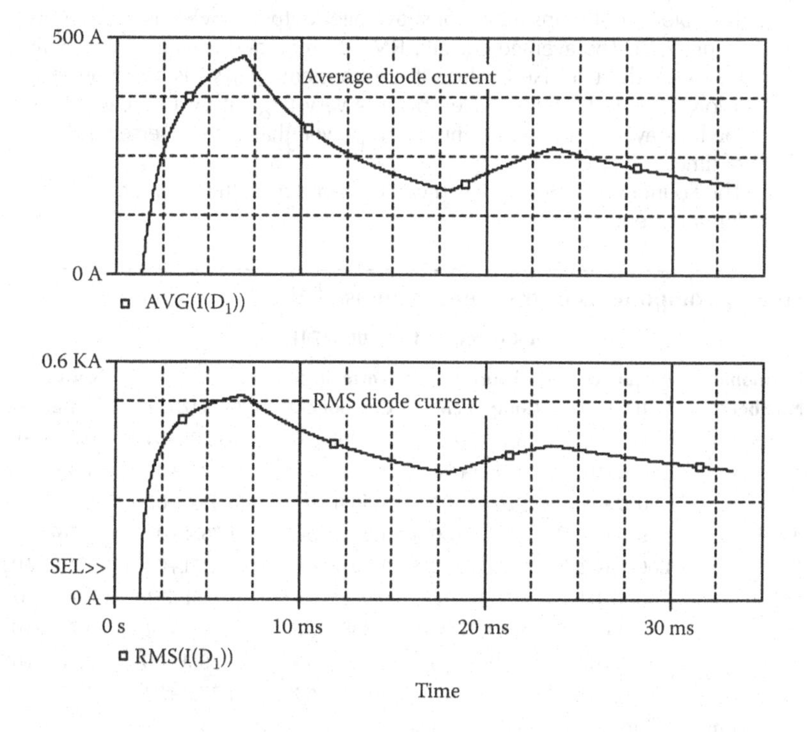

FIGURE 7.25 Plots of RMS and average currents through diode D1 for Example 7.5.

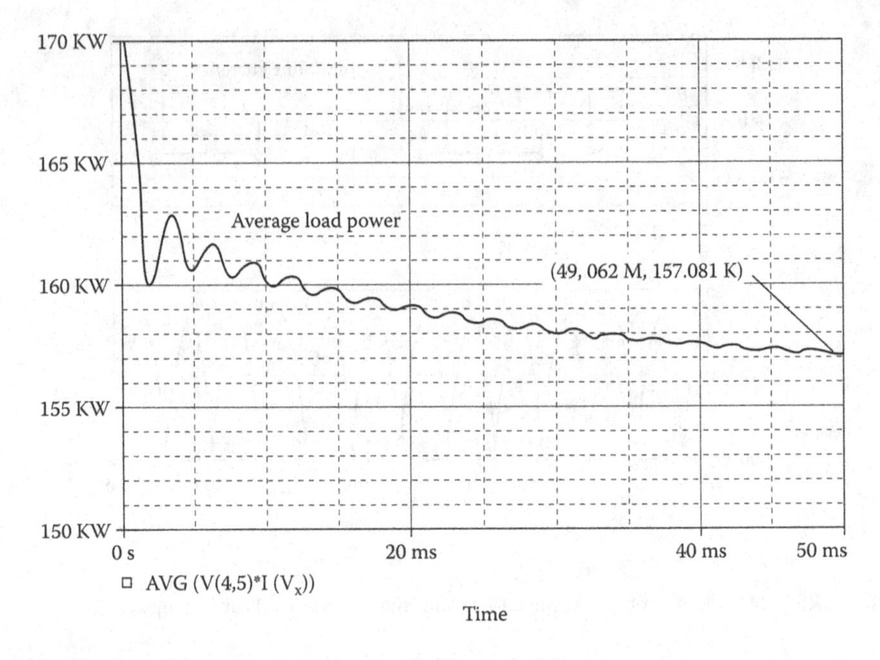

FIGURE 7.26 Instantaneous load power for Example 7.5.

c. The plot of the instantaneous average output power is shown in Figure 7.26. The average current, RMS current, and average power will reach steady-state fixed values if the transient analysis is continued for a longer period. The average power is very large at $t = 0$ because it is the time average of the instantaneous power, that is, an inverse function of time.
d. The Fourier coefficients, which can be found from the output file, of the input current are:

Fourier components of transient response I(V_y)

Dc Component = 2.066274E–01

Harmonic Number	Frequency (Hz)	Fourier Component	Normalized Component	Phase(Deg)	Normalized Phase(Deg)
1	6.000E+01	6.161E+02	1.000E+00	–8.420E–03	0.000E+00
2	1.200E+02	1.182E+00	1.919E–03	–1.692E+02	–1.692E+02
3	1.800E+02	9.265E–01	1.504E–03	–6.353E+00	–6.345E+00
4	2.400E+02	1.219E+00	1.979E–03	–1.767E+02	–1.767E+02
5	3.000E+02	1.227E+02	1.991E–01	1.797E+02	1.797E+02
6	3.600E+02	6.153E–02	9.987E–05	1.145E+02	1.145E+02
7	4.200E+02	2 8.839E+01	1.435E–01	–1.797E+02	–1.797E+02
8	4.800E+02	1.196E+00	1.941E+00	3.666E+00	3.675E+00
9	5.400E+02	9.152E–01	1.485E–03	1.779E+02	1.779E+02

Total harmonic distortion = 2.454718E+01%

From the Fourier components of $I(V_y)$, we get

THD of input current, $THD = 24.55\% = 0.2455$
Displacement angle, $\phi_1 \simeq 0°$
Displacement factor, $DF = \cos\phi_1 = \cos(0) \simeq 1$

Neglecting the DC input current $I_{in(DC)} = 0.207$ A, which is small relative to the fundamental component, we can find power factor PF from Equation 7.3 as

$$PF = \frac{1}{\left(1+0.2455^2\right)^{1/2}} \times 1 = 0.971 \text{ (lagging)}$$

Note: The power factor is significantly higher than that of a single-phase rectifier.

LTspice: The LTspice schematic for Example 7.5 is shown in Figure 7.27(a), the plots of instantaneous load current, load voltage and average load power in Figure 6.27(b) and the average and RMS load voltage in Figure 7.27(c), the average and RMS load current in Figure 7.27(d), and the average load power in Figure 7.27(e). The plots of the instantaneous input current and the input power are shown in Figure 7.27(f) and the average and RMS input current in Figure 7.27(g) and the average input power in Figure 7.27(h). We can note that the input power to the rectifier is $P_{in} = 15.608$ kW and the output load power is $P_{out} = 13.558$ kW. The power difference is the caused due to losses in the power diodes.

The power difference is the caused due to losses in the power diodes. We can find the power efficiency of the rectifier $\eta = \frac{P_{out}}{P_{in}} = \frac{79711}{92743} = 85.95\%$

And the input power factor as $PF = \frac{P_{in}}{3V_s I_s} = \frac{92743}{3\times120\times292.53} = 0.881$

Notes:

- For SPICE command menu, we need to select SPICE directive otherwise, the statement will be used as a comment, not as a directive.

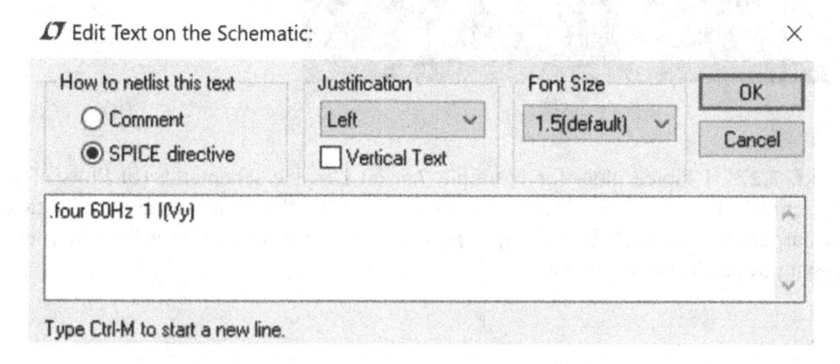

- For a .four command, the fundamental frequency is 60 Hz for the input current. However, the fundamental frequency of the output voltage and current is six times the supply frequency, that is, 360 Hz for a three-phase rectifier.

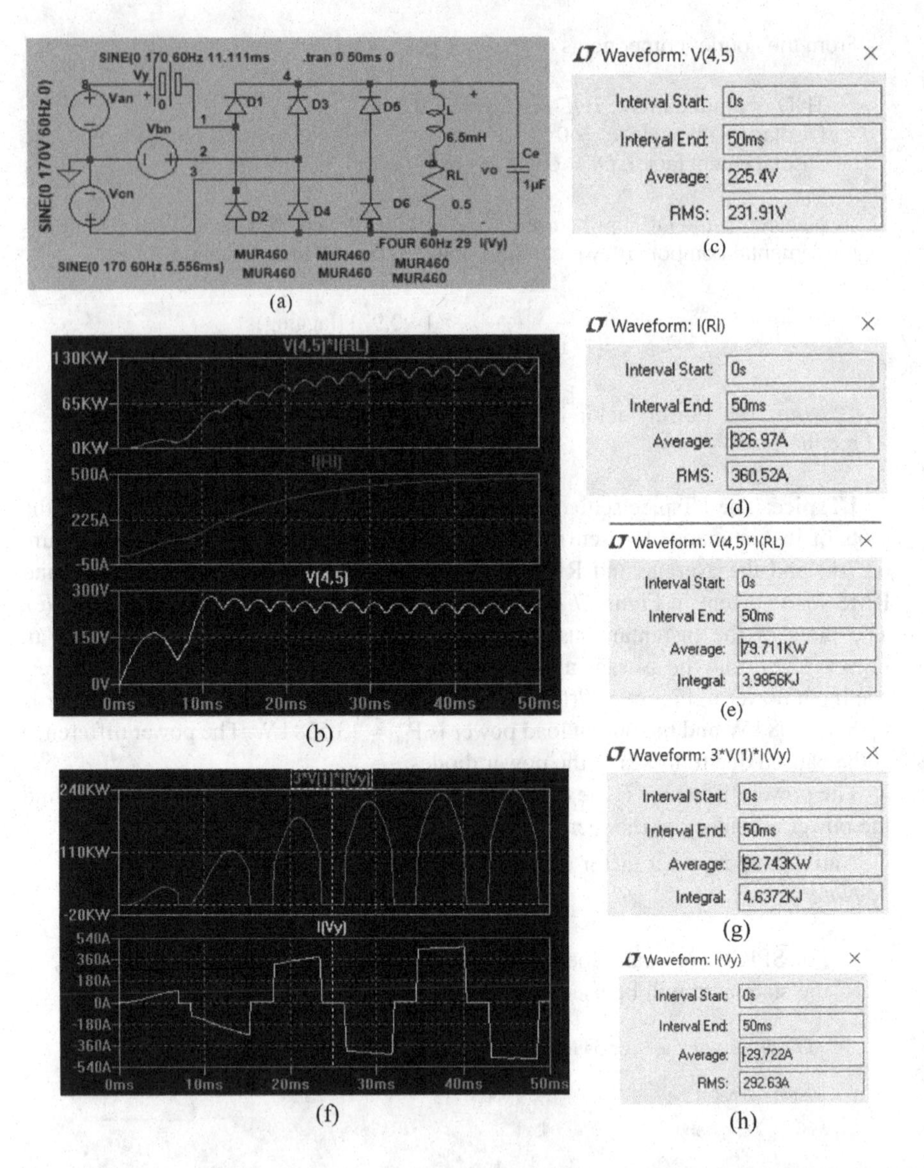

FIGURE 7.27 LTspice plots for Example 7.5. (a) LTspice schematic. (b) Plots of output power, current I® and voltage V(3,4). (c) Average and RMS output voltage. (d) Average and RMS output current. (e) Average output power. (f) Average input power and input current. (g) Input power. (h) Input current.

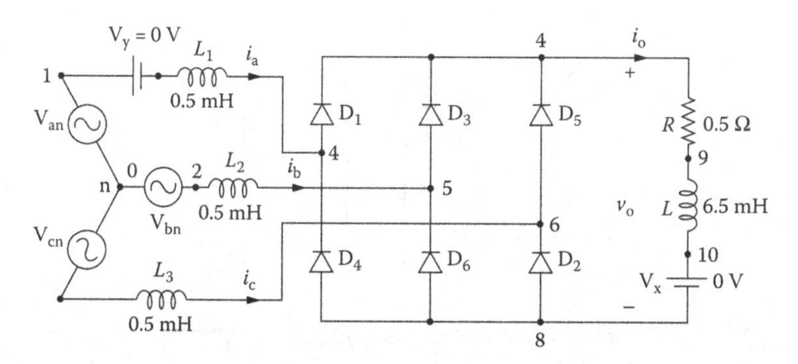

FIGURE 7.28 Three-phase bridge rectifier with source inductance.

Example 7.6:

Finding the Performance of a Three-Phase Bridge Rectifier with Line Inductances

A three-phase bridge rectifier with the inductances is shown in Figure 7.28. The rectifier is supplied from a balanced three-phase supply whose per-phase voltage has a peak of 169.7 V, 60 Hz. The load inductance L is 6.5 mH, and the load resistance R is 0.5 Ω. The line inductances are equal; $L_1 = L_2 = L_3 = 0.5$ mH. Use PSpice to plot the instantaneous line voltages v_{ac} and v_{bc}, and the instantaneous currents through diodes D_1, D_3, and D_5.

Plot the worst-case line current i_a if the resistances, inductances, and capacitances change by ±20%.

SOLUTION

The PSpice schematic is shown in Figure 7.29(a). The DC voltage V_y is used to measure the line current. The setup for worst case analysis is shown in Figure 7.29(b).

Peak phase voltages $V_m = 169.7$ V, and $f = 60$ Hz. The listing of the circuit file is as follows:

Three-Phase Bridge Rectifier with Source Inductances

```
SOURCE   ■    Van 1 0 SIN  (0 169.7V 60HZ)
CIRCUIT  ■■   L1 11 4 0.5MH
              Vbn 2  0 SIN  (0 169.7V 60HZ 0 0 120DEG)
              L2  2  5 0.5MH
              Vcn 3  0 SIN  (0 169.7V 60HZ 0 0 240DEG)
              Vy  1 11 DC
              L3  3  6 0.5MH
              R  7 9 0.5
              L  9 10 6.5MH
              VX 10  8 DC 0V ; Voltage source to measure the output
                               current
              D1  4  7 DMOD
              D3  5  7 DMOD
              D5  6  7 DMOD
```

(a)

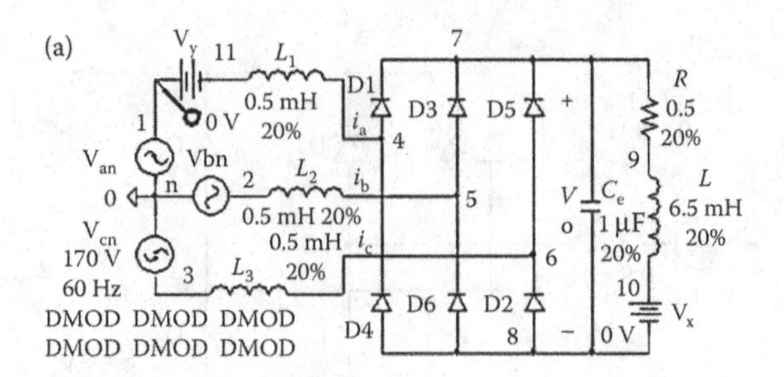

(b)

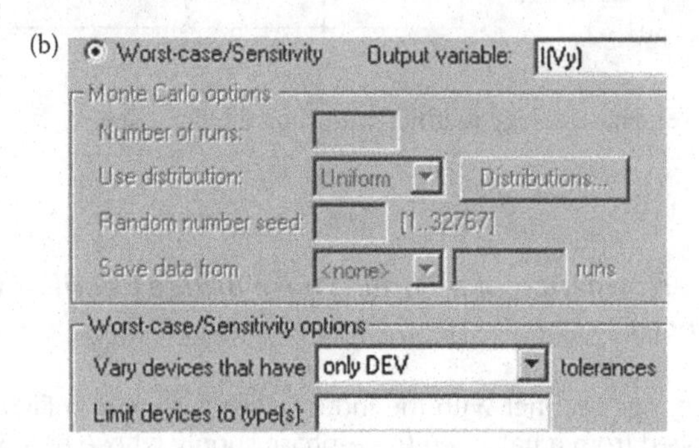

FIGURE 7.29 Three-phase bridge rectifier with line inductances for Example 7.6. (a) Schematic. (b) Worst case setup.

```
               D2 8 6 DMOD
               D6 8 5 DMOD
               D4 8 4 DMOD
               .MODEL DMOD D(IS=2.22E-15 BV=1200V IBV=13E-3
               CJO=2PF TT=1US)
ANALYSIS ■■■   .TRAN 10US 50MS 33.3333MS 10US ; Transient analysis
               .PROBE                         ; Graphics
                                                post-processor
               . OPTIONS ABSTOL = 0.0NRELTOL = 0.01VNTOL = 1.0M
                                    ITL5 = 10000 ; Convergence
.END
```

The PSpice plots of the instantaneous currents through diode D_1 I(D1), through diode D_3, I(D3), and diode D_5, I(D5), and the line voltages V(1,3) and V(2,3) are shown in Figure 7.30. Because of the source inductances, a commutation interval exists. During this interval, the current through the incoming diode rises, and that through the outgoing diode falls. The sum of these currents must equal the load current.

The worst case line current is shown in Figure 7.31.

Note: Because of the line inductances, there is a transition time for switching the line currents from one diode to another diode as shown in Figure 7.30. This causes a drop in the output voltage because of the commutation of the currents [1]. Due to the presence of the line inductances, there are no transient spikes in the line currents and the transition of current from one pair of diodes to another pair is slow and smooth.

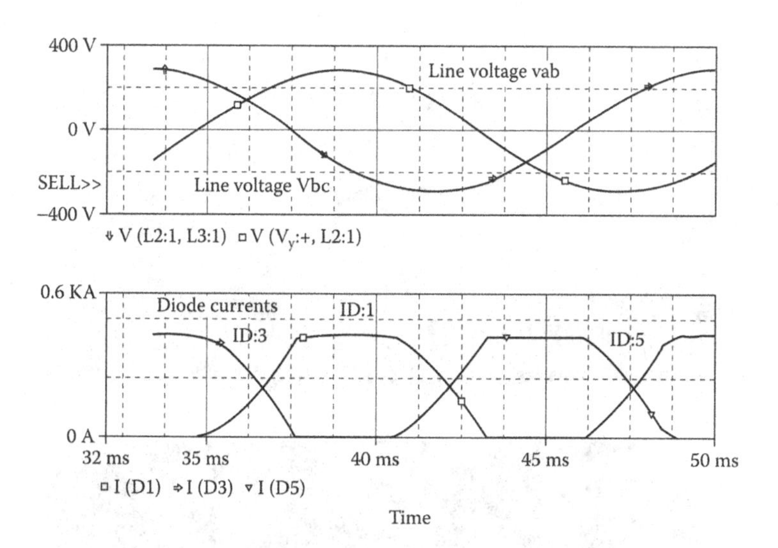

FIGURE 7.30 Plots of line voltages and diode currents for Example 7.6.

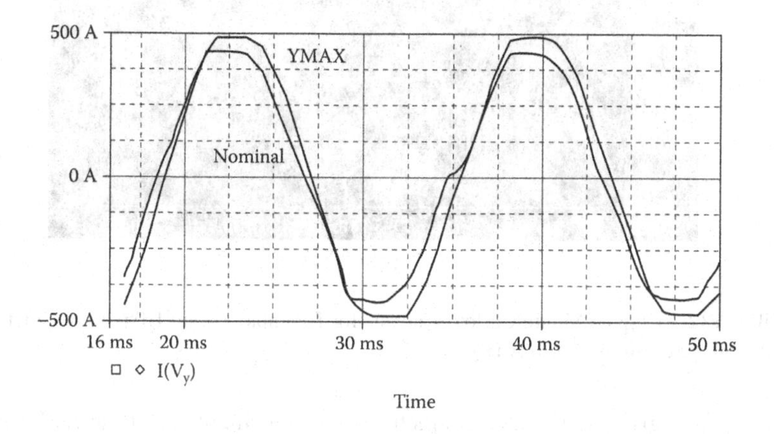

FIGURE 7.31 Worst case plot of the line current for Example 7.6.

7.7 LTspice MONTE CARLO ANALYSIS

The element values are described as variables having a nominal value (nom) and tolerance (tol), e.g., {mc(nom, tol)}. The user can specify the tolerance values as shown in Figure 7.32(a) or could define as a parameter, e.g., .paran tol = 0.1; 10%. The output voltage of Monte Carlos analysis is shown in Figure 7.32(b) for a diode rectifier shown in Figure 7.18(a) for 0 to 10 run with an increment of 4. Run is a dummy parameter to cycle Monte Carlo runs. First run gives the nominal values. Under the steady-state condition, the output voltage is a random variable depending on the run. For Monte-Carlo, the simulator will automatically change the components based on your Tolerance field. We need to run enough cases to get a statistically significant result, and we can calculate how many runs we need for a given confidence. Monte Carlo analysis randomly varies parameters (within their user-defined limits) for each new

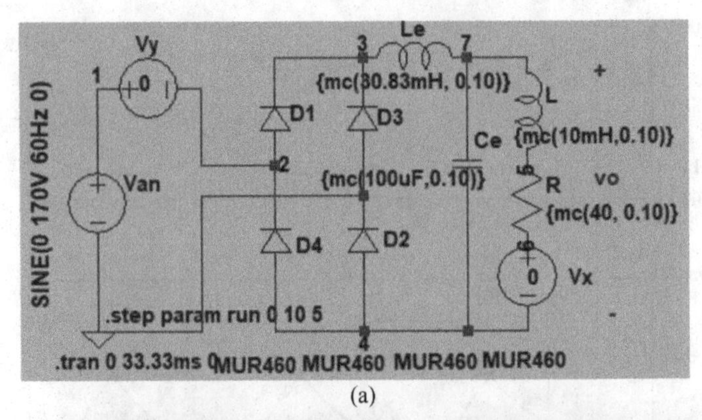

(a)

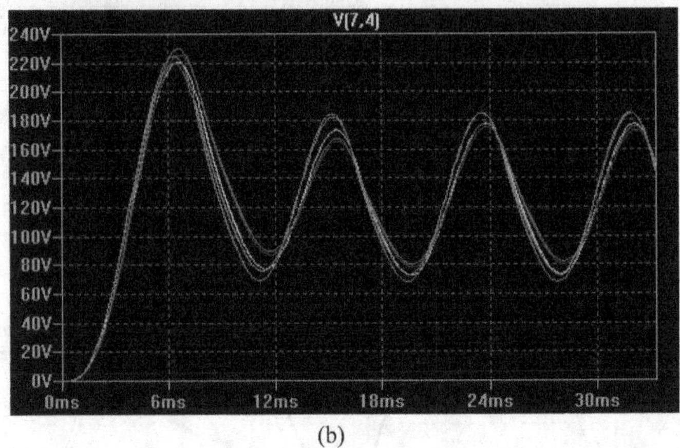

(b)

FIGURE 7.32 LTspice Monte Carlos Analysis for Example 7.4. (a) LTspice schematic. (b) Output voltage of Monte Carlo Analysis.

analysis run to give the user a useful picture of actual circuit performance. It can tell us what the performance is at these limits, if the circuit contains many components. But, it can take quite a lot of runs before its random selection of parameter values happens to simultaneously correspond to the worst-case limits of all of the components.

7.8 LTspice WORST CASE ANALYSIS

Similar to Monte Carlos analysis, the element values are described as variables having a nominal value (nom) and tolerance (tol), e.g., {wc(nom, tol)}. The user can specify the tolerance values as shown in Figure 7.33(a) or could define as a parameter, e.g., .paran tola = 0.1, paran tolb = 0.1, .paran tolc = 0.2. A function statement is necessary for each tolerance.

The function statement for worst case analysis for LTspice:

```
.function wc_x(nom, tola) if (run == 1, nom,
if(flat(1)>0,nom*(1+tolx),nom*(1-tolx)))
```

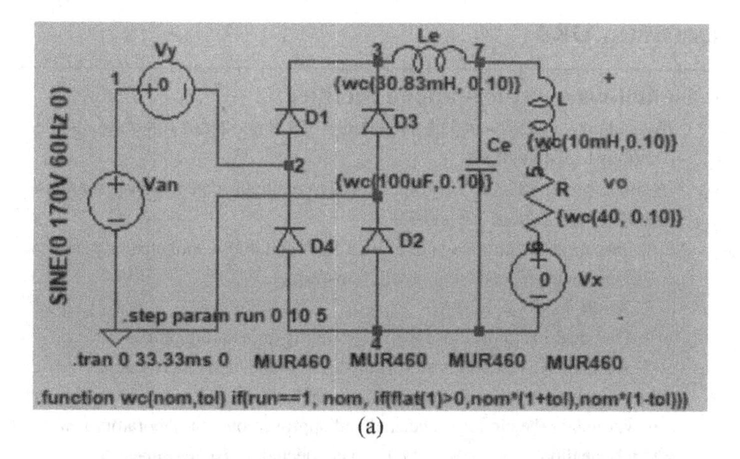

(a)

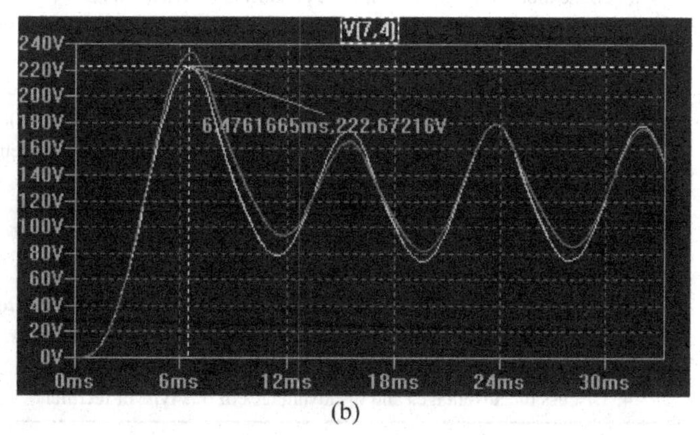

(b)

FIGURE 7.33 LTspice Worst Case Analysis for Example 7.4. (a) LTspoice schematic. (b) Output voltage of Worst case Analysis.

Since we have the same tolerance for all components, we need only one function statement as shown in Figure 3.33(a). The output voltage of worst case analysis is shown in Figure 7.33(b). Run is a dummy parameter to cycle the worst case runs. First run gives the nominal value. With the tolerance, the output voltage varies between the high and low values.

7.9 LABORATORY EXPERIMENTS

It is possible to develop many experiments for demonstrating the operation and characteristics of diode rectifiers. The following three experiments are suggested:

Single-phase full-wave center-tapped rectifier
Single-phase bridge rectifier
Three-phase bridge rectifier

7.9.1 Experiment DR.1

Single-phase full-wave center-tapped rectifier

Objective — To study the operation and characteristics of a single-phase full-wave rectifier under various load conditions

Applications — A single-phase full-wave rectifier is used as an input stage in power supplies, etc.

Textbook — See Ref. [1], Sections 3.8 and 3.9

Apparatus
1. Two diodes with ratings of at least 50 A and 400 V, mounted on heat sinks
2. One center-tapped (step-down) transformer
3. An *RL* load
4. One dual-beam oscilloscope with floating or isolating probes
5. AC and DC voltmeters and ammeters and one noninductive shunt

Warning — Before making any circuit connection, switch off the AC power. Do not switch on the power unless the circuit is checked and approved by your laboratory instructor. Do not touch the diode heat sinks, which are connected to live terminals

Experimental procedure
1. Set up the circuit as shown in Figure 7.34. Use the load resistance *R* only
2. Connect the measuring instruments as required
3. Observe and record the waveforms of the load voltage v_o and the load current i_o
4. Measure the average load voltage $V_{o(DC)}$, the RMS load voltage V_{oRMS}), the average load current $I_{o(DC)}$, the RMS load current $I_{O(RMS)}$, the RMS input current $I_{S(RMS)}$, the RMS input voltage $V_{S(RMS)}$, and the average load power P_L
5. Repeat Steps 2–4 with the load inductance *L* only
6. Repeat Steps 2–4 with both load resistance *R* and load inductance *L*

Report
1. Present all recorded waveforms and discuss all significant points
2. Compare the waveforms generated by SPICE with the experimental results, and comment
3. Compare the experimental results with the results predicted
4. Discuss the advantages and disadvantages of this type of rectifier

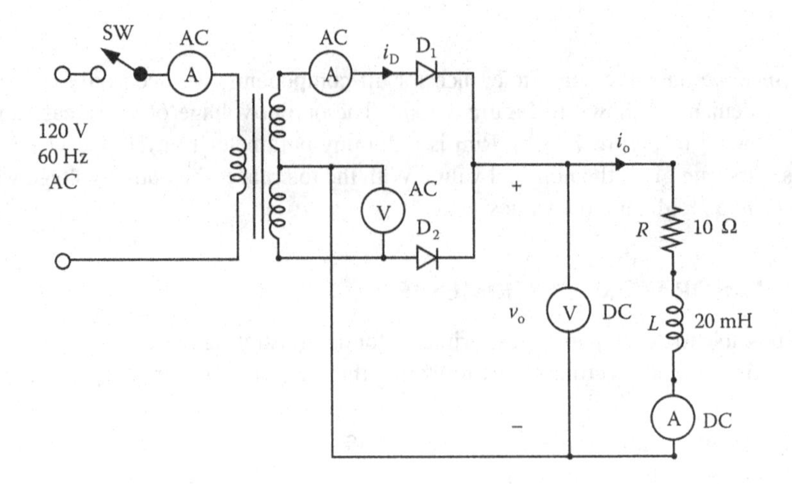

FIGURE 7.34 Single-phase full-wave rectifier.

7.9.2 Experiment DR.2

Single-phase bridge rectifier

Objective	To study the operation and characteristics of single-phase bridge-rectifier under various load conditions
Applications	A single-phase bridge rectifier is used as an input stage power supply, variable-speed AC/DC motor drives, etc.
Textbook	See Ref. [1], Sections 3.8 and 3.9
Apparatus	Same as Experiment DR.1, except that four diodes are required
Warning	See Experiment DR.1
Experimental procedure	Set up the circuit as shown in Figure 7.35 and follow the steps for Experiment fDR.1
Report	Repeat the steps of Experiment DR.1

7.9.3 Experiment DR.3

Three-phase bridge rectifier

Objective	To study the operation and characteristics of a three-phase bridge rectifier under various load conditions
Applications	A three-phase bridge rectifier is used as an input stage in variable-speed AC motor drives, etc.
Textbook	See Ref. [1], Sections 3.11 and 3.12
Apparatus	Same as Experiment DR.1, except that six diodes are required
Warning	See Experiment DR.1
Experimental procedure	Set up the circuit as shown in Figure 7.36 and follow the steps for Experiment DR.1
Report	Repeat the steps of Experiment DR.1

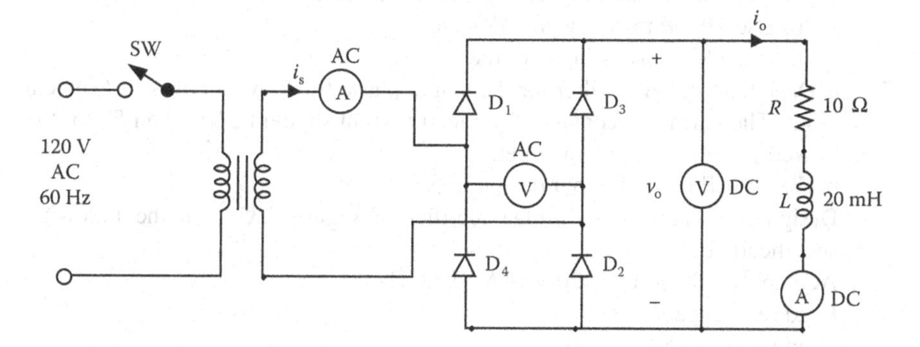

FIGURE 7.35 Single-phase bridge rectifier.

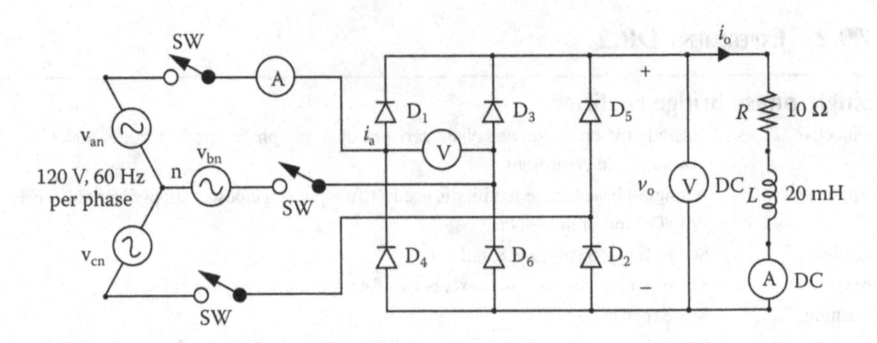

FIGURE 7.36 Three-phase bridge rectifier.

SUMMARY

The statements for diodes are as follows:

```
D<name> NA NK DNAME [(area) value].
MODEL DNAME TYPE(P1=B1 P2=B2 P3=B3 ... PN=BN)
```

I(D1)	Instantaneous current through diode D_1
RMS(I(D1))	RMS current of diode D_1
AVG(I(D1))	Average current of diode D_1
V(D1)	Instantaneous anode to cathode voltage of diode D_1

PROBLEMS

7.1 Design the single-phase full-wave rectifier of Figure 7.34 with the following specifications:
AC supply voltage $V_s = 120$ V (RMS), 60 Hz
Load resistance $R = 5\ \Omega$
Load inductance $L = 15$ mH
DC output voltage $V_{o(DC)} = 24$ V
  a. Determine the ratings of all components and devices.
  b. Use SPICE to verify your design.
  c. Provide a cost estimate of the circuit.

7.2 a. Design an output C filter for the single-phase full-wave rectifier of Problem 7.2. The harmonic content of the load current should be less than 5% of the average value without the filter.
  b. Use SPICE to verify your design in part (a).

7.3 Design the single-phase bridge rectifier of Figure 7.35 with the following specifications:
AC supply voltage $V_s = 120$ V (RMS), 60 Hz
Load resistance $R = 5\ \Omega$
Load inductance $L = 15$ mH
DC output voltage $V_{o(DC)} = 48$ V

a. Determine the ratings of all components and devices.
b. Use SPICE to verify your design.
c. Provide a cost estimate of the circuit.

7.4 a. Design an output LC filter for the single-phase bridge rectifier of Problem 7.4. The harmonic content of the load current should be less than 5% of the average value without the filter.
b. Use SPICE to verify your design in part (a).

7.5 a. Design an output C filter for the single-phase bridge rectifier of Problem 7.4. The harmonic content of the load voltage should be less than 5% of the value without the filter.
b. Use SPICE to verify your design in part (a).

7.6 The RMS input voltage to the single-phase bridge rectifier of Figure 7.35 is 120 V, 60 Hz, and it has an output LC filter. If the DC output voltage is V_{DC} = 48 V at I_{DC} = 25 A, determine the value of filter inductance L.

7.7 It is required to design the three-phase bridge rectifier of Figure 7.36 with the following specifications:
AC supply voltage per phase V_s = 120 V(RMS), 60 Hz
Load resistance $R = 5\ \Omega$
Load inductance L = 15 mH
DC output voltage $V_{o(DC)}$ = maximum possible value
a. Determine the ratings of all components and devices.
b. Use SPICE to verify your design.
c. Provide a cost estimate of the circuit.

7.8 a. Design an output LC filter for the three-phase bridge rectifier of Problem 7.7. The harmonic content of the load current should be less than 5% of the average value without the filter.
b. Use SPICE to verify your design in part (a).

7.9 a. Design an output C filter for the three-phase bridge rectifier of Problem 7.7. The harmonic content of the load voltage should be less than 5% of the average value without the filter.
b. Use SPICE to verify your design in part (a).

7.10 Repeat Example 7.2 with an antiparallel (or freewheeling) diode connected across terminals 2 and 0 of Figure 7.11. Complete Table P7.1.

TABLE P7.1

Rectifier	% THD of Input Current, THD_i	Input Power Factor, PF_i	% THD of Output Voltage, THD_{vo}	% THD of Load Current, THD_{io}
Half-wave				
Half-wave with a freewheeling diode				

7.11 Complete Table P7.2 for the single-phase half-wave rectifier in Figure 7.10, the single-phase bridge rectifier in Figure 7.13, and the three-phase bridge rectifier in Figure 7.20.

TABLE P7.2

Rectifier	% THD of Input Current, THD_i	Input Power Factor, PF_i	% THD of Output Voltage, THD_{vo}	% THD of Load Current, THD_{io}
Single-phase half-wave rectifier				
Single-phase bridge rectifier				
Three-phase bridge rectifier				

7.12 Complete Tables P7.3–P7.5 to find the effects of output filter capacitances on the performance of the single-phase half-wave rectifier in Figure 7.10, the single-phase bridge rectifier in Figure 7.13, and the three-phase bridge rectifier in Figure 7.20.

TABLE P7.3

Filter Capacitance Ce = 1 μF	% THD of Input Current, THD_i	Input Power Factor, PF_i	% THD of Output Voltage, THD_{vo}	% THD of Load Current, THD_{io}
Single-phase half-wave rectifier				
Single-phase bridge rectifier				
Three-phase bridge rectifier				

TABLE P7.4

Filter Capacitance Ce = 300 μF	% THD of Input Current, THD_i	Input Power Factor, PF_i	% THD of Output Voltage, THD_{vo}	% THD of Load Current, THD_{io}
Single-phase half-wave rectifier				
Single-phase bridge rectifier				
Three-phase bridge rectifier				

TABLE P7.5

Filter Capacitance Ce = 600 μF	% THD of Input Current, THD_i	Input Power Factor, PF_i	% THD of Output Voltage, THD_{vo}	% THD of Load Current, THD_{io}
Single-phase half-wave rectifier				
Single-phase bridge rectifier				
Three-phase bridge rectifier				

7.13 It is required to design a single-phase diode rectifier as shown in Figure 7.37 to provide an average output voltage to a load of $R = 10\ \Omega$ through a step-down input transformer of 10:1 turn ratio from a phase-input supply of $V_s = 120$ V (RMS).

1. Output parameters
    - Average output voltage V_{avg} =
    - RMS output voltage $V_{o(RMS)}$ =
    - Output RMS Power $P_{o(RMS)}$ =
    - Ripple factor of output voltage, $RF_{o(out)}$ =
    - Average output current $I_{o(av)}$ =
    - RMS output current $I_{o(RMS)}$ =
    - Output ripple voltage $V_{oripple}$=
    - Rectification efficiency, η =
2. Diode parameters
    - Peak diode current $I_{D(peak)}$ =
    - Average diode current $I_{D(av)}$ =
    - RMS diode current $I_{D(av)}$ =
    - Diode Peak Reverse voltage V_{peak} =
3. Input Supply parameters
    - Average input current $I_{s(av)}$ =
    - RMS input current $I_{s(RMS)}$ =
    - Input power factor PF_i =
    - Transformer utilization factor TUF

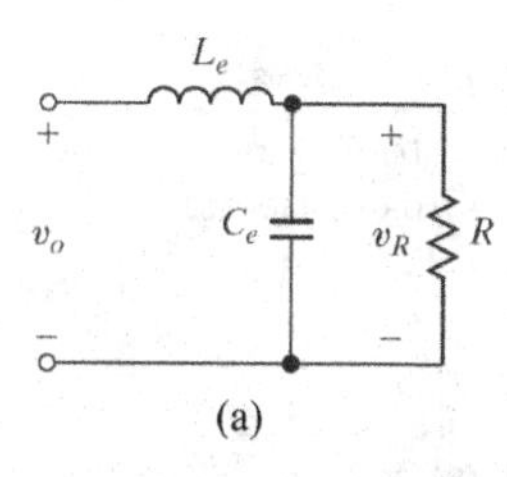

(a)

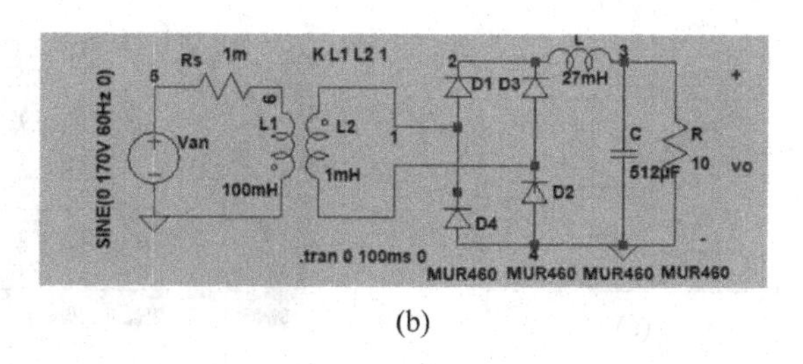

(b)

FIGURE 7.37 Single-phase rectifier with an output filter. (a) LC filter. (b) Diode rectifier with an output filter.

4. Neglecting the effect of inductor L_e, find the approximate value of C_e to keep the peak-to-peak ripple voltage to 5% of the average output voltage V_{avg} =
5. Neglecting the effect of C_e, find the approximate value of L_e to keep the peak-to-peak current to 5% of the average output current $I_{o(RMS)}$ =
6. Determine the values of L_e and C_e so that the ripple factor $RF_{o(out)}$ of the output voltage is ≤ 5% of the average output voltage.
7. Use Multisim/LTspice/Orcad to run the simulation from 0 to 50 ms so that the steady-state condition is reached.
8. Make modifications to meet the specifications if necessary.
   a. Plot the instantaneous output voltage $v_o(t)$ across the load, between the terminals 3 and 4.
   b. Plot the instantaneous current through diode D_1, $i_{D1}(t)$
   c. Plot the instantaneous current through supply $i_s(t)$, that is, -I(van)
   d. Include .four command to find the THD of the outcome current with 29th harmonic contents at the output frequency of 120Hz.
9. Complete Table P7.6 for simulated values.

TABLE P7.6

	V_{avg} (V)	$V_{o(RMS)}$ (V)	$V_{o(ripple)}$ (V)	% $RF_{o(output)}$
Calculated				
Simulated				

10. Estimated component costs: https://www.jameco.com/
11. Safety considerations if you would be building the product:
12. Risk factors of the design if you would be building the product:
13. What trade-offs have you considered?
14. Lessons Learned from the design assignment:

7.14 Design of a Three-Phase Diode Rectifier

It is required to design a three-phase diode rectifier as shown in Figure 7.38 to provide an average output voltage to a load of R = 10 Ω from a Y-connected 3-phase input supply of phase voltage V_s = 120 V (rms).

1. Output parameters
   - Average output voltage V_{avg} =
   - RMS output voltage $V_{o(RMS)}$ =

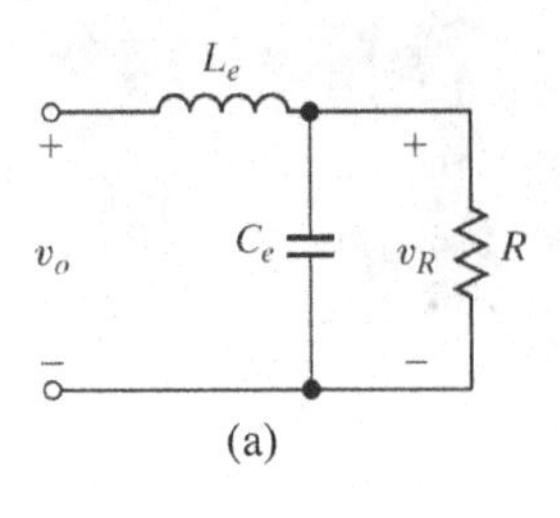

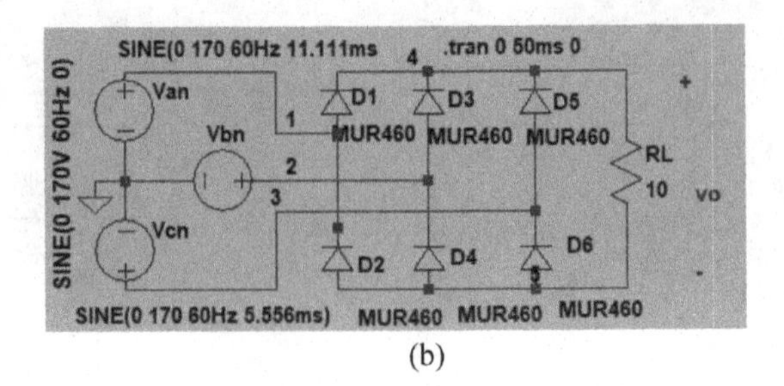

FIGURE 7.38 Three-phase rectifier. (a) LC filter. (b) Diode rectifier with an output filter.

- Output RMS Power $P_{o(RMS)}$ =
- Ripple factor of output voltage, $RF_{o(out)}$ =
- Average output current $I_{o(av)}$ =
- RMS output current $I_{o(RMS)}$ =
- Output ripple voltage $V_{oripple}$ =
- Rectification efficiency, η =

2. Diode parameters
   - Peak diode current $I_{D(peak)}$ =
   - Average diode current $I_{D(av)}$ =
   - RMS diode current $I_{D(av)}$ =
   - Diode Peak Reverse voltage V_{peak} =
3. Input supply parameters
   - Average input current $I_{s(av)}$ =
   - RMS input current $I_{s(RMS)}$ =
   - Input power factor PF_i =
   - Transformer utilization factor TUF
4. Neglecting the effect of inductor L_e, find the approximate value of C_e to keep the peak-to-peak ripple voltage to 5% of the average output voltage V_{avg} =
5. Neglecting the effect of $C_{e,}$ find the approximate value of L_e to keep the peak-to-peak current to 5% of the average output current $I_{o(RMS)}$ =
6. Determine the values of L_e and C_e so that the ripple factor $RF_{o(out)}$ of the output voltage is ≤ 5% of the average output voltage.
7. Use Multisim/LTspice/Orcad to run the simulation from 0 to 50 ms so that the steady-state condition is reached.
8. Make modifications to meet the specifications if necessary.
   a. Plot the instantaneous output voltage $v_o(t)$ across the load, between terminals 3 and 4.
   b. Plot the instantaneous current through diode D_1, $i_{D1}(t)$
   c. Plot the instantaneous current through supply $i_s(t)$, that is, -I(van)
   d. Include .four command to find the THD of the outcome current with 29th harmonic contents at the output frequency of 360 Hz.
9. Complete Table P7.7 for simulated values.

TABLE P7.7

	V_{avg} (V)	$V_{o(RMS)}$ (V)	$V_{o(ripple)}$ (V)	% $RF_{o(output)}$
Calculated				
Simulated				

10. Estimated component costs: https://www.jameco.com/
11. Safety considerations if you would be building the product:
12. Risk factors of the design if you would be building the product:
13. What trade-offs have you considered?
14. Lessons learned from the design assignment.

SUGGESTED READING

1. M.H. Rashid, *Power Electronics: Circuit, Devices and Applications*, Third Edition, Englewood Cliffs, NJ: Prentice-Hall, 2003, Chapters 2 and 3.
2. M.H. Rashid, *Introduction to PSpice Using OrCAD for Circuits and Electronics*, Third Edition, Englewood Cliffs, NJ: Prentice-Hall, 2003, Chapter 7.
3. M.H. Rashid. *SPICE for Power Electronics and Electric Power*, Englewood Cliffs, NJ: Prentice-Hall, 1993.
4. P. Antognetti, *Power Integrated Circuits*, New York: McGraw-Hill, 1986.
5. MicroSim Corporation, *PSpice Manual*, Irvine, CA, 1988.
6. The data sheets for electronic components http://www.alldatasheet.com/
7. P. Antognetti, and G. Massobri, *Semiconductor Device Modeling with SPICE*, New York: McGraw-Hill Book Co., 1988.
8. A. Laha, and D. Smart, A Zener diode model with application to SPICE2, *IEEE Journal of Solid-State Circuits*, *SC-16*(1), 21–22, 1981.
9. M. H. Rashid, *Introduction to PSpice Using OrCAD for Circuits and Electronics*, Englewood, N. J.: Prentice-Hall, Inc., 2003 (3/e).
10. M. H. Rashid, *Electronics Analysis and Design Using Electronics Workbench*, Boston, MA: PWS Publishing, 1997.
11. M. H. Rashid, *Microelectronic Circuits: Analysis and Design*, Boston, MA: Cengage Publishing, 2017.
12. M. H. Rashid, *SPICE for Power Electronics and Electric Power*, Boca Raton, FL: CRC Pres, 2012 (2/e).
13. *LTspice IV Manual*, Milpitas, CA: Linear Technology Corporation, 2013.
14. PSpice Models from Vendors. http://www.pspice.com/models/links.asp
15. Books on PSpice. http://www.pspice.com/publications/books.asp
16. R. Jacob Baker, *SPICE Software, MOSFET Models, and MOSIS Information. CMOS Circuit Design, Layout, and Simulation*, Third Edition, New York, NY: Wiley-IEEE Press, 2010.

8 DC–DC Converters

After completing this chapter, students should be able to do the following:

- Model BJTs, MOSFETs, and IGBTs in SPICE and specify their model parameters.
- Model PWM control in SPICE and as a hierarchy block.
- Perform transient analysis of DC–DC converters.
- Evaluate the performance of DC–DC converters.

8.1 INTRODUCTION

In a DC–DC converter, both input and output voltages are DC. It uses a power semiconductor device as a switch to turn on and off the DC supply to the load. The switching action can be implemented by a BJT, a MOSFET, or an IGBT. A DC–DC converter with only one switch is often known as a DC chopper.

A DC-to-DC converter converts a fixed direct current (DC) voltage to a variable direct current (DC) voltage. It uses normally one transistor acting as a switch and one or more diodes to transfer energy from the source to the load. The amount of energy transfer depends on the switching on-time of the transistor commonly known as the duty cycle of the transistor switch. The converter is normally operated under closed-loop control to vary the duty cycle to give the desired output voltage. Thus, the voltage and the current on the input side and the output side are of DC types. Table 8.1 lists the performance parameters of DC-to-DC converter.

TABLE 8.1
DC–DC Converter Performance Parameters

Input Side Parameters	**Output Side Parameters**
DC input voltage, V_S	Average output voltage, V_{dc}
Average input current, I_S	Average output current, I_{dc}
RMS input current, I_{rms}	The output dc power, $P_{dc} = V_{dc} I_{dc}$
Converter Power Efficiency, η	RMS output voltage, V_{rms}
The rms ripple content of the input current $I_{ripple} = \sqrt{I_{rms}^2 - I_S^2}$	RMS output current, I_{rms}
	The output ac power, $P_{ac} = V_{rms} I_{rms} = I_{rms}^2 R_L$ for a load resistance of R_L
	The rms ripple content of the output voltage $V_{ac} = \sqrt{V_{rms}^2 - V_{dc}^2}$
	The *ripple factor, RF* $= \frac{V_{ac}}{V_{dc}}$
	The rectification *efficiency or ratio,* $\eta = \frac{P_{dc}}{P_{ac}}$
	The *form factor, FF* $= \frac{V_{rms}}{V_{dc}}$

DOI: 10.1201/9781003284451-8

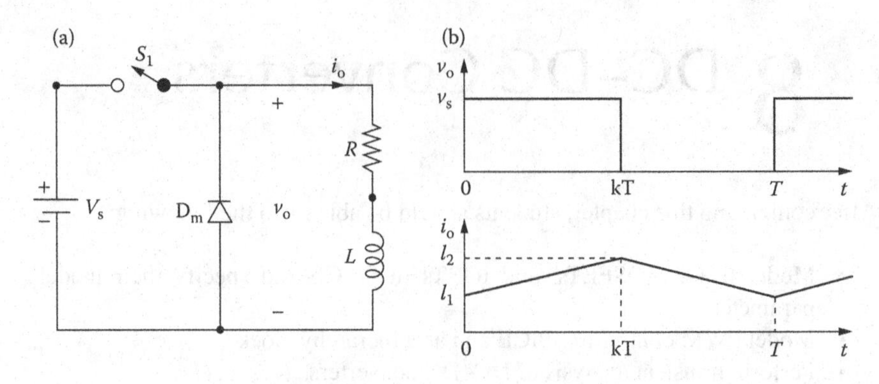

FIGURE 8.1 DC switch chopper. (a) Circuit. (b) Output voltage and current.

8.2 DC SWITCH CHOPPER

A chopper switch is shown in Figure 8.1(a). If switch S_1 is turned on, the supply voltage V_S is connected to the load. If the switch is turned off, the inductive load current i_o is forced to flow through diode D_m. The output voltage and load current are shown in Figure 8.1(b). As the output voltage is turned on and off by turning switch S_1 on and off, the load current rises from I_1 to I_2 and then falls to I_1 under steady-state conditions. The parameters of the switch can be adjusted to model the voltage drop of the chopper. We use the switch parameters RON = 1M, ROFF = 10E+6, VON = 1V, and VOFF = 0V and the diode parameters IS = 2.22E–15, BV = 1200V, CJO = 0F, and TT = 0. The diode parasitics are neglected; however, they will affect the transient behavior.

Note: The modeling of voltage-controlled and current-controlled switches is covered in Section 5.7.

Example 8.1:

Finding the Performance of a Step-Down DC–DC Converter with a Voltage-Controlled Switch

A DC chopper switch is shown in Figure 8.2(a). The DC input voltage is V_S = 220 V. The load resistance R is 5 Ω, and the load inductance L = 7.5 mH. The chopping frequency is f_o = 1 kHz, and the duty cycle of the chopper is k = 50%. The control voltage is shown in, Figure 8.2(b). Use PSpice to (a) plot the instantaneous output voltage v_o, the load current i_o, and the diode current i_{Dm}, (b) calculate the Fourier coefficients of the load current i_o, and (c) calculate the Fourier coefficients of the input current i_s.

SOLUTION

a. The DC supply voltage V_S = 220 V, k = 0.5, f_o = 1 kHz, $T = 1/f_o$ = 1 ms, and $t_{on} = k \times T = 0.5 \cdot 1$ ms = 0.5 ms.

   The corresponding PSpice schematic is shown in Figure 8.3(a). Comparing a triangular signal Vref with a carrier signal Vcr generates a PWM waveform. The PWM generator is implemented as a descending

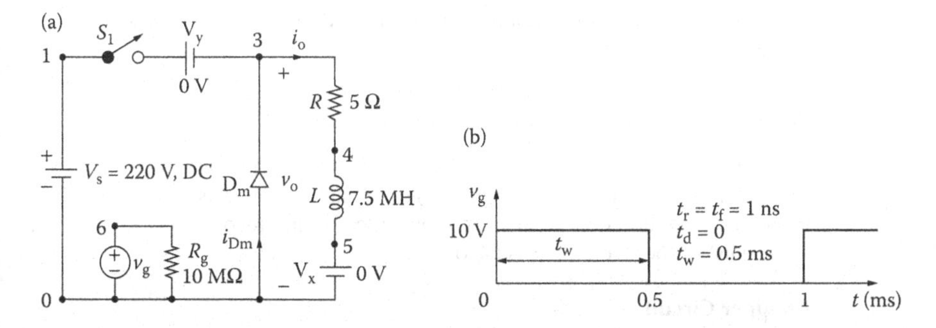

FIGURE 8.2 DC chopper for PSpice simulation. (a) Circuit. (b) Gate voltage.

hierarchy as shown in Figure 8.3(b). An ABM2 device that compares the two signals produces a square wave output between 0 and 1 V to drive the voltage control switch. Since we are comparing a DC voltage of 0–1 V with a triangular reference signal with a peak of 1 V, the assigned DC voltage corresponds to the duty cycle. Varying the voltage V_Duty_Cycle can vary the duty cycle of the switch and the average

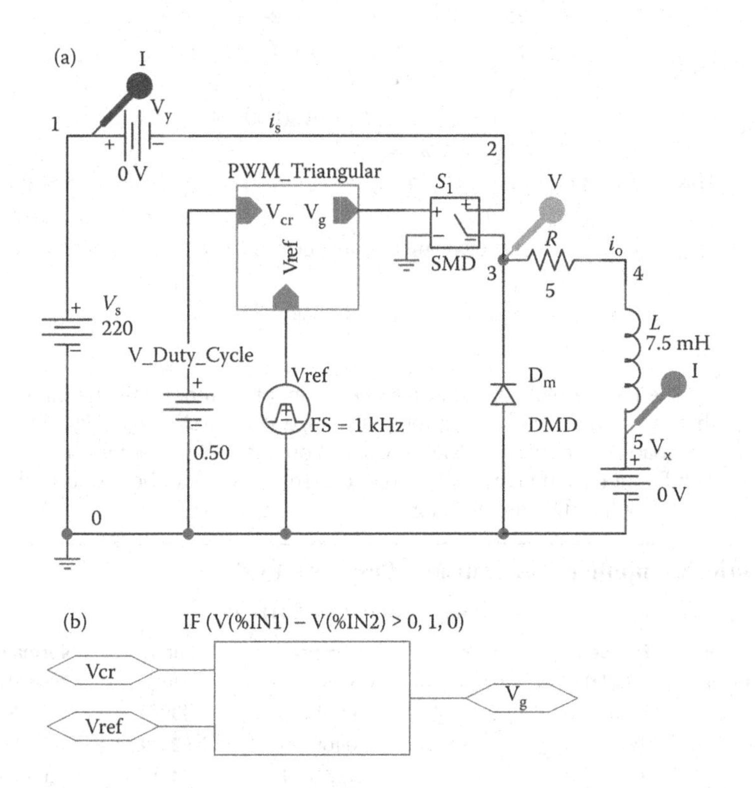

FIGURE 8.3 PSpice schematic for Example 8.1. (a) Schematic. (b) Descending hierarchy comparator.

output voltage. The model parameters for the switch and the freewheeling diode are as follows:

```
.MODEL SMD VSWITCH (RON=1M ROFF=10E6 VON=1V VOFF=0V)
.MODEL DMD D(IS=2.22E-15 BV=1200V CJO=1PF TT=0US))
```

Note: The descending hierarchy blocks can be assigned directly from the edit menu. V_x and V_y, as shown in Figure 8.3(a), are used as ammeters to monitor the load current and the input side current, respectively.

The list of the circuit files is as follows:

Chopper Circuit

```
SOURCE    ■    VS 1 0 DC 220V
               Vg 6 0 PULSE (0V 10V 0 1NS 1NS 0.5MS 1MS)
               Rg 6 0 10MEG
CIRCUIT   ■■   R  3 4 5
               L  4 5 7.5MH
               VX 5 0 DC 0V ; Load battery voltage
               VY 2 3 DC 0V ; Voltage source to measure
               chopper current
               DM 0 3 DMOD               ; Freewheeling diode
               .MODEL DMOD D(IS=2.22E-15 BV=1200V
               CJO=0PF TT=0)        ; Diode model
               S1 1 2 6 0 SMOD           ; Switch
               .MODEL SMOD VSWITCH (RON=1M ROFF=10E+6
               VON=1V VOFF=0V)
ANALYSIS  ■■■  .TRAN 1US 10MS 8MS        ; Transient analysis
               .PROBE                    ; Graphics post-processor
               .OPTIONS ABSTOL = 1.00N RELTOL = 0.01 VNTOL = 0.1
                                               ITL5 = 40000
               .FOUR 1KHZ I(VX) I(VY)    ; Fourier analysis
        .END
```

The PSpice plots of the instantaneous input current I(VY), the current through diode I(DM), and the output voltage V(3) are shown in Figure 8.4. The load current rises when the switch is on and falls when it is off.

b. The Fourier coefficients of the load current, which can be found in the PSpice output file, are as follows:

Fourier Components of Transient Response I(vx)

DC Component = 2.189331E+01

Harmonic Number	Frequency (Hz)	Fourier Component	Normalized Component	Phase (Deg)	Normalized Phase (Deg)
1	1.000E+03	2.955E+00	1.000E+00	–8.439E+01	0.000E+00
2	2.000E+03	2.378E–03	8.046E–04	–1.523E+02	–6.789E+01
3	3.000E+03	3.344E–01	1.132E–01	–8.743E+01	–3.040E+00
4	4.000E+03	5.029E–03	1.702E–03	–1.868E+01	6.570E+01
5	5.000E+03	1.147E–01	3.882E–02	–9.128E+01	–6.896E+00

(Continued)

Fourier Components of Transient Response I(vx) *(Continued)*

DC Component = 2.189331E+01

Harmonic Number	Frequency (Hz)	Fourier Component	Normalized Component	Phase (Deg)	Normalized Phase (Deg)
6	6.000E+03	5.937E–03	2.009E–03	1.115E+02	1.959E+02
7	7.000E+03	6.417E–02	2.171E–0201	–9.042E+01	–6.036E+00
8	8.000E+03	2.062E–03	6.979E–0401	–1.602E+02	–7.578E+01
9	9.000E+03	3.747E–02	1.268E–0201	–9.349E+01	–9.108E+00

Total harmonic distortion = 1.222766E+01%

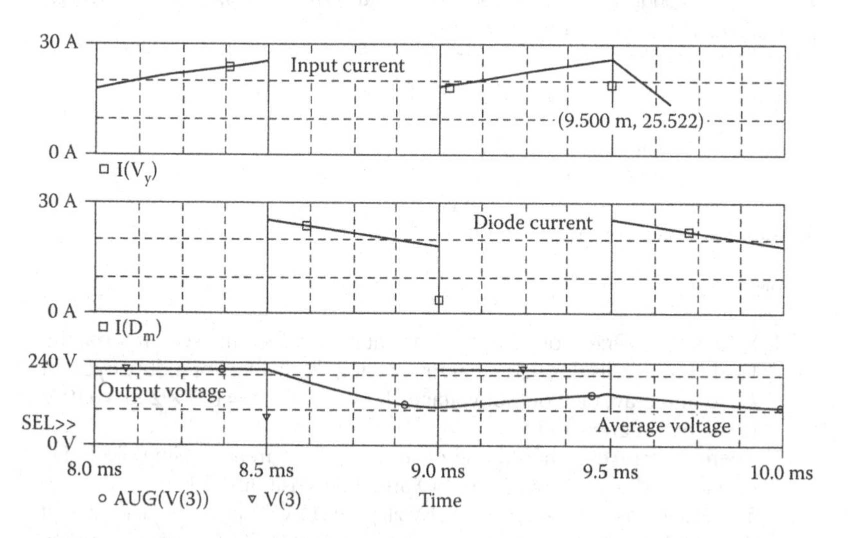

FIGURE 8.4 Plots for Example 8.1.

From the Fourier components of the load current, we get

DC component current = 21.89 A
Peak fundamental component at 1 kHz = 2.95 A
Fundamental phase delay = –84.39°
Total harmonic distortion = 12.22%

c. The Fourier coefficients of the input current, which can be found in the PSpice output file, are as follows:

Fourier Component of Transient Response I(vy)

DC Component = 1.113030E+01

Harmonic Number	Frequency (Hz)	Fourier Component	Normalized Component	Phase (Deg)	Normalized Phase (Deg)
1	1.000E+03	1.414E+01	1.000E+00	–5.237E+00	0.000E+00
2	2.000E+03	1.167E+00	8.254E–02	1.739E+02	1.792E+02
3	3.000E+03	4.652E+00	3.289E–01	1.861E–01	5.423E+00

(Continued)

Fourier Component of Transient Response I(vy) ***(Continued)***

DC Component = 1.113030E+01

Harmonic Number	Frequency (Hz)	Fourier Component	Normalized Component	Phase (Deg)	Normalized Phase (Deg)
4	4.000E+03	6.127E–01	4.333E–02	1.647E+02	1.699E+02
5	5.000E+03	2.784E+00	1.969E–01	2.486E+00	7.722E+00
6	6.000E+03	4.410E–01	3.118E–02	1.563E+02	1.615E+02
7	7.000E+03	1.995E+00	1.411E–01	4.250E+00	9.487E+00
8	8.000E+03	3.622E–01	2.561E–02	1.509E+02	1.561E+02
9	9.000E+03	1.552E+00	1.098E–01	5.976E+00	1.121E+01

Total harmonic distortion = 4.350004E+01%

From the Fourier components of the load current, we get

DC component current = 11.13 A
Peak fundamental component at 1 kHz = 14.14 A
Fundamental phase delay = –5.23°
Total harmonic distortion = 43.5%

Notes:

1. When the switch is on, the load current rises. When the switch is off, the load current falls through the freewheeling diode D_m. For a duty cycle of $k = 0.5$, the average output voltage [1] is $V_{o(av)} = kV_s = 0.5 \times 220 = 110$ V (PSpice also gives 110 V).
2. There is a spike at the turning off instant of the freewheeling diode D_m due to the diode capacitance CJO and the transit time TT.
3. The instantaneous average output voltage AVG(V(3)) has a higher value at the start. It will take a certain time before arriving at a final steady-state value.

LTspice: LTspice schematic for Example 8.1 is shown in Figure 8.5(a) and the output voltage and the load current are shown in Figure 8.5(b) for a duty cycle of 50%. The component attribute is shown as

Attribute	Value	Vis.
Prefix	S	
InstName	S1	X
SpiceModel		
Value	smod	X
Value2		

The general statement of a Voltage Controlled Switch is as follows:

```
Symbol Names: SW
Syntax: Sxxx n1 n2 nc+ nc- <model> [on,off]
```

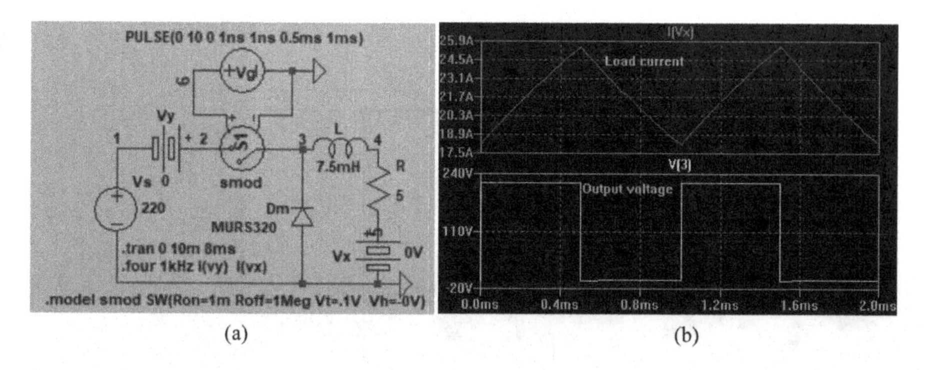

FIGURE 8.5 Ltspice schematic for Examples 8.1. (a) LTspice schematic. (b) Output voltage and load current.

Example:

```
S1 out 0 in 0 MySwitch
.model MySwitch SW(Ron=.1 Roff=1Meg Vt=0 Vh=-.5 Lser=10n Vser=.6)
```

where Ron – On resistance, and Roff – Off resistance, Vt – Threshold voltage defaulty to 0, and Vh – Hysteresis voltage, default to 0

8.3 BJT SPICE MODEL

SPICE generates a complex model of BJTs. The model equations that are used by SPICE are described in Refs. [1,2]. If a complex model is not necessary, many model parameters can be ignored by the users, and PSpice assigns default values to them. The PSpice model, which is based on the integral charge-control model of Gummel and Poon [1,2], is shown in Figure 8.6(a). The static (DC) model that is generated by PSpice is shown in Figure 8.6(b).

The model statement for NPN transistors has the general form

```
.MODEL QNAME NPN (P1=B1 P2=B2 P3=B3 ... PN=BN)
```

and the general form for PNP transistors is

```
.MODEL QNAME PNP (P1=B1 P2=B2 P3=B3 ... PN=BN)
```

where QNAME is the name of the BJT model. NPN and PNP are the type symbols for NPN and PNP transistors, respectively. QNAME, which is the model name, can begin with any character, and its word size is normally limited to eight characters. P1, P2, … and B1, B2, … are the parameters and their values, respectively. Table 8.2 shows the model parameters of BJTs. If certain parameters are not specified, PSpice assumes the simple model of Ebers and Moll [3], which is shown in Figure 8.6(c).

The area factor is used to determine the number of equivalent parallel BJTs of the model specified. The model parameters, which are affected by the area factor,

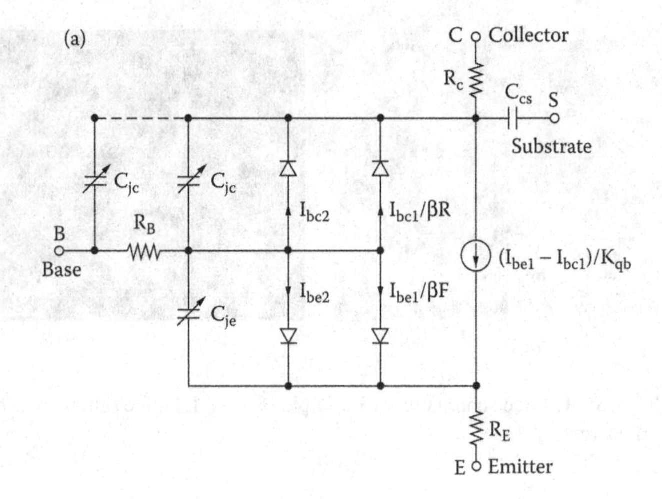

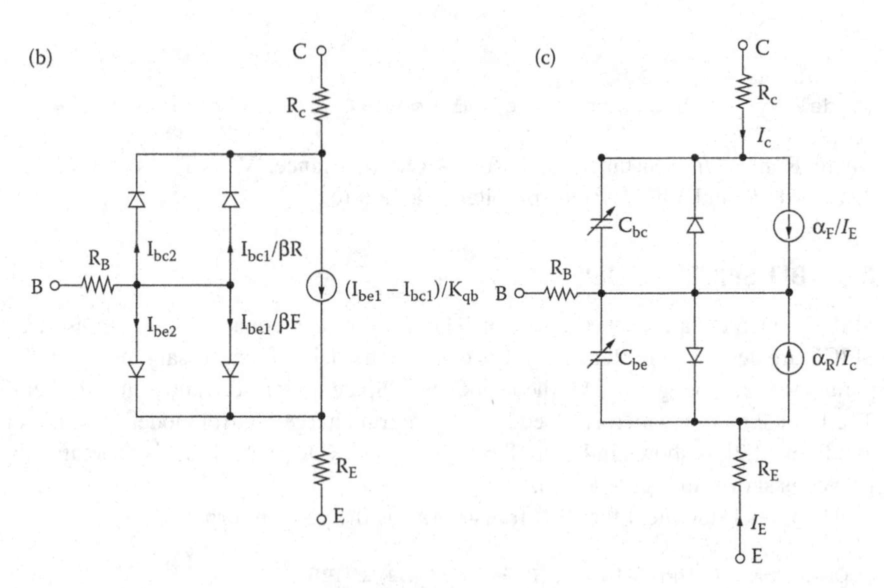

FIGURE 8.6 PSpice BJT model. (a) Gummel and Poon model. (b) DC model. (c) Ebers and Moll model.

are marked with an asterisk (*) in Table 8.2. A bipolar transistor is modeled as an intrinsic transistor with ohmic resistances in series with the collector (RC/area), the base (RB/area), and the emitter (RE/area). `[(area) value]` is the relative device area, defaults to 1. For those parameters that have alternative names such as VAF and VA (the alternative name is indicated in parentheses), either name may be used.

The parameters ISE (C2) and ISC (C4) may be set to be greater than 1. In this case, they are interpreted as multipliers of IS instead of absolute currents, that is, if ISE > 1, it is replaced by ISE*IS, and similarly for ISC. The DC model is defined by (1) parameters BF, C2, IK, and NE, which determine the forward current gain,

TABLE 8.2
Model Parameters of BJTs

Name	Area	Model Parameter	Unit	Default	Typical
IS	*	*p–n* Saturation current	A	1E–16	1E–16
BF		Ideal maximum forward beta		100	100
NF		forward current emission coefficient		1	1
VAF(VA)		forward early voltage	V	•	100
IKF(IK)		Corner for forward beta high-current roll-off	A	•	10M
ISE(C2)		Base–emitter leakage saturation current	A	0	1000
NE		Base–emitter leakage emission coefficient		1.5	2
BR		Ideal maximum reverse beta		1	0.1
NR		Reverse current emission coefficient		1	
VAR(VB)		Reverse early voltage	V	•	100
IKR	*	Corner for reverse beta high-current roll-off	A	•	100M
ISC(C4)		Base–collector leakage saturation current	A	0	1
NC		Base–collector leakage emission coefficient		2	2
RB	*	Zero-bias (maximum) base resistance	W	0	100
RBM		Minimum base resistance	W	RB	100
IRB		Current at which RB falls halfway to RBM	A	•	
RE	*	Emitter ohmic resistance	W	0	1
RC	*	Collector ohmic resistance	W	0	10
CJE	*	Base-emitter zero-bias *p–n* capacitance	F	0	2P
VJE(PE)		Base–emitter built-in potential	V	0.75	0.7
MJE(ME)		Base–emitter *p–n* grading factor	0.33	0.33	
CJC	*	Base–collector zero-bias *p–n* capacitance	F	0	1P
VJC(PC)		Base-collector built-in potential	V	0.75	0.5
MJC(MC)		Base–collector *p–n* grading factor	0.33	0.33	
XCJC		Fraction of C_{bc} connected internal to R_b		1	
CJS(CCS)		Collector–substrate zero-bias *p–n* capacitance	F	0	2PF
VJS(PS)		Collector–substrate built-in potential	V	0.75	
MJS(MS)		Collector–substrate *p–n* grading factor		0	
FC		forward-bias depletion capacitor coefficient		0.5	
TF		Ideal forward transit time	s	0	0.1NS
XTF		Transit-time bias dependence coefficient		0	
VTF		Transit-time dependency on V_{bc}	V	•	
ITF		Transit-time dependency on I_c	A	0	
PTF		Excess phase at $1/(2\pi \times \text{TF})$ Hz	degree	0	30°
TR		Ideal reverse transit time	s	0	10NS
EG		Band-gap voltage (barrier height)	eV	1.11	1.11
XTB		forward and reverse beta temperature coefficient		0	
XTI(PT)		IS temperature effect exponent		3	
KF		Flicker noise coefficient		0	6.6E–16
AF		Flicker noise exponent		1	1.2

(2) BR, C4, IKR, and VC, which determine the reverse current gain characteristics, (3) VA and VB, which determine the output conductance for forward and reverse regions, and (4) the reverse saturation current IS.

Base-charge storage is modeled by (1) forward and reverse transit times TF and TR, and nonlinear depletion-layer capacitances, which are determined by CJE, PE, and ME for a base-emitter junction, and (2) CJC, PC, and MC for a base-collector junction. CCS is a constant collector–substrate capacitance. The temperature dependence of the saturation current is determined by the energy gap EG and the saturation current temperature exponent PT.

The parameters that affect the switching behavior of a BJT are the most important ones for power electronics applications: IS, BF, CJE, CJC, and TF. The symbol of a bipolar junction transistor (BJT) is Q. The name of a bipolar transistor must start with Q and it takes the general form

```
Q <name> NC NB  NE NS QNAME [(area)  value]
```

where NC, NB, NE, and NS are the collector, base, emitter, and substrate nodes, respectively. QNAME could be any name of up to eight characters. The substrate node is optional: if not specified, it defaults to ground. Positive current is the current that flows into a terminal, that is, the current flows from the collector node, through the device, to the emitter node for an NPN BJT. That is, the sum of all currents must equal to zero, $I_E + I_B + I_C = 0$.

8.4 BJT PARAMETERS

The data sheet for power transistor 2N6546 is shown in Figure 8.7. SPICE parameters are not available from the data sheet. Some versions of SPICE (e.g., PSpice) support device library files. The software PARTS of PSpice can generate SPICE models from the data sheet parameters of transistors and diodes. Although PSpice allows one to specify many parameters, we shall use only those parameters that significantly affect the output of a power converter [8,9]. From the data sheet we get

$$I_{C(\text{rated})} = 10\text{A}$$
$$V_{BE} = 0.8\text{V} \quad \text{at } I_C = 2\text{A}$$

Assuming that $n = 1$ and $V_T = 25.8$ mV, we can apply the diode in Equation 7.1 to find the saturation current I_s:

$$I_C = I_s(e^{V_{BE}/nV_T} - 1) \tag{8.1}$$
$$2 = I_s(e^{0.8/(25.8\times10^{-3})} - 1)$$

which gives I_s = 2.33E–27 A. DC current gain at 10 A is h_{FE} = 6 to 30. Taking the geometric mean gives $\text{BF} = \text{BR} = \sqrt{6\times30} \approx 13$.

(a)

SWITCHMODE SERIES
NPN SILICON POWER TRANSISTORS

The 2N6546 and 2N6547 transistors are designed for high-voltage, high-speed, power switching in inductive circuits where fall time is critical. They are particularly suited for 115 and 220 volt line operated switch-mode applications such as:

- Switching regulators
- PWM inverters and motor controls
- Solenoid and relay drivers
- Defelection circuits

Specification features:-
High temperature performance specified for:
Reversed biased SOA with inductive loads
Stitching times with inductive loads
Saturation voltages
Leakage currents

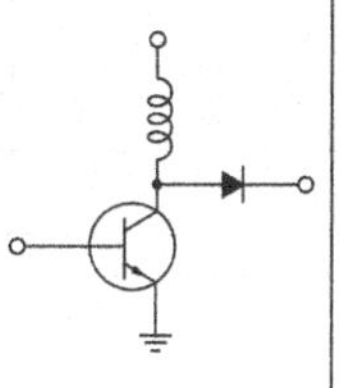

15 AMPERE
NPN SILICON
POWER TRANSISTORS

300 and 400 VOLTS
175 WATTS

Designer's data for
"Worst Case" Conditions

The designers data sheet permits the design of most cirucuits entirely from the information presented. Limit data–representing device characteristics boundaries–are given to facilitate "worst case" design

*MAXIMUM RATINGS

Rating	Symbol	2N6546	2N6547	Unit
Collector-emitter voltage	$V_{CEO(sus)}$	300	400	Vdc
Collector-emitter voltage	$V_{CEX(sus)}$	350	450	Vdc
Collector-emitter voltage	V_{CEV}	650	850	Vdc
Collector base voltage	V_{EB}	9.0		Vdc
Collector current – Continuous – Peak (1)	I_C I_{CM}	15 30		Adc
Base current – Continuous – Peak (1)	I_B I_{BM}	10 20		Adc
Emitter current – Continuous – Peak (1)	I_E I_{EM}	25 50		Adc
Total power dissipation @ T_C = 25°C @ T_C = 100°C Derate above 25°C	P_D	175 100 1.0		Watts W/°C
Operating and storage junction temperature range	T_J, T_{stg}	–65 to +200		°C

THERMAL CHARACTERISTICS

characteristic	Symbol	Max	Unit
Thermal resistance, junction to case	$R_{\theta JC}$	1.0	°C/W
Maximum lead temperature for soldering purposes: 1/8" from case for 5 seconds	T_L	275	°C

*Indicates JEDEC registered data
(1) Pulse test: Pulse width = 5.0 ms, Duty cycle ≤ 10%

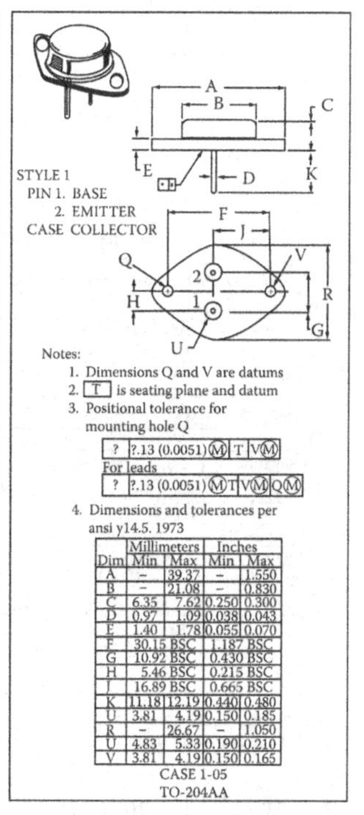

Dim	Millimeters Min	Millimeters Max	Inches Min	Inches Max
A	–	39.37	–	1.550
B	–	21.08	–	0.830
C	6.35	7.62	0.250	0.300
D	0.97	1.09	0.038	0.043
E	1.40	1.78	0.055	0.070
F	30.15 BSC		1.187 BSC	
G	10.92 BSC		0.430 BSC	
H	5.46 BSC		0.215 BSC	
J	16.89 BSC		0.665 BSC	
K	11.18	12.19	0.440	0.480
U	3.81	4.19	0.150	0.185
R	–	26.67	–	1.050
U	4.83	5.33	0.190	0.210
V	3.81	4.19	0.150	0.165

CASE 1-05
TO-204AA

FIGURE 8.7 Data sheet for transistor 2N6546. (a) Datasheet. (b) Datasheet (continued). (c) Characteristics. (d) Characteristics (continued). (Courtesy of Motorola, Inc.). *(Continued)*

The input capacitance at the base-emitter junction is very small, and its typical value is 0.2 to 1 pF. Let us assume that C_{je} = CJE = 1 pF. Output capacitance, C_{obo} = 125 to 500 pF at $V_{CB} = 10$ V, $I_E = 0$ (reverse biased). Taking the geometric mean gives $C_{obo} = C_{\mu} = \sqrt{125 \times 500} \approx 250$ pF. $C_{\mu o}$ can be found from [19]

$$C_{\mu} = \frac{C_{\mu o}}{(1 + V_{CB}/V_0)^m} \tag{8.2}$$

where $m = \frac{1}{3}$ and $V_0 = 0.75$ V. From Equation 8.2, $C_{\mu o}$ = CJC = 607.3 pF at $V_{CB} = 10$ V.

(b) *ELECTRICAL CHARACTERISTICS (T_C = 25°C unless otherwise noted.)

Characteristic		Symbol	Min	Max	Unit
OFF CHARACTERISTICS (1)					
Collector-Emitter sustaining voltage (I_C = 100 mA, I_B = 0)	2N6546 2N6547	$V_{CEO(sus)}$	300 400	— —	Vdc
Collector-Emitter sustaining voltage (I_C = 8.0 A, V_{clamp} = Rated V_{CEX}, T_C = 100°C) (I_C = 15 A, V_{clamp} = Rated V_{CEO} − 100 V, T_C = 100°C)	2N6546 2N6547 2N6546 2N6547	$V_{CEX(sus)}$	350 450 200 300	— — — —	Vdc
Collector cutoff current (V_{CEV} = Rated value, $V_{BE(off)}$ = 1.5 Vdc) (V_{CEV} = Rated value, $V_{BE(off)}$ = 1.5 Vdc, T_C = 100°C)		I_{CEV}	— —	1.0 4.0	mAdc
Collector cutoff current (V_{CE} = Rated V_{CEV}, R_{BE}= 50 Ω, T_C = 100°C)		I_{CER}	—	5.0	mAdc
Emitter cutoff current (V_{EB} = 9.0 Vdc, I_C = 0)		I_{EBO}	—	1.0	mAdc
SECOND BREAKDOWN					
Second breakdown collector current with base forward biased t = 1.0 s (non-repetitive) (V_{CE} = 100 Vdc)		$I_{S/b}$	0.2	—	Adc
ON CHARACTERISTICS (1)					
DC current gain (I_C = 5.0 Adc, V_{CE} = 2.0 Vdc) (I_C = 10 Adc, V_{CE} = 2.0 Vdc)		h_{FE}	12 6.0	60 30	—
Collector-Emitter saturation voltage (I_C = 10 Adc, I_B = 2.0 Adc) (I_C = 15 Adc, I_B = 3.0 Adc) (I_C = 10 Adc, I_B = 2.0 Adc, T_C = 100°C)		$V_{CE(sat)}$	— — —	1.5 5.0 2.5	Vdc
Base-Emitter saturation voltage (I_C = 10 Adc, I_B = 2.0 Adc) (I_C = 10 Adc, I_B = 2.0 Adc, T_C = 100°C)		$V_{BE(sat)}$	— —	1.6 1.6	Vdc
DYNAMIC CHARACTERISTICS					
Current-gain–Bandwidth product (I_C = 500 mAdc, V_{CE} = 10 Vdc, f_{test} = 1.0 MHz)		f_T	6.0	28	MHz
Output capacitance (V_{CB} = 10 Vdc, I_E = 0, f_{test} = 1.0 MHz)		C_{ob}	125	500	pF
SWITCHING CHARACTERISTICS					
Resistive load					
Delay time	(V_{CC} = 250 V, I_C = 10 A, I_{B1} = I_{B2} = 2.0 A, t_p = 100 µs, Duty cycle ≤ 2.0%)	t_d	—	0.05	µs
Rise time		t_r	—	1.0	µs
Storage time		t_s	—	4.0	µs
Fall time		t_f	—	0.7	µs
Inductive load, clamped					
Storage time	(I_C = 10 A(pk), V_{clamp} = Rated V_{CEX}, I_{B1} = 2.0 A, $V_{BE(off)}$ = 5.0 Vdc, T_C = 100°C)	t_s	—	5.0	µs
Fall time		t_f	—	1.5	µs
			Typical		
Storage time	(I_C = 10 A(pk), V_{clamp} = Rated V_{CEX}, I_{B1} = 2.0 A, $V_{BE(off)}$ = 5.0 Vdc, T_C = 25°C)	t_s	2.0		µs
Fall time		t_f	0.09		µs

*Indicates JEDEC registered data
(1) Puse test: Pulse width = 300 µs, Duty cycle = 2%.

FIGURE 8.7 *(Continued)*

The transition frequency $f_{T(min)}$ = 6 MHz at V_{CE} = 10 V, I_C = 500 mA. The transition period is $\tau_T = 1/2\pi f_T = 1/(2\pi \cdot 6\text{ MHz})$ = 26,525.8 ps. Thus, $V_{CB} \approx V_{CE} - V_{BE}$ = 10–0.7 = 9.3 V, and Equation 8.2 gives C_μ = 255.7 pF. The transconductance g_m is given by [19]

$$g_m = \frac{I_C}{V_T} = \frac{500\text{mA}}{25.8\text{mV}} = 19.38\text{A/V} \tag{8.3}$$

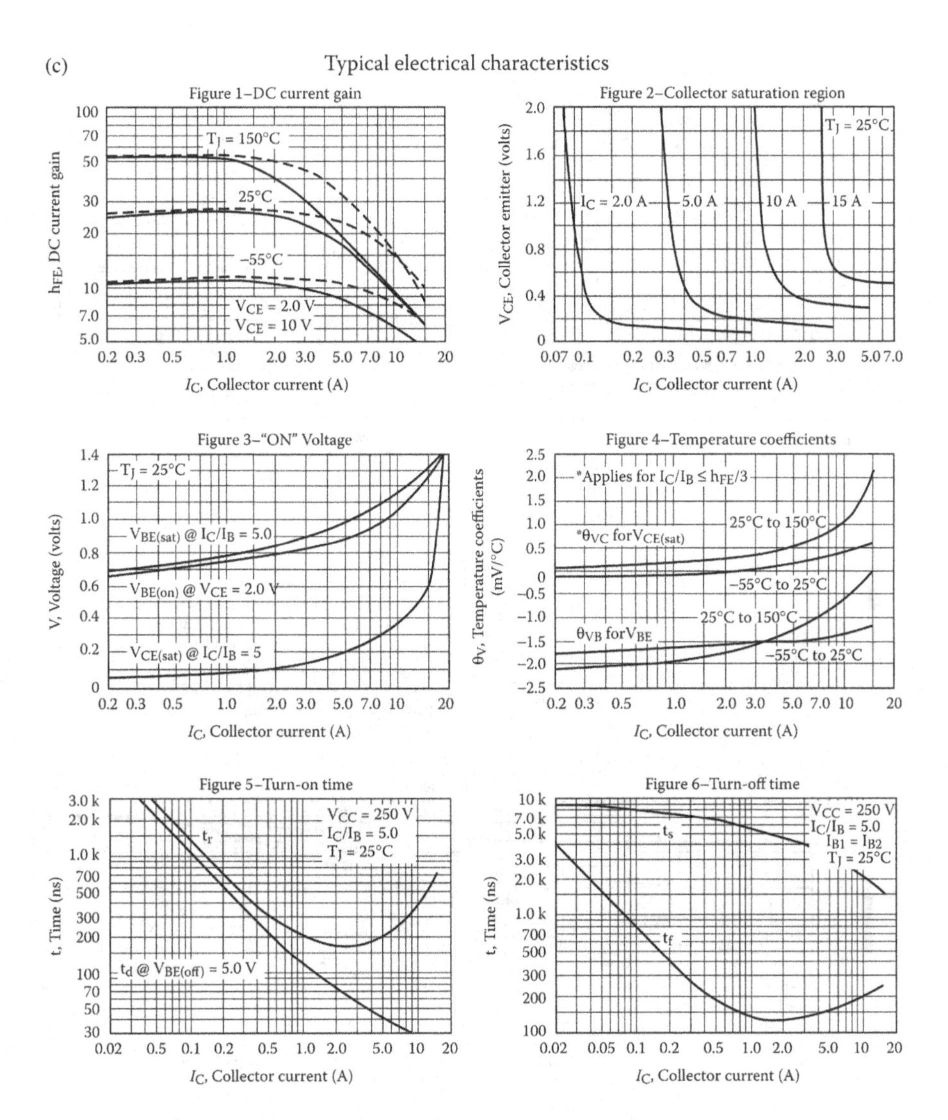

FIGURE 8.7 *(Continued)*

The transition period τ_T is related to forward transit time τ_F by

$$\tau_T = \tau_F + \frac{C_{Je}}{g_m} + \frac{C_n}{g_m} \tag{8.4}$$

or

$$26{,}525.8\,\text{ps} = \tau_F + \frac{1\text{pF}}{19.38} + \frac{255.7\text{pF}}{19.38}$$

(d)

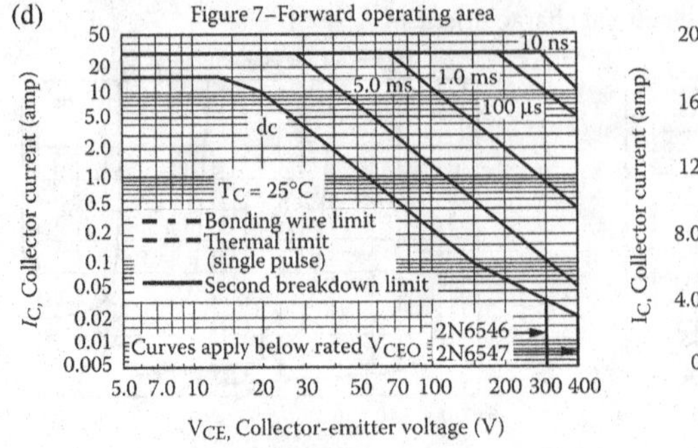

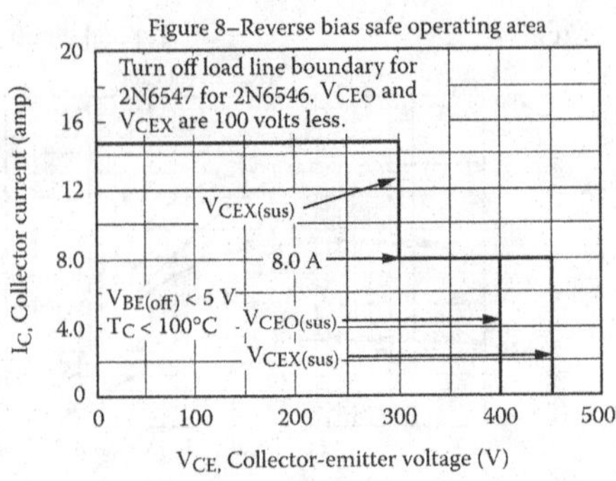

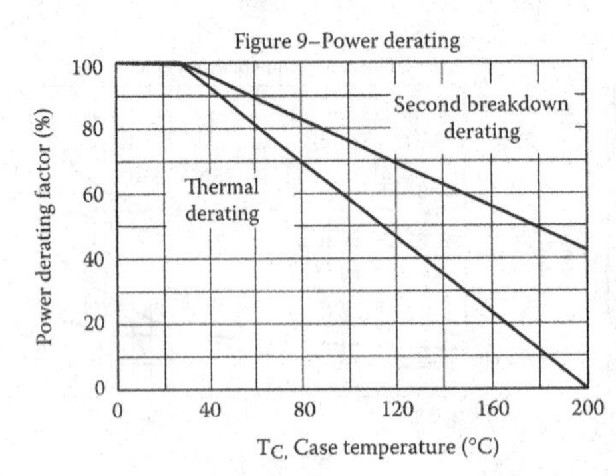

There are two limitations on the power handling ability of a transistor: Average junction temperature and second breakdown. Safe operating area curves indicate I_C–V_{CE} limits of the transistor that must not be subjected to greater dissipation than the curves indicate.

The data of figure 7 is based on T_C = 25°C; $T_{J(pk)}$ is variable depending on power level. Second breakdown pulse limits are valid for duty cycles to 10% but must be derated when $T_C \geq 25$°C. Second breakdown limitations do not derate the same as thermal limitations. Allowable current at the voltages shown on figure 7 may be found at any case temperature by using the appropriate curve on figure 9.

$T_{J(pk)}$ may be calculated from the data in figure 10. At high case temperatures, thermal limitations will reduce the power that can be handled to values less than the limitations imposed by second breakdown.

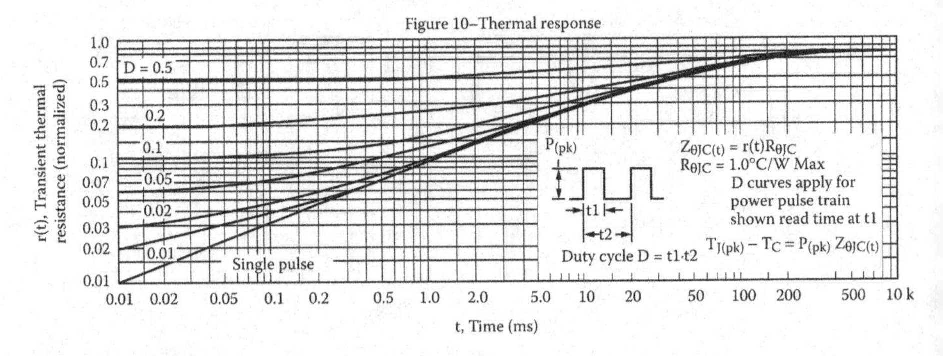

FIGURE 8.7 *(Continued)*

which gives $\tau_F = 26{,}512.6$ ps. Thus, the PSpice model statement for transistor 2N6546 is

```
.MODEL 2N6546 NPN (IS=6.83E-14 BF=13 CJE=1 PF CJC=607.3
PF TF=26.5NS)
```

This model can be used to plot the characteristics of the MOSFET. It may be necessary to modify the parameter values to conform to the actual characteristics.

Note: It is often necessary to adjust the base resistance RB or base (control) voltage V_g so that the transistor is driven into saturation.

LTspice BJT Characteristics: The schematic for plotting the output characteristics of a BJT is shown in Figure 8.8(a). A BJT is a current-dependent device, and

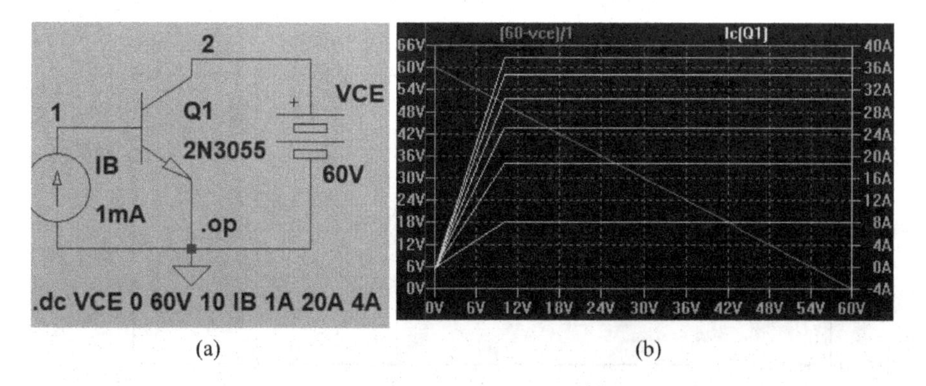

(a) (b)

FIGURE 8.8 LTspice schematic BJT characteristics. (a) BJT biasing. (b) BJT output characteristics.

its collector current depends on the base current. As the base current is increased, the collector current is also increased. As the collector-emitter voltage increases, the collector current increases until the collector current reaches the saturation region and remains practically constant. For operating the BJT as a switching device, its base current must be high enough to keep the collector-emitter low for a specific load line as shown in Figure 8.8(b) for a load resistance of 1 ohm.

8.5 EXAMPLES OF BJT DC-DC CONVERTERS

The applications of the SPICE BJT model are illustrated by some examples.

Example 8.2:

Finding the Performance of a Step-Down DC–DC Converter with a BJT Switch

A BJT buck chopper is shown in Figure 8.9(a). The DC input voltage is V_S = 12 V. The load resistance R is 5 Ω. The filter inductance is L = 145.84 μH, and the filter capacitance is C = 200 μF. The chopping or switching frequency is f_c = 25 kHz, and the duty cycle of the chopper is k = 42%. The control voltage is shown in Figure 8.9(b). Use PSpice to (a) plot the instantaneous load current i_o, the input current i_s, the diode voltage v_D, and the output voltage v_C and (b) calculate the Fourier coefficients of the input current i_s. Plot the frequency response of the converter output voltage from 10 kHz to 10 MHz and find the resonant frequency.

SOLUTION

The DC supply voltage $V_S = 12$ V, $k = 0.42$, $f_c = 25$ kHz, $T = 1/f_c = 40$ μs, and $t_{on} = k \times T = 16.7$ μs. An initial value for C is assigned to reach steady state faster.

The PSpice schematic is shown in Figure 8.10. A voltage-controlled voltage source with a voltage gain of 30 drives the BJT switch. The PWM generator is

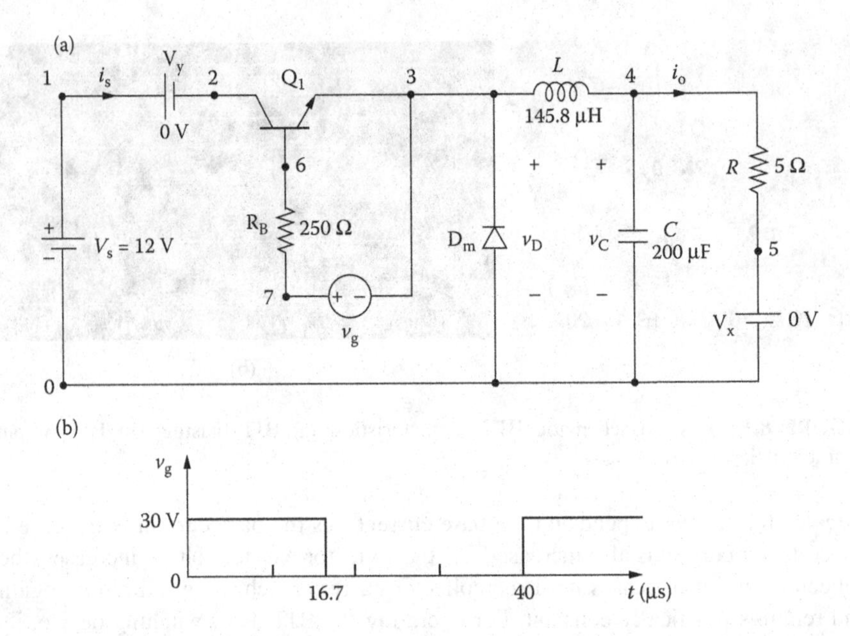

FIGURE 8.9 BJT buck chopper for PSpice simulation. (a) Circuit (b) Control voltage.

implemented as a descending hierarchy as shown in Figure 8.3(b). Note that E1 in Figure 8.10 provides the isolation between the power circuit and the low signal gating circuit. The model parameters for the BJT switch and the freewheeling diode are as follows:

```
.MODEL QMOD NPN(IS=6.83E-14 BF=13 CJE=1pF CJC=607.3PF TF=2 6.5NS)
.MODEL DMD D(IS=2.22E-15  BV=1200V CJO=1PF TT=0US))
```

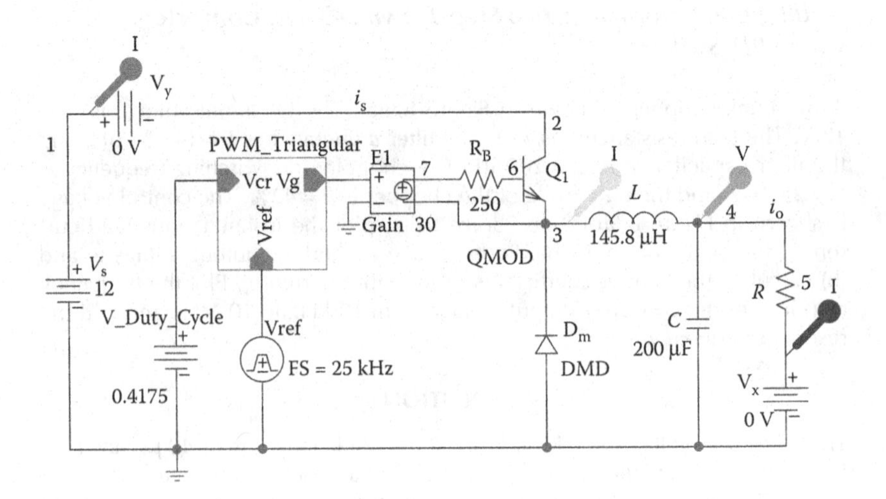

FIGURE 8.10 PSpice schematic for Example 8.2.

The list of the circuit file is as follows:

BJT Buck Chopper

```
SOURCE    ■     VS 1 0 DC 12V
                VY 1 2 DC 0V ; Voltage source to measure
                input current
                Vg 7 3 PULSE (0V 30V 0 0.1NS 0.1NS
                16.7US 40US)
CIRCUIT   ■■    RB  7 6 250 ; Transistor base resistance
                R   4 5 5
                L   3 4 145.8UH
                C   4 0 200UF IC=0V ; Initial voltage
                VX  5 0 DC 0V       ; To measure load current
                DM  0 3 DMOD        ; Freewheeling diode
                .MODEL DMOD D(IS=2.22E-15 BV=1200V CJO=0 TT=0)
                Q1  2 6 3 3 2N6546 ; BJT switch
                .MODEL 2N6546 NPN (IS=6.83E-14 BF=13 CJE=1PF
                CJC=607.3PF TF=26.5NS)
ANALYSIS  ■■■   .TRAN  0.1US 2.1MS 2MS UIC  ; Transient analysis
                .PROBE                      ; Graphics post-processor
                .OPTIONS ABSTOL =1.00N RELTOL = 0.01
                VNTOL = 0.1 ITL5 = 40000
                .FOUR 25KHZ I(VY) ; Fourier analysis
          .END
```

Note the following:

1. The PSpice plots of the instantaneous load current I(VX), the input current I(VY), the diode voltage V(3), and the output voltage V(5) are shown in Figure 8.11(a). The inductor current rises and falls with a minimum and a maximum value, whereas the capacitor voltage varies from a negative to a positive value, that is, the capacitor charges and discharges.
2. The Fourier coefficients of the input current, which can be found in the PSpice output file, are as follows:

Fourier Components of Transient Response I(Vy)

DC Component = 3.938953E–01

Harmonic Number	Frequency (Hz)	Fourier Component	Normalized Component	Phase (Deg)	Normalized Phase (Deg)
1	2.500E+04	5.919E–01	1.000E+00	1.792E+02	0.000E+00
2	5.000E+04	2.021E–01	3.415E–01	–1.106E+02	–2.898E+02
3	7.500E+04	1.545E–01	2.611E–01	–1.260E+02	–3.053E+02
4	1.000E+05	1.299E–01	2.195E–01	–6.211E+01	–2.413E+02
5	1.250E+05	6.035E–02	1.020E–01	–6.698E+01	–2.462E+02
6	1.500E+05	9.647E–02	1.630E–01	–2.055E+01	–1.998E+02

(Continued)

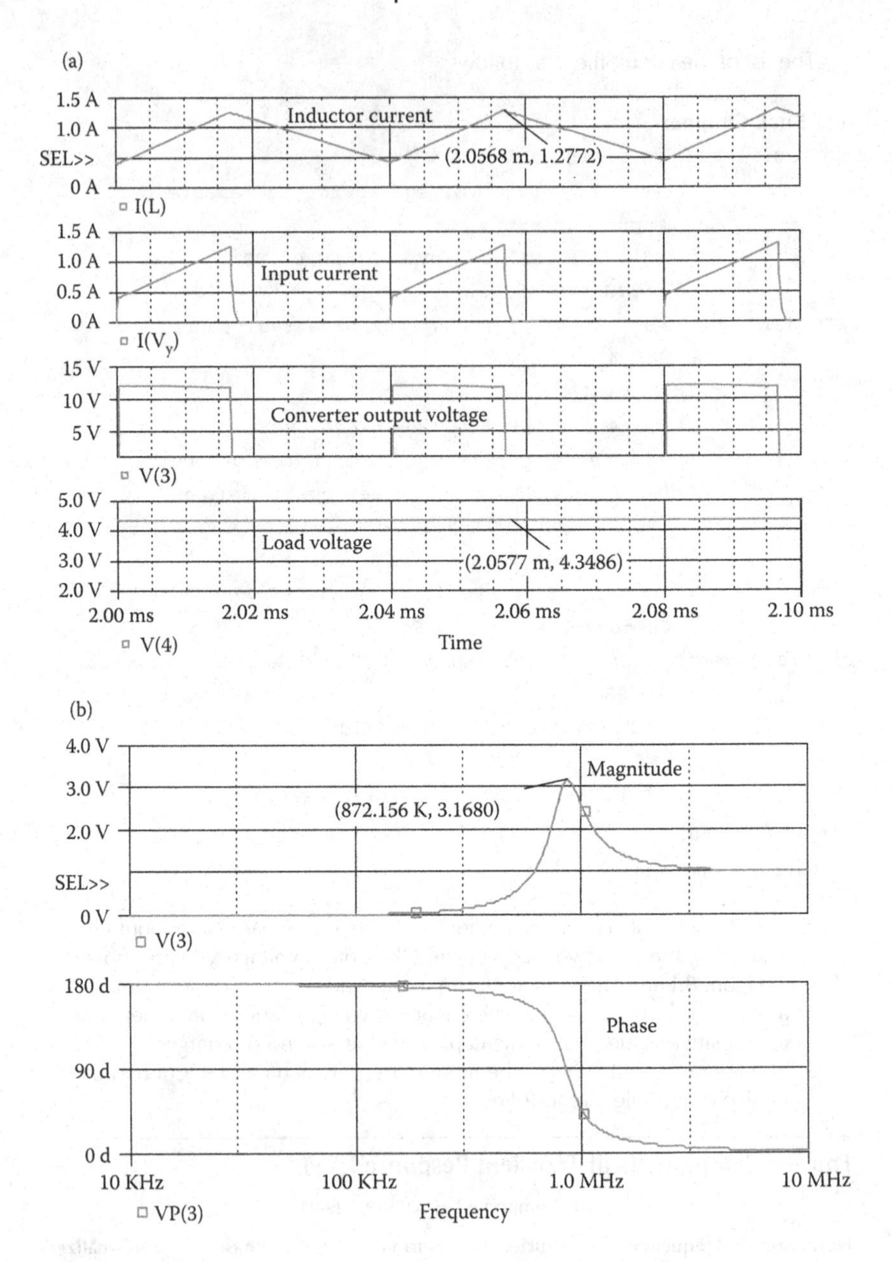

FIGURE 8.11 Plots for Example 8.2. (a) Transient plots. (b) Frequency response.

Fourier Components of Transient Response I(Vy) ***(Continued)***

DC Component = 3.938953E–01

Harmonic Number	Frequency (Hz)	Fourier Component	Normalized Component	Phase (Deg)	Normalized Phase (Deg)
7	1.750E+05	4.065E–02	6.869E–02	1.750E+01	–1.617E+02
8	2.000E+05	6.771E–02	1.144E–01	1.779E+01	–1.614E+02
9	2.250E+05	4.600E–02	7.772E–02	7.940E+01	–9.984E+01

Total harmonic distortion = 5.420088E+01%

From the Fourier component, we get

DC component current = 0.3938 A
Peak fundamental component at 250 kHz = 0.5919 A
Fundamental phase delay = 179.2°(expected 180°)
Total harmonic distortion = 54.2%

Figure 8.11(b) gives the resonant frequency of f_{res} = 872.15 Hz at a resonant peak gain of $A_{o(peak)}$ = 3.16.

Note: The base resistance RB is needed to limit the base drive current of the BJT switch. The *LC* output filter is needed to reduce the ripples on the load voltage and current. for a duty cycle of $k = 0.42$, the average voltage at the output of the converter [12] is $V_{o(av)} = kV_s = 0.42 \times 12 = 5.04$ V.

LTspice: LTspice schematic for Example 8.2 is shown in Figure 8.12(a). The plots of the input power, the input current, the output voltage, and the output power voltage are shown in Figure 8.12(b). The rms and average values of the input current and the output voltage are shown in Figure 8.12(d) and (f). The input power as shown in Figure 8.12(c) is 7.114 W and the output power as shown in Figure 8.12(e) is 4.2292 W. As expected the voltage across the converter V(3) is pulsed and V(3) is connected to the source when the transistor is turned on and is connected to the ground through the free-wheeling diode as shown in Figure 8.12(b). The power difference is caused due to losses in the transistor and the free-wheeling diode. We can find the power efficiency of the converter as $\eta = \frac{P_{out}}{P_{in}} = \frac{4.2292}{7.114} = 48.45\%$

Note: For a low output power, the losses in the semiconductor devices are significant compared to the output power and the efficiency becomes low.

Example 8.3:

Finding the Performance of a Buck–Boost DC–DC Converter with a BJT Switch

A BJT buck–boost chopper is shown in Figure 8.13(a). The DC input voltage is V_s = 12 V. The load resistance R is 5 Ω. The inductance is L = 150 μH, and the filter capacitance is C = 200 πF. The chopping frequency is f_c = 25 kHz, and the duty

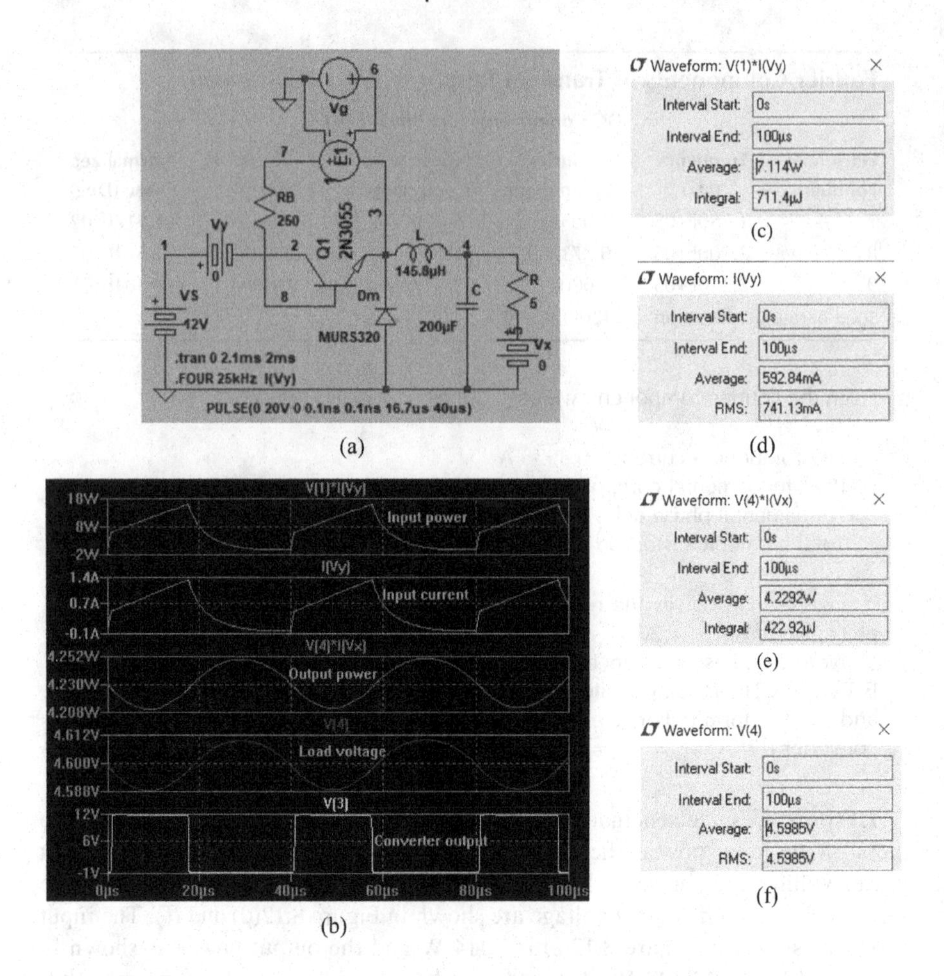

FIGURE 8.12 LTspice schematic for Example 8.2. (a) LTspice schematic. (b) Output waveforms. (c) Input power. (d) Input current. (e) Output power. (f) Output voltage.

cycle of the chopper is k = 25%. The control voltage is shown in Figure 8.13(b). Use PSpice to plot the instantaneous output voltage v_c, the capacitor current i_c, the inductor current i_L, and the inductor voltage v_L. Plot the frequency response of the converter output voltage from 100 kHz to 10 MHz and find the resonant frequency.

SOLUTION

The DC supply voltage $V_S = 12$ V, $k = 0.25$, $f_c = 25$ kHz, $T = 1/f_c = 40$ µs, and $t_{on} = k \times T = 0.25 \times 40 = 10$ µs.

The PSpice schematic is shown in Figure 8.14. A voltage-controlled voltage source with a voltage gain of 40 drives the BJT switch. The PWM generator is implemented as a descending hierarchy as shown in Figure 8.3(b). The model parameters for the BJT switch and the freewheeling diode are as follows:

```
.MODEL QMOD NPN(IS=6.83E-14 BF=13 CJE=1pF CJC=607.3PF TF=2 6.5NS)
.MODEL DMD D(IS=2.22E-15 BV=1200V CJO=1PF TT=0US))
```

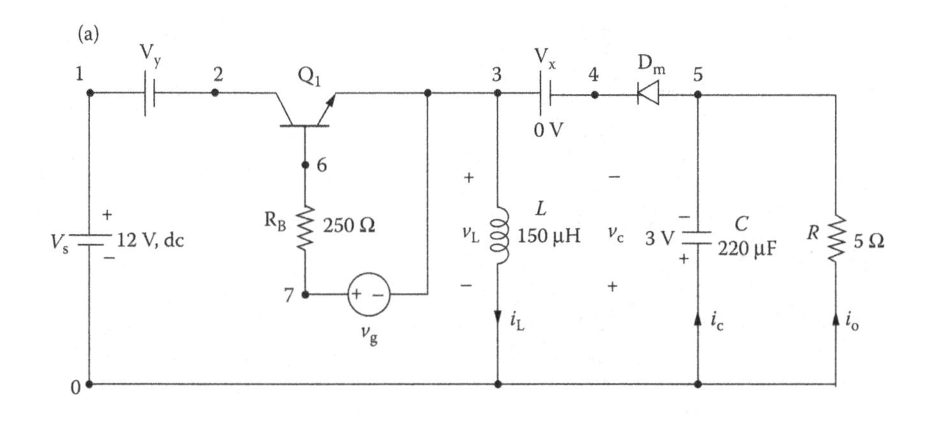

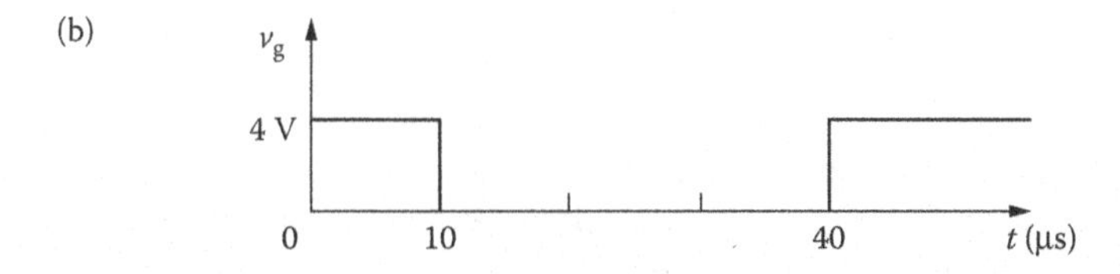

FIGURE 8.13 BJT buck–boost chopper for PSpice simulation. (a) Schematic circuit. (b) Gate control voltage.

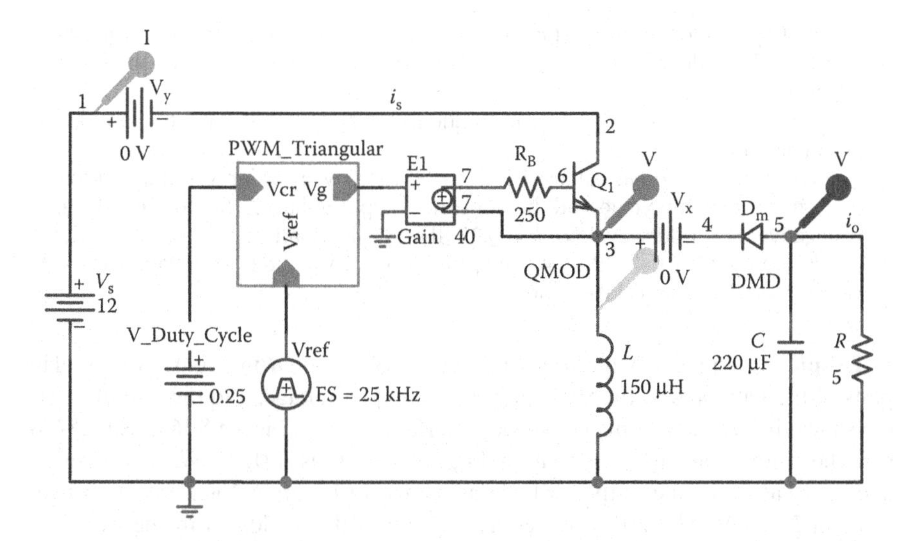

FIGURE 8.14 Boring/mixing tool for the TSM process. Finished TSM wall with inserted H beams.

The list of the circuit file is as follows:

BJT Buck-Boost Chopper

```
SOURCE    ■ VS  1 0 DC 12V
            VY 1 2 DC 0V ; Voltage source to measure input
            current
            Vg 7 3 PULSE (0V 40V 0 0.1NS 0.1NS 10US 40US)
CIRCUIT  ■■ RB 7 6 250 ; Transistor base resistance
             R 5 0 5
             L 3 0 150UH
             C 5 0 200UF IC=0V ; Initial voltage on
             capacitor
             VX 3 4 DC 0V ; Voltage source to measure diode
             current
             DM 5 4 DMOD ; Freewheeling diode
             .MODEL DMOD D(IS=2.22E-15 BV=1200V IBV=13E-3
             CJO=0 TT=0)                  ; Diode model
             Q1 2 6 3 3 2N6546            ; BJT switch
             .MODEL 2N6546 NPN (IS=6.83E-14 BF=13 CJE=1PF
             CJC=607.3PF TF=26.5NS)
ANALYSIS ■■■ .TRAN 0.1US 1MS 750US UIC ; Transient analysis
              .PROBE                     ; Graphics
              post-processor
              .OPTIONS ABSTOL=1.00N RELTOL=0.01 VNTOL=0.1
              ITL5=40000
              .FOUR 25KHZ I(VY)          ; Fourier analysis
         .END
```

The PSpice plots of the instantaneous output voltage V(5), the capacitor current I(C), the inductor current I(L), and the inductor voltage V(3) are shown in Figure 8.15(a).

Figure 8.15(b) gives the resonant frequency of f_{res} = 872.15 Hz at a resonant peak gain of $A_{o(peak)} = 3.2$.

Note: The base drive voltage is increased to 40 V in order to reduce distortion on the switch voltage and the converter output voltage. The output voltage is negative. For a duty cycle of $k = 0.25$, the average load voltage is $V_{o(av)} = kV_s/(1 - k) = 0.25 \times 12/(1 - 0.25) = 4$ V (PSpice gives 3.5 V). The load voltage has not yet reached the steady-state value.

LTspice: LTspice schematic for Example 8.3 is shown in Figure 8.16(a). The plots of the gate pulses, the input power, the output voltage, and the output power are shown in Figure 8.16(b). The input power as shown in Figure 8.16(c) is 6.257 W and the output power as shown in Figure 8.16(d) is 1.0793 W. The average and rms values of the output voltage as shown in Figure 8.16(e) are almost the same at 2.323 V. The power difference is caused due to losses in the transistor and the free-wheeling diode. We can find the power efficiency of the converter as $\eta = \frac{P_{out}}{P_{in}} = \frac{1.0793}{6.2573} = 17.25\%$

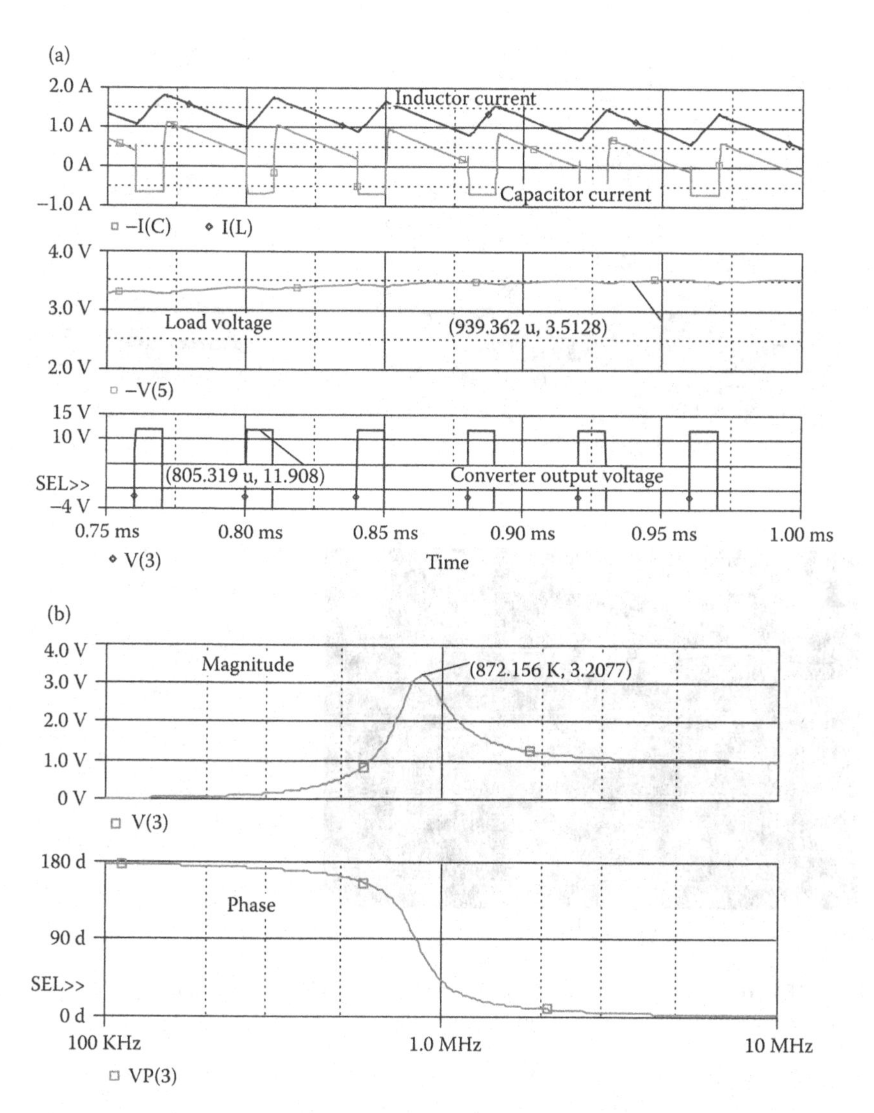

FIGURE 8.15 Boring/mixing tool for the TSM process. Finished TSM wall with inserted H beams. (a) Transient plots. (b) Frequency response.

Note: For a low output power, the losses in the semiconductor devices are significant compared to the output power and the efficiency becomes low.

Example 8.4:

Finding the Performance of Cuk DC–DC Converter with a BJT Switch

A BJT Cuk chopper is shown in Figure 8.17(a). The DC input voltage is V_S = 12 V. The load resistance R is 5 Ω. The inductances are L_1 = 200 μH and L_2 = 150 μH. The

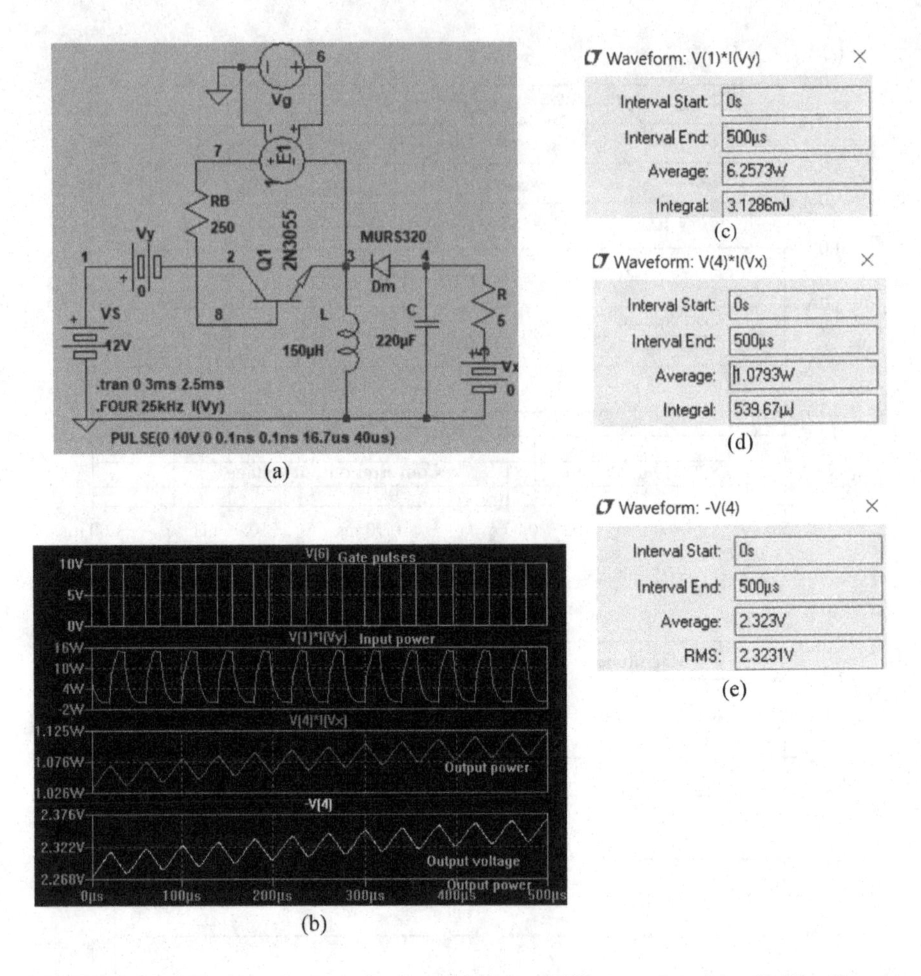

FIGURE 8.16 LTspice schematic for Example 8.3. (a) LTspice schematic. (b) Waveforms. (c) Input power. (d) Output power. (e) Output voltage.

capacitance are C_1 = 200 µF and C_2 = 220 µF. The chopping frequency is f_c = 25 kHz, and the duty cycle of the chopper is k = 25%. The control voltage is shown in Figure 8.17(b). Use PSpice to plot the instantaneous capacitor current i_{C1}, the capacitor current i_{C2}, the inductor current i_{L1}, the inductor current i_{L2}, and the transistor voltage v_T. Plot the frequency response of the converter output voltage from 100 Hz to 4 kHz and find the resonant frequency.

The PSpice schematic is shown in Figure 8.18. A voltage-controlled voltage source with a voltage gain of 30 drives the BJT switch. The PWM generator is implemented as a descending hierarchy as shown in Figure 8.3(b). The model parameters for the BJT switch and the freewheeling diode are as follows:

```
.MODEL  QMOD  NPN(IS=6.83E-14  BF=13  CJE=1pF  CJC=607.3PF  TF=2
6.5NS)
.MODEL  DMD  D(IS=2.22E-15  BV=1200V  CJO=1PF  TT=0US))
```

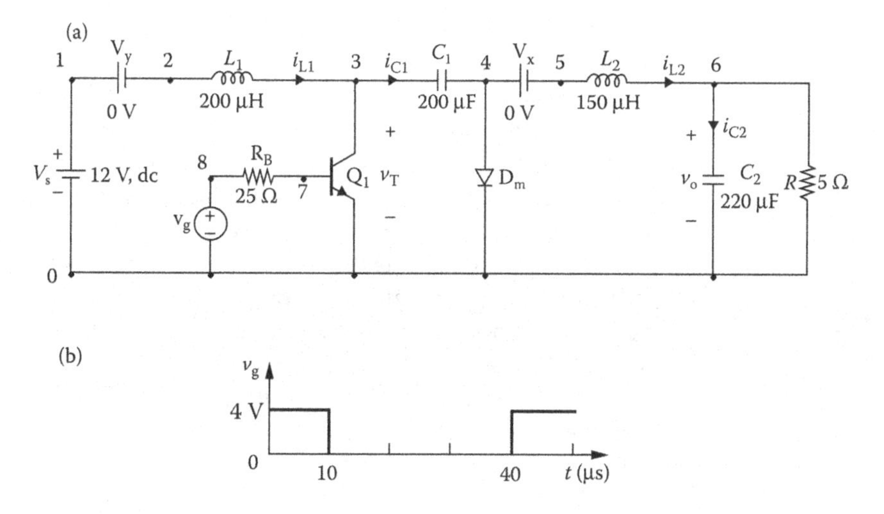

FIGURE 8.17 BJT Cuk chopper for PSpice simulation. (a) Schematic circuit. (b) Gate control voltage.

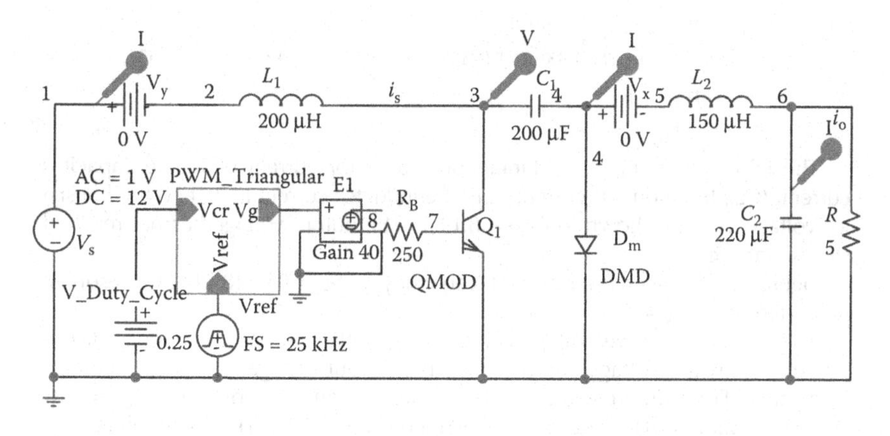

FIGURE 8.18 PSpice schematic for Example 8.4.

SOLUTION

The DC supply voltage $V_S = 12$ V. An initial capacitor voltage $V_C = 3$ V (assigned for reaching steady-state faster), $k = 0.25$, $f_c = 25$ kHz, $T = 1/f_c = 40$ μs, and $t_{on} = k \times T = 0.25 \times 40 = 10$ μs. The list of the circuit file is as follows:

BJT Cuk Chopper

```
SOURCE    ■ VS 1 0 DC 12V
            VY 1 2 DC 0V ; Voltage source to measure
            input current
            Vg 8 0 PULSE (0V 40V 0 0.1NS 0.1NS 10US 40US)
```

```
CIRCUIT    ■■ RB 8 7 25 ; Transistor base resistance
              R  6 0 5
              L1 2 3 200UH
              C1 3 4 200UF
              L2 5 6 150UF
              C2 6 0 220UF IC=0V      ; Initial conditions
              VX 4 5 DC   ; Voltage source to measure
              current of L2 DM 4 0 DMOD
              .MODEL DMOD D(IS=2.22E-15 BV=1200V
              IBV=13E-3 CJO=0 TT=0)  ; Diode model
              Q1 3 7 0 2N6546             ; BJT switch
              .MODEL 2N6546 NPN (IS=6.83E-14 BF=13
              CJE=1PF CJC=607.3PF TF=26.5NS)
ANALYSIS ■■■  .TRAN 0.1US 1MS 600US 750US ; Transient
              analysis
              .PROBE                          ; Graphics
              post-processor
              .OPTIONS ABSTOL = 1.00N RELTOL = 0.01
              VNTOL = 0.1
              ITL5 = 40000
              .FOUR 25KHZ I(VY)               ; Fourier analysis
          .END
```

The PSpice plots of the instantaneous capacitor current I(C1), the capacitor current I(C2), the inductor current I(L1), the inductor current I(L2), and the transistor voltage V(3) are shown in Figure 8.19(a). The output voltage has not reached steady state yet.

Figure 8.19(b) gives the resonant frequency of f_{res} = 833.28 Hz at a resonant peak gain of $A_{o(peak)} = 6.8$.

Note: The base drive voltage is increased to 40 V in order to reduce distortion on the switch voltage and the converter output voltage. The output voltage is negative. For a duty cycle of $k = 0.25$, the average load voltage is $V_{o(av)} = kV_s/(1 - k) = 0.25 \times 12/(1 - 0.25) = 4$ V (PSpice gives 3.5 V). The load voltage has not yet reached the steady-state value. If we run the simulation for a longer time as shown in Figure 8.20, we can find out the oscillation of the converter output voltage at its natural frequency.

LTspice: LTspice schematic for Example 8.4 is shown in Figure 8.21(a). The plots of the gate pulses, the input power, the output voltage, and the output power are shown in Figure 8.21(b). The input power as shown in Figure 8.21(c) is 7.3619 W and the output power as shown in Figure 8.21(d) is 5.5078 W. The power difference is caused due to losses in the transistor and the free-wheeling diode. We can find the power efficiency of the converter as $\eta = \frac{P_{out}}{P_{in}} = \frac{5.5078}{7.3619} = 74.75\%$

Note: For a low output power, the losses in the semiconductor devices are significant compared to the output power and the efficiency becomes low.

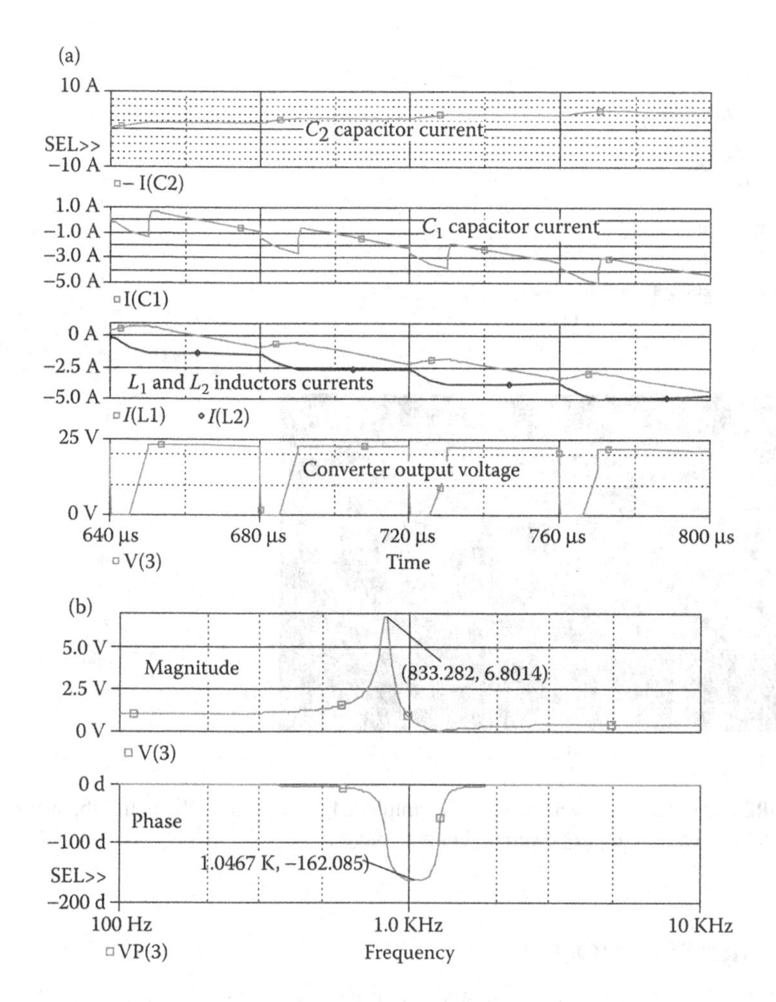

FIGURE 8.19 Plots for Example 8.4. (a) Transient plots. (b) Frequency response.

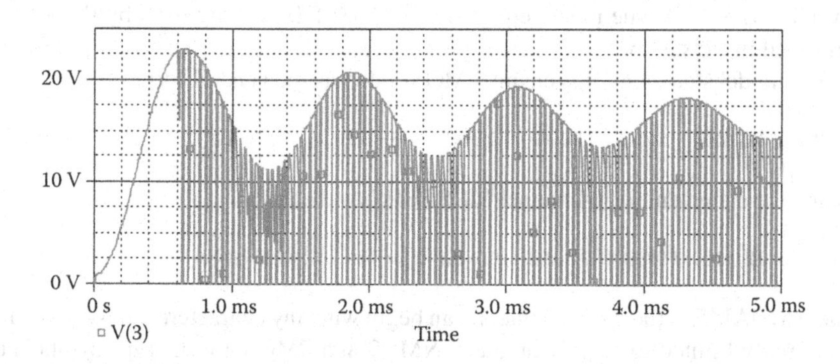

FIGURE 8.20 Oscillation on the output voltage.

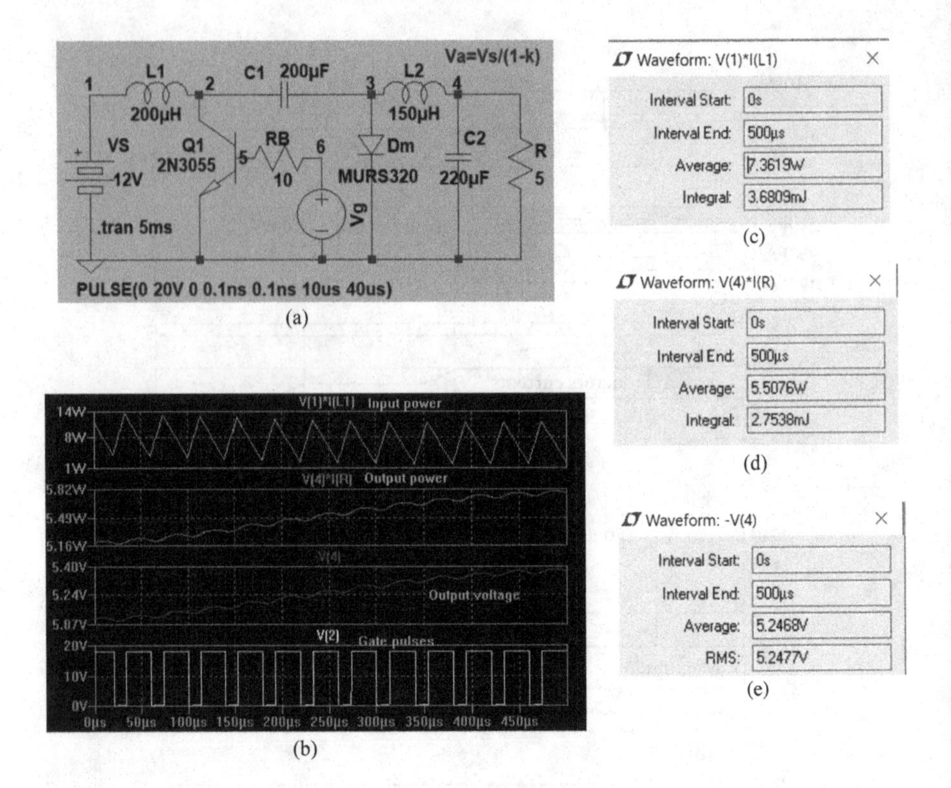

FIGURE 8.21 LTspice schematic for Example 8.4. (a) LTspice schematic. (b) Waveforms. (c) Input power. (d) Output power. (e) Output voltage.

8.6 MOSFET CHOPPERS

The PSpice model of an *n*-channel MOSFET [4–6] is shown in Figure 8.22(a). The static (DC) model that is generated by PSpice is shown in Figure 8.22(b). The model parameters for a MOSFET device and the default values assigned by PSpice are given in Table 8.3. The model equations of MOSFETs that are used by PSpice are described in Refs. [4,6].

The model statement of *n*-channel MOSFETs has the general form

```
.MODEL MNAME  NMOS  (P1=B1  P2=B2  P3=B3  ...  PN=BN)
```

and the statement for *p*-channel MOSFETs has the form

```
.MODEL MNAME PMOS  (P1=B1  P2=B2  P3=B3  ...  PN=BN)
```

where MNAME is the model name. It can begin with any character, and its word size is normally limited to eight characters. NMOS and PMOS are the type symbols of *n*-channel and *p*-channel MOSFETs, respectively. P1, P2, … and B1, B2, … are the parameters and their values, respectively.

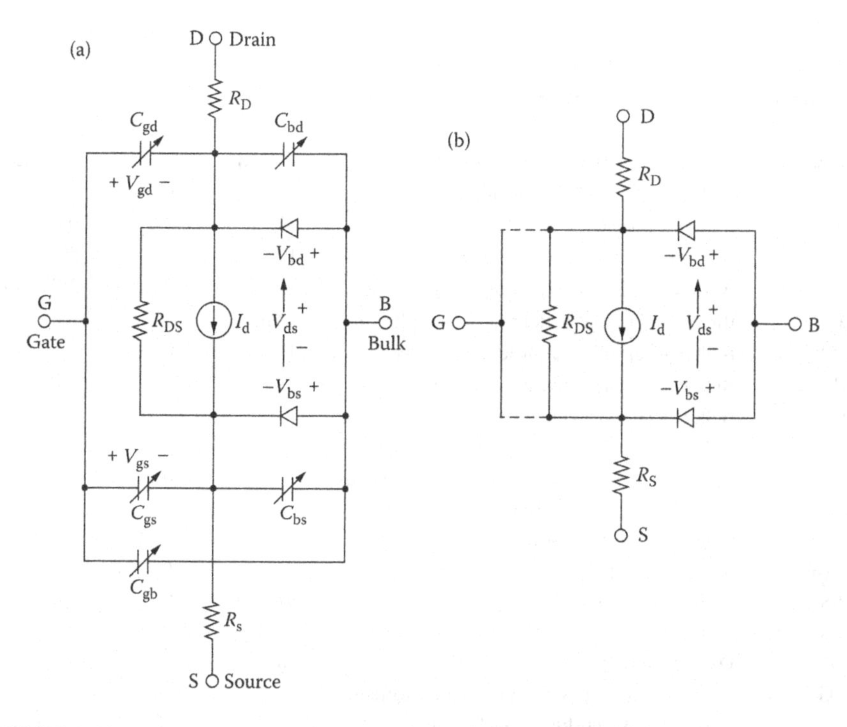

FIGURE 8.22 PSpice n-channel MOSFET model. (a) PSpice model. (b) DC model.

TABLE 8.3
Model Parameters of MOSFETs

Name	Model Parameter	Unit	Default	Typical
LEVEL	Model type (1, 2, or 3)		1	
L	Channel length	m	DEFL	
W	Channel width	m	DEFW	
LD	Lateral diffusion (length)	m	0	
WD	Lateral diffusion (width)	m	0	
VTO	Zero-bias threshold voltage	V	0	0.1
KP	Transconductance	A/V^2	2E–5	2.5E–5
GAMMA	Bulk threshold parameter	$V^{1/2}$	0	0.35
PHI	Surface potential	V	0.6	0.65
LAMBDA	Channel-length modulation (LEVEL = 1 or 2)	V^{-1}	0	0.02
RD	Drain ohmic resistance	W	0	10
RS	Source ohmic resistance	W	0	10
RG	Gate ohmic resistance	W	0	1
RB	Bulk ohmic resistance	W	0	1
RDS	Drain–source shunt resistance	W	•	
RSH	Drain–source diffusion sheet resistance	Ω/square	0	20
IS	Bulk *p–n* saturation current	A	1E–14	1E–13;15

(Continued)

TABLE 8.3 *(Continued)*
Model Parameters of MOSFETs

Name	Model Parameter	Unit	Default	Typical
JS	Bulk *p–n* saturation current/area	A/m^2	0	1E–8
PB	Bulk *p–n* potential	V	0.8	0.75
CBD	Bulk–drain zero-bias *p–n* capacitance	F	0	5PF
CBS	Bulk–source zero-bias *p–n* capacitance	F	0	2PF
CJ	Bulk *p–n* zero-bias bottom capacitance/length	F/m^2	0	
CJSW	Bulk *p–n* zero-bias perimeter capacitance/length	F/m	0	
MJ	Bulk *p–n* bottom grading coefficient	0.5		
MJSW	Bulk *p–n* sidewall grading coefficient		0.33	
FC	Bulk *p–n* forward-bias capacitance coefficient		0.5	
CGSO	Gate–source overlap capacitance/channel width	F/m	0	
CGDO	Gate–drain overlap capacitance/channel width	F/m	0	
CGBO	Gate–bulk overlap capacitance/channel length	F/m	0	
NSUB	Substrate doping density	1/cm^3	0	
NSS	Surface state density	1/cm^2	0	
NFS	Fast surface state density	1/cm^2	0	
TOX	Oxide thickness	m	•	
TPG	Gate material type: +1 = opposite of substrate; –1 = same as substrate; 0 = aluminum		+1	
XJ	Metallurgical junction depth	m	0	
UO	Surface mobility	cm^2/V ≥ s	600	
UCRIT	Mobility degradation critical field (LEVEL = 2)	V/cm	1E4	
UEXP	Mobility degradation exponent (LEVEL = 2)		0	
UTRA	*(not used)* Mobility degradation transverse field coefficient			
VMAX	Maximum drift velocity	m/s	0	
NEFF	Channel charge coefficient (LEVEL = 2)		1	
XQC	Fraction of channel charge attributed to drain		1	
DELTA	Width effect on threshold		0	
THETA	Mobility modulation (LEVEL = 3)	V^{-1}	0	
ETA	Static feedback (LEVEL = 3)		0	
KAPPA	Saturation field factor (LEVEL = 3)		0.2	
KF	Flicker noise coefficient		0	1E–26
AF	Flicker noise exponent		1	1.2

L and *W* are the channel length and width, respectively. AD and AS are the drain and source diffusion areas. *L* is decreased by twice LD to get the effective channel length. *W* is decreased by twice WD to get the effective channel width. *L* and *W* can be specified on the device, the model, or on the .OPTION statement. The value on the device supersedes the value on the model, which supersedes the value on the .OPTION statement.

AD and AS are the drain and source diffusion areas. PD and PS are the drain and source diffusion perimeters. The drain–bulk and source–bulk saturation currents can be specified either by JS, which is multiplied by AD and AS, or by IS, which is an absolute value. The zero-bias depletion capacitance can be specified by CJ, which is multiplied by AD and AS, and by CJSW, which is multiplied by PD and PS. Alternatively, these capacitances can be set by CBD and CBS, which are absolute values.

A MOSFET is modeled as an intrinsic MOSFET with ohmic resistances in series with the drain, source, gate, and bulk (substrate). There is also a shunt resistance (RDS) in parallel with the drain-source channel. NRD, NRS, NRG, and NRB are the relative resistivities of the drain, source, gate, and substrate in squares. These parasitic (ohmic) resistances can be specified by RSH, which is multiplied by NRD, NRS, NRG, and NRB, respectively, or, alternatively, the absolute values of RD, RS, RG, and RB can be specified directly.

PD and PS default to 0, NRD and NRS default to 1, and NRG and NRB default to 0. Defaults for *L*, *W*, AD, and AS may be set in the .OPTIONS statement. If AD or AS defaults are not set, they also default to 0. If *L* or *W* defaults are not set, they default to 100 μm.

The DC characteristics are defined by parameters VTO, KP, LAMBDA, PHI, and GAMMA, which are computed by PSpice using the fabrication-process parameters NSUB, TOX, NSS, NFS, TPG, and so on. The values of VTO, KP, LAMBDA, PHI, and GAMMA, which are specified on the model statement, supersede the values calculated by PSpice based on fabrication-process parameters. VTO is positive for enhancement-type *n*-channel MOSFETs and for depletion-type *p*-channel MOSFETs. VTO is negative for enhancement-type *p*-channel MOSFETs and for depletion-type *n*-channel MOSFETs.

PSpice incorporates three MOSFET device models. The LEVEL parameter selects between different models for the intrinsic MOSFET. If LEVEL = 1, the Shichman–Hodges model [5] is used. If LEVEL = 2, an advanced version of the Shichman–Hodges model, which is a geometry-based analytical model and incorporates extensive second-order effects [6], is used. If LEVEL = 3, a modified version of the Shichman–Hodges model, which is a semiempirical short-channel model [7], is used.

The LEVEL-1 model, which employs fewer fitting parameters, gives approximate results. However, it is useful for a quick, rough estimate of circuit performance and is normally adequate for the analysis of power electronics circuits. The LEVEL-2 model, which can take into consideration various parameters, requires considerable CPU time for the calculations and could cause convergence problems. The LEVEL-3 model introduces a smaller error than the LEVEL-2 model, and the CPU time is also ~25% less. The LEVEL-3 model is designed for MOSFETs with a short channel.

The parameters that affect the switching behavior of a BJT in power electronics applications are

```
L W VTO KP CGSO CGDO
```

The symbol of a metal-oxide silicon field-effect transistor (MOSFET) is *M*. The name of a MOSFET must start with *M* and take the general form

```
M <name>  ND    NG    NS    NB    MNAME
+       [L=<value]   [W=<value>]
+       [AD=<value>]       [AS=<value>]
+       [PD=<value>]       [PS=<value>]
+       [NRD=<value>]      [NRS=<value>]
+       [NRG=<value>]      [NRB=<value>]
```

where ND, NG, NS, and NB are the drain, gate, source, and bulk (or substrate) nodes, respectively. MNAME is the model name and can begin with any character; its word size is normally limited to eight characters. Positive current is the current that flows into a terminal, that is, the current flows from the drain node, through the device, to the source node for an *n*-channel MOSFET.

8.7 MOSFET PARAMETERS

The data sheet for an *n*-channel MOSFET of type IRF150 is shown in Figure 8.23. The library file of the student version of PSpice supports a model of this MOSFET as follows:

```
.MODEL  IRF150  NMOS  (TOX=100N  PHI=.6  KP=20.53U  W=.3
+  L=2UV  TO=2.831  RD=1.031M  RDS=444.4K  CBD=3.229N  PB=.8 MJ=.5
+  CGSO=9.027N  CGDO=1.679N  RG=13.89  IS=194E-18  N=1  TT=288N)
```

However, we shall generate approximate values of some parameters [8, 9]. From the data sheet we get $I_{DSS} = 250$ μA at $V_{CS} = 0$ V, $V_{DS} = 100$ V, $V_{Th} = 2\text{–}4$ V. Geometric mean, $V_{Th} = \text{VTO} = \sqrt{2 \times 4} = 2.83\text{V}$. The constant K_p can be found from

$$I_D = K_p(V_{GS} - V_{Th})^2 \tag{8.5}$$

for $I_D = I_{DSS} = 250$ μA, and $V_{Th} = 2.83$ V, Equation 8.5 gives $K_p = 250\ \mu\text{A}/2.83^2 = 31.2\ \mu\text{A/V}^2$. K_p is related to channel length L and channel width W by Equation 8.6

$$K_p = \frac{\mu_a C_o}{2}\left(\frac{W}{L}\right) \tag{8.6}$$

where C_o is the capacitance per unit area of the oxide layer, a typical value for a power MOSFET being 3.5–10^{-8} F/cm² at a thickness of 0.1 μm, and μ_a is the surface mobility of electrons, 600 cm²/(*V* > *s*).

The ratio *W*/*L* can be found from

$$\frac{W}{L} = \frac{2K_p}{\mu_a C_o} = \frac{2 \times 31.2 \times 10^{-6}}{600 \times 3.5 \times 10^{-8}} = 3$$

Let $L = 1$ μm and $W = 3000$ μm = 3 mm $C_{rss} = 350\text{–}500$ pF at $V_{GS} = 0$, and $V_{DS} = 25$ V. The geometric mean, $C_{rss} = C_{gd} = \sqrt{350 \times 500} = 418.3\text{pF}$ at $V_{DG} = 25$ V.

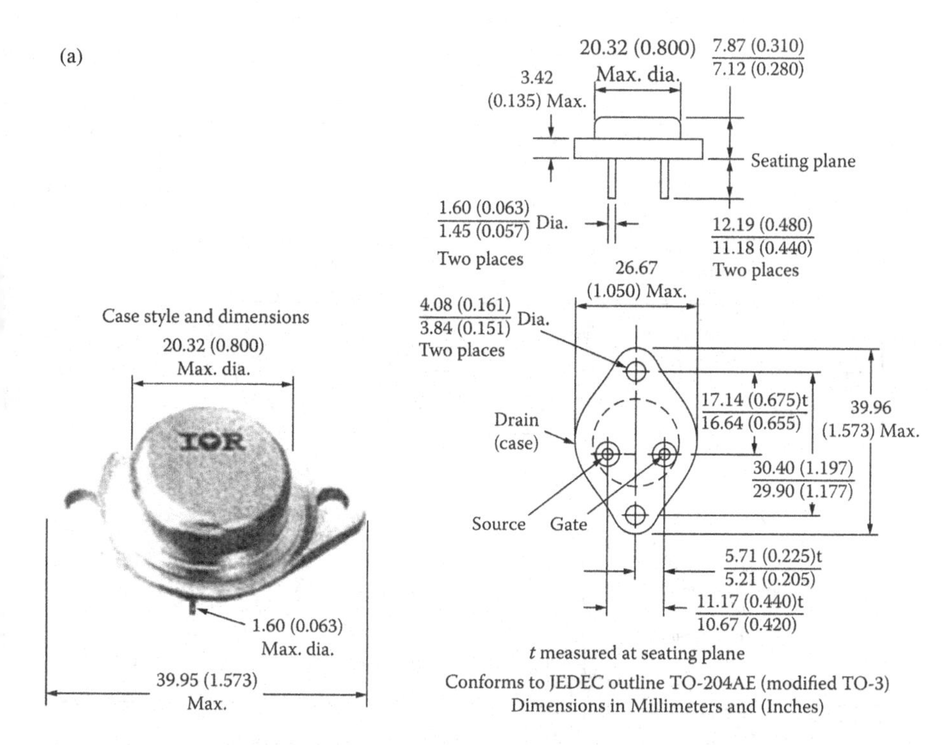

FIGURE 8.23 (a) Data sheetfor MOSFET of type IRF150. Datasheet. (b) Datasheet (continued). (c) Datasheet (continued). (d) Characteristics. (e) Characteristics (continued). (f) Characteristics (continued). (Courtesy of International Rectifier.) *(Continued)*

For a MOSFET, the values of C_{gs} and C_{gd} remain relatively constant with changing V_{GS} or V_{DS}. They are determined mainly by the thickness and type of the insulating oxide. Although the curves of the capacitances versus drain-source voltage show some variations, we will assume constant capacitances. Thus, $C_{gdo} = 418.3$ pF and $C_{iss} = 2000$ to 3000 pF. The geometric mean $C_{iss} = \sqrt{2000 \times 3000} = 2450\,\text{pF}$. Because C_{iss} is measured at $V_{GS} = 0$ V, $C_{gs} = C_{gso}$. That is,

$$C_{iss} = C_{gso} + C_{gd}$$

which gives $C_{gso} = C_{iss} – C_{sd} = 2450–418.3 = 2032$ pF = 2.032 nF. Thus, the PSpice model statement for MOSFET IRF150 is

```
.MODEL  IRF150  NMOS  (VTO=2.83  KP=31.2U  L=1U  W=30M
CGDO=0.418N  CGSO=2.032N)
```

The model can be used to plot the characteristics of the MOSFET. It may be necessary to modify the parameter values to conform with the actual characteristics. It should be noted that the parameters differ from those given in the PSpice library because their values are dependent on the constants used in derivations. Students are encouraged to run the following circuit file with the PSpice library model and compare the results.

(b) Absolute maximum ratings

	Parameter	IRF 150	IRF 151	IRF 152	IRF 153	Units
V_{DS}	Drain-Source voltage ①	100	60	100	60	V
V_{DGR}	Drain-Gate voltage (R_{GS} = 20 kΩ) ①	100	60	100	60	V
I_D @ T_C = 25°C	Continuous drain current	40	40	33	33	A
I_D @ T_C = 100°C	Continuous drain current	25	25	20	20	A
I_{DM}	Pulse drain current ③	160	160	132	132	A
V_{GS}	Gate - Source voltage	±20				V
P_D @ T_C = 25°C	Max. power dissipation	150 (See Fig. 14)				W
	Linear derating factor	1.2 (See Fig. 14)				W/K
I_{LM}	Inductive current, clamped	(See Fig. 15 and 16) L = 100 μH 160	160	132	132	A
T_J T_{stg}	Operating junction and Storage temperature range	−55 to 150				°C
	Lead temperature	300 (0.063 in. (1.6 mm) from case for 10s)				°C

Electrical characteristics @ T_C = 25°C (Unless otherwise specified)

	Parameter	Type	Min.	Typ.	Max.	Units	Test conditions
BV_{DSS}	Drain-Source breakdown voltage	IRF 150 IRF 152	100	–	–	V	V_{GS} = 0 V
		IRF 151 IRF 153	60	–	–	V	I_D = 250 μA
$V_{GS(th)}$	Gate threshold voltage	All	2.0	–	4.0	V	V_{DS} = V_{GS}, I_D = 250 μA
I_{GSS}	Gate-Source leakage forward	All	–	–	100	nA	V_{GS} = 20 V
I_{GSS}	Gate-Source leakage reverse	All	–	–	−100	nA	V_{GS} = 0.20 V
I_{DSS}	Zero gate voltage drain current	All	–	–	250	μA	V_{DS} = Max. rating, V_{GS} = 0 V
			–	–	1000	μA	V_{DS} = Max. rating × 0.8, V_{GS} = 0 V, T_C = 125°C
$I_{D(on)}$	On-State drain current ②	IRF 150 IRF 151	40	–	–	A	$V_{DS} > I_{D(on)} \times R_{DS(on)max.}$, V_{GS} = 10 V
		IRF 152 IRF 153	33	–	–	A	
$R_{DS(on)}$	Static Drain-Source on-state resistance ②	IRF 150 IRF 151	–	0.045	0.055	Ω	V_{GS} = 10 V, I_D = 20 A
		IRF 152 IRF 153	–	0.06	0.08	Ω	
g_{fs}	Forward transconductance ②	All	9.0	11	–	S(℧)	$V_{DS} > I_{D(on)} \times R_{DS(on)max.}$, I_D = 20 A
C_{iss}	Input capacitance	All	–	2000	3000	pF	V_{GS} = 0 V, V_{DS} = 25 V, f = 1.0 MHz See Fig. 10
C_{oss}	Output capacitance	All	–	1000	1500	pF	
C_{rss}	Reverse transfer capacitance	All	–	350	500	pF	
$t_{d(on)}$	Turn-on delay time	All	–	–	35	ns	V_{DD} = 24 V, I_D = 20 A, Z_o = 4.7 Ω See figure 17. (MOSFET switching times are essentially independent of operating temperature)
t_r	Rise time	All	–	–	100	ns	
$t_{d(off)}$	Turn-off delay time	All	–	–	125	ns	
t_f	Fall time	All	–	–	100	ns	
Q_g	Total gate charge (Gate-Source plus Gate-Drain)	All	–	63	120	nC	V_{GS} = 10 V, I_D = 50 A, V_{DS} = 0.8 max. rating. See Fig. 18 for test circuit. (Gate charge is essentially independent of operating temperature)
Q_{gs}	Gate-Source charge	All	–	27	–	nC	
Q_{gd}	Gate-Drain ("Miller") charge	All	–	36	–	nC	
L_D	Internal drain inductance	All	–	5.0	–	nH	Measured between the contact screw on header that is closer to source and gate pins and center of die — Modified MOSFET symbol showing the internal device inductances
L_S	Internal source inductance	All	–	12.5	–	nH	Measured from the source pin, 6 mm (0.25 in.) from header and source bonding pad

Thermal resistance

R_{thJC}	Junction-to-case	All	–	–	0.83	K/W	
R_{thCS}	Case-to-sink	All	–	0.1	–	K/W	Mounting surface flat, smooth, and greased
R_{thJA}	Junction-to-Ambient	All	–	–	0.83	K/W	Free air operation

FIGURE 8.23 *(Continued)*

(c) Source-Drain diode ratings and charecteristics

Symbol	Parameter	Type	Min	Typ	Max	Units	Test Conditions
I_S	Continuous source current (Body diode)	IRF 150 IRF 151	–	–	40	A	Modified MOSFET symbol showing the integral reverse P–N junction rectifier (D, G, S)
		IRF 152 IRF 153	–	–	33	A	
I_{SM}	Pulse source current (Body diode) ③	IRF 150 IRF 151	–	–	160	A	
		IRF 152 IRF 153	–	–	132	A	
V_{SD}	Diode forward voltage ②	IRF 150 IRF 151	–	–	2.5	V	$T_C = 25°C$, $I_S = 40$ A, $V_{GS} = 0$ V
		IRF 152 IRF 153	–	–	2.3	V	$T_C = 25°C$, $I_S = 33$ A, $V_{GS} = 0$ V
t_{rr}	Reverse recovery time	All	–	600	–	ns	$T_J = 150°C$, $I_F = 40$ A, $dI_F/dt = 100$ A/μs
Q_{RR}	Reverse recovered charge	All	–	3.3	–	μC	$T_J = 150°C$, $I_F = 40$ A, $dI_F/dt = 100$ A/μs
t_{on}	Forward Turn-on time	All	Intrinsic turn-on time is negligible. Turn-on speed is substantially controlled by $L_S + L_D$				

① Tj = 25°C to 150°C. ② Pulse test: Pulse width ≤ 300 μs, Duty cycle ≤ 2%. ② Repititive rating: Pulse width limited by max. junction temperature. See transient thermal impedance curve (Fig. 5).

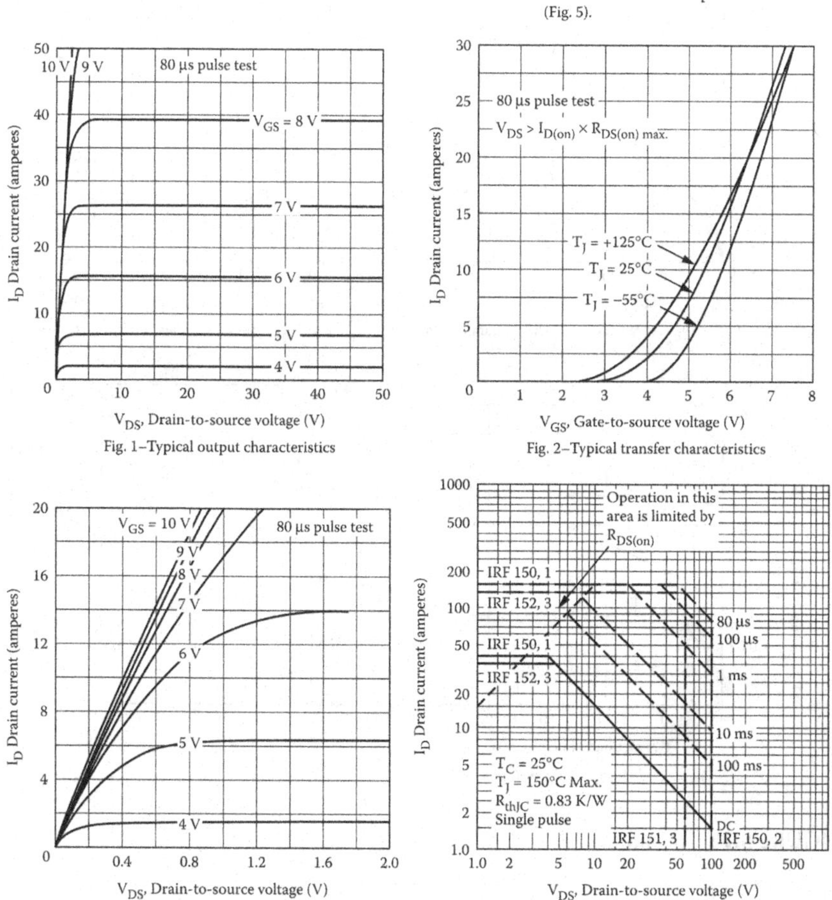

Fig. 1–Typical output characteristics

Fig. 2–Typical transfer characteristics

Fig. 3–Typical saturation characteristics

Fig. 4–Maximum safe operating area

FIGURE 8.23 *(Continued)*

(d)

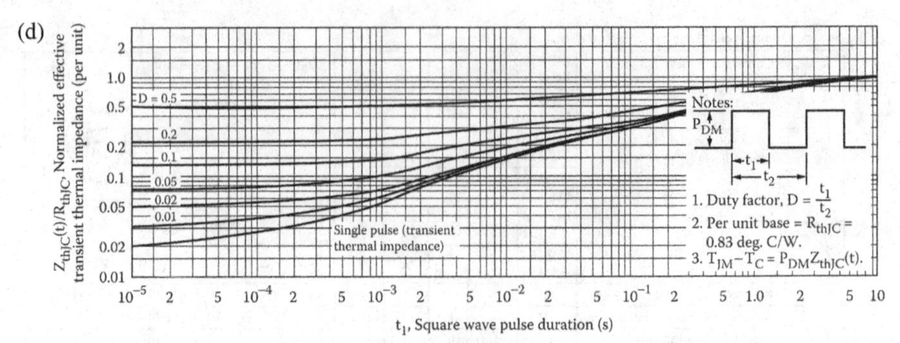

Fig. 5–Maximum effective transient thermal impedance, junction-to-case vs. pulse duration

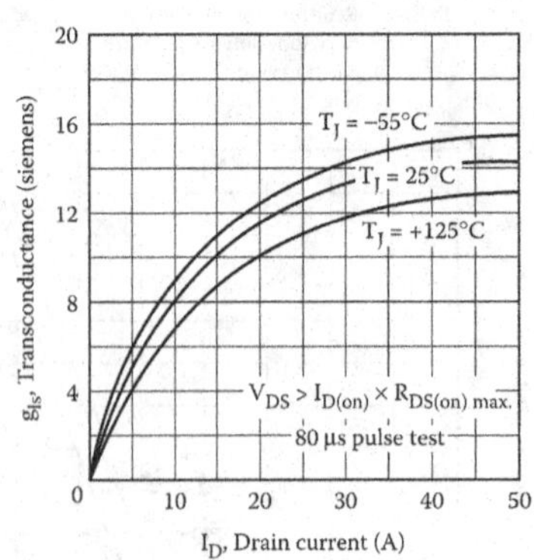

Fig. 6–Typical transconductance vs. drain current

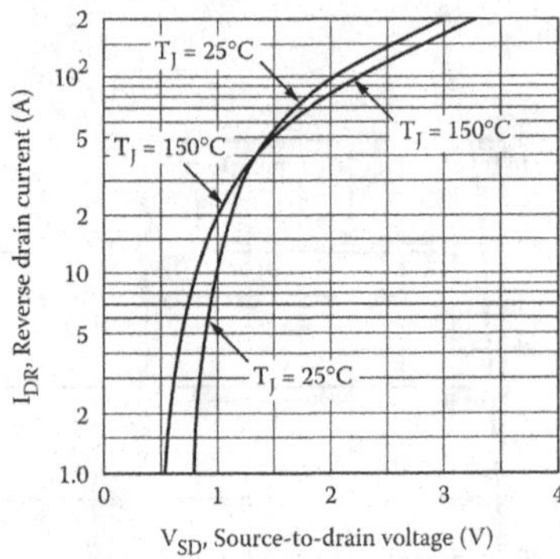

Fig. 7–Typical source-drain diode forward voltage

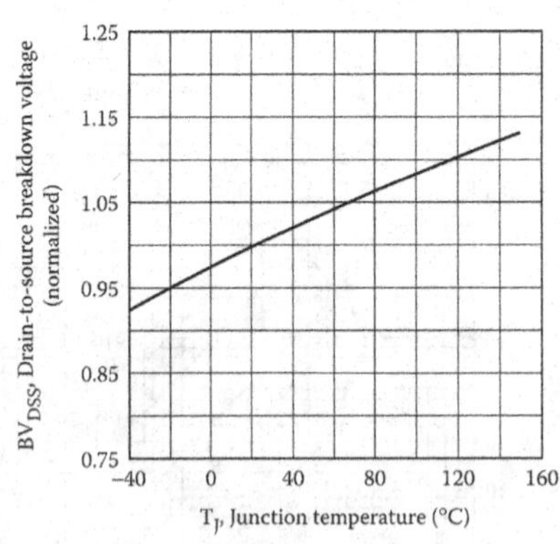

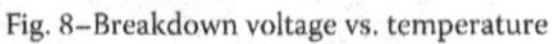

Fig. 8–Breakdown voltage vs. temperature

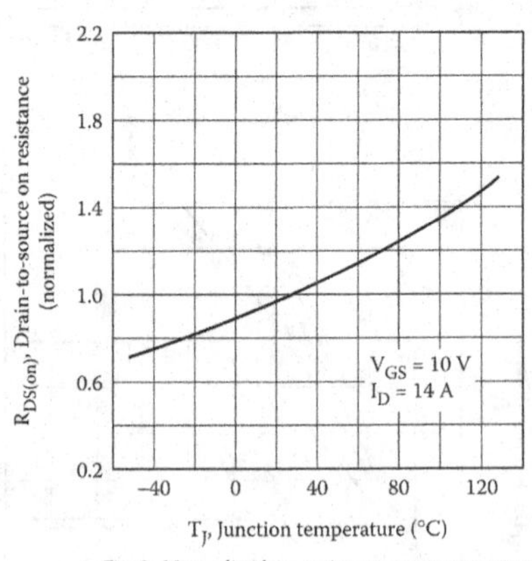

Fig. 9–Normalized on-resistance vs. temperature

FIGURE 8.23 *(Continued)*

(e)

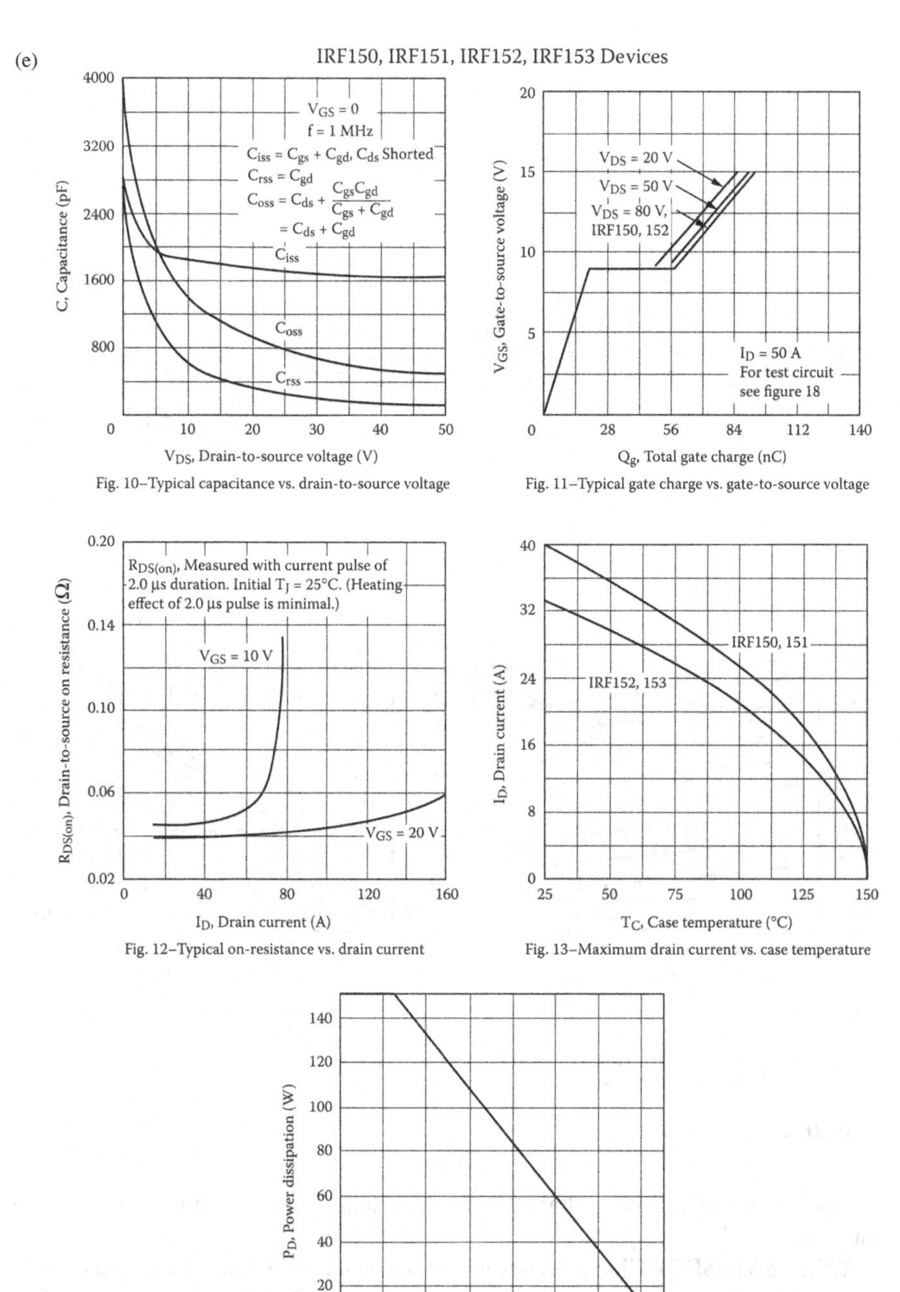

Fig. 10–Typical capacitance vs. drain-to-source voltage

Fig. 11–Typical gate charge vs. gate-to-source voltage

Fig. 12–Typical on-resistance vs. drain current

Fig. 13–Maximum drain current vs. case temperature

FIGURE 8.23 *(Continued)*

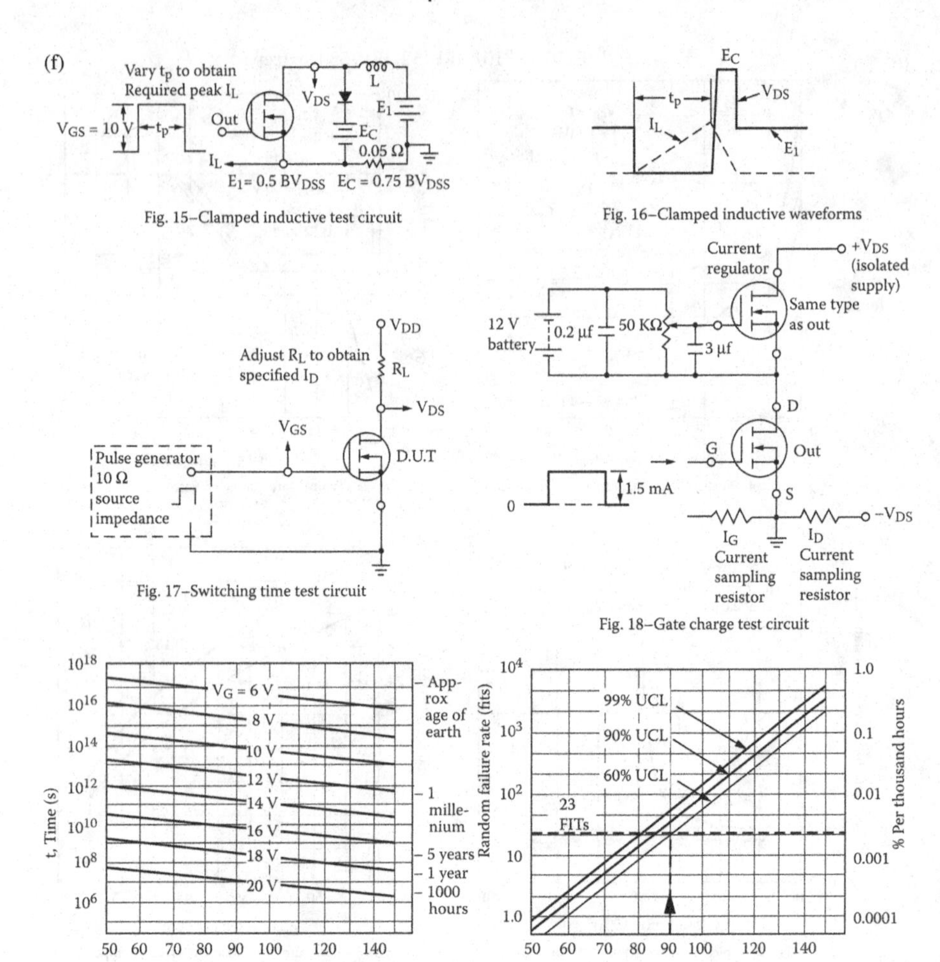

FIGURE 8.23 *(Continued)*

Note: The gate (control) voltage Vg should be adjusted to drive the MOSFET into saturation.

LTspice MOSFET Characteristics: The schematic for plotting the output characteristics of a MOSFET is shown in Figure 8.24(a). An MOSFET is a voltage-dependent device, and its drain current depends on the gate-source voltage. As the gate-source voltage is increased, the drain current is also increased. As the drain-source voltage increases, the drain current increases until the drain current reaches the saturation region and remains practically constant. For operating the MOSFET as a switching device, its gate-source voltage must be high enough to keep the drain-source voltage low for a specific load line as shown in Figure 8.24(b) for a load resistance of 1 ohm.

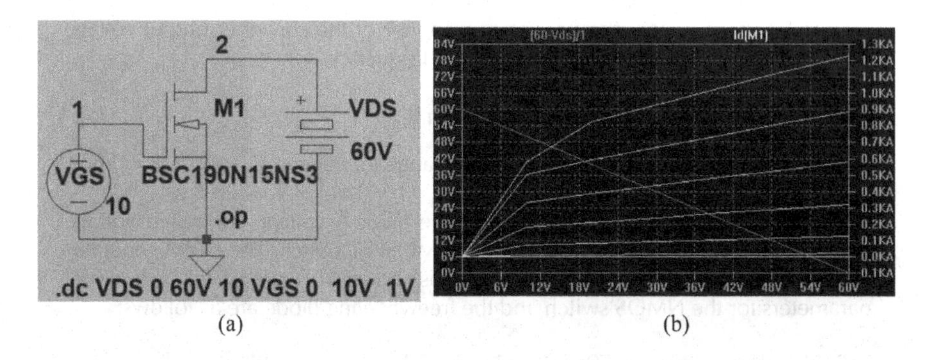

FIGURE 8.24 LTspice schematic MOSFET characteristics. (a) MOSFET biasing. (b) MOSFET output characteristics.

8.8 EXAMPLES OF MOSFET DC–DC CONVERTERS

Example 8.5:

Finding the Performance of Boos Cuk DC–DC Converter with an NMOS Switch

A MOSFET Boost Chopper is shown in Figure 8.25(a). The DC input voltage is $V_S = 5$ V. The load resistance R is 100 Ω. The inductance is $L = 150$ μH, and the filter capacitance is $C = 220$ μF. The chopping frequency is $f_c = 25$ kHz, and the duty cycle of the chopper is $k = 66.7\%$. The control voltage is shown in Figure 8.25(b). Use PSpice to plot the instantaneous output voltage v_c, the input current i_s, and the

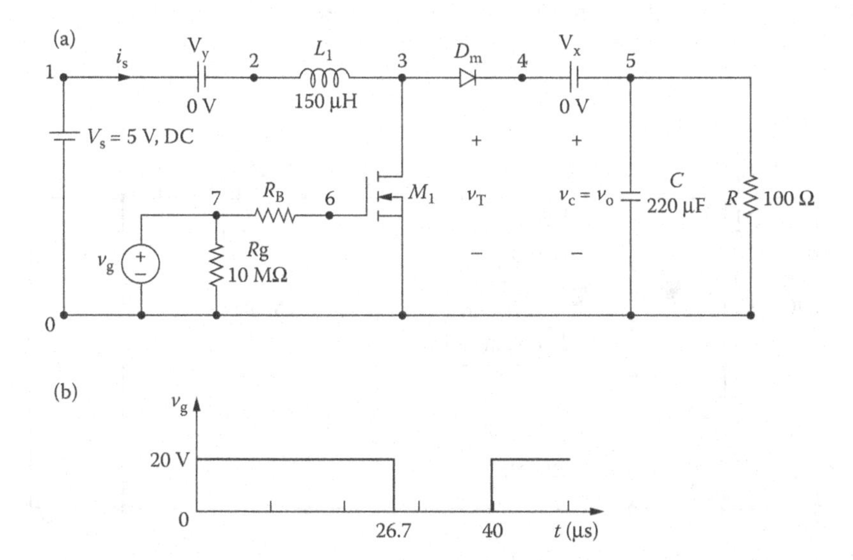

FIGURE 8.25 MOSFET boost chopper for PSpice simulation. (a) Circuit. (b) Gate voltage.

MOSFET voltage v_T. Plot the frequency response of the converter output voltage from 10 Hz to 10 kHz and find the resonant frequency.

SOLUTION

The DC supply voltage V_S = 5 V. An initial capacitor voltage V_C = 12 V, k = 0.667, f_c = 25 kHz, and $T = 1/f_c$ = 40 μs, and $t_{on} = k \times T = 0.667 \times 40$ = 26.7 μs

The PSpice schematic is shown in Figure 8.26. A voltage-controlled voltage source with a voltage gain of six drives the MOSFET switch. The PWM generator is implemented as a descending hierarchy as shown in Figure 8.3(b). The model parameters for the NMOS switch and the freewheeling diode are as follows:

```
.MODEL  NMOD  NMOS  (VTO=2.83  KP=31.2U  L=1U  W=30M
CGDO=0.418N  CGSO=2.032N)
.MODEL  DMD  D(IS=2.22E-15  BV=1200V  CJO=1PF  TT=0US))
```

Note: Rg is connected at the gate of the MOSFET to a DC path because the gate-source of a MOSFET is ideally infinity. E1 in Figure 8.26 provides the isolation between the power circuit and the low signal gating circuit.

The list of the circuit is as follows:

MOSFET Boost Chopper

```
SOURCE    ■ VS 1 0 DC 5V
            VY 1 2 DC 0V    ; Voltage source to measure input
            current
            Vg 7 0 PULSE (0V 6V 0 0.1NS 0.1NS 26.7US 40US)
            Rg 7 0 10MEG
CIRCUIT  ■■ RB 7 6 10k     ; Transistor base resistance
             R  5 0 100
             C  5 0 220UF IC=12V ; With initial condition
             L  2 3 150UH
             VX 4 5 DC 0V   ; Voltage source to inductor
             current
```

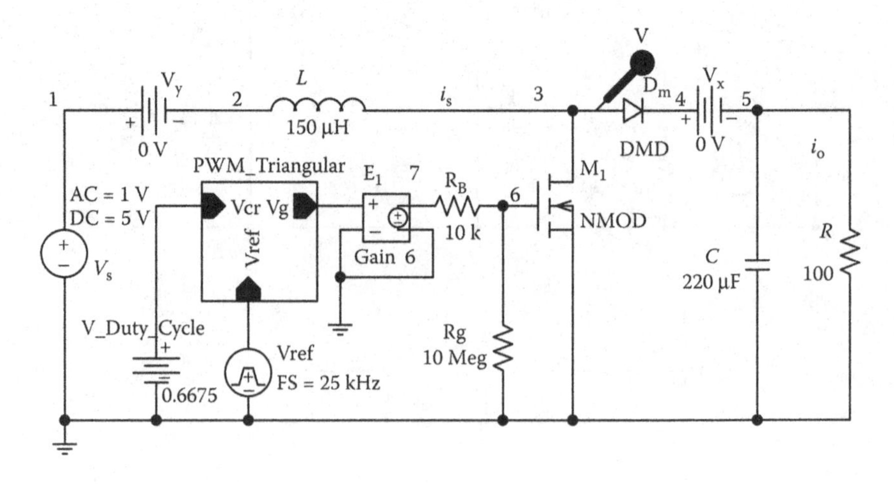

FIGURE 8.26 PSpice schematic for Example 8.5.

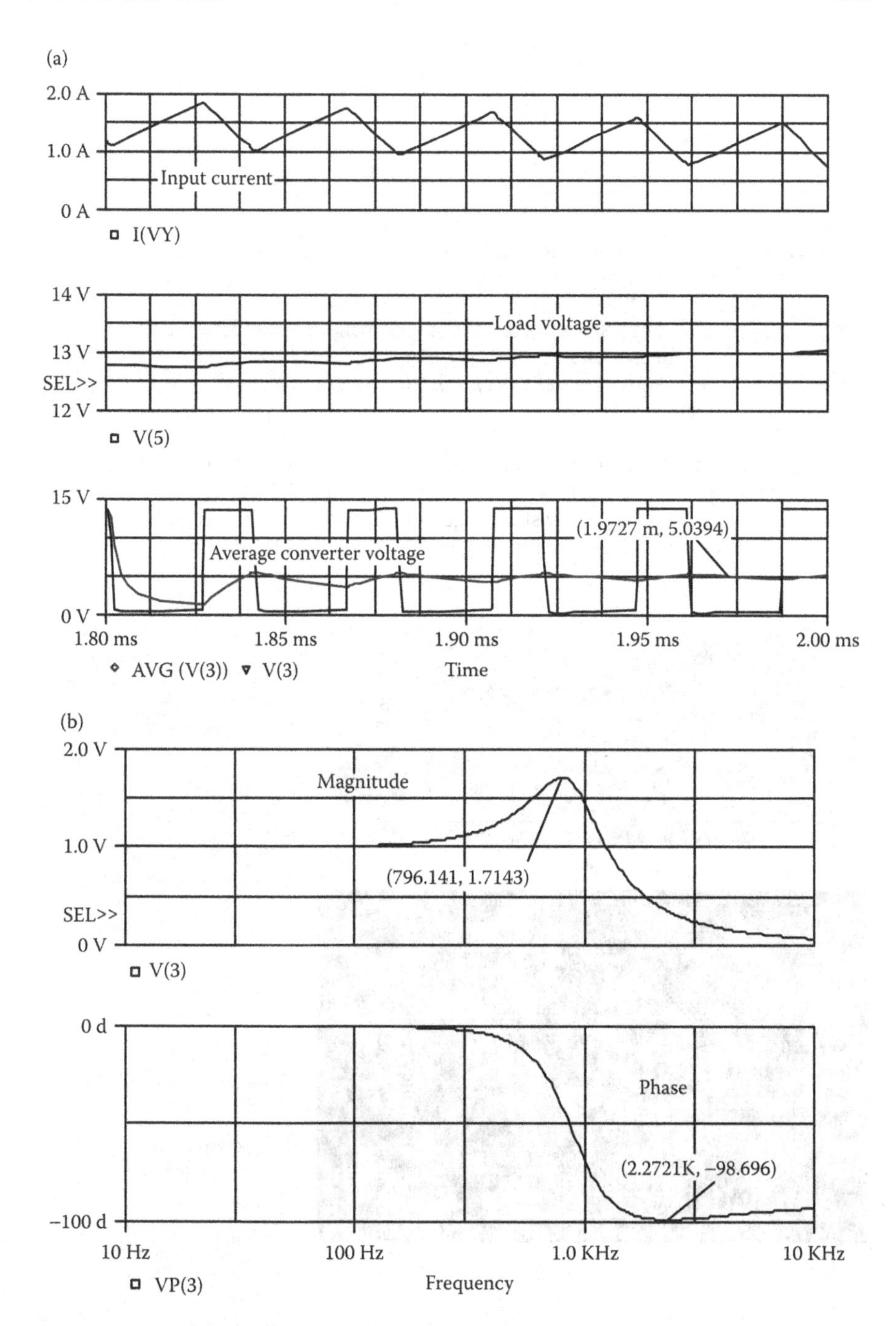

FIGURE 8.27 Plots for Example 8.5. (a) Transient plots. (b) Frequency response.

```
            DM 3 4 DMOD              ; Freewheeling diode
            .MODEL DMOD D(IS=2.22E-15 BV=1200V
            IBV=13E-3 CJO=0 TT=0) ;Diode model
            M1 3 6 0 0 IRF150        ; MOSFET switch
            .MODEL IRF150 NMOS (VTO=2.83 KP=31.2U
            L=1U W=30M+CGDO=0.418N CGSO=2.032N)
                                ; MOSFET parameters
ANALYSIS ■■■ .TRAN 0.1US 2MS 1.8MS UIC    ; Transient
            analysis with initial conditions
            .PROBE               ; Graphics post-processor
            .OPTIONS ABSTOL=1.00N RELTOL=0.01 VNTOL=0.1
            ITL5=40000
            .FOUR 25KHZ I(VY)    ; Fourier analysis
          .END
```

The PSpice plots of the instantaneous MOSFET voltage V(3), the input current I(VY), and the output voltage V(5) are shown in Figure 8.27(a). The output voltage has not reached the steady state. Figure 8.27(b) gives the resonant frequency of f_{res} = 796.14 Hz at a resonant peak gain of $A_{o(peak)}$ = 1.71.

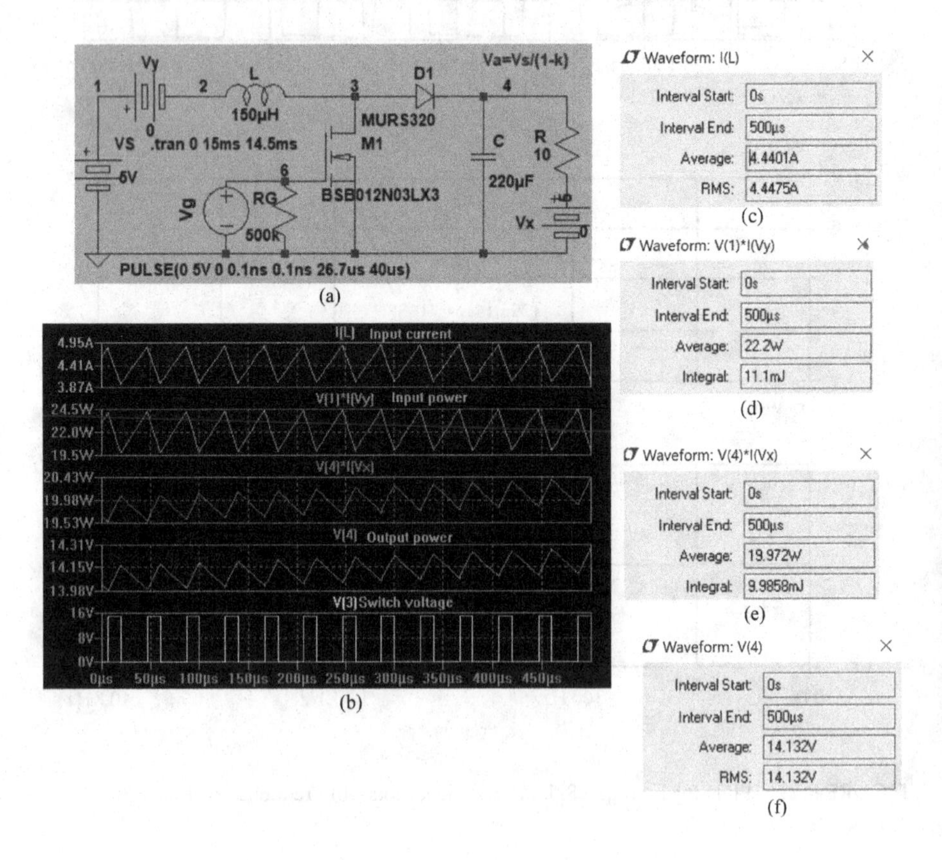

FIGURE 8.28 LTspice schematic for Example 8.5. (a) LTspice schematic. (b) Waveforms. (c) Input current. (d) Input power. (e) Output power. (f) Output voltage.

Note: The base drive voltage is reduced to 60 V. For a duty cycle of $k = 0.6675$, the average load voltage [1] is $V_{o(av)} = V_s/(1-k) = 5/(1 - 0.6675) = 15$ V (PSpice gives 13.0 V). The load voltage has not yet reached the steady-state value.

LTspice: LTspice schematic for Example 8.5 is shown in Figure 8.28(a). The plots of the input current, input power, the output voltage, and the output power are shown in Figure 8.28(b). The input current as shown in Figure 8.28(c) is 4.44 A and the output voltage as shown in Figure 8.28(f) is 14.32 A. The input power as shown in Figure 8.28(d) is 22.2 W and the output power as shown in Figure 8.28(e) is 14.32 W. The power difference is caused due to losses in the transistor and the forward diode. We can find the power efficiency of the converter as $\eta = \frac{P_{out}}{P_{in}} = \frac{19.97}{22.2} = 89.95\%$

Note: For a low output power, the losses in the semiconductor devices are significant compared to the output power and the efficiency becomes low. We should note that the selection of the switching device and the diode would contribute to the losses and the converter efficiency.

8.9 IGBT MODEL

The IGBT shown in Figure 8.29(a) behaves as a MOSFET from the input side and as a BJT from the output side. The modeling of an IGBT is very complex [8].

There are two main ways to model IGBT in SPICE: (1) composite model and (2) equation model. The composite model connects the existing BJT and MOSFET models of PSpice in a Darlington configuration and uses their built-in equations. The equivalent circuit of the composite model is shown in Figure 8.29(b). This model computes quickly and reliably, but it does not model the behavior of the IGBT accurately.

$$C_{dg} = \frac{\varepsilon_{si} C_{oxd}}{\sqrt{\frac{2\varepsilon_{si} V_{dg}}{qN_B}} C_{oxd} + A_{dg}\varepsilon_{si}}$$

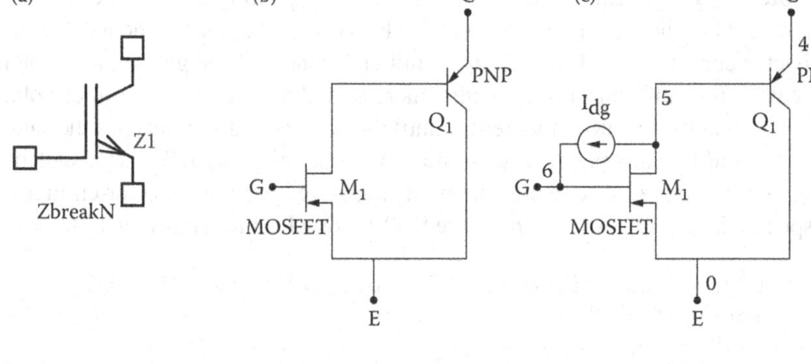

FIGURE 8.29 Equivalent circuits of IGBT SPICE models. (a) *n*-type IGBT. (b) Composite model. (c) Sheng PSpice model.

The equation model [16,17] implements the physics-based equations and models the internal carrier and charge to simulate the circuit behavior of the IGBT accurately. This model is complicated, often unreliable, and computationally slow because the equations are derived from the complex semiconductor physics theory. Simulation times can be over 10 times longer than for the composite model.

There are numerous papers on the SPICE modeling of IGBTs. Sheng [18] compares the merits and limitations of various models. Figure 8.29(c) shows the equivalent circuit of Sheng's model [15], which adds a current source from the drain to the gate. The major inaccuracy in dynamic electrical properties [15] is associated with the modeling of the drain-to-gate capacitance of the *n*-channel MOSFET. During high-voltage switching, the drain-to-gate capacitance C_{dg} changes by two orders of magnitude due to any changes in drain-to-gate voltage V_{dg}, that is, C_{dg} is expressed by

$$C_{dg} = \frac{\varepsilon_{si} C_{oxd}}{\sqrt{\frac{2\varepsilon_{si} V_{dg}}{qN_B}} C_{oxd} + A_{dg}\varepsilon_{si}}$$

PSpice Schematics does not incorporate a capacitance model involving the square root, which models the space charge layer variation for a step junction. The PSpice model can implement the equations describing the highly nonlinear gate-drain capacitance into the composite model by using the analog behavioral modeling function of PSpice.

The student version of the PSpice Schematics or OrCAD library comes with one breakout IGBT device, ZbreakN, and one real device, IXGH40N60. Although a complex model is needed for accurately simulating the behavior of an IGBT circuit, these simple PSpice IGBT models can simulate the behavior of a converter for most applications. The model parameters of the IGBT of IXGH40N60 are as follows:

```
.MODEL  IXGH4 0N6 0  NIGBT  (TAU=287.56E-9  KP=50.034
AREA=37.500E-6
+ AGD=18.750E-6  VT=4.1822  KF=.36047  CGS=31.94 2E-9
COXD=53.188E-9  VTD=2.6570
```

LTspice IGBT Characteristics: The schematic for plotting the output characteristics of an IGBT is shown in Figure 8.30(a). An IGBT is a voltage-dependent device, and its collector current depends on the gate-emitter voltage. As the gate-emitter voltage is increased, the collector current is also increased. As the collector-emitter voltage increases, the collector current increases until the collector current reaches the saturation region and remains practically constant. For operating the IGBT as a switching device, its gate-emitter voltage must be high enough to keep the collector-emitter low for a specific load line as shown in Figure 8.30(b) for a load resistance of 1 ohm.

```
The list of circuit file for Figure 8.30 is as follows:
** C:\Users\mrashid\  ; file location
VDS 2 0 60V
VGS 1 0 10
XFGW50XS65 2 1 0 FGW50XS65
.dc VDS 0 60V 10 VGS 0  10V  1V
```

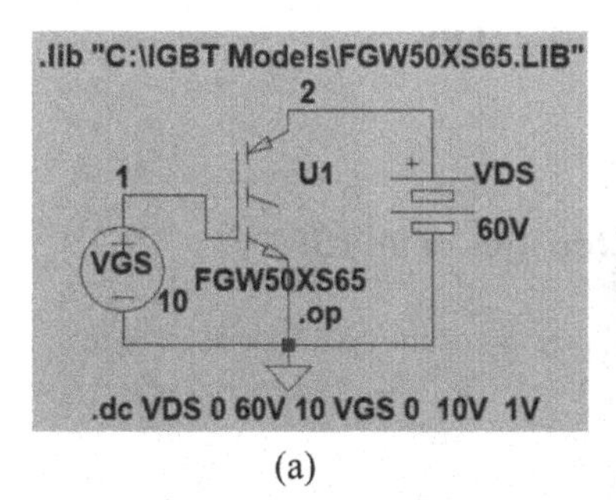

(a)

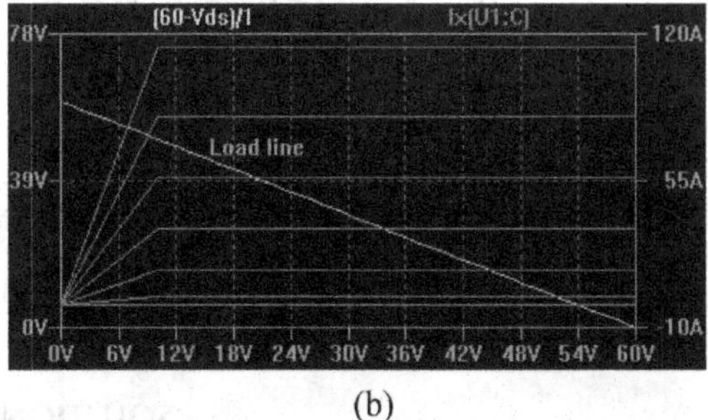

(b)

FIGURE 8.30 Characteristics of Fuji Electric IGBT FGW50XS65 [27]. (a) IGBT Test circuit. (b) v-I output characteristics.

```
.op
.lib "C:\IGBT Models\FGW50XS65.LIB"
.backanno
.end
*****BODY DIODE SPECIAL MODEL*****
.SUBCKT DFGW50XS65C_sp 1 2
V_V1          N1 N2 DC 0Vdc AC 0Vac
E_E2          N14 0 VALUE { LIMIT(10/(V(N11)+0.001), 0, 5) }
G_ABMI12      0 N9 VALUE { I(V_Vsns) }
V_Vsns        1 N1 DC 0Vdc AC 0Vac
C_C1          N8 2  0.45n
C_C2          N11 0 0.01n
D_D1          N2 2  DFGW50XS65C_s
R_R2          N11 0 10MEG
E_E7          N3 0 VALUE { LIMIT(I(V_V1)*1000, 0, 5) }
S_S1          N7 N8 N2 2 SW_S1
RS_S1         N2 2 1G
.MODEL        SW_S1 VSWITCH Roff=10e6 Ron=0.1m Voff=-0.0V
Von=0.5V
E_E5          0 N10 VALUE { LIMIT(V(N9, 0), -500, 0) }
E_E3          N8 N13 VALUE { V(N13,2)*(V(N12)-1) }
L_L1          0 N9  0.01nH
S_S2          N10 N11 N3 0 SW_S2
RS_S2         N3 0 1G
.MODEL        SW_S2 VSWITCH Roff=10e6 Ron=0.1m Voff=4.0V
Von=4.5V
R_R1          N13 2  10
R_R3          N3 0  10MEG
G_ABMI1       2 N1 VALUE { V(N7,N8) }
C_C3          N12 0  0.01n
E_E1          N7 2 VALUE { I(V_V1) }
R_R4          N14 N12  100
.MODEL DFGW50XS65C_s D
+ IS=500.00E-6 N=3.9704 RS=6.1058E-3 IKF=5.8231
+ CJO=1.2533E-9 M=0.47774 VJ=34.556E-3 ISR=0
+ BV=670 IBV=100.00E-6 TT=37.20E-9
.ENDS DFGW50XS65C_sp
```

8.10 EXAMPLES OF IGBT DC–DC CONVERTERS

Example 8.6:

Plotting the Output and Transfer Characteristics of an IGBT

Use PSpice to plot the output characteristics (V_{CE} vs. I_C) of the IGBT for V_{CE} = 0 to 100 V and V_{GS} = 4 to 210 V.

SOLUTION

The PSpice schematic of an *n*-type IGBT is shown in Figure 8.31(a). The output characteristics I_C versus V_{CE} are shown in Figure 8.31(b). The transfer

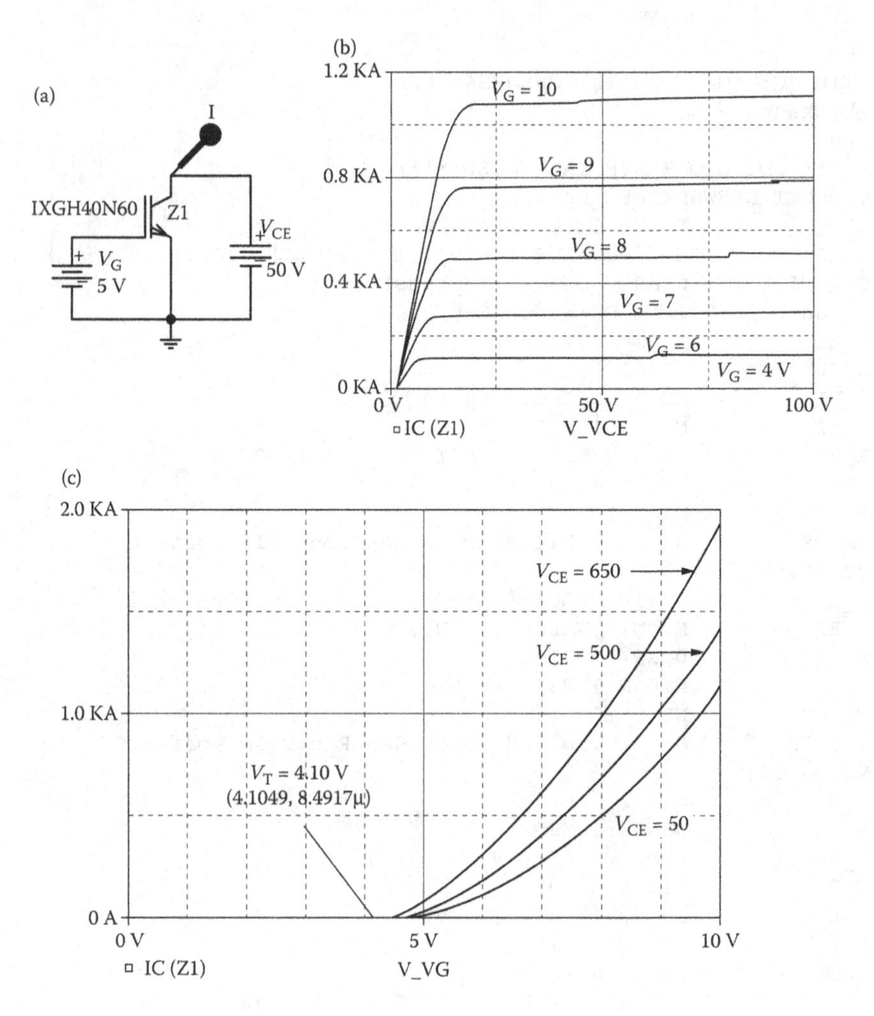

FIGURE 8.31 Transfer characteristic of a typical *n*-type IGBT. (a) Schematic. (b) Output characteristics. (c) Transfer characteristics.

characteristics I_C versus V_G are shown in Figure 8.31(c), which gives the threshold voltage V_T = 4.1049 V. An Zbreak N device will permit adjustment of the model parameters. If gate voltage is less than the threshold voltage, the device remains off.

LTspice IGBT Characteristics: The schematic for plotting the output characteristics of an IGBT is shown in Figure 8.32(a). An IGBT is a voltage-dependent device, and its collector current depends on the gate-emitter voltage. As the gate-emitter voltage is increased, the collector current is also increased. As the collector-emitter voltage increases, the collector current increases until the collector current reaches the saturation region and remains practically constant. For operating the IGBT as a switching device, its gate-emitter voltage must be high enough to keep the collector-emitter low for a specific load line as shown in Figure 8.32(b) for a load resistance of 1 ohm.

The list of the circuit file for IGBT FGW50XS65 and the subcircuit MFGW50XS65_p is as follows:

```
* LTspice File location for "Figure 8-24-IGBT" for Figure 8.32
VDS 2 0 60V
VGS 1 0 10
XU1 2 1 0 FGW50XS65
.dc VDS 0 60V 10 VGS 0 10V 1V
.op
.lib "C:\IGBT Models\FGW50XS65.LIB"
.backanno
.end
*$
* PART NUMBER: FGW50XS65
* MANUFACTURER: Fuji Electric
* VCES=650V, IC=50A @100C
* (C) Fuji Electric Co., Ltd. All Rights Reserved.
.SUBCKT FGW50XS65 C G E;    C - collector, G - Gate and
E - Emitter
X_U1 C G E  MFGW50XS65_p
.ENDS FGW50XS65
```

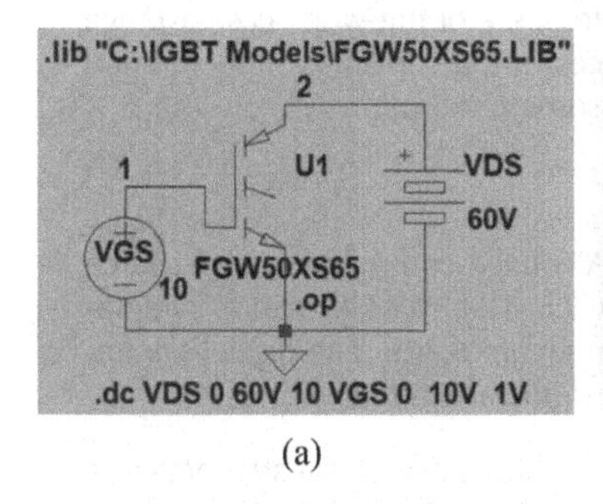

(a)

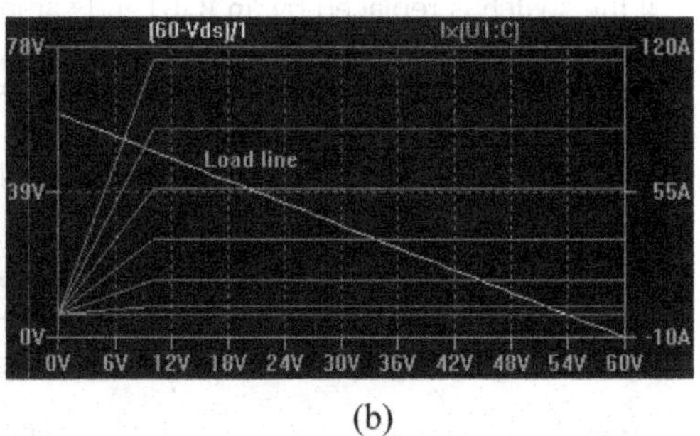

(b)

FIGURE 8.32 Characteristics of Fuji Electric IGBT FGW50XS65 [26]. IGBT Test circuit. (b) v-I output characteristics.

```
*$
.SUBCKT MFGW50XS65_p C G E
E_E1  C2 C1 VALUE {(-1)*LIMIT(0.041331+0.0071217*I(V_VC)
+ +3.7146E-5*I(V_VC)*I(V_VC),-10,10)}
R_RE  83 E    10u
M_M1  81 82 83 83 MFIN
D_DDS 83 81       DO
C_CGE 82 83    6.77n
R_RG  G 82  3.2
D_D1  2 81        DGD
V_VC  C C2    0Vdc
R_R1  1 82  100MEG
M_S2  2 81 82 82  MNSW
R_R2  81 2  10MEG
R_RC  C1 85  10u
M_S1  1 82 81 81  MNSW
D_DBE 85 81  DE
Q_Q1  83 81 85    QOUT
C_CGD 1 82    4.0n
.MODEL DGD D (CJO=1.0938E-9 M={0.4451*0.6} VJ=16.973E-3)
.MODEL MFIN NMOS (LEVEL=3 L=1U W=1U LEVEL=3
+ VTO=4.0 KP=4.5 THETA=70m)
.MODEL DO D (IS=3.79p)
.MODEL DE D (IS=3.79p N=2)
.MODEL QOUT PNP (IS=3.79p NF=1.20 BF=1
+ XTB=1.3 CJC=0.01n CJE=0.01n VAF=100k TF=0.1n)
.MODEL MNSW NMOS Vto=-0.0 KP=5 N=1Meg Rds=1e12
.ENDS MFGW50XS65_p
*$
```

Example 8.7:

Finding the Performance of Boost DC–DC Converter with an IGBT switch

If the switch is replaced by an IGBT in Example 8.5, plot the worst-case average and instantaneous load voltages if the tolerances of the capacitor, inductor, and resistor are ±20% (with standard deviation). Assume $C = 5\ \mu F$.

SOLUTION

The PSpice schematic is shown in Figure 8.33. A voltage-controlled voltage source with a voltage gain of six drives the MOSFET switch. The PWM generator is implemented as a descending hierarchy as shown in Figure 8.3(b). The model parameters for the IGBT switch and the freewheeling diode are as follows:

```
.MODEL IXGH4 0N60 NIGBT (TAU=287.56E-9 KP=50.034 AREA=37.500E-6
+ AGD=18.750E-6 VT=4.1822 KF=.36047 CGS=31.942E-9 COXD=53.188E-9
VTD=2.6570
.MODEL DMD D(IS=2.22E-15 BV=1200V CJO=1PF TT=0US)
```

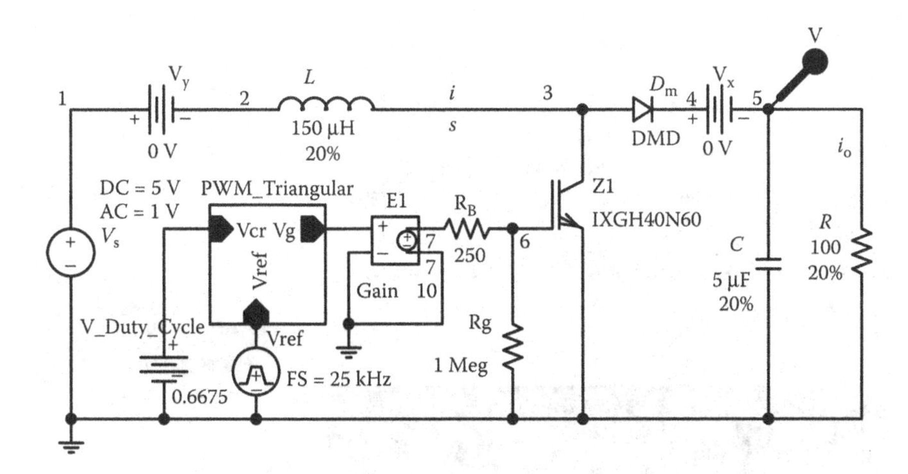

FIGURE 8.33 PSpice schematic for Example 8.7.

The plots of the average and instantaneous output voltages are shown in Figure 8.34 under nominal and maximum worst-case conditions. With the specified tolerances, the difference is approximately +0.5 V. For the output files, the sensitivity of each component on the output could be found.

LTspice: LTspice schematic for Example 8.7 is shown in Figure 8.35(a). The plots of the switch voltage, the input power, input current, the output voltage, and the output power are shown in Figure 8.35(b). The input current as shown in Figure 8.35(c)

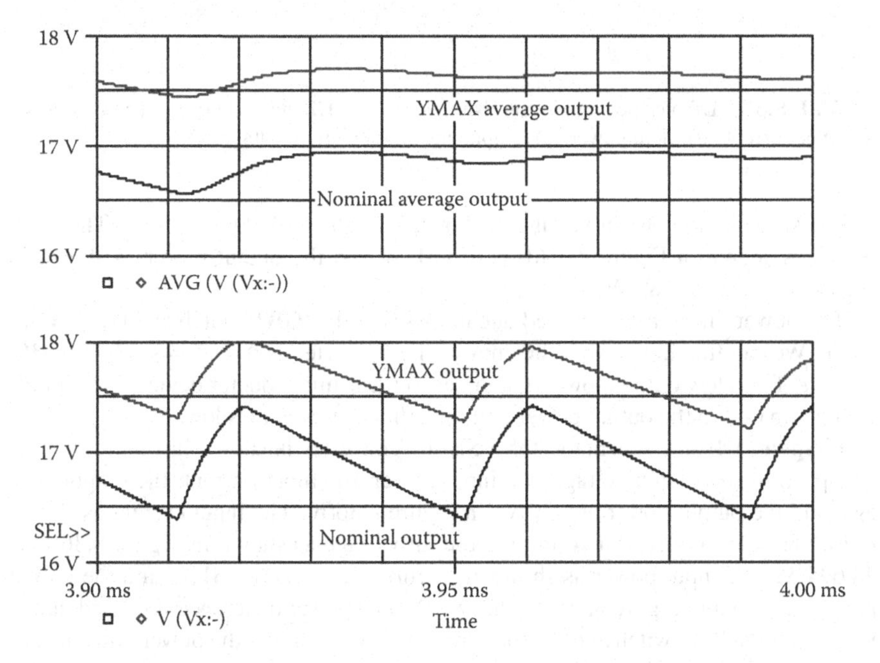

FIGURE 8.34 Worst-case and nominal output voltages for Example 8.7.

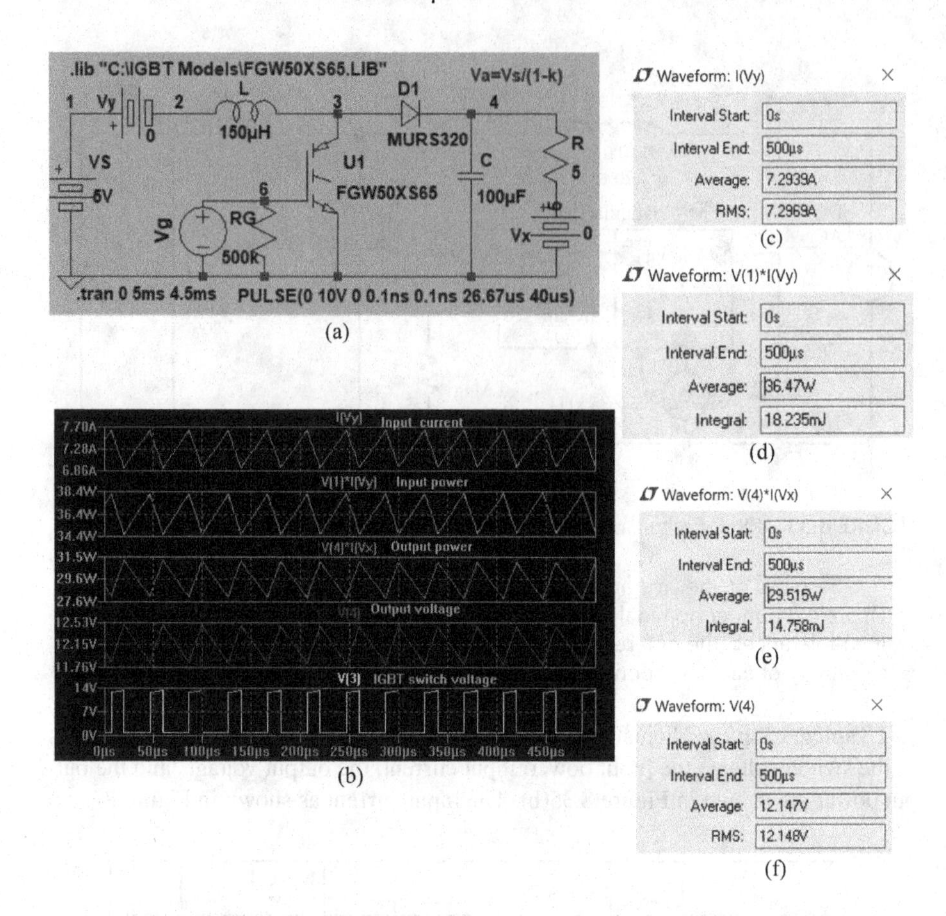

FIGURE 8.35 LTspice schematic for Example 8.7. (a) LTspice schematic. (b) Waveform. (c) Input current. (d) Input power. (e) Output power. (f) Output voltage.

is 70.994 A and the output voltage as shown in Figure 8.35(f) is 78.69 V. The input power as shown in Figure 8.35(d) is 851.81 W and the output power as shown in Figure 8.35(e) is 621.89 W.

The power difference is caused due to losses in the IGBT switch and the forward diode. We can find the power efficiency of the converter as $\eta = \frac{P_{out}}{P_{in}} = \frac{29.515}{36.47} = 80.93\%$

Note: For a low output power, the losses in the semiconductor devices are significant compared to the output power and the efficiency becomes low.

LTspice: LTspice schematic for a SEPIC converter is shown in Figure 8.36(a). The plots of the switch voltage, the input power, the input current, the output voltage, and the output power are shown in Figure 8.36(b). The input current as shown in Figure 8.36(c) is 70.984 A and the output voltage as shown in Figure 8.36(f) is 78.694 V. The input power as shown in Figure 8.36(d) is 851.81 W and the output power as shown in Figure 8.36(e) is 621.89 W. The power difference is caused due to losses in the IGBT switch and the forward diode. We can find the power efficiency of the converter as $\eta = \frac{P_{out}}{P_{in}} = \frac{521.89}{851.81} = 61.23\%$

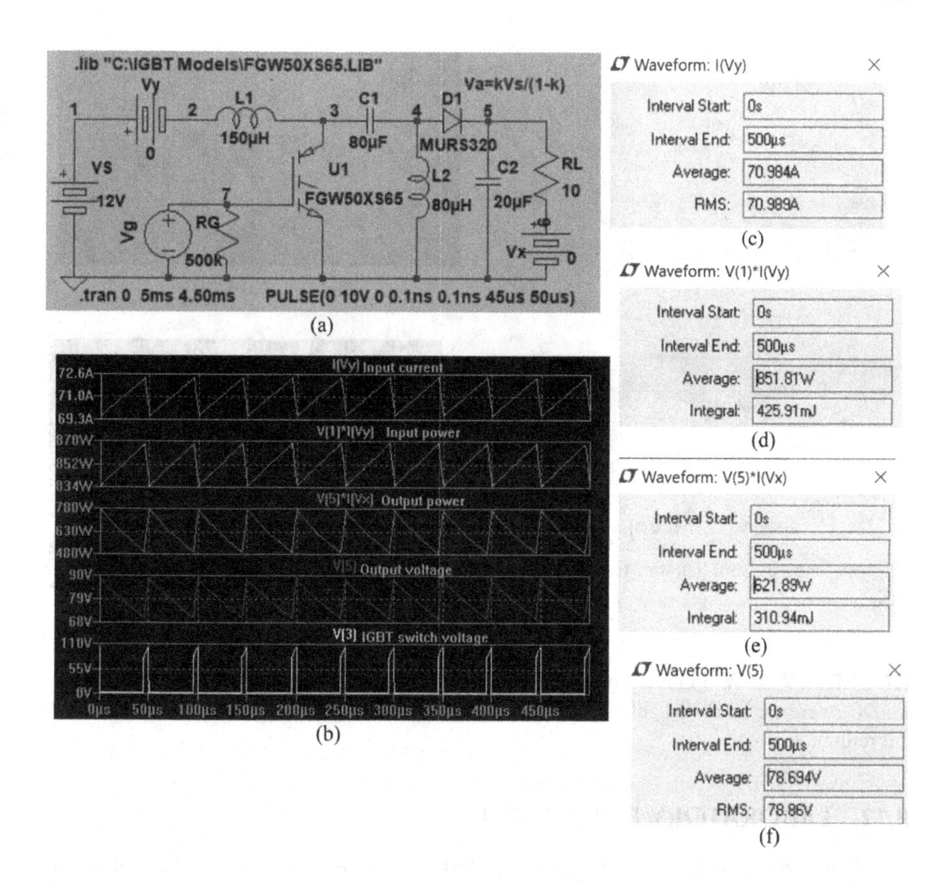

FIGURE 8.36 LTspice schematic for Example 8.8. (a) LTspice schematic. (b) Waveforms. (c) Input current. (d) Input power. (e) Output power. (f) Output voltage.

Note: For a low output power, the losses in the semiconductor devices are significant compared to the output power and the efficiency becomes low.

8.11 PWM GENERATOR

The output voltage of the DC–DC converters depends on the duty cycle k of the PWM gating signal. The gating signal is normally from a PWM generator and the duty cycle depends on the desired output voltage requirement of the converter. A PWM waveform can be generated by op-amp circuit by comparing a triangular carrier waveform of a desired frequency with a DC reference signal. As the reference voltage is varied, the width of the PWM pulse is varied.

LTspice schematics for PWM generator are shown in Figure 8.37(a) for DC reference and (c) for sinusoidal wave reference. The PWM waveform is generated by comparing a triangular waveform with a DC reference signal as shown in Figure 8.37(b) or a sinusoidal reference signal shown in Figure 8.37(d). DC reference is generally used for DC–DC converters and sinewave reference is normally used for DC–AC converters known as an inverter.

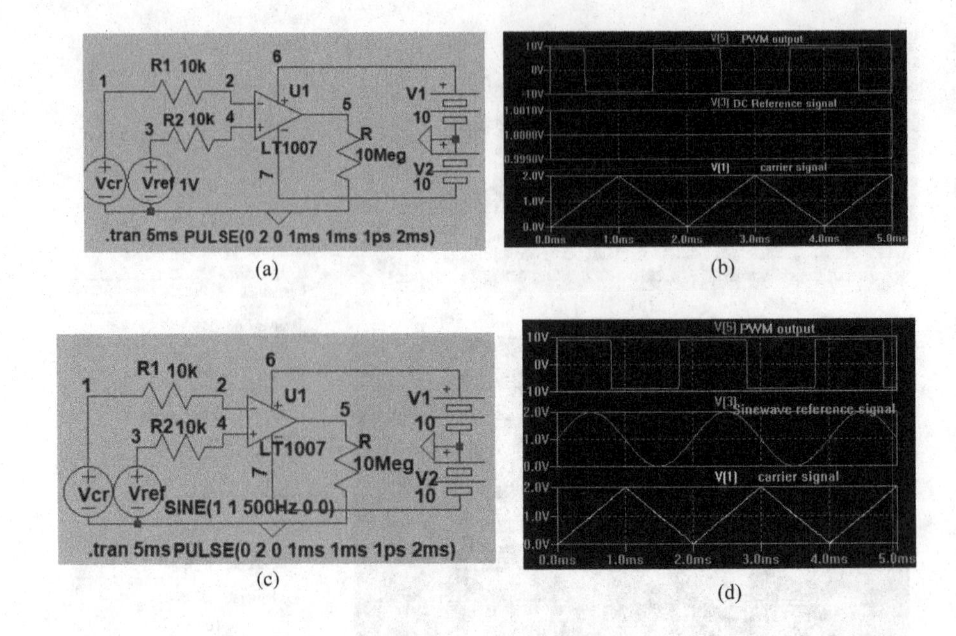

(a) (b)

(c) (d)

FIGURE 8.37 PWM Generator. (a) PWM generator for DC reference. (b) PWM waveforms for DC reference (c) PWM generator forsinewave reference. (d) PWM waveforms for sinusoidal reference.

8.12 LABORATORY EXPERIMENT

The following two experiments are suggested to demonstrate the operation and characteristics of DC choppers:

DC buck chopper
DC boost chopper

8.12.1 Experiment TP.1

DC Buck Chopper

Objective	To study the operation and characteristics of a DC buck chopper under various load conditions
Applications	The DC buck (step-down) chopper is used to control power flow in power supplies, DC motor control, input stages to inverters, etc.
Textbook	See Ref. [12] and Sections 5.3 and 5.7
Apparatus	1. One BJT/MOSFET with ratings of at least 50 A and 500 V, mounted on a heat sink 2. One fast-recovery diode with ratings of at least 50 A and 500 V, mounted on a heat sink 3. A firing pulse generator with isolating signals for gating the BJT 4. An *RL* load 5. One dual-beam oscilloscope with floating or isolating probes 6. DC voltmeters and ammeters and one noninductive shunt

(Continued)

DC Buck Chopper *(Continued)*

Warning	Before making any circuit connection, switch the DC power off. Do not switch on the power unless the circuit is checked and approved by your laboratory instructor. Do not touch the transistor heat sinks, which are connected to live terminals
Experimental procedure	1. Set up the circuit as shown in Figure 8.38. Use a load resistance R only 2. Connect the measuring instruments as required 3. Set the chopping frequency to $fc = 1$ kHz and the duty cycle to $k = 50\%$ 4. Connect the firing pulse to the BJT/MOSFET. 5. Observe and record the waveforms of the load voltage V_o, the load current i_o, and the input current i_s 6. Measure the average load voltage $V_{o(DC)}$, the average load current $I_{o(DC)}$, the RMS transistor current $I_{T(RMS)}$, the average input current Is(DC), and the load power P_L 7. Repeat steps 2–6 with a load inductance L only 8. Repeat steps 2–6 with both load resistance R and load inductance L
Report	1. Present all recorded waveforms and discuss all significant points 2. Compare the waveforms generated by SPICE with the experimental results, and comment 3. Compare the experimental results with the predicted results 4. Calculate and plot the average output voltage $V_{o(DC)}$ against the duty cycle 5. Discuss the advantages and disadvantages of this type of chopper

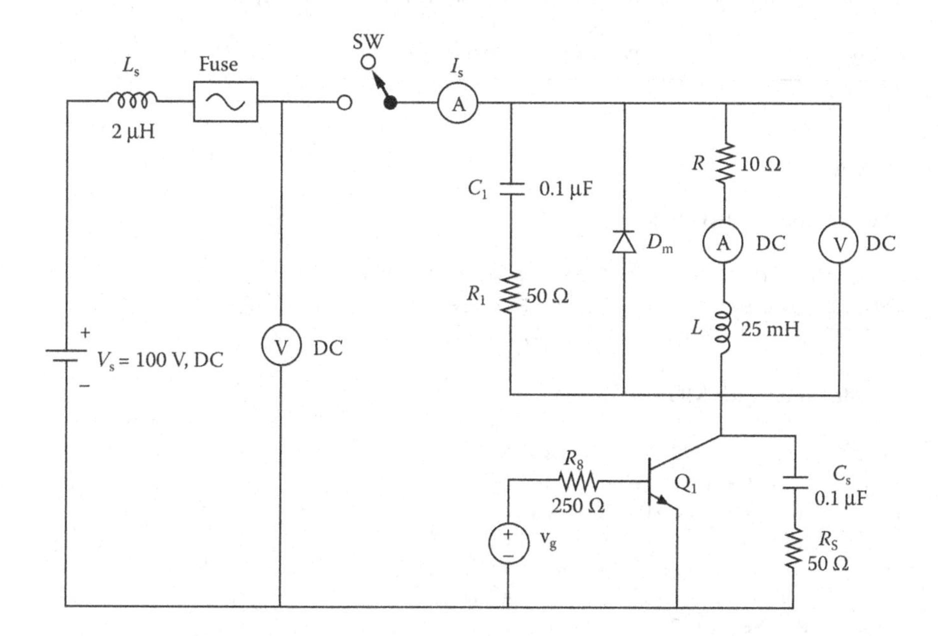

FIGURE 8.38 BJT step-down chopper.

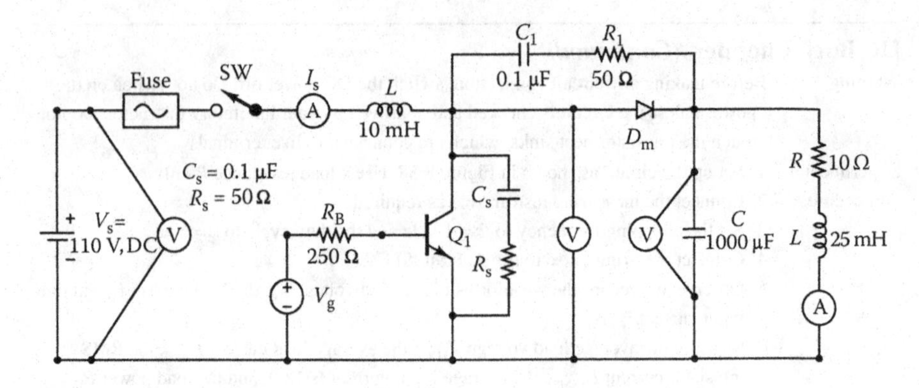

FIGURE 8.39 BJT step-up chopper.

8.12.2 Experiment TP.2

DC Boost Chopper

Objective	To study the operation and characteristics of a DC boost chopper under various load conditions
Applications	The DC boost (step-up) chopper is used to control power flow in power supplies, DC motor control, input stages to inverters, etc
Textbook	See Ref. [12], Sections 5.4 and 5.7
Apparatus	See Experiment TP.1
Warning	See Experiment TP.1
Experimental procedure	Set up the circuit as shown in Figure 8.39. Repeat the steps of Experiment TP.1
Report	See Experiment TP.1

SUMMARY

The statements of BJTs are

```
Q<name>  NC NB NE NS QNAME [(area) value]
.MODEL QNAME NPN (P1=B1 P2=B2 P3=B3 ... PN=VN
.MODEL QNAME NPN (P1=B1 P2=B2  P3=B3 ... PN=VN
```

The statements for MOSFETs are

```
M <name> ND NG NS NB MNAME
+       [L=<value>] [W=<value>]
+       [AD=<value>] [AS=<value>]
+       [PD=<value>] [PS=<value>]
+       [NRD=<value>] [NRS=<value>]
+       [NRG=<value>] [NRB=<value>]
.MODEL MNAME NMOS (P1=V1 P2=V2 P3=V3 ... PN=VN)
.MODEL MNAME PMOS (P1=V1 P2=V2 P3=V3 ... PN=VN)
```

The statements for JGBTs are

```
Z1 NC NG NE ZNAME
.MODEL ZNAME NIGBT (P1 = B1 P2 = B2 ... PN=BN) for N-type
MODEL ZNAME PIGBT (P1=B1 P2=B2 ... PN=BN) for P-type
```

PROBLEMS

8.1 It is required to design the step-up chopper of Figure 8.25 with the following specifications:
DC supply voltage, $V_s = 110$ V
Load resistance, $R = 5\ \Omega$
Load inductance, $L = 15$ mH
DC output voltage $V_{o(DC)}$ is 80% of the maximum permissible value Peak-to-peak load ripple current should be less than 10% of the load average value, $I_{o(DC)}$
a. Determine the ratings of all devices and components under worst-case conditions.
b. Use SPICE to verify your design.
c. Provide a cost estimate of the circuit.

8.2 1. Design an input LC filter for the chopper of Problem 8.1. The harmonic content of the input current should be less than 10% of the value without the filter.
2. Use SPICE to verify your design in part (a).

8.3 It is required to design a step-up chopper of Figure 8.25 with the following specifications:
DC supply voltage, $V_s = 24$ V
Load resistance, $R = 5\ \Omega$
Load inductance, $L = 15$ mH
DC output voltage $V_{o(DC)}$ is 150% of the maximum permissible value
Average load current, $I_{o(DC)} = 2.5$ A
Peak-to-peak load ripple current should be less than 10% of the load average value, $I_{o(DC)}$
a. Determine the ratings of all devices under worst-case conditions.
b. Use SPICE to verify your design.
c. Provide a cost estimate of the circuit.

8.4 1. Design an output C filter for the chopper of Problem 8.3. The harmonic content of the output voltage should be less than 10% of the value without the filter.
2. Use SPICE to verify your design in part (a).

8.5 It is required to design a buck regulator with the following specifications:
DC input voltage, $V_s = 15$ V
Average output voltage, $V_{o(DC)} = 10$ V
Average load voltage, $I_{o(DC)} = 2.5$ A
Peak-to-peak output ripple voltage should be less than 50 mV
Switching frequency, $f_c = 20$ kHz
Peak-to-peak inductor ripple current should be less than 0.5 A

a. Determine the ratings of all devices and components under worst-case conditions.
b. Use SPICE to verify your design.
c. Provide a cost estimate of the circuit.

8.6 It is required to design a boost regulator with the following specifications:
DC input voltage, $V_s = 12$ V
Average output voltage, $V_{o(DC)} = 24$ V
Average load current, $I_{o(DC)} = 2.5$ A
Switching frequency, $f_c = 20$ kHz
Peak-to-peak output ripple voltage should be less than 50 mV
Peak-to-peak inductor ripple current should be less than 0.5 A
a. Determine the ratings of all devices and components under worst-case conditions.
b. Use SPICE to verify your design.
c. Provide a cost estimate of the circuit.

8.7 It is required to design a buck-boost regulator with the following specifications:
DC input voltage, $V_s = 15$ V
Average output voltage, $V_{o(DC)} = 10$ V
Average load current, $I_{o(DC)} = 2.5$ A
Switching frequency, $f_c = 20$ kHz
Peak-to-peak output ripple voltage should be less than 50 mV
Peak-to-peak ripple current of inductor should be less than 0.2 A
a. Determine the ratings of all devices and components under worst-case conditions.
b. Use SPICE to verify your design.
c. Provide a cost estimate of the circuit.

8.8 It is required to design a Cuk regulator with the following specifications:
DC input voltage, $V_s = 15$ V
Average output voltage, $V_{o(DC)} = 10$ V
Average load current, $I_{o(DC)} = 2.5$ A
Switching frequency, $f_c = 20$ kHz
Peak-to-peak ripple voltages of capacitors should be less than 50 mV
Peak-to-peak ripple currents of inductors should be less than 0.2 A
a. Determine the ratings of all devices and components under worst-case conditions.
b. Use SPICE to verify your design.
c. Provide a cost estimate of the circuit.

8.9 Assuming ±20% tolerances with standard deviations for all capacitors, inductors, and resistors, plot the worst-case average and instantaneous load voltages for Problem 8.1.

8.10 Assuming ±20% tolerances with standard deviations for all capacitors, inductors, and resistors, plot the worst-case average and instantaneous load voltages for Problem 8.2.

8.11 Assuming ±20% tolerances with standard deviations for all capacitors, inductors, and resistors, plot the worst-case average and instantaneous load voltages for Problem 8.3.

8.12 Assuming ±20% tolerances with standard deviations for all capacitors, inductors, and resistors, plot the worst-case average and instantaneous load voltages for Problem 8.4.

8.13 Assuming ±20% tolerances with standard deviations for all capacitors, inductors, and resistors, plot the worst-case average and instantaneous load voltages for Problem 8.5.

8.14 Assuming ±20% tolerances with standard deviations for all capacitors, inductors, and resistors, plot the worst-case average and instantaneous load voltages for Problem 8.6.

8.15 Assuming ±20% tolerances with standard deviations for all capacitors, inductors, and resistors, plot the worst-case average and instantaneous load voltages for Problem 8.7.

8.16 Assuming ±20% tolerances with standard deviations for all capacitors, inductors, and resistors, plot the worst-case average and instantaneous load voltages for Problem 8.8.

8.17 The SEPIC converter as shown in Figure P8.17 has an input voltage of $Vs = 12$ V. The required average output voltage is $V_a = 98$ V at a load current of Ia = 0.5 A. The peak-to-peak ripple current of inductors L1 and L2 is limited to 5% of its average current. The peak-to-peak output ripple voltage is 5% of the average output voltage. The switching frequency is 20 kHz.

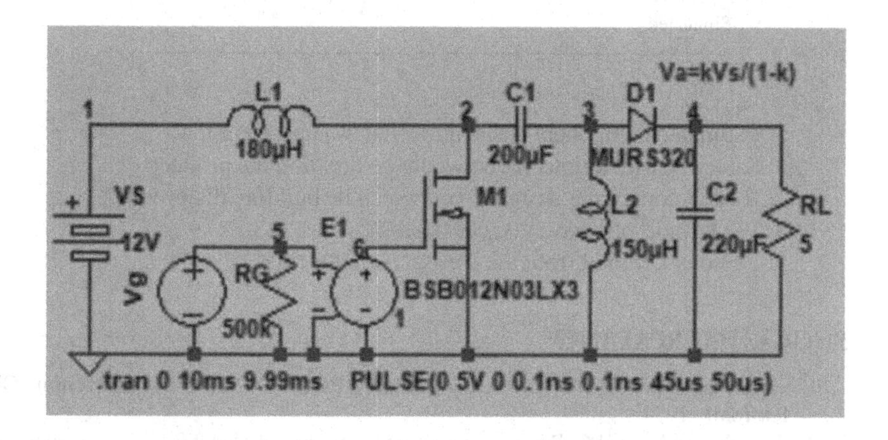

FIGURE P8.17 SEPIC converter.

1. Calculate the performance parameters and complete Table P8.17. Assume an ideal transistor switch.
   a. the duty cycle k
   b. the average current through inductor I_{L1}
   c. the filter inductance L_1
   d. the average current through inductor I_{L2}
   e. the filter inductance L_2
   f. the filter capacitor C_1

g. the critical value of inductor L_{C1}
h. the critical value of inductor L_{C2}
i. the critical value of capacitor C_2

2. Use LTspice/Orcad.
3. Apply a pulse pulse-voltage of 5V (peak) between the gate and the source terminals through a voltage-controlled voltage source as an isolating unit. Use MOSFET Type: BSB012N03LX3 and Power Diode type: MURS320.
   Run the simulation from 0 to 10 ms so that the steady-state condition is reached. Make sure that the transistor is turned on and operated as a switch.
   a. Plot the instantaneous output voltage $v_o(t)$
   b. Plot the instantaneous voltage across diode D_1, $v_{d1}(t)$
   c. Plot the instantaneous inductor current $i_{L1}(t)$
   d. Include .four command to find the THD of the outcome current with 29th harmonic contents at the output freqeuncy of 20 kHz.
4. Complete Table P8.17 for simulated values.

TABLE P8.17

	V_a (V)	ΔV_c (mv)	I_{L1} (A)	ΔI_{L1} (A)	I_{L2} (A)	ΔI_{L2} (A)
Calculated	98					
Simulated						

5. Estimated component costs: https://www.jameco.com/
6. Safety considerations if you would be building the product:
7. Risk factors of the design if you would be building the product:
8. What trade-offs have you considered?
9. Lessons Learned from the Design Assignment:

SUGGESTED READING

1. I. Getreu, *Modeling the Bipolar Transistor*, Part # 062-2841-00, Beaverton, OR: Tektronix, 1979.
2. H.K. Gummel and H.C. Poon. An integral charge control model for bipolar transistors, *Bell Systems Technical Journal*, *49*, 1970, 115–120.
3. J.J. Ebers, and J.J. Moll, Large signal behavior of junction transistors, *Proceedings of the IRE*, 42, 1954,1161–1172.
4. H. Schichman and D.A. Hodges, Modeling and simulation of insulated gate field effect transistor switching circuits, *IEEE Journal of Solid-State Circuits*, SC*3*, 1968, 285–289.
5. J.F. Meyer, MOS models and circuit simulation, *RCA Review*, *32*, 1971, 42–63.
6. A. Vladimirescu and S. Liu, *The Simulation of MOS Integrated Circuits Using SPICE2*, Memorandum M80/7, Berkeley: University of California, February 1980.
7. S. Hangeman, Behavioral modeling and PSpice simulation SMPS control loops: Parts I and II, *PCIM Magazine*, April–May 1990.
8. P.O. Lauretzen, *Power semiconductor device models for use in circuit simulation*, *Conference Proceedings of the IEEE-IAS Annual Meeting*, 1990, pp. 1559–1560.

9. S. Natarajan, An effective approach to obtain model parameters for BJTs and FETs from data books, *IEEE Transactions on Education*, *35*(2), 1992, 164–169.
10. G. Fay and J. Sutor, Power FET models are easy and accurate, *Powertechnics Magazine*, 1987, 16–21.
11. C.F. Wheatley, H.R. Jr. Ronan, and G.M. Dolny, Spicing up SPICE II software for power MOSFET modeling, *Powertechnics Magazine*, 1987, 28–32.
12. M.H. Rashid, *Power Electronics: Circuits, Devices, and Applications*, Third Edition, Englewood Cliffs, NJ: Prentice-Hall, 2004, Chapter 6.
13. M.H. Rashid, *Introduction to PSpice Using OrCAD for Circuits and Electronics*, Third Edition, Englewood Cliffs, NJ: Prentice-Hall, 2003, Chapters 8 and 9.
14. B.J. Baliga, *Power Semiconductor Devices*, Boston, MA: PWS Publishing, 1996.
15. K. Sheng, S.J. Finney and B.W. Williams, Fast and accurate IGBT model for PSpice, *Electronics Letters*, *32*(*25*), 1996, 2294–2295.
16. A.G.M. Strollo, A new IGBT circuit model for SPICE simulation, *Power Electronics Specialists Conference*, June 1997, *1*, 133–138.
17. K. Sheng, S.J. Finney and B.W. Williams, A new analytical IGBT model with improved electrical characteristics, *IEEE Transactions on Power Electronics*, *14*(*1*), 1999, 98–107.
18. K. Sheng, B.W. Williams and S.J. Finney, A review of IGBT models, *IEEE Transactions on Power Electronics*, *15*(*6*), 2000, 1250–1266.
19. M.H. Rashid, *Microelectronic Circuits: Analysis and Design*, Florence, KY: Cengage Publishing, 2011, Chapters 7 and 8.
20. F. Petrie and C. Hymowitz, A SPICE Model for IGBTs. Intusoft, CA.
21. C.E. Hymowitz, Intusoft Newsletter, "Intusoft Modeling Corner", Intusoft, June 1992, San Pedro, CA 90731.
22. International Rectifier. *Insulated Gate Bipolar Transistor Designer's Manual*, International Rectifier, El Segundo, CA 90245.
23. Infineon: Introduction to Infineon's simulation models for IGBTs and silicon diodes in discrete packages, AN 2020-17, Infineon.
24. Infineon-IKW30N65EL5-DataSheet-v02_02-EN. https://www.infineon.com/dgdl/Infineon-IGW30N65L5-DataSheet-v02_02-EN.pdf?fileId=5546d4624b0b249c014b11cd55583ac9
25. IGBT Spice Model, https://www.globalspec.com/industrial-directory/igbt_spice_model
26. Fujielectric. Discrete IGBT P-Spice Models, https://www.fujielectric.com/products/semiconductor/model/igbt/technical/design_tool.html

9 Pulse Width-Modulated Inverters

After completing this chapter, students should be able to do the following:

- Model BJTs and IGBTs in SPICE, and specify their mode parameters.
- Model pulse-width modulation (PWM) control in SPICE as a hierarchy block.
- Perform transient analysis of voltage-source and current-source inverters.
- Evaluate the performance of voltage-source and current-source inverters.
- Evaluate the performance of DC-linked inverters.
- Evaluate the performance of multilevel inverters.

9.1 INTRODUCTION

A DC–AC converter is commonly known as an inverter. The input to an inverter is DC, and the output is AC. The power semiconductor devices perform the switching action, and the desired output is obtained by varying their turn-on and turn-off times. They must have controllable turn-on and turn-off characteristics. The commonly used devices are BJTs, MOSFETs, IGBTs, GTOs, MCTs, and forced-commutated thyristors. We use PSpice switches, BJTs, and IGBTs to simulate the characteristics of the following inverters:

- Voltage-source inverters
- Current-source inverters

Thus, the voltage and the current on the input side are of DC types and the voltage and current on the output side are of AC types. Table 9.1 lists the performance parameters of inverters.

9.2 VOLTAGE-SOURCE INVERTERS

Two types of control are commonly used for varying the output of an inverter: PWM and sinusoidal pulse-width modulation (SPWM). The number of pulses per half-cycle usually ranges from 1 to 15.

In PWM control, the conduction angles of power semiconductor devices are generated by comparing a reference signal v_r with a carrier signal v_c, as shown in Figure 9.1. The pulse width δ can be varied by changing the carrier voltage v_c. This technique can be implemented by a descending hierarchy as shown in Figure 8.3(b). The input voltages to the comparator are v_r and v_c, and its output

DOI: 10.1201/9781003284451-9

TABLE 9.1
Inverter Performance Parameters for an Ideal No-load Inverter

Input side parameters	Output side parameters
DC input voltage, V_S	RMS output voltage, V_o
Average input current, I_S	RMS output current, I_o
RMS input current, I_{RMS}	The output power, $P_o = V_o I_o = I_o^2 R_L$ for a load resistance of R_L
Converter Power Efficiency, η	RMS value of the fundamental output voltage, V_{o1}
The RMS ripple content of the input current $I_{ripple} = \sqrt{I_{rms}^2 - I_S^2}$	The RMS ripple content of the output voltage $V_{ac} = \sqrt{V_o^2 - V_{o1}^2}$
	The *ripple factor*, $RF_o = \frac{V_{ac}}{V_{o1}}$
	Total harmonic distortion (THD) of the output voltage, %

is the conduction angle δ for which the switch remains on. The modulation index M is defined by

$$M = \frac{A_c}{A_r} \tag{9.1}$$

where A_r is the peak value of the reference signal v_r, and A_c is the peak value of the carrier signal v_c.

The PWM modulator can be used as a subcircuit to generate control signals for a triangular reference voltage of one or more pulses per half-cycle and a DC carrier signal. The PSpice schematic for generating PWM control signals is shown in Figure 9.2(b). V_{ref} produces a pulse signal at a reference (or switching) frequency of $f_s = p \times f_o$, where p is the number of pulses per half-cycle and f_o is the frequency of the output voltage. A

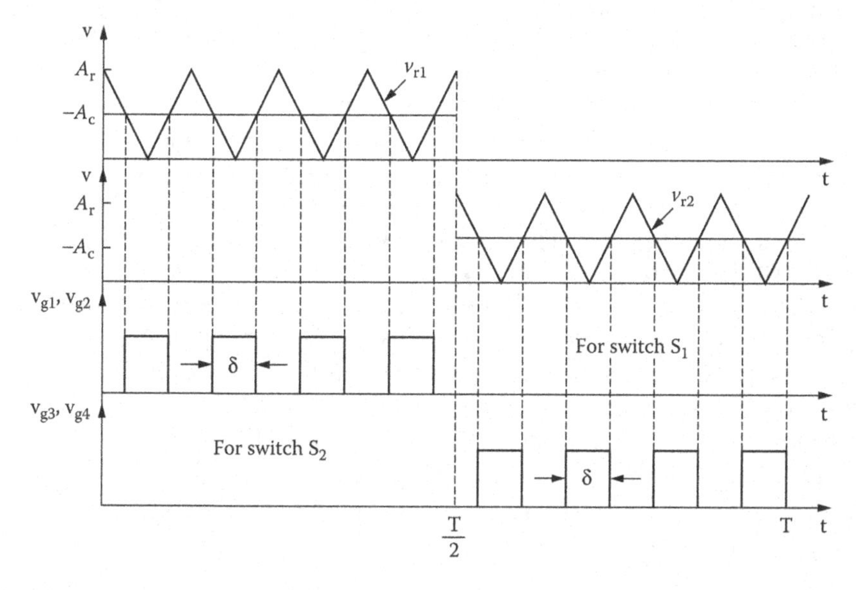

FIGURE 9.1 PWM control.

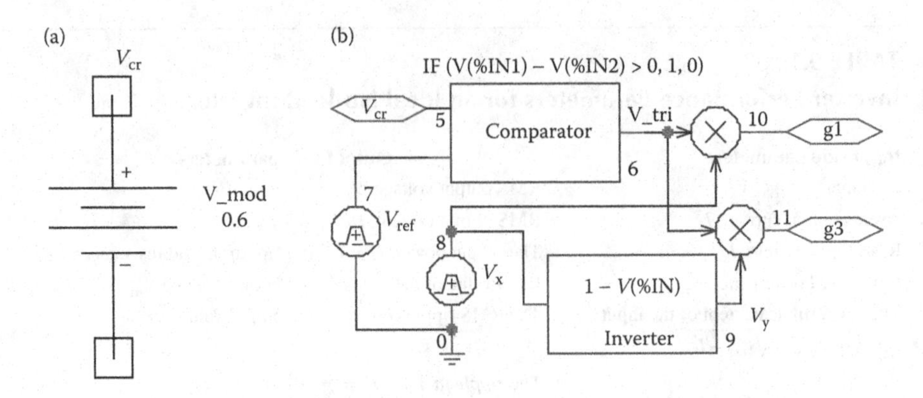

FIGURE 9.2 Schematics for PWM signals. (a) Carrier signal. (b) Comparator.

DC carrier signal V_{cr} as shown in Figure 9.2(a) is compared with the reference signal V_{ref} by an ABM2 device as in Figure 9.2(b). V_{ref} is a triangular signal at the carrier frequency of f_c, V_x is a pulse signal at an output frequency of f_o at a 50% duty cycle, and it releases gating signals during the first half-cycle of the output voltage. V_y is the logical inverse of V_x, that is, $V_y = \bar{V}_x$, and it releases gating signals during the second half-cycle of the output voltage. g_1 and g_3 are the gating signals for switches S_1 and S_3.

Note: V_x and V_y are multiplied by the output of the comparator to produce the gate signals for switch pairs: S_1 and S_3 and S_2 and S_4.

The subcircuit definition for the modulator model PWM can be described as follows:

```
* Subcircuit for PWM Control
.SUBCKT PWM     7       5          10             11 PARAMS:
* fout = 60Hz  p = 4
*        Model  ref.    carrier  +g1 control  +g3 control
*        Name   signal signal    voltage      voltage
* Where fout is the output frequency and p is the number of
* pulses per half cycle
Vref 7 0 PULSE (1 0 0 {1/(2*{2*{p}*{fout}})}
{1/(2*{2*{p}*{fout}})-1ns}1ns {1/{2*{p}*
+ {fout}}})             ; Reference voltage at output frequency
V_mod 5 0 0.6     ; Modulation index of less than 1.0
Vx 8 0 PULSE (0 1 0 1ns 1ns {1/(2*{fout})-
2ns} {1/{fout}}) ; pulse at the output frequency
E_ABM21  6 0 VALUE {IF(V(5)-V(7) > 0, 1, 0)}
E_ABM12  9 0 VALUE {1-V(8)}       ; Inverter
E_MULT1 10 0 VALUE {V(8)*V(6)} ; Multiplier 1
E_MULT2 11 0 VALUE {V(9)*V(6)} ; Multiplier 2
.ENDS  PWM   ; Ends subcircuit definition
```

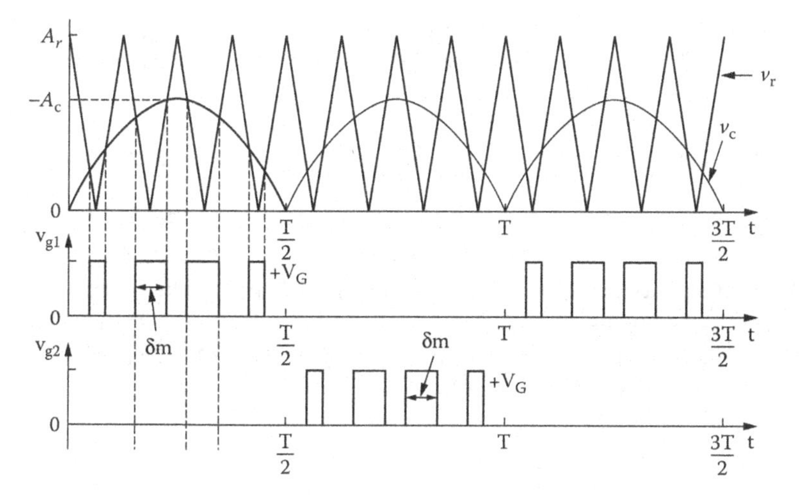

FIGURE 9.3 Gate voltages with an SPWM control.

In SPWM control, the carrier signal is a rectified sine wave as shown in Figure 9.3. PSpice generates only a sine wave. Thus, we can use a precision rectifier to convert a sine wave signal to rectified sine wave pulses and a comparator to generate a PWM waveform. This can be implemented using the PSpice Schematic as shown in Figure 9.4. Sinusoidal PWM signals are generated by replacing the DC carrier signal V_{cr} in Figure 9.2(a) with a rectified sinusoidal signal $v_{cr} = |M \sin(2\pi f_o)|$, where M is the modulation index and f_o is the frequency of the output voltage. An ABS unit as shown in Figure 9.4(a) performs the rectification function and produces the sinusoidal carrier signal. V_{ref} is a triangular signal at the switching frequency of f_s. V_x is a pulse signal at an output frequency of f_o at a 50% duty cycle, and it releases gating signals during the first half-cycle of the output voltage. V_y is the logical inverse of V_x, that is, $V_y = \overline{V_x}$, and it releases gating signals during the second half-cycle of the output voltage, g_1 and g_3 as shown in Figure 9.4(b) are generated by the comparator and are the gating signals for switches S_1 and S_3.

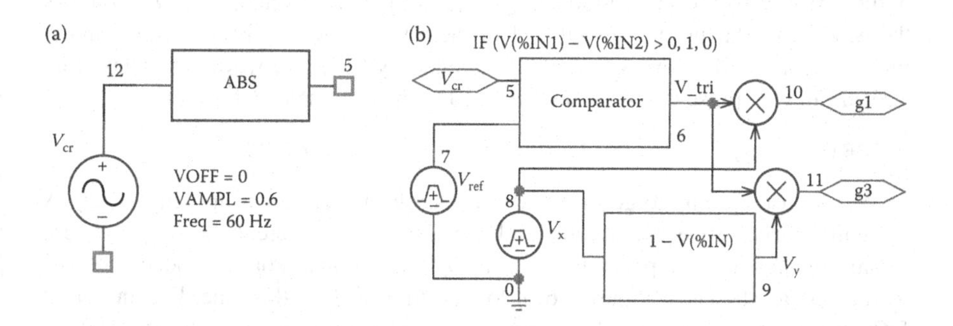

FIGURE 9.4 PSpice schematic for SPWM signals. (a) Sinusoidal carrier signal. (b) Comparator.

We can use a subcircuit SPWM for the generation of control signals. The subcircuit definition for the modulator model SPWM can be described as follows:

```
* Subcircuit for Sinusoidal PWM Control
.SUBCKT SPWM  7       5       10            11   PARAMS: fout =
60Hz p = 4
*         Model ref.    carrier  +g1 control  +g3 control
*         Name  signal signal   voltage        voltage
* Where fout is the output frequency and p is the number
* of pulses per half cycle
Vref 7 0 PULSE 1 0 0 {1/(2*{2*{p}*{fout}})} {1/(2*{2*{p}*
{fout}})-1ns} 1ns {1/{2*{p}*
+ {fout}}} ;  Reference voltage at output frequency
V_mod 12 0 SIN (0 0.6 60Hz)       ; Modulation index
                                    of less than 1.0
V_mod 12 5
Vx 8 0 PULSE (0 1 0 1ns 1ns {1/(2*{fout})-2ns}
{1/{fout}}) ; pulse           at the output frequency
E_ABM21  6 0 VALUE {IF(V(5)-V(7) > 0, 1, 0)}
E_ABM12  9 0 VALUE {1-V(8)}       ; Inverter
E_MULT1 10 0 VALUE {V(8)*V(6)} ; Multiplier 1
E_MULT2 11 0 VALUE {V(9)*V(6)} ; Multiplier 2
.ENDS SPWM
```

LTspice: In Section 9.2, we have applied behavioral functions of OrCAD to generate gating pulses for user defined output functions. For LTspice, we can use op-amp to generate PWM waveforms as shown in Figure 9.5(a). A triangular carrier wave V(2) is compared with a pulse reference wave V(1) with 50% on and 50% off. These signal voltages are compared through an op-amp comparator to produce PWM waveforms as shown in Figure 9.5(b) for the first 50% of the time. The {amp} is the amplitude of the reference signal and is defined as a variable so that we can vary the duty cycle of the pulses. For an output frequency of $f = 60$ Hz, the frequency of the carrier waveform is $8f$ (480 Hz) for 4 pulses per half-cycle. Figure 9.5(c) shows the waveforms during the second half of the reference wave. The reference signal for the second half of the waveform is obtained by the pulse waveform delayed by 8.33 ms and the pulse description is changed to the statement as follows.

```
PULSE(0 {amp} 8.33ms 0.1ns 0.1ns 8.33ms 16.67ms)
```

We can use op-amp to generate SPWM waveforms as shown in Figure 9.6(a). A triangular carrier wave V(2) is compared with a sinusoidal reference wave V(1). These signal voltages are compared through an op-amp comparator to produce SPWM waveforms as shown in Figure 9.6(b) for the first 50% of the time. For an output frequency of $f = 60$ Hz, the frequency of the carrier waveform is 8f (480 Hz) for 4 pulses per half-cycle. Figure 9.6(c) shows the waveforms during the second half of

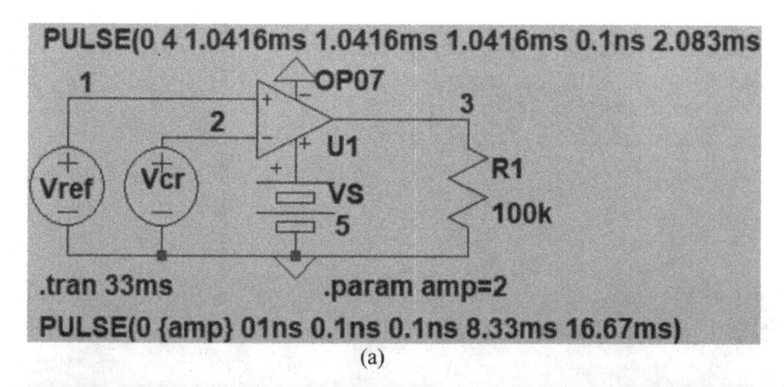

(a)

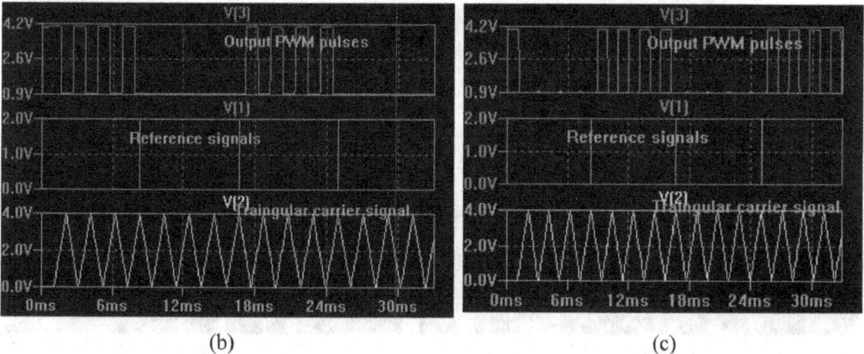

(b) (c)

FIGURE 9.5 PWM Generator for 4 pulses per half-cycle. (a) LTspice schematic. (b) PWM pulse for the first half-cycle. (c) PWM pulse for the second half-cycle.

the reference wave. The {amp} is the amplitude of the reference signal and is defined as a variable so that we can vary the duty cycle of the pulses. The reference signal for second half of the waveform is obtained by the sinusoidal reference waveform delayed by 180° and the reference sinewave description is changed to the statement as follows.

```
SINE(0 {amp} 60Hz 0 0 180 0)
```

9.2.1 Examples of Single-Phase PWM Inverters

Example 9.1:

Finding the Performance of a Single-Phase Inverter with Voltage-Controlled Switches

A single-phase bridge inverter is shown in Figure 9.7. The DC input voltage is 100 V. It is operated at an output frequency of $f_o = 60$ Hz with PWM control and four pulses per half-cycle. The modulation index $M = 0.6$. The load is purely resistive with $R = 2.5\ \Omega$. Use PSpice to (a) plot the instantaneous output voltage v_o, the instantaneous carrier and reference voltages and (b) calculate the Fourier coefficients of output voltage v_o. Use voltage-controlled switches to perform the switching action.

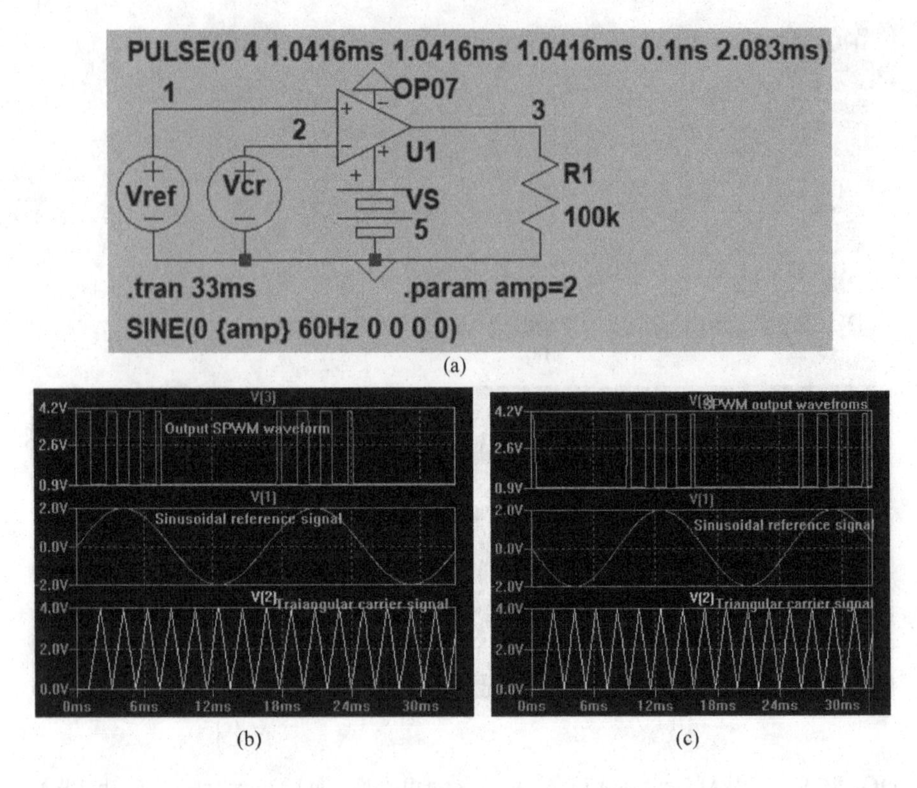

FIGURE 9.6 SPWM Generator for 4 pulses per half-cycle. (a) LTspice schematic. (b) PWM pulse for the first half-cycle. (c) PWM pulse for the second half-cycle.

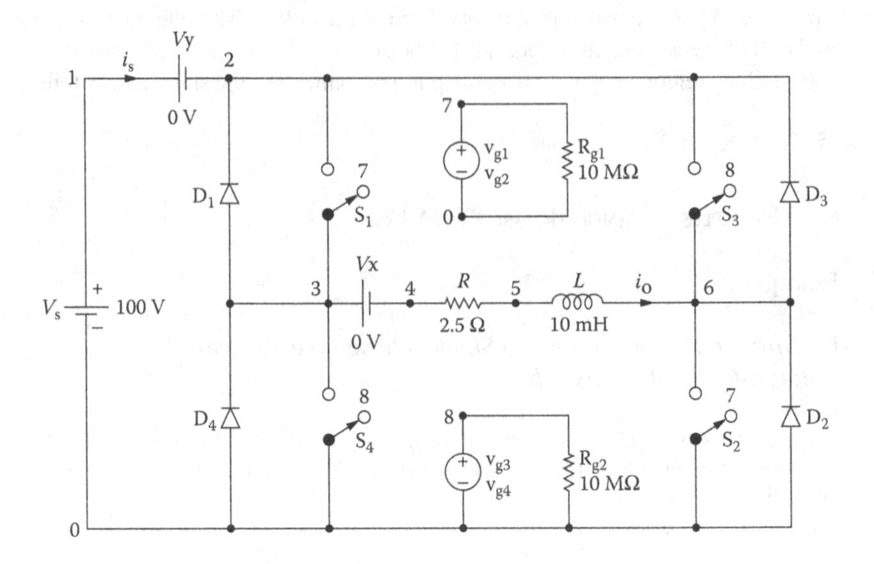

FIGURE 9.7 Single-phase bridge inverter.

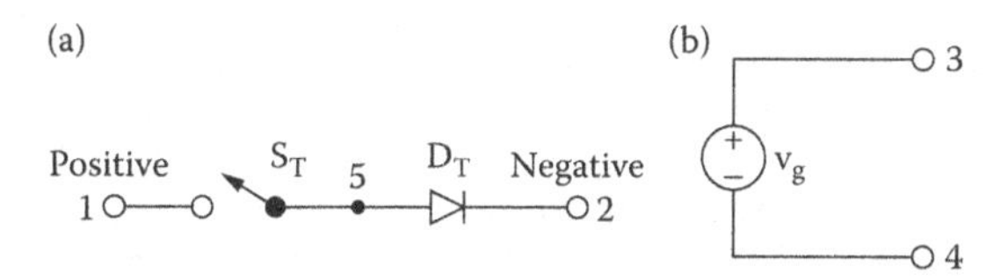

FIGURE 9.8 Switched transistor model. (a) Switch. (b) Control voltage.

SOLUTION

$P = 4$, $M = 0.6$. Assuming that $A_r = 10$ V, $A_c = 10 \times 0.6 = 6$ V, we can model a power transistor device as a voltage-controlled switch as shown in Figure 9.8. Switching characteristics are modeled by a voltage-controlled switch and the diode DT blocks any reverse current flow. The subcircuit definition for the switched transistor model (STMOD) can be described as follows:

```
* Subcircuit for switched transistor model:
.SUBCKT  STMOD    1       2         3          4
*        model   anode  cathode   +control   -control
*        name                     voltage    voltage
DT    5    2    DMOT                  ; Switch diode
ST    1    5    3    4    SMOD        ; Switch
.MODEL DMOT D(IS=2.2E-15 BV=1200V CJO=0 TT=0)  ; Diode model parameters
.MODEL SMOD VSWITCH RON = 0.01 ROFF = 10E+6 VON = 10V VOFF=5V)
.ENDS STMOD                           ; Ends subcircuit definition
```

We can use the subcircuit PWM in Figure 9.1 for the generation of control signals. The PSpice schematic is shown in Figure 9.9(a). By varying the modulation voltage V_mod (or carrier signal V_{cr}) from 0.01 to 0.99 V, the on-time of a switch and the output voltage can be changed. The comparator subcircuit for the PWM block is shown in Figure 9.9(b). The model parameters for the switch and the free-wheeling diode are as follows:

```
.MODEL SMD VSWITCH (RON=1M ROFF=10E6 VON=1V VOFF=0V)
for switches.
MODEL DMD D(IS=2.22E-15 BV=1200V CJO=1PF TT=0US))
for diodes
```

The listing of the circuit file for the inverter is as follows:

Single-Phase Inverter with PWM Control

```
Vs   4   0   100V   ; DC input voltage
Vy   4   3    0V    ; Monitors input current
V_mod  5   0    0.6    ; Modulation index of less than 1.0
.PARAM fout = 60Hz p = 4 ; parameters
* Parameters: fout  =  output frequency and p  =  #
* of pulses per half cycle
S1   3   1   10   0   SMD ; Voltage-controlled switches with SMD
```

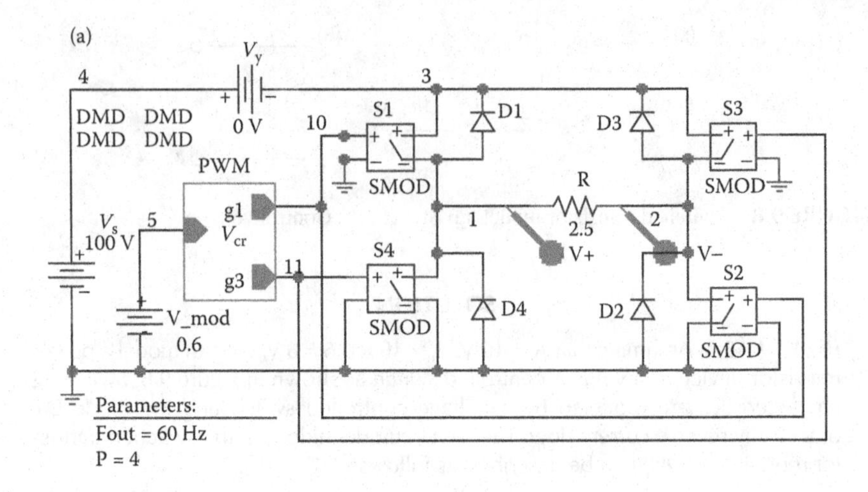

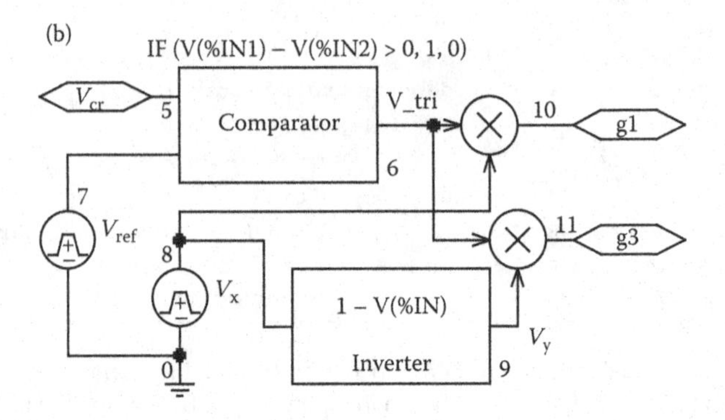

FIGURE 9.9 PSpice schematic for Example 9.1. (a) Schematic. (b) Descending hierarchy comparator.

```
S2  2  0  10  0  SMD
S3  3  2  11  0  SMD
S4  1  0  11  0  SMD
D1  1  3  DMD ; Diodes with model DMD
D2  0  2  DMD
D4  0  1  DMD
D3  2  3  DMD
R   1  2  2.5
.MODEL DMD D(IS=2.2E-15 BV=1800V TT=0) ; Diode
* model parameters
* Switch model parameters
.MODEL SMD VSWITCH (RON = 0.01 ROFF=10E+6 VON =
* 1V VOFF = 0V)
E_ABM21 6 0 VALUE {IF(V(5)-V(7) > 0, 1, 0)}
Vx 8 0 PULSE (0 1 0 1ns 1ns {1/(2*{fout})-2ns}
{1/({fout}}) ; pulse at the output frequency Vref 7 0
```

```
+PULSE (1 0 0 {1/(2*{2*{p}*{fout}})} {1/(2*{2*{p}*
{fout}})-1ns} 1ns {1/{2*{p}*
+ {fout}}}) ; Reference voltage at output frequency
E_ABM12 9     0 VALUE {1-V(8)}      ; Inverter
E_MULT1 10    0 VALUE {V(8)*V(6)} ; Multiplier 1
E_MULT2 11    0 VALUE {V(9)*V(6)} ; Multiplier 2
.TRAN   1US  16.67MS 0 0.1US       ; Transient Analysis
.FOUR   60Hz 10 V(1,2)             ; Fourier Analysis
.OPTIONS ABSTOL=1UA RELTOL=0.1 VNTOL=0.1
.PROBE
.END
```

The conduction angles are generated from two reference signals as shown in Figure 9.10. Assume that the reference voltage $V_r = 1$ V. for $M = 0.4$, the carrier voltage is $V_c = MV_r = 0.4$ V.

Note the following:

a. The PSpice plots of the instantaneous output voltage V(1,2), the instantaneous carrier voltages, and the instantaneous reference voltage are shown in Figure 9.10. The output voltage is symmetrical and the widths of the output pulses are uniform as expected.
b. The Fourier coefficients of the output voltage, which can be found in the PSpice output file, are as follows:

Fourier Components of Transient Response v(3,6)

DC Component = 8.339770E−04

Harmonic Number	Frequency (Hz)	Fourier Component	Normalized Component	Phase (Deg)	Normalized Phase (Deg)
1	6.000E+01	7.449E+01	1.000E+00	−2.945E−02	0.000E+00
2	1.200E+02	1.720E−03	2.310E−05	1.596E+02	1.597E+02
3	1.800E+02	2.871E+01	3.855E−01	−1.179E−01	−8.846E−02
4	2.400E+02	1.852E−03	2.486E−05	−1.307E+02	−1.307E+02
5	3.000E+02	2.472E+01	3.318E−01	2.240E−02	5.185E−02
6	3.600E+02	1.994E−03	2.677E−05	−5.965E+01	−5.962E+01
7	4.200E+02	4.608E+01	6.186E−01	6.028E−01	6.323E−01
8	4.800E+02	2.076E−03	2.787E−05	1.365E+01	1.368E+01
9	5.400E+02	3.132E+01	4.205E−01	−1.791E+02	−1.790E+02

Total harmonic distortion = 9.044873E+01%

From the Fourier component, we get

DC component = 0.83 mV
Peak fundamental component at 60 Hz = 74.49 V
Fundamental phase delay = – 0.0294°
Total harmonic distortion = 90.44%

Note: Because voltage-controlled switches will allow bidirectional current flow, they will produce the correct output voltage with a resistive load only.

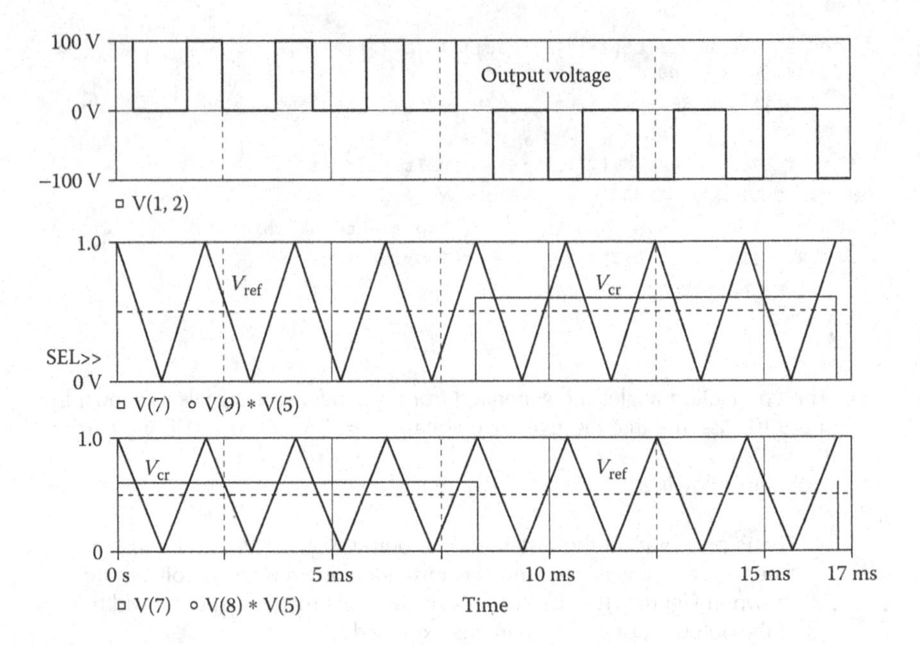

FIGURE 9.10 Plots for Example 9.1.

LTspice: LTspice schematic for Example 9.1 is shown in Figure 9.11(a). The plots of the input power, the output current, the output voltage, and the output power voltage are shown in Figure 9.11(b). The RMS and average values of the output voltage are shown in Figure 9.11(f). The input power as shown in Figure 9.11(c) is 1836.8 W and the output power as shown in Figure 9.11(d) is 1203.3W. As expected the voltage across the converter V(3,6) is pulsed and V(3,6) is connected to the source when the switch is turned on and is connected to the ground through the free-wheeling diodes as shown in Figure 9.11(a). The power difference is caused due to losses in the switches and the free-wheeling diode. We can find the power efficiency of the converter as $\eta = \frac{P_{out}}{P_{in}} = \frac{1293.3}{1836.8} = 70.41\%$

Note: For a low output power, the losses in the semiconductor devices are significant compared to the output power and the efficiency becomes low.

Example 9.2:

Finding the Performance of a Single-Phase PWM Inverter with an RL Load

The PWM inverter in Example 9.1 has a load resistance $R = 2.5\ \Omega$, and a load inductance $L = 10$ mH. Use PSpice to (a) plot the instantaneous output voltage v_o, the instantaneous output current i_o, and the instantaneous supply current i_s and (b) calculate the Fourier coefficients of the output voltage v_o and the output current i_o.

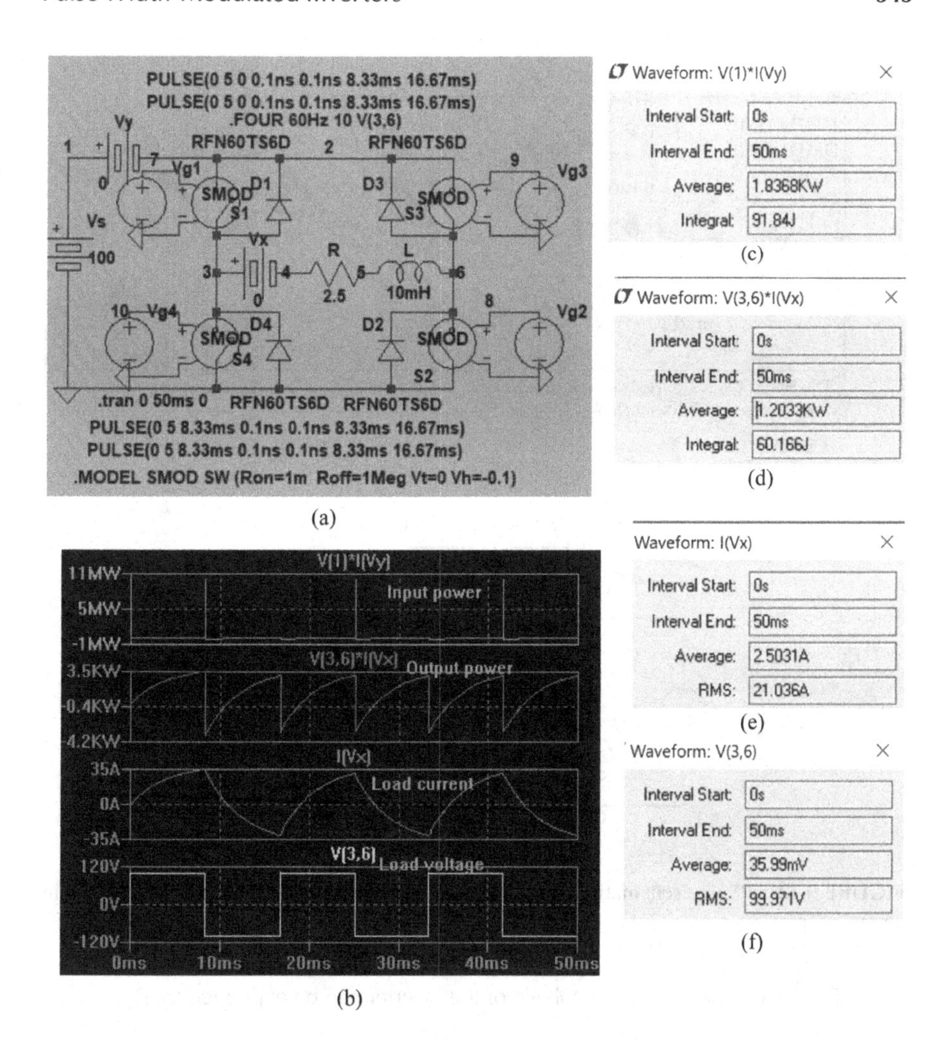

FIGURE 9.11 LTspice schematic for Example 9.1. (a) LTspice schematic. (b) Output waveforms. (c) Input power. (d) Output power. (e) Load current. (f) Output voltage.

SOLUTION

Because a load inductance will cause a bidirectional current flow through the load, we will use unidirectional switching devices such as IGBTs. The PSpice schematic is shown in Figure 9.12(a). By varying the modulation voltage V_mod from 0.01 to 0.99 V, the conduction time of the IGBTs and the output voltage can be changed. The comparator subcircuit for the PWM block is shown in Figure 9.12(b). The model parameters for IGBTs and freewheeling diodes are as follows:

```
.MODEL IXGH40N60 (TAU=287.56NS KP=50.034 AREA=37.5U
AGD=18.75U
+ VT=4.1822 KF=.36047 CGS=31.942NF COXD=53.188NF
VTD=2.6570) for IGBTs.
MODEL DMD D(IS=2.22E-15 BV=1200V CJO = 1PF
TT=0US))      for diodes
```

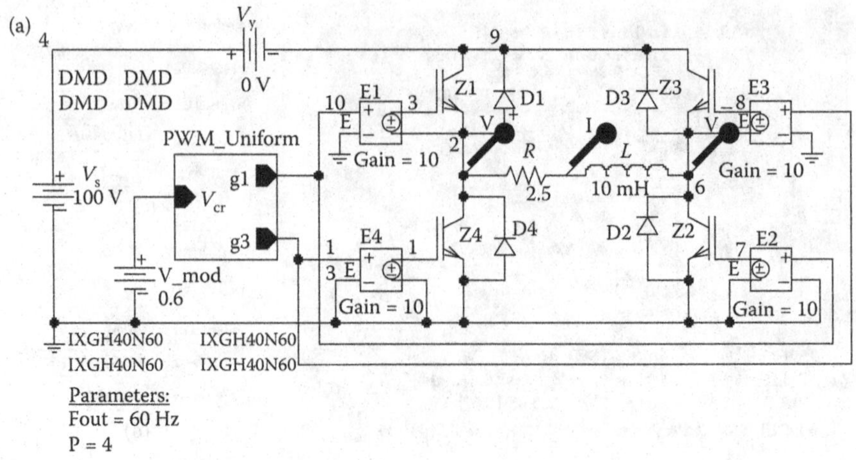

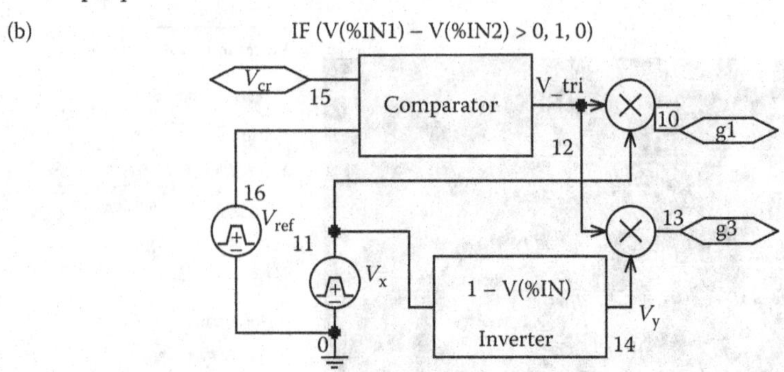

FIGURE 9.12 PSpice schematic for Example 9.2. (a) Schematic. (b) Descending hierarchy comparator.

The instantaneous output voltage of the inverter can be expressed by [1]

$$V_o = \sum_{n=1,3,5\ldots}^{\infty} \frac{4V_S}{n\pi} \sin n\omega t$$

From which we can get the RMS value of the fundamental line voltage for $n = 1$ as

$$V_{ol} = \frac{4V_s}{\sqrt{2}\pi} = 0.90V_S$$

The listing of the circuit file is changed as follows:

Single-Phase Inverter with PWM Control

```
Vs    4     0  100V    ; DC input voltage
Vy    4     9  0V      ; Monitors input current
V_mod 15 0  0.6        ; Modulation index of less than 1.0
```

```
.PARAM   Fout=60Hz   P=4
* Parameters: fout=output frequency and p=#
* of pulses per half cycle
Z1   9  3  2 IXGH40N60 ; IGBTs with a model IXGH40N60
Z2   6  7  0 IXGH40N60
Z3   9  8  6 IXGH40N60
Z4   2  1  0 IXGH40N60
.MODEL IXGH40N60 NIGBT (TAU=287.56ns KP=50.034 AREA=
* 37.5umAGD=18.75um
+ VT=4.1822 KF=.36047 CGS=31.942nf COXD=53.188nf
+ VTD=2.6570)
D1   2  9  DMD          ; Diodes with model DMD
D2   0  6  DMD
D3   6  9  DMD
D4   0  2  DMD
.MODEL DMD D(IS=2.2E-15 BV=1800V TT=0) ; Diode
* model parameters
E1   3  2  10  0  10   ; Voltage controlled voltage source
E2   7  0  10  0  10
E3   8  6  13  0  10
E4   1  0  13  0  10
R    2  5  2.5          ; Load resistance
L    5  6  10mH         ; Load inductance
Vref 16 0  PULSE (1  0 0 {1/(2*{2*{p}*{fout}})} {1/(2*{2*{p}*
{fout}})-1ns} 1ns + {1/{2*{p}*{fout}}})
Vx   11 0  PULSE (0 1 0 1ns 1ns {1/(2*{fout})-2ns} {1/{fout}})
E_MULT1 10   0 VALUE {V(11)*V(12)}          ; Multiplier 1
E_MULT2 13   0 VALUE {V(14)*V(12)}          ; Multiplier 2
E_ABM21 12   0 VALUE {IF(V(15)-V(16) > 0, 1, 0)} ; Comparator
E_ABM12 14   0 VALUE {1-V(11)}              ; Inverter
.TRAN   1US  16.67MS 0 0.1US                ; Transient Analysis
.FOUR   60Hz 10 V(1,2) I(Vy)                ; Fourier Analysis
.OPTIONS ABSTOL=1uA CHGTOL=0.1uC RELTOL=0.1 VNTOL=0.1
.PROBE
.END
```

Note the following:

a. The PSpice plots of the instantaneous output voltage V(2,6), the output current I(R), and the supply current I(VY) are shown in Figure 9.13. The output voltage is pulsed shape AC whereas the output current tries to be continuous due to each half-cycle. When the output voltage is positive, the load current rises; when the output is disconnected from the DC supply, the load current continues to flow through the freewheeling diodes and the load current falls.

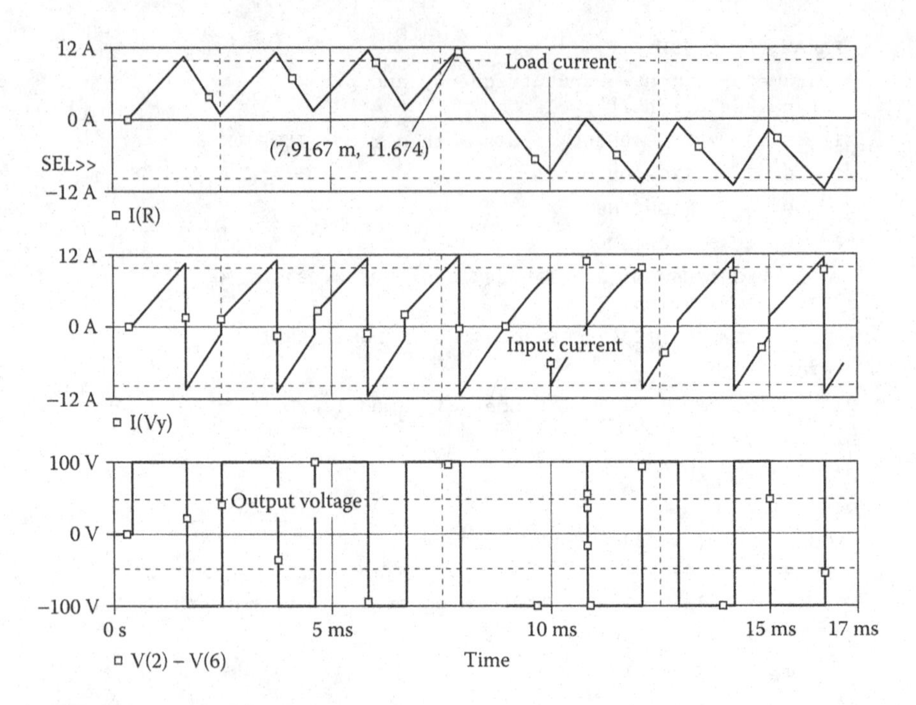

FIGURE 9.13 Plots for Example 9.2.

b. The Fourier coefficients of the output voltage and the output current, which can be found in the PSpice output file, are as follows:

Fourier Components of Transient Response V(3,6)

DC Component = −7.146290E−02

Harmonic Number	Frequency (Hz)	Fourier Component	Normalized Component	Phase (deg)	Normalized Phase (Deg)
1	6.000E+01	3.267E+01	1.000E+00	6.813E+01	0.000E+00
2	1.200E+02	1.267E−01	3.879E-03	−1.364E+02	−2.045E+02
3	1.800E+02	2.885E+01	8.831E-01	1.298E+02	6.168E+01
4	2.400E+02	1.155E-01	3.535E-03	1.703E+02	1.021E+02
5	3.000E+02	3.777E+01	1.156E+00	1.736E+02	1.055E+02
6	3.600E+02	1.202E−01	3.679E−03	1.140E+02	4.591E+01
7	4.200E+02	8.850E+01	2.709E+00	−1.471E+02	−2.153E+02
8	4.800E+02	1.341E−01	4.103E-03	6.515E+01	−2.979E+00
9	5.400E+02	6.730E+01	2.060E+00	6.753E+01	−5.927E−01

Total harmonic distortion = 3.700917E + 02%

From the Fourier component, we get

DC component = −71.46 mV
Peak fundamental component at 60 Hz = 32.67 V
Fundamental phase delay = 68.13°
Total harmonic distortion = 370.09% (very rich in harmonic components)

Fourier Components of Transient Response I(R)

DC Component = 3.288993E–01

Harmonic Number	Frequency (Hz)	Fourier Component	Normalized Component	Phase (deg)	Normalized Phase (Deg)
1	6.000E+01	7.469E+00	1.000E+00	–1.394E+01	0.000E+00
2	1.200E+02	2.804E–01	3.754E–02	4.549E+01	5.942E+01
3	1.800E+02	2.535E+00	3.395E–01	–2.891E+01	–1.497E+01
4	2.400E+02	1.832E–01	2.453E–02	4.369E+01	5.762E+01
5	3.000E+02	1.946E+00	2.605E–01	–4.735E+01	–3.341E+01
6	3.600E+02	1.489E–01	1.994E–02	4.922E+01	6.316E+01
7	4.200E+02	3.264E+00	4.370E–01	–7.088E+01	–5.694E+01
8	4.800E+02	1.471E–01	1.969E–02	5.020E+01	6.414E+01
9	5.400E+02	2.088E+00	2.796E–01	8.048E+01	9.442E+01

Total harmonic distortion = 6.745586E+01%

From the Fourier component, we get

DC component = 328 mA
Peak fundamental component at 60 Hz = 7.46 A
Fundamental phase delay = –13.94°
Total harmonic distortion = 67.45%

Note: With an inductive load, the output voltage V(2,3) switches between $-V_S$ to $+V_S$ to adjust to the load current I(R), which rises and falls.

LTspice: LTspice schematic for Example 9.2 is shown in Figure 9.14(a). The plots of the inverter output voltage for L = 1uH, the PWM reference voltages, the carrier signals, and the gating signals for 4 pulses per half-cycle are shown in Figure 9.14(b). The reference signals are compared with the carrier signal through op-amp comparators. The gating signals for transistors M1 and M3 are isolated from the ground through voltage-dependent voltage sources vg1 and vg2. The inverter output voltage is shown with a small load inductor. If we run the simulation with a load inductor of L = 10 mH, we can observe high-frequency transients and spikes due to the internal capacitances, the storage time, and the reverse recovery time of the transistors and diodes. However, if we plot the load current, the load power, and the input current, we should observe these parameters as expected. Students are encouraged to run the simulation and observe the effects of device parameters.

9.2.2 Examples of Single-Phase SPWM Inverters

Example 9.3:

Finding the Performance of a Single-Phase SPWM Inverter

Repeat Example 9.1 with SPWM control.

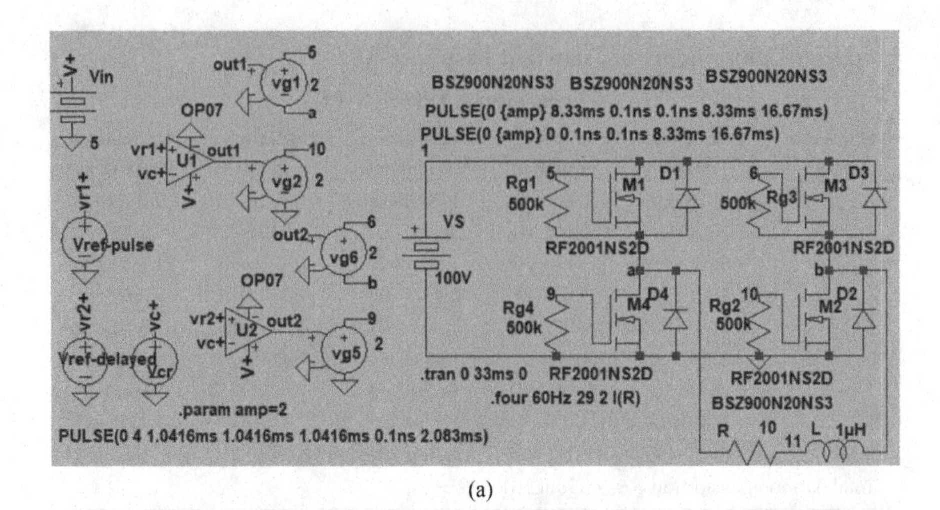

(a)

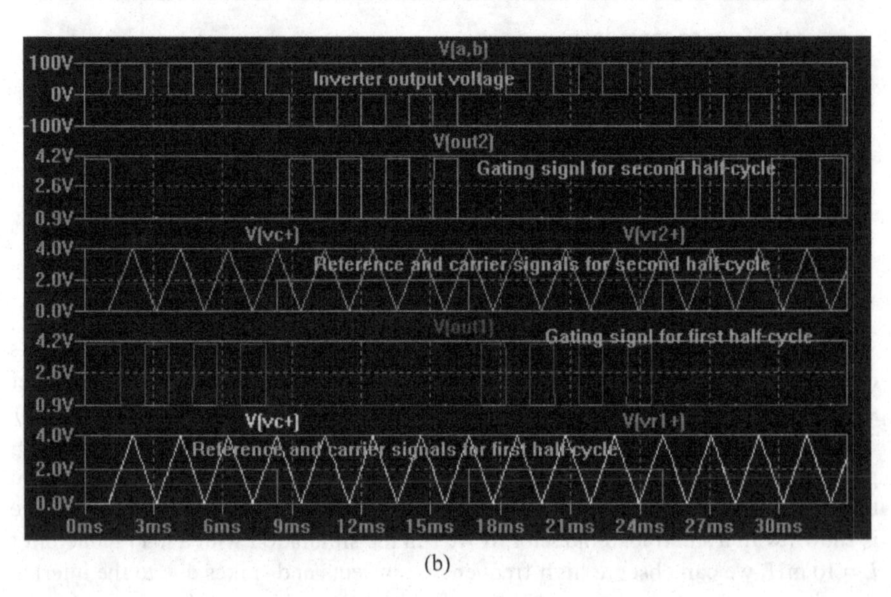

(b)

FIGURE 9.14 LTspice schematic for Example 9.2. (a) LTspice schematic. (b) Waveforms for inverter output voltage and gating signals.

SOLUTION

$P = 4$, $M = 0.6$. Let us assume that $A_r = 10$ V. Also, $A_c = 10 \times 0.6 = 6$ V. The two reference voltages are generated with a PWL function as shown in Figure 9.10. A precision rectifier can also convert a sine wave input signal to two half-sine-wave pulses, and a comparator generates the PWM waveform. This is shown in Figure 9.16. We can use the subcircuit SPWM for the generation of the control signals. IGBTs are used as switching devices in the PSpice schematic shown in Figure 9.15, which is similar to Figure 9.12(a). The carrier signal is a rectified sinusoidal waveform, as shown in Figure 9.4(a), of the form $v_{cr} = |M \sin(2\pi f_o)|$, where M is the modulation index and

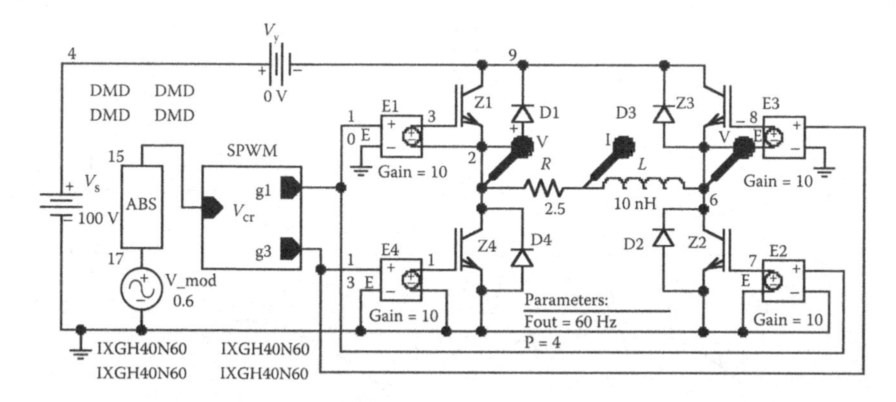

FIGURE 9.15 PSpice schematic for Example 9.3.

f_o is the frequency of the output voltage. The listing of the circuit file for the inverter in Example 9.2 can be changed to modify the modulating signal as follows:

```
V_mod   17  0   AC  0   SIN (0 {M} {fout} 0 0 0)
E_ABS   15  0   VALUE {ABS(V(17))}
```

Note the following:

a. The PSpice plots of the instantaneous output voltage V(2,6), the carrier voltages, and the reference voltages are shown in Figure 9.16. The output voltage pulses become wider toward the center and become narrower toward the sides.
b. The Fourier coefficients of the output voltage, which can be found in the PSpice output file, are as follows:

Fourier Components of Transient Response v(3,6)

DC Component = −4.386737E−03

Harmonic Number	Frequency (Hz)	Fourier Component	Normalized Component	Phase (Deg)	Normalized Phase (Deg)
1	6.000E+01	5.734E+01	1.000E+00	2.895E+01	0.000E+00
2	1.200E+02	1.233E−02	2.150E−04	−5.973E+01	−8.868E+01
3	1.800E+02	2.328E−02	4.061E−04	−1.778E+01	−4.673E+01
4	2.400E+02	1.630E−02	2.842E−04	−5.811E+01	−8.706E+01
5	3.000E+02	6.256E+00	1.091E−01	1.427E+02	1.137E+02
6	3.600E+02	1.604E−02	2.798E−04	−6.504E+01	−9.399E+01
7	4.200E+02	3.599E+01	6.278E−01	−1.583E+02	−1.873E+02
8	4.800E+02	1.197E−02	2.087E−04	−7.286E+01	−1.018E+02
9	5.400E+02	3.616E+01	6.307E−01	7.955E+01	5.060E+01

Total harmonic distortion = 8.965353E+01%

From the Fourier component, we get

DC component = −4.38 mV
Peak fundamental component at 60 Hz = 57.34 V

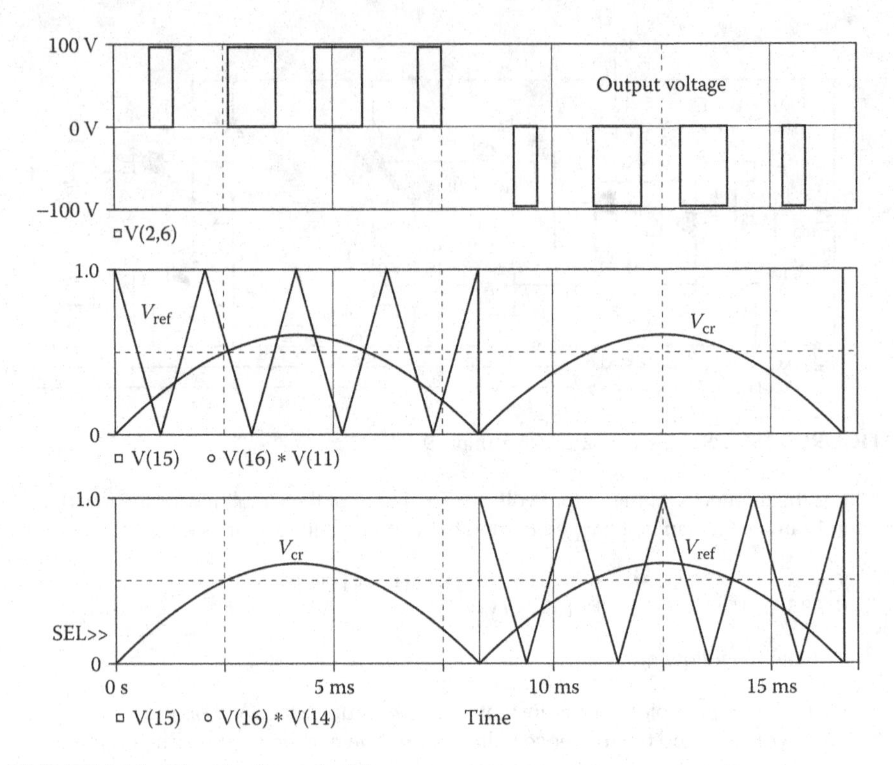

FIGURE 9.16 Plots for Example 9.3.

Fundamental phase delay = 28.95°
Total harmonic distortion = 89.65%

Note: With an inductive load, the output voltage will contain switching transients due to the switching times of the IGTBs and diodes. By varying the modulation voltage V_{cr}, or modulation index M, from 0.01 to 0.99 V, the conduction time of the IGBTs and the output voltage can be changed.

LTspice: LTspice schematic for Example 9.3 is shown in Figure 9.17(a). The plots of the inverter output voltage for L = 1uH, the PWM reference voltages, the carrier signals, and the gating signals for 4 pulses per half-cycle are shown in Figure 9.17(b). The reference signals are compared with the carrier signal through op-amp comparators. The gating signals for transistors M1 and M3 are isolated from the ground through voltage-dependent voltage sources vg1 and vg2. The inverter output voltage is shown with a small load inductor. If we run the simulation with a load inductor of L = 10 mH, we can observe high-frequency transients and spikes due to the internal capacitances, the storage time, and the reverse recovery time of the transistors and diodes. However, if we plot the load current, the load power, and the input current, we should observe these parameters as expected. Students are encouraged to run the simulation and observe the effects of device parameters.

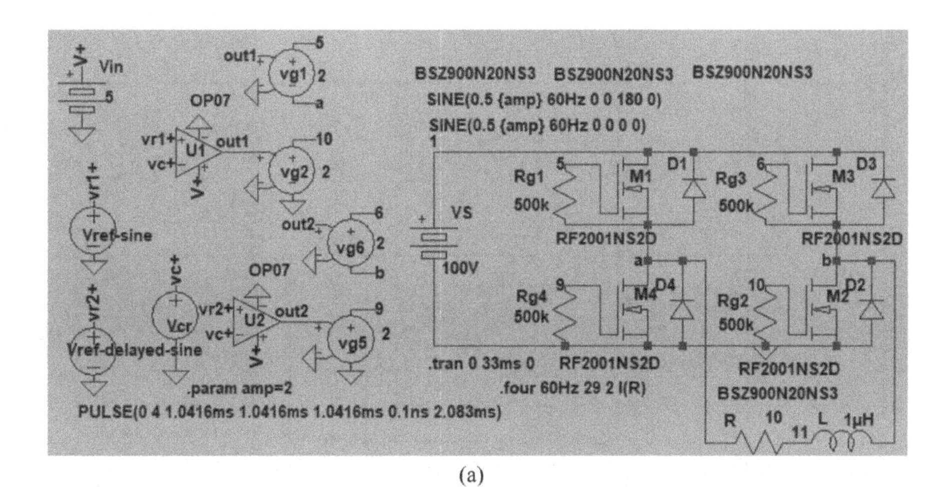

(a)

(b)

FIGURE 9.17 LTspice schematic for Example 9.3. (a) LTspice schematic. (b) Waveforms for inverter output voltage and gating signals.

9.2.3 Examples of Three-Phase PWM Inverters

Example 9.4:

Finding the Performance of a Three-Phase PWM Inverter

A three-phase bridge inverter is shown in Figure 9.18(a). The DC input voltage is 100 V. The control voltages are shown in Figure 9.18(b). The output frequency is

(a)

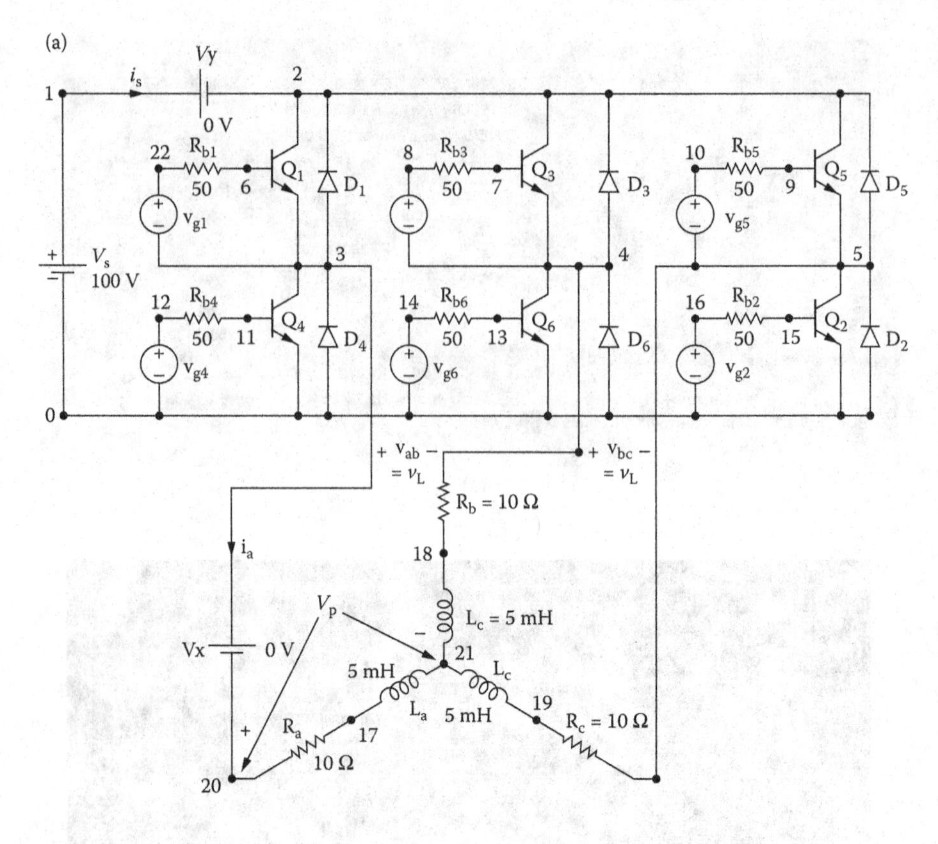

(b)

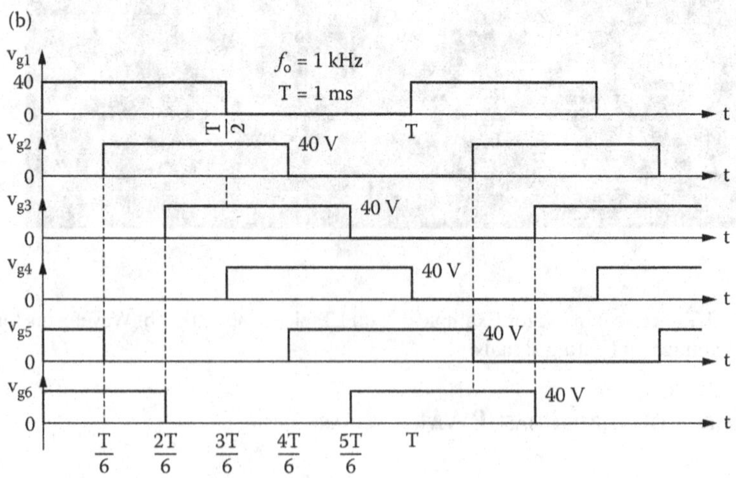

FIGURE 9.18 Three-phase bridge inverter. (a) Circuit. (b) Control voltage.

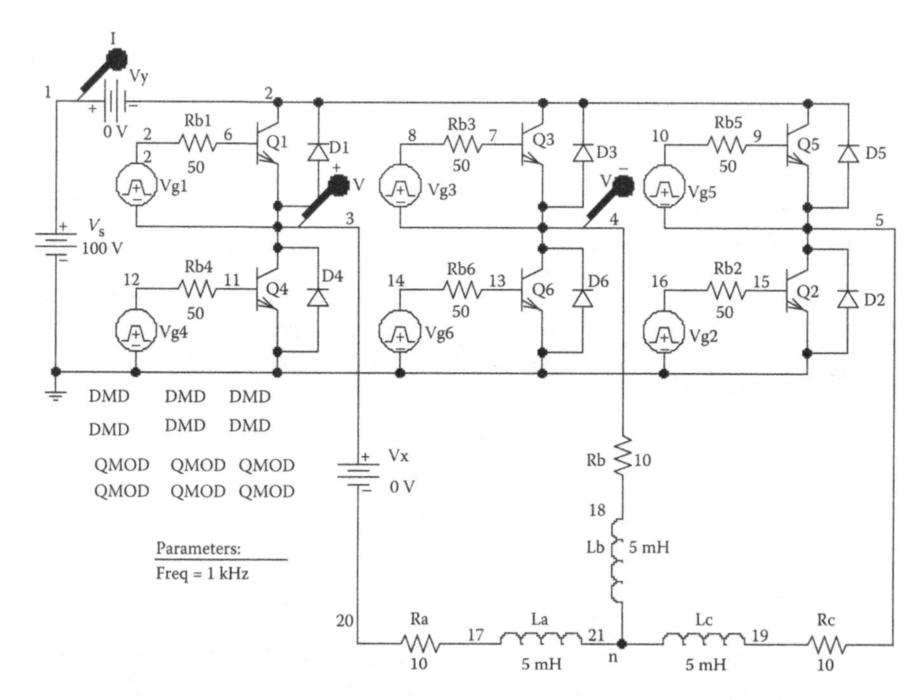

FIGURE 9.19 PSpice schematic for Example 9.4.

f_o = 1 kHz. The load resistance is R = 10 Ω and the load inductance is L = 5 mH. Use PSpice to (a) plot the instantaneous output line–line voltage v_L, the instantaneous output phase voltage v_p, and the instantaneous output current i_a for phase a and (b) calculate the Fourier coefficients of the phase voltage v_p and the phase current i_a. The model parameters of the BJTs are IS = 2.33E27, BF = 13, CJE = 1PF, CJC = 607.3PF, and TF = 26.5NS.

SOLUTION

BJTs are used as switching devices in the PSpice schematic shown in Figure 9.14. A base resistor (R_{b1}) must be inserted as shown in Figure 9.19 at the base of a transistor to limit the base current, and its value should be such that the transistor is driven into saturation, that is, the collector-emitter voltage in saturation $V_{CE(sat)}$ should be low.

The listing of the circuit file for the inverter is as follows:

Three-Phase Inverter with PWM Control

```
.PARAM Freq=1kHz ; * Parameters: Freq = output frequency
Vs   15 0 100V ; DC input voltage
Vy   15 6 0V ; Monitors input current
Vx   5 19 0V ; Monitors output phase current
Rb1 22 21 50
Vg1 22 5 PULSE (0 40 0 1ns 1ns {1/(2*{Freq})-2ns} {1/
{Freq}})
```

```
Rb2 9 12 50
Vg2 9 0 PULSE (0 40 {1/(6*{Freq})} 1ns 1ns {1/(2*{Freq})-
2ns} {1/{Freq}})
Rb3 1 11 50
Vg3 1 2 PULSE (0 40 {2/(6*{Freq})} 1ns 1ns {1/(2*{Freq})-
2ns} {1/{Freq}})
Rb4 7 13 50
Vg4 7 0 PULSE (0 40 {3/(6*{Freq})} 1ns 1ns {1/(2*{Freq})-
2ns} {1/{Freq}})
Rb5 3 10 50
Vg5 3 4 PULSE (0 40 {4/(6*{Freq})} 1ns 1ns {1/(2*{Freq})-
2ns} {1/{Freq}})
Rb6 8 14 50
Vg6 8 0 PULSE (0 40 {5/(6*{Freq})} 1ns 1ns {1/(2*{Freq})-
2ns} {1/{Freq}})
D1  5 6 DMD ; Diodes with model DMD
D2  0 4 DMD
D3  2 6 DMD
D4  0 5 DMD
D5  4 6 DMD
D6  0 2 DMD
.MODEL DMD D(IS=2.2E-15 BV=1800V TT=0) ; Diode model
 parameters
Q1  6 21 5 QMOD ; BJTs with model QMOD
Q5  6 10 4 QMOD
Q3  6 11 2 QMOD
Q2  4 12 0 QMOD
Q4  5 13 0 QMOD
Q6  2 14 0 QMOD
.MODEL QMOD NPN(IS=6.83E-14 BF=13 CJE=1pF CJC=607.3PF
TF=26.5NS
La 18 17 5mH ; Load inductance for phase a
Ra 19 18 10 ; Load resistance for phase a
Rb 16 2 10
Lb 17 16 5mH
Rc 20 4 10
Lc 17 20 5mH
.TRAN 0.1US  2.5MS 1MS 0.1e-6 ; Transient Analysis
.OPTIONS ABSTOL=1uA CHGTOL=0.01nC ITL2=100 ITL4=150
RELTOL=0.1 VNTOL=0.1
.PROBE
.END
```

Note the following:

a. The PSpice plots of the instantaneous output line–line voltages V(3,4), V(4,5), and V(5,3) are shown in Figure 9.20. The output phase voltages

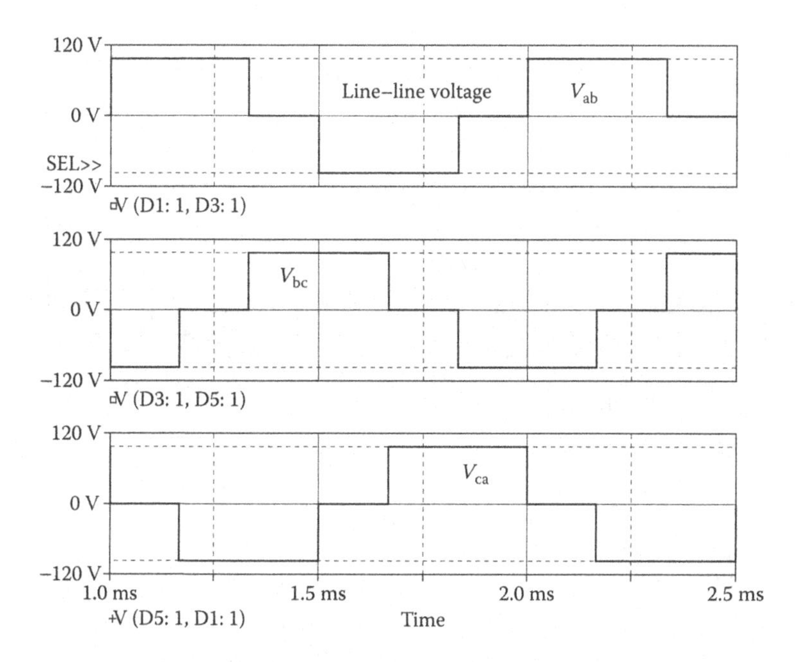

FIGURE 9.20 Line-to-line output voltage plots for Example 9.4.

V(3,21) and V(4,21) and the output phase current I(VX) are shown in Figure 9.21. The phase current is almost sinusoidal with THD = 4.72%.

b. The Fourier coefficients of the output phase voltage and phase current, which can be found in the PSpice output file, are as follows:

Fourier Components of Transient Response V(3,21)

DC Component = −3.689844E−03

Harmonic Number	Frequency (Hz)	Fourier Component	Normalized Component	Phase (Deg)	Normalized Phase (Deg)
1	1.000E+03	6.349E+01	1.000E+00	1.796E+02	0.000E+00
2	2.000E+03	2.118E−02	3.336E−04	1.098E+02	−6.981E+01
3	3.000E+03	6.643E−01	1.046E−02	8.981E+01	−8.979E+01
4	4.000E+03	2.138E−02	3.3683E−04	−6.049E+01	−2.401E+02
5	5.000E+03	1.306E+01	2.057E−01	1.775E+02	−2.135E+00
6	6.000E+03	6.534E−03	1.029E−04	1.216E+02	−5.799E+01
7	7.000E+03	9.015E+00	1.420E−01	1.759E+02	−3.683E+00
8	8.000E+03	1.352E−02	2.129E−04	−1.845E+01	−1.980E+02
9	9.000E+03	6.873E−01	1.083E−02	8.654E+01	−9.305E+01

Total harmonic distortion = 2.503806E+01%

From the Fourier component, we get

DC component = −3.68 mV
Peak fundamental component at 1 kHz = 63.49 V

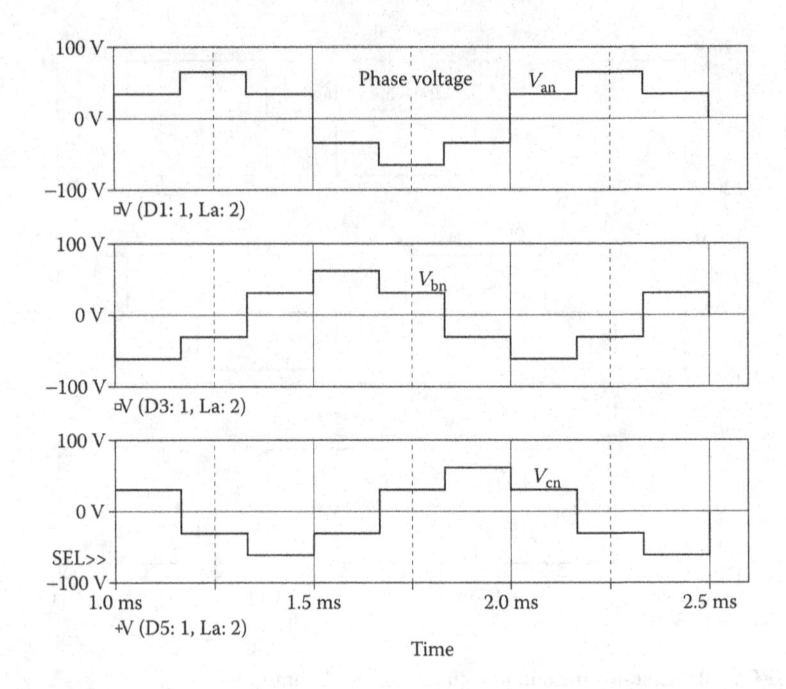

FIGURE 9.21 Line-to-neutral output voltage plots for Example 9.4.

Fundamental phase delay = 179.6°
Total harmonic distortion = 25.03%

Fourier Components of Transient Response I(VX)

DC Component = −4.969840E−03

Harmonic Number	Frequency (Hz)	Fourier Component	Normalized Component	Phase (Deg)	Normalized Phase (Deg)
1	1.000E+03	1.931E+00	1.000E+00	1.078E+02	0.000E+00
2	2.000E+03	8.719E−04	4.516E−04	−1.507E+02	−2.586E+02
3	3.000E+03	4.645E−04	2.406E−04	1.384E+02	3.053E+01
4	4.000E+03	8.019E−04	4.153E−04	1.741E+02	6.625E+01
5	5.000E+03	8.075E−02	4.182E−02	9.345E+01	−1.439E+01
6	6.000E+03	4.428E−04	2.293E−04	−1.476E+02	−2.555E+02
7	7.000E+03	4.190E−02	2.170E−02	9.405E+01	−1.380E+01
8	8.000E+03	2.887E−04	1.495E−04	1.744E+02	6.659E+01
9	9.000E+03	2.040E−04	1.057E−04	1.227E+02	1.488E+01

Total harmonic distortion = 4.711973E+00%

From the Fourier component, we get

DC component = −4.96 mA
Peak fundamental component at 1 kHz = 1.93 A
Fundamental phase delay = 107.8°
Total harmonic distortion = 4.71%

Note: The peak value of the line voltage varies between $+V_s$ to $-V_s$, and that of the phase voltage varies between $+2V_s/3$ to $-2V_s/3$. Each phase is shifted by 120° from the other.

LTspice: LTspice schematic for Example 9.4 is shown in Figure 9.22(a). The plots of the output current, the line-line output voltages, the line-neutral voltage, and the line-current are shown in Figure 9.22(b). The RMS values of the line-line is almost the same close to 80 V are shown in Figure 9.22(c, d, e).

The RMS values of the line-neutral are almost the same close to 45.769 V are shown in Figure 9.22(f) and are related to the line-line voltage 80V by a factor of $\sqrt{3}$. As expected, the output voltages across the inverter are PWM waveforms. The line "a" is connected to the source when the switch M1 is turned on and is connected to the ground through the free-wheeling diodes as shown in Figure 8.22(a). Due to the inductive load, the line current I(Ra) is a continuous AC waveform, and its RMS value of 4.356 A. We can find the output power as

$$P_{\text{out}} = 3I_o^2R = 3 \times 4.356^2 \times 10 = 570.3\,W$$

Note: Power is dissipated or lost in the resistor and there is power loss in the inductor. The factor 3 is included for a three-phase output voltages.

Example 9.5:

Finding the Performance of a Three-Phase Uniform Pulse PWM Inverter

Plot the output voltage and the gating signals g_1 and g_3 of the three-phase inverter in Figure 9.18(a) with two pulses per half-cycle, $p = 2$, and a modulation index of $M = 0.6$. Assume uniform pulse modulation.

SOLUTION

IGBTs are used as switching devices in the PSpice schematic as shown in Figure 9.23(a). The gating signal g_1 is generated by comparing triangular reference signal V_{ref} of +1 V to −1 V with a carrier pulse signal of magnitude {+M} to {−M} having a 50% duty cycle at the frequency f_o of the output voltage. The switching frequency f_s is related to the output frequency f_o by $f_s = 2 \times p \times f_o$. The signal g_3 is generated from V_{cb}, which is phase-shifted by 120° from V_{ca}. The PSpice schematic for the implementation of the PWM waveform is shown in Figure 9.23(b).

The generation of the gate signals g_1 and g_3 is shown in Figure 9.18. The instantaneous line-to-line output voltages are $V_{ab} = V_s(g_1 - g_3)$, $V_{bc} = V_s(g_3 - g_5)$, and $V_{ca} = V_s(g_5 - g_1)$. The line-line voltage V_{ab} is also shown in Figure 9.24, which is $V_{ab} = V_s(g_1 - g_3)$.

Note: The presence of g_3 limits the duration of g_1, thereby causing notches in the output voltage. For example (Figure 9.18), we have specified two pulses per half-cycle ($p = 2$), but the output voltage has three pulses per half-cycle as shown

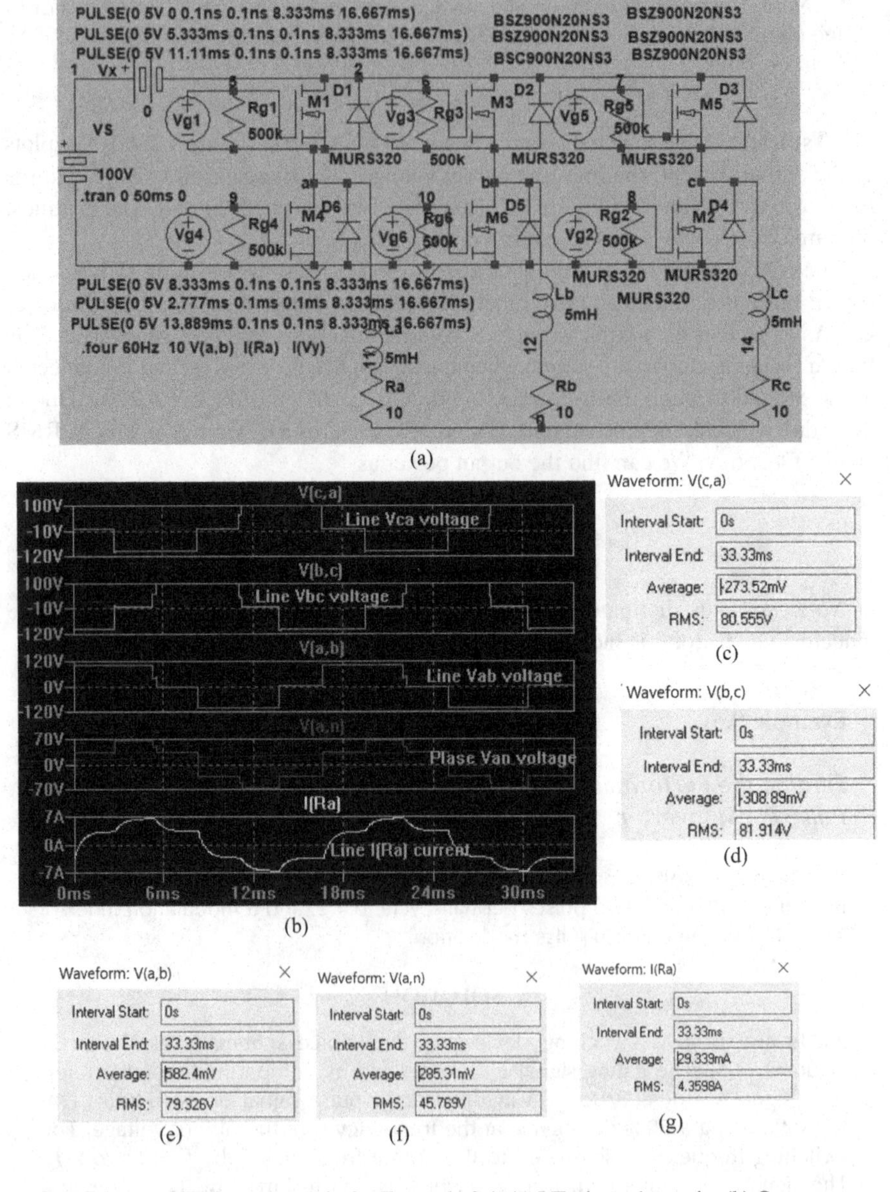

FIGURE 9.22 LTspice schematic for Example 9.4. (a) LTspice schematic. (b) Output waveforms. (c) Line c-a voltage. (d) Line b-c voltage. (e) Line a-b voltage. (f) Line to neutral voltage. (g) Line "a" current.

in Figure 9.24. The line-to-line output voltage v_{ab} (between lines a and b) can be expressed as [1]

$$V_{ab} = \sum_{n=1,3,5,\ldots}^{\infty} \frac{4V}{n\pi} \sin\frac{n\pi}{2} \sin\frac{n\pi}{3} n\left(\omega t + \frac{\pi}{6}\right)$$

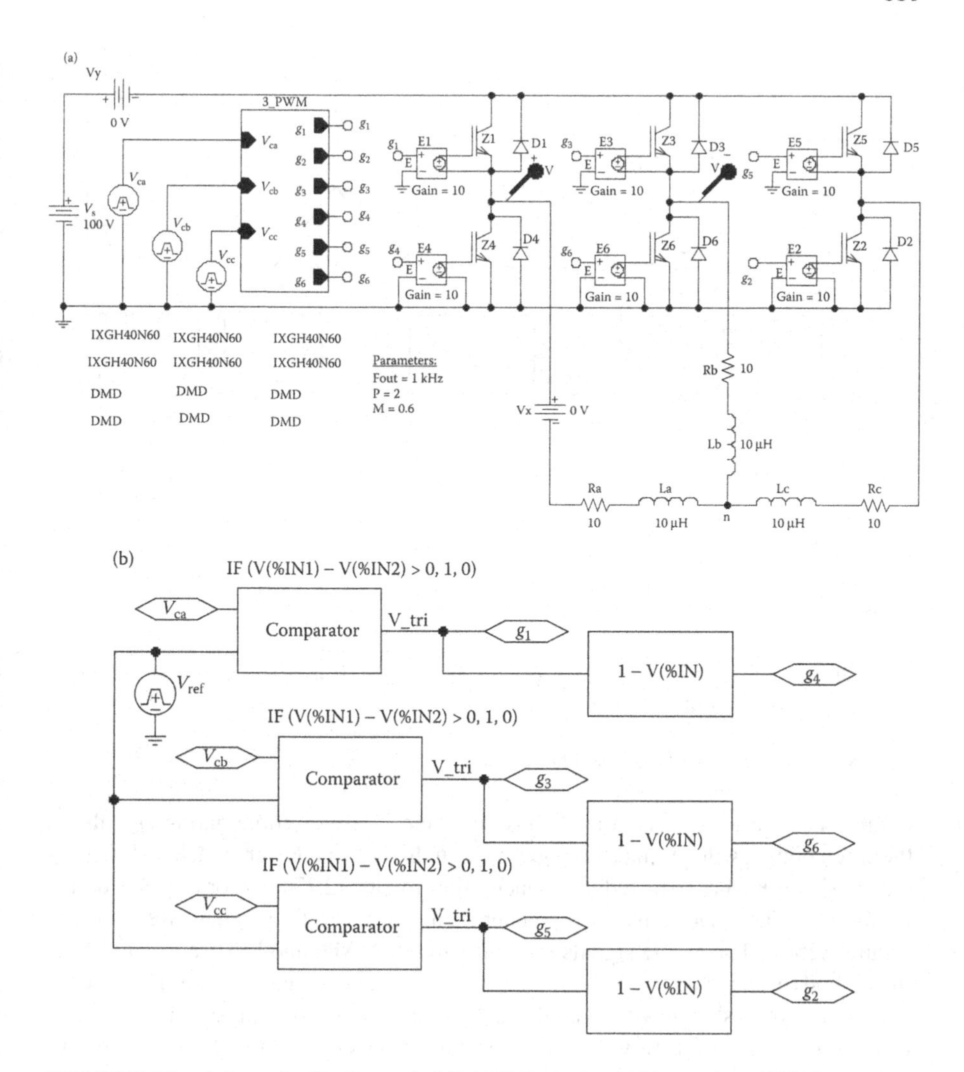

FIGURE 9.23 Schematics for Example 9.5. (a) Schematic. (b) Three-phase PWM generator.

which gives the RMS value of the fundamental line voltage for $n = 1$ as

$$V_l = \sqrt{\frac{2}{3}}V_S = 0.8165V_S$$

and the RMS value of the line-to-neutral voltage is

$$V_p = \frac{V_L}{\sqrt{3}} = 0.4714V_S$$

LTspice: LTspice schematic for Example 9.5 is shown in Figure 9.25(a). The plots of the inverter output voltage for L = 1uH, the PWM reference voltages, the carrier signals, and the gating signals for 4 pulses per half-cycle are shown in Figure 8.13(b). The reference signals are compared with the carrier signal through

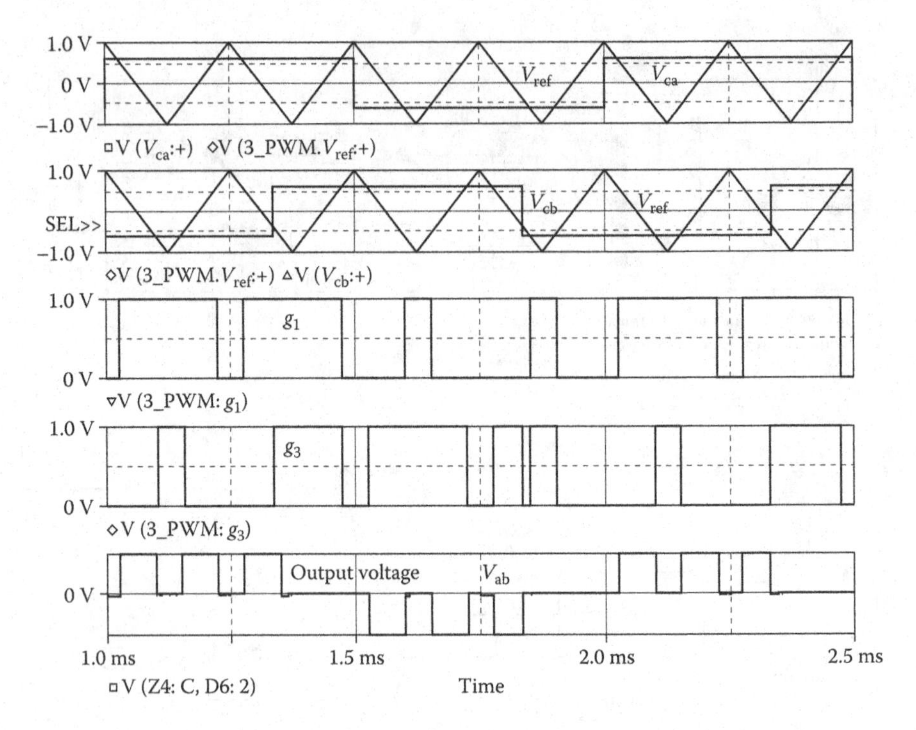

FIGURE 9.24 Plots for Example 9.5.

op-amp comparators. For a three-phase inverter, there are three gating signals for the three upper switches and three gating signals for three lower switches. Each reference signal is phase-shifted from each other by 60° (2.778 ms for a 60 Hz output frequency). The waveforms for the generation of three gating signals are shown in Figure 9.25(b). The gating signals for transistors M1, M3, and M5 are isolated from the ground through voltage-dependent voltage sources vg1, vg3, and vg5. The inverter output voltage is shown with a small load inductor. If we run the simulation with a load inductor of $L = 5$ mH, we can observe high-frequency transients and spikes due to the internal capacitances, the storage time, and the reverse recovery time of the transistors and diodes. However, if we plot the load current, the load power, and the input current, we should observe these parameters as expected. Students are encouraged to run the simulation and observe the effects of device parameters.

9.2.4 Examples of Three-Phase SPWM Inverters

Example 9.6:

Finding the Performance of a Three-Phase PWM Inverter

Plot the output voltage and the gating signals g_1 and g_3 of the three-phase inverter in Figure 9.18(a) with two pulses per half-cycle, $p = 2$, and a modulation index of $M = 0.6$. Assume sinusoidal pulse modulation.

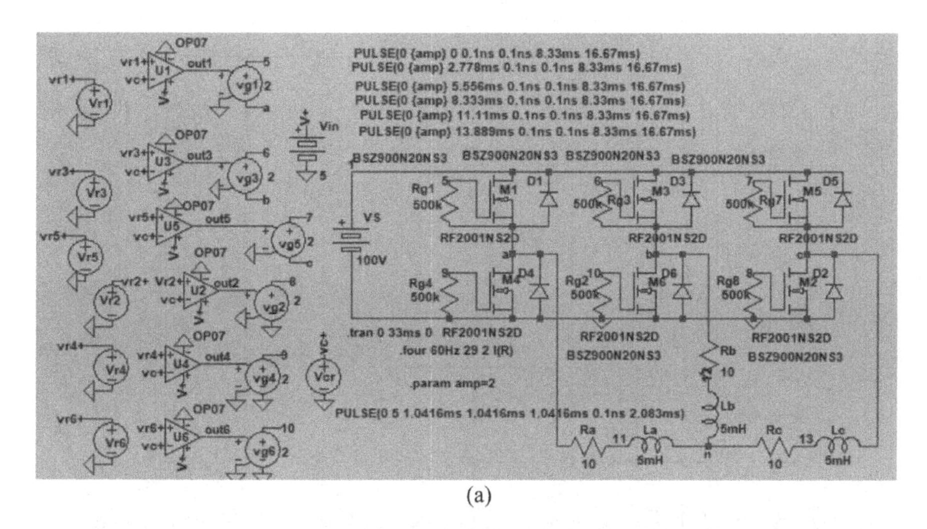

(a)

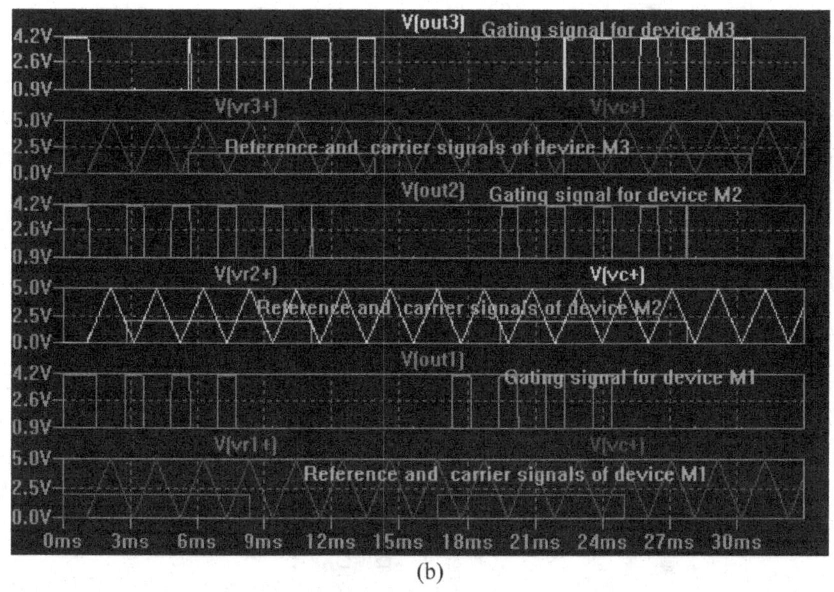

(b)

FIGURE 9.25 LTspice schematic for Example 9.5. (a) LTspice schematic. (b) Waveforms for PWM gating signals generation.

SOLUTION

IGBTs are used as switching devices in the PSpice schematic as shown in Figure 9.26. The gating signal g_1 is generated by comparing triangular reference signal V_{ref} of +1 V to −1 V with a sinusoidal carrier signal V_{ca} of magnitude {+M} at the frequency f_o of the output voltage such that $v_{cr} = M \sin(2\pi f_o)$. The switching frequency f_s is related to the output frequency f_o by $f_s = 2 \times p \times f_o$. The signal g_3 is generated from V_{cb}, which is phase-shifted by 120° from V_{ca}. Note that E1 in Figure 9.23 provides the isolation between the power circuit and the low signal gating

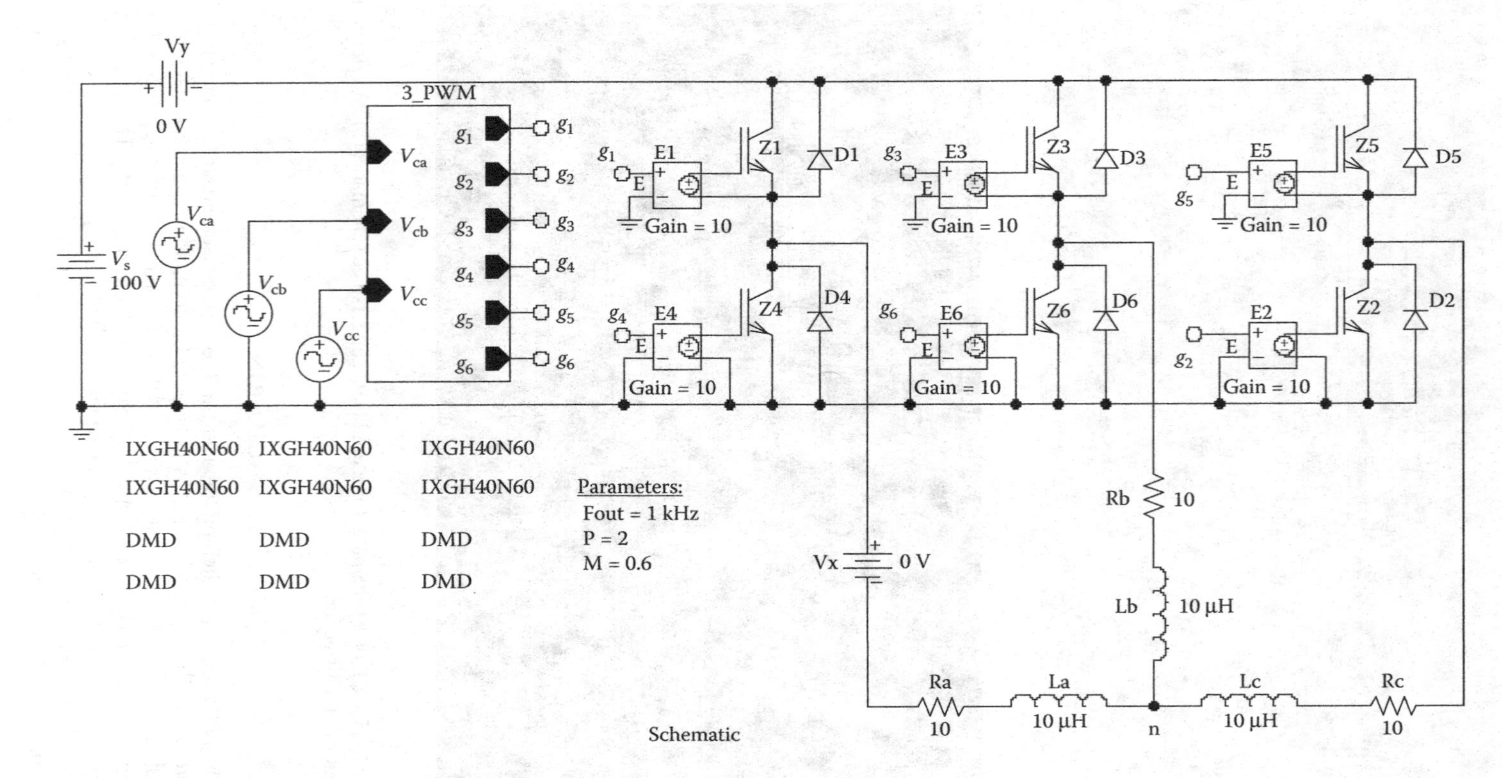

FIGURE 9.26 Schematics for Example 9.6.

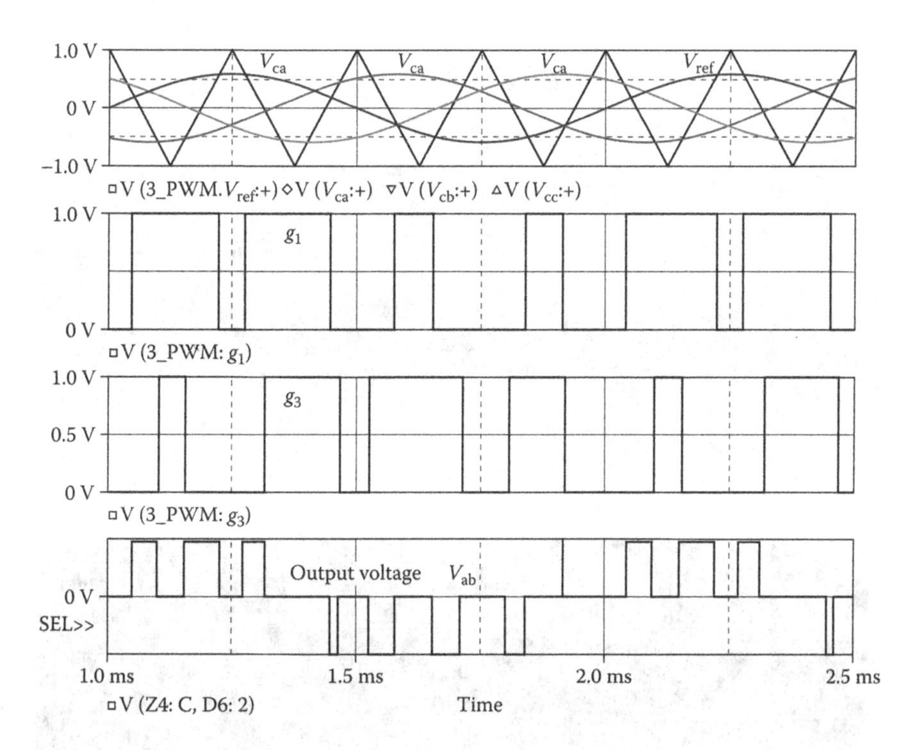

FIGURE 9.27 Plots for Example 9.6.

circuit. The PSpice schematic for the implementation of the PWM waveform is shown in Figure 9.23(b).

The generation of the gate signals g_1 and g_3 is shown in Figure 9.27. The instantaneous line-to-line output voltages are $V_{ab} = V_s(g_1 - g_3)$, $V_{bc} = V_s(g_3 - g_5)$, and $V_{ca} = V_s(g_5 - g_1)$. The line-line voltage V_{ab} is also shown in Figure 9.27 and is given by $V_{ab} = V_s(g_1 - g_3)$.

Note: Because of the sinusoidal PWM, the middle pulse is the widest. The gate signal g_3 produces notches. We have specified two pulses per half-cycle (p = 2), but the output voltage has four pulses per half-cycle as shown in Figure 9.27.

LTspice: LTspice schematic for Example 9.6 is shown in Figure 9.28(a). The plots of the inverter output voltage for L = 1uH, the sinusoidal reference voltages, the carrier signals, and the gating signals for 4 pulses per half-cycle are shown in Figure 8.28(b). The sinusoidal reference signals are compared with the PWM carrier signal through op-amp comparators. For a three-phase inverter, there are three gating signals for the three upper switches and three gating signals for three lower switches. Each sinusoidal reference signal is phase-shifted from each other by 60°. However, for a three-phase supply, the line-line voltages are phase-shifted by 30° with respect to the phase voltages. Taking a phase voltage as the reference, the sinusoidal reference voltages as shown in Figure 9.28(b) are phase-shifted by 30°. The waveforms for the generation of three gating signals are shown in

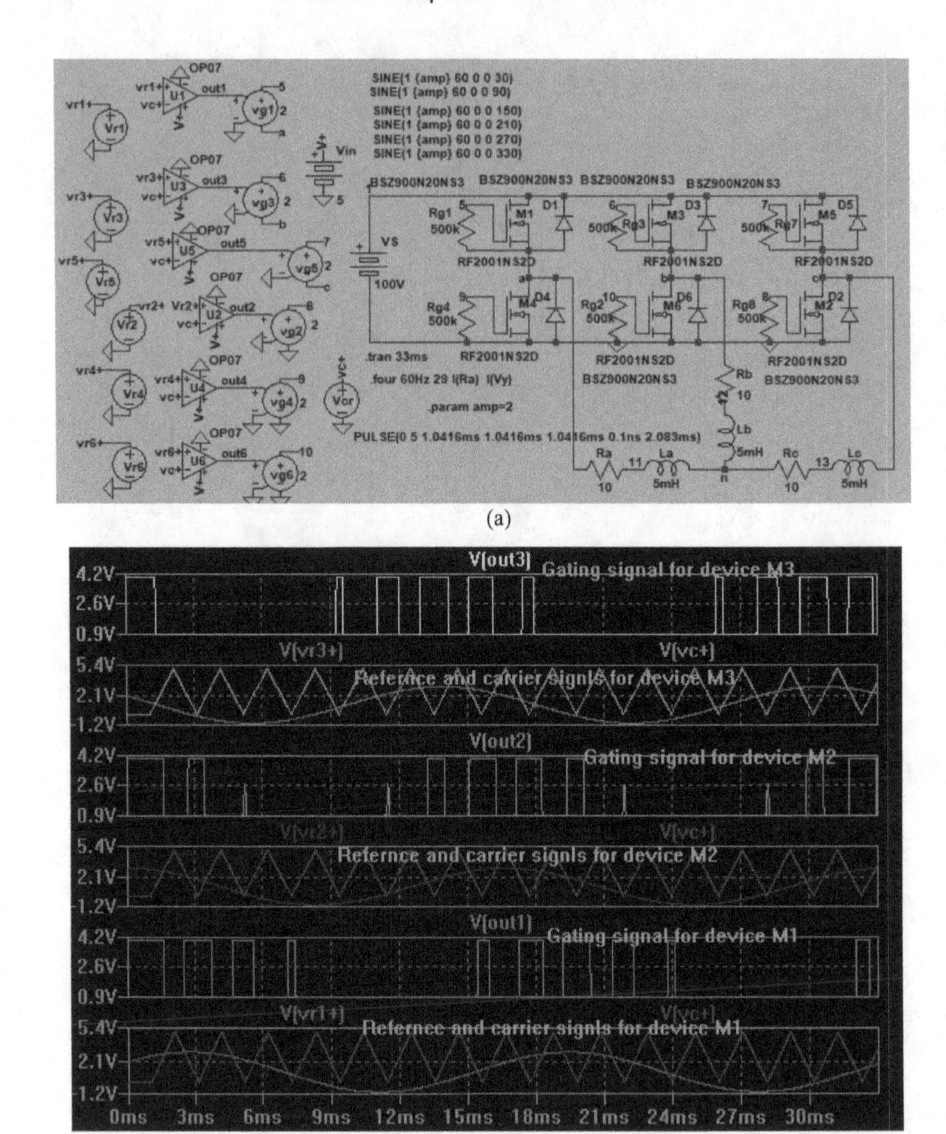

FIGURE 9.28 LTspice schematic for Example 9.6. (a) LTspice schematic. (b) Waveforms for PWM gating signals generation.

Figure 9.28(b). The gating signals for transistors M1, M3, and M5 are isolated from the ground through voltage-dependent voltage sources vg1, vg3, and vg5. The inverter output voltage is shown with a small load inductor. If we run the simulation with a load inductor of $L = 5$ mH, we can observe high-frequency transients and spikes due to the internal capacitances, the storage time, and the reverse recovery time of the transistors and diodes. However, if we plot the load current, the load power, and the input current, we should observe these parameters as expected. Students are encouraged to run the simulation and observe the effects of device parameters.

9.3 CURRENT-SOURCE INVERTERS

The input current of a current-source inverter is maintained approximately constant by having a large inductor at the input side. The magnitude of this current is normally varied by a chopper with an output filter. The time during which the input source current flows through the load is controlled by varying the turn-on and turn-off times of the inverter switches. The control can also use PWM, SPWM, and other advanced modulation techniques.

9.3.1 EXAMPLE OF CURRENT-SOURCE INVERTER

Example 9.7:

Finding the Performance of a Single-Phase Current-Source Inverter

A single-phase current-source inverter is shown in Figure 9.29(a). The control voltages are shown in Figure 9.29(b). The DC input voltage is 100 V. The output

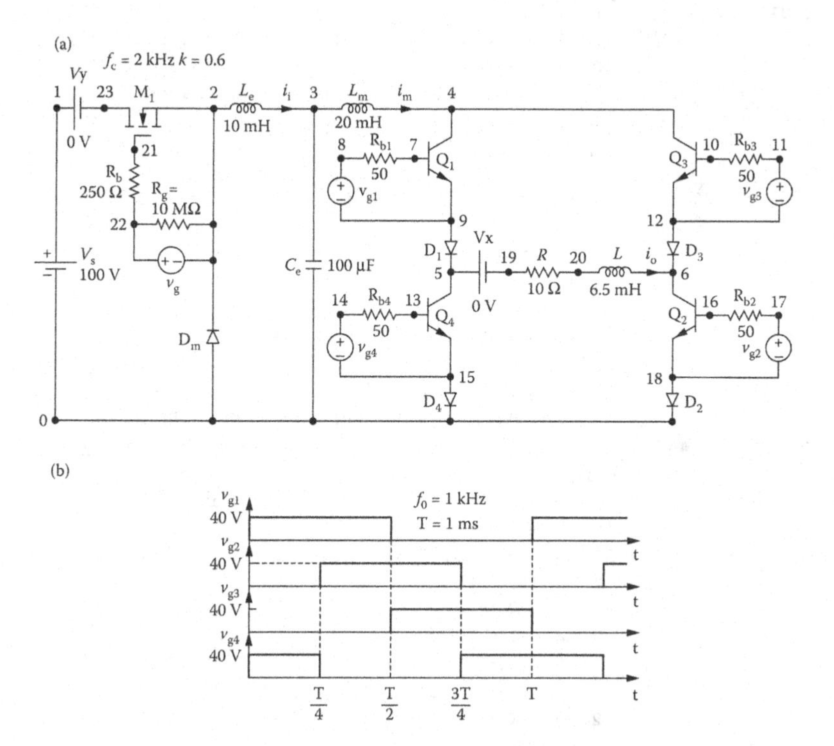

FIGURE 9.29 Single-phase current-source inverter. (a) Circuit. (b) Control voltage.

frequency is $f = 1$ kHz. The chopping frequency is $f_s = 2$ kHz, and its duty cycle is $k = 0.6$. The load resistance is $R = 10\ \Omega$, and the load inductance is $L = 6.5$ mH. Use PSpice to (a) plot the instantaneous output current i_o, the instantaneous source current i_s, and the instantaneous current i_1 through inductor L_e and (b) calculate the Fourier coefficients of the output current i_o. The model parameters of the BJTs are IS = 2.33E27, BF = 13, CJE = 1PF, CJC = 607.3PF, and TF = 26.5NS. The model parameters of the MOSFETs are VTO = 2.83, KP = 31.2U, L = 1U, W = 3.0M, CGDO = 1.359N, and CGSO = 2.032N.

SOLUTION

$f_o = 1$ kHz, $f_s = 2$ kHz, and $k = 0.6$. BJTs are used as switching devices in the PSpice schematic shown in Figure 9.30. The PWM generator varies the duty cycle of the DC–DC converter, which can change its output voltage in order to vary the current-source inductor current I_m.

Note: If you are building this circuit in Figure 9.29 or Figure 9.30, you will need to isolate the gate signals v_{g1} and v_{g3} from the power circuit through a pulse transformer or an opto-coupler. You cannot connect the ground terminal of gate signal v_{g1} directly to the emitter terminal of Q_1 and the ground terminal of gate signal v_{g3} directly to the emitter terminal of Q_3.

The listing of the circuit file for the inverter is as follows:

Single-Phase Current-Source Inverter

```
SOURCE    ■ VS    1   0 DC 100V
            Vg   22   2 PULSE (0V 40V 0 1NS 1NS 300US 500US)
            Rg   22   2 10MEG
            RB   22 21 250 ; Transistor base resistance
            Rb1   8   7 50
            Rg1   8   9 10MEG
            Vg1   8   9 PULSE (0 40V 01NS 1NS 0.5MS 1MS)
            Rb2 17 16 50
            Rg2 17 18 10MEG
            Vg2 17 18 PULSE (0 40V 250US 1NS 1NS 0.5MS 1MS)
            Rb3 11 10 50
            Rg3 11 12 10MEG
            Vg3 11 12 PULSE (0 40V 500US 1NS 1NS 0.5MS 1MS)
            Rb4 14 13 50
            Rg4 14 15 10MEG
            Vg4 14 15 PULSE (0 40V 750US 1NS 1NS 0.5MS 1MS)
CIRCUIT   ■■ VY  1 23 DC 0V  ; Voltage source to measure supply
             current
             VX   5 19 DC 0V   ; Measures load current
             R   19 20 10
             L   20   6 6.5MH  ; L is included
             Le   2   3 10MH IC=1A
             Ce   3   0 100UF
             Lm   3   4 20MH IC=3A
             D1   9   5 DMOD ;   Diode
             D2 18   0 DMOD   ; Diode
             D3 12   6 DMOD   ; Diode
             D4 15   0 DMOD   ; Diode
             DM   0   2 DMOD   ; Diode
```

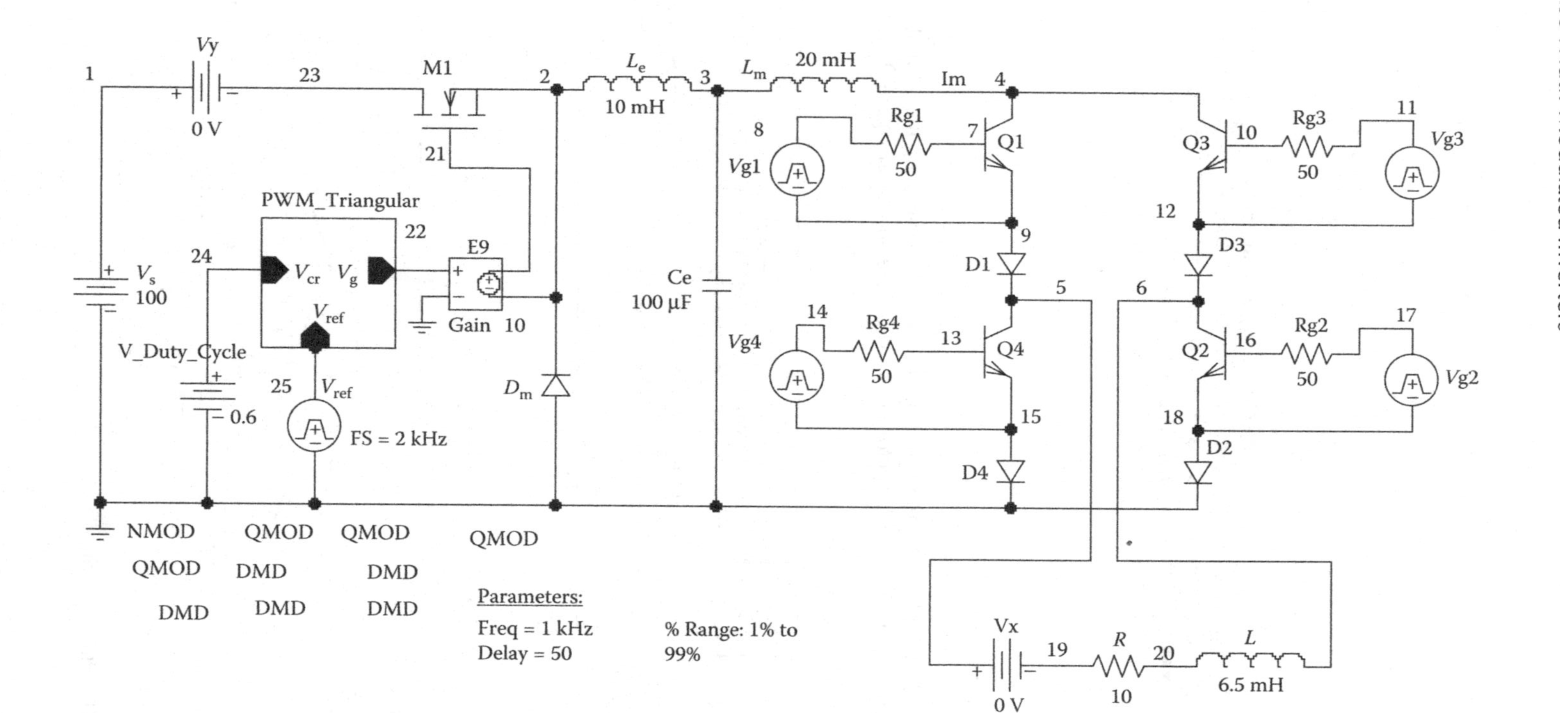

FIGURE 9.30 Schematics for Example 9.7.

```
            .MODELDMODD (IS=2.22E-15 BV-1200V
            IBV=13E-3 CJO=1PF TT=0)    ; Diode model
            M1 23 21 2 2 IRF150 ; MOSFET switch
            MODEL IRF150 NMOS(VTO=2.83 KP=31.2U L=1U
            W=3.0M +    CGDO=1.359N
            CGSO=2.032N)              ; MOSFET parameters
            Q1  4  7  9  9 2N6546     ; BJT switch
            Q2  6 16 18 18 2N6546     ; BJT switch
            Q3  4 10 12 12 2N6546     ; BJT switch
            Q4  5 13 15 15 2N6546     ; BJT switch
            .MODEL 2N6546 NPN(IS = 2.33E-27 BF=13
            CJE=1PF CJC=607.3PF TF=26.5NS)
ANALYSIS ■■■ .TRAN 10US 5MS 3MS UIC ; Transient analysis
             .PROBE                 ; Graphics post-processor
             .OPTIONS ABSTOL = 1.00U RELTOL = 0.02
             VNTOL = 0.1 ITL5 = 50000
             .FOUR 1KHZ I(VX)       ; Fourier analysis
        .END
```

Note the following:

a. The PSpice plots of the instantaneous output current I(VX), the current source I(Lm), and the inductor current I(Le) are shown in Figure 9.31.

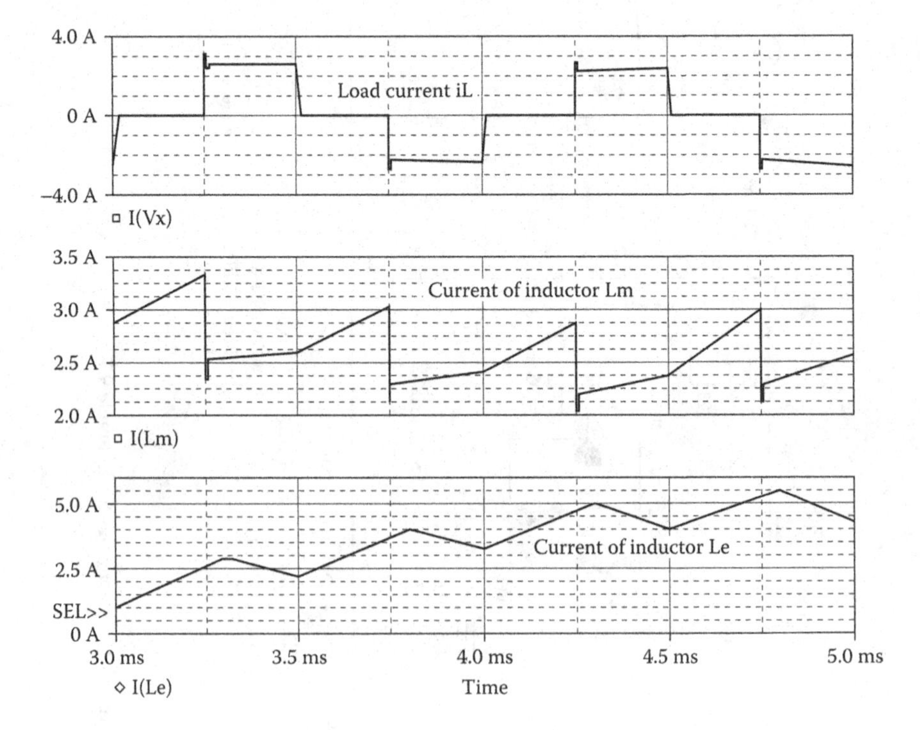

FIGURE 9.31 Plots for Example 9.7.

The output voltage is a square wave, as expected. It should be noted that the currents have not reached a steady state. We need to run the simulation for a longer time to see the steady-state conditions and the spikes in the output current are due to the diode capacitance CJO and transit time TT.

b. The Fourier coefficients of the output current are as follows:

Fourier Components of Transient Response I(VX)

Dc Component = 9.959469E−02

Harmonic Number	Frequency (Hz)	Fourier Component	normalized Component	Phase (Deg)	Normalized Phase (Deg)
1	1.000E+03	3.009E+00	1.000E+00	−4.814E+01	0.000E+00
2	2.000E+03	1.437E−01	4.775E−02	−1.759E+02	−1.278E+02
3	3.000E+03	9.189E−01	3.054E−01	3.520E+01	8.334E+01
4	4.000E+03	2.512E−02	8.347E−03	−1.064E+02	−5.831E+01
5	5.000E+03	6.507E−01	2.162E−01	−5.957E+01	−1.143E+01
6	6.000E+03	5.224E−02	1.736E−02	−1.633E+02	−1.152E+02
7	7.000E+03	3.570E−01	1.186E−01	2.344E+01	7.158E+01
8	8.000E+03	2.363E−02	7.853E−03	−1.062E+02	−5.803E+01
9	9.000E+03	3.878E−01	1.289E−01	−7.128E+01	−2.314E+01

Total harmonic distortion = 4.164380E+01%

From the Fourier components, we get

DC component = −99.59 mA
Peak fundamental component at 1 kHz = 3.09 A
Fundamental phase delay = −48.14°
Total harmonic distortion = 41.64%

Note: During the switching of the output current, the load voltage will exhibit high transients with an inductive load because of the rapid change of the current. The series diodes D_1 to D_4 are used to protect the transistors from these transients.

LTspice: LTspice schematic for Example 9.7 is shown in Figure 9.32(a). The plots of the inverter load current I(Vx), the current through the DC filter inductor I(Le), the DC current through the source inductor I(Lm), the output voltage of the DC–DC converter V(4) are shown in Figure 9.32(b). The duty cycle and the output of the DC–DC converter can be varied by varying the on-time of the transistor M5. The gating signals for transistors M1 and M3 are normally isolated from the ground through voltage-dependent voltage sources vg1 and vg2. The inverter output voltage is shown with a load inductor of L = 6.5 mH. We can observe that the waveform for voltage V(4) and the current I(Le) oscillate due to the resonant frequency of filter inductor Le and filter capacitor Ce.

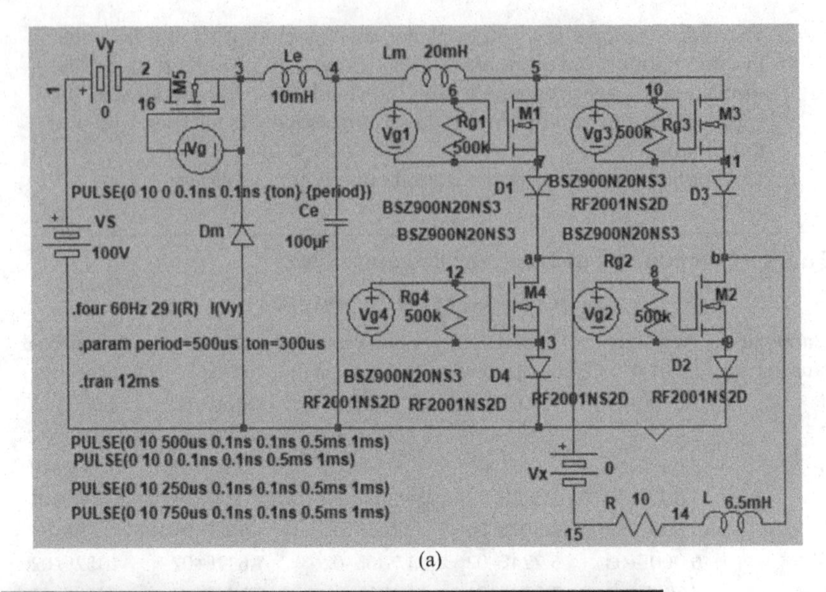

(a)

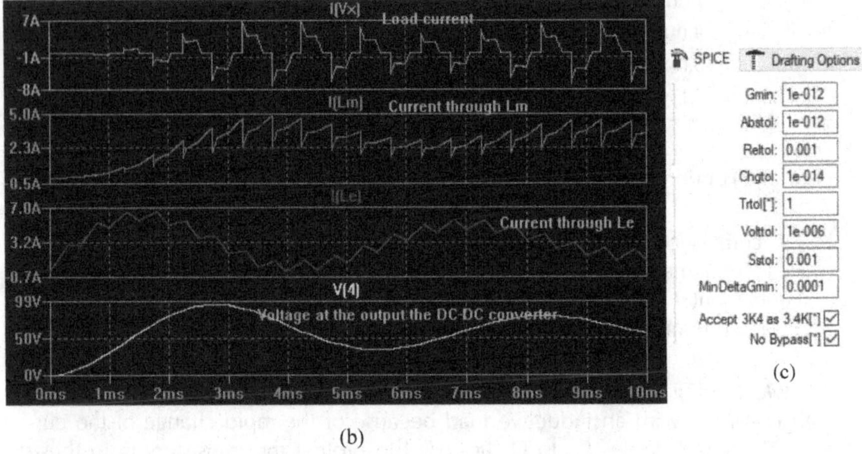

(b)

(c)

FIGURE 9.32 LTspice schematic for Example 9.7. (a) LTspice schematic. (b) Waveforms for inverter output voltage and gating signals. (c) Control option.

Note: We often need to adjust the tolerances with a. options command as shown in Figure 9.32(c).

9.4 DC LINK INVERTERS

The output voltage of a PWM inverter is varied by varying the modulation index M and the total harmonic distortion is also a function of the modulation index. As a result, the filtering requirements also vary. In a DC link inverter, the shape of the output voltage is kept fixed and the DC link voltage is varied by varying the duty cycle of a DC–DC converter at the input side.

That is, a DC link inverter consists of a DC–DC converter as the input stage as shown in the current-source inverter of Figure 9.30, and a DC–AC converter (inverter) operating at a fixed output voltage as shown in the three-phase inverter of Figure 9.19.

9.4.1 Example of DC Link Three-Phase Inverter

Example 9.8:

Finding the Performance of a DC Link Three-Phase Inverter

A DC link three-phase inverter is shown in Figure 9.33. The control voltages are the same as those shown in Figure 9.18(b). The DC input voltage is 100 V. The output frequency is $f_o = 1$ kHz. The chopping or switching frequency of the DC link converter is $f_s = 2$ kHz, and its duty cycle is $k = 0.1$, 0.8. The load resistance for each phase is $R = 10\ \Omega$, and the load inductance for each phase is $L = 5$ mH. Use PSpice to (a) plot the instantaneous output line–line voltage v_L and the instantaneous output phase voltage v_P. The model parameters of the BJTs are IS = 2.33E27, BF = 13, CJE = 1PF, CJC = 607.3PF, and TF = 26.5NS. The model parameters of the MOSFETs are VTO = 2.83, KP = 31.2U, L = 1U, W = 3.0M, CGDO = 1.359N, and CGSO = 2.032N.

SOLUTION

$f_o = 1$ kHz and $f_s = 2$ kHz. BJTs are used as switching devices in the PSpice schematic shown in Figure 9.33. The PWM generator varies the duty cycle of the DC–DC converter, which can change its output voltage in order to vary the DC input of the inverter.

The listing of the circuit file for the inverter is as follows:

DC Link Three-Phase Inverter

```
SOURCE     ■ V_Vy    1   2   0V
             V_Vg3   5   6
             +PULSE  0   40  {2/(6*{Freq})} 1ns
             1ns {1/(2*{Freq})-2ns} {1/{Freq}}
             V_Vg5   7   8
             +PULSE 0 40 {4/(6*{Freq})} 1ns
             1ns {1/(2*{Freq})-2ns} {1/{Freq}}
             V_Vx    15  16  0V
             V_Vg1   18  15
             +PULSE 0 40 0 1ns 1ns {1/(2*{Freq})-2ns} {1/{Freq}}
             V_Vz    26  10  0V
             E_PWM_Triangular_ABM21 28 0 VALUE
             { IF(V(27)-V(29)>0, 1,+ 0) }
```

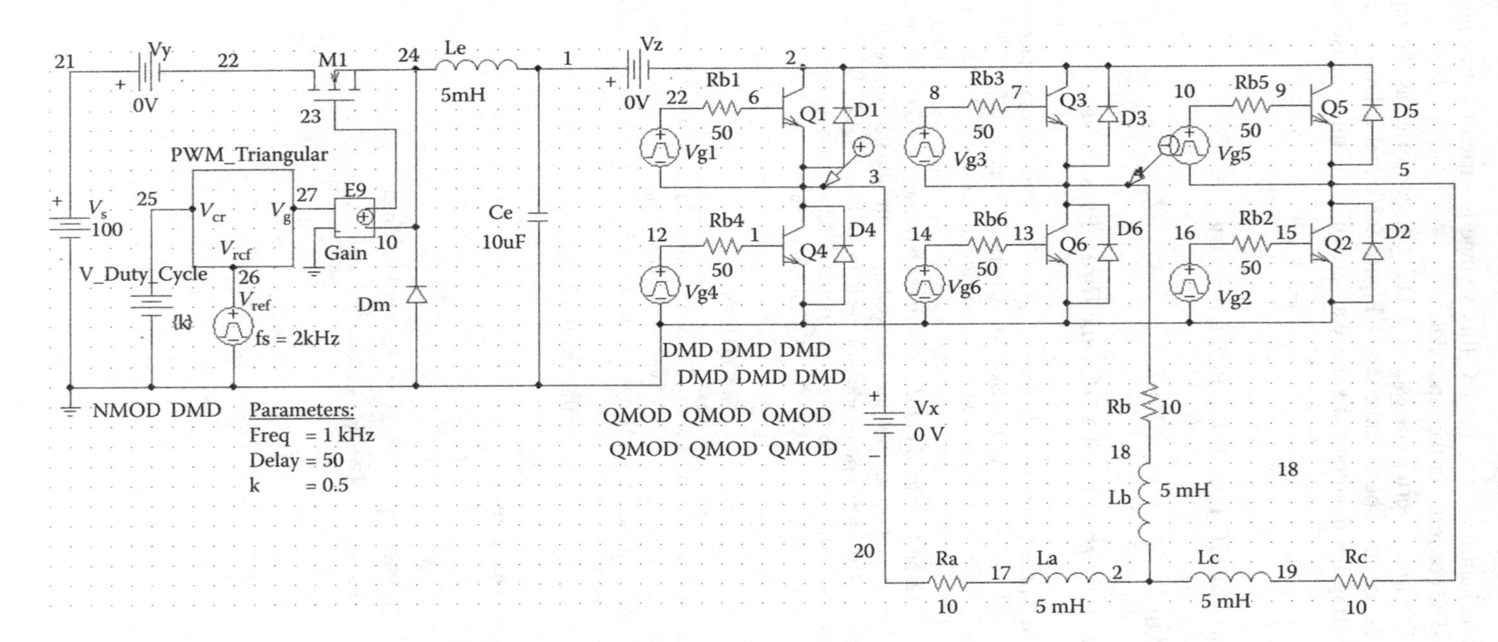

FIGURE 9.33 DC link three-phase inverter PSpice schematic for Example 9.8.

```
V_Vref  29  0
+PULSE 0 1 0 {1/2kHz-0.02us} 0.01u 0.01us {1/2kHz}
V_Vs   1  0  100
V_Vg4   21  0
+PULSE 0 40 {3/(6*{Freq})} 1ns
1ns {1/(2*{Freq})-2ns} {1/{Freq}}
V_Vg2   25  0
+PULSE 0 40 {1/(6*{Freq})} 1ns
1ns {1/(2*{Freq})-2ns} {1/{Freq}}
V_Vg6   23  0
+PULSE 0 40 {5/(6*{Freq})} 1ns
1ns {1/(2*{Freq})-2ns} {1/{Freq}}
V_V_Duty_Cycle    27    0    {k}
E_E9 4        3    28    0    10
```

CIRCUIT ■■

```
M_M1     2    4    3    3    NMOD L=1U W=0.03
Q_Q3     10        9    6    QMOD
L_Lb     11        12   5mH
L_La     13        11   5mH
L_Lc     11        14   5mH
R_Rb1    18   17   50
R_Rb3    5         9    50
R_Rb5    7         19   50
R_Rb4    21   20   50
R_Rb6    23   22   50
R_Rb2    25   24   50
R_Ra     16        13   10
R_Rc     14        8    10
Q_Q5     10        19   8    QMOD
D_Dm     0         3    DMD
Q_Q6     6         22   0    QMOD
Q_Q2     8         24   0    QMOD
Q_Q4     15        20   0    QMOD
D_D5     8         10   DMD
D_D6     0         6    DMD
D_D3     6         10   DMD
D_D1     15        10   DMD
D_D4     0         15   DMD
D_D2     0         8    DMD
Q_Q1     10        17   15   QMOD
L_Le     3         26   5mH
C_Ce     0         26   10uF
R_Rb     12        6    10
```

```
ANALYSIS ■■■ .PARAM Freq=1kHz   Delay=50 k=0.5
             .tran 0.     1us   10ms   8ms 0  .1e-6
             .four 1kHz   10    V([15],[6])
             .OPTIONS ABSTOL=1uA CHGTOL=0.01n CITL2=100
             ITL4=150 RELTOL=0.1 VNTOL=0.1
             .STEP PARAM k      LIST   0.1    0.8
             .PROBE
         .END
```

Note that the PSpice plots of the instantaneous line-to-line output voltage and the line-to-neutral voltage are shown in Figure 9.34. The line-to-line output voltage is a square wave, as expected, and the line-neutral output voltage is a stair-case wave, as expected. As the duty cycle is varied, the shape of the output

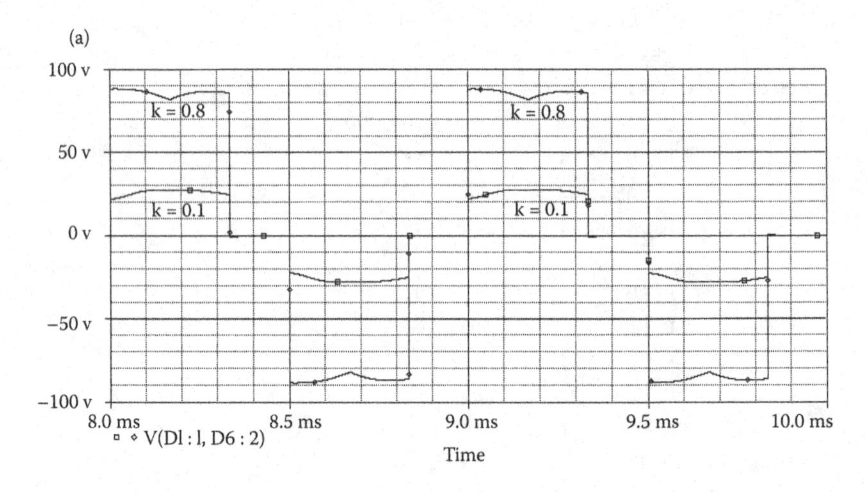

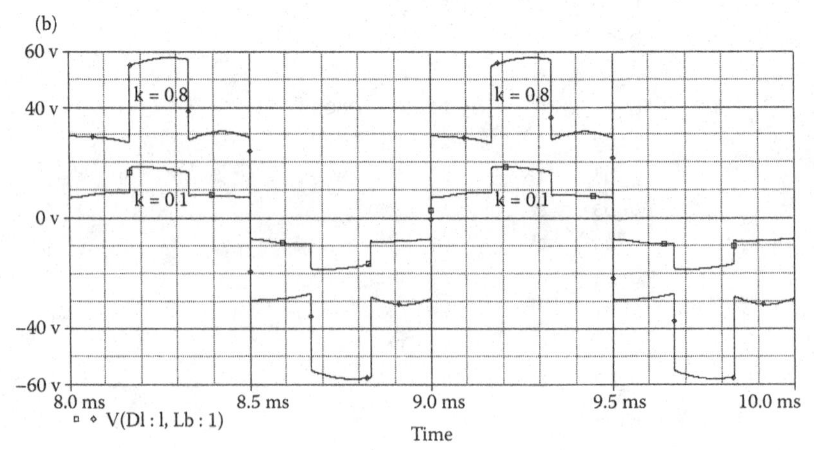

FIGURE 9.34 Output voltages of DC link three-phase inverter for Example 9.8. (a) Line-to-line voltage (V_{ab}). (b) Line-to-neutral voltage (V_{an}).

voltage remains the same, but the DC level varies with the duty cycle. The Fourier coefficients of the output voltages, which are obtained from the PSpice output file, are as follows:

Fourier Components of Transient Response V(15,6)

Duty Cycle, *k* = 0.1; Dc Component = −1.303634E–02

Harmonic Number	Frequency (Hz)	Fourier Component	Normalized Component	Phase (Deg)	Normalized Phase (Deg)
1	1.000E+03	2.935E+01	1.000E+00	2.897E+01	0.000E+00
2	2.000E+03	1.542E−02	5.253E−04	1.199E+02	6.192E+01
3	3.000E+03	1.327E+00	4.520E−02	−1.086E+02	−1.955E+02
4	4.000E+03	2.474E−03	8.428E−05	−1.134E+02	−2.292E+02
5	5.000E+03	5.505E+00	1.875E−01	−3.484E+01	−1.797E+02
6	6.000E+03	2.538E−03	8.647E−05	7.842E+00	−1.660E+02
7	7.000E+03	3.438E+00	1.171E−01	3.274E+01	−1.700E+02
8	8.000E+03	2.364E−03	8.052E−05	1.127E+02	−1.190E+02
9	9.000E+03	2.543E−01	8.664E−03	−1.742E+02	−4.349E+02
10	1.000E+04	1.489E−03	5.073E−05	−1.151E+02	−4.048E+02

Total harmonic distortion = 2.258544E+01%

Fourier Components of Transient Response V(15,6)

DC Component = 1.477034E–02

Harmonic Number	Frequency (Hz)	Fourier Component	Normalized Component	Phase (Deg)	Normalized Phase (Deg)
1	1.000E+03	9.455E+01	1.000E+00	3.023E+01	0.000E+00
2	2.000E+03	1.423E−02	1.505E−04	7.117E+00	−5.335E+01
3	3.000E+03	1.331E+00	1.408E−02	7.766E+01	−1.304E+01
4	4.000E+03	4.206E−03	4.448E−05	2.476E+01	−9.618E+01
5	5.000E+03	2.020E+01	2.137E−01	−2.948E+01	−1.807E+02
6	6.000E+03	2.991E−03	3.163E−05	8.036E+00	−1.734E+02
7	7.000E+03	1.329E+01	1.406E−01	2.980E+01	−1.818E+02
8	8.000E+03	3.797E−03	4.016E−05	4.615E+01	−1.957E+02
9	9.000E+03	1.728E−01	1.827E−03	5.576E+01	−2.164E+02
10	1.000E+04	2.549E−03	2.696E−05	8.097E+01	−2.214E+02

Total harmonic distortion = 2.561788E+01%

Note: THD, which is a measure of the quality of the line-to-neutral output voltage, is 22.58% for $k = 0.1$ and 25.58% for $k = 0.8$.

LTspice: LTspice schematic for Example 9.8 is shown in Figure 9.35(a). The plots of the inverter load current I(Ra), the output voltage of the DC–DC converter V(2), the line-line voltage V_{ab}, and the gating signal of the DC–DC converter

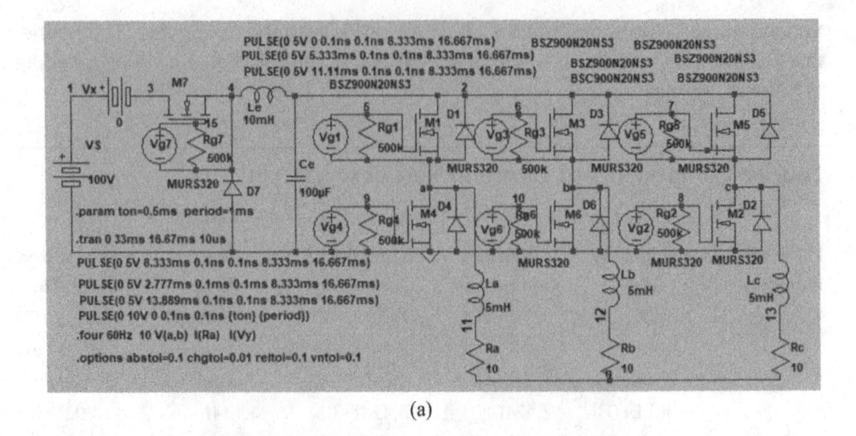

(a)

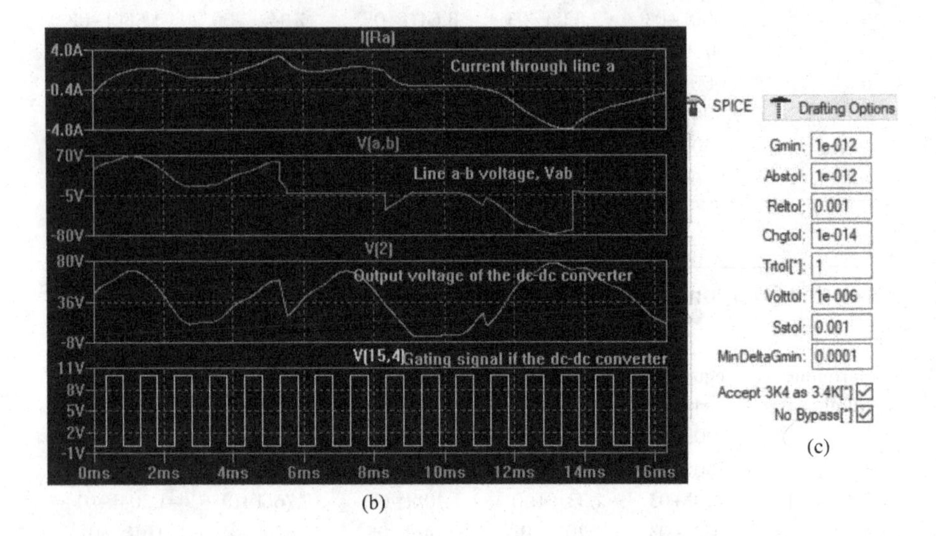

(b)

(c)

FIGURE 9.35 LTspice schematic for Example 9.8. (a) LTspice schematic. (b) Waveforms for inverter output voltage and gating signals. (c) Control options.

V(15,4) are shown in Figure 9.32(b). The duty cycle and the output of the DC–DC converter can be varied by varying the on-time of the transistor M7. The gating signals for transistors M1, M3, and M5 are normally isolated from the ground through voltage-dependent voltage sources vg1, vg3, and vg3. The inverter output voltage is shown with a load inductor of L = 5 mH. We can observe that the waveform for voltage V(2) varies significantly and continues. The filter consisting of Le and Ce should designed to keep the voltage V(2) within certain limiting values. Voltage V(2) oscillates due to the resonant frequency of filter inductor Le and filter capacitor Ce.

Note: We often need to adjust the tolerances with a. options command as shown in Figure 9.32(c).

9.5 LABORATORY EXPERIMENTS

It is possible to develop many experiments to demonstrate the operation and characteristics of inverters. The following experiments are suggested:

Single-phase half-bridge inverter
Single-phase full-bridge inverter
Single-phase full-bridge inverter with PWM control
Single-phase full-bridge inverter with SPWM control
Three-phase bridge inverter
Single-phase current-source inverter
Three-phase current-source inverter

Warning: The gate signal circuits in all the following experiments must be isolated from the power circuit by isolating transformers or optocouplers.

9.5.1 Experiment PW.1

Single-Phase Half-Bridge Inverter

Objective	To study the operation and characteristics of a single-phase half-bridge (transistor) inverter under various load conditions
Applications	The single-phase half-bridge inverter is used to control power flow in AC and DC power supplies, input stages of other converters, etc.
Textbook	See Ref. [1], Section 6.2
Apparatus	1. Two BJTs or MOSFETs with ratings of at least 50 A and 400 V, mounted on heat sinks 2. Two fast-recovery diodes with ratings of at least 50 A and 400 V, mounted on heat sinks 3. A firing pulse generator with isolating signals for gating transistors 4. An *RL* load 5. One dual-beam oscilloscope with floating or isolating probes 6. AC and DC voltmeters and ammeters and one noninductive shunt
Warning	Before making any circuit connection, switch the DC power off. Do not switch on the power unless the circuit is checked and approved by your laboratory instructor. Do not touch the transistor or diode heat sinks, which are connected to live terminals
Experimental procedure	1. Set up the circuit as shown in Figure 9.36. Use the load resistance *R* only 2. Connect the measuring instruments as required 3. Connect the firing pulses to the appropriate transistors 4. Set one pulse per half-cycle with a duty cycle of $k = 0.5$ 5. Observe and record the waveforms of the load voltage v_o and the load current i_o 6. Measure the RMS load voltage $V_{o(RMS)}$, the RMS load current $I_{o(RMS)}$, the average input current $I_{s(DC)}$, the DC input voltage $V_{s(DC}$, and the THD of the output voltage and current 7. Measure the conduction angles of transistor Q_1 and diode D_1 8. Repeat steps 2–7 with both load resistance *R* and load inductance *L*

(Continued)

Single-Phase Half-Bridge Inverter ***(Continued)***

Report	1. Present all recorded waveforms and discuss all significant points Single-Phase Half-Bridge Inverter 2. Compare the waveforms generated by SPICE with the experimental results, and comment 3. Compare the experimental results with the predicted results 4. Discuss the advantages and disadvantages of this type of inverter 5. Discuss the effects of the diodes on the performance of the inverter

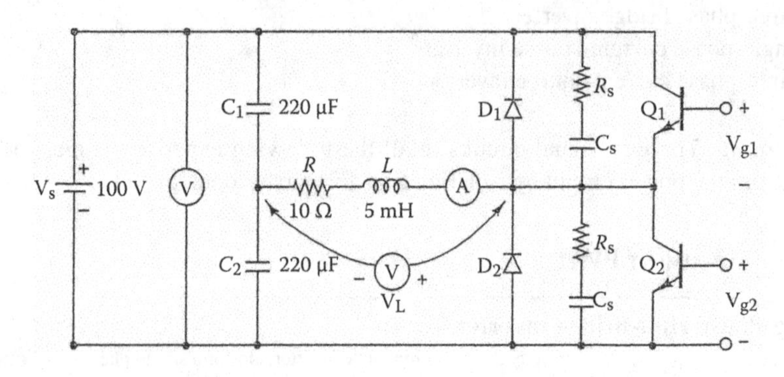

FIGURE 9.36 Single-phase half-bridge inverter.

9.5.2 Experiment PW.2

Single-Phase Full-Bridge Inverter

Objective	To study the operation and characteristics of a single-phase full-bridge (transistor) inverter under various load conditions
Applications	See Experiment PW.1
Textbook	See Ref. [1], Section 6.4
Apparatus	Similar to Experiment PW.1, except that four BJTs or MOSFETs with four fast-recovery diodes are required as shown in Figure 9.27
Warning	See Experiment PW.1
Experimental procedure	See Experiment PW.1
Report	See Experiment PW.1

Warning: The ground terminal of v_{g1} cannot be connected directly to the emitter terminal of transistor Q_1. The gate signal v_{g1} must be isolated from the power circuit by an isolating transformer or an opto-coupler.

9.5.3 Experiment PW.3

Single-Phase Full-Bridge Inverter with PWM Control

Objective	To study the effects of PWM control on the THD of the output voltage in a single-phase full-bridge (transistor) inverter with a resistive load
Applications	Similar to Experiment PW.1
Textbook	See Ref. [1], Sections 6.4 and 6.6

(Continued)

Single-Phase Full-Bridge Inverter with PWM Control *(Continued)*

Apparatus	Similar to Experiment PW.1, except that four BJTs or MOSFETs and four fast-recovery diodes are required
Warning	See Experiment PW.1
Experimental procedure	1. Set up the circuit as shown in Figure 9.37. Use the load resistance R only 2. Set one pulse per half-cycle, $p = 1$, and the modulation index $M = 0.5$ 3. Measure the RMS load voltage $V_{o(RMS)}$ and the THD of the output voltage 4. Repeat step 3 for $p = 2, 3, 4$, and 5 5. Repeat steps 2 and 4 for $M = 0.1$ to 1 with an increment of 0.1
Report	1. Plot the RMS output voltage and the THD of the output voltage against the modulation M for various values of p 2. Use SPICE or MathCAD to predict the RMS output voltage and the THD 3. Compare the experimental results with the predicted results, and comment

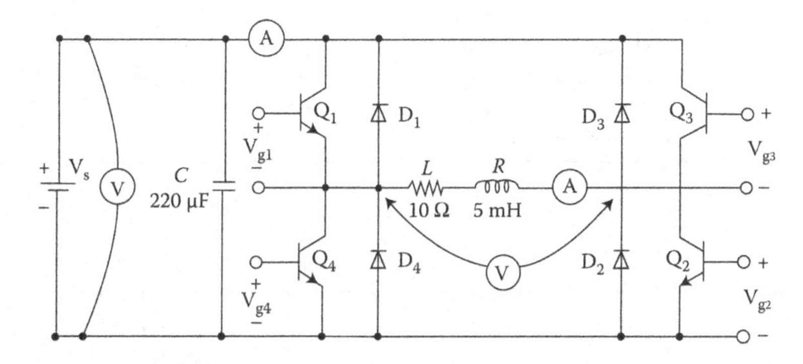

FIGURE 9.37 Single-phase full-bridge inverter.

9.5.4 Experiment PW.4

Single-Phase Full-Bridge Inverter with SPWM Converter

Objective	To study the effects of SPWM control on the THD of the output voltage for a single-phase full-bridge (transistor) inverter under a resistive load Applications See Experiment PW.1
Textbook	See Ref. [1], Sections 6.4 and 6.6
Apparatus	See Experiment PW.2
Warning	See Experiment PW.1
Experimental procedure	See Experiment PW.3
Report	See Experiment PW.3

9.5.5 Experiment PW.5

Three-Phase Bridge Inverter

Objective	To study the operation and characteristics of a three-phase bridge (transistor) inverter under various load conditions
Applications	The three-phase bridge inverter is used to control power flow in AC power supplies, AC motor drives, etc.

(Continued)

Three-Phase Bridge Inverter *(Continued)*

Textbook	See Ref. [1], Sections 6.5 and 6.6
Apparatus	Similar to Experiment PW.1, except that six BJTs or MOSFETs and six fast-recovery diodes are required
Warning	See Experiment PW.1
Experimental procedure	1. Set up the circuit as shown in Figure 9.38. Use a wye-connected resistive load *R* only 2. Connect the measuring instruments as required 3. Connect the firing pulses to the appropriate transistors 4. Set one pulse per half-cycle with a duty cycle of $k = 0.5$ 5. Observe and record the waveforms of the output phase voltage, the output line–line voltage, and the load phase current i_o 6. Measure the RMS load phase voltage $V_{p(RMS)}$, the RMS load phase current $I_{p(RMS)}$, the average input current $I_{s(DC)}$, the average input voltage $V_{s(DC)}$, and the THD of the output phase voltage and phase current 7. Measure the conduction angles of transistor Q_1 and diode D_1 8. Repeat steps 2–7 with both load resistance *R* and load inductance *L* 9. Repeat steps 1–8 with a delta-connected load
Report	See Experiment PW.1

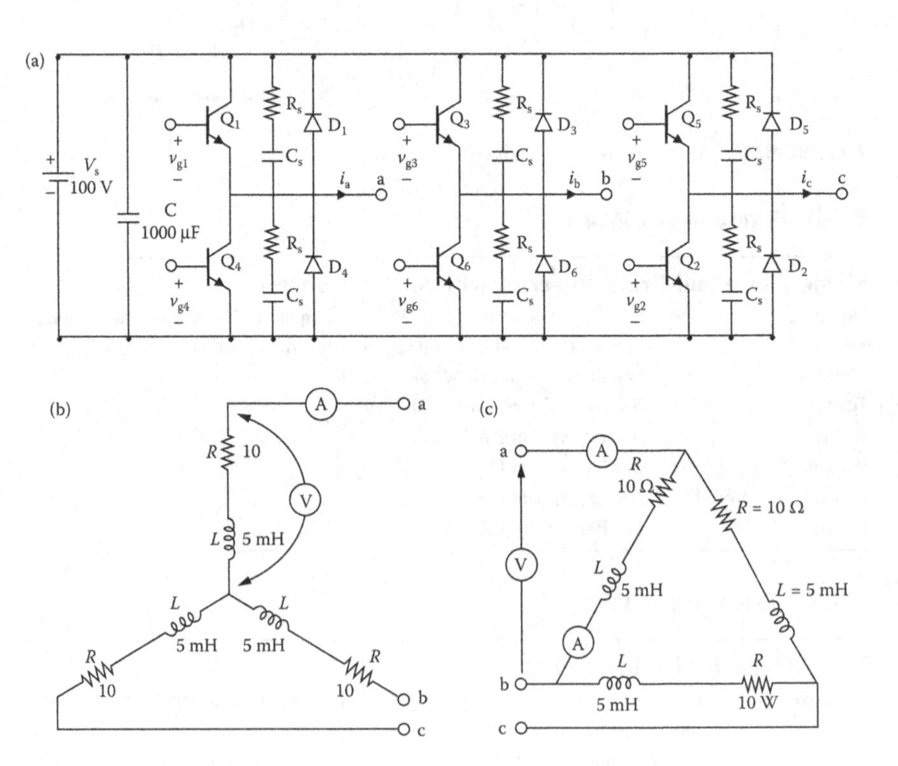

FIGURE 9.38 Three-phase bridge inverter. (a) Circuit. (b) Wye-load. (c) Delta-load.

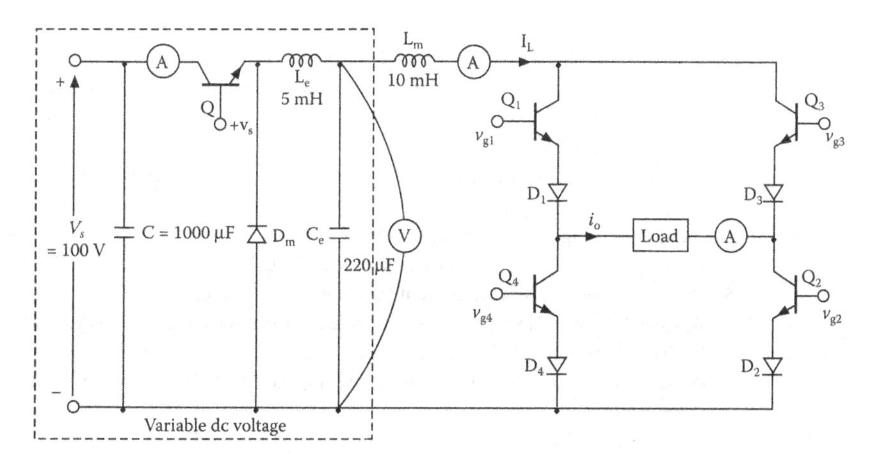

FIGURE 9.39 Single-phase current-source inverter.

9.5.6 Experiment PW.6

Single-Phase Current-Source Inverter

Objective — To study the operation and characteristics of a single–phase current-source(transistor) inverter under various load conditions

Applications — The current-source inverter is used to control power flow in AC power supplies, etc.

Textbook — See Ref. [1], Section 6.11

Apparatus — Similar to Experiment PW.1, except that four BJTs or MOSFETs and four fast-recovery diodes are required

Warning — See Experiment PW.1

Experimental procedure —

1. Set up the circuit as shown in Figure 9.39. Use a resistive load *R* only
2. Connect the measuring instruments as required
3. Connect the firing pulses to the appropriate transistors
4. Set the duty cycle of the chopper at $k = 0.5$. Set the inverter to one pulse per half-cycle with a duty cycle of $k = 0.5$
5. Observe and record the waveforms of the output voltage and the output current
6. Measure the RMS load voltage $V_{o(RMS)}$, the RMS load current $I_{o(RMS)}$, the average input current $I_{S(DC)}$, the DC input voltage $V_{S(DC)}$, and the THD of the output current
7. Measure the conduction angles of the transistor Q_1
8. Repeat steps 2–7 with both load resistance *R* and load inductance *L*

Report — See Experiment PW.1

9.5.7 Experiment PW.7

Three-Phase Current-Source Inverter

Objective — To study the operation and characteristics of a three-phase current-source (transistor) inverter under resistive load

Applications — The three-phase current-source inverter is used to control power flow in AC power supplies, AC motor drives, etc.

Textbook — See Ref. [1], Section 6.11

(Continued)

Three-Phase Current-Source Inverter *(Continued)*

Apparatus	Similar to Experiment PW.1, except that six BJTs or MOSFETs and six fast-recovery diodes are required
Warning	See Experiment PW.1
Experimental procedure	1. Set up the circuit as shown in Figure 9.40. Use a wye-connected resistive load *R* only 2. Connect the measuring instruments as required 3. Connect the firing pulses to the appropriate transistors 4. Set one pulse per half-cycle with a duty cycle of $k = 0.5$ 5. Observe and record the waveforms of the output phase current i_o 6. Measure the RMS load phase current, average input current, DC input voltage, and the THD of the output phase current 7. Measure the conduction angles of the transistor Q_1 and diode D_1 Report See Experiment PW.1

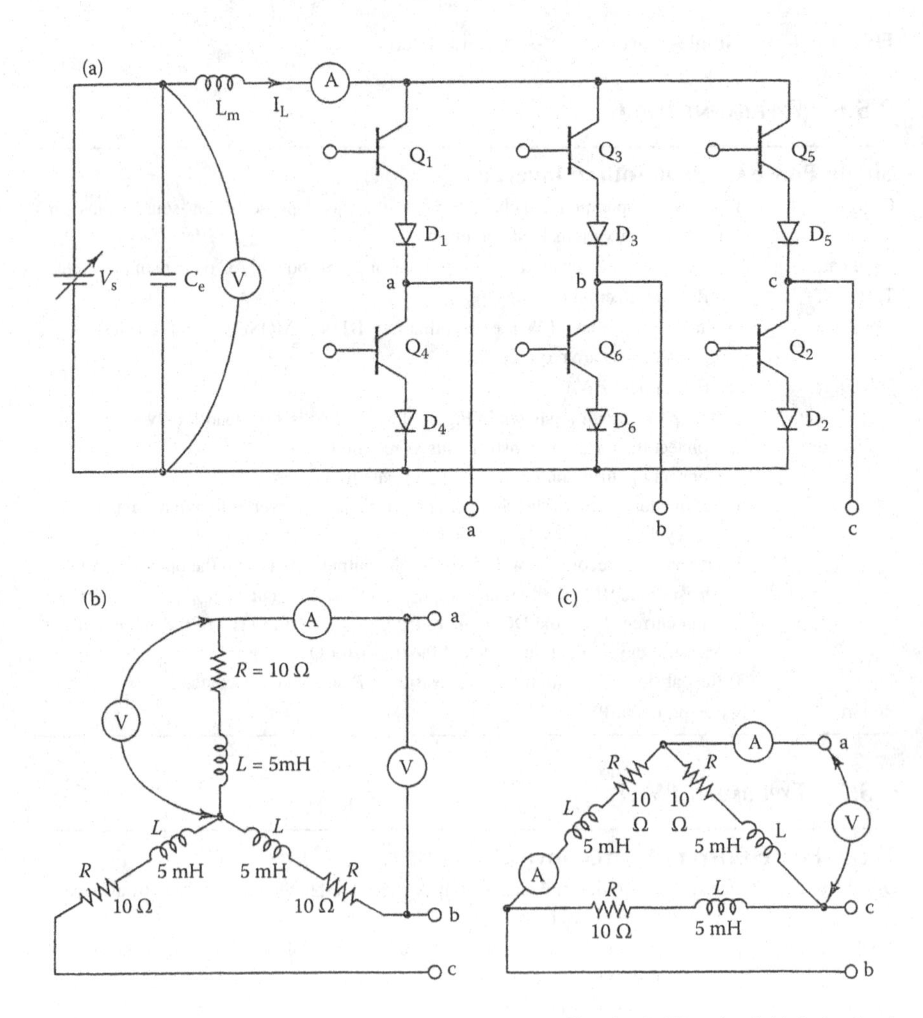

FIGURE 9.40 Three-phase current-source inverter. (a) Circuit. (b) Wye-load. (c) Delta-load.

SUMMARY

The statements for an AC thyristor are as follows:

```
*Subcircuit call for switched transistor model:
XT1      +N          -N          +NG          -NG          QM
     positive    negative    +control     -control     model
*                             voltage      voltage     name
*Subcircuit call for PWM control:
XPWM    VR       VC          +NG         -NG        PWM
*      ref     carrier    +control    -control    model
*     input     input     voltage     voltage     name
*Subcircuit call for sinusoidal PWM control:
XSPWM   VR       VS          +NG         -NG           VC            SPWM
*       ref   sine wave   +control  -control      rectified       model
*      input    input      voltage  voltage carrier sine wave name
```

DESIGN PROBLEMS

9.1 It is required to design the single-phase half-bridge inverter of Figure 9.36 with the following specifications:
DC supply voltage, $V_s = 100$ V
Load resistance, $R = 5\ \Omega$
Load inductance, $L = 5$ mH
Output frequency, $f_o = 1$ kHz
  a. Determine the ratings of all components and devices under worst-case conditions.
  b. Use SPICE to verify your design.
  c. Provide a cost estimate of the circuit.

9.2 It is required to design the single-phase full-bridge inverter of Figure 9.37 with the following specifications:
DC supply voltage, $V_s = 100$ V
Load resistance, $R = 5$ Q
Load inductance, $L = 5$ mH

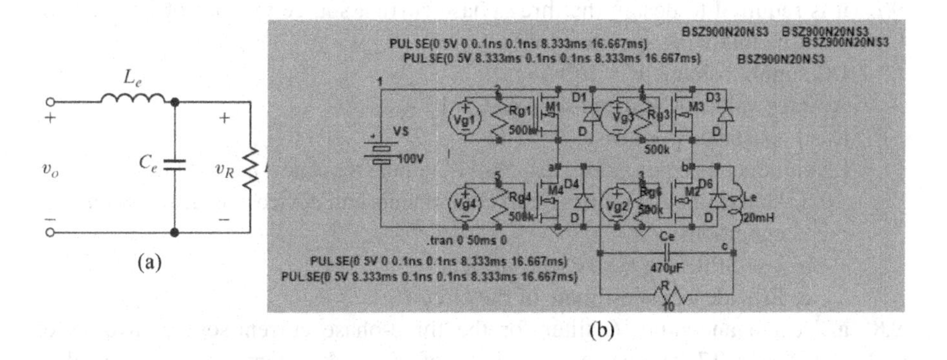

FIGURE P9.2 (a) LC-filter. (b) Inverter with an output filter.

Output frequency, $f_o = 1$ kHz

a. Determine the ratings of all components and devices under worst-case conditions.
b. Use SPICE to verify your design.
c. Provide a cost estimate of the circuit.

9.3 a. Design an output C filter for the single-phase full-bridge inverter of Problem 9.2. The harmonic content of the load current should be less than 5% of the value without the filter.
b. Use SPICE to verify your design in part (a).

9.4 It is required to design the three-phase bridge inverter of Figure 9.38 with the following specifications:
DC supply voltage, $V_s = 100$
Load resistance per phase, $R = 5\ \Omega$
Load inductance per phase, $L = 5$ mH
Output frequency, $f_o = 1$ kHz

a. Determine the ratings of all components and devices under worst-case conditions.
b. Use SPICE to verify your design.
c. Provide a cost estimate of the circuit.

9.5 It is required to design the single-phase current-source inverter of Figure 9.39 with the following specifications:
DC supply voltage, $V_s = 100$ V
Average value DC current source, $I_s = 10$ A
Load resistance, $R = 5\ \Omega$
Load inductance, $L = 5$ mH Output frequency, $f_o = 400$ Hz

a. Determine the ratings of all components and devices under worst-case conditions.
b. Use SPICE to verify your design.
c. Provide a cost estimate of the circuit.

9.6 a. Design an output C filter for the single-phase current-source inverter of Problem 9.5. The harmonic content of the load current should be less than 5% of the value without the filter.
b. Use SPICE to verify your design in part (a).

9.7 It is required to design the three-phase current-source inverter of Figure 9.40 with the following specifications:
DC supply voltage, $V_s = 100$ V
Average value DC current source, $I_s = 10$ A
Load resistance per phase, $R = 5\ \Omega$
Load inductance per phase, $L = 5$ mH Output frequency, $f_o = 400$ Hz

a. Determine the ratings of all components and devices under worst-case conditions.
b. Use SPICE to verify your design.
c. Provide a cost estimate of the circuit.

9.8 a. Design an output C filter for the three-phase current-source inverter of Problem 9.7. The harmonic content of the load current should be less than 5% of the value without the filter.
b. Use SPICE to verify your design in part (a).

9.9 Use PSpice to find the THD of the output voltage for a single-phase PWM inverter. Complete the following table for a modulation index of $M = 0.05$ to 0.95 and different pulses per half-cycle, $p = 1$ to 8. Assume a resistive load of $R = 10\ \Omega$, a DC input voltage $V_s = 100$ V, and an output frequency of $f_o = 60$ Hz. Use any suitable switching devices.

	p/m	0.05	0.1	0.2	0.3	0.4	0.5	0.6	0.7	0.8	0.9	0.95
THD% of output voltage	1											
	2											
	4											
	5											
	6											
	7											
	8											

9.10 Use PSpice to find the THD of the output voltage for a single-phase SPWM inverter. Complete the table of Problem 9.9 with a modulation index of $M = 0.05$ to 0.95 and different pulses per half-cycle, $p = 1$ to 8. Assume a resistive load of $R = 10\ \Omega$, a DC input voltage $V_s = 100$ V, and an output frequency of $f_o = 60$ Hz. Use any suitable switching devices.

9.11 Use PSpice to find the THD of the wye-connected output phase voltage for a three-phase PWM inverter. Complete the table of Problem 9.9 with a modulation index of $M = 0.05$ to 0.95 and different pulses per half-cycle, $p = 1$ to 8. Assume a resistive load of $R = 10\ \Omega$, a DC input voltage $V_s = 100$ V, and output frequency of $f_o = 60$ Hz. Use any suitable switching devices.

9.12 Use PSpice to find the THD of the wye-connected output phase voltage for a three-phase sinusoidal PWM inverter. Complete the table of Problem 9.9 with a modulation index of $M = 0.05$ to 0.95 and different pulses per half-cycle, $p = 1$ to 8. Assume a resistive load of $R = 10\ \Omega$, a DC input voltage $V_s = 100$ V, and an output frequency of $f_o = 60$ Hz. Use any suitable switching devices.

9.13 Use PSpice to find the THD of the output current for a single-phase current-source inverter for different delay angles from 0 to 180° with an increment of 20°. Assume a resistive load of $R = 10\ \Omega$, a DC input voltage $V_s = 100$ V, and an output frequency of $f_o = 60$ Hz. Use any suitable switching devices.

9.14 Use PSpice to (a) plot the instantaneous output line–line voltage v_L and the instantaneous output phase voltage v_p of the DC link three-phase inverter shown in Figure 9.24. The control voltages are the same as thos shown in Figure 9.18(b). The DC input voltage is 120 V. The output frequency is $f_o = 60$ Hz. The chopping or switching frequency of the DC link converter is $f_s = 2$ kHz, and its duty cycle is $k = 0.0.5$. The load resistance for each phase is $R = 10$ Q, and the load inductance for each phase is $L = 1.5$ mH. The model parameters of the BJTs are IS = 2.33E27, BF = 13, CJE = 1PF, CJC = 607.3PF, and TF = 26.5NS. The model parameters of the MOSFETs are VTO = 2.83, KP = 31.2U, L = 1U, W = 3.0M, CGDO = 1.359N, and CGSO = 2.032N.

9.15 The single-phase full-bridge inverter as shown in Figure P9.24 supplies a load of $R = 10\ \Omega$, The DC input voltage is $V_s = 108$ V and the inverter frequency is fo = 60 Hz. An output LC filter as shown is used to eliminate third and higher order harmonics.
1. Determine the performance parameters
   a. Total RMS output voltage, V_o
   b. RMS value of the fundamental output voltage, V_{o1}
   c. Total output power, Po
   d. Average current from the battery, Is
   e. Average current through each transistor. I_Q
   f. RMS current of each transistor, Ir
   g. THD of the output voltage, %
2. Determine the values of Le and Ce so that the THD of the output current through R is ≤ 5%.
   a. Find the value of Ce
   b. Find the value of Le
3. Use Multisim/LTspice/Orcad to run the simulation from 0 to 50 ms so that the steady-state condition is reached.
4. Make modifications to meet the specifications if necessary to meet the specifications
   a. Plot the instantaneous load current through R, $i_o(t)$.
   b. Plot the instantaneous current through supply $i_s(t)$, that is, –I(Vs)
   c. Run the Fourier analysis and find the THD of the load current for 2 cycles, up to 29th harmonics.
5. Complete Table P9.15 for simulated values.

TABLE P9.15

	$I_{o(RMS\text{-}fundamental)}$ (A)	THD of the current through R, io(t)
Calculated		≤ 5%
Simulated		

6. Estimated component costs: https://www.jameco.com/
7. Safety considerations if you would be building the product.
8. Risk factors of the design if you would be building the product.
9. What trade-offs have you considered?
10. Lessons Learned from the design assignment.

SUGGESTED READING

1. M.H. Rashid, *Power Electronics: Circuit, Devices and Applications*, Fourth Edition, Englewood Cliffs, NJ: Prentice-Hall, 2014, Chapter 6.
2. M.H. Rashid, *Introduction to PSpice Using OrCAD for Circuits and Electronics*, Third Edition, Englewood Cliffs, NJ: Prentice-Hall, 2003, Chapter 7.

3. M.H. Rashid, *SPICE for Power Electronics and Electric Power*, Englewood Cliffs, NJ: Prentice-Hall, 1993.
4. N. Mohan, T.M. Undeland and W.P. Robbins, *Power Electronics: Converters, Applications and Design*, New York: John Wiley & Sons, 2003.
5. v. Voperian,R. Tymerski and F.C.Y. Lee, Equivalent circuit models for resonant and PWM switches, *IEEE Transactions on Power Electronics*, *4*(2), 1990, 205–214.
6. M.H. Rashid (editor), *Power Electronics Handbook*, Butterworth Heinemann. 4/e, 2017.

10 Resonant-Pulse Inverters

After completing this chapter, students should be able to do the following:

- Model BJTs and IGBTs in SPICE, and specify their mode parameters.
- Model current- and voltage-controlled switches.
- Evaluate the performance of voltage-source and current-source resonant-pulse inverters.
- Perform worst-case analysis of resonant-pulse inverters for parametric variations of model parameters and tolerances.

10.1 INTRODUCTION

The input to a resonant inverter is a DC voltage or current source, and the output is a voltage or current resonant pulse. Power semiconductor devices perform the switching action, and the desired output is obtained by varying their turn-on and turn-off times. The commonly used devices are BJTs, MOSFETs, IGBTs, MCTs, GTOs, and SCRs. We shall use PSpice switches, IGBTs, and BJTs to simulate the characteristics of the following inverters:

Resonant-pulse inverter
Zero-current switching converter (ZCSC)
Zero-voltage switching converter (ZVSC)

Thus, the voltage and the current on the input side are of DC types, and the voltage and current on the output side are of AC types. Table 10.1 lists the performance parameters of inverters.

TABLE 10.1
Resonant-Pulse Inverter Performance Parameters for an Ideal No-Load Inverter

Input side parameters	Output side parameters
DC input voltage, V_S	RMS output voltage, V_o
Average input current, I_S	RMS output current, I_o
RMS input current, I_{RMS}	The output power, $P_o = V_o I_o = I_o^2 R_L$ for a load resistance of R_L
Converter Power Efficiency, η	RMS value of the fundamental output voltage, V_{o1}
The RMS ripple content of the input current $I_{ripple} = \sqrt{I_{rms}^2 - I_S^2}$	The RMS ripple content of the output voltage $V_{ac} = \sqrt{V_o^2 - V_{o1}^2}$
	The *ripple factor*, $RF_o = \frac{V_{ac}}{V_{o1}}$
	Total harmonic distortion (THD) of the output voltage, %

DOI: 10.1201/9781003284451-10

Tips for SPICE/LTspcie Simulations:

- After completing the drawing of the circuit file, Run the Simulation.
- Plot the gating signals and ensure the desired gating signals.
- Plot the voltages across the switching devices to ensure that switching devices are turned on and off according to the gating signals.
- Increase the voltage level of the gating signals if the switching devices are not fully turned on.

10.2 RESONANT-PULSE INVERTERS

The switches of resonant inverters are turned on to initiate resonant oscillations in an underdamped *RLC* circuit and are maintained in an on-state condition to complete the oscillations. The output waveform depends mainly on the circuit parameters and the input source. The on-time and switching frequency of power devices must match the resonant frequency of the circuit.

10.2.1 Examples of Resonant-Pulse Inverters

Example 10.1:

Finding The Performance of a Half-Bridge Resonant-Pulse Inverter With BJT Switches

A half-bridge resonant inverter is shown in Figure 10.1(a). The control voltages are shown in Figure 10.1(b) The DC input voltage is 100 V. The output frequency is $f_o = 5$ kHz. The load resistance is $R = 1\ \Omega$ and the load inductance is $L = 50\ \mu$H. Use PSpice to (a) plot the instantaneous output current i_o and the instantaneous input supply current i_s and (b) calculate the Fourier coefficients of the output current i_0. The BJT parameters are IS = 2.33E–27, BF = 13, CJE = 1PF, CJC = 607.3PF, and TF = 26.5NS.

Notes:

1. Vx and Vy, as shown in Figure 10.1(a), are used as ammeters to monitor the load current and the input side current, respectively.
2. A base resistor (R_{b1}) must be inserted as shown in Figure 10.1(a) at the base of each transistor to limit the base current and its value should be such that the transistor is driven into saturation, that is, the collector-emitter voltage in saturation $V_{CE(sat)}$ should be low.

SOLUTION

The values of gate voltage V_g and base resistance R_b must be such that the transistors are driven into saturation at the expected load current. The PSpice schematic with BJTs is shown in Figure 10.2. Varying the duty cycle can change the output voltage. The switching frequency {FREQ} and the duty cycle {DUTY_CYCLE} are

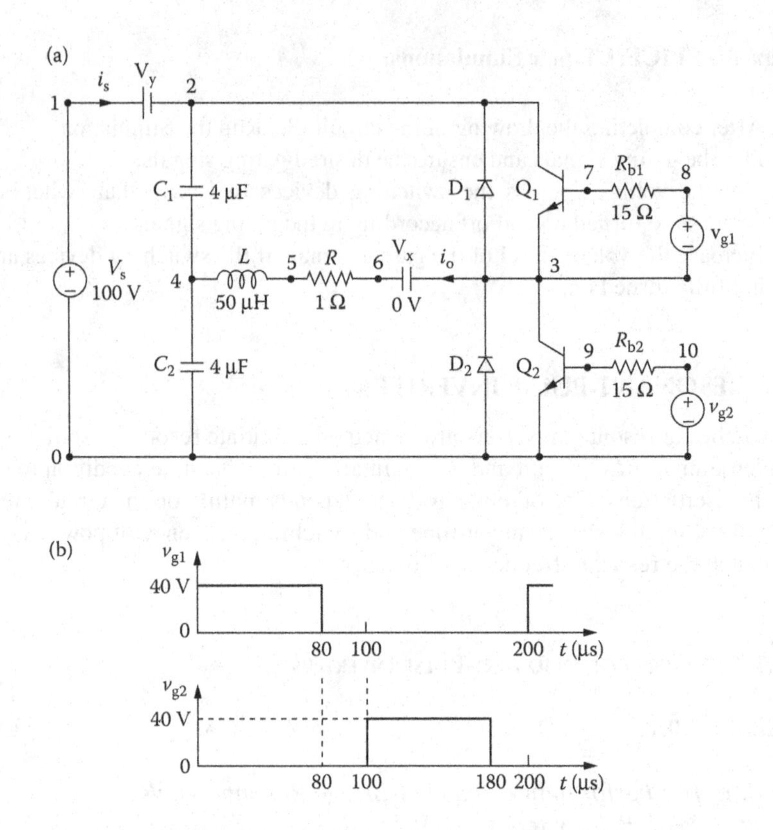

FIGURE 10.1 Half-bridge resonant inverter. (a) Circuit. (b) Control voltages.

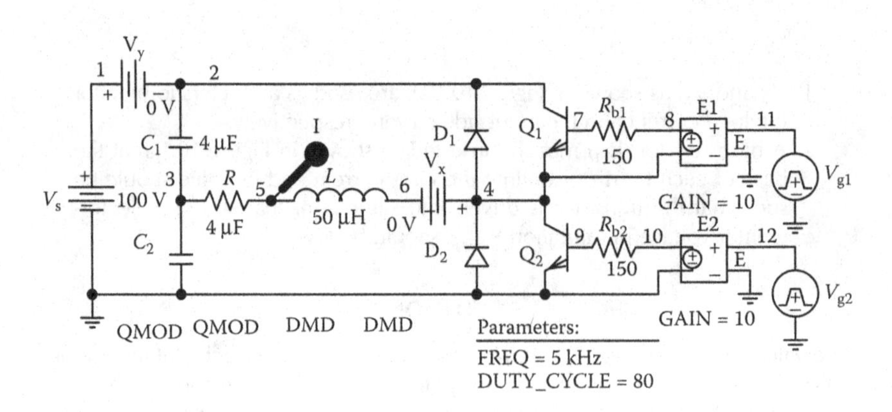

FIGURE 10.2 PSpice schematic for Example 10.1.

defined as variables. The model parameters for the BJTs and the freewheeling diodes are as follows:

```
.MODEL QMOD NPN(IS = 6.83E-14 BF = 13 CJE = 1pF CJC = 607.3PF
TF = 26.5NS) for BJTs
.MODEL DMD D(IS = 2.22E-15 BV = 1200V CJO = 0PF TT = 0US)) for BJTs
```

The listing of the circuit file is as follows:

Half-Bridge Resonant Inverter

```
SOURCE    ■     VS    1 0 DC 100V
                .PARAM DUTY_CYCLE=80 FREQ=5kHz
                Vg1  8 3 PULSE (0 40 0 1ns 1ns {{Duty_Cycle}/(200*
                {Freq})-2ns} {1/{Freq}})
                Vg2 10 0 PULSE 0 40 {1/(2*{Freq})} 1ns 1ns
                {{Duty_Cycle}/(200*{Freq})-2ns} {1/{Freq}}
CIRCUIT   ■■    Rb1 8 7 150
                Rb2 10 9 150
                VY   1 2 DC 0V ; Voltage source to measure
                supply  current
                VX   4 6 DC 0V        ; Measures load current
                R    3 5 1
                L    5 6 50UH
                C1   2 4 4UF
                C2   4 0 4UF
                D1   3 2 DMOD         ; Diode
                D2   0 3 DMOD         ; Diode
                .MODEL DMOD D(IS=2. 2E-15 BV=1200V CJO=0 TT=0) ;
                Diode
                                              model parameters
                Q1   2 7 3 3 2N6546        ; BJT switch
                Q2   3 9 0 0 2N6546        ; BJT switch
                .MODEL 2N6546 NPN (IS=2.33E-27 BF=13 CJE=1PF
                CJC=607.3PF TF=26.5NS)
ANALYSIS  ■■■   .TRAN 1US 400US         ; Transient analysis
                .PROBE ; Graphics post-processor
                .OPTIONS ABSTOL=1.00N RELTOL=0.01 VNTOL 0.1
                ITL5=50000
                .FOUR 5KHZ I(VX)        ; Fourier analysis
          .END
```

Note the following:

a. The PSpice plots of the instantaneous output current I(VX) and the current source I(VY) for $f_o = 5$ kHz are shown in Figure 10.3. For $f_o = 4$ kHz, the switching period is changed to 250 μs, and the plots are shown in Figure 10.4. As expected, the output voltage can be varied by changing the switching frequency. The completion of the full-resonant cycle was interrupted by the switching of the second transistor switch, that is, the second switch was turned on before the completion of the resonant cycle.
b. The Fourier coefficients of the output current, which can be found in the PSpice output file, are as follows:

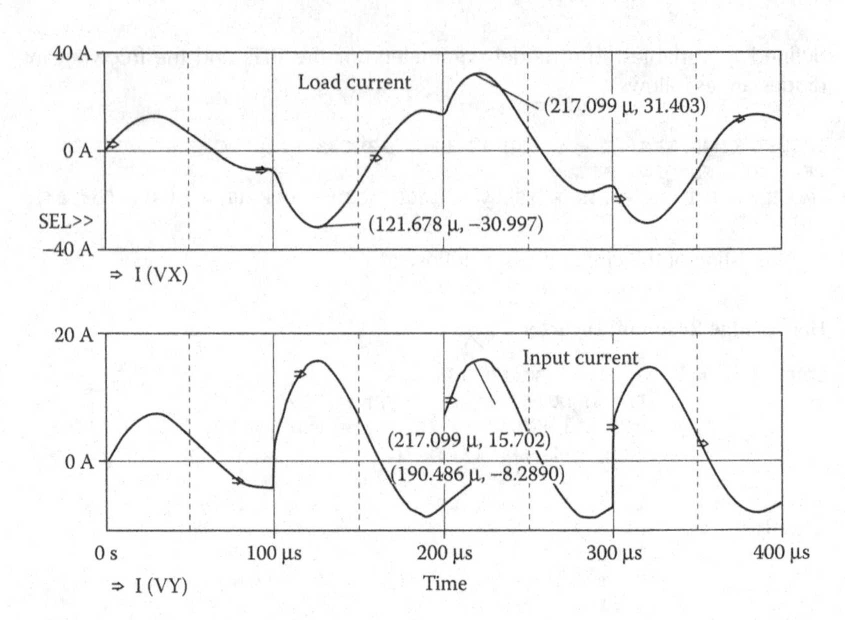

FIGURE 10.3 Plots at $f_o = 5$ kHz for Example 10.1.

Fourier Components of Transient Response I(VX)

DC Component = 2.185143E–01

Harmonic Number	Frequency (Hz)	Fourier Component	Normalized Component	Phase (Deg)	Normalized Phase (Deg)
1	5.000E+03	2.566E+01	1.000E+00	6.723E+01	0.000E+00
2	1.000E+04	9.637E–01	3.756E–02	1.020E+01	–5.703E+01
3	1.500E+04	6.027E+00	2.349E–01	–7.231E+01	–1.395E+02
4	2.000E+04	2.216E–01	8.636E–03	–3.107E–01	–6.754E+01
5	2.500E+04	1.781E+00	6.943E–02	–7.669E+01	–1.439E+02
6	3.000E+04	1.362E–01	5.308E–03	–5.628E+00	–7.286E+01
7	3.500E+04	8.531E–01	3.325E–02	–7.579E+01	–1.430E+02
8	4.000E+04	1.193E–01	4.649E–03	1.046E+00	–6.619E+01
9	4.500E+04	5.075E–01	1.978E–02	–7.477E+01	–1.420E+02

Total harmonic distortion = 2.510610E+01%

From the Fourier component, we get

DC component = –218.5 mA
Peak fundamental component at 5 kHz = 25.66 A
Fundamental phase delay = 67.23°
Total harmonic distortion = 25.10%

Note: With a lower frequency, there may not be enough time for the resonant oscillation to complete the cycle and the peak current is lower. The resonant current may also be discontinuous.

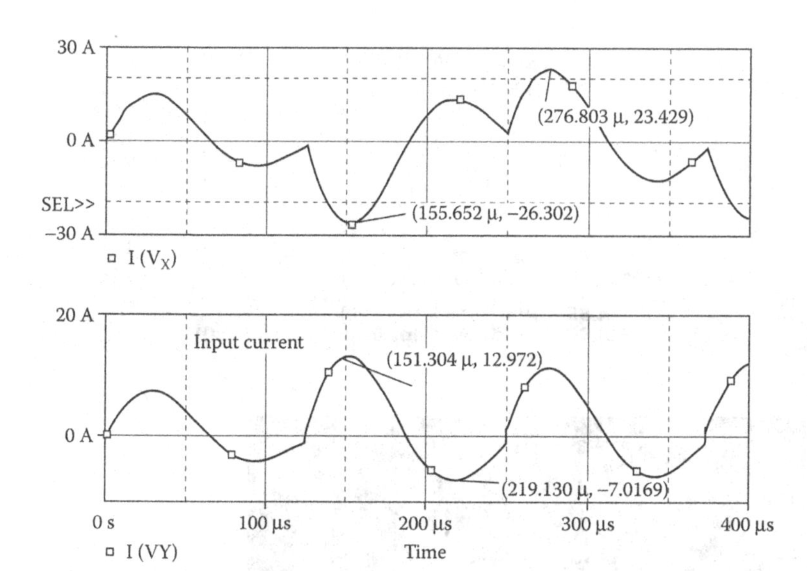

FIGURE 10.4 Plots at $f_o = 4$ kHz for Example 10.1.

The resonant current of an *RLC* circuit can be found in [1]

$$i_o(t) = \frac{V_s + V_c}{\omega_r L} e^{-\alpha t} \sin(\omega_r t)$$

where V_c is the initial capacitor voltage and

$$\alpha = \frac{R}{2L} \quad \omega_r = \sqrt{\frac{1}{L(C_1 + C_s)} - \frac{R^2}{4L^2}}$$

$$V_c = \frac{V_s}{e^{\alpha\pi/\omega r} - 1}$$

LTspice: Figure 10.5 shows the gating signals with a pulse width of 80 μs and delayed by 80 μs at a period of 200 μs.

LTspice schematic for Example 10.1 is shown in Figure 10.6(a). The plots of the load current, the load voltage, and the load power are shown in Figure 10.6(b). The

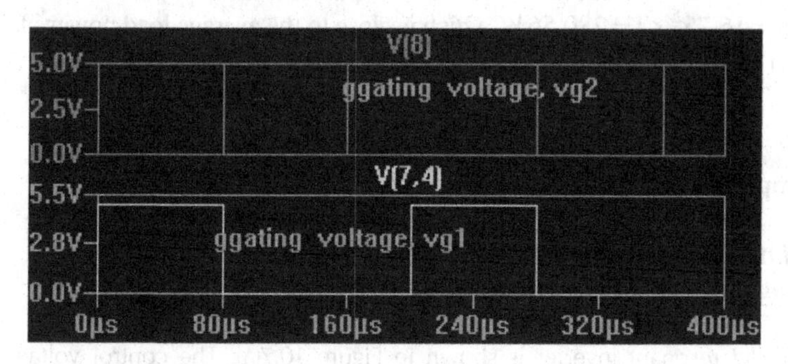

FIGURE 10.5 -LTspice gating signals for Example 10.1

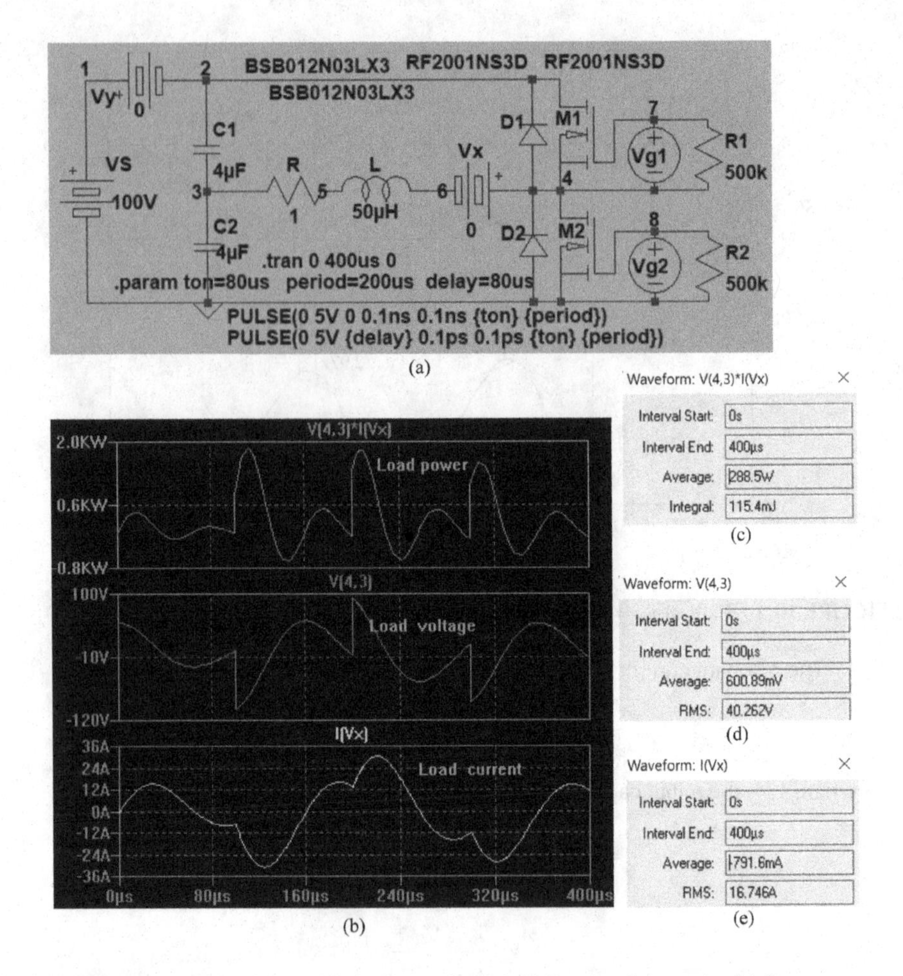

FIGURE 10.6 LTspice schematic for Example 10.1. (a) Ltspice Schematic. (b) Output waveforms. (c) Output power. (d) Output voltage. (e) Output current.

RMS value of the load voltage V(4,3) is 40.26 V. The RMS value of the load current I(Vx) is 16.75 A. We can find the power dissipated in the load resistor of R = 1 Ω as $P_L = I^2 R = 16.75^2 \times 1 = 280.56\,W$ which is close to the average load power of 288.5 W as shown in Figure 10.6(c).

Note: There is no power loss in the load inductor.

Example 10.2:

Finding the Performance of a Parallel Resonant Inverter with Voltage-Controlled Switches

A parallel resonant inverter is shown in Figure 10.7(a). The control voltages are shown in Figure 10.7(b). The DC input voltage is 100 V. The output frequency is f_o = 29.3 kHz. Use PSpice to (a) plot the instantaneous output current i_o and the

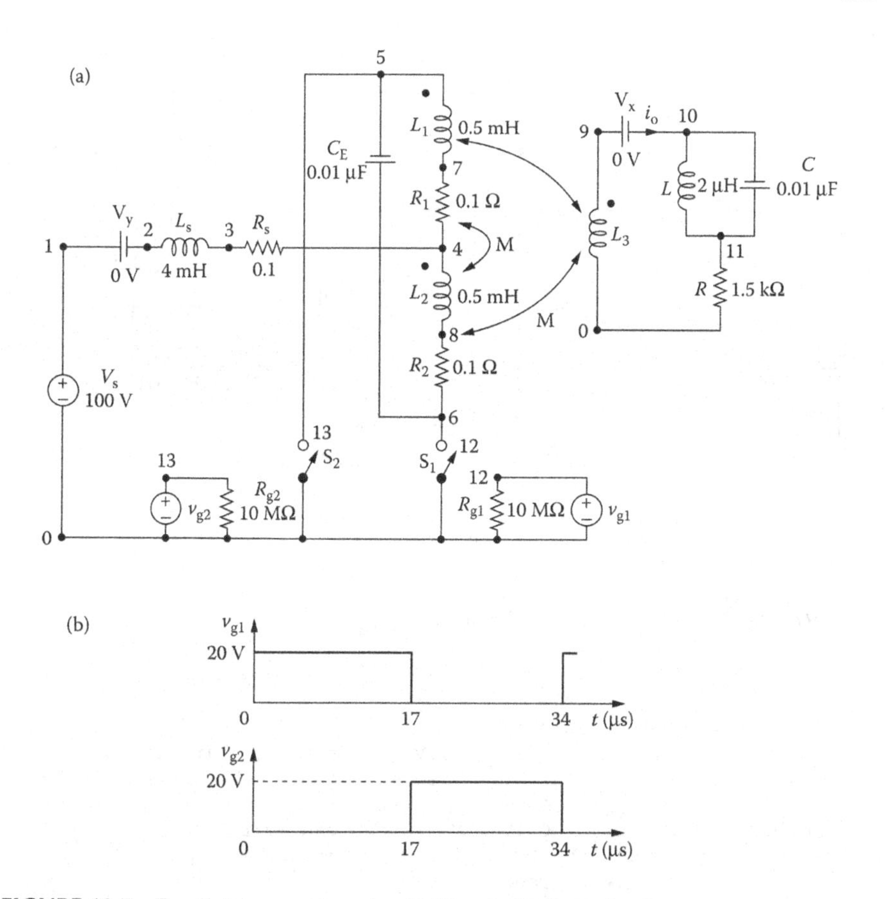

FIGURE 10.7 Parallel resonant inverter. (a) Circuit. (b) Control voltages.

instantaneous input supply current i_s and (b) calculate the Fourier coefficients of the output current i_o. Use voltage-controlled switches to perform the switching action.

SOLUTION

The inductor L_s acts as a current source. It is generally necessary to adjust the on-time of the switches to match the resonant frequency of the circuit. The PSpice schematic is shown in Figure 10.8. Note that the inductors should be connected with proper dot signs. The switching frequency can be varied by changing the parameter {Freq}. The model parameters for the switch are as follows:

```
.MODEL SMD VSWITCH (RON = 1M ROFF = 10E6 VON = 1V VOFF = 0V)
```

The listing of the circuit file is as follows:

Push-Pull Resonant Inverter

```
SOURCE  ■   VS   1   0 DC 100V
            .PARAM Freq=29.4kHz
            Vg1 12 0 PULSE (0 10V 0 1ns 1ns {1/(2*{Freq})-2ns}
            {1/{Freq}})
```

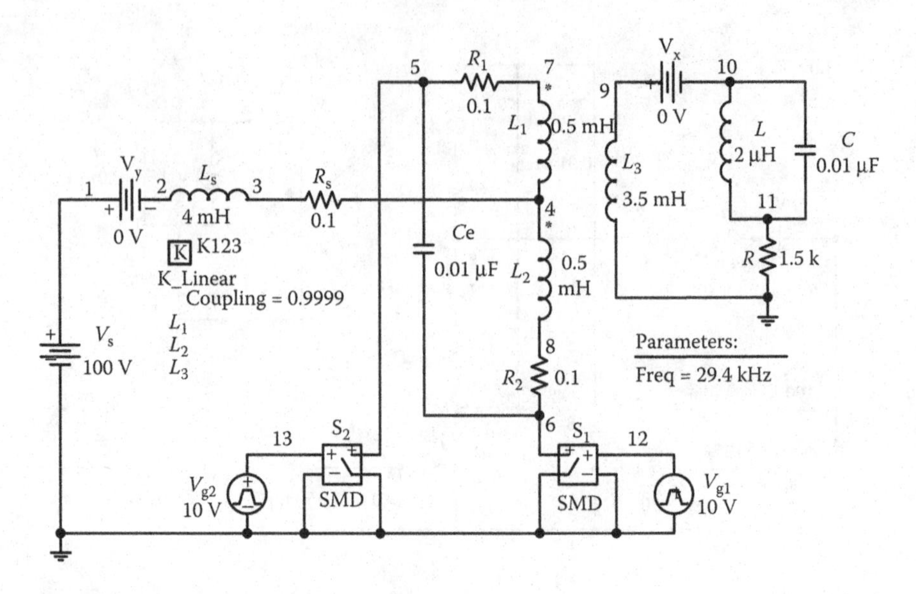

FIGURE 10.8 PSpice schematic for Example 10.2.

```
                Rg1 13 0 10MEG
                Vg2 13 0 PULSE (0 10V {1/(2*{Freq})} 1ns 1ns {1/(2*
                {Freq})-2ns} {1/{Freq}})
                Rg2 13 0 10MEG
CIRCUIT  ■■     VX   9 10 DC 0V ; Measures load current
                VY   1 2 DC 0V ; Voltage source to measure
                supply current
                RS   3 4 0.1
                LS   2 3 4MH
                CE   5 6 0.01UF
                L1   5 7 0.5MH
                R1   7  4 0.1
                L2   4  8 0.5MH
                R2   8  6 0.1
                L3   9  0 3.5MH
                K12 L1 L2 0.9999
                K14 L1 L4 0.9999
                K24 L2 L4 0.9999
                L    10 11 2UH
                C    10 11 0.01UF
                R    11 0  1.5K
                S1   6  0  12 0 SMOD ; Voltage-controlled switch
                S2   5  0  13 0 SMOD ; Voltage-controlled switch
                .MODEL SMOD VSWITCH (RON=0.01 ROFF=10E+6
                VON=1V VOFF=0MV)
ANALYSIS ■■■    .TRAN 0.1US 120US ; Transient analysis
                .PROBE ; Graphics post-processor .OPTIONS
                ABSTOL = 1.00N RELTOL=0.01 VNTOL=0.1
                ITL5=50000
                .FOUR 29.3KHZ I(VX) ; Fourier analysis
           .END
```

Note the following:

a. The PSpice plots of the instantaneous output current I(VX) and the current source I(VY) are shown in Figure 10.9. The output voltage and the output current oscillate as a sine wave while the input current is maintained as a DC value with ripples on it.
b. The Fourier coefficients of the output current are as follows:

Fourier Components of Transient Response I(VX)

DC Component = 9.019181E–05

Harmonic Number	Frequency (Hz)	Fourier Component	Normalized Component	Phase (Deg)	Normalized Phase (Deg)
1	2.930E+04	7.433E–02	1.000E+00	5.243E+01	0.000E+00
2	5.860E+04	2.302E–02	3.098E–01	1.345E+02	8.202E+01
3	8.790E+04	8.330E–03	1.121E–01	–4.895E+01	-1.014E+02
4	1.172E+05	4.527E–03	6.091E–02	1.299E+02	7.745E+01
5	1.465E+05	2.641E–03	3.553E–02	–4.305E+01	-9.548E+01
6	1.758E+05	1.825E–03	2.456E–02	1.424E+02	8.993E+01
7	2.051E+05	1.300E–03	1.749E–02	–2.515E+01	-7.759E+01
8	2.344E+05	1.083E–03	1.457E–02	1.535E+02	1.011E+02
9	2.637E+05	8.259E–04	1.111E–02	–1.467E+01	-6.710E+01

Total harmonic distortion = 3.387136E+01%

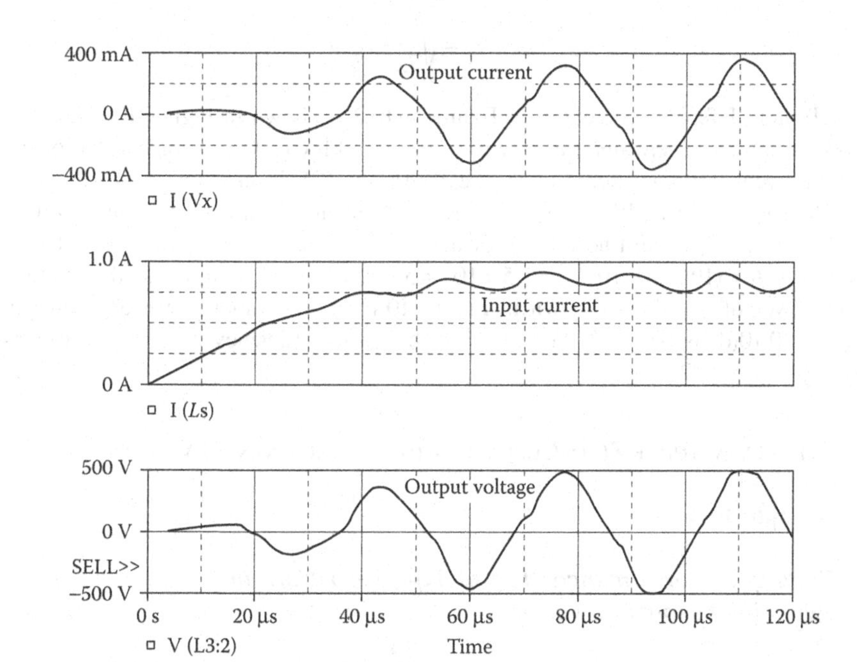

FIGURE 10.9 Plots for Example 10.2.

From the Fourier components, we get

DC component = 90.19 μA
Peak fundamental component at 29.3 kHz = 74.33 mA
Fundamental phase delay = 52.43°
Total harmonic distortion = 33.87%

Note: Because of the constant-current inductor L_s, the input current remains approximately constant within a certain amount of ripples. The output voltage V(L_3:L_1) will depend on the turns ratio $N_s/N_p = \sqrt{L_3/L_1}$.

10.3 ZERO-CURRENT SWITCHING CONVERTERS

A power device of a ZCSC is turned on and off at zero current by using an *LC*-resonant circuit. The device remains on and provides a path for completing the resonant oscillation. When the device is off, it has to withstand the peak voltage at zero current, thereby reducing the switching loss of the device. The output waveforms depend primarily on the circuit parameters and the input supply voltage. The switching period must be long enough to complete the resonant oscillation. The resonant current through the capacitor *C* can be found in [1]

$$i_r(t) = V_s\sqrt{\frac{C}{L}}\sin(\omega_o t)$$

where

$$\omega_o = \sqrt{\frac{1}{LC}}$$

LTspice: LTspice schematic for Example 10.2 is shown in Figure 10.10(a). The plots of output waveforms are shown in Figure 10.10(b). The average value of the input current I(Vy) as shown in Figure 10.10(c) is 625.56 mA and the RMS value is. 638.8 mA. The RMS value of the load current I(Vx) shown in Figure 10.10(f) is 196.48 mA. We can find the power dissipated in the load resistor of R = 1.5kΩ as $P_L = I^2R = (196.48\times10^{-3})^2\times1.5\times10^3 = 57.88\,\text{W}$ which is close to the average load power of 57.91W as shown in Figure 10.10(e). The input power as shown in Figure 10.10(d) is 62.56 W. We can find the power efficiency of the converter as $\eta = \frac{P_{out}}{P_{in}} = \frac{57.91}{62.56} = 92.56\%$

10.3.1 Examples of Zero-Current Switching Resonant Inverters

Example 10.3:

Finding the Performance of an M-Type Zero-Current Switching Resonant Inverter

An M-type ZCSC is shown in Figure 10.11(a). The control voltage is shown in Figure 10.11(b). The DC input voltage is 15 V. The output frequency is f_o = 8.33 kHz.

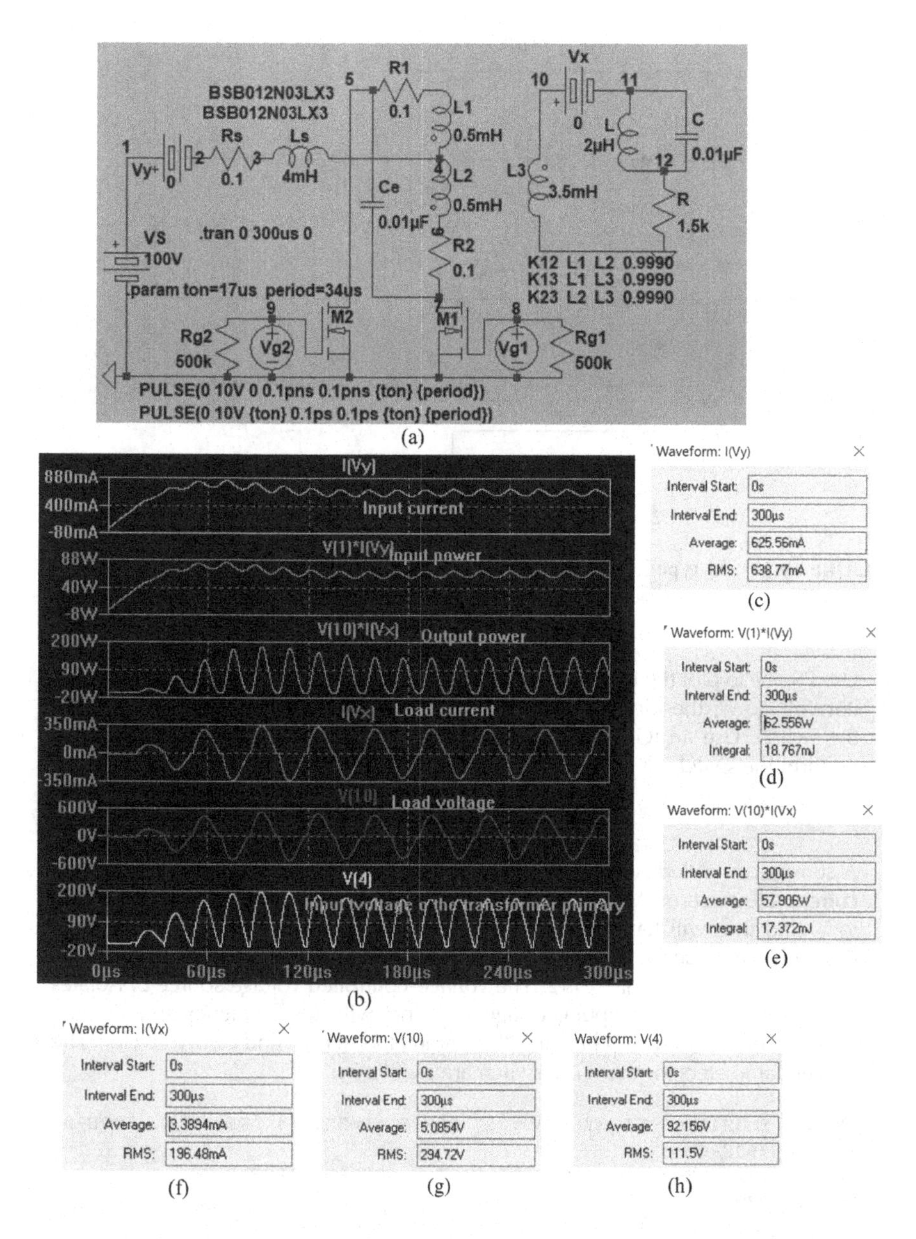

FIGURE 10.10 LTspice schematic for Example 10.2. (Ltspice Schematic. (b) Output waveforms. (c) Input current. (d) Input power. (e) Output Power. (f) Output current. (g) Output voltage. (h) Transfomer input voltage.

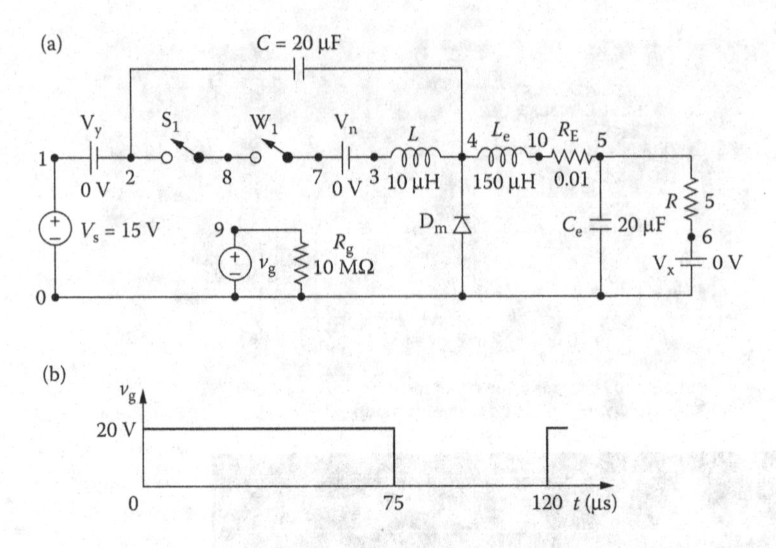

FIGURE 10.11 M-type ZCSC. (a) Circuit. (b) Control voltage.

Use PSpice to plot the instantaneous capacitor current i_c, the instantaneous capacitor voltage v_c, the diode voltage v_{Dm}, and the instantaneous power loss across the switch. Use an IGBT to replace voltage- and current-controlled switches to perform the switching action.

SOLUTION

A current-controlled switch W_1 is required to break the circuit when the resonant current falls to zero. When the supply is turned on, the capacitor will be charged to V_s and thus will have an initial voltage. The current-controlled switch W_1 in Figure 10.11(a) can be replaced by a diode. The PSpice schematic with an IGBT switch is shown in Figure 10.12. The voltage-controlled voltage source E_1 isolates the IGBT gate from the input gate signal V_g. The switching frequency and the duty cycle can be varied by changing the parameters {Freq} and {Duty_Cycle}. The model parameters for the IGBT switch are as follows:

```
.MODEL IXGH40N60 NIGBT (TAU=287.56E-9 KP=50.034 AREA=3 7. 500E-6
AGD=18.750E-6
+ VT=4.1822 KF=.36 047 CGS=31.942E-9 COXD=53.188E-9
VTD=2.6570)
```

The listing of the circuit file with an implementation by switches S_1 and W_1 rather than an IGBT is as follows:

ZCSC

```
SOURCE  ■   VS  1 0 DC 15V
            .PARAM Freq=8.33kHz Duty_Cycle=62.5
            Rg  9 0 10MEG ; Control voltage
            Vg  9 0 PULSE (0 1 0 1ns 1ns {{Duty_Cycle}/(100*
            {Freq})-2ns} {1/{Freq}})
```

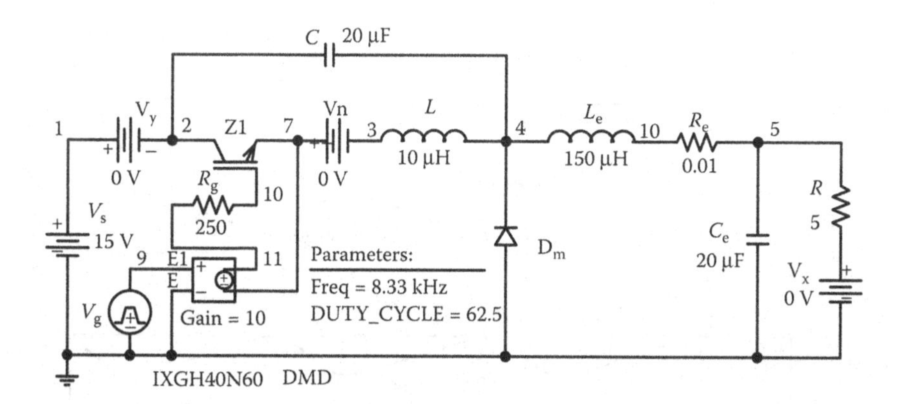

FIGURE 10.12 PSpice schematic for Example 10.3.

```
CIRCUIT ■■    VY 1 2 DC 0V ; Voltage source to measure supply
              current
              VX 6 0 DC 0V ; Measures load current
              VN 7 3 DC 0V ; Measures the current-controlled
              switch
              R  5 6 55
              LE 4 10 150UH
              RE 10 5 0.01
              CE 5 0 20UF
              L  3 4 10UH
              C  2 4 20UF   ; Initial condition
              DM 0 4 DMOD   ; Diode
              .MODEL DMOD D (IS=2.2E-15BV=1200VCJO=0TT=0) ;
              Diode model parameters
              S1 2 8 9 0 SMOD ; Voltage-controlled switch
              .MODEL SMOD VSWITCH (RON=0.001ROFF=10E+6VON=
              10VVOFF=5V)
              W1 8 7 VN IMOD ; Current-controlled switch
              .MODEL IMOD ISWITCH (RON=1E+6 ROFF=0.01 ION=0
              IOFF=1UA)        ; Model parameters
ANALYSIS ■■■  .TRAN 0.1US 400US ; Transient analysis
              .PROBE ; Graphics post-processor
              .OPTIONS ABSTOL=1.00NRELTOL=0.1VNTOL=
              0.1ITL5=50000
         .END
```

The PSpice plots of the gate voltage V(9), the instantaneous capacitor voltage V(2,4), the capacitor current I(C), the diode voltage V(4), and the instantaneous power of the IGBT switch are shown in Figure 10.13, where its peak instantaneous power is 13.06 W.

LTspice: LTspice schematic for Example 10.3 is shown in Figure 10.14(a). The plots of output waveforms are shown in Figure 10.14(b). The RMS value of the output voltage V(8) as shown in Figure 10.14(c) is 10.55 V and the average value is 10.12 V. The average output power as shown in Figure 10.14(e) is 22.27 W. The input power

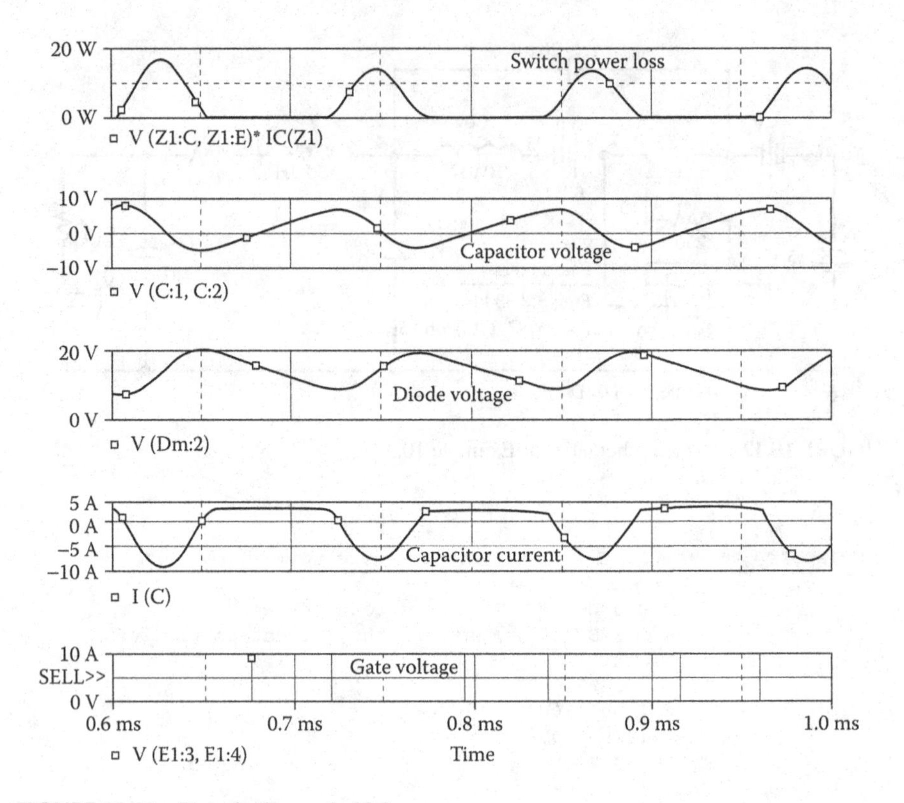

FIGURE 10.13 Plots for Example 10.3.

as shown in Figure 10.14(f) is 24.26 W. We can find the power efficiency of the converter as $\eta = \frac{P_{out}}{P_{in}} = \frac{22.27}{24.26} = 91.8\%$

Note: For a low output power, the losses in the semiconductor devices could be significant compared to the output power and the efficiency becomes low.

Example 10.4:

Finding the Performance of an L-Type Zero-Current Switching Resonant Inverter

Repeat Example 10.3 for the L-type converter shown in Figure 10.15(a). The control voltage is shown in Figure 10.15(b).

SOLUTION

The PSpice schematic is shown in Figure 10.16, which is similar to that of Figure 10.10.

The circuit file is similar to that of Example 10.3, except that the capacitor *C* is connected across the diode D_m. The statement for *C* is changed to

```
C 4 0 20UF; No initial condition
```

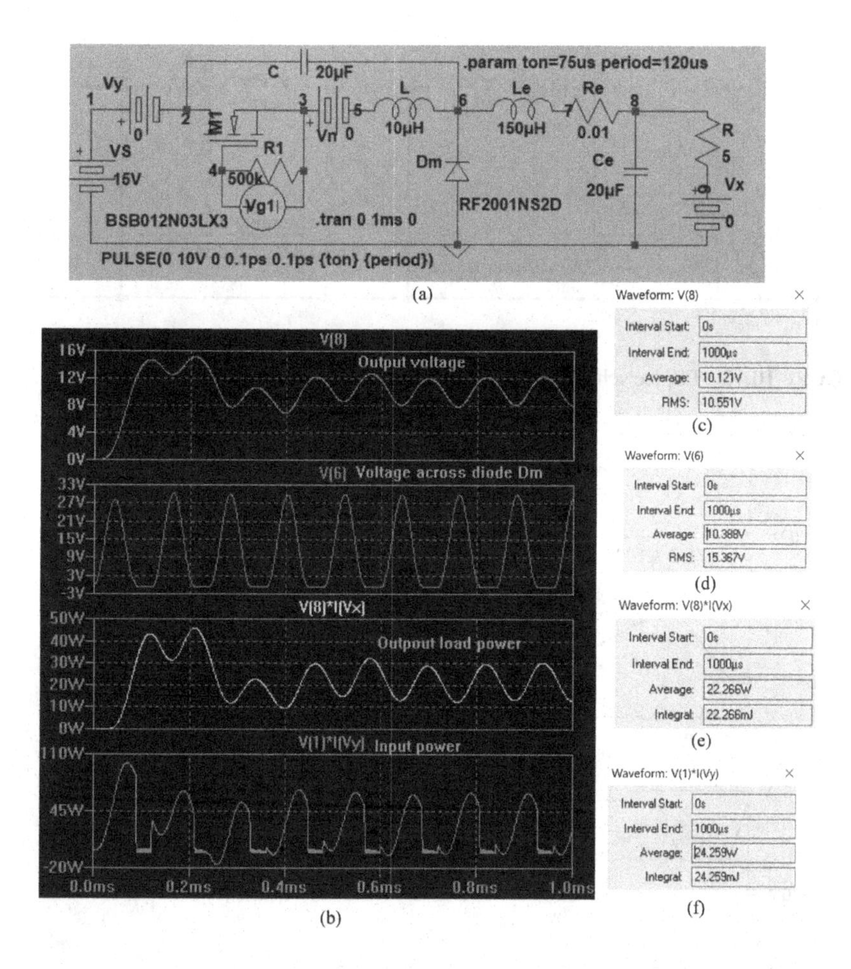

FIGURE 10.14 LTspice schematic for Example 10.3. (a) Ltspice Schematic. (b) Output Waveforms. (c) Output voltage. (d) Diode Voltage. (e) Output power. (f) Input power.

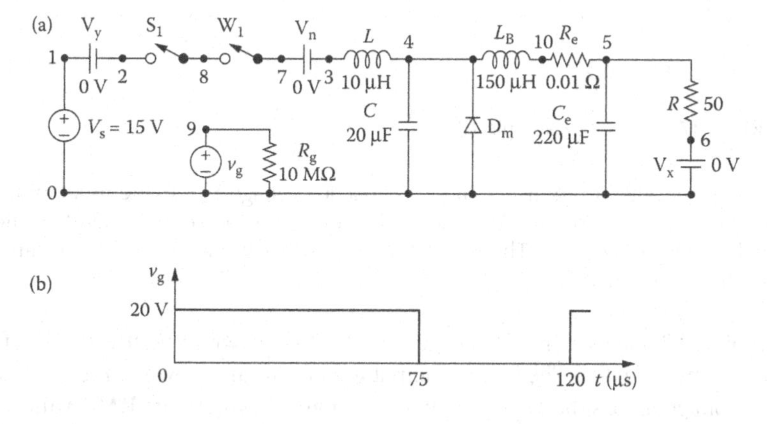

FIGURE 10.15 L-type ZCSC. (a) Circuit. (b) Control voltage.

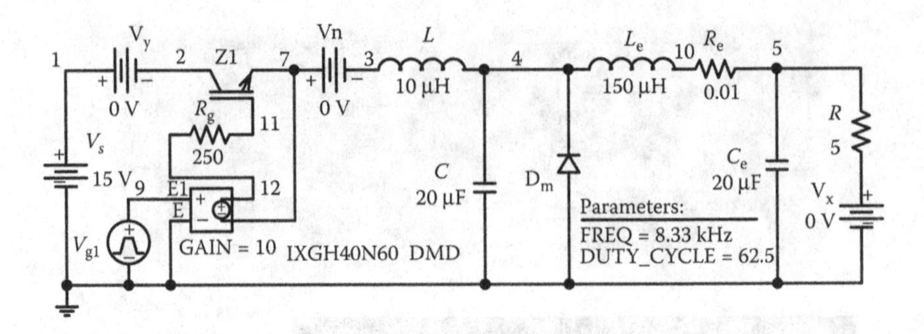

FIGURE 10.16 PSpice schematic for Example 10.4.

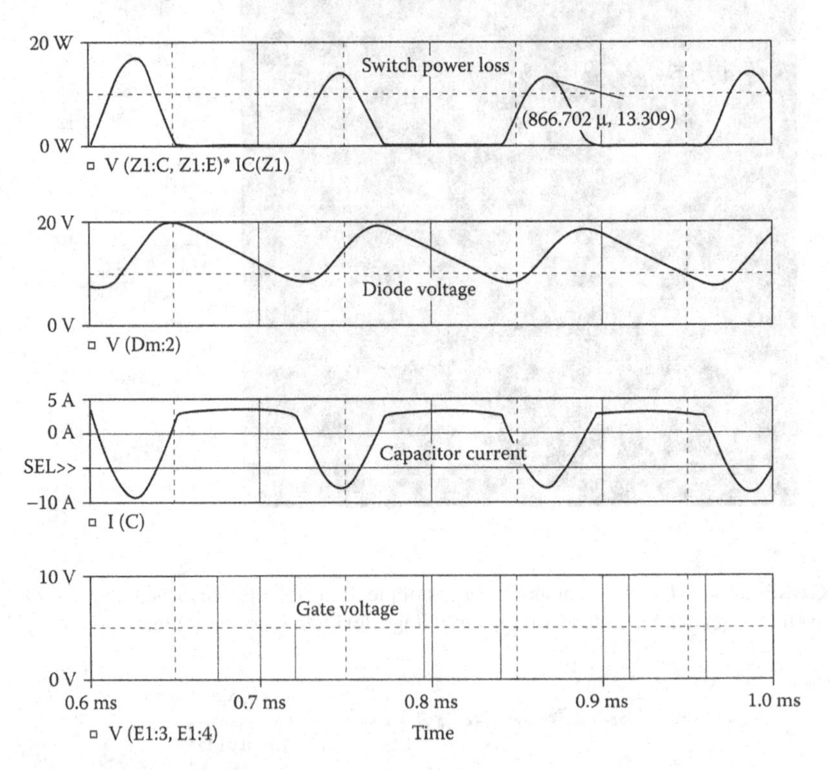

FIGURE 10.17 Plots for Example 10.4.

The PSpice plots of the instantaneous capacitor voltage V(2,4), the inductor current I(L), the diode voltage V(4), and the instantaneous power of the IGBT switch are shown in Figure 10.17. The instantaneous power is the same as that of Example 10.3, that is, 13.06 W.

LTspice: LTspice schematic for Example 10.4 is shown in Figure 10.18(a) for an IGBT switching device. The plots of output waveforms are shown in Figure 10.18(b).

The voltage across diode D_m is shown in Figure 10.18(c). The RMS value of the output voltage V(8) as shown in Figure 10.18(d) is 14.42 V and the average value is

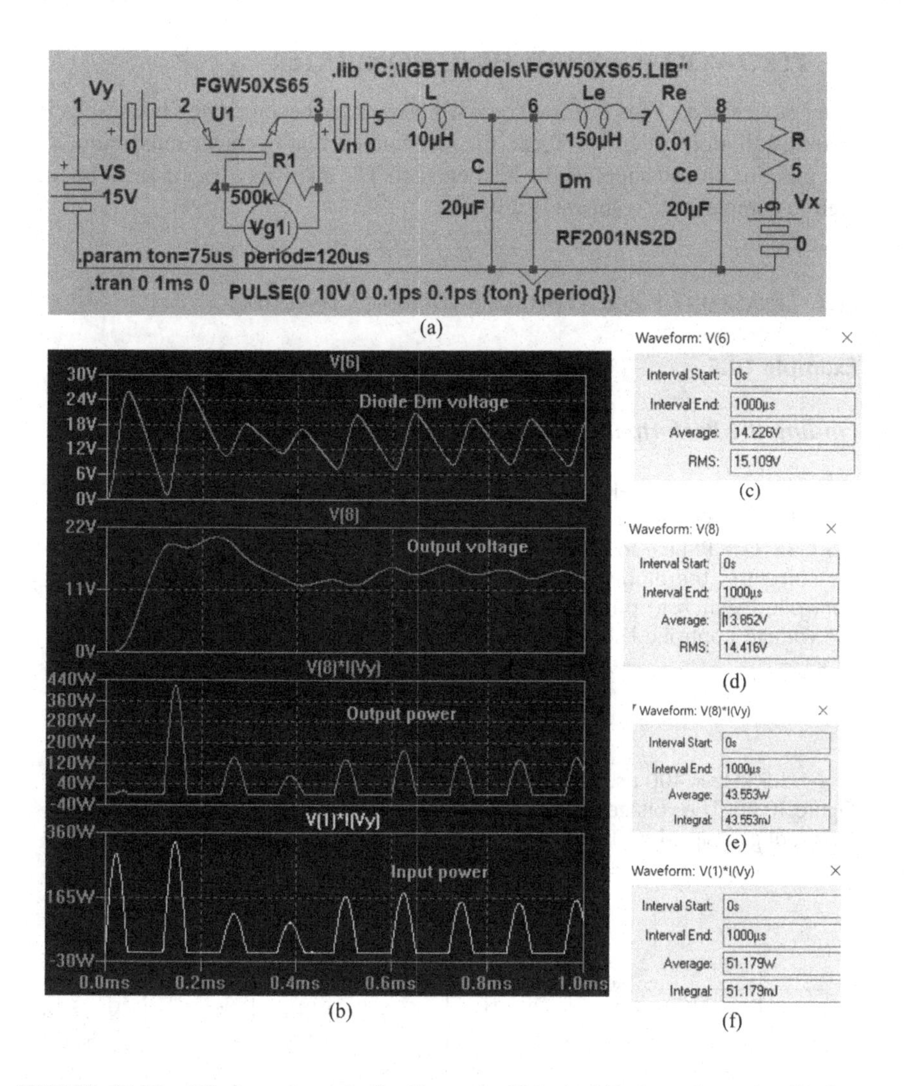

FIGURE 10.18 LTspice schematic for Example 10.4. (a) LTspice schematic, (b) Output Waveforms. (c) Diode voltage. (d) Output voltage. (e) Output power. (f) Input power.

13.85 V. The average input power as shown in Figure 10.18(f) is 51.18 W. The average output power as shown in Figure 10.18(e) is 43.55 W. We can find the power efficiency of the converter as $\eta = \frac{P_{out}}{P_{in}} = \frac{43.55}{51.18} = 85.1\%$

We can also find the output power from the output voltage as

$$P_{out} = \frac{V(8)^2}{R} = \frac{14.42^2}{5} = 41.58\text{W}$$

which is close to 43.55 W.

Note: For a low output power, the losses in the semiconductor devices could be significant compared to the output power and the efficiency becomes low.

10.4 ZERO-VOLTAGE SWITCHING CONVERTER

A power device of a ZVSC is turned on when its voltage becomes zero because of the resonant oscillation. At zero voltage, the resonant current becomes maximum. The device remains on and supplies the load current. The switching period must be long enough to complete the resonant oscillation.

10.4.1 EXAMPLES OF ZERO-VOLTAGE SWITCHING CONVERTERS

Example 10.5:

Finding the Performance of a Zero-Voltage Switching Inverter

A ZVSC is shown in Figure 10.19(a). The control voltage is shown in Figure 10.19(b). The DC input voltage is V_s = 15 V. The switching frequency is f_s = 2.5 kHz. Use PSpice to plot the instantaneous capacitor voltage v_c, the inductor current i_L, the diode current v_{Dm}, and the load voltage v_L. Use a voltage-controlled switch to perform the switching action. The switch parameters of the voltage-controlled switch are RON = 0.01, ROFF = 10E + 6, VON = 1V, and VOFF = 0V.

SOLUTION

V_s = 1/2.5 kHz = 400 μs. The PSpice schematic with an IGBT switch is shown in Figure 10.20. The voltage-controlled voltage source E_1 isolates the IGBT gate from the input gate signal V_g. The switching frequency and the duty cycle can be varied

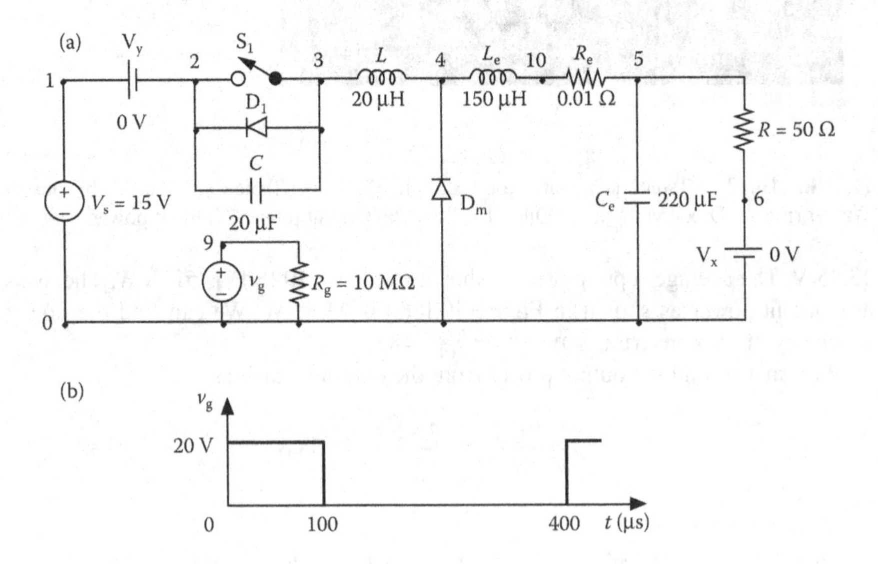

FIGURE 10.19 ZVSC. (a) Circuit. (b) Control voltage.

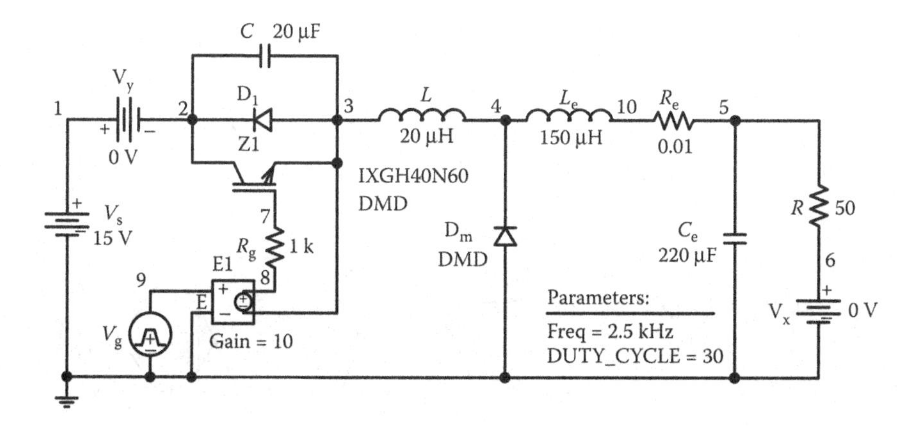

FIGURE 10.20 PSpice schematic for Example 10.5.

by changing the parameters {Freq} and {Duty_Cycle}. The model parameters for the IGBT switch are as follows:

```
.MODEL IXGH40N60 NIGBT (TAU=287.56E-9 KP=50.034
AREA=37.500E-6 AGD=18.750E-6
+ VT=4.1822 KF=.36 047 CGS=31.942E-9 COXD=53.188E-9
VTD=2.6570) for IGBTs
```

The listing of the circuit file with an implementation by switch S_1 is as follows:

ZVSC

```
SOURCE   ■    VS 1 0 DC 15V
              .PARAM Freq = 2.5kHz Duty_Cycle = 25
              Rg 9 0 10MEG ; Control voltage
              Vg 9 0 PULSE (0 1 0 1ns 1ns {{Duty_Cycle}/(100*
              {Freq})-2ns} {1/{Freq}})
CIRCUIT  ■■   VY  1  2 DC 0V ; Voltage source to measure
              supply current
              VX  6  0 DC 0V ; Measures load current
              R   5  6 50
              LE  4 10 150UH
              RE 10  5 0.01
              CE  5  0 220UF
              L   3  4 20UH
              C   2  3 20UF
              D1  3  2 DMOD ; Diode
              DM  0  4 DMOD ; Diode
              .MODEL DMOD D (IS=2.2E-15 BV=1200V CJO=0 TT=0) ;
              Diode
              model parameters
              S1  2  3  9  0  SMOD ; Voltage-controlled switch
              .MODEL SMOD VSWITCH (RON = 0.01 ROFF = 10E+6
              VON = 1V VOFF = 0V)
ANALYSIS ■■■  .TRAN 1US 1.6MS 0.40MS ; Transient analysis
              .PROBE ; Graphics post-processor
              .OPTIONS ABSTOL = 1.00 NRELTOL = 0.1 VNTOL = 0.1I
              TL5 = 50000
          .END
```

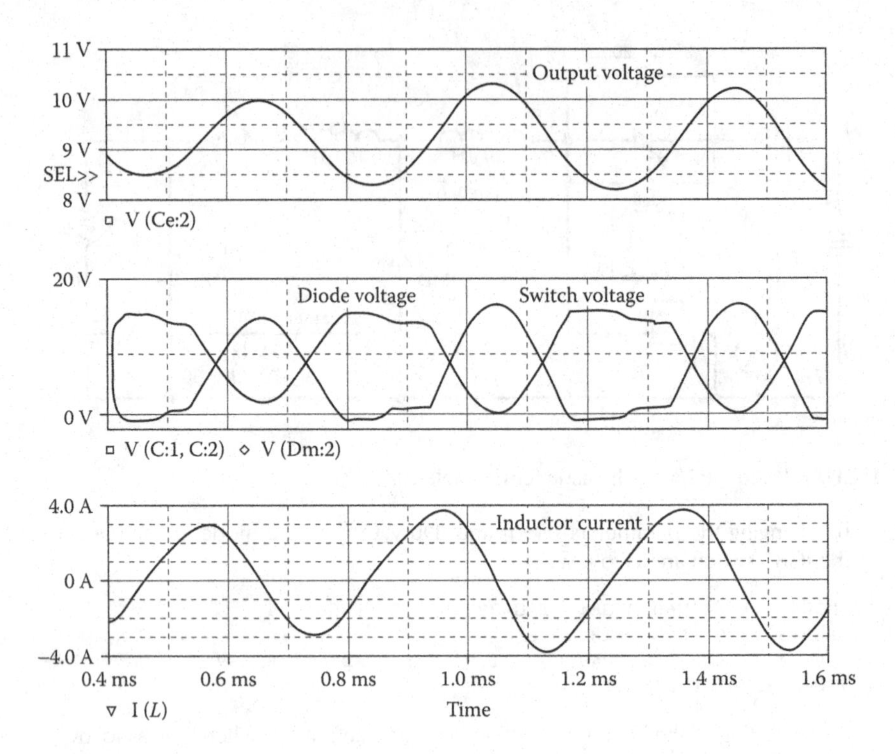

FIGURE 10.21 Plots for Example 10.5.

The PSpice plots of the instantaneous capacitor voltage V(2,4), the inductor current I(L), the diode voltage V(4), and the load voltage V(5) are shown in Figure 10.21. Note that the switch and diode voltages vary inversely, that is, when the diode voltage is opposite of the switch voltage. The inductor current oscillates and goes through zero when the switch voltage reaches its peak.

LTspice: LTspice schematic for Example 10.5 is shown in Figure 10.22(a) for an IGBT switching device. The plots of output waveforms are shown in Figure 10.22(b). The input current is shown in Figure 10.22(c). The RMS value of the output voltage V(8) as shown in Figure 10.22(d) is 6.38 V and the average value is 6.13 V. The average output power as shown in Figure 10.22(e) is 8.13 W. The input power as shown in Figure 10.22(f) is 13.36 W. We can find the power efficiency of the converter as $\eta = \frac{P_{out}}{P_{in}} = \frac{8.13}{13.36} = 60.85\%$

We can also find the output power from the output voltage as

$$P_{out} = \frac{V(8)^2}{R} = \frac{6.38^2}{5} = 8.15\text{W}$$

which is close to 8.13 W.

Note: For a low output power, the losses in the semiconductor devices could be significant compared to the output power and the efficiency becomes low.

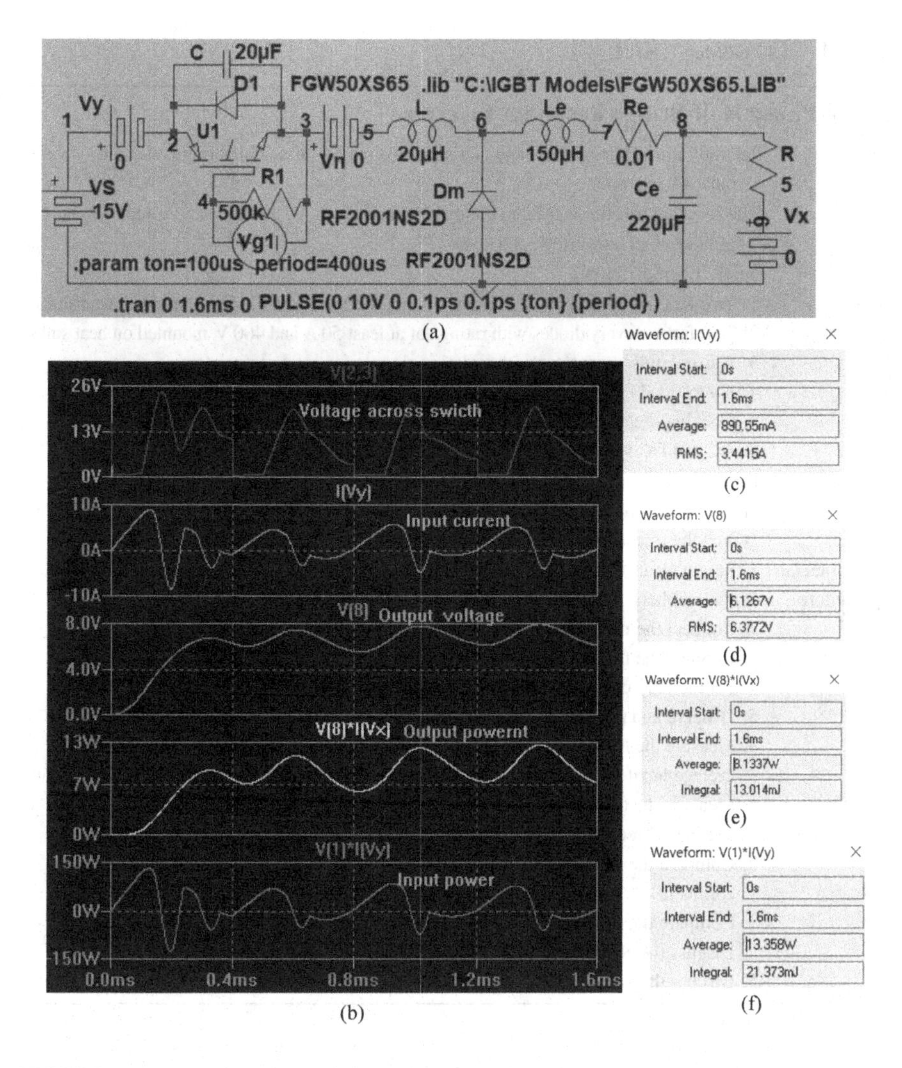

FIGURE 10.22 LTspice schematic for Example 10.5. (a) LTspice schematic. (b) Output Waveforms. (c) Input current. (d) Output voltage. (e) Output power. (f) Input power.

10.5 LABORATORY EXPERIMENTS

It is possible to develop many experiments to demonstrate the operation and characteristics of inverters. The following experiments are suggested:

Single-phase half-bridge resonant inverter
Single-phase full-bridge resonant inverter
Push-pull inverter
Parallel resonant inverter
ZCSC
ZVSC

10.5.1 Experiment RI.1

Single-Phase Half-Bridge Resonant Inverter

Objective	To study the operation and characteristics of a single-phase half-bridge resonant (transistor) inverter
Applications	The resonant inverter is used to control power flow in AC and DC power supplies, input stages of other converters, and so on
Textbook	See Ref. 1, Section 8.2
Apparatus	1. Two BJTs or MOSFETs with ratings of at least 50 A and 400 V, mounted on heat sinks 2. Two fast-recovery diodes with ratings of at least 50 A and 400 V, mounted on heat sinks 3. A firing pulse generator with isolating signals for gating transistors 4. An *RL* load 5. One dual-beam oscilloscope with floating or isolating probes 6. AC and DC voltmeters and ammeters and one noninductive shunt
Warning	Before making any circuit connection, switch the DC power off. Do not switch on the power unless the circuit is checked and approved by your laboratory instructor. Do not touch the transistor or diode heat sinks, which are connected to live terminals
Experimental procedure	1. Set up the circuit as shown in Figure 10.23. Use an RLC load. Design suitable values for snubbers 2. Connect the measuring instruments as required 3. Connect the firing pulses to the appropriate transistors 4. Set the duty cycle to $k = 0.5$ 5. Observe and record the waveforms of the load voltage v_o and the load current i_o 6. Measure the RMS load voltage, the RMS load current, the average input current, the average input voltage, and the total harmonic distortion of the output voltage and current 7. Measure the conduction angles of the transistor Q_1 and diode D_1
Report	1. Present all recorded waveforms and discuss all significant points 2. Compare the waveforms generated by SPICE with the experimental results, and comment 3. Compare the experimental results with the predicted results 4. Discuss the advantages and disadvantages of this type of inverter 5. Discuss the effects of the diodes on the performance of the inverter

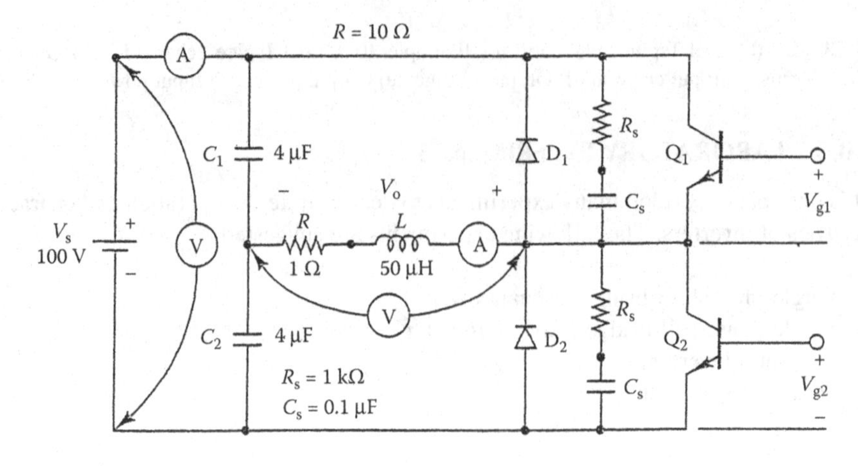

FIGURE 10.23 Single-phase half-bridge resonant inverter.

10.5.2 EXPERIMENT RI.2

Single-Phase Full-Bridge Resonant Inverter

Objective	To study the operation and characteristics of a single-phase full-bridge resonant (transistor) inverter
Applications	The resonant inverter is used to control power flow in high-frequency applications, AC and DC power supplies, input stages of other converters, and so on
Textbook	See Ref. 1, Section 8.2
Apparatus	See Experiment RI.1
Warning	See Experiment RI.1
Experimental procedure	Set up the circuit as shown in Figure 10.24. Repeat the steps in Experiment RI.1
Report	See Experiment RI.1

10.5.3 EXPERIMENT RI.3

Push-Pull Inverter

Objective	To study the operation and characteristics of a push–pull (transistor) inverter
Applications	The push–pull inverter is used to control power flow in high-frequency applications, AC and DC power supplies, input stages of other converters, and so on
Textbook	See Ref. 1, Section 8.2
Apparatus	Experiment RI.1
Warning	Experiment RI.1
Experimental procedure	Set up the circuit as shown in Figure 10.25. Repeat the steps of Experiment RI.1
Report	See Experiment RI.1

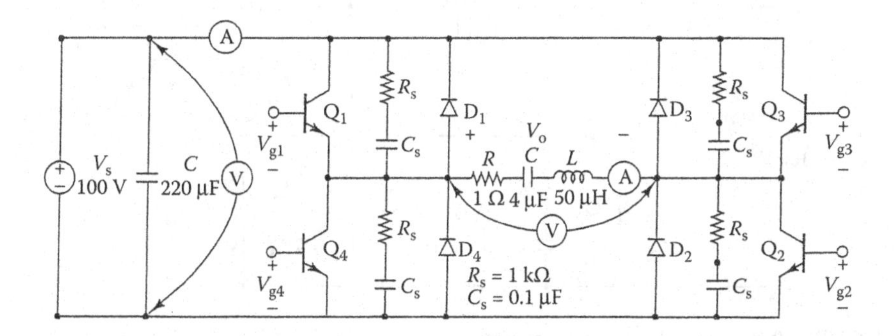

FIGURE 10.24 Single-phase full-bridge resonant inverter.

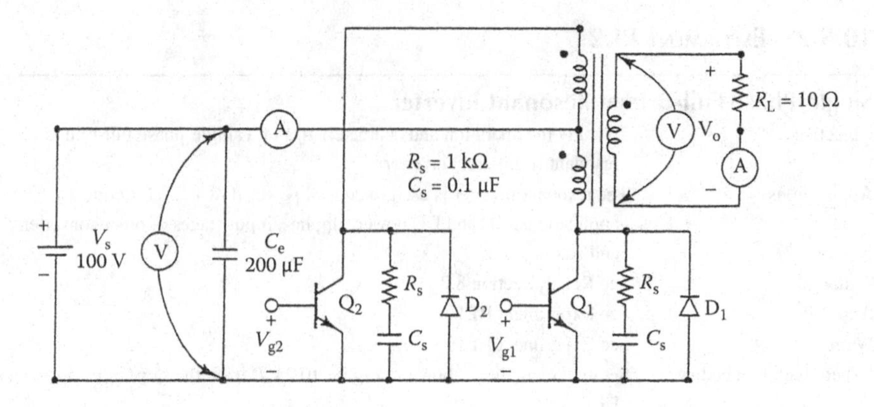

FIGURE 10.25 Push–pull inverter.

10.5.4 Experiment RI.4

Parallel Resonant Inverter

Objective	To study the operation and characteristics of a single-phase push–pull parallel resonant (transistor) inverter
Applications	The parallel resonant inverter is used to control power flow in high-frequency applications, AC and DC power supplies, input stages of other converters, and so on
Textbook	See Ref. 1, Section 8.3
Apparatus	See Experiment RI.1
Warning	See Experiment RI.1
Experimental procedure	Set up the circuit as shown in Figure 10.26. Repeat the steps of Experiment RI.1
Report	See Experiment RI.1

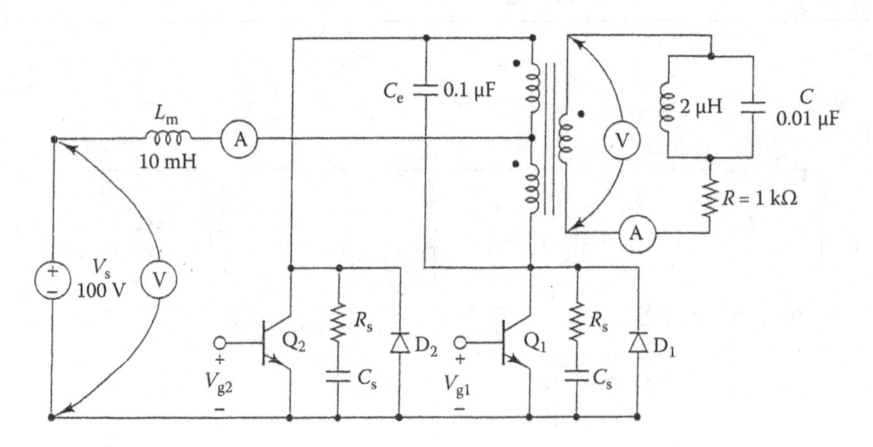

FIGURE 10.26 Parallel resonant inverter.

10.5.5 Experiment RI.5

ZCSC

Objective	To study the operation and characteristics of a ZCSC
Applications	The ZCSC is used to control power flow in AC and DC power supplies, and so on
Textbook	See Ref. 1, Section 8.6
Apparatus	1. One BJT or MOSFET with ratings of at least 50 A and 400 V, mounted on a heat sink 2. One fast-recovery diode with ratings of at least 50 A and 400 V, mounted on a heat sink 3. A firing pulse generator with isolating signals for gating transistor 4. An *R* load, capacitors, and chokes 5. One dual-beam oscilloscope with floating or isolating probes 6. AC and DC voltmeters and ammeters and one noninductive shunt
Warning	See Experiment RI.1
Experimental procedure	1. Set up the circuit as shown in Figure 10.27 2. Connect the measuring instruments as required 3. Connect the firing pulses to the appropriate transistors 4. Set the duty cycle to $k = 0.5$ 5. Observe and record the waveforms of the load voltage v_o, the currents through *C*, *L*, and Q_1, and the voltage across diode D_m 6. Measure the average output voltage, the average output current, the average input current, and the average input voltage 7. Measure the conduction angles of transistor Q_1 8. Repeat steps 2–7 for the circuit of Figure 10.28
Report	See Experiment RI.1

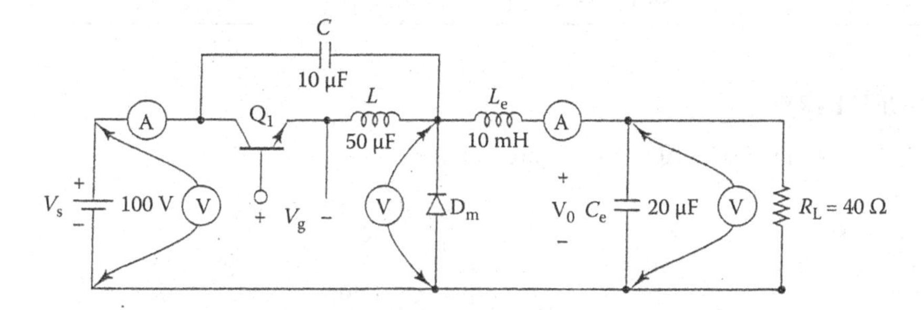

FIGURE 10.27 First ZCSC circuit for Experiment RI.5.

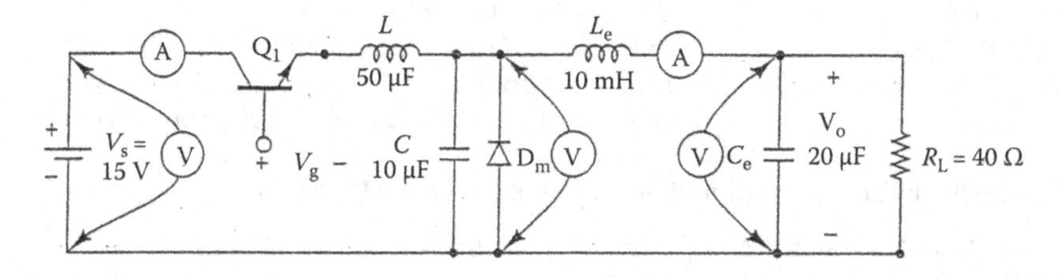

FIGURE 10.28 Second ZCSC circuit for Experiment RI.5.

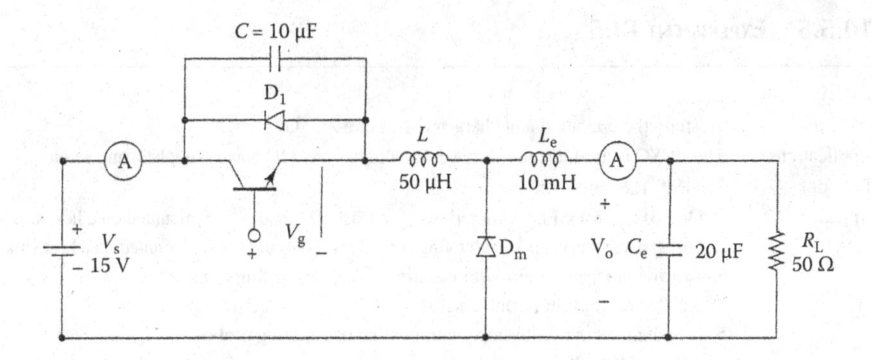

FIGURE 10.29 ZVSC.

10.5.6 Experiment RI.6

ZVSC

Objective	To study the operation and characteristics of a ZVSC
Applications	The ZVSC is used to control power flow in AC and DC power supplies, and so on
Textbook	See Ref. 1, Section 8.7
Apparatus	See Experiment RI.5
Warning	See Experiment RI.1
Experimental procedure	Set up the circuit as shown in Figure 10.29 See Experiment RI.5
Report	See Experiment RI.1

SUMMARY

The statements for an AC thyristor are as follows:

```
* Subcircuit call for switched transistor model:
XT1  +N          -N        +NG        -NG        QM
    positive  negative  +control   - control  model
*                       voltage    voltage    name
* Subcircuit call for PWM control:
XPWM   VR    VC       +NG        -NG        PWM
*     ref. carrier  +control  -control   model
*    input  input   voltage   voltage    name
* Subcircuit call for sinusoidal PWM control:
XSPWM  VR       VS        +NG       -NG          VC           SPWM
*  ref.   sine-wave  +control  -control    rectified        model
*  input    input   voltage  voltage   carrier sine wave   name
```

The model statement of n-channel IGBT type IXGH40N60

```
.MODEL IXGH40N60 NIGBT (TAU = 287.56E-9 KP = 50.034
AREA = 37.500E-6 AGD = 18.750E-6 VT = 4.1822 KF = .36047
CGS = 31.942E-9 COXD = 53.188E-9 VTD = 2.6570)
```

The typical model statements of a voltage controlled switch S1 with a model of SMOD is

```
S1      2    3        9     0      SMOD
.MODEL SMOD VSWITCH (RON=0.01 ROFF=10E+6 VON=1V VOFF=0V)
```

DESIGN PROBLEMS

10.1 It is required to design the single-phase half-bridge resonant inverter of Figure 10.23 with the following specifications:
DC supply voltage, $V_s = 100$ V
Load resistance, $R = 1\ \Omega$
Load inductance, $L = 100\ \mu$H
Output frequency should be as high as possible
  a. Determine the ratings of all components and devices under worst-case conditions.
  b. Use SPICE to verify your design.
  c. Provide a component cost estimate of the circuit.

10.2 It is required to design the single-phase full-bridge resonant inverter of Figure 10.24 with the following specifications:
DC supply voltage, $V_s = 100$ V
Load resistance, $R = 1\Omega$
Load inductance, $L = 50$ pF
Load capacitance, $C = 4\ \mu$F
Output frequency should be as high as possible
  a. Determine the ratings of all components and devices under worst-case conditions.
  b. Use SPICE to verify your design.
  c. Provide a component cost estimate of the circuit.

10.3 It is required to design the push–pull inverter of Figure 10.25 with the following specifications:
DC supply voltage, $V_s = 100$ V
Load resistance, $R = 100\ \Omega$
Peak value of load voltage, $V_p = 140$ V
Output frequency, $f_o = 1$ kHz
  a. Determine the ratings of all components and devices under worst-case conditions.
  b. Use SPICE to verify your design.
  c. Provide a component cost estimate of the circuit.

10.4 It is required to design the parallel resonant inverter of Figure 10.26 with the following specifications:
DC supply voltage, $V_s = 100$ V
Load resistance, $R = 1$ kΩ
Load inductance, $L = 2$ pF
Load capacitance, $C = 0.1$ pF
Peak value of load voltage, $V_p = 140$ V

a. Determine the ratings of all components and devices under worst-case conditions.
b. Use SPICE to verify your design.
c. Provide a component cost estimate of the circuit.

10.5 It is required to design the ZCSC of Figure 10.27 with the following specifications:
DC supply voltage, $V_s = 20$ V
Load resistance, $R = 100\ \Omega$
Average output voltage, $V_{(DC)} = 10$ V with ±5% ripple
a. Determine the ratings of all components and devices under worst-case conditions.
b. Use SPICE to verify your design.
c. Provide a component cost estimate of the circuit.

10.6 It is required to design the ZCSC of Figure 10.28 with the following specifications:
DC supply voltage, $V_s = 20$ V
Load resistance, $R = 100\ \Omega$
Average output voltage, $V_{(DC)} = 10$ V with ±5% ripple
a. Determine the ratings of all components and devices under worst-case conditions.
b. Use SPICE to verify your design.
c. Provide a component cost estimate of the circuit.

10.7 It is required to design the ZVSC of Figure 10.29 with the following specifications:
DC supply voltage, $V_s = 20$ V
Load resistance, $R = 100\ \Omega$
Average output voltage, $V_{(DC)} = 10$ V with ±5% ripple
a. Determine the ratings of all components and devices under worst-case conditions.
b. Use SPICE to verify your design.
c. Provide a component cost estimate of the circuit.

10.8 Use PSpice to find the worst-case minimum and maximum DC output voltages ($V_{o(min)}$ and $V_{o(max)}$) for Problem 10.1. Assume uniform tolerances of ±15% for all passive elements and an operating temperature of 25°C.

10.9 Use PSpice to find the worst-case minimum and maximum DC output voltages ($V_{o(min)}$ and $V_{o(max)}$) for Problem 10.2. Assume uniform tolerances of ±15% for all passive elements and an operating temperature of 25°C.

10.10 Use PSpice to find the worst-case minimum and maximum DC output voltages ($V_{o(min)}$ and $V_{o(max)}$) for Problem 10.3. Assume uniform tolerances of ±15% for all passive elements and an operating temperature of 25°C.

10.11 Use PSpice to find the worst-case minimum and maximum DC output voltages ($V_{o(min)}$ and $V_{o(max)}$) for Problem 10.4. Assume uniform tolerances of ±15% for all passive elements and an operating temperature of 25°C.

10.12 Use PSpice to find the worst-case minimum and maximum DC output voltages ($V_{o(min)}$ and $V_{o(max)}$) for Problem 10.5. Assume uniform tolerances of ±15% for all passive elements and an operating temperature of 25°C.

10.13 Use PSpice to find the worst-case minimum and maximum DC output voltages ($V_{o(min)}$ and $V_{o(max)}$) for Problem 10.6. Assume uniform tolerances of ±15% for all passive elements and an operating temperature of 25°C.

10.14 Use PSpice to find the worst-case minimum and maximum DC output voltages ($V_{o(min)}$ and $V_{o(max)}$) for Problem 10.7. Assume uniform tolerances of ±15% for all passive elements and an operating temperature of 25°C.

10.15 Design of a Class E-Inverter. The class E inverter as shown in Figure P10.15 operates at resonance and has $Vs = 48$ V and $R = 8\ \Omega$. The switching frequency is $fs = 25$ kHz. Assume a quality factor $Q = 7$. Use the values in the problem, not as shown in Figure P10.15.

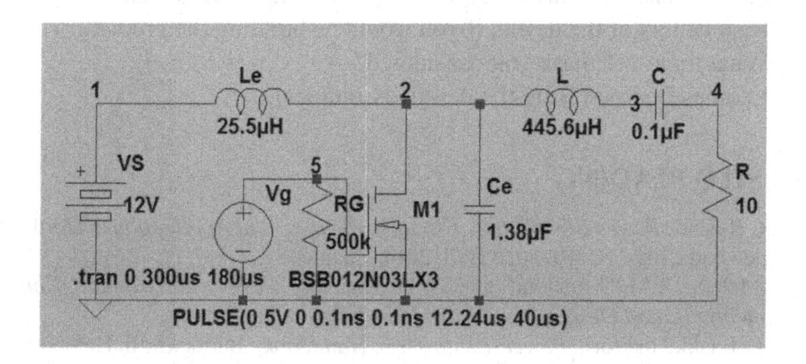

FIGURE P10.15 Class E inverter.

1. Calculate the parameters and complete Table 10.1. Assume an ideal transistor switch.
   a. The optimum values on inductor L
   b. The optimum value of capacitor C
   c. The optimum values on inductor $_{Le}$
   d. The optimum value of capacitor C_e
   e. The damping factor δ
   f. The peak output voltage V_o for $V_i = 12$ V from

$$G(j\omega) = \frac{V_o}{V_i}(j\omega) = \frac{1}{1 + j\omega L/R - j/(\omega CR)}$$

2. Use Multisim/LTspice/Orcad. Apply a pulse voltage of 5V (peak) with a duty cycle of $k = 0.304$.

   Use any MOSFET (e.g., 2N6756) or IGBT that meets the voltage requirements.
3. Run the simulation from 0 to 1.5 ms so that the steady-state condition is reached. Make sure that the transistor is turned on and operated as a switch.
   a. Plot the instantaneous output voltage $v_o(t)$.
   b. Plot the instantaneous voltage across the transistor switch $v_T(t)$.

4. Complete Table P10.2

TABLE P10.2

	$V_{o(peak)}$ (V)	f_o (kHz)	$I_{M1(avg)}$ (A)	$I_{M1(RMS)}$ (A)
Calculated		25		
Simulated				

5. Estimated component costs: https://www.jameco.com/
6. Safety considerations if you would be building the product:
7. Risk factors of the design if you would be building the product.
8. What trade-offs have you considered?
9. Lessons Learned from the design assignment.

SUGGESTED READING

1. M.R. Rashid, *Power Electronics: Circuits, Devices, and Applications*, Third Edition, Englewood Cliffs, NJ: Prentice Hall, 2014, Chapter 8.
2. N. Mohan, T.M. Undeland and W.P. Robbins, *Power Electronics: Converters, Applications, and Design*, New York: John Wiley & Sons, 2003.
3. M.H. Rashid (editor), *Power Electronics Handbook*, Butterworth Heinemann. 4/e, 2017.
4. Company report: Infineon, Introduction to Infineon's simulation models for IGBTs and silicon diodes in discrete packages, AN 2020-17, Infineon.
5. Infineon-IKW30N65EL5-DataSheet-v02_02-EN. https://www.infineon.com/dgdl/Infineon-IGW30N65L5-DataSheet-v02_02-EN.pdf?fileId=5546d4624b0b249c014b11cd55583ac9
6. IGBT Spice Model, https://www.globalspec.com/industrial-directory/igbt_spice_model
7. Discrete IGBT P-Spice Models, https://www.fujielectric.com/products/semiconductor/model/igbt/technical/design_tool.html

11 Controlled Rectifiers

After completing this chapter, students should be able to do the following:

- Model SCRs and specify their model parameters.
- Model voltage-controlled switches.
- Perform transient analysis of controlled rectifiers.
- Evaluate the performance of controlled rectifiers.
- Perform worst-case analysis of resonant-pulse inverters for parametric variations of model parameters and tolerances.

11.1 INTRODUCTION

A thyristor can be turned on by applying a pulse of short duration. Once the thyristor is on, the gate pulse has no effect, and it remains on until its current is reduced to zero. It is a latching device. PWM techniques can be applied to controlled rectifiers with bidirectional switches in order to improve the input power factor of the converters.

11.2 AC THYRISTOR MODEL

There are a number of published AC thyristor models [3–6]. We shall use a very simple model that can obtain the various waveforms of controlled rectifiers. Let us assume that the thyristor (see Figure 11.1(a)) is operated from an AC supply. This thyristor should exhibit the following characteristics:

1. It should switch to the on-state with the application of a small positive gate voltage, provided that the anode-to-cathode voltage is positive.

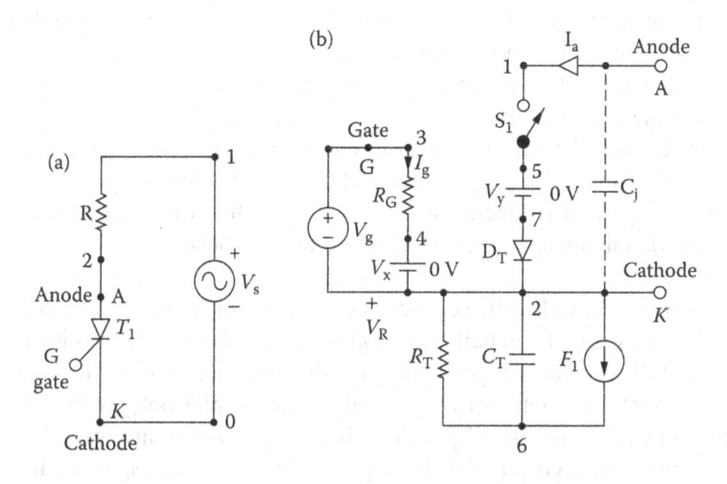

FIGURE 11.1 AC thyristor model. (a) Thyristor circuit. (b) Thyristor Model.

DOI: 10.1201/9781003284451-11

2. It should remain in the on-state for as long as the anode current flows.
3. It should switch to the off-state when the anode current goes through zero in the negative direction.

The switching action of the thyristor can be modeled by a voltage-controlled switch and a polynomial current source [3]. This is shown in Figure 11.1(b). The turn on process can be explained by the following steps:

1. For a positive gate voltage V_g between nodes 3 and 2, the gate current is $I_g = \text{I(VX)} = V_g/R_G$.
2. The gate current I_g activates the current-controlled current source F_1 and produces a current of value $F_g = P_1 I_g = P_1 \times \text{I(VX)}$, such that $F_1 = F_g + F_a$.
3. The current source F_1 produces a rapidly rising voltage V_R across resistance R_T.
4. As the voltage V_R increases above zero, the resistance R_S of the voltage-controlled switch S_1 decreases from R_{OFF} toward R_{ON}.
5. As the switch resistance R_S decreases, the anode current $I_a = \text{I(VY)}$ increases, provided that the anode-to-cathode voltage is positive. This increasing anode current I_a produces a current $F_a = P_2 I_a = P_2 \times \text{I(VY)}$. This causes the value of voltage V_R to increase.
6. This then produces a regenerative condition with the switch rapidly being driven into low resistance (the on-state). The switch remains on if the gate voltage V_g is removed.
7. The anode current I_a continues to flow as long as it is positive, and the switch remains in the on-state.

During turn-off, the gate current is off, and $I_g = 0$. That is, $I_g = 0$ and $F_g = 0$, $F_1 = F_g + F_a = F_a$. The turn-off operation can be explained by the following steps:

1. As the anode current I_a goes negative, the current F_1 reverses provided that the gate voltage V_g is no longer present.
2. With a negative F_1, the capacitor C_T discharges through the current source F_1 and the resistance R_T.
3. With the fall of voltage V_R to a low level, the resistance R_S of switch S_1 increases from a low (R_{ON}) value to a high (R_{OFF}) value.
4. This is, again, a regenerative condition with the switch resistance being driven rapidly to an R_{OFF} value as the voltage V_R becomes zero.

This model works well with a converter circuit in which the thyristor current falls to zero itself, for example, in half-wave controlled rectifiers and AC voltage controllers. But in full-wave converters with a continuous load current, the current of a thyristor is diverted to another thyristor, and this model may not give the true output. This problem can be remedied by adding diode D_T as shown in Figure 11.1(b). The diode prevents reverse current flow through the thyristor resulting from the firing of another thyristor in the circuit.

Let us consider the thyristor whose data sheets are shown in Figure 11.2. Suitable values of the model parameters can be chosen to satisfy the characteristics of a particular thyristor by the following steps:

1. The switch parameters V_{ON} and V_{OFF} can be chosen arbitrarily. Let $V_{ON} = 1$ V and $V_{OFF} = 0$ V.
2. R_T can also be chosen arbitrarily. Let $R_T = 1\ \Omega$.

(a)

S18CF SERIES
1200–1000 VOLTS RANGE
STANDARD TURN-OFF TIME 16 μs
110 AMP RMS, CENTER AMPLIFYING GATE
INVERTER TYPE STUD MOUNTED SCRs

Voltage ratings

Voltage code (1)	V_{RRM}, V_{DRM} – (V) Max. rep. peak reverse and off-state voltage	V_{RSH} – (V) Max. non-rep. peak· reverse voltage $t_p \le 5$ ms	Notes
	T_J = –40° to 125°C	T_J = 25° to 125°C	
12	1200	1300	Gate open
10	1000	1100	

Maximum allowable ratings

Parameter		Value	Units	Notes	
T_J	Junction temperature	–40 to 125	°C		
T_{stg}	Storage temperature	–40 to 150	°C		
$I_{T(AV)}$	Max. ev, current	70	A	180° half sine wave	
	Max. T_C	85	°C		
$I_{T(RMS)}$	Max. RMS current	110	A		
I_{TSM}	Max. peak non-repetitive surge current	1910	A	50 Hz half cycle sine wave	Initial T_J = 125°C, rated V_{RRM} applied after surge.
		2000		60 Hz half cycle sine wave	
		2270		50 Hz half cycle sine wave	Initial T_J = 125°C, no voltage applied after surge.
		2380		60 Hz half cycle sine wave	
$I^2\sqrt{t}$	Max. I^2t capability	18	$kA^2 8$	t = 10 ms Initial T_J = 125°C, rated V_{RRM} applied after surge.	
		17		t = 8.3 ms	
		26		t = 10 ms Initial T_J = 125°C, on voltage applied after surge.	
		24		t = 8.3 ms	
$I^2\sqrt{t}$	Max. $I^2\sqrt{t}$ capability	258	$kA^2\sqrt{v}$	Initial T_J = 125°C, on voltage applied after surge. I^2t for time $t_x = I^2\sqrt{t}\cdot\sqrt{t_x}$ $0.1 \le t_x \le 10$ ms.	
di/dt	Max. non-repetitive rate-of rise of current	800	A/μs	T_J = 125°C, $V_D = V_{DRM}$, I_{TM} = 1600 A, gate pules: 20 V, 20 Ω, 10 μs, 0.5 μs rise time. Max. repetitive di/dt is approximately 40% of non-repetitive value.	
P_{GH}	Max. peak gate power	10	W	$t_p \le 5$ ms	
$P_{G(AV)}$	Max. ev. gate power	2	W		
I_{GM}	Max. peak gate current	3	A	$t_p \le 5$ ms	
$-V_{GH}$	Max. peak negative gate voltage	15	V		
T	Mounting tor[illegible]ue	15.5 (137) ± 10%	N·m	Non-lubricated threads	
		14 (120) ± 10%	(tbf–in)	Lubricated threads	

(1) To complete the part number, refer to the ordering information table.

FIGURE 11.2 Data S\sheets for IR thyristors type S18. (a) Datasheet. (b) Datasheet (continued). (c) Characteristics. (d) Characteristics (continued). (d) Characteristics (continued). (Courtesy Of International Rectifier.) *(Continued)*

(b)

S18CF SERIES
1200–1000 VOLTS RANGE

Characteristics

	Parameter	Min.	Typ.	Max.	Units	Test conditions
V_{TM}	Peak on-state voltage	—	1.95	2.06	V	Initial T_J = 25°C, 50–60 Hz half sine, I_{peak} = 220 A.
$V_{T(T0)1}$	Low-level threshold	—	—	1.28	V	T_J = 125°C Av. power = $V_{T(TO)} \cdot I_{T(AV)} + rT \cdot [I_{T(RHS)}]^2$ Use low level values for $I_{TH} \leq \pi$ rated $I_{T(AV)}$
$V_{T(T0)2}$	High-level threshold	—	—	1.61		
rT_1	Low-level threshold	—	—	3.54	mΩ	
rT_2	High-level threshold	—	—	2.27		
I_L	Latching current	—	270	—	ma	T_C = 25°C, 12 V anode. gate pulse: 10 V, 20 Ω, 100 μs.
I_H	Holding current	—	90	500	ma	T_C = 25°C, 12 V anode. Initial I_T = 3A.
t_d	Delay time	—	0.5	1.5	μs	T_C = 25°C, V_D = rated V_{DRH}, 50 A resugtive load. gate pulse: 10 V, 20 Ω, 10 μs, 1 μs rise time
t_q	Turn-off time				μs	T_J = 125°C, I_{TM} = 200 A, di_R/dt = 10 A/μs, V_R = 50 V, dv/dt = 200 V/μs line to 80% rated V_{DRM}. Gate: 0 V, 100 Ω.
	"A" suffix	—	—	16		
	"B" suffix	—	—	20		
$t_{q(diode)}$	Turn-off time with feedback diode				μs	T_J = 125°C, I_{TM} = 200 A, di_R/dt = 10 A/μs, V_R = 1 V, dv/dt = 600V/μs line to 40% rated V_{DRM}. Gate: 0 V, 100 Ω.
	"A" suffix	—	—	20		
	"B" suffix	—	—	25		
$I_{RM(REC)}$	Recovery current	—	57	—	A	T_J = 125°C. I_{TM} = 400 A, di_R/dt = 50 A/μs.
O_{RR}	Recovery charge	—	58	—	μc	
dv/dt	Critical rate-of-rise of off-state voltage	500	700	—	V/μs	T_J = 125°C. Exp. to 100% or lin. Higher dv/dt values to 80% V_{DRM}, gets open, available.
		1000	—	—		T_J = 125°C. Exp. to 67% V_{DRH}, gets open.
I_{RH} I_{DM}	Peak reverse and off-state current	—	10	20	mA	T_J = 125°C, Rated V_{RRM} and V_{DRM}, gets open.
I_{GT}	DC gate voltage to trigger	—	—	300	mA	T_C = −40°C +12 V anode-to-cathode. For recommended gate drive see "Gate Characteristic" figure.
		25	20	150		T_C = 25°C
V_{GT}	DC gate current to trigger	—	—	3.3	V	T_C = −40°C
		—	1.2	2.5		T_C = 25°C
V_{GD}	DC gate voltage not to trigger	—	—	0.3	V	T_C = 125°C. Max. value which will not trigger with rated V_{DRM} anode-to-cathode.
R_{thJC}	Thermal resistance, junction-to-case	—	—	0.250	°C/W	DC operation
		—	—	0.291	°C/W	180° sine wave
		—	—	0.302	°C/W	120° rectangular wave
R_{thCS}	Thermal resistance, junction-to-case	—	—	0.100	°C/W	Mtg. surface smooth, flat and greased.
wt	weight	—	100(3.5)	—	g(oz.)	
	Case style		To–209AC (To–94)		JEDEC	

FIGURE 11.2 *(Continued)*

3. R_{ON} should be chosen to model the on-state resistance of the thyristor so that $R_{on} = V_{TM}/I_{TM}$ at 25°C. From the on-state characteristic of the data sheet, $V_{TM} = 2V$ at $I_{TM} = 190$ A. Thus, $R_{ON} = 2/190 = 0.0105\Omega$.
4. R_{OFF} should be chosen to model the off-state resistance of the thyristor so that $R_{OFF} = V_{RRM}/I_{DRM}$ at 25°C. From the data sheet, $V_{RRM} = 1200V$ at $I_{DRM} =$ 10 mA. Thus, $R_{OFF} = 1200/(10 \times 10^{-3}) = 120$ kΩ.
5. The switch resistance R_s can be found from

$$R_s = R_{ON} \times \frac{R_{OFF}}{R_{ON}}\left[0.5 + 2\left(V_R - 0.5\frac{V_{ON}+V_{OFF}}{V_{ON}-V_{OFF}}\right)^3 - 1.5\left(V_R - 0.5\frac{V_{ON}+V_{OFF}}{V_{ON}-V_{OFF}}\right)\right] \tag{11.1}$$

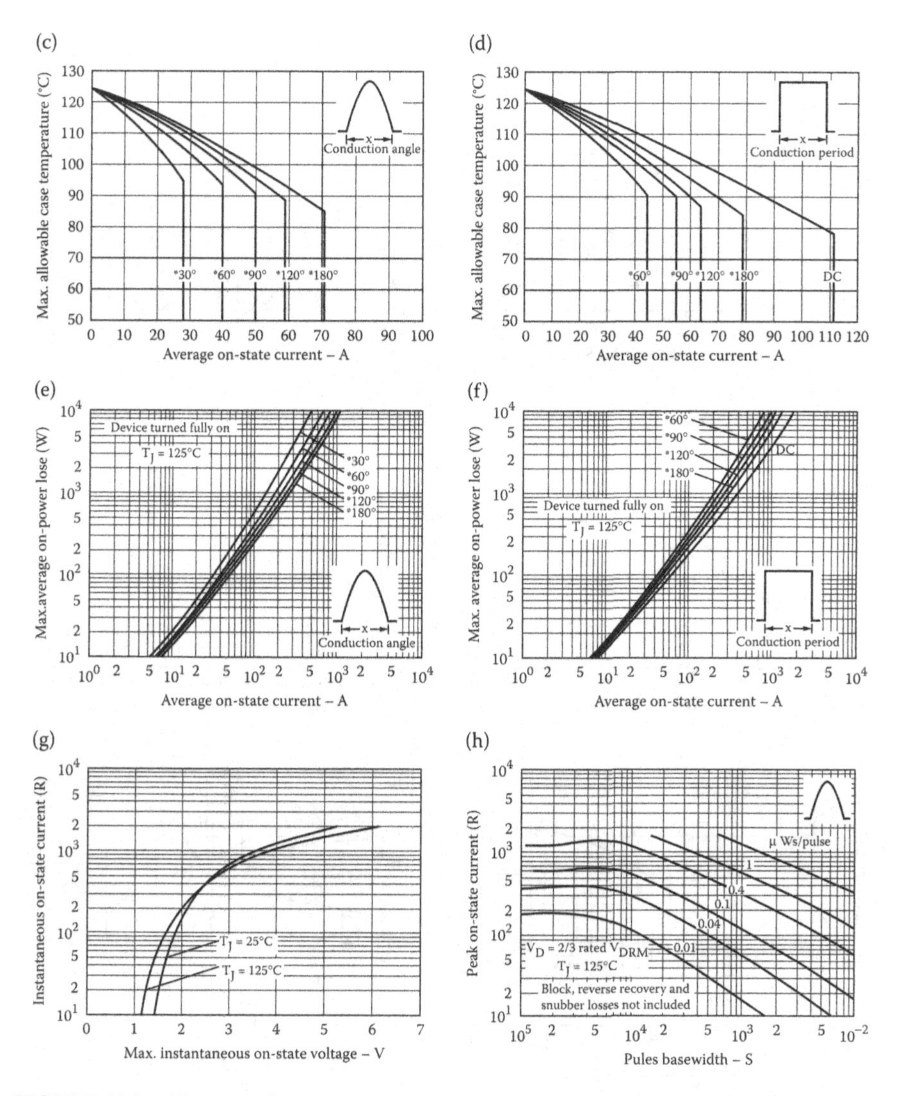

FIGURE 11.2 *(Continued)*

For $V_R = 0$, $R_S = R_{OFF}$, and for $V_R = 1$, $R_S = R_{ON}$. As V_R varies from 0 to 1 V.

6. The value of P_2 must be such that the device turns off (with zero gate current $I_g = 0$) at the maximum anode-to-cathode voltage V_{DRM}. If V_{R1} is 15% of V_R, that is, $V_{R1} = 0.15V_R$, the switch will essentially be off. The voltage V_{R1} due to the anode current only is given by

$$V_{R1} < P_2 I_a = P_2 \frac{V_{DRM}}{R_{S1}}$$
$$P_2 > \frac{V_{R1} R_{S1}}{V_{DRM}} \tag{11.2}$$

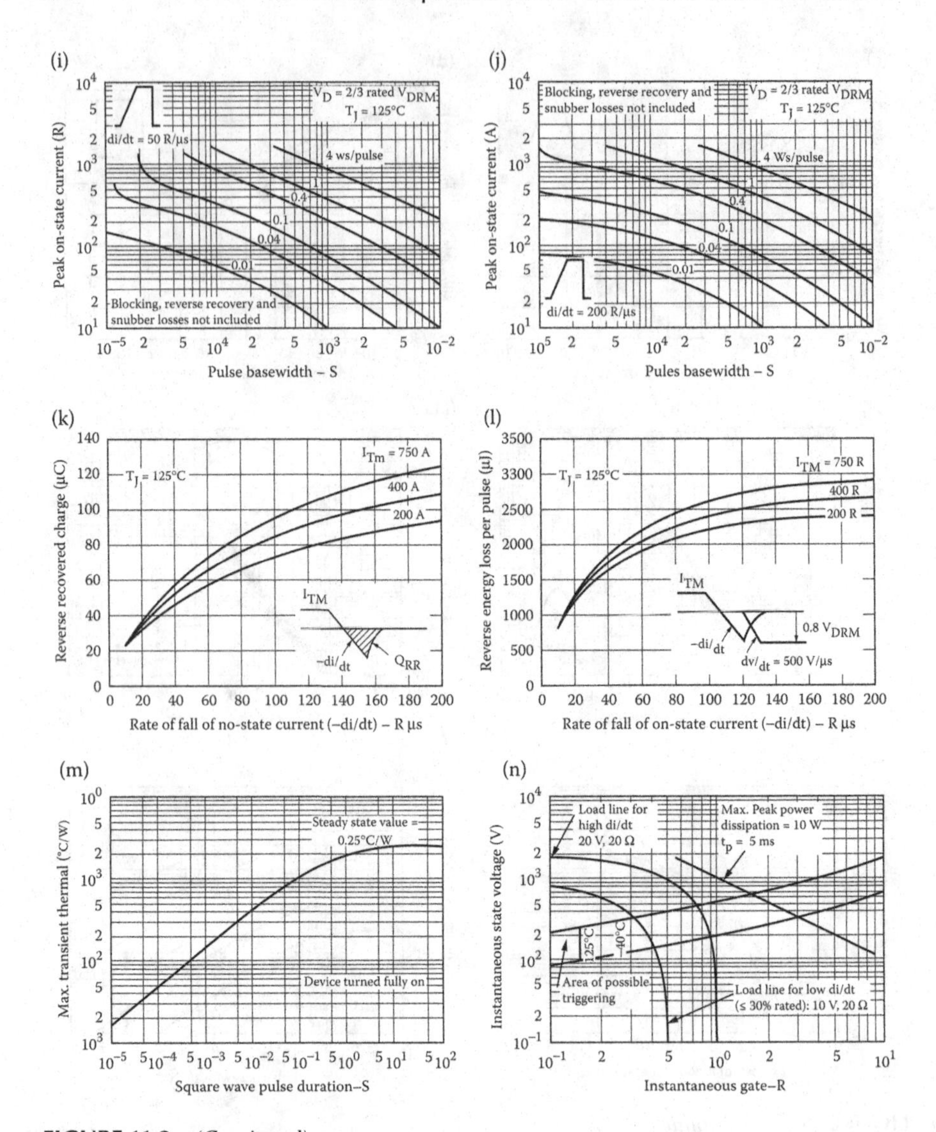

FIGURE 11.2 *(Continued)*

At the latching current I_L, the switch must be fully on, $V_R = 1$ V. That is, $P_2 > V_{R1}/I_L$. Thus, the value of P_2 must satisfy the condition.

$$\frac{V_R}{I_L} < P_2 < \frac{V_{R1} R_{S1}}{V_{DRM}} \tag{11.3}$$

From the data sheet, $I_L = 270$ mA. That is, (1/270 mA) < P_2 > (0.15 × 120 kΩ/1200) or 3.7 < P_2 > 15. Let us choose $P_2 = 11$.

7. The capacitor C_T is introduced primarily to facilitate SPICE convergence. If C_T is too small, the thyristor will go off before the anode current becomes

(o)

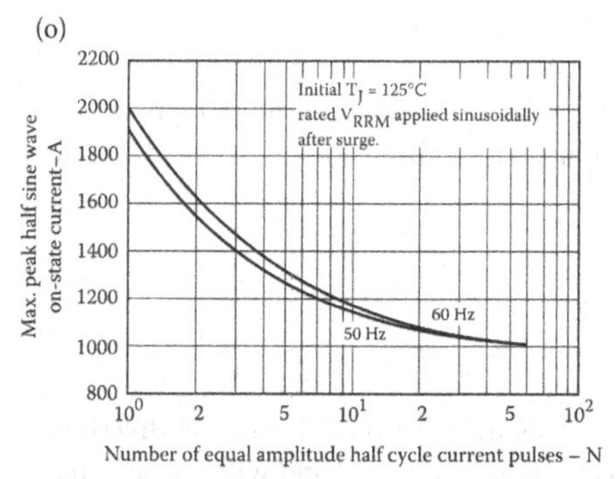

(p)

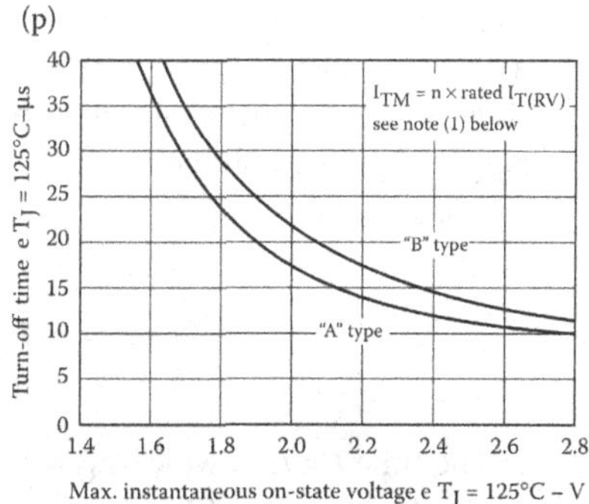

(1) These curve are intended as a guideline. To specify non-standard t_q/v_{TM} contact factory.

Ordering information

Type	Package (1)		Fast	Temperature		Voltage		Turn-off		Leads &-terminals	
	Code	Description		Code	Max. T_J	Code	V_{DRM}	Code	Max. T_q	Code	Description
S18	C	1/2" stud, ceramic housing.	F	—	125°C	12	1200 V	A	16 µs	0	Flexible leads, ayelet terminals. standard in USA. (fig. 1)
						10	1000 V	B	20 µs		
										1	Flexible leads, fast-on terminals. standard in europe. (fig. 2)

(q)

(1) Other packages are also available:
– Supplied with flag terminals.
For further details contact factory.

For a device with standerd USA case, max T_J = 125°C, V_{DRM} = 1200 V, max. t_q = 16 µm, order as: S1BCF12AD.

FIGURE 11.2 *(Continued)*

zero. If C_T has a sufficiently large value, the thyristor will continue to conduct beyond the zero crossing of supply voltage, and C_R can be chosen large enough to model the turn-off time t_{off} of the thyristor. The time constant C_TR_T should be much smaller than the period T of the supply voltage. This condition is generally satisfied by the relation $C_TR_T \le 0.01T$. Let us choose $C_T = 10\ \mu\text{F}$.

8. As far as the model is concerned, R_G can be chosen arbitrarily. From the data sheet, $V_g = 5$ V at $I_g = 100$ mA. Therefore, we get $R_G = V_g/I_g = 5/0.1 = 50\ \Omega$.
9. For a known value of R_G, the multiplier P_1 can be determined from $V_{ON} = R_TP_1I_g$, which gives

$$P_1 = \frac{V_{ON}}{R_TI_g} = \frac{V_{ON}}{\dfrac{R_TV_{g(peak)}}{R_G}} = \frac{V_{ON}R_G}{R_TV_{g(peak)}}$$

$$= \frac{1\times 50}{1\times 5} = 10 \tag{11.4}$$

where $V_{g(peak)}$ is the peak voltage of the gate pulse. At the leading edge of the gate pulse, most of the controlling current F_1 flows through C_T. Thus, the gate pulse must be applied for a sufficient time to cause on-triggering. The voltage v_{R2} due to gate triggering only (with zero anode current) is given by

$$v_{R_2} = R_T P_1 I_g \left(1 - e^{\frac{-t}{R_T C_T}}\right) \quad (11.5)$$

The time $t = t_n$ at which $v_{R2} = V_{ON}$ gives the turn-on delay of the thyristor. Due to the presence of C_T, the voltage v_{R2} will be delayed and will not be equal to $V_{ON} = R_T P_1 I_g$ instantly. A higher value of P_1 is required to turn on the thyristor. Taking five times more than the value given by Equation 11.4, let $P_1 = 50$.

This thyristor model can be used as a subcircuit. The switch S_1 is controlled by the controlling voltage V_R connected between nodes 6 and 2. The switch and the diode parameters can be adjusted to yield the desired on-state drop of the thyristor. In the following examples, we shall use a superdiode with parameters IS = 2.2E15, BV = 1200V, TT = 0, and CJO = 0, and the switch parameters RON = 0.0105, ROFF = 10E+5, VON = 0.5V, and VOFF = 0V. The subcircuit definition for the thyristor model SCR can be described as follows:

```
* Subcircuit for AC thyristor model
.SUBCKT  SCRMOD      1        3          2
*         model    anode  +control   cathode
*         name              voltage
S1    1    5    6    2    SMOD        ; Switch
RG    3    4    50
VX    4    2    DC   0V
VY    5    7    DC   0V
DT    7    2    DMOD                  ; Switch diode
RT    6    2    1
CT    6    2    10UF
F1    2    6    POLY(2)  VX   VY   0       50       11
.MODEL SMOD VSWITCH (RON=0.0105 ROFF=10E+5 VON=0.5V
                          VOFF=0V)  ; Switch model
.MODEL DMOD D(IS=2.2E-15 BV=1200V  TT=0  CJO=0)
                                      ; Diode model parameters
.ENDS SCRMOD                          ; Ends subcircuit definition
```

11.3 CONTROLLED RECTIFIERS

A controlled rectifier converts a fixed AC voltage to a variable DC voltage and uses thyristors as switching devices. Ideally, the output voltage of an ideal rectifier should be a pure sine wave and contain no harmonics. That is, the *THD* of the input

TABLE 11.1
Controlled Rectifier Performance Parameters

Input Side Parameters	**Output Side Parameters**
RMS input voltage, Vs	Average output voltage, V_{dc}
RMS input current, Is	Average output current, I_{dc}
The *transformer utilization factor, TUF* $= \frac{P_{dc}}{V_s I_s}$	The output dc power, $P_{dc} = V_{dc} I_{dc}$
The input power, $PF = \frac{P_{ac}}{V_s I_s}$	RMS output voltage, V_{rms}
	RMS output current, I_{rms}
	The output ac power, $P_{ac} = V_{rms} I_{rms} = I_{rms}^2 R_L$ for a load resistance of R_L
	The rms ripple content of the output voltage $V_{ac} = \sqrt{V_{rms}^2 - V_{dc}^2}$
	The *ripple factor, RF* $= \frac{V_{ac}}{V_{dc}}$
	The rectification *efficiency or ratio,* $\eta = \frac{P_{dc}}{P_{ac}}$
	The *form factor, FF* $= \frac{V_{rms}}{V_{dc}}$

current and output voltage should be zero, and the input power factor should be unity. However, the *THD* and *PF* of a controlled rectifier will differ from the ideal values. Table 11.1 lists the performance parameters of controlled rectifiers.

11.4 EXAMPLES OF CONTROLLED RECTIFIERS

Let us apply the thyristor model of Figure 11.1(b) to the circuit of Figure 11.3(a). Thyristor T_1 is turned on by the voltage V_g connected between the gate and cathode voltage. The gate voltage V_g is shown in Figure 11.3(b). The listing of the PSpice circuit file for determining the transient response is as follows:

```
SOURCE  ■   VS 1 0 SIN (0 169.7V 60HZ)
            Vg 4 0 PULSE (0V 10V 2777.8US 1NS 1NS 100US 16666.7US)
```

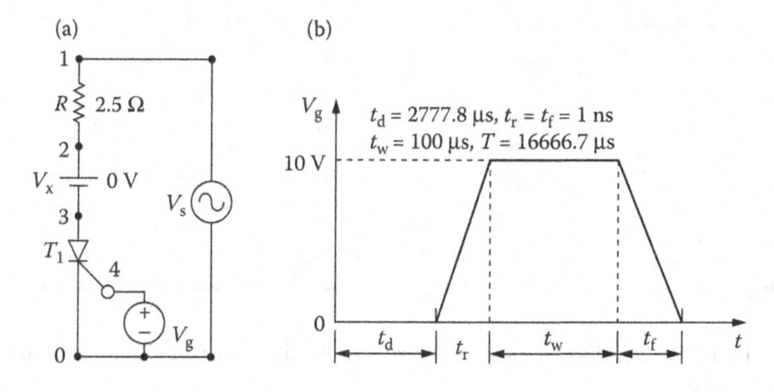

FIGURE 11.3 AC thyristor circuit. (a) Circuit. (b) Gate voltage.

```
CIRCUIT ■■  R     1 2 2.5
            VX  2 3 DC 0V ; Voltage source to measure the
            load current
            * Subcircuit call for thyristor model:
            XT1 3 0 4 0 SCRMOD ; Thyristor T1
ANALYSIS ■■■ .TRAN  1US 5MS     ; Transient analysis from 0 to 5 ms
            .PROBE              ; Graphics post-processor
        .END
```

11.4.1 Examples of Single-Phase Controlled Rectifiers

Example 11.1:

Finding the Performance of a Single-Phase Half-Wave Controlled Rectifier

A single-phase half-wave rectifier is shown in Figure 11.4(a). The input has a peak voltage of 169.7 V, 60 Hz. The load inductance L is 6.5 mH, and the load resistance R is 0.5 Ω. The delay angle is $\alpha = 60°$. The gate voltage is shown in Figure 11.4(b). Use PSpice to (a) plot the instantaneous output voltage v_o and the load current i_o and (b) calculate the Fourier coefficients of the input current i_s and the input power factor PF.

SOLUTION

The peak supply voltage $V_m = 169.7$ V. for $\alpha = 60°$, time delay $t_1 = (60/360) \times (1000/60\ \text{Hz}) \times 1000 = 2777.78\ \mu s$. The PSpice schematic with an SCR is shown in Figure 11.5. The model name of the SCR 2N1595 is changed to SCRMOD, the subcircuit definition of which is listed in Section 11.2. Varying the delay cycle can vary the output voltage. The supply frequency {FREQ} and the duty cycle {DELAY_ANGLE} are defined as variables. The listing of the circuit file is as follows:

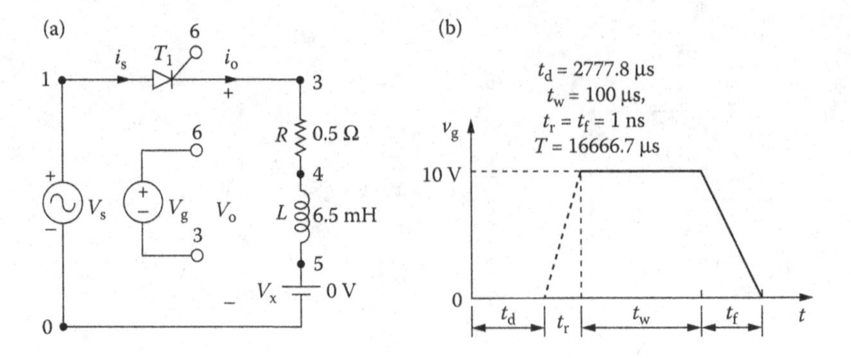

FIGURE 11.4 Single-phase half-wave controlled rectifier for PSpice simulation. (a) Circuit. (b) Gate voltage.

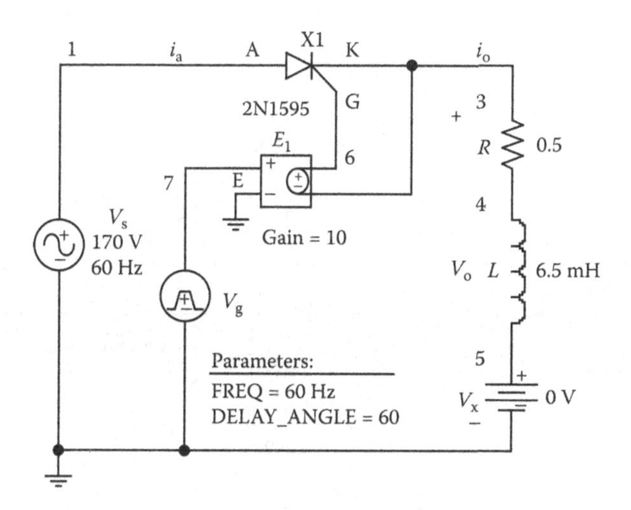

FIGURE 11.5 PSpice schematic for Example 11.1.

Single-Phase Half-Wave Controlled Rectifier

```
SOURCE    ■    VS 1 0 SIN (0169.7V60HZ)
               .PARAM Freq  =  60Hz Delay_Angle  =  60
               Vg 6 3 PULSE (0 1 {{Delay_Angle}/
               {360*{Freq}}} 1ns 1ns 100us {1/{Freq}})
CIRCUIT   ■■   R  3 4 0.5
               L  4 5 6.5MH
               VX 5 0 DC 0V ; Voltage source to measure
               the load current
               * Subcircuit call for thyristor model:
               XT1 1 6 3 SCRMOD ; Thyristor T1
               * Subcircuit SCR, which is missing,
               * must be inserted.
ANALYSIS ■■■   .TRAN 1US 50.0MS 33.33MS ; Transient analysis
               .PROBE ; Graphics post-processor
               .OPTIONS ABSTOL=1.0N RELTOL =1.0M VNTOL =1.0M
               ITL5 =10000                 ; Convergence
               .FOUR 60HZI (VX)            ; Fourier analysis
          .END
```

Note the following:

a. The PSpice plots of the instantaneous output voltage V(3) and the load current I(VX) are shown in Figure 11.6. The thyristor T_1 turns off when its current falls to zero, but not when the input voltage becomes zero. Note that the frequency of the output ripple voltage is the same as the supply frequency. That is, $f_o = f = 60$ Hz.
b. To find the input power factor, we need to find the Fourier series of the input current, which is the same as the current through source VX.

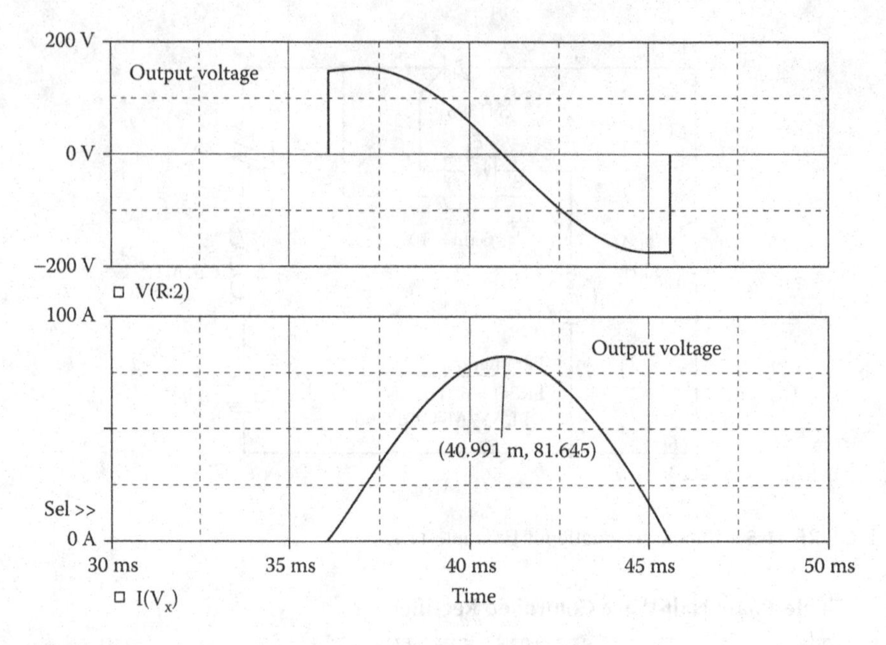

FIGURE 11.6 Plots for Example 11.1 for delay angle = 60.

Fourier Components of Transient Response I(VX)

DC Component = 2.912470E+01

Harmonic Number	Frequency (Hz)	Fourier Component	Normalized Component	Phase (Deg)	Normalized Phase (Deg)
1	6.000E+01	4.266E+01	1.000E+00	–7.433E+01	0.000E+00
2	1.200E+02	1.315E+01	3.082E–01	1.200E+02	1.943E+02
3	1.800E+02	2.897E+00	6.792E–02	1.448E+02	2.191E+02
4	2.400E+02	1.845E+00	4.325E–02	–2.775E+01	4.657E+01
5	3.000E+02	1.510E+00	3.540E–02	–3.585E+00	7.074E+01
6	3.600E+02	3.159E–01	7.406E–03	1.761E+02	2.504E+02
7	4.200E+02	8.563E–01	2.007E–02	–1.511E+02	–7.681E+01
8	4.800E+02	1.373E–01	3.218E–03	–1.137E+02	–3.936E+01
9	5.400E+02	4.678E–01	1.097E–02	6.119E+01	1.355E+02

Total harmonic distortion = 3.214001E+01%

From the Fourier components, we get

DC input current $I_{\text{in(DC)}} = 29.12$ A
RMS fundamental input current $I_{1(\text{RMS})} = 42.66\sqrt{2} = 30.17A$
Total harmonic distortion of input current $THD = 32.14\% = 0.3214$
RMS harmonic current $I_{\text{h(RMS)}} = I_{1(\text{RMS})} \times THD = 30.17 \times 0.3214 = 9.69$

$$\text{RMS input current, } I_s = \left[I^2_{\text{in(DC)}} + I^2_{(\text{RMS})} + I^2_{\text{h(RMS)}}\right]^{1/2}$$

$$= \left(29.12^2 + 30.17^2 + 9.69^2\right)^{1/2} = 43.4$$

Displacement angle, $\phi_1 = -74.33°$
Displacement factor, $DF = \cos\phi_1 = \cos(-74.33) = 0.27$ (lagging)
The input power factor is

$$PF = \frac{I_{1(RMS)}}{I_s}\cos\phi_1 = \frac{30.17}{43.4} \times 0.27 = 0.1877 \text{(lagging)}$$

Note: For a .FOUR command, TSTART and TSTOP values should be a multiple of the period of input voltage (e.g., for 60 Hz, a multiple of 16.667 ms). The Fourier coefficients will vary slightly with TMAX because it sets the number of samples in a period. The average output voltage of a single-phase half-wave controlled rectifier (for a delay angle of α) is given by [1]

$$v_{O(av)} = \frac{V_m}{2\pi}(1+\cos\alpha)$$

LTspice: LTspice schematic for Example 11.1 is shown in Figure 11.7(a). The plots of the load current, the input power, and the load power are shown in Figure 10.7(b). The average input power as shown in Figure 11.7(c) is 833.41 W, and the average output power as shown in Figure 11.7(d) is 760.12 W. The rms value of the load current I(Vx) is shown in Figure 11.7(e). We can find the power dissipated in the load resistor of $R = 0.5\ \Omega$ as $P_L = I^2R = 38.95^2 \times 0.5 = 758.55\,W$ which is close to the average load power of 760.52 W as shown in Figure 11.7(d). We can find the power efficiency of the converter as $\eta = \frac{P_{out}}{P_{in}} = \frac{760.12}{833.41} = 91.2\%$

Note: For a low output power, the losses in the semiconductor devices are significant compared to the output power and the efficiency becomes low. There is no power loss in the load inductor.

```
The list of circuit file for Figure 11.7 is as follows:
* C:\Users\mrashid\- file location
Vs 1 0 SINE(0 170V 60Hz 0 0 0)
Vg 5 0 PULSE(0 10V {delay} 0.1ns 0.1ns 100us {period})
XU1 1 6 2 2N5171
R 2 3 0.5
L 3 4 6.5mH
Vx 4 0 0
E1 6 2 5 0 1
.param delay=2777.8us period=16.667ms
.tran 0 50ms 30ms
.lib "C:\SCR Models\2N5171.txt"
.four 60Hz 2 I(vx)
.options abstol=1.0n reltol=1.0m vntol=1.0m
.backanno
.end
* ============= SPICE Model =================
.SUBCKT 2N5171 10 30 20
* 10-A, 30-G, 20-K
.MODEL DGAT D (IS=1.0e-12 N=1 RS=0.001)
.MODEL DMOD D (IS=1.0e-12 N=0.001)
.MODEL DON D (IS=1.000e-012 N=1.000e+000 RS=3.534e-002
BV=7.200e+002)
```

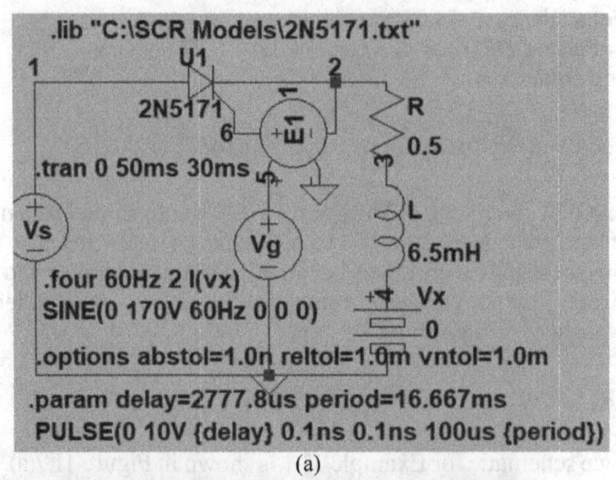

(a)

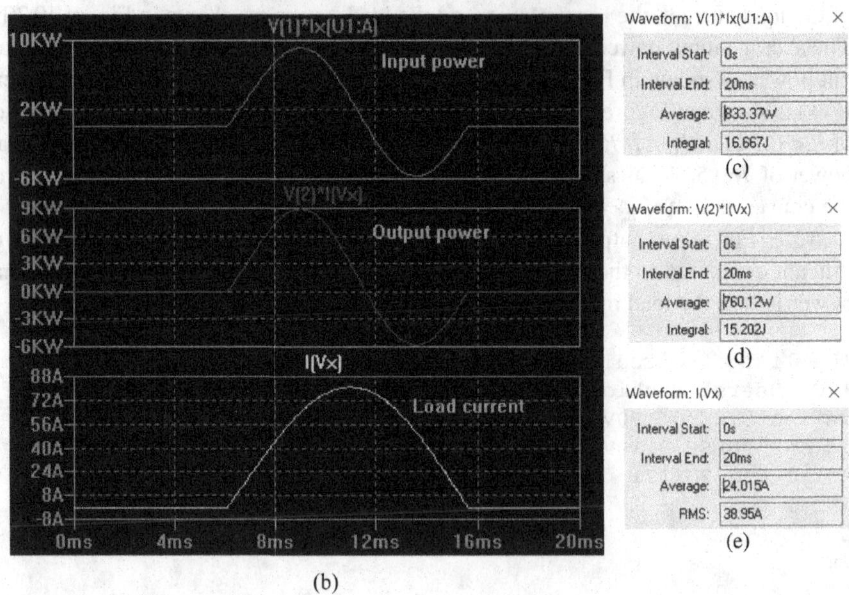

(b) (c) (d) (e)

FIGURE 11.7 LTspice schematic for Example 11.1. (a) LTspice Schematic. (b) Output waveforms. (c) Input power. (d) Output power. (e) Load current.

```
.MODEL DBREAK D (IS=1.000e-012 N=9.404e+002 BV=7.200e+002)
V1 10 14 DC 0
DON1 14 222 DON
VV 222 22 DC 0
E1 22 20 poly(2) 10 20 3 20 0 0 0 0 1
DBRK1 14 27 DBREAK
DBRK2 20 27 DBREAK
RLEAK 14 20 6.000e+007
CRISE 14 20 1.000e-009
FC1 3 20 poly(2) VGD V1 -4.000e-002 1 8.000e-001
```

```
CON 3 20 4.000e-008 IC=1.5
DS1 3 31 DMOD
DS2 20 3 DMOD
VW 31 20 DC 1
DGATE 30 7 DGAT
VGD 7 20 DC 8.677e-001
.ENDS
```

Example 11.2:

Finding the Performance of a Single-Phase Semiconverter

A single-phase semiconverter is shown in Figure 11.8(a). The input voltage has a peak of 169.7 V, 60 Hz. The load inductance L is 6.5 mH, and the load resistance R is 0.5 Ω. The load battery voltage is $V_x = 10$ V. The delay angle is $\alpha = 60°$. The gate voltages are shown in Figure 11.8(b). Use PSpice to (a) plot the instantaneous output voltage, the input current i_s, and the load current i_o and (b) calculate the Fourier coefficients of the input current i_s and the input power factor PF.

SOLUTION

The peak supply voltage $V_m = 169.7$ V. For $\alpha = 60°$,

$$\text{time delay } t_1 = \frac{60}{360} \times \frac{1000}{60\text{Hz}} \times 1000 = 2777.78 \mu\text{s}$$

$$\text{time delay } t_2 = \frac{240}{360} \times \frac{1000}{60\text{Hz}} \times 1000 = 11{,}111.1 \mu\text{s}$$

The PSpice schematic with SCRs is shown in Figure 11.9. Note that the voltage-controlled voltage sources E_1 and E_2 are connected to provide the isolation between the low signal gating signals from the power circuit. The model name of the SCR 2N1595 is changed to SCRMOD, whose subcircuit definition is listed in Section 11.2. Varying the delay cycle can vary the output voltage. The supply

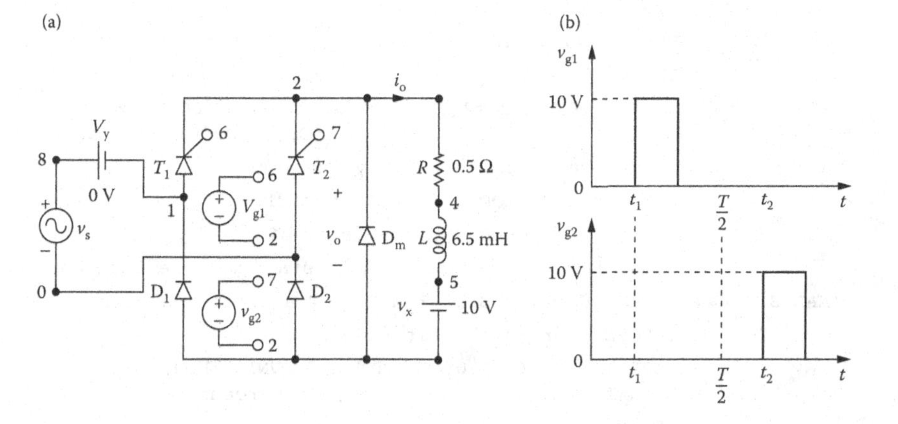

FIGURE 11.8 Single-phase semiconverter for PSpice simulation. (a) Circuit. (b) Gate voltages.

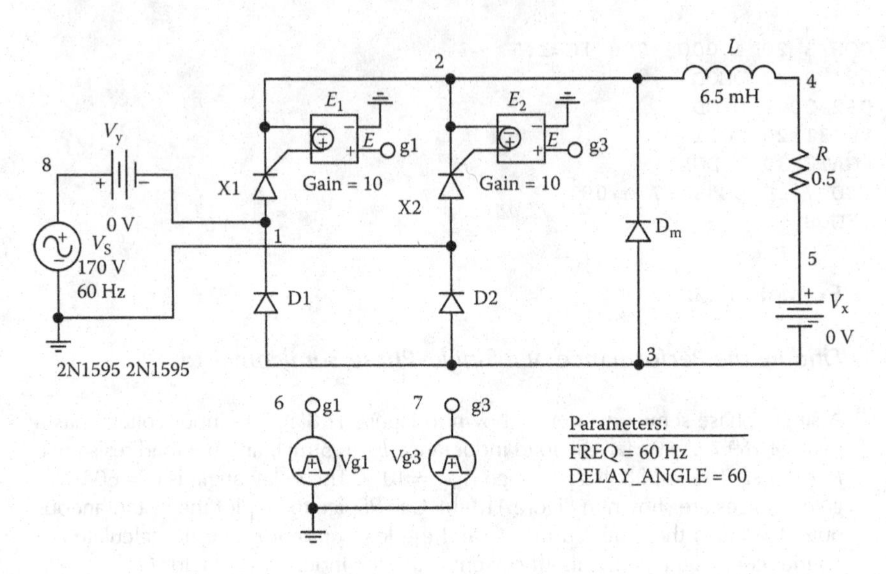

FIGURE 11.9 PSpice schematic for Example 11.2.

frequency {FREQ} and the duty cycle {DELAY_ANGLE} are defined as variables. The model parameters for the freewheeling diode are as follows:

```
SOURCE    ■ VS   8 0 SIN(0169.7V60HZ)
           .PARAM DELAY_ANGLE=60 FREQ=60Hz
             Vg1 6 2 PULSE (0 1 {{Delay_Angle}/{360*{Freq}}}
             1ns 1ns 100us {1/{Freq}})
             Vg3 7 2 PULSE (0 1 {{{Delay_Angle} + 180}/
             {360*{Freq}}} 1ns 1ns 100us {1/{Freq}})
CIRCUIT ■■   R    2 4 0.5
             L 4 5 6.5MH
             VX5 3 DC 10V   ; Load battery voltage
             VY8 1 DC 0V    ; Voltage source to supply current
             D13 1 DMOD
             D23 0 DMOD
             DM3 2 DMOD
             .MODEL DMOD D(IS=2.2E-15 BV=1200V TT=0 CJO=0)
                                          ; Diode model parameters
             * Subcircuit calls for thyristor model:
             XT1 1 6 2 SCRMOD    ; Thyristor T1
             XT2 0 7 2 SCRMOD    ; Thyristor T2
             * Subcircuit SCR, which is missing, must be inserted.
ANALYSIS ■■■ .TRAN 1US 50.0MS 33.33MS    ; Transient analysis
             .PROBE ; Graphics post-processor
             .OPTIONS ABSTOL=1.00N RELTOL=1.0M VNTOL=1.0M
             ITL5=10000                  ; Convergence
             .FOUR 60HZ I(VY)
         .END
```

The listing of the circuit file is as follows:

Single-Phase Semiconverter

```
SOURCE    ■ VS   8 0 SIN(0169.7V60HZ)
            .PARAM DELAY_ANGLE=60 FREQ=60Hz
```

Note the following:

a. The PSpice plots of the instantaneous output voltage V(2, 3), the input current I(VY), and the load current I(VX) are shown in Figure 11.10. The load current is continuous as expected, and the effects of load current ripples can be noticed on the input current.
b. To find the input power factor, we need to find the Fourier series of the input current, which is the same as the current through source VY.

Fourier Components of Transient Response I(VY)

DC Component = 4.587839E+00

Harmonic Number	Frequency (Hz)	Fourier Component	Normalized Component	Phase (Deg)	Normalized Phase (Deg)
1	6.000E+01	1.436E+01	1.000E+00	2.960E+01	0.000E+00
2	1.200E+02	1.144E+01	7.966E–02	1.008E+02	1.303E+02
3	1.800E+02	1.730E+01	1.204E–01	7.189E+01	1.015E+02
4	2.400E+02	1.035E+01	7.204E–02	1.200E+02	1.496E+02
5	3.000E+02	2.854E+01	1.987E–01	4.806E+01	7.766E+01
6	3.600E+02	1.037E+01	7.218E–02	1.353E+02	1.649E+02
7	4.200E+02	9.400E+00	6.545E–02	–7.113E+00	2.249E+01
8	4.800E+02	9.306E+00	6.480E–02	1.495E+02	1.791E+02
9	5.400E+02	7.692E+00	5.356E–02	1.414E+02	1.710E+02

Total harmonic distortion = 2.865256E+01%

From the Fourier components, we get

Total harmonic distortion of input current $THD = 28.65\% = 0.2865$
Displacement angle $\phi_1 = -29.6°$
Displacement factor $DF = \cos\phi_1 = \cos(-29.6) = 0.87$ (lagging)
From Equation 7.3, the input power factor:

$$PF = \frac{1}{(1+THD^2)^{1/2}}\text{COS}\phi_1 = \frac{1}{(1+0.2865^2)^{1/2}} \times 0.87 = 0.836(\text{lagging})$$

Note: The freewheeling diode D_m reduces the ripple content of the load current, thereby improving the input power factor of the converter. The average output voltage of a single-phase semiconverter (for a delay angle of α) is given by [1]

$$V_{o(av)} = \frac{V_m}{\pi}(1+\cos\alpha)$$

LTspice: LTspice schematic for Example 11.2 is shown in Figure 11.11(a). The plots of the load current, the input power, the output voltage, and the load power are

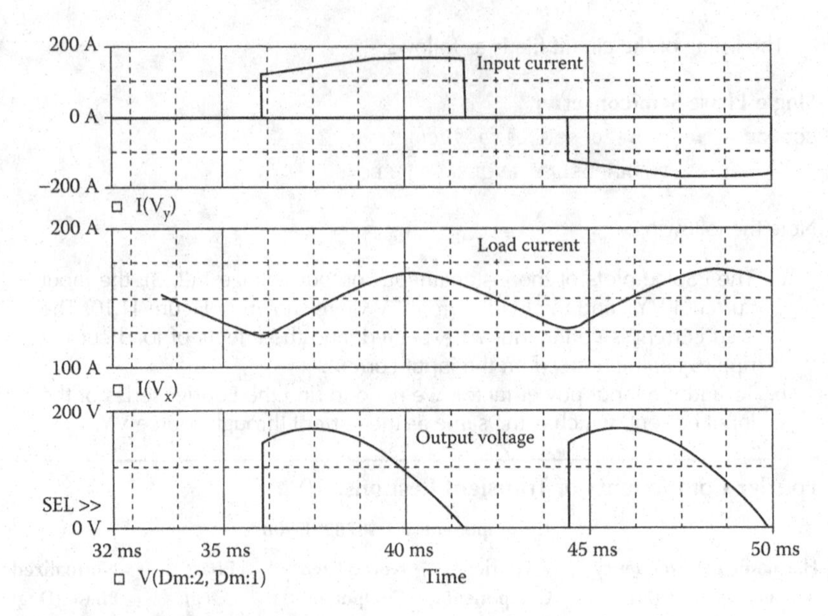

FIGURE 11.10 Plots for Example 11.2.

shown in Figure 11.11(b). The average input power as shown in Figure 11.11(c) is 10.3 kW, and the average output power as shown in Figure 11.11(d) is 9.18 kW. The rms value of the load current I(Vx) is 123 A and the rms value of the output voltage is 98.94 V. We can find the power efficiency of the converter as $\eta = \frac{P_{out}}{P_{in}} = \frac{9.18k}{10.3k} = 89.12\%$

Note: Diode Dm acting as a free-wheeling diode should be of recovery time. For a low output power, the losses in the semiconductor devices are significant compared to the output power and the efficiency becomes low. There is no power loss in the load inductor.

Example 11.3:

Finding the Performance of a Single-Phase Full Converter

A single-phase full converter is shown in Figure 11.12(a). The input voltage has a peak of 169.7 V, 60 Hz. The load inductance L is 6.5 mH, and the load resistance R is 0.5 Ω. The load battery voltage is $V_x = 10$ V. The delay angle is $\alpha = 60°$. The gate voltages are shown in Figure 11.12(b). Use PSpice to (a) plot the instantaneous output voltage v_o, the input current i_s, and the load current i_o and (b) calculate the Fourier coefficients of the input current i_s and the input power factor PF.

SOLUTION

The peak supply voltage $V_m = 169.7$ V. for $\alpha_1 = 60°$,

$$\text{time delay } t_1 = \frac{60}{360} \times \frac{1000}{60\text{Hz}} \times 1000 = 2777.78 \mu\text{s}$$

$$\text{time delay } t_2 = \frac{240}{360} \times \frac{1000}{60\text{Hz}} \times 1000 = 11{,}111.1 \mu\text{s}$$

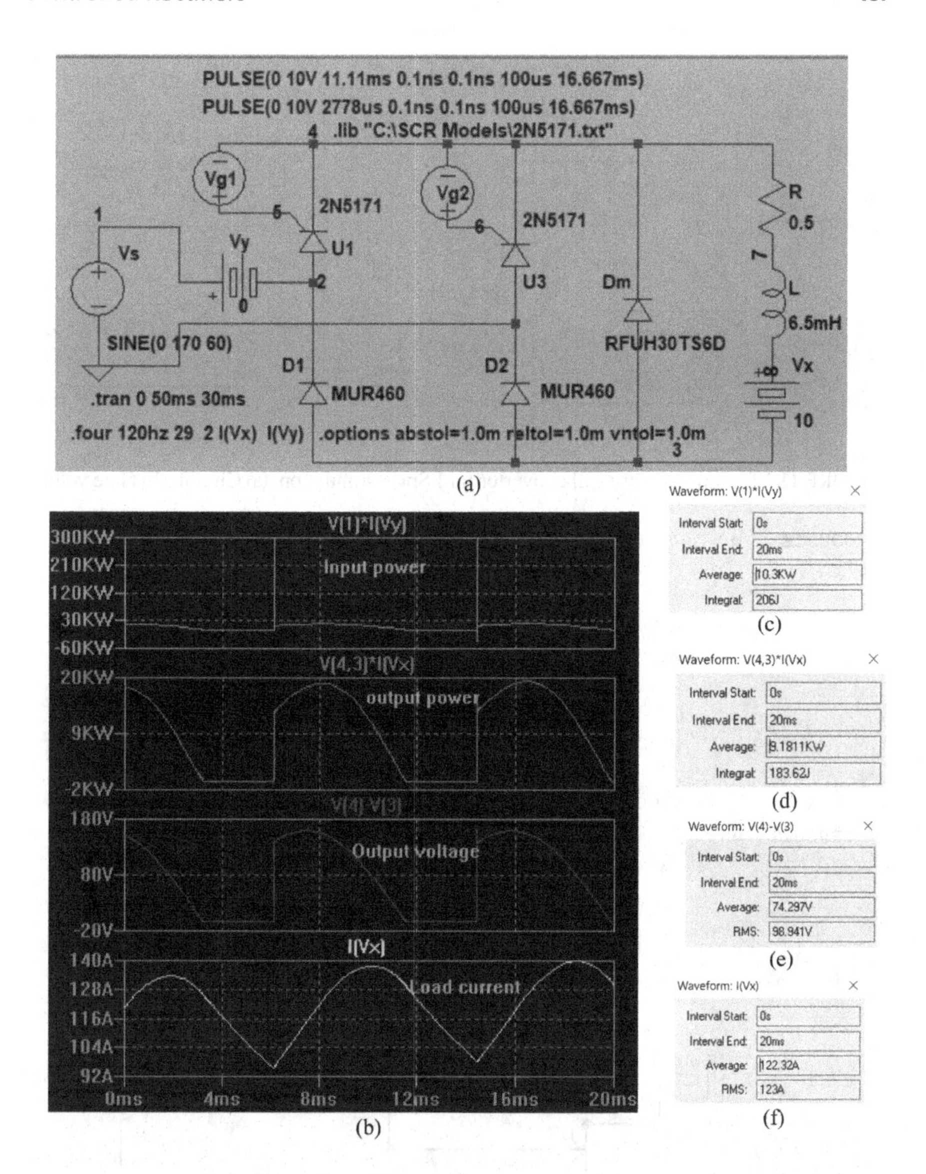

FIGURE 11.11 LTspice schematic for Example 11.2. (a) Tspice Schematic. (b) Output waveforms. (c) Input power. (d) Output power. (e) Load voltage. (f) Load current.

The PSpice schematic with SCRs is shown in Figure 11.13. Note that the voltage-controlled voltage sources E_1, E_2, E_3, and E_4 are connected to provide the isolation between the low signal gating signals from the power circuit. The model name of the SCR 2N1595 is changed to SCRMOD whose subcircuit definition is listed in Section 11.2. Varying the delay cycle can vary the output voltage. The supply frequency {FREQ} and the duty cycle {DELAY_ANGLE} are defined as variables. The model parameters for the freewheeling diode are as follows:

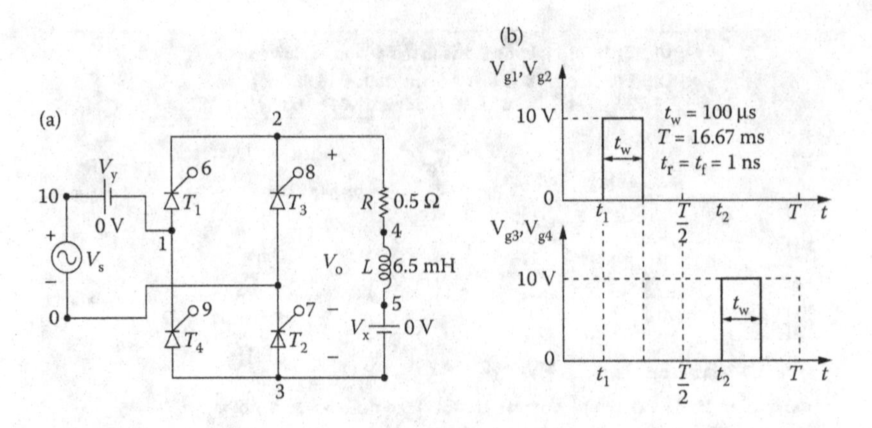

FIGURE 11.12 Single-phase full converter for PSpice simulation. (a) Circuit. (b) Gate voltage.

```
SOURCE    ■    VS 10 0 SIN(0169.7V60HZ)
               .PARAM DELAY_ANGLE=60 FREQ=60Hz
               Vg1 6 2 PULSE (0 1 {{Delay_Angle}/
               {360*{Freq}}} 1ns 1ns 100us {1/{Freq}})
               Vg2 7 0 PULSE (0 1 {{Delay_Angle}/
               {360*{Freq}}} 1ns 1ns 100us {1/{Freq}})
               Vg3 8 2 PULSE (0 1 {{{Delay_Angle} + 180}/
               {360*{Freq}}} 1ns 1ns 100us {1/{Freq}})
               Vg4 9 1 PULSE (0 1 {{{Delay_Angle} + 180}/
               {360*{Freq}}} 1ns 1ns 100us {1/{Freq}})
CIRCUIT   ■■   R  2 4 0.5
               L4 5 6.5MH
               VX5 3 DC 10V ; Load battery voltage
```

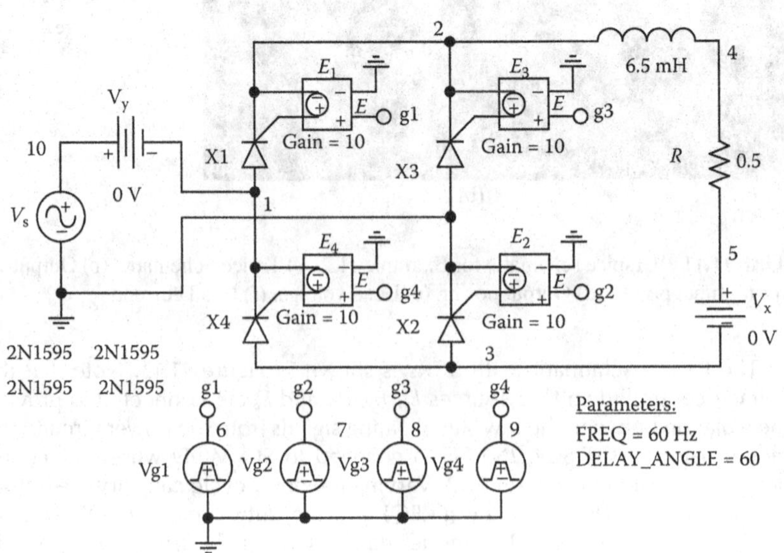

FIGURE 11.13 PSpice schematic for Example 11.3.

```
VY10 1 DC 0V   ; Voltage source to measure
supply current
*Subcircuit calls for thyristor model:
XT1 1 6 2 SCRMOD    ; Thyristor T1
```

The listing of the circuit file is as follows:

Single-Phase Full-Bridge Converter

```
          XT3 0 8 2 SCRMOD    ; Thyristor T2
          XT2 3 7 0 SCRMOD    ; Thyristor T3
          XT4 3 9 1 SCRMOD    ; Thyristor T4
          *Subcircuit SCR, which is missing, must; be inserted
ANALYSIS ■■■ .TRAN 1US 50MS 33.33MS   ;Transient analysis
          .PROBE                 ;Graphics post-processor
          .OPTIONS ABSTOL=1.00N RELTOL=1.0M VNTOL=0.01
          ITL5=20000
          .FOUR 60HZ I(VY)
     .END
```

Note the following:

a. The PSpice plots of the instantaneous output voltage V(2, 3), the input current I(VY), and the load current I(VX) are shown in Figure 11.14.The instantaneous output voltage can be negative, but the load current is always positive. Note that the frequency of the output ripple voltage is two times the supply frequency. That is, $f_o = 2f = 120$ Hz. The output current is a DC waveform, while the input current is an AC waveform. The input current can be derived from

$$i_{\text{in}}(t) = i_o(\omega t = 0 \text{ to } \pi) - i_o(\omega t = \pi \text{ to } 2\pi)$$

b. To find the input power factor, we need to find the Fourier series of the current through the voltage source VY.

Fourier Components of Transient Response I(VY)

DC Component = 1.200052E–02

Harmonic Number	Frequency (Hz)	Fourier Component	Normalized Component	Phase (Deg)	Normalized Phase (Deg)
1	6.000E+01	1.113E+02	1.000E+00	–6.211E+01	0.000E+00
2	1.200E+02	8.525E–01	7.662E–03	2.244E+01	8.456E+01
3	1.800E+02	1.633E+01	1.468E–01	–1.744E+02	–1.123E+02
4	2.400E+02	3.016E–01	2.711E–03	–6.781E+01	–5.696E+00
5	3.000E+02	1.007E+01	9.046E–02	6.171E+01	1.238E+02
6	3.600E+02	2.470E–01	2.220E–03	9.033E+01	1.524E+02
7	4.200E+02	7.179E+00	6.452E–02	–5.831E+01	3.806E+00
8	4.800E+02	3.580E–01	3.218E–03	–3.425E+00	5.869E+01
9	5.400E+02	5.454E+00	4.902E–02	1.796E+02	2.417E+02

Total harmonic distortion = 1.907363E+01%

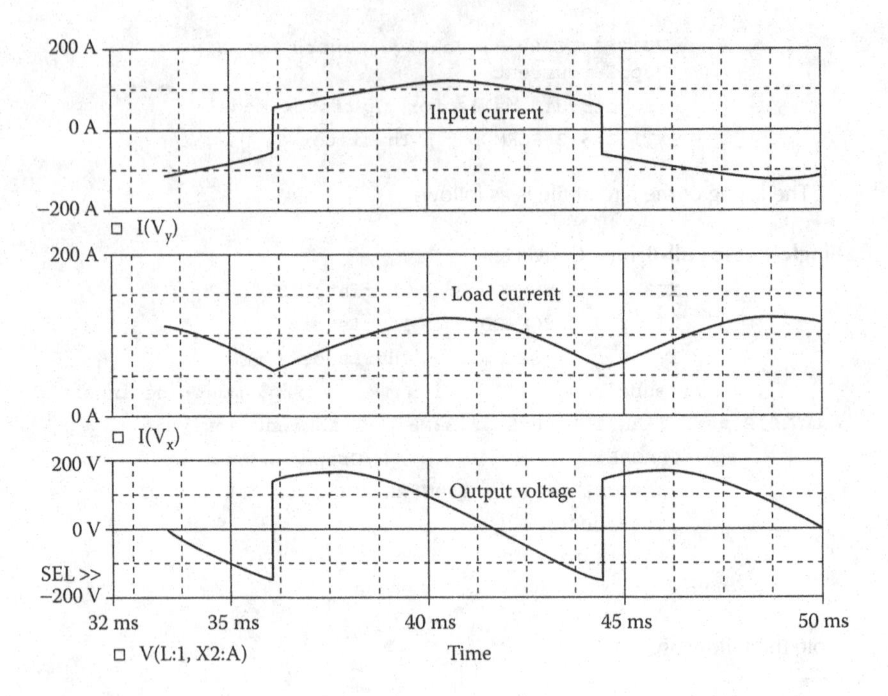

FIGURE 11.14 Plots for Example 11.3 for delay angle = 60.

From the Fourier components, we get

Total harmonic distortion of input current THD = 19.07% = 0.1907
Displacement angle ϕ_1 = −62.1°
Displacement factor DF = cos = cos(−62.1) = 0.468 (lagging)
From Equation 7.3, the input power factor is

$$PF = \frac{1}{(1+THD^2)^{1/2}} \cos\phi_1 = \frac{1}{(1+0.1907^2)^{1/2}} \times 0.468 = 0.46\,(\text{lagging})$$

Note: for an inductive load, the output voltage can be either positive or negative. The input power factor, however, is lower than that of the semiconverter at the same delay angle $\alpha = 60°$.

The average output voltage of a single-phase controlled rectifier (for a delay angle of α) is given by [1]

$$V_{o(av)} = \frac{2V_m}{\pi}\cos\alpha$$

LTspice: LTspice schematic for Example 11.3 is shown in Figure 11.15(a). The plots of the load current, the input power, the output voltage, and the load power are shown in Figure 10.15(b). The average input power as shown in Figure 11.15(c) is 5793.5W, and the average output power as shown in Figure 11.15(d) is 5018.8 W.

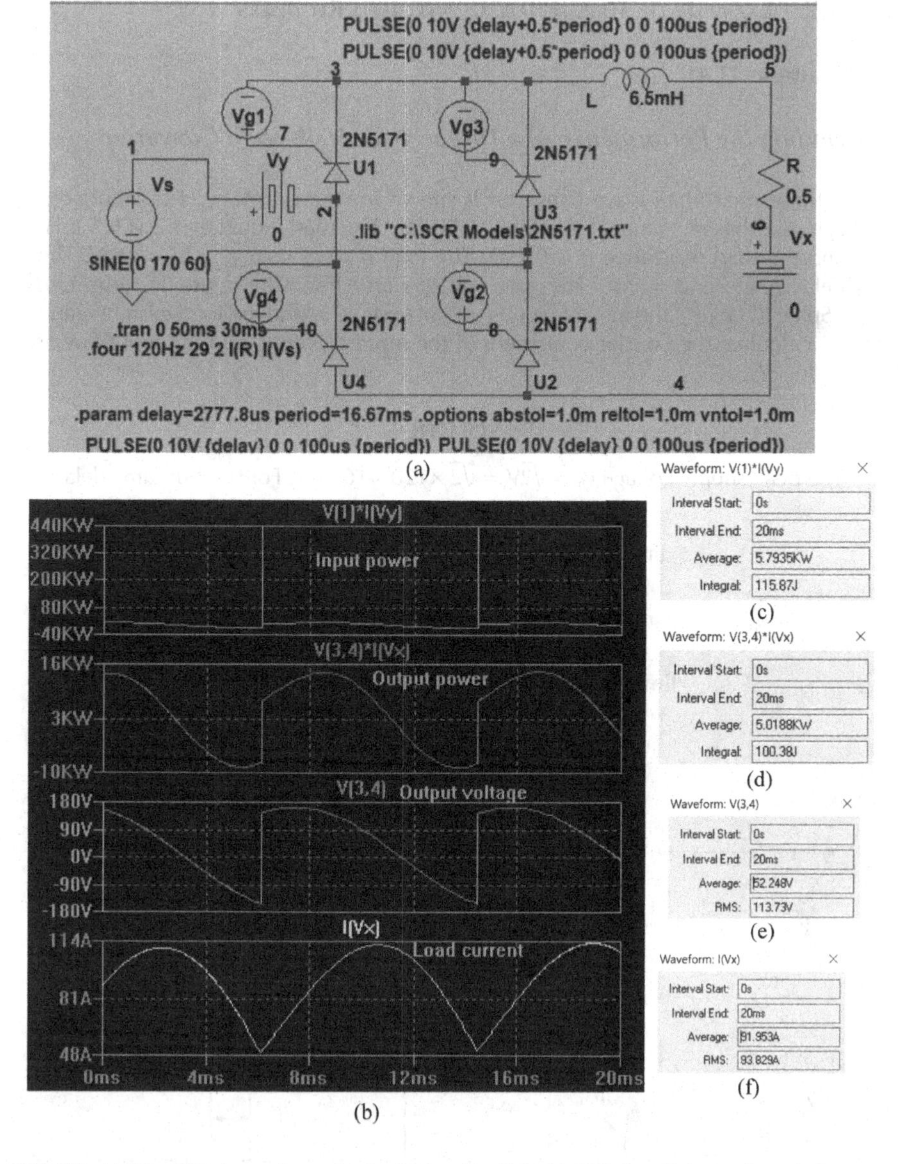

FIGURE 11.15 LTspice schematic for Example 11.3. (a) LTspice Schematic. (b) Output waveforms. (c) Input power. (d) Output power. (e) Load voltage. (f) Load current.

The rms value of the load current I(Vx) is 113.73A and the rms value of the output voltage is 93.83 V. We can find the power efficiency of the converter as $\eta = \frac{P_{out}}{P_{in}} = \frac{5018.8}{5718.9} = 87.76\%$

Note: For a low output power, the losses in the semiconductor devices are significant compared to the output power and the efficiency becomes low. There is no power loss in the load inductor.

11.4.2 Examples of Three-Phase Controlled Rectifiers

Example 11.4:

Finding the Performance of a Three-Phase Half-Wave Converter

A three-phase half-wave converter is shown in Figure 11.16(a). The input voltage per phase has a peak of 169.7 V, 50 Hz. The load inductance L is 6.5 mH, and the load resistance R is 0.5 Ω. The load battery voltage is V_x = 10 V. The delay angle is α = 90°. The gate voltages are shown in Figure 11.16(b). Use PSpice to (a) plot the instantaneous output voltage v_o and the load current i_o and (b) calculate the Fourier coefficient of the input current i_s and the input power factor PF.

SOLUTION

The peak supply voltage $V_m = \sqrt{2}V_s = \sqrt{2} \times 120 = 169.7\text{V}$. For α = 90°, time delay

$$\text{time delay}\, t_1 = \frac{90}{360} \times \frac{1000}{60\text{Hz}} \times 1000 = 4166.7\,\mu\text{s}$$

$$\text{time delay}\, t_2 = \frac{210}{360} \times \frac{1000}{60\text{Hz}} \times 1000 = 9722.2\,\mu\text{s}$$

$$\text{time delay}\, t_3 = \frac{330}{360} \times \frac{1000}{60\text{Hz}} \times 1000 = 15,277.8\,\mu\text{s}$$

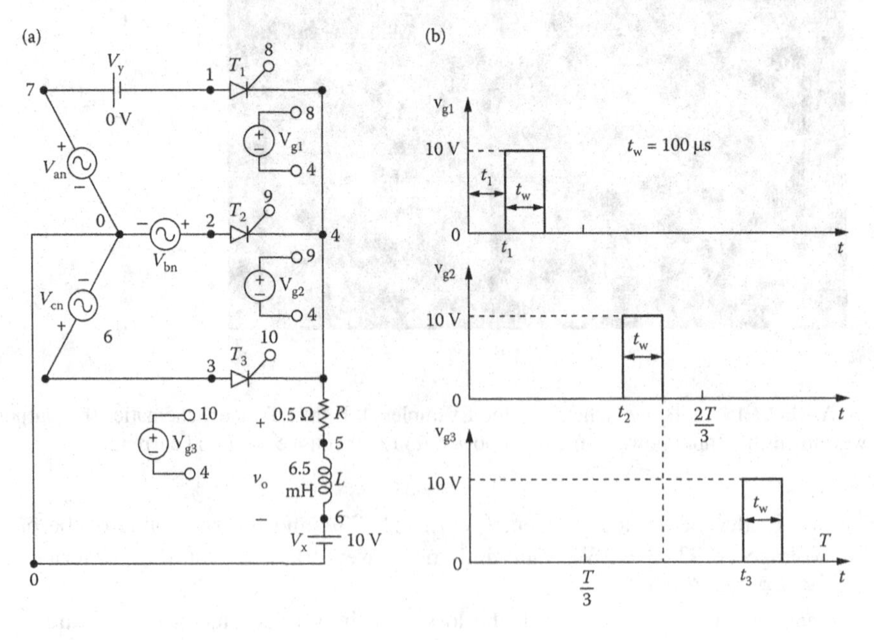

FIGURE 11.16 Three-phase half-wave converter for PSpice simulation. (a) Circuit. (b) Gate voltage.

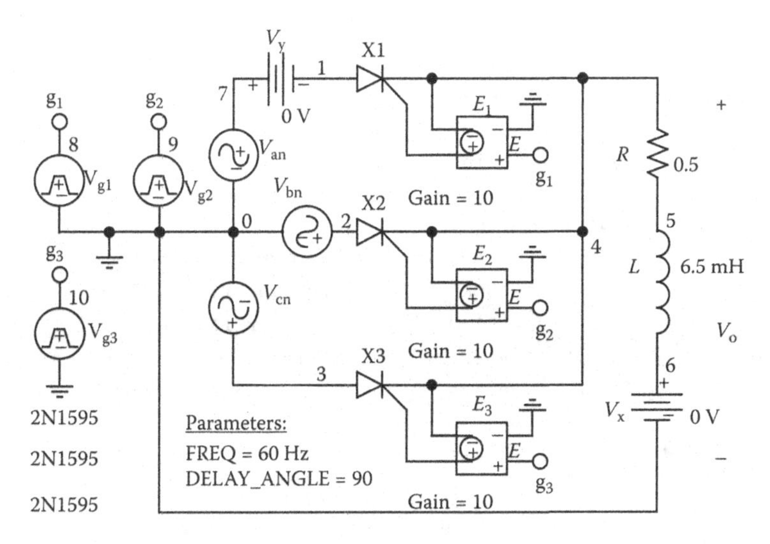

FIGURE 11.17 PSpice schematic for Example 11.4.

The PSpice schematic with SCRs is shown in Figure 11.17. Note that the voltage-controlled voltage sources E_1, E_2, and E_3 are connected to provide the isolation between the low signal gating signals from the power circuit. The model name of the SCR 2N1595 is changed to SCRMOD whose subcircuit definition is listed in Section 11.2. Varying the delay cycle can vary the output voltage. The supply frequency {FREQ} and the duty cycle {DELAY_ANGLE} are defined as variables. The model parameters for the freewheeling diode are as follows:

```
.MODEL DMD D(IS=2.22E-15 BV=1200V CJO=0PF TT=0US)) for BJTs
```

The listing of the circuit file is as follows:

Three-Phase Half-Wave Converter

```
SOURCE  ■  Van 7  0 SIN(0 169.7V 60HZ)
           Vbn 2  0 SIN(0 169.7V 60HZ 00-120DEG)
           Vcn 3  0 SIN(0 169.7V 60HZ 00-240DEG)
           Vg1 8  4 PULSE (0V 10V  4166.7US 1NS
           1NS 100US 16666.7US)
           Vg2 9  4 PULSE (0V 10V  9722.2US 1NS
           1NS 100US 16666.7US)
           Vg3 10 4 PULSE (0V 10V 15277.8US 1NS
           1NS 100US 16666.7US)
           .PARAM  Freq=60Hz Delay_Angle=90
           Vg1 8  4 PULSE (0 1 {{Delay_Angle}/
           {360*{Freq}}} 1ns 1ns 100us {1/{Freq}})
           Vg2 9  4 PULSE (0 1 {{{Delay_Angle} + 120}/
           {360*{Freq}}} 1ns 1ns 100us {1/{Freq}})
```

```
          Vg3 10 4 PULSE (0 1 {{{Delay_Angle} + 240}/
          {360*{Freq}}} 1ns 1ns 100us {1/{Freq}})
CIRCUIT ■■ R   4 5 0.5
          L   5 6 6.5MH
          VX  6 0 DC 10V    ; Load battery voltage
          VY  7 1 DC 0V     ; Voltage source to measure
          supply current
          * Subcircuit calls for thyristor model:
          XT1 1 8 4 SCRD      ; Thyristor T1
          XT2 2 9 4 SCRD      ; Thyristor T3
          XT3 3 10 4 SCRD     ; Thyristor T2
          * Subcircuit SCRD, which is missing, must be
          * inserted.
ANALYSIS ■■■ .TRAN 100US 50MS 33.33MS  ; Transient analysis
          .PROBE                 ; Graphics post-processor
          .OPTIONS ABSTOL=1.00N RELTOL=1.0M
          VNTOL=1.0M ITL5=20000      ; Convergence
          .FOUR 60HZ I(VY)
       .END
```

Note the following:

a. The PSpice plots of the instantaneous output voltage V(4) and the load current I(VX) are shown in Figure 11.18. The average load current is greater than that of a single-phase converter. Note that the frequency of the output ripple voltage is three times the supply frequency. That is, $f_o = 3f = 180$ Hz.

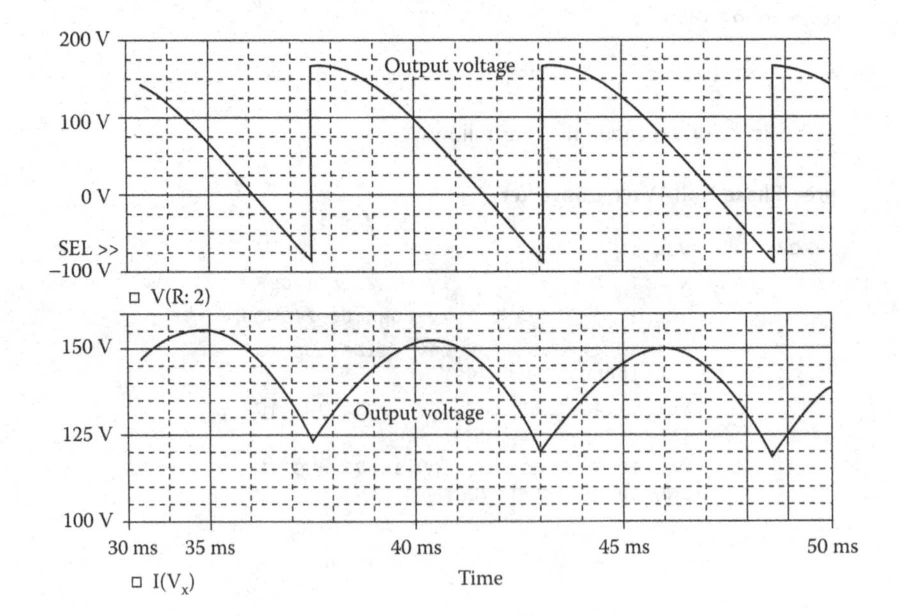

FIGURE 11.18 Plots for Example 11.4 for delay angle α = 90°.

b. The Fourier series of the current through voltage source VY is as follows:

Fourier Components of Transient Response I(VY)

DC Component = 4.048497E+01

Harmonic Number	Frequency (Hz)	Fourier Component	Normalized Component	Phase (Deg)	Normalized Phase (Deg)
1	6.000E+01	6.793E+01	1.000E+00	−6.001E+01	0.000E+00
2	1.200E+02	3.650E+01	5.373E−01	1.499E+02	2.099E+02
3	1.800E+02	4.317E+00	6.354E−02	−2.866E+00	5.715E+01
4	2.400E+02	1.315E+01	1.936E−01	3.128E+01	9.129E+01
5	3.000E+02	1.216E+01	1.790E−01	−1.188E+02	−5.882E+01
6	3.600E+02	1.108E+00	1.631E−02	9.453E+01	1.545E+02
7	4.200E+02	7.661E+00	1.128E−01	1.206E+02	1.806E+02
8	4.800E+02	7.409E+00	1.091E−01	−2.886E+01	3.116E+01
9	5.400E+02	5.800E−01	8.538E−03	−1.720E+02	−1.120E+02

Total harmonic distortion = 6.222461E+01%

From the Fourier components, we get

DC input current $I_{in(DC)} = 40.48$ A
RMS fundamental input current $I_{1(RMS)} = I_{1(RMS)} = 67.93/\sqrt{2} = 48.03A$
Total harmonic distortion of input current $THD = 62.22\% = 0.6222$
RMS harmonic current $I_{h(RMS)} = I_{1(RMS)} \times THD = 48.03 \times 0.6222 = 29.88$ A
RMS input current $I_s = [I_{in(DC)}^2 + I_{1(RMS)}^2 + I_{h(RMS)}^2]^{1/2}$
$= (40.48^2 + 48.03^2 + 29.88^2)^{1/2} = 69.56A$
Displacement factor $\phi_1 = -60.01°$
Displacement factor $DF = \cos\phi_1 = \cos(-60.01) = 0.5$ (lagging)
Thus, the input power factor is

$$PF = \frac{I_{1(RMS)}}{I_s}\cos\phi_1 = \frac{48.03}{69.56} \times 0.5 = 0.3452(\text{lagging})$$

Note: The input current from each phase contains a DC component, thereby causing a low power factor and DC saturation problem for the input-side transformer.

LTspice: LTspice schematic for Example 11.4 is shown in Figure 11.19(a). The plots of the load current, the input power, the output voltage, and the load power are shown in Figure 10.19(b). The average input power as shown in Figure 11.19(c) is 7222.4 W, the average output power as shown in Figure 11.19(d) is 7621.6. The rms value of the load current I(Vx) is 122.41A and the rms value of the output voltage is 104.36 V. We can find the power dissipated in the load resistor of $R = 0.5\ \Omega$ as $P_L = I^2R = 122.41^2 \times 0.5 = 7492.1W$ which is close to the average load power of 7621.6 W as shown in Figure 11.19(d)

Note: Note that the input power is lower than the output power which is not practical. It could be due to the averaging instantaneous power over the initial transient period. For a low output power, the losses in the semiconductor devices are significant compared to the output power and the efficiency becomes low. There is no power loss in the load inductor.

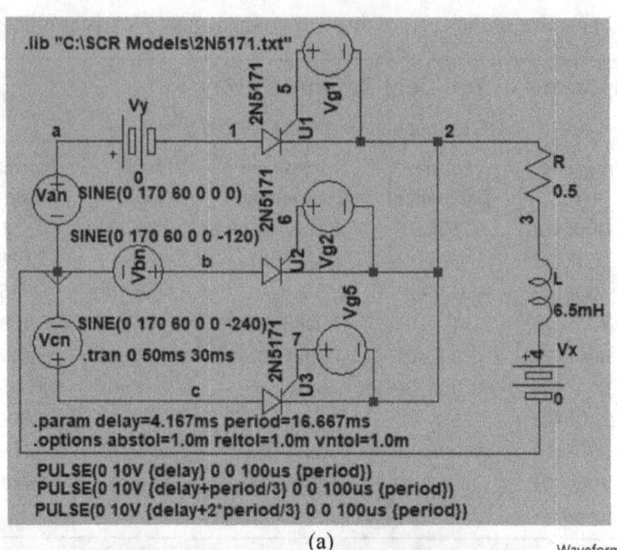

(a)

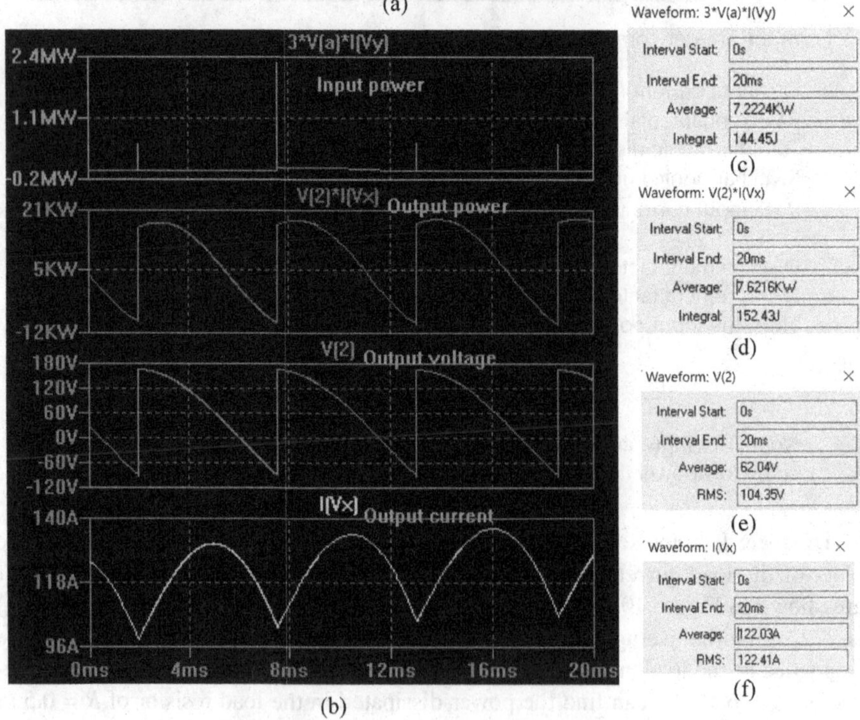

(b)

(c)

(d)

(e)

(f)

FIGURE 11.19 LTspice schematic for Example 11.4. (a) LTspice Schematic. (b) Output waveforms. (c) Input power. (d) Output power. (e) Load voltage. (f) Load current.

Example 11.5:

Finding the Performance of a Three-Phase Semiconverter

A three-phase semiconverter is shown in Figure 11.20(a). The input voltage per phase has a peak of 169.7 V, 60 Hz. The load inductance L is 6.5 mH, and the load resistance R is 0.5 Ω. The load battery voltage is $V_x = 10$ V. The delay angle is $\alpha = 90°$. The gate voltages are shown in Figure 11.20(b). Use PSpice to (a) plot the instantaneous output voltage v_o, the load current i_o, and the input current i_s and (b) calculate the Fourier coefficients of the input current i_s and the input power factor *PF*.

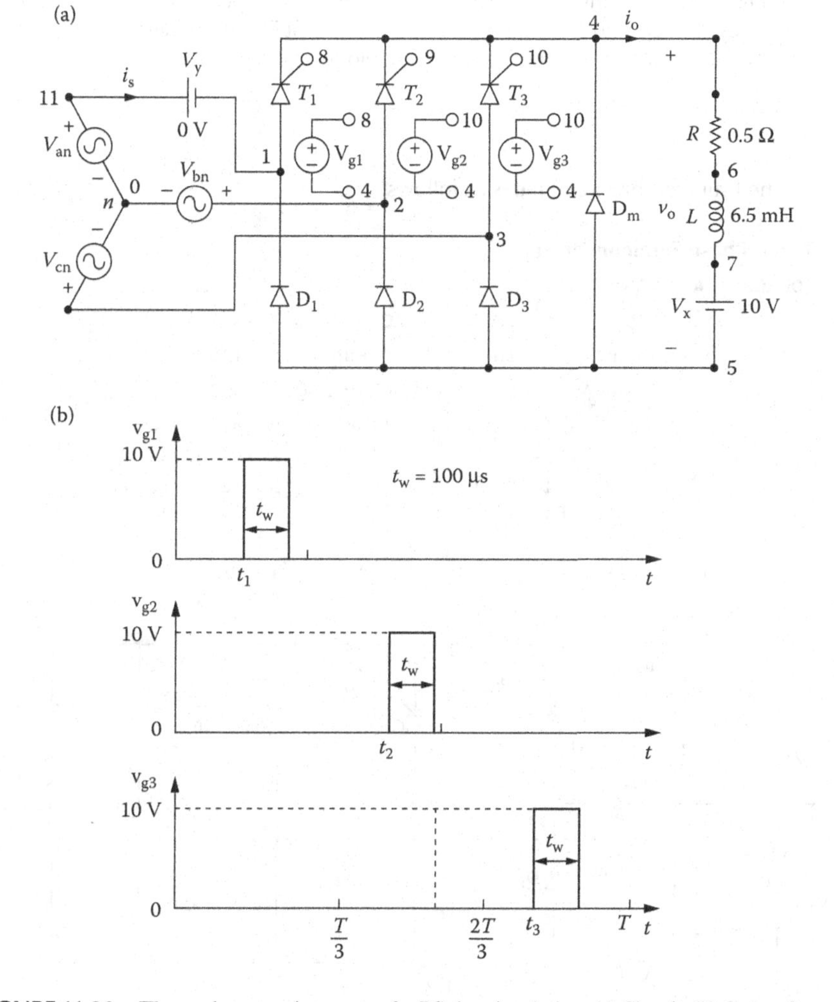

FIGURE 11.20 Three-phase semiconverter for PSpice simulation. (a) Circuit. (b) Gate voltages.

SOLUTION

The peak supply voltage $V_m = 169.7$ V. FOR $\alpha = 90°$,

$$\text{time delay } t_1 = \frac{90}{360} \times \frac{1000}{60\text{Hz}} \times 1000 = 4166.7\,\mu s$$

$$\text{time delay } t_2 = \frac{210}{360} \times \frac{1000}{60\text{Hz}} \times 1000 = 9722.2\,\mu s$$

$$\text{time delay } t_3 = \frac{330}{360} \times \frac{1000}{60\text{Hz}} \times 1000 = 15,277.8\,\mu s$$

The PSpice schematic with SCRs is shown in Figure 11.21. Note that the voltage-controlled voltage sources E_1, E_2, and E_3 are connected to provide the isolation between the low signal gating signals from the power circuit. The model name of the SCR 2N1595 is changed to SCRMOD, whose subcircuit definition is listed in Section 11.2. Varying the delay cycle can vary the output voltage. The supply frequency {FREQ} and the duty cycle {DELAY_ANGLE} are defined as variables. The model parameters for the freewheeling diode are as follows:

```
.MODEL DMD D(IS=2.22E-15 BV=1200V CJO=0PF TT=0US)) for
Diodes
```

The listing of the circuit file is as follows:

Three-Phase Semiconverter

```
SOURCE  ■   Van 11 0 SIN(0 169.7V 60HZ)
            Vbn  2 0 SIN(0 169.7C 60HZ00 – 120DEG)
            Vcn  3 0 SIN(0 169.7C 60HZ00 – 240DEG)
            .PARAM Freq = 60Hz Delay_Angle = 90
            Vg1  8 4 PULSE (0 1 {{Delay_Angle}/
            {360*{Freq}}} 1ns 1ns 100us {1/{Freq}})
            Vg2  9 4 PULSE (0 1 {{{Delay_Angle} + 120}/
            {360*{Freq}}} 1ns 1ns 100us {1/{Freq}})
            Vg3 10 4 PULSE (0 1 {{{Delay_Angle} + 240}/
            {360*{Freq}}} 1ns 1ns 100us {1/{Freq}})
```

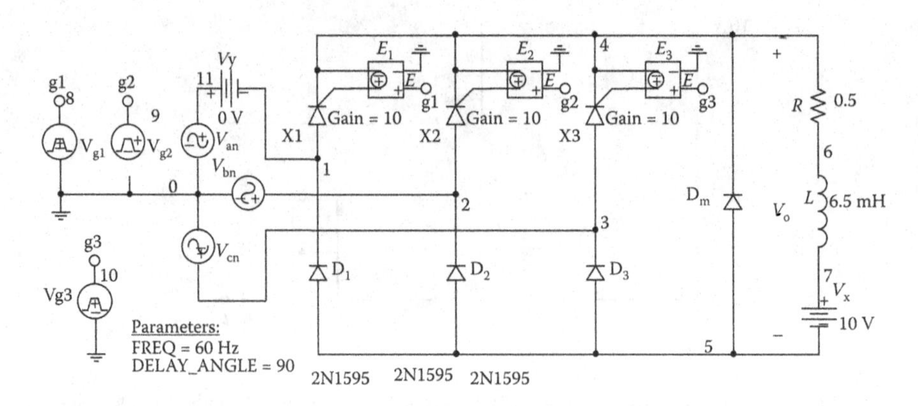

FIGURE 11.21 PSpice schematic for Example 11.5.

```
CIRCUIT ■■   R    4 6 0.5
             L    6 7 6.5MH
             VX   7 5 DC 10V ; Load battery voltage
             VY 11 1 DC 0V   ; Voltage source to measure
             supply current
             D1   5 1 DMOD
             D2   5 2 DMOD
             D3   5 3 DMOD
             DM   5 4 DMOD
             .MODEL DMOD D(IS = 2.2E - 15 BV = 1200V TT = 0
             CJO = 1PF)   ; Diode model parameters
             *Subcircuit calls for thyristor model:
             XT1 1 8 4 SCRMOD   ; Thyristor T1
             XT2 2 9 4 SCRMOD   ; Thyristor T3
             XT3 3 10 4 SCRMOD  ; Thyristor T2
             *Subcircuit SCR, which is missing, must be
             *inserted.
ANALYSIS ■■■ .TRAN 1US 50MS 33.33MS ; Transient analysis
             .PROBE                 ; Graphics post-processor
             .OPTIONS ABSTOL = 100.U RELTOL = 0.01
             VNTOL = 0.01 ITL5 = 20000
             .FOUR 60HZ I(VY)
         .END
```

Note the following:

a. The PSpice plots of the instantaneous output voltage V(4,5), the load current I(VX), and the input current I(VY) are shown in Figure 11.22. The ripple contents on the output voltage and output current are lower than those of a single-phase converter.
b. The Fourier series of the input current is as follows:

Fourier Components of Transient Response I(VY)

DC Component = 8.208448E–01

Harmonic Number	Frequency (Hz)	Fourier Component	Normalized Component	Phase (Deg)	Normalized Phase (Deg)
1	6.000E+01	3.753E+02	1.000E+00	–3.022E+01	0.000E+00
2	1.200E+02	1.943E+02	5.177E–01	1.195E+02	1.498E+02
3	1.800E+02	1.153E–01	3.073E–04	8.665E+01	1.169E+02
4	2.400E+02	8.585E+01	2.287E–01	6.213E+01	9.235E+01
5	3.000E+02	7.424E+01	1.978E–01	–1.483E+02	–1.180E+02
6	3.600E+02	5.945E–02	1.584E–04	1.676E+02	1.978E+02
7	4.200E+02	4.864E+01	1.296E–01	1.519E+02	1.822E+02
8	4.800E+02	4.660E+01	1.242E–01	–5.781E+01	–2.759E+01
9	5.400E+02	4.144E–02	1.104E–04	–1.095E+02	–7.932E+01

Total harmonic distortion = 6.258354E+01%

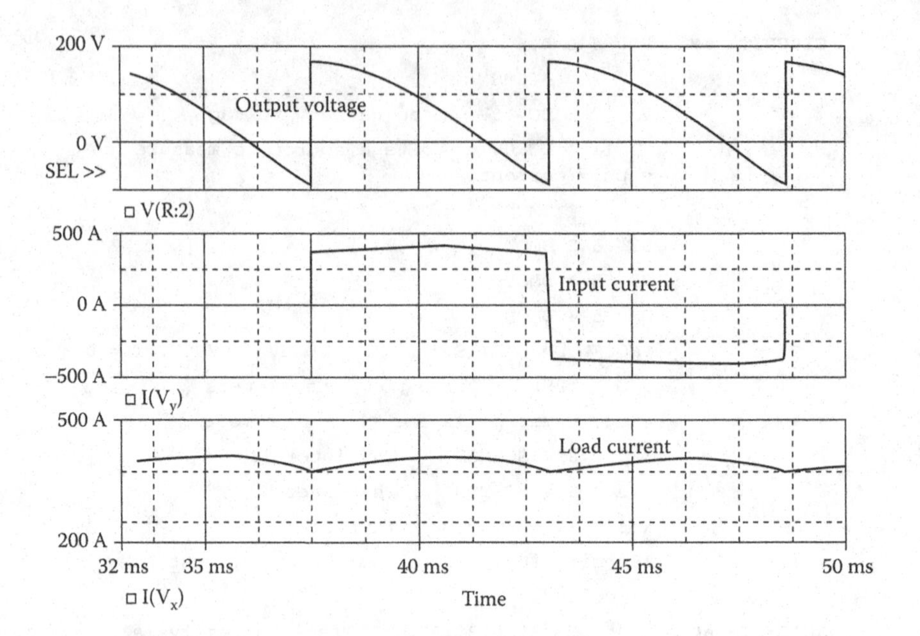

FIGURE 11.22 Plots for Example 11.5 for delay angle = 90°.

From the Fourier components, we get

Total harmonic distortion of input current $THD = 62.58\% = 0.6258$
Displacement angle $\phi_1 = -30.2°$
Displacement factor $DF = \cos\phi_1 = \cos(-30.2) = 0.864$ (lagging)

The input power factor is

$$PF = \frac{1}{(1+THD^2)^{1/2}}\cos\phi_1 = \frac{1}{(1+0.6258^2)^{1/2}} \times 0.864 = 0.732(\text{lagging})$$

Note: The freewheeling diode D_m reduces the ripple of the load current, thereby improving the input power factor of the converter.

LTspice: LTspice schematic for Example 11.5 is shown in Figure 11.23(a). The plots of the load current, the input power, the output voltage, and the load power are shown in Figure 10.23(b). The average input power as shown in Figure 11.23(c) is 78124 W, and the average output power as shown in Figure 11.23(d) is 61194 W. The rms value of the load current I(Vx) is 334.38A and the rms value of the output voltage is 204.3 V. We can find the power efficiency of the converter as $\eta = \frac{P_{out}}{P_{in}} = \frac{61194}{78124} = 78.32\%$

Note: For a low output power, the losses in the semiconductor devices are significant compared to the output power and the efficiency becomes low. There is no power loss in the load inductor.

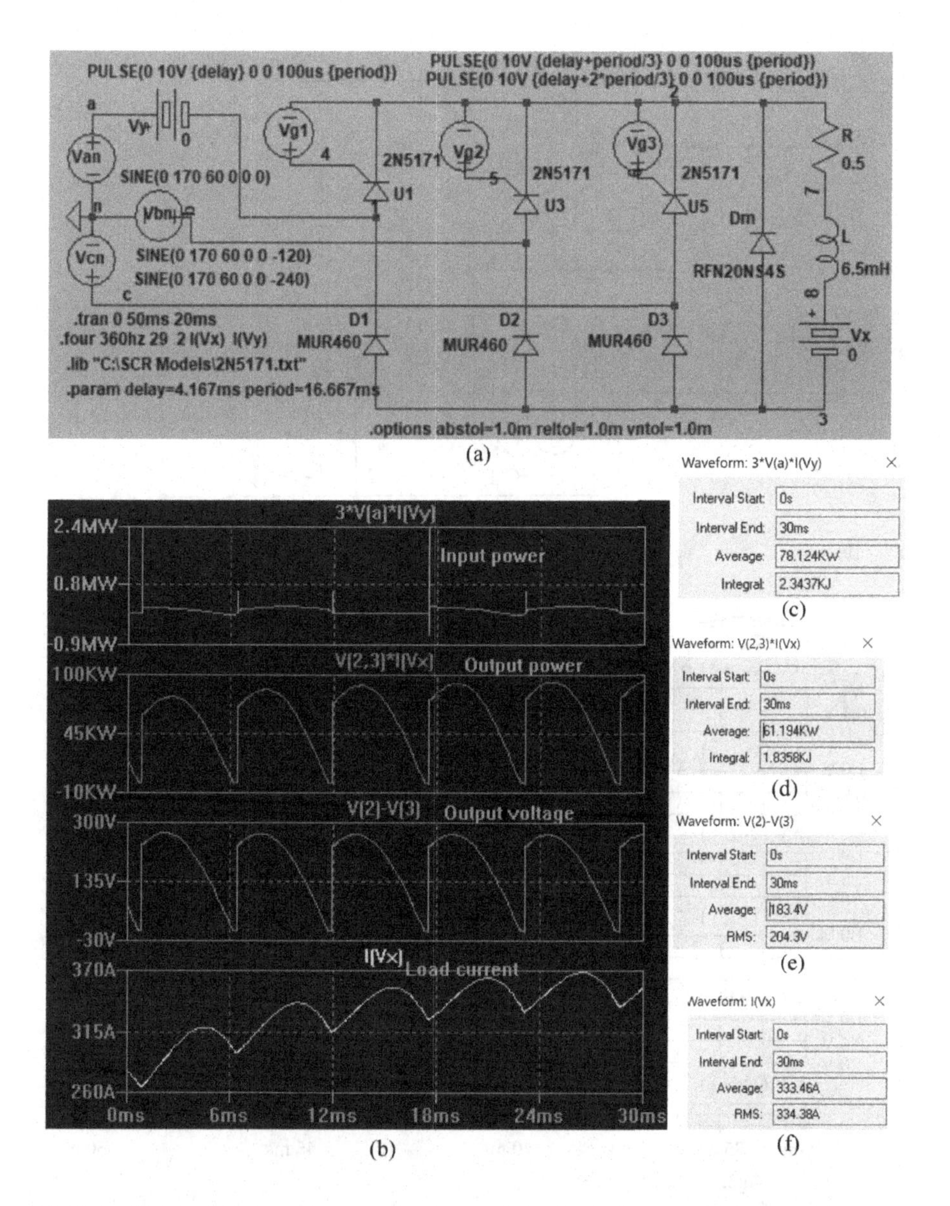

FIGURE 11.23 LTspice schematic for Example 11.5. (a) LTspice Schematic. (b) Output waveforms. (c) Input power. (d) Output power. (e) Load voltage. (f) Load current.

Example 11.6:

Finding the Performance of a Three-Phase Full-Bridge Converter

A three-phase full-bridge converter is shown in Figure 11.24(a). The input voltage per phase has a peak of 169.7 V, 60 Hz. The load inductance L is 6.5 mH, and the

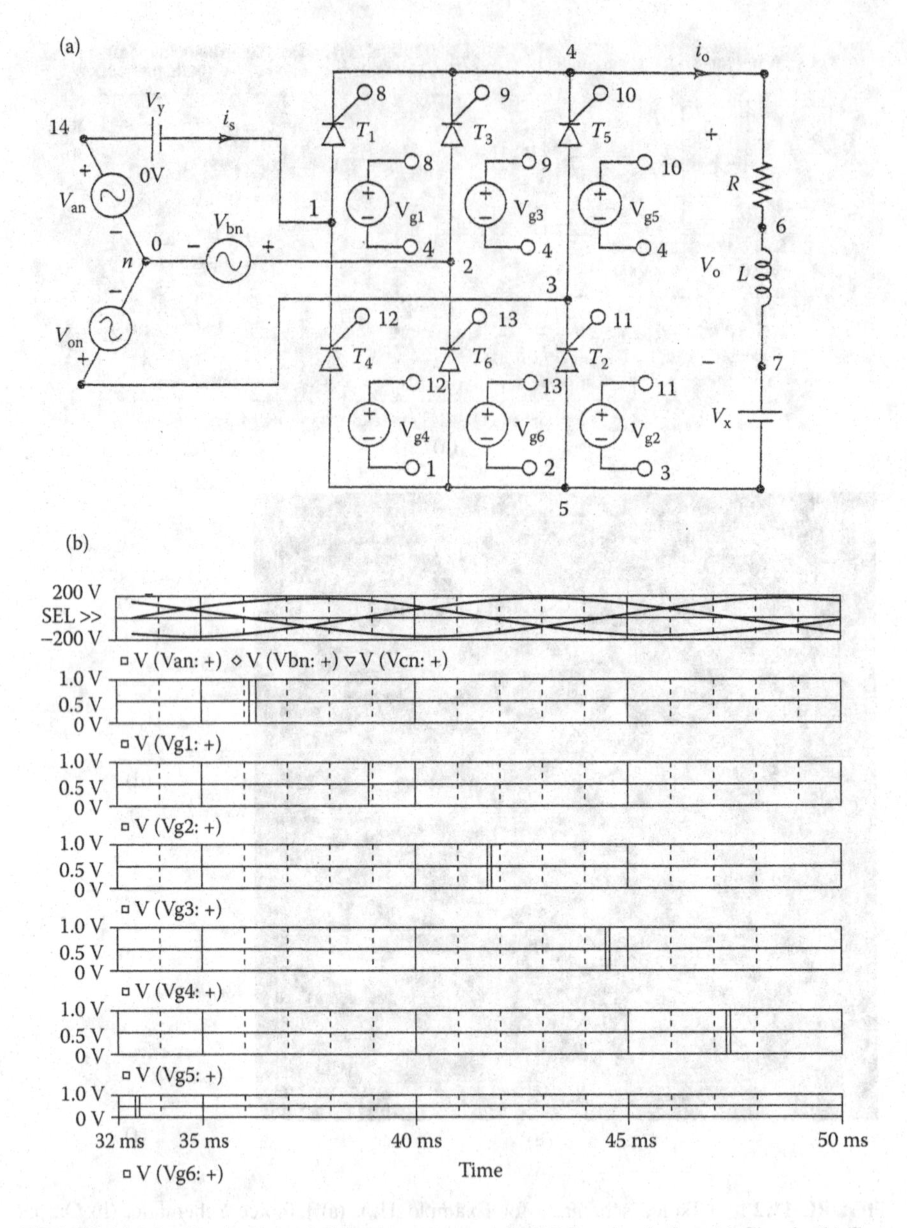

FIGURE 11.24 Three-phase full-bridge converter for PSpice simulation. (a) Circuit. (b) Gate voltages.

load resistance R is 0.5 Ω. The load battery voltage is $V_x = 10$ V. The delay angle is $\alpha = 90°$. The gate voltages are shown in Figure 11.24(b). Use PSpice to (a) plot the instantaneous output voltage v_o, the load current i_o, and the input current i_s and (b) calculate the Fourier coefficients of the input current i_s and the input power factor PF.

SOLUTION

For $\alpha = 90°$,

$$\text{time delay } t_1 = \frac{90}{360} \times \frac{1000}{60\text{Hz}} \times 1000 = 4166.7\,\mu\text{s}$$

$$\text{time delay } t_3 = \frac{210}{360} \times \frac{1000}{60\text{Hz}} \times 1000 = 9722.2\,\mu\text{s}$$

$$\text{time delay } t_5 = \frac{330}{360} \times \frac{1000}{60\text{Hz}} \times 1000 = 15{,}277.8\,\mu\text{s}$$

$$\text{time delay } t_2 = \frac{150}{360} \times \frac{1000}{60\text{Hz}} \times 1000 = 6944.4\,\mu\text{s}$$

$$\text{time delay } t_4 = \frac{270}{360} \times \frac{1000}{60\text{Hz}} \times 1000 = 12{,}500\,\mu\text{s}$$

$$\text{time delay } t_6 = \frac{30}{360} \times \frac{1000}{60\text{Hz}} \times 1000 = 1388.9\,\mu\text{s}$$

The PSpice schematic with SCRs is shown in Figure 11.25 Note that the voltage-controlled voltage sources E_1, E_2, E_3, E_4, E_5, and E_6 are connected to provide the isolation between the low signal gating signals from the power circuit. The model name of the SCR 2N1595 is changed to SCRMOD, whose subcircuit definition is listed in Section 11.2. Varying the delay cycle can vary the output voltage.

The supply frequency {FREQ} and the duty cycle {DELAY_ANGLE} are defined as variables. The model parameters for the freewheeling diode are as follows:

```
.MODEL DMD D(IS=2.22E-15 BV=1200V CJO=0PF TT=0US)) for
Diodes
```

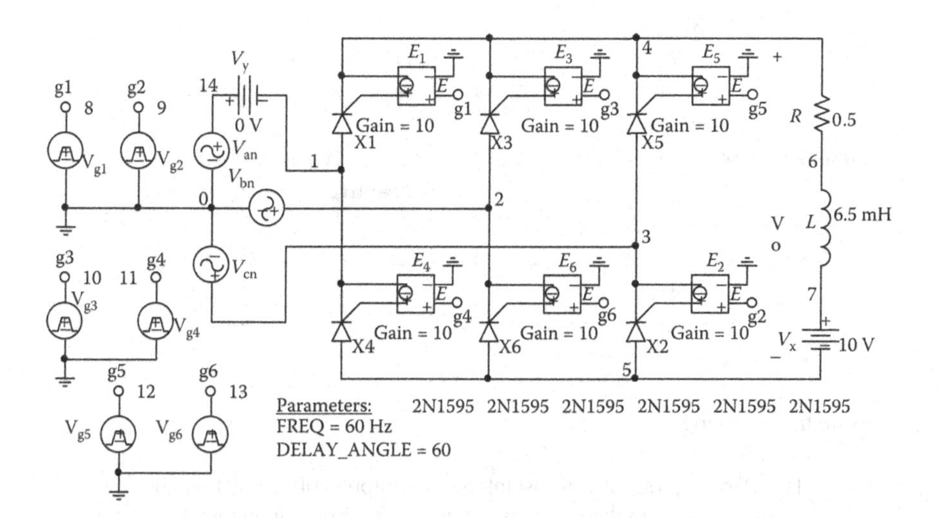

FIGURE 11.25 PSpice schematic for Example 11.6.

The listing of the circuit file is as follows:

Three-Phase Full-Bridge Converter

```
SOURCE   ■     Van 14 0 SIN(0 169.7V 60HZ)
               Vbn  2 0 SIN(0 169.7C 60HZ00 - 120DEG)
               Vcn  3 0 SIN(0 169.7C 60HZ00 - 240DEG)
               .PARAM DELAY_ANGLE = 60  FREQ = 60Hz
               Vg1  8 4 PULSE (0 1 {{Delay_Angle}/
               {360*{Freq}}} 1ns 1ns 100us {1/{Freq}})
               Vg3  9 4 PULSE (0 1 {{{Delay_Angle} + 120}/
               {360*{Freq}}} 1ns 1ns 100us {1/{Freq}})
               Vg5 10 4 PULSE (0 1 {{{Delay_Angle} + 240}/
               {360*{Freq}}} 1ns 1ns 100us {1/{Freq}})
               Vg2 11 3 PULSE (0 1 {{{Delay_Angle} + 60}/
               {360*{Freq}}} 1ns 1ns 100us {1/{Freq}})
               Vg4 12 1 PULSE (0 1 {{{Delay_Angle} + 180}/
               {360*{Freq}}} 1ns 1ns 100us {1/{Freq}})
               Vg6 13 2 PULSE (0 1 {{{Delay_Angle} + 300}/
               {360*{Freq}}} 1ns 1ns 100us {1/{Freq}})
CIRCUIT  ■■■   R    4  6 0.5
               L    6  7 6.5MH
               VX   7  5 DC 10V      ; Load battery voltage
               VY 14   1 DC 0V       ; Voltage source to measure
               supply current
               *Subcircuit calls for thyristor model:
               XT1 1   8 4 SCRMOD    ; Thyristor T1
               XT3 2   9 4 SCRMOD    ; Thyristor T3
               XT5 3 10 4 SCRMOD     ; Thyristor T5
               XT2 5 11 3 SCRMOD     ; Thyristor T2
               XT4 5 12 1 SCRMOD     ; Thyristor T4
               XT6 5 13 2 SCRMOD     ; Thyristor T6
               *Subcircuit SCR, which is missing, must be
               *inserted.
ANALYSIS ■■■   .TRAN 1US 50MS 33.33MS      ; Transient analysis
               .PROBE                    ; Graphics post-processor
               .FOUR 60HZ I(VY)
               .OPTIONS ABSTOL=1.00N RELTOL=0.01 VNTOL=0.01
               *ITL5=20000
               ;Convergence
          .END
```

Note the following:

a. The PSpice plots of the instantaneous output voltage V(4,5), the load current I(VX), and the input current I(VY) are shown in Figure 11.26. The

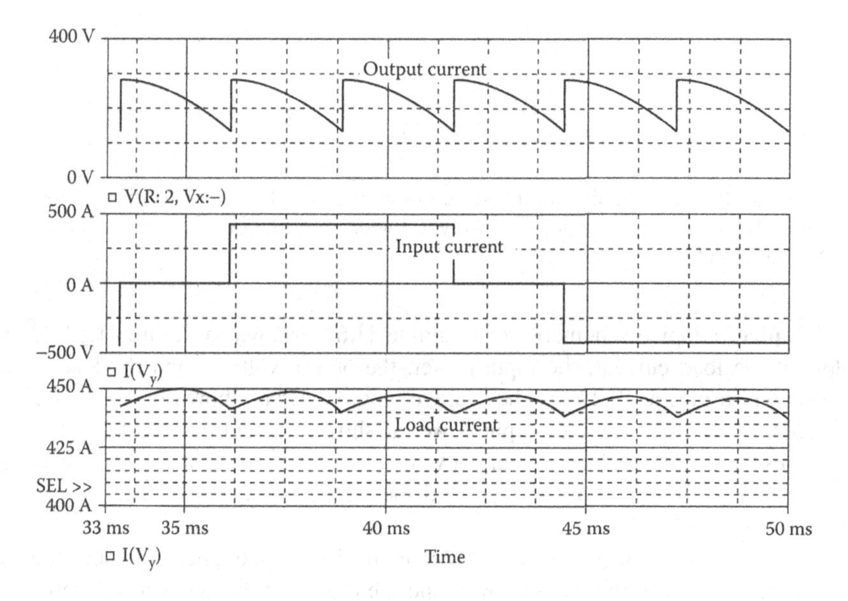

FIGURE 11.26 Plots for Example 11.6 for delay angle $\alpha = 90°$.

load current has not reached the steady-state condition. Note that the frequency of the output ripple voltage is six times the supply frequency. That is, $f_o = 6f = 360$ Hz.

b. The Fourier series of the input current is as follows:

Fourier Components of Transient Response I(VY)

DC Component = 7.862956E–01

Harmonic Number	Frequency (Hz)	Fourier Component	Normalized Component	Phase (Deg)	Normalized Phase (Deg)
1	6.000E+01	2.823E+02	1.000E+00	–6.021E+01	0.000E+00
2	1.200E+02	2.387E+00	8.453E–03	–1.651E+02	–1.048E+02
3	1.800E+02	1.885E+00	6.676E–03	–1.540E+02	–9.383E+01
4	2.400E+02	1.952E+00	6.915E–03	–1.020E+02	–4.179E+01
5	3.000E+02	6.049E+01	2.143E–01	–1.211E+02	–6.088E+01
6	3.600E+02	1.836E+00	6.503E–03	–1.761E+02	–1.159E+02
7	4.200E+02	3.774E+01	1.337E–01	1.216E+02	1.818E+02
8	4.800E+02	1.473E+00	5.216E–03	–1.451E+02	–8.488E+01
9	5.400E+02	6.411E–01	2.271E–03	1.662E+02	2.264E+02

Total harmonic distortion = 2.530035E+01%

Total harmonic distortion of input current $THD = 25.3\% = 0.253$
Displacement angle $\phi_1 = -60.21°$
Displacement factor $DF = \cos \phi_1 = \cos(-60.21) = 0.497$ (lagging)

The input power factor is

$$PF = \frac{1}{(1+THD^2)^{1/2}}\cos\phi_1 = \frac{1}{(1+0.253^2)^{1/2}} \times 0.497 = 0.482(\text{lagging})$$

Note: For an inductive load, the output voltage can be either positive or negative. The input power factor is, however, lower than that of the semiconverter at the same delay angle $\alpha = 90°$.

LTspice: LTspice schematic for Example 11.6 is shown in Figure 11.27(a). The plots of the load current, the input power, the output voltage, and the load power are shown in Figure 10.27(b). The average input power as shown in Figure 11.27(c) is 42536 W, and the average output power as shown in Figure 11.27(d) is 31148 W. The rms value of the load current I(Vx) is 244.39 A and the rms value of the output voltage is 127.63 V. We can find the power efficiency of the converter as $\eta = \frac{P_{out}}{P_{in}} = \frac{31148}{42536} = 73.22\%$

Note: For a low output power, the losses in the semiconductor devices are significant compared to the output power and the efficiency becomes low. There is no power loss in the load inductor.

11.5 SWITCHED THYRISTOR DC MODEL

Forced-commutated converters are being used increasingly to improve the input power factor. These converters operate power devices as switches. The switches are turned on or off at a specified time. We do not need latching characteristics. Rather, we need to operate them as on-and-off switches, similar to gate turn-off (GTO) thyristors. We shall model a power device as the voltage-controlled switch shown in Figure 11.28. This model, called a *switched thyristor DC model*, is also used in Chapter 10. The subcircuit definition of SSCR can be described as follows:

```
* Subcircuit for switched thyristor model:
.SUBCKT   SSCR        1         2          3          4
*        model     anode     cathode    +control    -control
*         name                          voltage     voltage
DT     5    2    DMOD                 ; Switch diode
ST     1    5    3    4    SMOD       ; Switch
.MODEL DMOD D (IS=2.2E - 15 BV=1200V TT=0 CJO=0) ; Diode
model parameters
.MODEL SMOD VSWITCH (RON=0.01 ROFF=10E + 6 VON=10V VOFF=5V)
.ENDS SSCR ; Ends subcircuit definition
```

11.6 GTO THYRISTOR MODEL

There are dedicated PSpice GTO models [1,7–9]. The AC thyristor model of Figure 11.1(b) can, however, be used as a GTO. The turn-on is similar to a normal SCR. However, the turn-off gate voltage must be negative with appropriate magnitude and

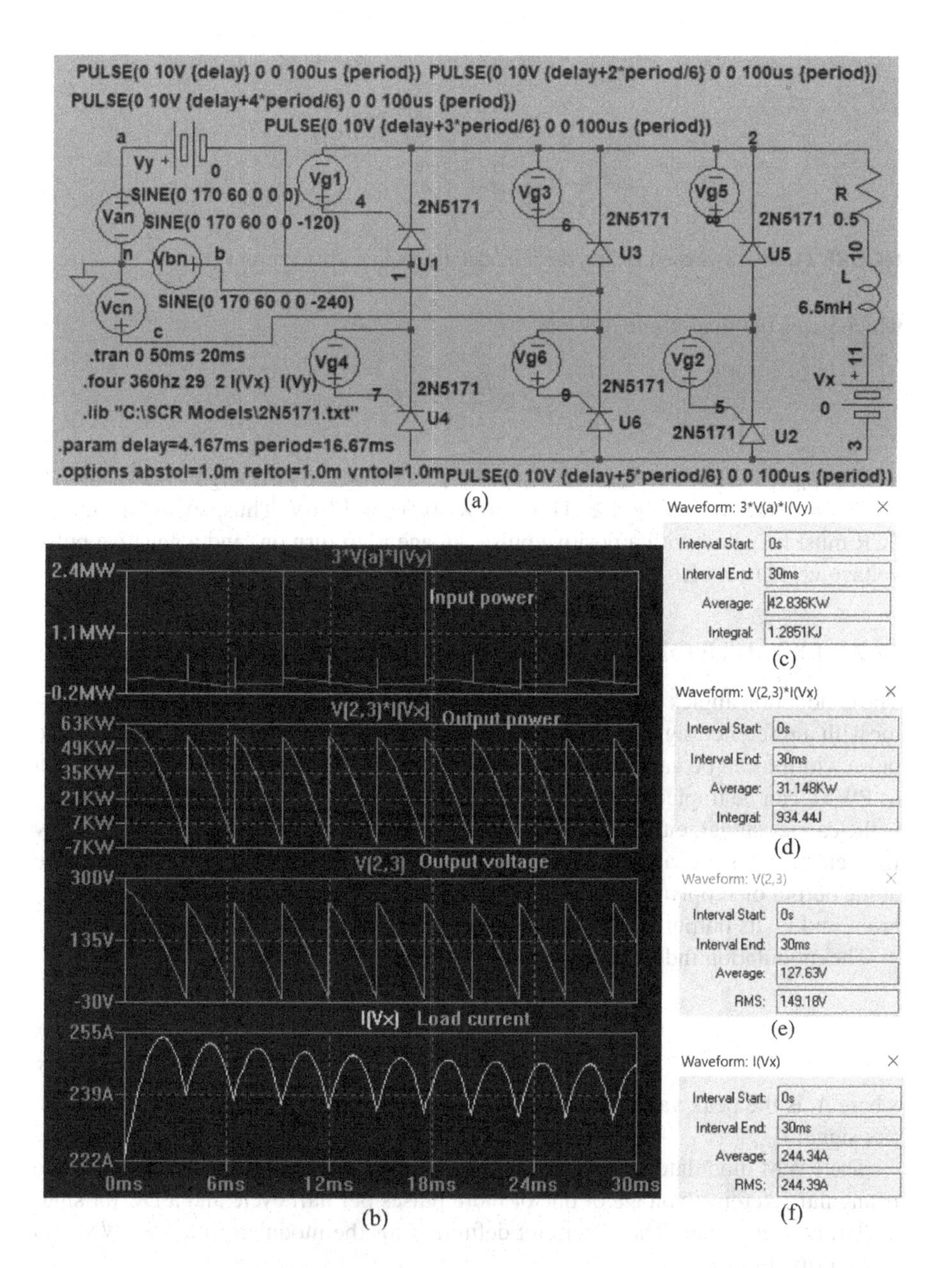

FIGURE 11.27 LTspice schematic for Example 11.6. (Ltspice Schematic. (b) Output waveforms. (c) Input power. (d) Output power. (e) Load voltage. (f) Load current.

be capable of turning off at the maximum possible current. Therefore, the turn-off gate voltage v_{gn} must satisfy the condition

$$P_1 I_g + P_2 I_{a(\max)} \le 0 \quad \text{or} \quad P_1 \frac{v_{gn}}{R_G} + P_2 I_{a(\max)} \le 0$$

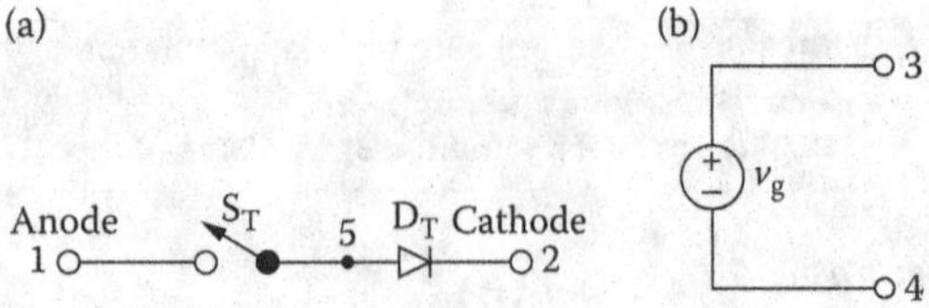

(a) Anode 1, S_T, 5, D_T Cathode 2

(b) 3, v_g, 4

FIGURE 11.28 Switched DC thyristor model. (a) Switch. (b) Control voltage.

which gives the magnitude of v_{gn} as

$$|v_{gn}| \geq \frac{P_2 I_{a(max)}}{P_1} R_G$$

For $I_{T(RMS)} = 110$ A, $I_{a(max)} = \sqrt{2} I_{T(RMS)} = \sqrt{2} \times 110 = 156$V. Using $P_1 = 50$, $P_2 = 11$, and $R_G = 50\ \Omega$, we get $|V_{gn}| \geq (11 \times 156 \times 50/50) = 1716$V. Thus, subcircuit model SCR must be gated with a positive pulse voltage v_g to turn on, and a negative pulse voltage v_{gn} to turn off.

11.7 EXAMPLE OF FORCED-COMMUTATED RECTIFIERS

The conduction angles of one or more switches can be generated by using a comparator with a reference signal v_r and a carrier signal v_c as shown in Figure 11.29(a). The pulse width δ can be varied by varying the carrier voltage v_c. This technique, known as PWM, can be implemented in PSpice Schematics as shown in Figure 11.29(b). V_{cr} is the carrier signal, either DC or a rectifier sinusoidal signal at the output or supply frequency. V_{ref} is a triangular signal at the switching frequency, and V_p is the pulse at the output or supply frequency with a 50% duty cycle. The inputs to the amplifier are v_r and v_c. Its output is the conduction angle δ during which a switch remains on.

The modulation index M is defined by

$$M = \frac{A_c}{A_r}$$

where A_r is the peak value of a reference signal v_r, and A_c is the peak value of a carrier signal v_c.

The PWM modulator can be used as a subcircuit to generate control signals for a triangular reference voltage of one or more pulses per half cycle and a DC (or sinusoidal) carrier signal. The subcircuit definition for the modulator model PWM can be described as follows:

```
* Subcircuit for generating PWM control signals
.SUBCKT    PWM       1         2          3          4
*        model   ref.    carrier     +g1        +g2
*         name   input    input    voltage   voltage
E_ABM1 6 0 VALUE {IF(V(2) -V(1)>0, 1, 0)}       ; Comparator
Vp 5 0 PULSE (0 1 0 1ns 1ns {1/(2*{fout}) -2ns} {1/{fout}}); Pulse
```

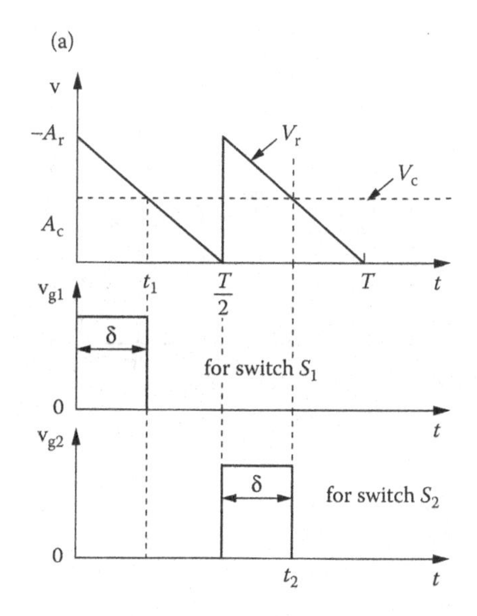

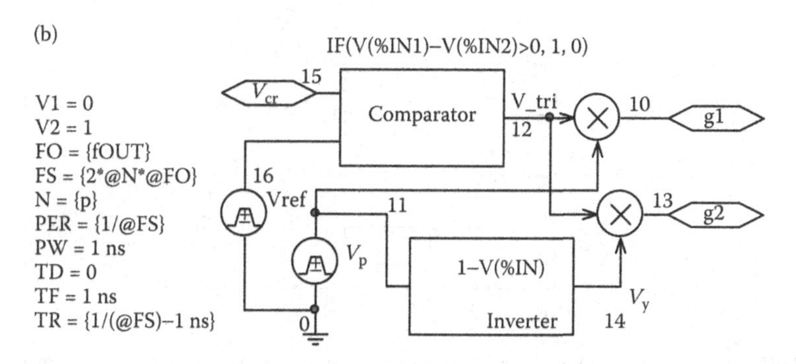

FIGURE 11.29 PWM control. (a) Gate signals. (b) PSpice comparator.

```
E_ABM2 7 0 VALUE {1-V(5)} ; Inverter
E_MULT1 3 0 VALUE {V(5)*V(6)}   ; Multiplier 1
E_MULT2 4 0 VALUE {V(7)*V(6)}   ; Multiplier 2
.ENDS PWM                       ; Ends subcircuit definition
```

Example 11.7:

Finding the Performance of a Single-Phase Converter with an Extinction Angle Control

A single-phase converter with an extinction angle control is shown in Figure 11.30(a). The input voltage has a peak of 169.7 V, 60 Hz. The load inductance L is 6.5 mH,

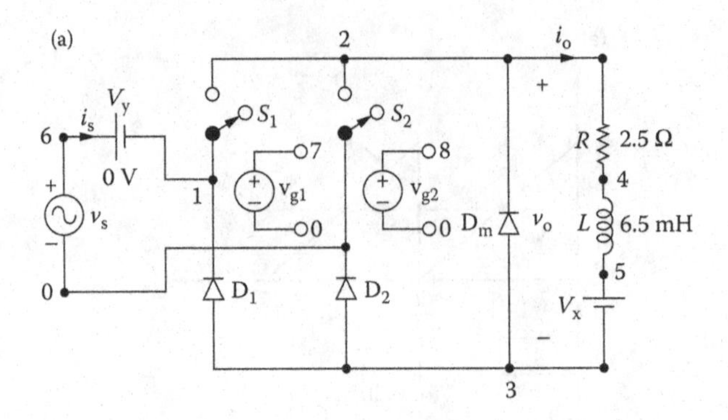

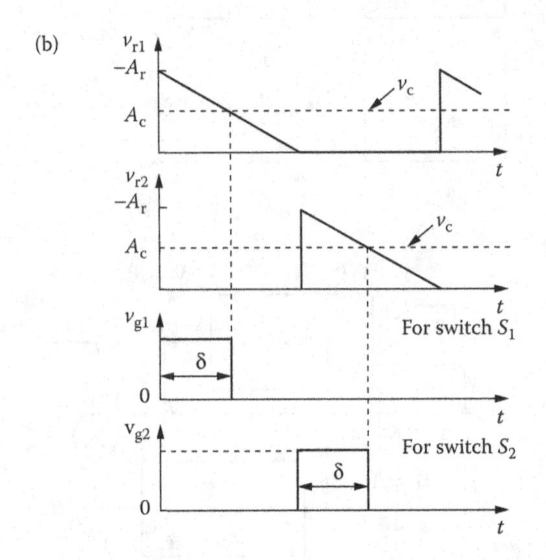

FIGURE 11.30 Single-phase converter with extinction angle control. (a) Circuit. (b) Gate voltages.

and the load resistance R is 0.5 Ω. The load battery voltage is $V_x = 10$ V. The extinction angle is $\beta = 60°$. Use PSpice to (a) plot the instantaneous output v_o, the load current i_o, and the input current i_s and (b) calculate the Fourier coefficients of the input current i_s and the input power factor PF.

SOLUTION

The conduction angles are generated with two carrier voltages as shown in Figure 11.30(b). Let us assume that $A_r = 10$ V. for $\beta = 60°$, $A_c = 10 \times 60/360 = 3.33$ V, and $M = 3.33/10 = 0.333$. v_c is generated by a PWL waveform, and v_r by a PULSE waveform.

The PSpice schematic with IGBTs is shown in Figure 11.31(a). Varying the modulation voltage V_mod can vary the output voltage. The output frequency {FOUT} of the modulating signal and the number of pulses P are defined as variables. The model parameters for the IGBTs and the freewheeling diode are as follows:

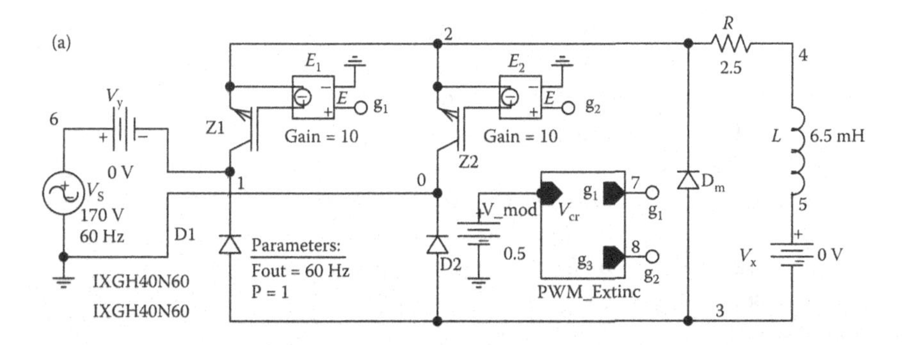

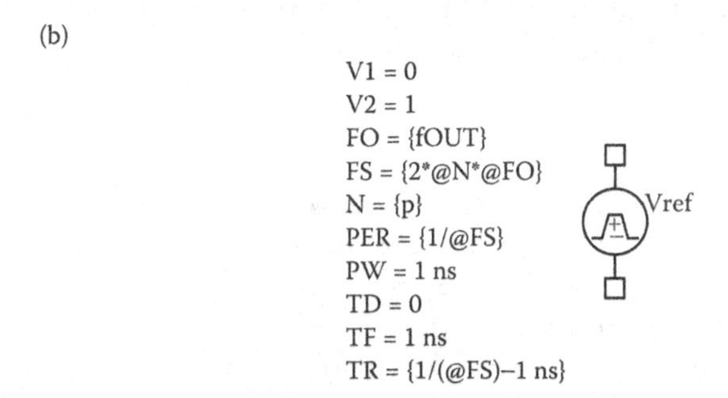

FIGURE 11.31 PSpice schematic for Example 11.7. (a) Schematic. (b) Schematic parameters.

```
.MODEL IXGH40N60 NIGBT (TAU=287.56ns KP=50.034
AREA=37.5um AGD=18.75um

+ VT=4.1822 KF=.36047 CGS=31.942nf COXD=53.188nf
VTD=2.6570) for IGBTs
.MODEL DMD D(IS=2.22E - 15 BV=1200V CJO=0PF TT=0US)) for
Diodes
```

The parameters of the reference pulse signal can be adjusted to produce the triangular wave as shown in Figure 11.30(b). These parameters, as shown in Figure 11.31(b), are related to the supply frequency f and the number of pulse per half cycle p ($p = 1$ for an extinction angle control) as follows: delay time $t_d = 0$, switching frequency $f_s = 2pf$ switching period $T_s = 1/f_s$, pulse width $t_w = 1$ ns, rise time $t_r = (1/f_s) - 1$ ns, and fall time $t_f = 1$ ns.

The listing of the circuit file for the converter is as follows:

Single-Phase Semiconverter with Extinction Angle Control

```
VS 6  0 AC  0  SIN (0 170V 60Hz 0 0 0)
.PARAM FOUT=60Hz  P=1  M=0.6
* Parameters: fout = output frequency and p = # of pulses
* per half cycle
Vref 10  0 PULSE (0 1 0 {1/({2*{p}*{fout}}) - 1ns} 1ns
```

```
+ 1ns {1/{2*{p}*{fout}}}) ; Reference Signal
V_mod 11 0 DC {M} ; Modulation signal
E_ABS 12 0 VALUE {ABS(V(11))} ; Carrier signal
R  2 4 2.5          ; Load Resistance
L  4 5 6.5mH        ; Load Inductance
Vx 5 3 0V           ; Measues the load current
Vy 6 1 0V           ; Measues the input current
D1 3 1 DMD          ; Diodes with model DMD
D2 3 0 DMD
Dm 3 2 DMD
.MODEL DMD D(IS=1E-25 BV=1000V) ; Diode model parameters
Z1 1 7 2 IXGH40N60 ; IGBTs with a model IXGH40N60
Z2 0 8 2 IXGH40N60
.MODEL IXGH40N60 NIGBT (TAU=287.56ns KP=50.034 AREA=37.5um
AGD=18.75um
+ VT=4.1822 KF=.36047 CGS=31.942nf COXD=53.188nf VTD=2.6570)
E1 7 2  g1 0 10         ; Voltage-controlled voltage source
E2 8 2  g2 0 10         ; Voltage-controlled voltage source
XPWM 10 12 g1 g2 PWM   ; Control voltages g1 and g3
* Subcircuit PWM, which is missing, must be inserted
.TRAN 1US 50MS 33.33MS 0.1E-6    ; Transient analysis
.FOUR 60 10 I(Vy)   ; Fourier analysis
.OPTIONS ABSTOL= 10u CHGTOL= 0.01nC RELTOL= 0.1 TNOM= 1m
VNTOL= 0.1
.PROBE    ; Graphics post-processor
.END
```

Note the following:

a. The PSpice plots of the instantaneous output voltage V(2,3), the load current I(VX), and the input current I(VY) are shown in Figure 11.32. The input current is not symmetrical about the 0-axis, and its shape depends on the load current.
b. The Fourier series of the input current is as follows:

Fourier Components of Transient Response I(VY)

DC Component = –8.990786E–02

Harmonic Number	Frequency (Hz)	Fourier Component	Normalized Component	Phase (Deg)	Normalized Phase (Deg)
1	6.000E+01	3.078E+01	1.000E+00	1.338E+01	0.000E+00
2	1.200E+02	1.281E–01	4.163E–03	1.912E+01	5.742E+00
3	1.800E+02	1.057E+01	3.434E–01	1.779E+02	1.645E+02
4	2.400E+02	1.335E–01	4.338E–03	1.697E+02	1.564E+02
5	3.000E+02	6.870E+00	2.232E–01	–4.495E+01	–5.833E+01
6	3.600E+02	1.798E–01	5.841E–03	–8.377E+01	–8.377E+01
7	4.200E+02	5.066E+00	1.646E–01	5.811E+01	4.473E+01

(Continued)

Fourier Components of Transient Response I(VY) *(Continued)*

DC Component = −8.990786E−02

Harmonic Number	Frequency (Hz)	Fourier Component	Normalized Component	Phase (Deg)	Normalized Phase (Deg)
8	4.800E+02	1.225E−01	3.980E−03	2.827E+01	1.490E+01
9	5.400E+02	2.582E+00	8.390E−02	−1.671E+02	−1.805E+02

Total harmonic distortion = 4.493854E+01%

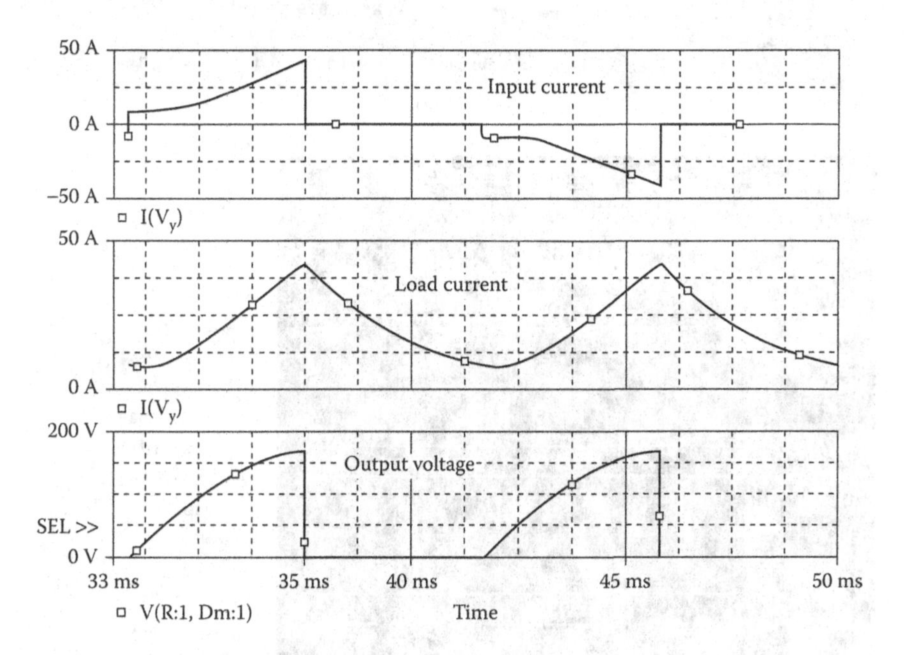

FIGURE 11.32 Plots for Example 11.7 for delay angle $\beta = 60°$.

From the Fourier components, we get

Total harmonic distortion of input current $THD = 44.93\% = 0.4493$
Displacement angle $\phi_1 = 13.38$
Displacement factor $DF = \cos \phi_1 = \cos(13.38) = 0.9728$ (leading)

The input power factor is

$$PF = \frac{1}{(1+THD^2)^{1/2}} \cos\phi_1 = \frac{1}{(1+0.4493^2)^{1/2}} \times 0.9728 = 0.8873 \text{(leading)}$$

Note: The power factor is high due to the low displacement angle.

LTspice: LTspice schematic for Example 11.7 is shown in Figure 11.33(a). The plots of the input current, the input power, the output voltage, and the load power

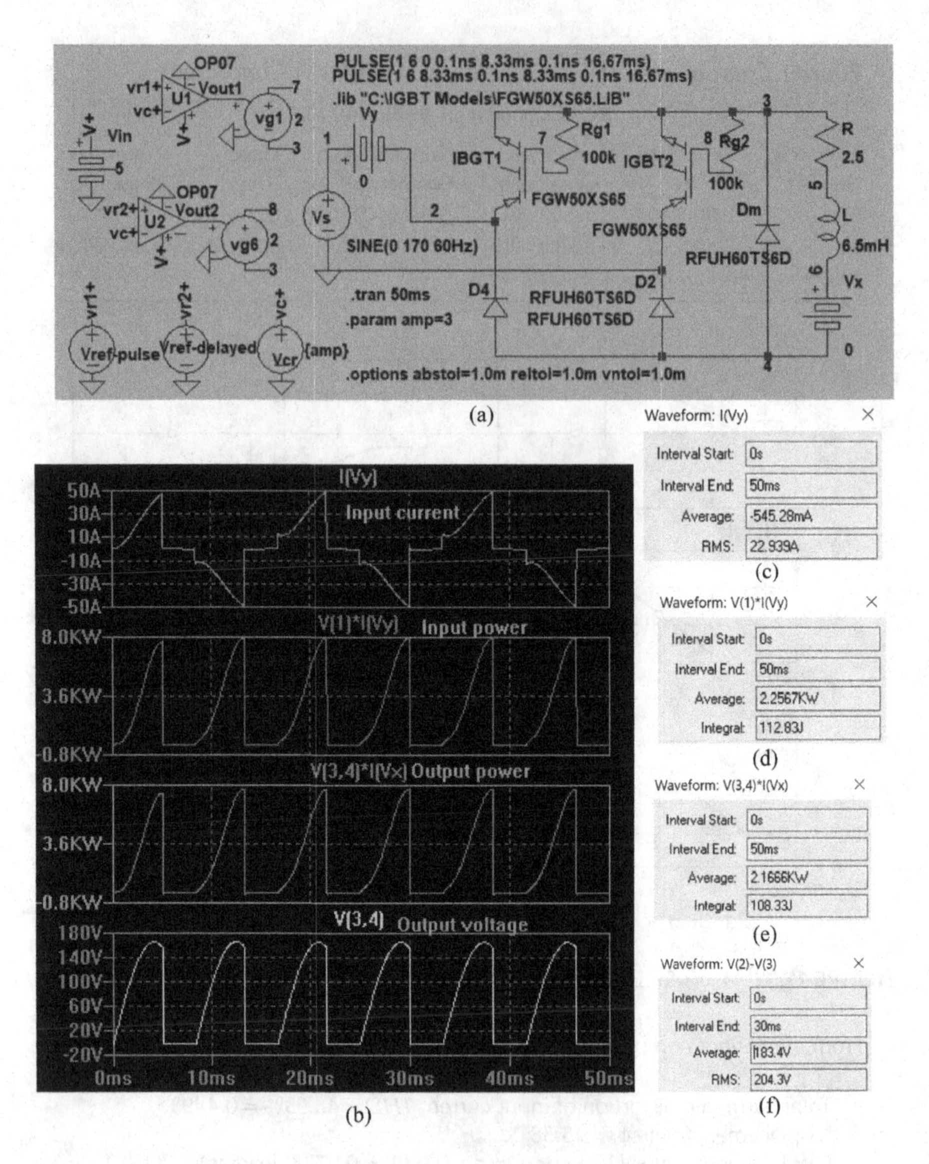

FIGURE 11.33 LTspice schematic for Example 11.7. (a) LTspice Schematic. (b) Output waveforms. (c) Input current. (d) Input power. (e) Output power. (f) Load voltage.

are shown in Figure 10.33(b). The rms input current as shown in Figure 11.33(c) is 22.94 A, the average input power as shown in Figure 11.33(d) is 2256.7 W, and the average output power as shown in Figure 11.33(e) is 2166.6 W. The rms value of the output voltage is 204.3 V. We can find the power efficiency of the converter as $\eta = \frac{P_{out}}{P_{in}} = \frac{2166.6}{2256.7} = 96\%$

We can find the input power factor as $PF_i = \frac{P_{in}}{V_s I_s} = \frac{2166.6}{120 \times 22.94} = 0.78$

Note: For a low output power, the losses in the semiconductor devices are significant compared to the output power and the efficiency becomes low. There is no power loss in the load inductor.

Example 11.8:

Finding the Performance of a Single-Phase Converter with a Symmetrical Angle Control

A symmetrical angle control is applied to the converter of Figure 11.30(a). The input voltage has a peak of 169.7 V, 60 Hz. The load inductance L is 6.5 mH, and the load resistance R is 2.5 Ω. The load battery voltage is $V_x = 10$ V. The conduction angle is $\beta = 60°$. Use PSpice to (a) plot the instantaneous output voltage v_o, the load current i_o, and the input current i_s, and (b) calculate the Fourier coefficients of the input current i_s and the input power factor PF.

SOLUTION

The conduction angles can be generated with two carrier voltages as shown in Figure 11.34. Let us assume that $A_r = 10$ V. For $\beta = 60°$, $A_c = 10 \times 60/360 = 3.33$ V, and $M = 3.33/10 = 0.333$. The subcircuit PWM is used to generate control signals. v_r is generated by a PULSE generator with a very small pulse width, say, TW = 1NS. v_c is generated by a PWL generator.

The PSpice schematic is the same as that in Figure 11.31(a). The parameters of the reference pulse signal can be adjusted to give the triangular wave as shown in Figure 11.34. These parameters, as shown in Figure 11.35, are related to the supply frequency f and the number of pulse per half cycle p ($p = 1$ for a symmetrical angle

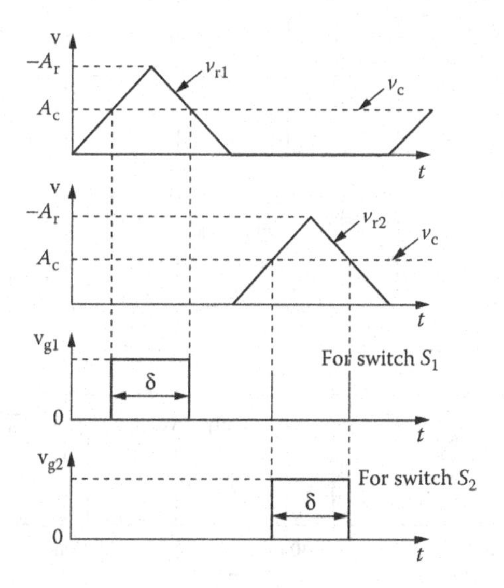

FIGURE 11.34 Gate voltages with symmetrical angle control.

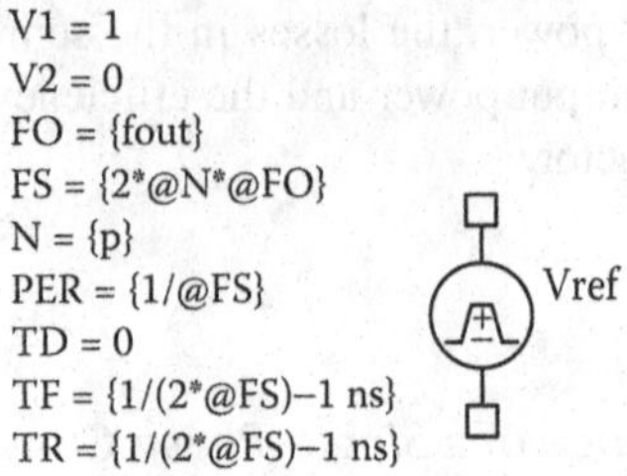

V1 = 1
V2 = 0
FO = {fout}
FS = {2*@N*@FO}
N = {p}
PER = {1/@FS}
TD = 0
TF = {1/(2*@FS)–1 ns}
TR = {1/(2*@FS)–1 ns}

Vref

FIGURE 11.35 Model parameters of the reference pulse signal for symmetrical angle control.

control) as follows: delay time $t_d = 0$, switching frequency $f_s = 2pf$ switching period $T_s = 1/f_s$, pulse width $t_w = 1$ ns, rise time $t_r = (1/2f_s) - 1$ ns, and fall time $t_f = (1/2f_s) - 1$ ns.

The listing of the circuit file is the same as that for Example 11.7, except that the statement for the reference signal is changed as follows:

```
Vref 10 0 PULSE ( 1 0 1ns {1/{2*{2*{p}*{fout}}}} {1/
{2*2*{p}*{fout}}} 1ns {1/{2*
+ {p}*{fout}}} ) ; Reference SignalExample 11.8 Single-phase
semiconverter with symmetrical angle control
```

Note the following:

a. The PSpice plots of the instantaneous output voltage V(2,3), the load current I(VX), and the input current I(VY) are shown in Figure 11.36. The output voltage is symmetrical, as expected.

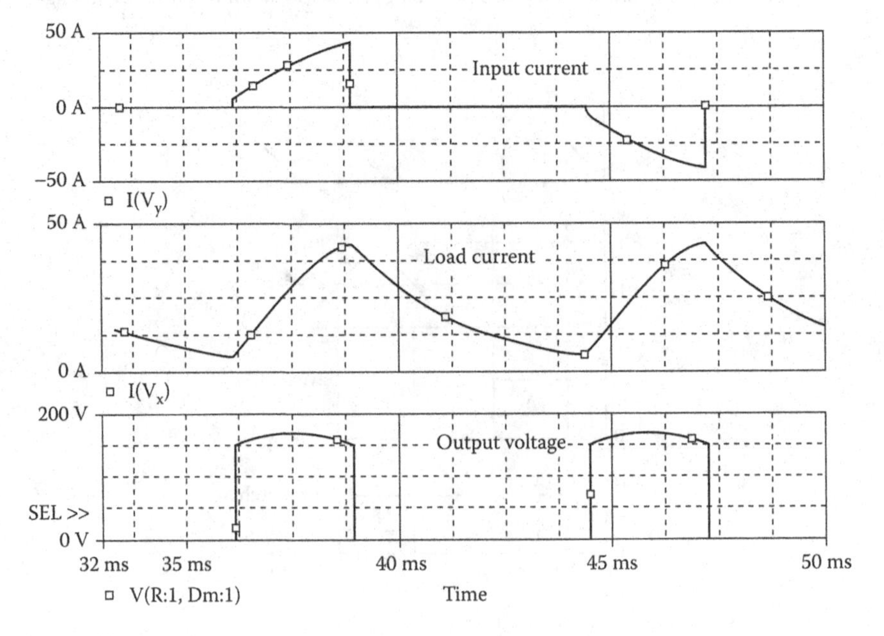

FIGURE 11.36 Plots for Example 11.8 for delay angle $\beta = 60^{\circ\circ}$.

b. The Fourier series of the input current is as follows:

Fourier Components of Transient Response I(VY)

DC Component = –1.888986E–02

Harmonic Number	Frequency (Hz)	Fourier Component	Normalized Component	Phase (Deg)	Normalized Phase (Deg)
1	6.000E+01	4.063E+01	1.000E+00	–1.143E+01	0.000E+00
2	1.200E+02	4.407E–02	1.085E–03	1.524E+02	1.639E+02
3	1.800E+02	9.826E+00	2.419E–01	1.115E+02	1.230E+02
4	2.400E+02	4.161E–02	1.024E–03	–1.490E+02	–1.376E+02
5	3.000E+02	7.352E+00	1.810E–01	–1.630E+02	1.744E+02
6	3.600E+02	4.074E–02	1.003E–03	–8.388E+01	–7.245E+01
7	4.200E+02	4.407E+00	1.085E–01	–1.558E+02	–1.444E+02
8	4.800E+02	4.312E–02	1.061E–03	–2.285E+01	–1.142E+01
9	5.400E+02	2.517E+00	6.195E–02	–7.257E+01	–6.113E+01

Total harmonic distortion = 3.214001E+0%

From the Fourier components, we get

Total harmonic distortion of input current $THD = 32.69\% = 0.3269$
Displacement angle $\phi_1 = -11.43°$
Displacement factor $DF = \cos\phi_1 = \cos(-11.434) = 0.98$ (lagging)

From Equation 7.3, the input power factor

$$PF = \frac{1}{(1+THD^2)^{1/2}}\cos\phi_1 = \frac{1}{(1+0.3269^2)^{1/2}} \times 0.98 = 0.9315(\text{lagging})$$

Note: The power factor of a converter with symmetrical angle control is better than that with extinction angle control for the same amount of pulse width or modulation index.

LTspice: LTspice schematic for Example 11.8 is shown in Figure 11.37(a). The plots of the input current, the input power, the output voltage, and the load power are shown in Figure 11.37(b). The rms input current as shown in Figure 11.37(c) is 30.8 A, the average input power as shown in Figure 11.37(d) is 3348.2 W, and the average output power as shown in Figure 11.37(e) is 3221.3 W. The rms value of the output voltage is 110.83 V. We can find the power efficiency of the converter as $\eta = \frac{P_{out}}{P_{in}} = \frac{3221.3}{3348.2} = 96.21\%$

We can find the input power factor as $PF_i = \frac{P_{in}}{V_s I_s} = \frac{3348.2}{120 \times 30.8} = 0.906$

Note: For a low output power, the losses in the semiconductor devices are significant compared to the output power and the efficiency becomes low. There is no power loss in the load inductor.

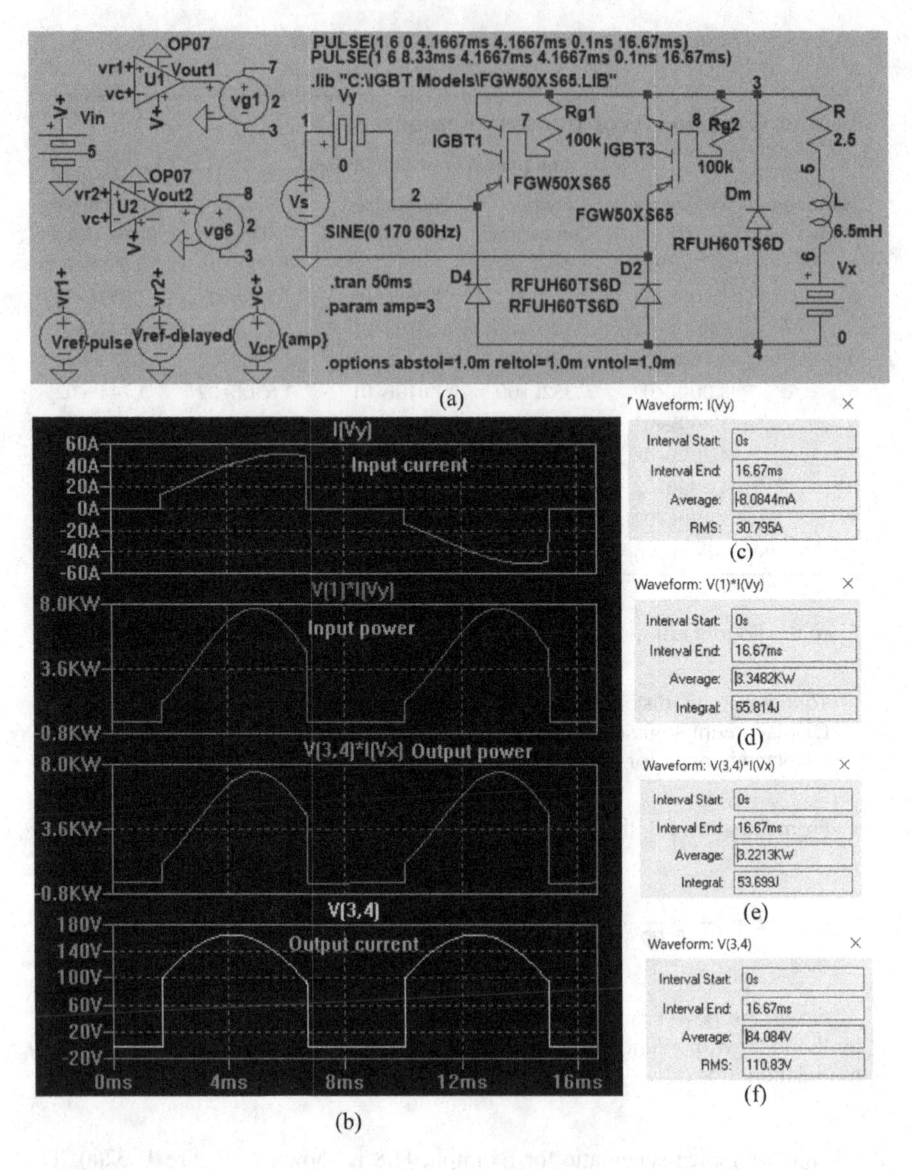

FIGURE 11.37 LTspice schematic for Example 11.8. (a) LTspice Schematic. (b) Output waveforms. (c) Input current. (d) Input power. (e) Output power. (f) Load voltage.

Example 11.9:

Finding the Performance of a Single-Phase Converter with a Uniform PWM Control

PWM control with four pulses per half cycle is applied to the converter of Figure 11.30(a). The input voltage has a peak of 169.7 V, 60 Hz. The load inductance L is 6.5 mH, and the load resistance R is 0.5 Ω. The load battery voltage is V_x = 10 V. Use PSpice to (a) plot the instantaneous output voltage v_o, the load current i_o, and the input current i_s and (b) calculate the Fourier coefficients of the input current i_s and the input power factor PF. Assume a modulation index of 0.4.

SOLUTION

The conduction angles are generated with two carrier voltages as shown in Figure 11.38. Assume that reference voltage V_r = 10 V, for M = 0.4, the carrier voltage $V_c = MV_r$ = 4 V. We can use the same subcircuit PWM for the generation of control signals. Note that the carrier voltages are generated with a PWL generator. Instead, we could use a PULSE generator with a very small pulse width, say, TW = 1 ns.

The PSpice schematic is the same as that in Figure 11.31(a). The parameters of the reference pulse signal can be adjusted to give the triangular wave as shown in Figure 11.38. These parameters, as shown in Figure 11.35, are related to the supply frequency f and the number of pulse per half cycle p as follows: delay time $t_d = 0$, switching frequency $f_s = 2pf$ switching period $T_s = 1/f_s$, pulse width t_w = 1 ns, rise time $t_r = 1/2f_s$, and fall time $t_f = (1/2f_s)$ – 1 ns.

The listing of the circuit file is the same as that for Example 11.8, except that the parameter p is changed to 4.

Note the following:

1. The PSpice plots of the instantaneous output voltage V(2,3), the load current I(VX), and the input current I(VY) are shown in Figure 11.39. The input current consists of four pulses per half cycle.

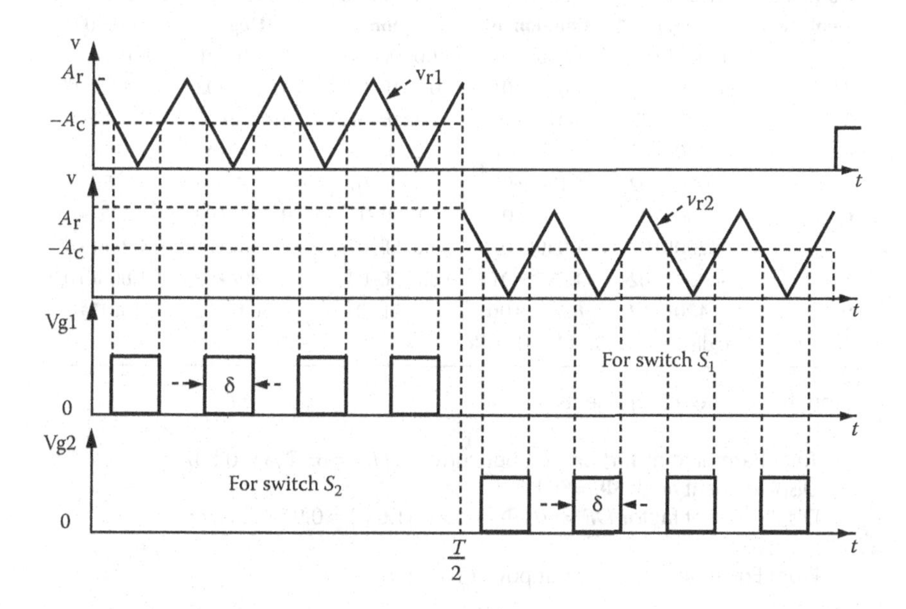

FIGURE 11.38 Gate voltages with PWM control.

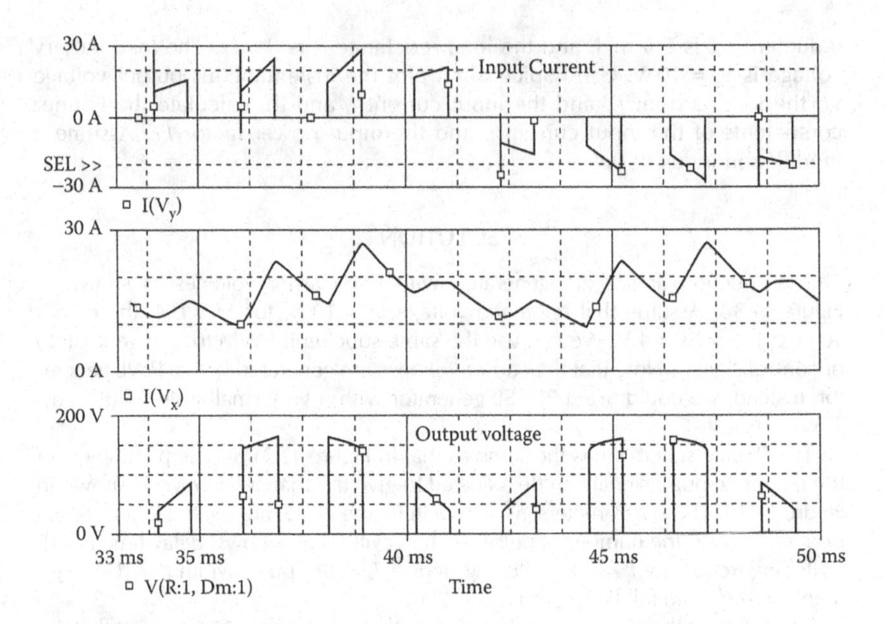

FIGURE 11.39 Plots for Example 11.9 for modulation index $M = 0.4$.

2. The Fourier series of the input current is as follows:

Fourier Components of Transient Response I(VX)

DC Component = −5.261076E−02

Harmonic Number	Frequency (Hz)	Fourier Component	Normalized Component	Phase (Deg)	Normalized Phase (Deg)
1	6.000E+01	1.768E+01	1.000E+00	−1.163E+01	0.000E+00
2	1.200E+02	1.070E−01	6.051E−03	1.324E+02	1.441E+02
3	1.800E+02	4.811E+00	2.721E−01	1.433E+01	2.596E+01
4	2.400E+02	2.142E−01	1.212E−02	1.416E+00	1.305E+01
5	3.000E+02	4.184E+00	2.367E−01	−4.615E+00	7.015E+00
6	3.600E+02	2.012E−01	1.138E−02	1.386E+02	1.502E+02
7	4.200E+02	1.039E+01	5.876E−01	−1.230E+00	1.286E+01
8	4.800E+02	1.529E−01	8.651E−03	1.689E+02	1.805E+02
9	5.400E+02	8.397E+00	4.750E−01	1.564E+02	1.681E+02

Total harmonic distortion = 3.214001E+0%

From the Fourier components, we get

Total harmonic distortion of input current $THD = 83.7\% = 0.837$
Displacement angle $\Phi_1 = -11.63°$
Displacement factor $DF = \cos \Phi_1 = \cos(-11.63°) = 0.979$ (lagging)

From Equation 7.3, the input power factor is

$$PF = \frac{1}{(1+THD^2)^{1/2}}\cos\phi_1 = \frac{1}{(1+0.837^2)^{1/2}} \times 0.979 = 0.75(\text{lagg})$$

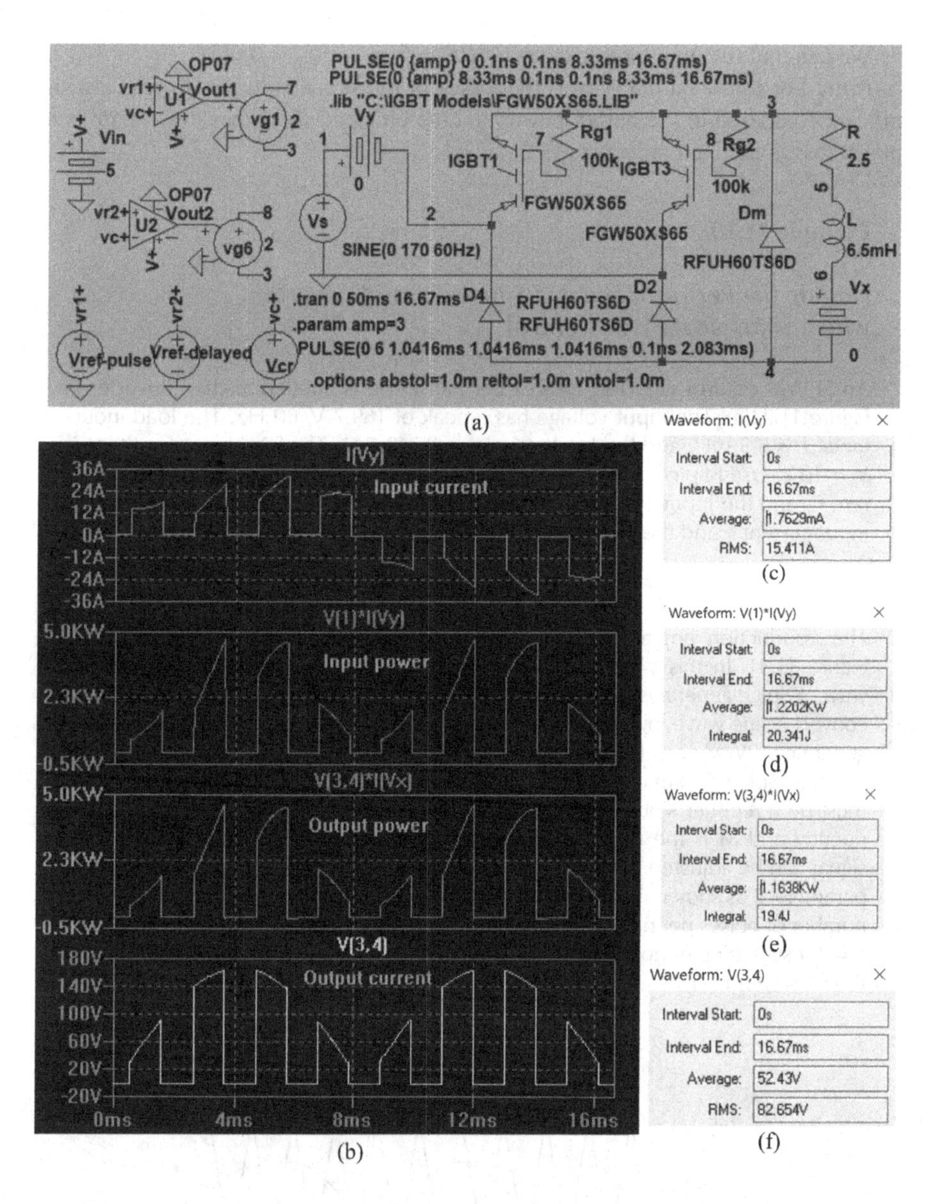

FIGURE 11.40 LTspice schematic for Example 11.9. (a) LTspice Schematic. (b) Output waveforms. (c) Input current. (d) Input power. (e) Output power. (f) Load voltage.

LTspice: LTspice schematic for Example 11.9 is shown in Figure 11.40(a). The plots of the input current, the input power, the output voltage, and the load power are shown in Figure 10.40(b). The rms input current as shown in Figure 11.40(c) is 15.41 A, the average input power as shown in Figure 11.40(d) is 1220.2 W, and the average output power as shown in Figure 11.40(e) is 1163.8 W. The rms value of the output voltage is 82.65 V. We can find the power efficiency of the converter as $\eta = \frac{P_{out}}{P_{in}} = \frac{1163.8}{1220.2} = 95.4\%$

We can find the input power factor as $PF_i = \frac{P_{in}}{V_s I_s} = \frac{1220.2}{120 \times 15.41} = 0.66$

Note: For a low output power, the losses in the semiconductor devices are significant compared to the output power and the efficiency becomes low. There is no power loss in the load inductor.

Example 11.10:

Finding the Performance of a Single-Phase Converter with a Sinusoidal PWM Control

An SPWM control with four pulses per half cycle is applied to the converter of Figure 11.24(a). The input voltage has a peak of 169.7 V, 60 Hz. The load inductance L is 6.5 mH, and the load resistance R is 2.5 Ω. The load battery voltage is $V_x = 10$ V. Use PSpice to (a) plot the instantaneous output voltage v_o, the load current i_o, and the input current i_s and (b) to calculate the Fourier coefficients of the input current i_s and the input power factor PF. Assume a modulation index of 0.4.

SOLUTION

The conduction angles are generated with two carrier voltages as shown in Figure 11.41. In this type of control, the carrier signal must be a rectified sine wave. PSpice generates only a sine wave. Thus, we can use a precision rectifier to convert a sine wave input signal to two sine wave pulses and use a comparator to generate a PWM waveform.

The PSpice schematic is the same as that in Figure 11.39(a). The carrier signal must be a rectifier sinusoidal signal v_{cr} = **M* sin(2*πft*)*, where f is the supply frequency and M is the modulation index. The parameters of the reference pulse signal can be adjusted to give the triangular wave as shown in Figure 11.41. These parameters, as shown in Figure 11.42, are related to the supply frequency f and the number of pulses per half cycle p as follows: delay time $t_d = 0$, switching frequency $f_s = 2pf$ switching period $T_s = 1/f_s$, pulse width $t_w = 1$ ns, rise time $t_r = 1/2f_s$, and fall time $t_f = (1/2f_s) - 1$ ns.

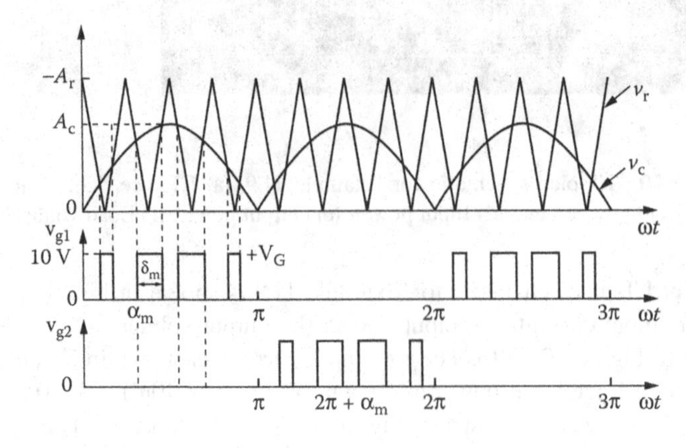

FIGURE 11.41 Gate voltages with SPWM control.

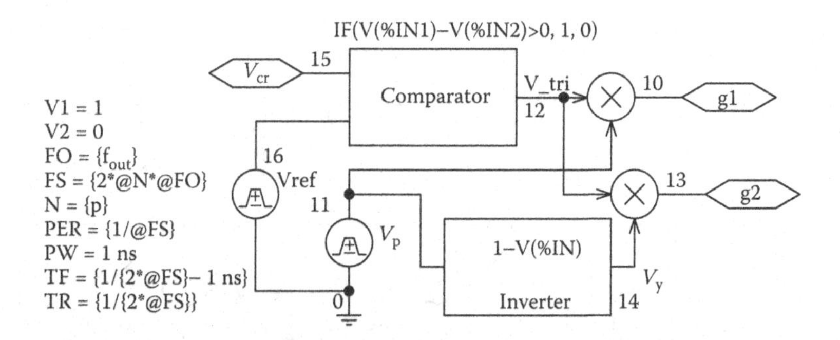

FIGURE 11.42 PSpice schematic SPWM signals.

We can use the subcircuit SPWM for the generation of the control signals. The subcircuit definition for the sinusoidal modulation model SPWM is the same as that for PWM, except that the carrier signal must be a rectified sinusoidal signal of magnitude *M* at the supply frequency *f*.

Let us assume a reference voltage $V_r = 10$ V, for $M = 0.4$, the carrier voltage $V_c = MV_r = 4$ V. The listing of the circuit file is the same as that for Example 11.9, except that the statements for the carrier signal will be changed as follows:

```
V_mod   11   0  AC   0   SIN (0 {M} {fout} 0 0 0)    ; Modulation
signal
E_ABS   12   0  VALUE   {ABS(V(11))}    ; Absolute value of the
carrier signal:
```

Note the following:

a. The PSpice plots of the instantaneous output voltage V(2,3), the load current I(VX), and the input current I(VY) are shown in Figure 11.43. The input current pulses follow a sinusoidal pattern.
b. The Fourier series of the input current is as follows:

Fourier Components of Transient Response I(VX)

DC Component = −3.111144E−02

Harmonic Number	Frequency (Hz)	Fourier Component	Normalized Component	Phase (Deg)	Normalized Phase (Deg)
1	6.000E+01	2.255E+01	1.000E+00	−1.127E+01	0.000E+00
2	1.200E+02	2.386E−01	1.058−02	9.025E+00	2.029E+01
3	1.800E+02	8.748E+00	3.880E−01	6.041E+00	1.731E+01
4	2.400E+ 02	2.635E−01	1.169E−022	−1.781E+02	−1.668E+02
5	3.000E+02	4.986E+00	2.211E−01	1.760E+01	2.887E+01
6	3.600E+02	9.425E−02	4.180E−03	3.283E+01	4.410E+01
7	4.200E+02	5.257E+00	2.332E−01	1.662E+02	1.775E+02
8	4.800E+02	1.603E−01	7.111E−03	1.688E+02	1.801E+02
9	5.400E+02	1.184E+01	5.252E−01	−1.120E+01	7.249E−02

Total harmonic distortion = 3.214001E+0%

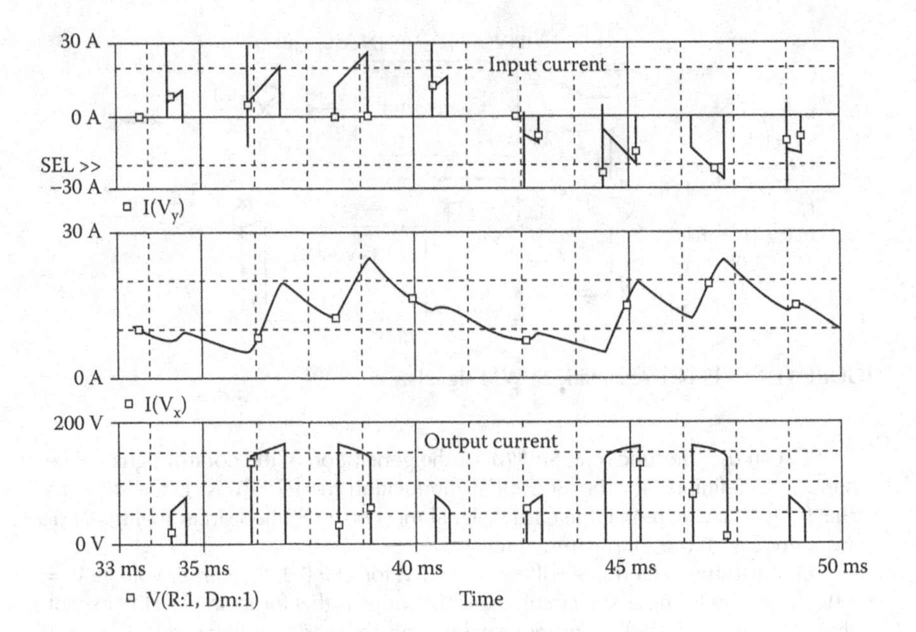

FIGURE 11.43 Plots for Example 11.10 for modulation index $M = 0.4$.

From the Fourier components, we get

Total harmonic distortion of input current $THD = 72.8\% = 0.728$
Displacement angle $\phi_1 = -11.27°$
Displacement factor $DF = \cos\phi_1 = \cos(-11.27) = 0.9807$ (lagging)

From Equation 7.3, the input power factor is

$$PF = \frac{1}{(1+THD^2)^{1/2}}\cos\phi_1 = \frac{1}{(1+0.728^2)^{1/2}} \times 0.9807 = 0.7928\,(\text{lagging})$$

Note: The power factor of a converter with sinusoidal PWM control is better than that with PWM control for the same amount of pulse width or modulation index.

LTspice: LTspice schematic for Example 11.10 is shown in Figure 11.44(a). The plots of the input current, the input power, the output voltage, and the load power are shown in Figure 11.44(b). The rms input current as shown in Figure 11.44(c) is 10.544 A, the average input power as shown in Figure 11.44(d) is 775.6 W, and the average output power as shown in Figure 11.44(e) is738 W. The rms value of the output voltage is 76.24 V. We can find the power efficiency of the converter as $\eta = \frac{P_{out}}{P_{in}} = \frac{738}{775.6} = 95.2\%$

We can find the input power factor as $PF_i = \frac{P_{in}}{V_s I_s} = \frac{775.6}{120\times 10.54} = 0.61$

Note: For a low output power, the losses in the semiconductor devices are significant compared to the output power and the efficiency becomes low. There is no power loss in the load inductor.

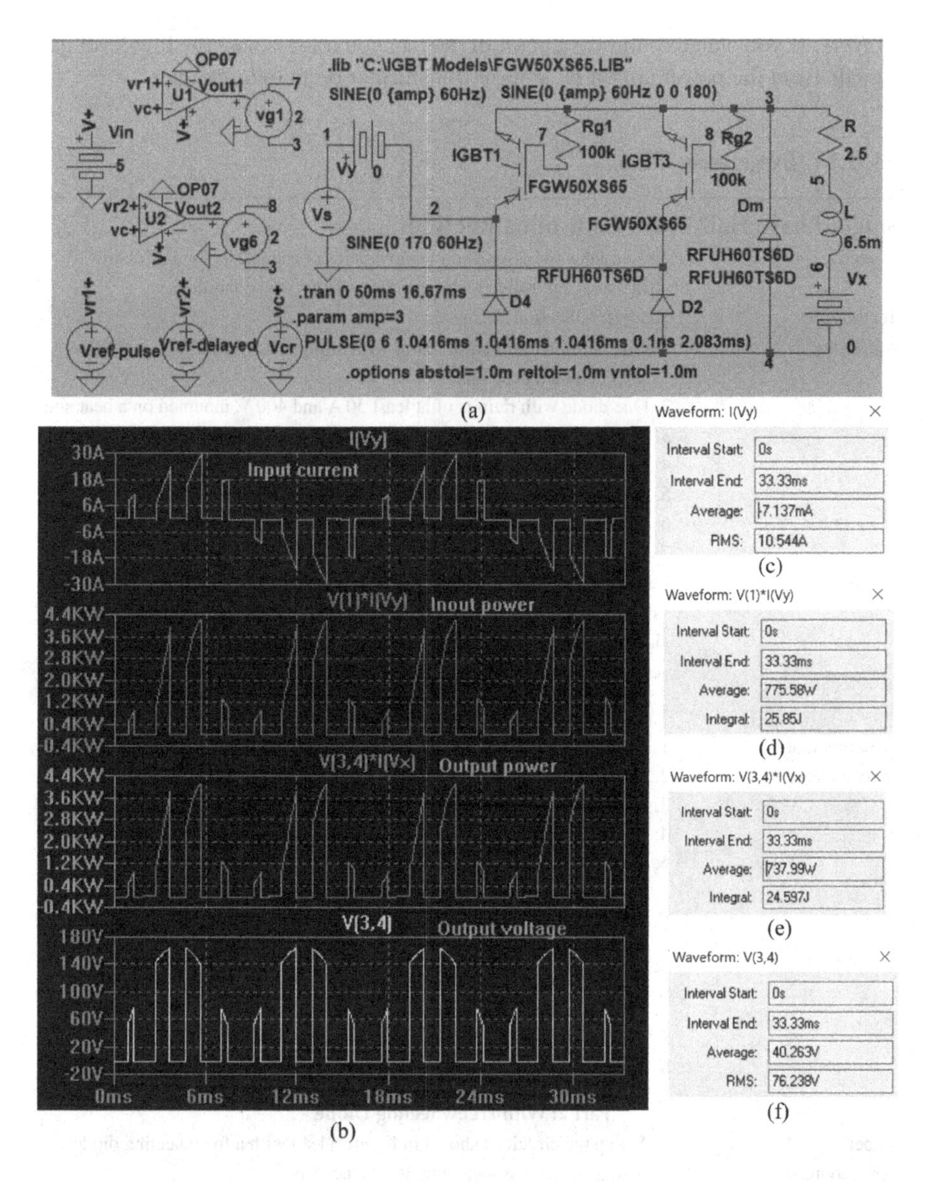

FIGURE 11.44 LTspice schematic for Example 11.10. (a) LTspice Schematic. (b) Output waveforms. (c) Input current. (d) Input power. (e) Output power. (f) Load voltage.

11.8 LABORATORY EXPERIMENTS

It is possible to develop many experiments for demonstrating the operation and characteristics of thyristor-controlled rectifiers. The following experiments are suggested:

Single-phase half-wave controlled rectifier
Single-phase full-wave controlled rectifier
Three-phase full-wave controlled rectifier

Note: If you plan to build the circuit in the lab, you must isolate the low-level gate signals from the power circuit by a pulse transformer or an opto-coupler.

11.8.1 Experiment TC.1

Single-Phase Half-Wave Controlled Rectifier

Objective — To study the operation and characteristics of a single-phase half-wave (thyristor) controlled rectifier under various load conditions.

Textbook — See Ref. 1, Section 10.2.

Apparatus
1. One phase-controlled thyristor with ratings of at least 50 A and 400 V mounted on a heat sink.
2. One diode with ratings of at least 50 A and 400 V, mounted on a heat sink.
3. A firing pulse generator with isolating signals for gating thyristors.
4. An *RL* load.
5. One dual-beam oscilloscope with floating or isolating probes.
6. AC and DC voltmeters and ammeters and one noninductive shunt.
7. One isolation transformer.

Warning — Before making any circuit connection, switch off the AC power. Do not switch on the power unless the circuit is checked and approved by your laboratory instructor. Do not touch the thyristor heat sinks, which are connected to live terminals.

Part 1: Without Freewheeling Diode

Experimental procedure
1. Set up the circuit as shown in Figure 11.35. Use the load resistance *R* only.
2. Connect the measuring instruments as required.
3. Connected the firing pulses to the appropriate thyristors.
4. Set the delay angle to $\alpha = \pi/3$.
5. Observe and record waveforms of the load voltage v_o and the load current i_0.
6. Measure the average load voltage $V_{o(DC)}$, the RMS load voltage $V_{o(RMS)}$, the average load current $I_{o(DC)}$, the RMS load current $I_{o(RMS)}$, the RMS input current $I_{s(RMS)}$, the RMS input voltage $V_{s(RMS)}$, and the load power P_L.
7. Measure the conduction angle of the thyristor T_1.
8. Repeat Steps 2–7 with the load inductance *L* only.
9. Repeat Steps 2–7 with both load resistance *R* and load inductance *L*.

Part 2: With Freewheeling Diode

Experimental procedure
1. Set up the circuit as shown in Figure 11.45 with a freewheeling diode across the load as shown by the dashed lines.
2. Repeat the steps in Part 1.

Report
1. Present all recorded waveforms and discuss all significant points.
2. Compare the waveforms generated by SPICE with the experimental results, and comment.
3. Compare the experimental results with the predicted results.
4. Calculate and plot the average output voltage $V_{o(DC)}$ and the input power factor *PF* against the delay angle α.
5. Discuss the advantages and disadvantages of this type of rectifier.
6. Discuss the effects of the freewheeling diode on the performance of the rectifier.

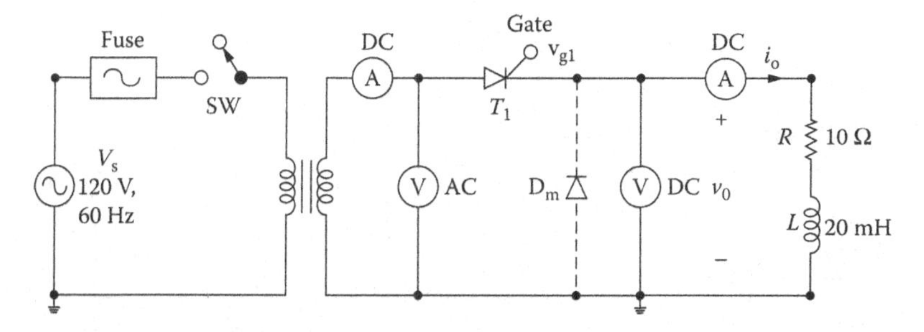

FIGURE 11.45 Single-phase half-wave controlled rectifier.

11.8.2 Experiment TC.2

Single-Phase Full-Wave Controlled Rectifier

Objective	To study the operation and characteristics of a single-phase full-wave (thyristor) controlled rectifier under various load conditions.
Applications	The single-phase full-wave controlled rectifier is used to control power flow, power supplies, variable-speed DC motor drives, input stages of other converters, etc.
Textbook	See Ref. 1, Sections 10.3 and 10.4.
Apparatus	1. Four phase-controlled thyristors with ratings of at least 50 A and 400 V, mounted on heat sinks. 2. One diode with ratings of at least 50 A and 400 V, mounted on a heat sink. 3. A firing pulse generator with isolating signals for gating thyristors. 4. An *RL* load. 5. One dual-beam oscilloscope with floating or isolating probes. 6. AC and DC voltmeters and ammeters and one noninductive shunt.
Warning	See Experiment TC.1.
Experimental procedure	Set up the circuit as shown in Figure 11.46. Repeat the steps in Parts 1 and 2 of Experiment TC.1 on the single-phase half-wave controlled rectifier.
Report	Repeat the steps in Experiment TC.1 for the single-phase half-wave controlled rectifier.

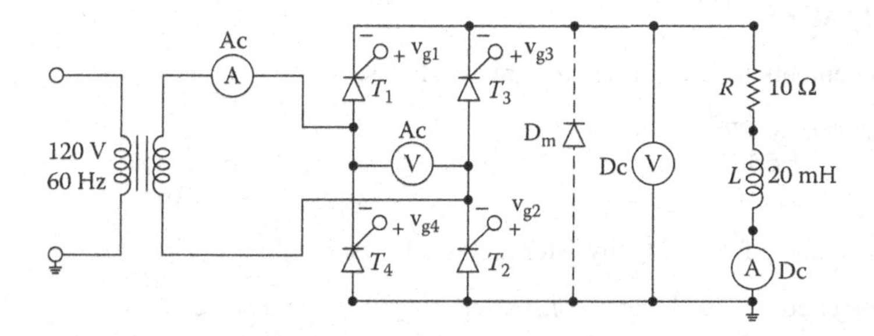

FIGURE 11.46 Single-phase full-wave controlled rectifier.

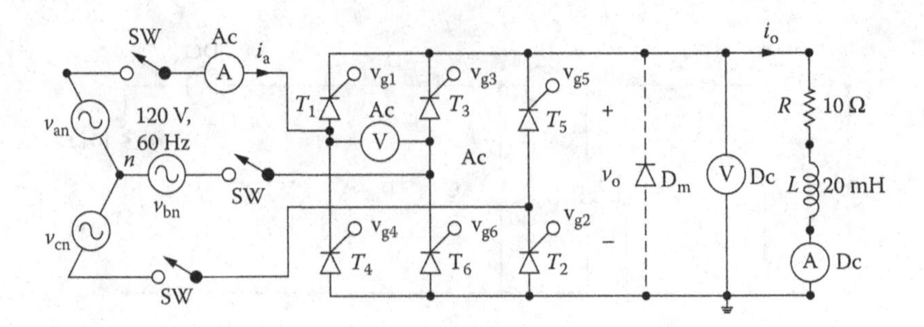

FIGURE 11.47 Three-phase full-wave controlled rectifier.

11.8.3 Experiment TC.3

Three-Phase Full-Wave Controlled Rectifier

Objective: To study the operation and characteristics of a three-phase full-wave (thyristor) controlled rectifier under various load conditions.

Applications: The three-phase full-wave controlled rectifier is used to control power flow, power supplies, variable-speed DC motor drives, input stages of other converters, etc.

Textbook: See Ref. 1, Sections 10.6 and 10.7.

Apparatus:

1. Six phase-controlled thyristors with ratings of at least 50 A and 400 V, mounted on heat sinks.
2. One diode with ratings of at least 50 A and 400 V, mounted on a heat sink.
3. A firing pulse generator with isolating signals for gating thyristors.
4. An *RL* load.
5. One dual-beam oscilloscope with floating or isolating probes.
6. AC and DC voltmeters and ammeters and one noninductive shunt.

Warning: See Experiment TC.1.

Experimental procedure: Set up the circuit as shown in Figure 11.47. Repeat the steps in Parts 1 and 2 for Experiment TC.1 on the single-phase half-wave controlled rectifier.

Report:

1. Present all recorded waveforms and discuss all significant points.
2. Compare the waveforms generated by SPICE with the experimental results, and comment.
3. Compare the experimental results with the predicted results.
4. Discuss the advantages and disadvantages of this type of rectifier.
5. Discuss the effects of the freewheeling diode on the performance of the rectifier.

SUMMARY

The statements for a DC thyristor are as follows:

```
.SUBCKT   SSCR     1       2          3          4
*        model  anode  cathode  +control   -control
*        name                    voltage   voltage
```

The statements for an AC thyristor are as follows:

```
* Subcircuit call for switched thyristor model:
XT1        NA        NK        +NG        −NG        SCR
         anode   cathode  +control   −control    model
```

```
*          voltage   voltage    name
* Subcircuit call for PWM control:
XPW      VR        VC          +NG         - NG        PWM
*       ref.   carrier  +control  -control   model
*      input    input     voltage   voltage    name
* Subcircuit calls for sinusoidal PWM control:
XSPW   VR         VS          + NG        - NG            VC            SPWM
*      ref.  sine-wave +control- control      rectified      model
*     input    input    voltage   voltage carrier sine wave name
```

DESIGN PROBLEMS

11.1 Design the single-phase semiconverter of Figure 11.8 with the following specifications:

AC supply voltage $V_s = 120$ V (RMS), 60 Hz

Load resistance $R = 5\ \Omega$

Load inductance $L = 15$ mH

DC output voltage $V_{o(DC)} = 80\%$ of the maximum permissible value.

a. Determine the ratings of all components and devices under worst-case conditions.

b. Use SPICE to verify your design.

c. Provide a component cost estimate of the circuit.

11.2 a. Design an output C filter for the single-phase semiconverter of Problem 11.1. The harmonic content of the load current should be less than 5% of the average value without the filter.

b. Use SPICE to verify your design in part (a).

11.3 Design the single-phase full converter of Figure 11.12(a) with the following specifications:

AC supply voltage $V_s = 120$ V (RMS), 60 Hz

Load resistance $R = 5\ \Omega$

Load inductance $L = 15$ mH

DC output voltage $V_{o(DC)} = 80\%$ of the maximum value

a. Determine the ratings of all components and devices under worst-case conditions.

b. Use SPICE to verify your design.

c. Provide a component cost estimate of the circuit.

11.4 a. Design an output C filter for the single-phase semiconverter of Problem 11.3. The harmonic content of the load current should be less than 5% of the average value without the filter.

b. Use SPICE to verify your design in part (a).

11.5 Design the three-phase semiconverter of Figure 11.20(a) with the following specifications:

AC supply voltage per phase, $V_s = 120$ V (RMS), 60 Hz

Load resistance $R = 5\ \Omega$

Load inductance $L = 15$ mH

DC output voltage $V_{o(DC)} = 80\%$ of the maximum value

a. Determine the ratings of all components and devices under worst-case conditions.
b. Use SPICE to verify your design.
c. Provide a component cost estimate of the circuit.

11.6 a. Design an output *C* filter for the three-phase semiconverter of Problem 11.5. The harmonic content of the load current should be less than 5% of the average value without the filter.
b. Use SPICE to verify your design in part (a).

11.7 Design the three-phase full-bridge converter of Figure 11.24(a) with the following specifications:
AC supply voltage per phase, $V_s = 120$ V (RMS), 60 Hz
Load resistance $R = 5\ \Omega$
Load inductance $L = 15$ mH
DC output voltage $V_{o(DC)} = 80\%$ of the maximum value
a. Determine the ratings of all components and devices under worst-case conditions.
b. Use SPICE to verify your design.
c. Provide a component cost estimate of the circuit.

11.8 a. Design an output C filter for the converter of Problem 11.7. The harmonic content of the load current should be less than 5% of the average value without the filter.
b. Use SPICE to verify your design in part (a).

11.9 Use PSpice to find the worst-case minimum and maximum output voltages ($V_{o(min)}$ and $V_{o(max)}$) for Problem 11.1. Assume uniform tolerances of ±15% for all passive elements and an operating temperature of 25°C.

11.10 Use PSpice to find the worst-case minimum and maximum output voltages ($V_{o(min)}$ and $V_{o(max)}$) for Problem 11.2. Assume uniform tolerances of ±15% for all passive elements and an operating temperature of 25°C.

11.11 Use PSpice to find the worst-case minimum and maximum output voltages ($V_{o(min)}$ and $V_{o(max)}$) and for Problem 11.3. Assume uniform tolerances of ±15% for all passive elements and an operating temperature of 25°C.

11.12 Use PSpice to find the worst-case minimum and maximum output voltages ($V_{o(min)}$ and $V_{o(max)}$) for Problem 11.4. Assume uniform tolerances of ±15% for all passive elements and an operating temperature of 25°C.

11.13 Use PSpice to find the worst-case minimum and maximum output voltages ($V_{o(min)}$ and $V_{o(max)}$) for Problem 11.5. Assume uniform tolerances of ±15% for all passive elements and an operating temperature of 25°C.

11.14 Use PSpice to find the worst-case minimum and maximum output voltages ($V_{o(min)}$ and $V_{o(max)}$) for Problem 11.6. Assume uniform tolerances of ±15% for all passive elements and an operating temperature of 25°C.

11.15 Use PSpice to find the worst-case minimum and maximum output voltages ($V_{o(min)}$ and $V_{o(max)}$) for Problem 11.7. Assume uniform tolerances of ±15% for all passive elements and an operating temperature of 25°C.

11.16 Use PSpice to find the worst-case minimum and maximum output voltages ($V_{o(min)}$ and $V_{o(max)}$) for Problem 11.8. Assume uniform tolerances of ±15% for all passive elements and an operating temperature of 25°C.

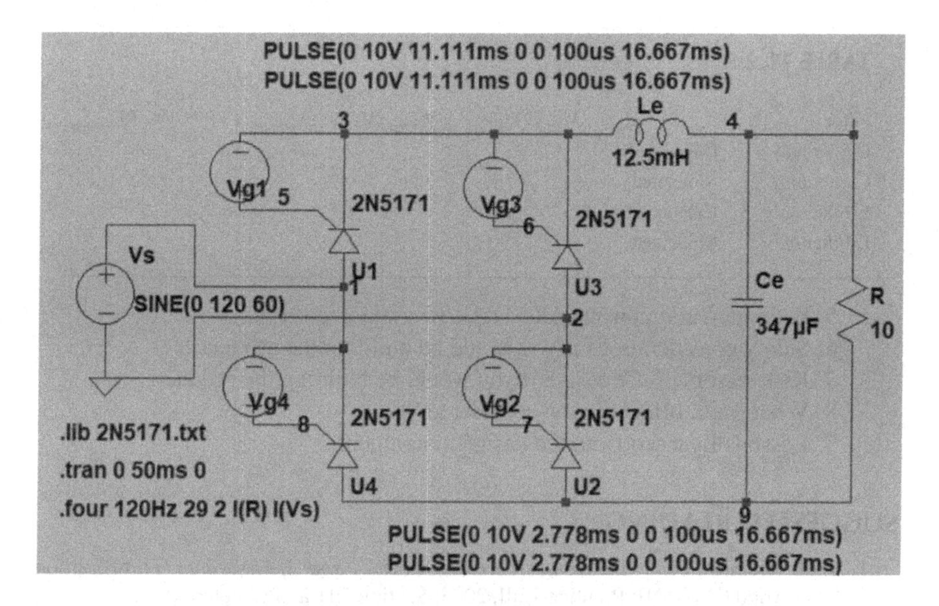

FIGURE 11.48 Single-phase full converter.

11.17 The single-phase full converter as shown in Figure 11.48 has a resistive load of 10 Ω. The input voltage is V_s = 120 V at (rms) 60 Hz. The delay angle is $\omega t = \alpha = 60$ degrees.

1. Calculate the performance parameters. Assume ideal thyristor switches.
   a. The average output current I_{dc}
   b. The steady-state initial load current I_{Lo} at the delay angle $\omega t = \alpha$
   c. The average output current I_{dc}
   d. The rms output current I_{rms}
   e. The average thyristor current I_A
   f. The rms thyristor current I_R,
   g. Find the value of series inductor Le so that the ripple on the load current is within 5% of the average value, assume Ce = 0
   h. Find the value of filter capacitor Ce so that its impedance is much smaller than the load resistance R so that most of the harmonic currents flow through the filter capacitor Ce
2. Use Multisim/LTspice/Orcad. Apply a pulse-voltage of 10V (peak) with a pulse width of 100 μs. Use the subcircuit for thyristor type 2N5171
3. Run the simulation from 0 to 50 ms so that the steady-state condition is reached. Make sure that the thyristors are turned on and operated as switches.
   a. Plot the instantaneous output voltage $v_o(t)$
   b. Plot the instantaneous input current from the supply $i_s(t)$
   c. Include .four commands to find the THD of the outcome current with 29th harmonic contents at the output frequency of 120Hz.
4. Compute Table 11.2

TABLE 11.2

Filter		V_{avg} (V)	$V_{o(rms)}$ (V)	$V_{o(ripple)}$ (V)	% $RF_{o(output)}$
C filter only	Calculated				
C filter only	Simulated				
L-Filter only	Calculated				
L-Filter only	Simulated				

5. Estimated component costs: https://www.jameco.com/
6. Safety considerations if you would be building the product:
7. Risk factors of the design if you would be building the product:
8. What trade-offs have you considered?
9. Lessons Learned from the design assignment:

SUGGESTED READING

1. M.H. Rashid, *Power Electronics Circuits, Devices and Applications*, Third Edition, Englewood Cliffs, NJ: Prentice-Hall, 2003, Section 7.11 and Chapter 10.
2. M.H. Rashid, *Introduction to PSpice Using OrCAD for Circuits and Electronics*, Third Edition, Englewood Cliffs, NJ: Prentice-Hall, 2003, Chapter 6.
3. L.J. Giacoletto, Simple SCR and TRIAC PSPICE computer models, *IEEE Transactions on Industrial Electronics, 36(3)*, 1989, Chapt 451–455.
4. G.L. Arsov, Comments on a nonideal macromodel of thyristor for transient analysis in power electronics, *IEEE Transactions on Industrial Electronics, 39*(2), 1992, 175–176.
5. F.J. Garcia, F. Arizti, F.J. Aranceta, A nonideal macromodel of thyristor for transient analysis in power electronics, *IEEE Transactions on Industrial Electronics, 37(6)*, 1990, 514–520.
6. R.L. Avant, F.C.Y. Lee, A nonideal macromodel of thyristor for transient analysis in power electronics, *IEEE Transactions on Industrial Electronics, 32(1)*, 1985, 1–12.
7. E.Y. Ho, P.C. Sen, Effect of gate drive on GTO thyristor characteristics, *IEEE Transactions on Industrial Electronics*, IE*33*(3), 1986, 325–331.
8. M.A.I. El-Amin, *GTO PSPICE Model and Its Applications, The Fourth Saudi Engineering Conference*, November 1995, Vol. III, 271–277.
9. G. Busatto, F. Iannuzzo and L. Fratelli, *Proceedings of Symposium on Power Electronics Electrical Drives, Advanced Machine Power Quality, The Fourth Saudi Engineering Conference*, June 3–5, 1998, Sorrento, Italy, Col. 1, pp. P2/5–10.
10. A.A. Zekry, G.T. Sayah and F.A. Soliman, SPICE model of thyristors with amplifying gate and emitter-shorts, *IET Power Electronics, 7*(3), 2014, 724–735. https://doi.org/10.1049/iet-pel.2013.0158
11. A. Barili, G. Cottafava and E. Dallago, A SPICE2 SCR model for power circuit analysis. *Computer-Aided Design, 16*(5), 1984, 279–284. https://doi.org/10.1016/0010-4485(84)90086-1
12. Y.C. Liang and V.J. Gosbell, Transient model for gate turn-off thyristor in power electronic simulations, *International Journal of Electronics, 70*(1), 1991, 85–99, DOI: 10.1080/00207219108921259
13. A.A. Zekry, G.T. Sayah, SPICE model of thyristors with amplifying gate and emitter-shorts, *IET Power Electronics, 7* (3), March 2014, 724–735, https://doi.org/10.1049/iet-pel.2013.0158

12 AC Voltage Controllers

After completing this chapter, students should be able to do the following:

- Model SCRs for AC applications and specify their model parameters.
- Model voltage-controlled switches.
- Perform transient analysis of AC voltage controllers.
- Evaluate the performance of AC voltage controllers.
- Perform worst-case analysis of AC voltage controllers for parametric variations of model parameters and tolerances.

12.1 INTRODUCTION

The input voltage and output current of AC voltage controllers pass through the value zero in every cycle. This simplifies the modeling of a thyristor. A thyristor can be turned on by applying a pulse of short duration, and it is turned off by natural commutation due to the characteristics of the input voltage and the current. PWM techniques can be applied to AC voltage controllers with bidirectional switches in order to improve the input power factor of the converters.

12.2 AC THYRISTOR MODEL

The load current of AC voltage controllers is AC, and the current of a thyristor always passes through zero. There being no need for the diode D_T of Figure 11.1(b), the thyristor model can be simplified to Figure 12.1. This model can be used as a subcircuit. Switch S_j is controlled by the controlling voltage V_R connected between

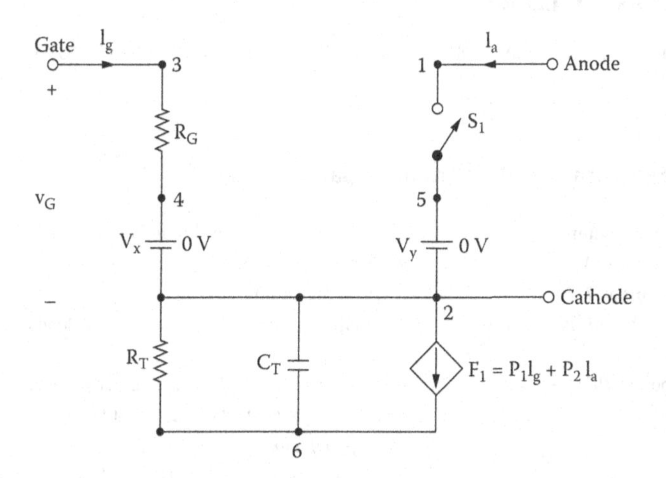

FIGURE 12.1 AC thyristor model.

DOI: 10.1201/9781003284451-12

nodes 6 and 2. The switch parameters can be adjusted to yield the desired on-state voltage drop across the thyristor. In the following examples, we will use the switch parameters RON = 0.01, ROFF = 10E+5, VON = 0.1 V, and VOFF = 0 V. The other parameters are discussed in Section 11.2.

The subcircuit definition for the thyristor model ASCR can be described as follows:

```
* Subcircuit for AC thyristor model
.SUBCKT   ASCR     1      2        3         2
*         model   anode  cathode  +control  -control
*         name                    voltage   voltage
S1    1    5     6    2   SMOD        ; Switch
RG    3    4     50
VX    4    2     DC   0V
VY    5    2     DC   0V
RT    2    6     1
CT    6    2     10UF
F1    2    6     POLY(2)    VX    VY    0    50   11
.MODEL SMOD VSWITCH (RON=0.01 ROFF=10E+5 VON=0.1V VOFF=0V)
.ENDS ASCR ; Ends subcircuit definition
```

12.3 PHASE-CONTROLLED AC VOLTAGE CONTROLLERS

The input to an AC voltage controller is a fixed AC voltage, and its output is a variable AC voltage. When the converter switches are turned on, the input voltage is connected to the load. The output voltage, which is varied by changing the conduction time of the switches, is discontinuous. The input power factor is low. Table 12.1 lists the performance parameters of AC voltage rectifiers.

12.4 EXAMPLES OF PHASE-CONTROLLED AC VOLTAGE CONTROLLERS

The applications of the AC thyristor model are illustrated by the following examples.

TABLE 12.1
AC Voltage Converter Performance parameters

Input Side Parameters	Output Side Parameters
RMS input voltage, V_s	RMS output voltage, V_o
RMS input current, I_s	RMS output current, I_o
The *transformer utilization factor,* $TUF = \frac{P_{dc}}{V_s I_s}$	The output power, $P_o = V_o I_o = I_o^2 R_L$ for a load resistance of R_L
The input power, $PF = \frac{P_{ac}}{V_s I_s}$	RMS value of the fundamental output voltage, V_{o1}
	The rms ripple content of the output voltage $V_{ac} = \sqrt{V_o^2 - V_{o1}^2}$
	The *ripple factor,* $RF_o = \frac{V_{ac}}{V_{o1}}$
	Total harmonic distortion (THD) of the output voltage, %

12.4.1 Examples of Single-Phase AC Voltage Controllers

Example 12.1:

Finding the Performance of a Single-Phase Full-Wave AC Voltage Controller

A single-phase full-wave AC voltage controller is shown in Figure 12.2(a). The input voltage has a peak of 169.7 V, 60 Hz. The load inductance L is 6.5 mH, and the load resistance R is 2.5 Ω. The delay angles are equal: $\alpha_1 = \alpha_2 = 90°$. The gate voltages are shown in Figure 12.2(b). Use PSpice to (a) plot the instantaneous output voltage v_o and the load current i_o and (b) calculate the Fourier coefficients of the input current i_s and the input power factor PF.

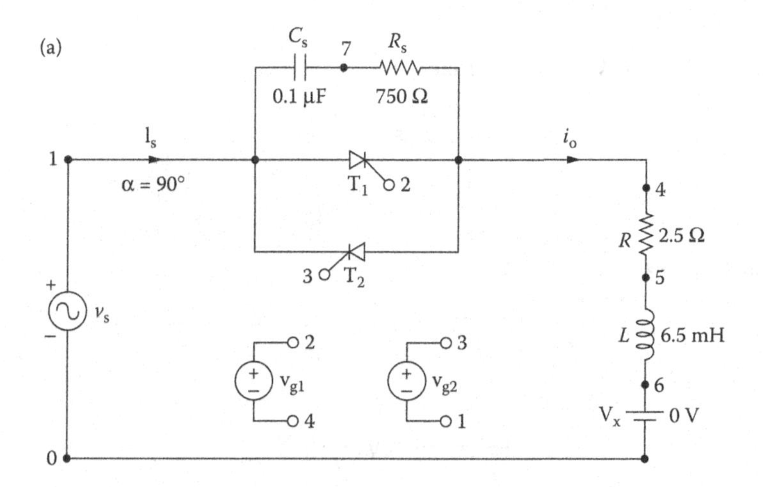

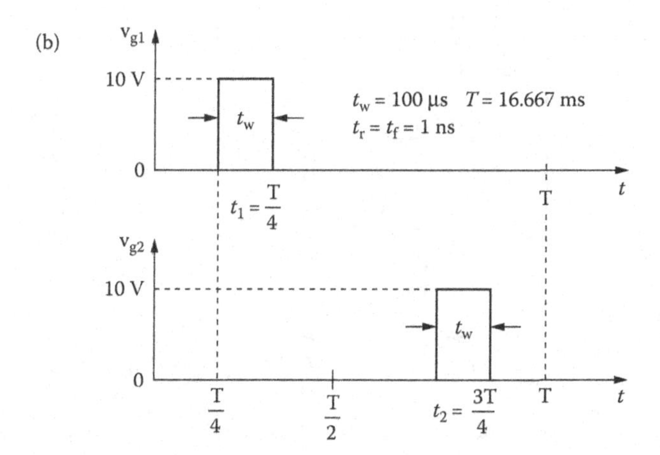

FIGURE 12.2 Single-phase full-wave AC voltage controller for PSpice simulation. (a) Circuit. (b) Gate voltages.

SOLUTION

The peak supply voltage V_m = 169.7 V, for $\alpha_1 = \alpha_2 = 90°$, time delay t_1 = (90/360) × (1000/60 Hz) × 1000 = 4166.7µs. A series snubber with C_s = 0.1 µF and R_s = 750Ω is connected across the thyristor to cope with the transient voltage due to the inductive load and also to protect the devices from high *dv/dt*.

The PSpice schematic with an SCR is shown in Figure 12.3. The model name of the SCR, 2N1595, is changed to ASCR, whose subcircuit definition is listed in Section 12.2. By varying the delay cycle, the output voltage can be changed. The supply frequency {FREQ} and the duty cycle {DELAY_ANGLE} are defined as variables.

Note that the voltage-controlled voltage sources E_1 and E_2 (as shown in Figure 12.3) are connected to provide the isolation between the low signal gating signals from the power circuit. This could be done in a practical circuit with a pulse transformer or an opto-coupler isolation.

The listing of the circuit file is as follows:

Single-Phase AC Voltage Controller

```
SOURCE   ■    VS 1 0 SIN (0 169.7V 60HZ)
              .PARAM DELAY_ANGLE=90 FREQ=60Hz
              Vg1 2 4 PULSE ( 0 10 {{Delay_Angle}/
              {360*{Freq}}} 1ns 1ns 100us {1/{Freq}})
              Vg2 3 1  PULSE ( 0 10 {{{Delay_Angle} + 180}/
              {360*{Freq}}} 1ns 1ns 100us {1/{Freq}})
CIRCUIT  ■■   R 4 5 2.5
              L 5 6 6.5MH
              VX 6 0 DC 0V ; Voltage source to measure the
              load current
```

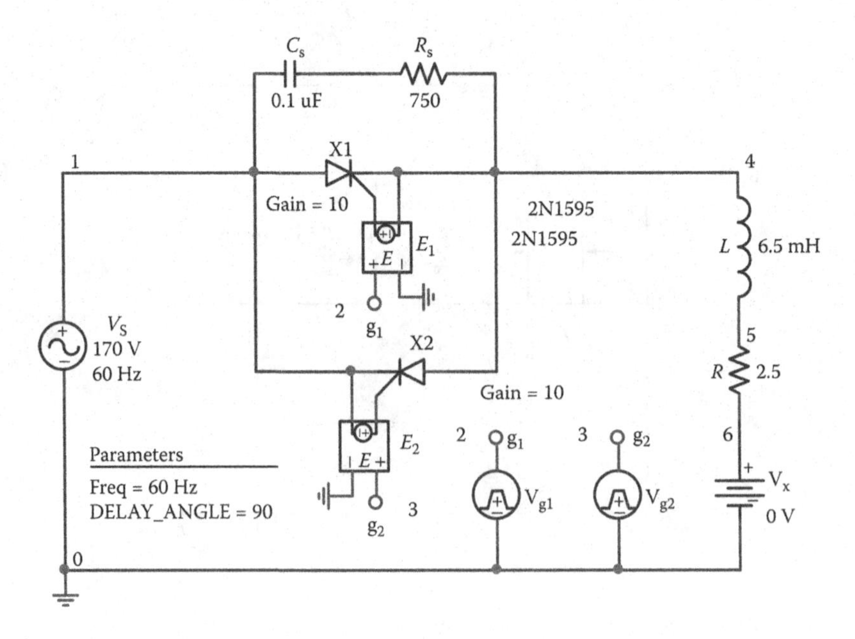

FIGURE 12.3 PSpice schematic for Example 12.1.

```
                CS 1 7 0.1UF
                RS 7 4 750
                * Subcircuit call for thyristor model:
                XT1 1 4 2 4 ASCR ; Thyristor T1
                XT2 4 1 3 1 ASCR ; Thyristor T2
                * Subcircuit ASCR, which is missing, must be *
                * inserted.
ANALYSIS ■■■    .TRAN 1US 33.33MS ; Transient analysis
                .PROBE ; Graphics post-processor
                .OPTIONS ABSTOL=1.00N RELTOL=1.0M VNTOL=1.0M
                ITL5=10000;
                *Convergence
                .FOUR 60HZ I(VX) ; Fourier analysis
           .END
```

Note the following:

a. The PSpice plots of instantaneous output voltage V(4) and load current I(VX) are shown in Figure 12.4. The output voltage and current are discontinuous as expected.
b. The Fourier series of the input current, which is the same as the current through source VX, is as follows:

Fourier Components of Transient Response I(VX)

DC Component = –1.31252104

Harmonic Number	Frequency (Hz)	Fourier Component	Normalized Component	Phase (Deg)	Normalized Phase (Deg)
1	6000E+01	2.883E+01	1.000E+00	–6.2444E+01	0.000E+00
2	1.200E+02	1.068E–04	3.704E–06	1.714E+01	7.958E+01
3	1.800E+02	7.966E+00	2.763E–01	–2.494E+00	5.995E+01
4	2.400E+02	9.049E–05	3.139E–06	–1.676E+02	–1.052E+02
5	3.000E+02	2.671E+00	9.265E–02	–1.472E+02	–8.473E+01
6	3.600E+02	3.556E–05	1.233E–06	3.215E+01	9.459E+01
7	4.200E+02	4.161E–01	1.443E–02	3.990E+01	1.023E+02
8	4.800E+02	3.673E–05	1.274E–06	1.672E+02	2.296E+02
9	5.400E+02	5.995E–01	2.079E–02	1.496E+02	2.120E+02

Total harmonic distortion = 2.925271E+01%

From the Fourier components, we get

Total harmonic distortion of input current $THD = 29.25\% = 0.2925$
Displacement angle $\phi_1 = -62.44°$
Displacement factor $DF = \cos(-62.44) = 0.4627$(lagging)

From Equation 7.3, the input power factor is

$$PF = \frac{1}{(1 + THD^2)^{1/2}} \cos\phi = \frac{1}{(1 + 0.2925^2)^{1/2}} \times 0.4627 = 0.444 \text{ (lagging)}$$

Note: The input power will decrease with the delay angle of the converter.

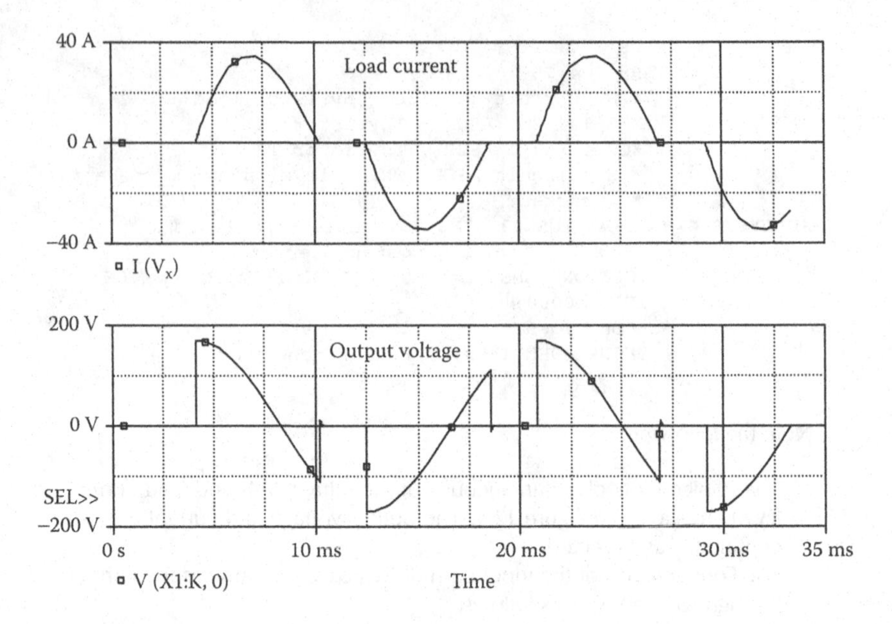

FIGURE 12.4 Plots for Example 12.1 for delay angle $\alpha = 90°$.

LTspice:. The gating signals for Example 12.1 are as shown in Figure 12.5 for an on-time width of 80 µs, and a period of 200 µs. The gating signal for T_1 is delayed by angle α and the delay angle for T_2 is delayed by angle of $(\pi+\alpha)$.

LTspice schematic for Example 12.1 is shown in Figure 12.6(a). The plots of the input current, the input power, the output voltage, and the load power are shown in Figure 12.6(b). The rms input current as shown in Figure 12.6(c) is 20.57 A, the average input power as shown in Figure 12.6(d) is 1172 W, and the average output power as shown in Figure 12.6(e) is 1145 W. As shown in Figure 5.6(f), the rms value of the output load voltage is 88.58 V. The rms value of the output voltage is 204.3 V. We can find the power efficiency of the converter as $\eta = \frac{P_{out}}{P_{in}} = \frac{1145}{1172} = 97.7\%$

We can find the input power factor as $PF_i = \frac{P_{in}}{V_s I_s} = \frac{1172}{120 \times 20.57} = 0.502$

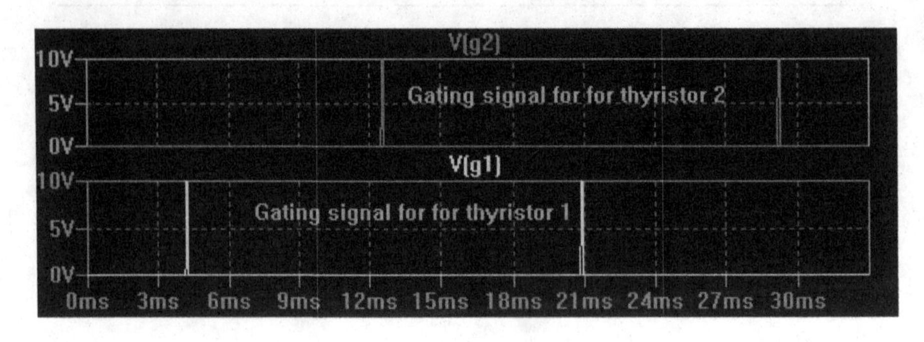

FIGURE 12.5 LTspice gating signals for Example 12.1.

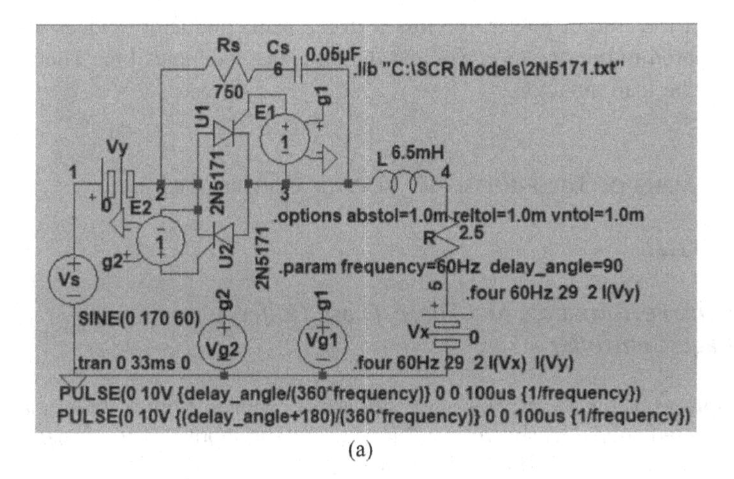

(a)

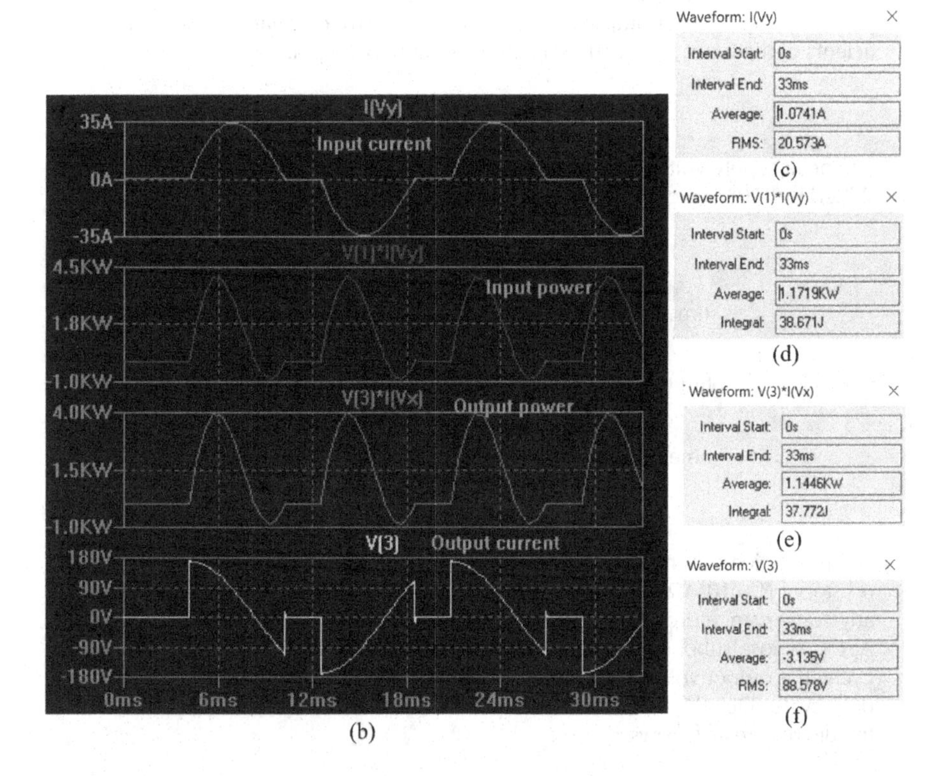

FIGURE 12.6 LTspice schematic for Example 12.1. (a) LTspice Schematic. (b) Output waveforms. (c) Input current. (d) Input power. (e) Output power. (f) Load voltage.

Note: For a low output power, the losses in the semiconductor devices are significant compared to the output power and the efficiency becomes low. There is no power loss in the load inductor.

12.4.2 Examples of Three-Phase AC Voltage Controllers

Example 12.2:

Finding the Performance of a Three-Phase Half-Wave AC Voltage Controller

A single-phase half-wave AC voltage controller is supplied from a three-phase wye-connected supply as shown in Figure 12.7(a). The input phase voltage has a peak of 169.7 V, 60 Hz. The load resistance per phase is $R = 2.5\ \Omega$. The delay angle is $\alpha = 60°$. The gate voltages are shown in Figure 12.7(b). Use PSpice to (a) plot the instantaneous output phase voltage v_o and (b) calculate the Fourier coefficients of the input phase current i_s and the input power factor *PF*.

SOLUTION

The peak supply voltage per phase $V_m = 169.7$ V. for $\alpha = 60°$, since each phase is shifted by 120°, we can get the delay times as

$$\text{time delay}\, t_1 = \frac{60}{360} \times \frac{1000}{60\text{Hz}} \times 1000 = 2777.78\,\mu\text{s}$$

$$\text{time delay}\, t_2 = \frac{180}{360} \times \frac{1000}{60\text{Hz}} \times 1000 = 8333.3\,\mu\text{s}$$

$$\text{time delay}\, t_3 = \frac{300}{360} \times \frac{1000}{60\text{Hz}} \times 1000 = 13{,}888.9\,\mu\text{s}$$

The PSpice schematic with SCRs is shown in Figure 12.8. Note that a large resistor of $R_s = 10\text{M}\Omega$ is connected across the SCRs to allow a DC path and reduce any convergence problems. The model name of the SCR, 2N1595, is changed to ASCR, whose subcircuit definition is listed in Section 12.2. By varying the delay cycle, the output voltage can be changed. The supply frequency {FREQ} and the duty cycle {DELAY_ANGLE} are defined as variables. The model parameters for the diodes are as follows:

```
.MODEL DMD D(IS=2.22E-15 BV=1200V CJO=0PF TT=0US) for
diodes
```

Note that the voltage-controlled voltage sources E_l, E_3, and E_5 (as shown in Figure 12.8) are connected to provide the isolation between the low signal gating signals from the power circuit. This could be done in a practical circuit with a pulse transformer or an opto-coupler isolation.

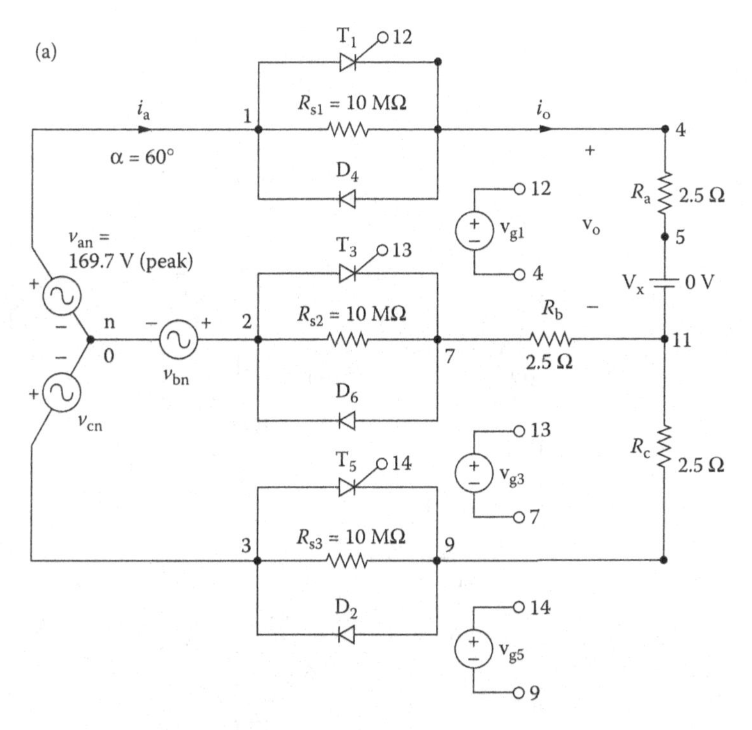

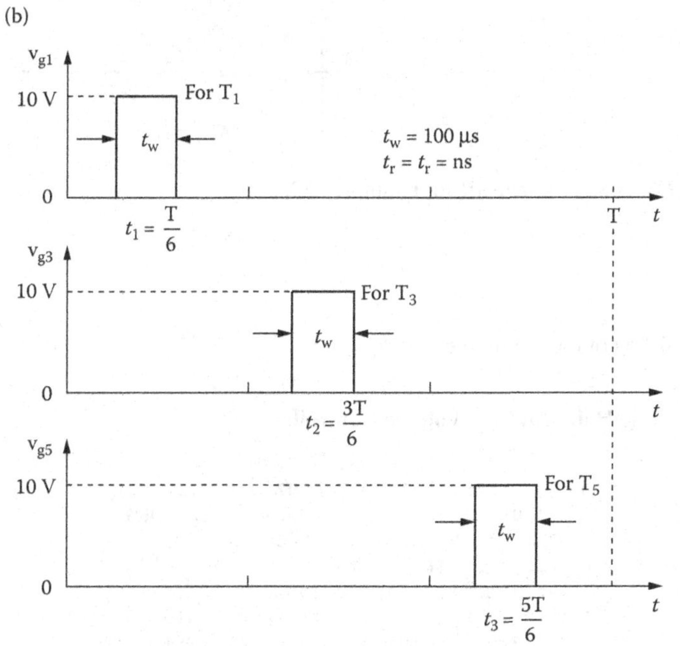

FIGURE 12.7 (p12.5) Three-phase half-wave AC controller for PSpice simulation. (a) Circuit. (b) Gate voltages.

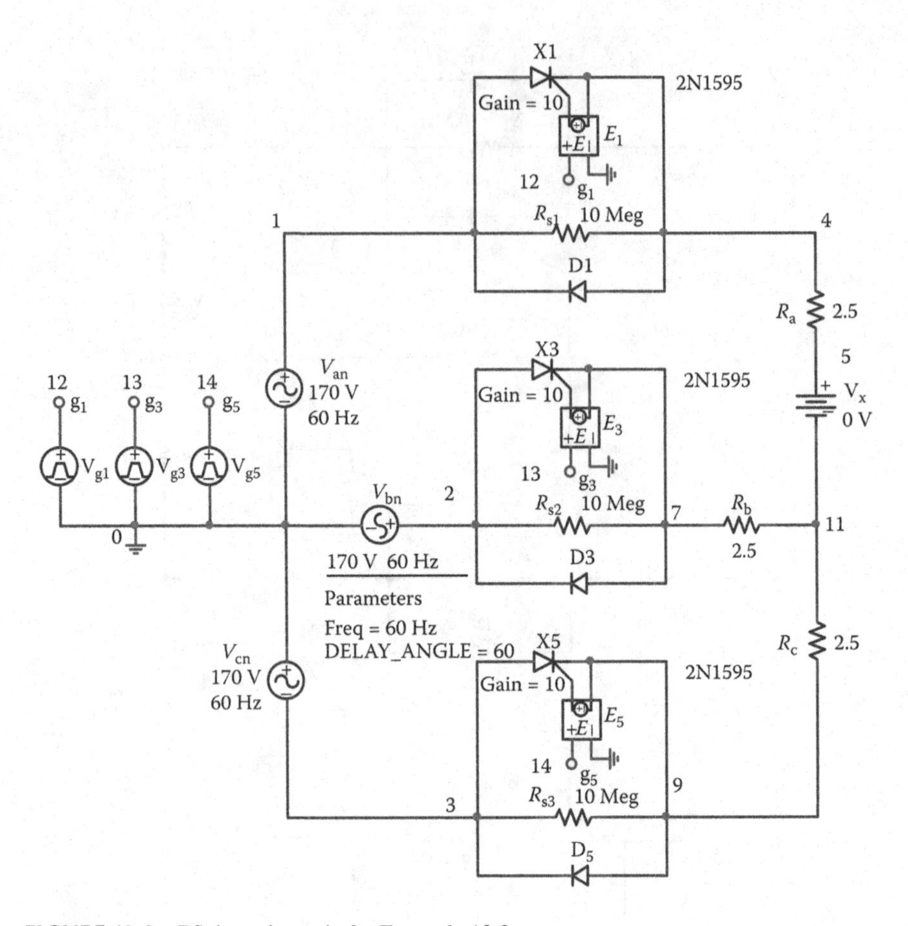

FIGURE 12.8 PSpice schematic for Example 12.2.

The listing of the circuit file is as follows:

Three-Phase Half-Wave AC Voltage Controller

```
SOURCE   ■  Van 1 0 SIN (0 169.7V 60HZ)
            Vbn 2 0 SIN (0 169.7V 60HZ 0 0 -120DEG)
            Vcn 3 0 SIN (0 169.7V 60HZ 0 0 -240DEG)
            .PARAM DELAY_ANGLE=60 FREQ=60Hz
            Vg1 12 4 PULSE ( 0 10 {{Delay_Angle}/
            {360*{Freq}}} 1ns 1ns 100us{1/{Freq}})
            Vg3 13 7 PULSE (0 10 {{{Delay_Angle} + 120}/
            {360*{Freq}}} 1ns 1ns 100us {1/{Freq}})
            Vg5 14 9 PULSE (0 10 {{{Delay_Angle} + 240}/
            {360*{Freq}}} 1ns 1ns 100us {1/{Freq}})
CIRCUIT  ■■ Rs1 1 4 10MEG
            Rs2 2 7 10MEG
```

```
             Rs3 3 9 10MEG
             Ra 4 5 2.5
             VX 5 11 DC 0V ; To measure load current
             Rb 7 11 2.5
             Rc 9 11 2.5
             * Subcircuit calls for thyristor model:
             XT1 1 4 12 4 ASCR ; Thyristor T1
             XT3 2 7 13 7 ASCR ; Thyristor T3
             XT5 3 9 14 9 ASCR ; Thyristor T5
             D2 9 3 DMOD ; Diode
             D4 4 1 DMOD ; Diode
             D6 7 2 DMOD ; Diode
             .MODEL DMOD D (IS=2.22E-15 BV=1200V IBV=13E-3
             CJ0=2PF TT=1US)
             * Subcircuit ASCR, which is missing, must be *
             * inserted.
ANALYSIS ■■■ .TRAN 10US 33.33MS 0 0.1MS ; Transient analysis
             .PROBE ; Graphics post-processor
             .OPTIONS  ABSTOL=1.00N RELTOL=0.01 VNTOL=0.01
             ITL5=10000;
             *                                  Convergence
             .FOUR 60HZ I(VX)
         .END
```

Note the following:

a. The PSpice plots of the instantaneous output voltage V(4,11) are shown in Figure 12.9. Note that the output voltage is discontinuous and has a high harmonic content.
b. The Fourier series of the input current is as follows:

Fourier Components of Transient Response I(VX)

DC Component = −1.573343E−02

Harmonic Number	Frequency (Hz)	Fourier Component	Normalized Component	Phase (Deg)	Normalized Phase (Deg)
1	6.000E+01	5.881E+01	1.000E+00	−1.195E+01	0.000E+00
2	1.200E+02	1.423E+01	2.419E−01	−1.695E+02	−1.576E+02
3	1.800E+02	3.855E−02	6.555E−04	8.972E+01	1.017E+02
4	2.400E+02	9.580E+00	1.629E−01	1.064E+02	1.184E+02
5	3.000E+02	6.952E+00	1.182E−01	5.972E+01	7.167E+01
6	3.600E+02	3.150E−02	5.356E−04	−9.084E+01	−7.889E+01
7	4.200E+02	3.487E+00	5.930E−02	−6.156E+01	−4.961E+01
8	4.800E+02	3.140E+00	5.339E−02	−1.305E+02	−1.185E+02
9	5.400E+02	3.854E−02	6.554E−04	8.912E+01	1.011E+02

Total harmonic distortion = 3.246548E+01%

From the Fourier components, we get

Total harmonic distortion of input current $THD = 32.47\% = 0.3247$
Displacement angle $\phi_1 = -11.95°$
Displacement factor $DF = \cos\phi_1 = \cos(-11.95) = 0.9783$(lagging)

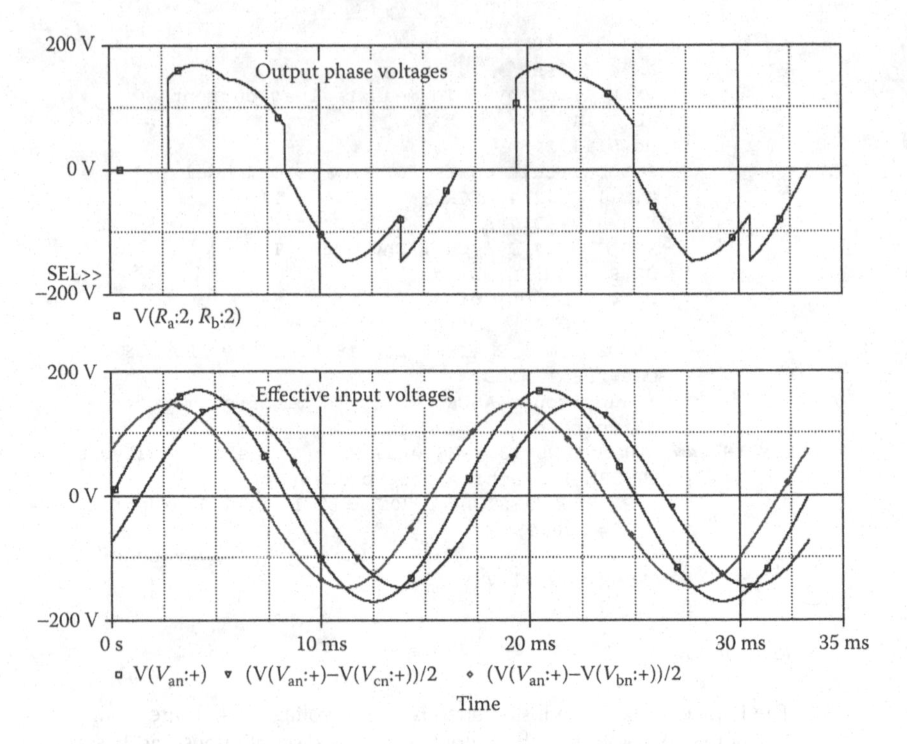

FIGURE 12.9 Plots for Example 12.2 for a delay angle of $\alpha = 60°$.

From Equation 7.3, the input power factor is as follows:

$$PF = \frac{1}{(1+THD^2)^{1/2}}\cos\phi = \frac{1}{(1+0.3247^2)^{1/2}} \times 0.9783 = 0.93 \text{ (lagging)}$$

Note: The instantaneous output voltage depends on the instantaneous effective input voltages as shown in Figure 12.9.

LTspice: The LTspice gating signals for Example 12.2 for a delay angle of 90° as shown are phase-shifted by 120°are shown in Figure 12.10.

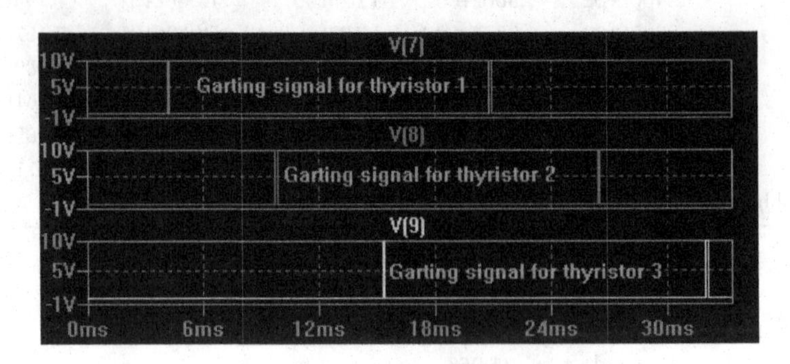

FIGURE 12.10 LTspice gating signals for Example 12.2.

LTspice schematic for Example 12.2 is shown in Figure 12.11(a).The plots of the input current, the input power, the output voltage, and the load power are shown in Figure 12.11(b). The rms input current as shown in Figure 12.11(c) is 37.45 A, the average input power as shown in Figure 12.11(d) is 10673 W, and the average output power as shown in Figure 12.11(e) is 10519 W. The rms value of the output voltage is 93.62 V, as shown in Figure 12.11(f). We can find the power efficiency of the converter as $\eta = \frac{P_{out}}{P_{in}} = \frac{10519}{10673} = 98.56\%$

We can find the input power factor as $PF_i = \frac{P_{in}}{3V_s I_s} = \frac{10673}{3 \times 120 \times 37.45} = 0.79$

Note: For a low output power, the losses in the semiconductor devices are significant compared to the output power and the efficiency becomes low. There is no power loss in the load inductor.

Example 12.3:

Finding the Performance of a Three-Phase Full-Wave AC Voltage Controller

A three-phase full-wave AC voltage controller is supplied from a wye-connected supply as shown in Figure 12.12(a). The input phase voltage has a peak of 169.7 V, 60 Hz. The load resistance per phase is $R = 2.5\Omega$. The delay angle is $\alpha = 60°$. The gate voltages are shown in Figure 12.12(b). Use PSpice to (a) plot the instantaneous output phase voltage v_o and (b) calculate the Fourier coefficients of the input current i_a and the input power factor *PF*.

SOLUTION

The peak supply voltage per phase $V_m = 169.7$ V. for $\alpha_1 = 60°$, since each phase is shifted by 120°, we can get the delay times as

$$\text{time delay}\, t_1 = \frac{60}{360} \times \frac{1000}{60\text{Hz}} \times 1000 = 2777.78\,\mu\text{s}$$

$$\text{time delay}\, t_3 = \frac{180}{360} \times \frac{1000}{60\text{Hz}} \times 1000 = 8333.3\,\mu\text{s}$$

$$\text{time delay}\, t_5 = \frac{300}{360} \times \frac{1000}{60\text{Hz}} \times 1000 = 13{,}888.9\,\mu\text{s}$$

$$\text{time delay}\, t_2 = \frac{120}{360} \times \frac{1000}{60\text{Hz}} \times 1000 = 5555.78\,\mu\text{s}$$

$$\text{time delay}\, t_4 = \frac{240}{360} \times \frac{1000}{60\text{Hz}} \times 1000 = 11{,}111.1\,\mu\text{s}$$

$$\text{time delay}\, t_6 = \frac{360}{360} \times \frac{1000}{60\text{Hz}} \times 1000 = 16{,}666.7\,\mu\text{s}$$

The PSpice schematic with SCRs is shown in Figure 12.13. The model name of the SCR, 2N1595, is changed to ASCR, whose subcircuit definition is listed in

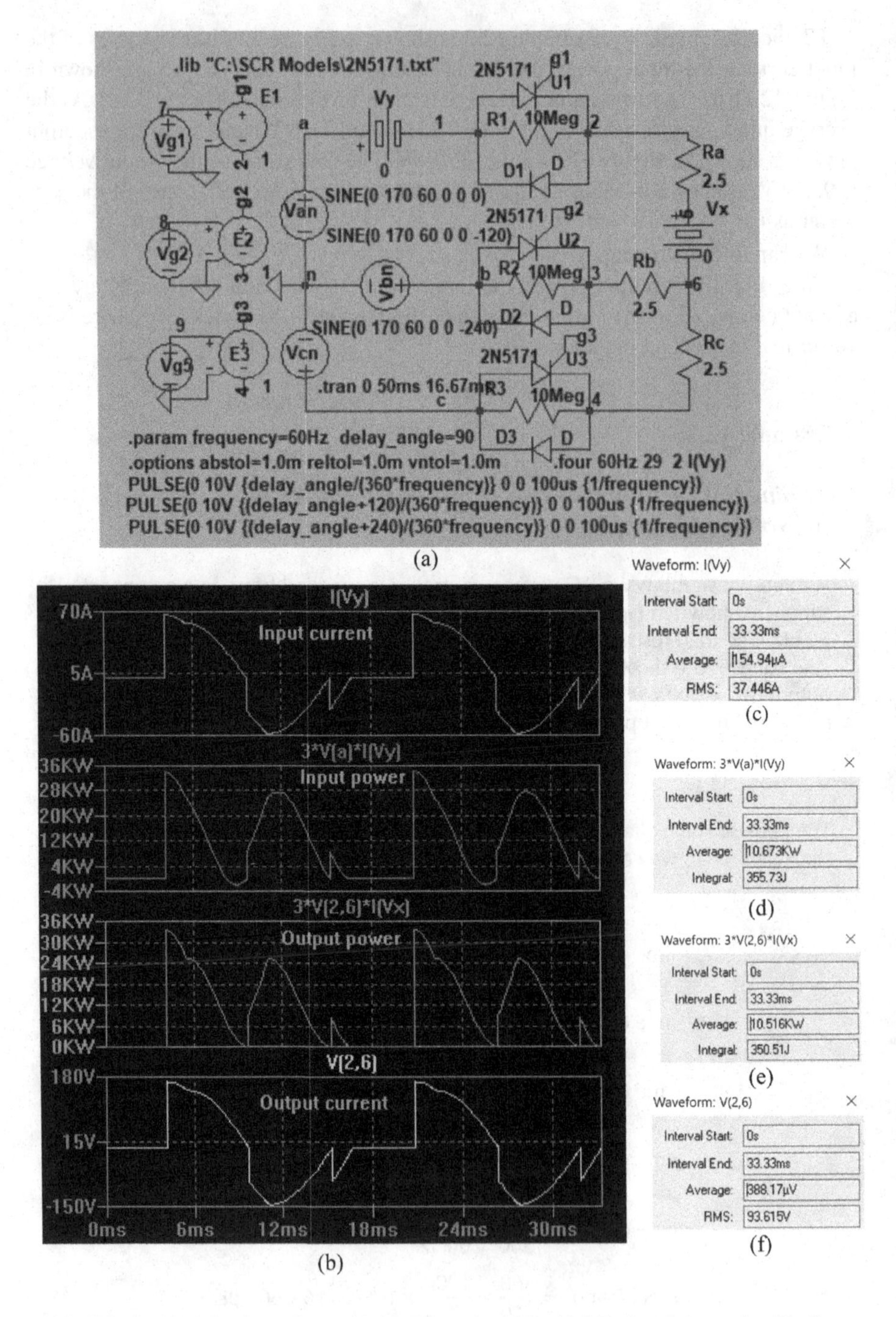

FIGURE 12.11 LTspice schematic for Example 12.2. (a) LTspice Schematic. (b) Output waveforms. (c) Input current. (d) Input power. (e) Output power. (f) Load voltage.

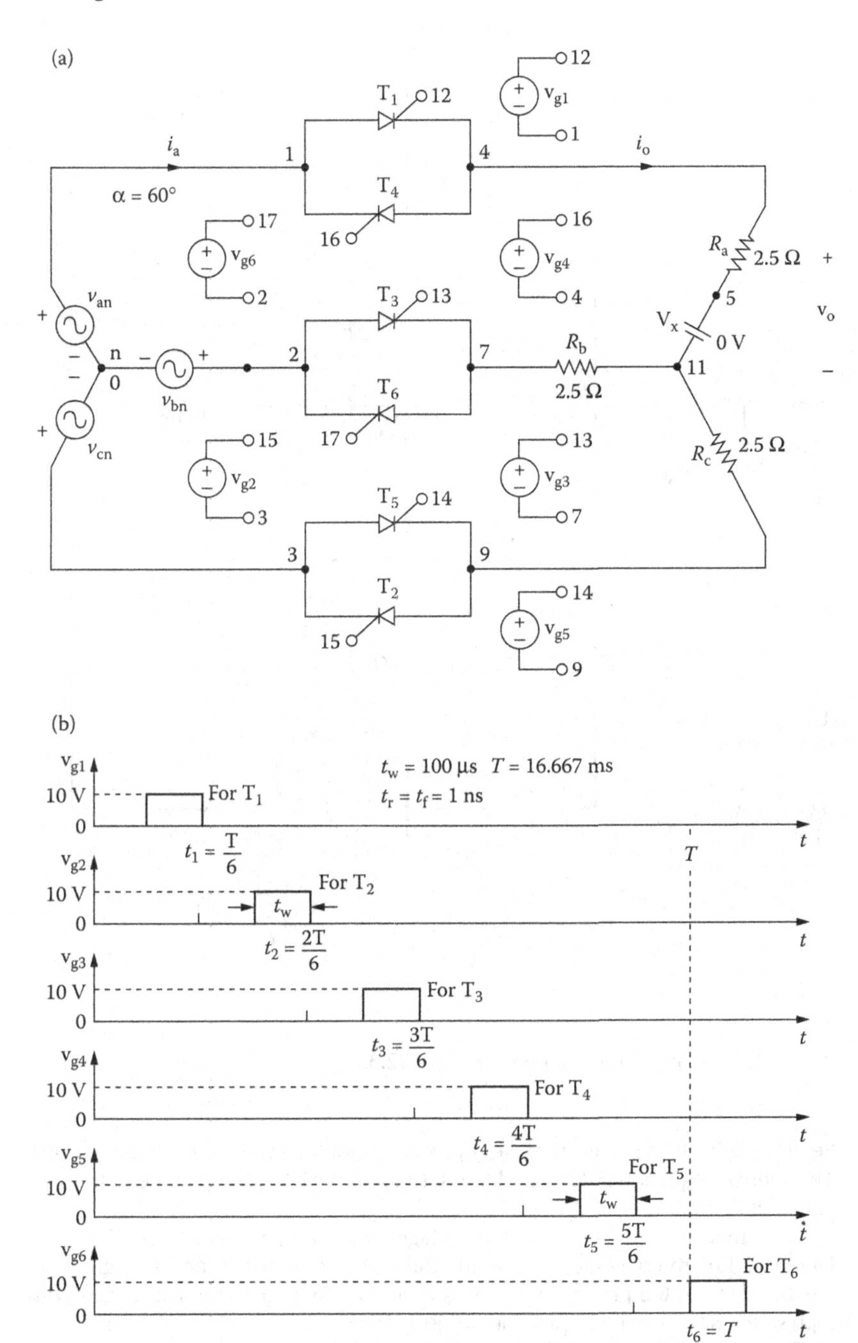

FIGURE 12.12 Three-phase full-wave AC controller for PSpice simulation. (a) Circuit. (b) Gate signal voltages.

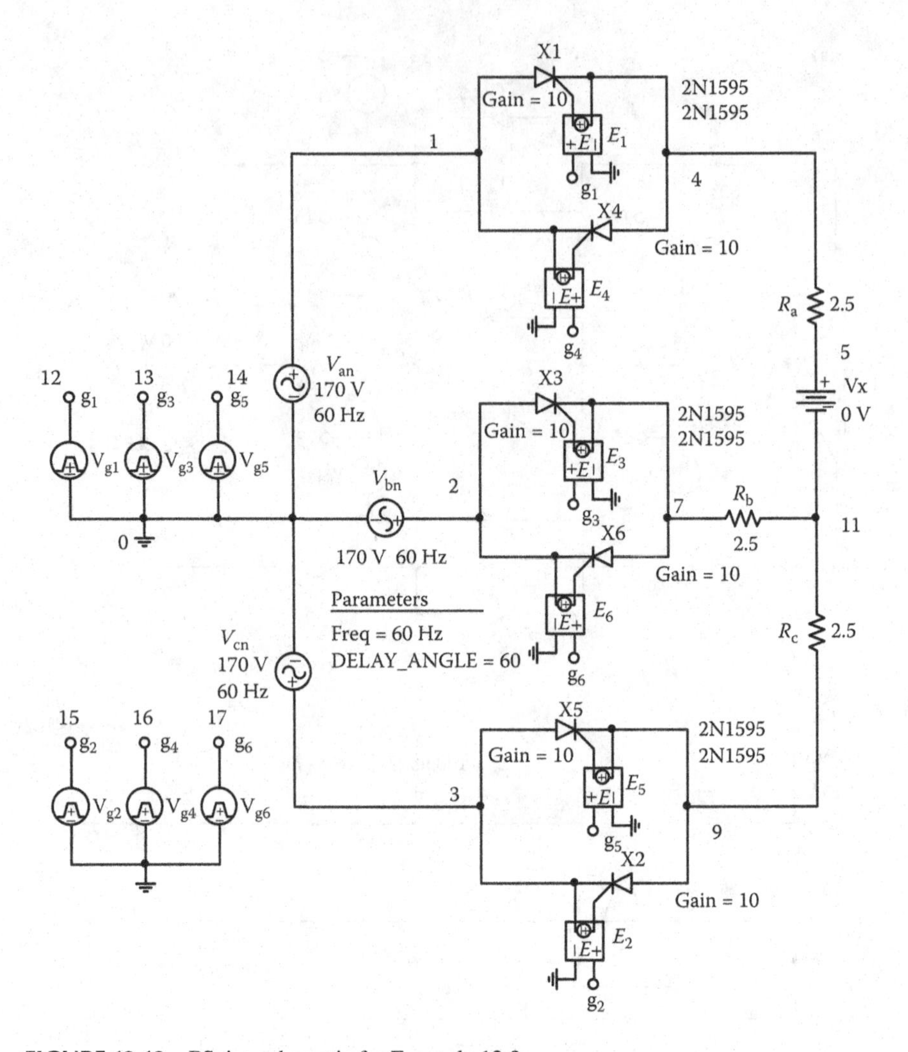

FIGURE 12.13 PSpice schematic for Example 12.3.

Section 12.2. By varying the delay cycle a, the output voltage can be changed. The supply frequency {FREQ} and the duty cycle {DELAY_ANGLE} are defined as variables.

Note that the voltage-controlled voltage sources E_1, E_3, and E_5 (as shown in Figure 12.13) are connected to provide the isolation between the low signal gating signals from the power circuit. This could be done in a practical circuit with a pulse transformer or an opto-coupler isolation.

The listing of the circuit file is as follows:

Three-Phase Full-Wave AC Voltage Controller

```
SOURCE   ■  Van 1 0 SIN (0 169.7V 60HZ)
            Vbn 2 0 SIN (0 169.7V 60HZ 0 0 -120DEG)
            Vcn 3 0 SIN (0 169.7V 60HZ 0 0 -240DEG)
            .PARAM DELAY_ANGLE=60 FREQ=60Hz
```

```
              Vg1 12 4 PULSE (0 10 {{Delay_Angle}/
              {360*{Freq}}} 1ns 1ns 100us {1/{Freq}})
              Vg2 13 7 PULSE (0 10 {({Delay_Angle}+60)/
              {360*{Freq}}} 1ns 1ns 100us {1/{Freq}})
              Vg3 14 9 PULSE (0 10 {{{Delay_Angle}+120}/
              {360*{Freq}}} 1ns 1ns 100us {1/{Freq}})
              Vg4 15 3 PULSE (0 10 {{{Delay_Angle}+180}/
              {360*{Freq}}} 1ns 1ns 100us {1/{Freq}})
              Vg5 16 1 PULSE (0 10 {{{Delay_Angle}+240}/
              {360*{Freq}}} 1ns 1ns 100us {1/{Freq}})
              Vg6 17 2 PULSE (0 10 {{{Delay_Angle}+300}/
              {360*{Freq}}} 1ns 1ns 100us {1/{Freq}})
CIRCUIT  ■■   Ra 4 5 2.5
              VX 5 11 DC 0V ; To measure load current
              Rb 7 11 2.5
              Rc 9 11 2.5
              * Subcircuit calls for thyristor model:
              XT1 1 4 12 4 ASCR ; Thyristor T1
              XT3 2 7 13 7 ASCR ; Thyristor T3
              XT5 3 9 14 9 ASCR ; Thyristor T5
              XT2 9 3 15 3 ASCR ; Thyristor T2
              XT4 4 1 16 1 ASCR ; Thyristor T4
              XT6 7 2 17 2 ASCR ; Thyristor T6
              * Subcircuit ASCR, which is missing, must be inserted.
ANALYSIS ■■■  .TRAN 0.1MS 33.33MS ; Transient analysis
              .PROBE ; Graphics post-processor
              .OPTIONS ABSTOL=1.00N RELTOL=0.01 VNTOL=0.01
              ITL5=10000;
              *                                   Convergence
              .FOUR 60HZ I(VX)
          .END
```

Note the following:

a. The PSpice plots of instantaneous output phase voltage V(4,11) and input voltages are shown in Figure 12.14. Note that the output voltage waveform is discontinuous, but it is a symmetrical AC waveform.
b. The Fourier series of the input phase current is as follows:

Fourier Components of Transient Response I(VX)

DC Component = 3.465652E–05

Harmonic Number	Frequency (Hz)	Fourier Component	Normalized Component	Phase (Deg)	Normalized Phase (Deg)
1	6.000E+01	5.354E+01	1.000E+00	–2.701E+01	0.000E+00
2	1.200E+02	6.775E–05	1.266E–06	9.195E+01	1.190E+02
3	1.800E+02	6.944E–02	1.297E–03	8.967E+01	1.167E+02
4	2.400E+02	6.785E–05	1.267E–06	8.835E+01	1.154E+02
5	3.000E+02	1.396E+01	2.608E–01	5.936E+01	8.637E+01
6	3.600E+02	6.873E–05	1.284E–06	8.983E+01	1.168E+02
7	4.200E+02	7.005E+00	1.308E–01	–6.106E+01	–3.404E+01
8	4.800E+02	6.862E–05	1.282E–06	9.200E+01	1.190E+02
9	5.400E+02	6.945E–02	1.297E–03	8.901E+01	1.160E+02

Total harmonic distortion = 2.917517E+01%

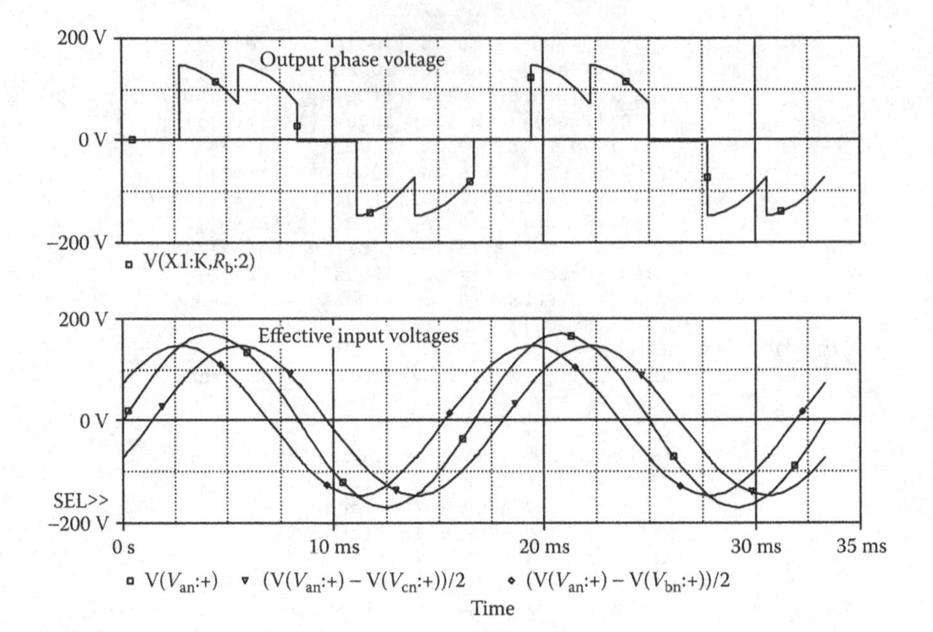

FIGURE 12.14 Plots for Example 12.3 for a delay angle of $\alpha = 60°$.

From the Fourier components, we get

Total harmonic distortion of input current $THD = 29.18\% = 0.2918$
Displacement angle $\phi_1 = -27.01°$
Displacement factor $DF = \cos\phi_1 = \cos(-27.01) = 0.891$(lagging)

From Equation 7.3, the input power factor is

$$PF = \frac{1}{(1+THD^2)^{1/2}}\cos\phi_1 = \frac{1}{(1+02918^2)^{1/2}} \times 0.891 = 0.855 \text{ (lagging)}$$

Note: However, the input power factor of the full-wave controller is lower than that of the half-wave controller for the same delay angle $\alpha = 60°$.

LTspice:. The LTspice gating signals for Example 12.3 for a delay angle of 90° as shown are phase-shifted by 60° from each other as shown n Figure 12.15.

LTspice schematic for Example 12.3 is shown in Figure 12.16(a). The plots of the net filed input current, the input power, the output voltage, and the load power are shown in Figure 12.16(b). The rms input current as shown in Figure 12.16(c) is 37.23 A, the average input power as shown in Figure 12.16(d) is 10610 W, and the average output power as shown in Figure 12.16(e) is 10393 W. The rms value of the output voltage as shown in Figure 12.16(f) is 93.06 V. We can find the power efficiency of the converter as $\eta = \frac{P_{out}}{P_{in}} = \frac{10393}{10610} = 98\%$

We can find the input power factor as $PF_i = \frac{P_{in}}{3V_sI_s} = \frac{10610}{3\times120\times37.23} = 0.792$

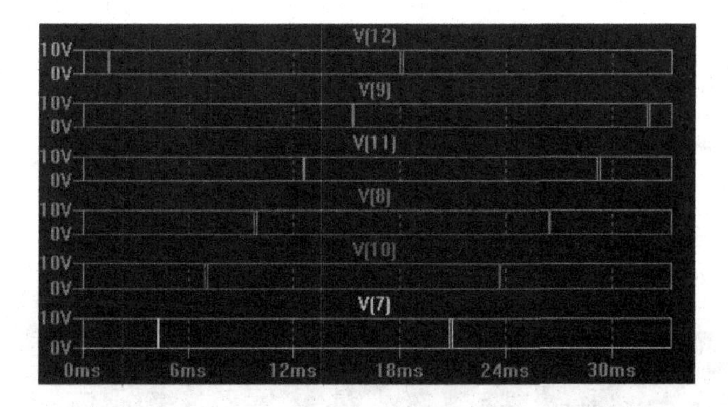

FIGURE 12.15 LTspice gating signals for Example 12.3.

Note: For a low output power, the losses in the semiconductor devices are significant compared to the output power and the efficiency becomes low. There is no power loss in the load inductor.

Example 12.4:

Finding the Performance of a Three-Phase Full-Wave Delta-Connected Controller

A three-phase full-wave delta-connected controller is supplied from a wye-connected three-phase supply as shown in Figure 12.17(a). The input phase voltage has a peak of 169.7V, 60 Hz. The load resistance per phase is $R = 2.5\Omega$. The delay angle is $\alpha = 60°$. The gate voltages are shown in Figure 12.17(b). Use PSpice to (a) plot the instantaneous output phase voltage v_o and the input line current i_a and (b) calculate the Fourier coefficients of the output phase current i_o.

SOLUTION

The PSpice schematic with SCRs is shown in Figure 12.18. The model name of the SCR, 2N1595, is changed to ASCR, whose subcircuit definition is listed in Section 12.2. By varying the delay cycle α, the output voltage can be changed. The supply frequency {FREQ} and the duty cycle {DELAY_ANGLE} are defined as variables.

The delay angles are the same as in Example 12.3. The listing of the circuit file for Figure 12.13(a) is as follows:

Three-Phase Full-Wave Delta-Connected AC Controller

```
SOURCE  ■ Van 1 0 SIN (0 169.7V 60HZ)
          Vbn 2 0 SIN (0 169.7V 60HZ 0 0 -120DEG)
          Vcn 3 0 SIN (0 169.7V 60HZ 0 0 -240DEG)
          .PARAM Freq=60Hz Delay_Angle=90
          Vg1 9 2 PULSE (0 10 {{Delay_Angle}/{360*{Freq}}}
          1ns 1ns 100us {1/{Freq}})
          Vg2 12 8 PULSE (0 10 {({Delay_Angle}+60)/
          {360*{Freq}}} 1ns 1ns 100us {1/{Freq}})
          Vg3 10 3 PULSE (0 10 {{{Delay_Angle}+120}/
          {360*{Freq}}} 1ns 1ns 100us {1/{Freq}})
```

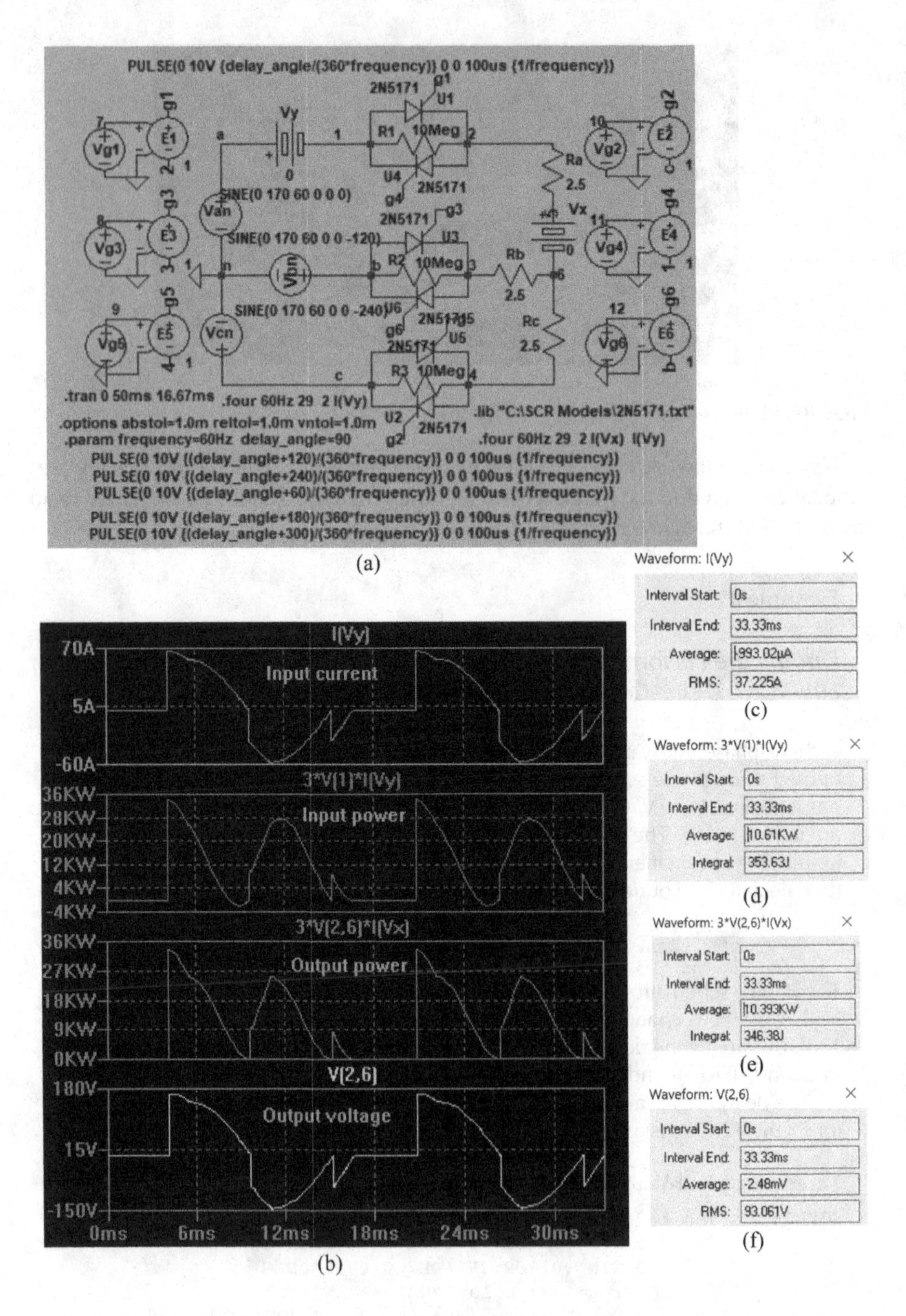

FIGURE 12.16 LTspice schematic for Example 12.3. (a) LTspice schematic. (b) Output waveforms. (c) Input current. (d) Input power. (e) Output power. (f) Load voltage.

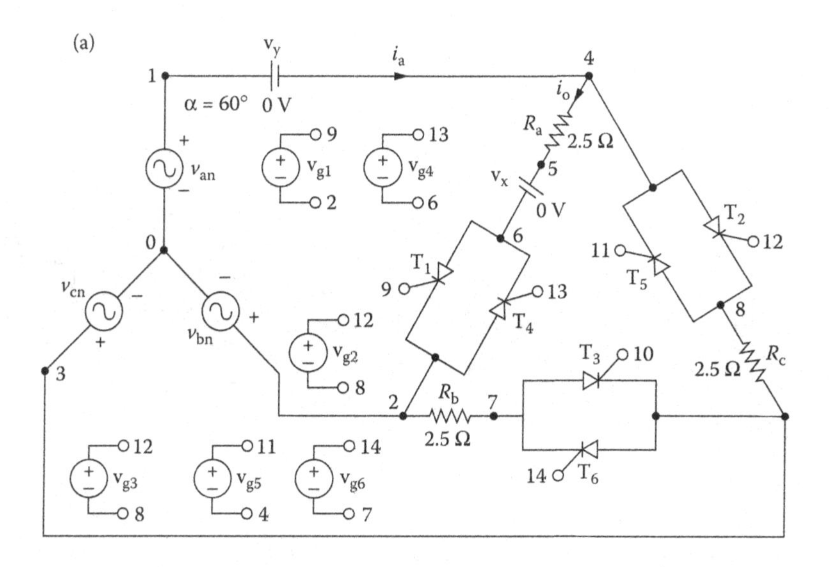

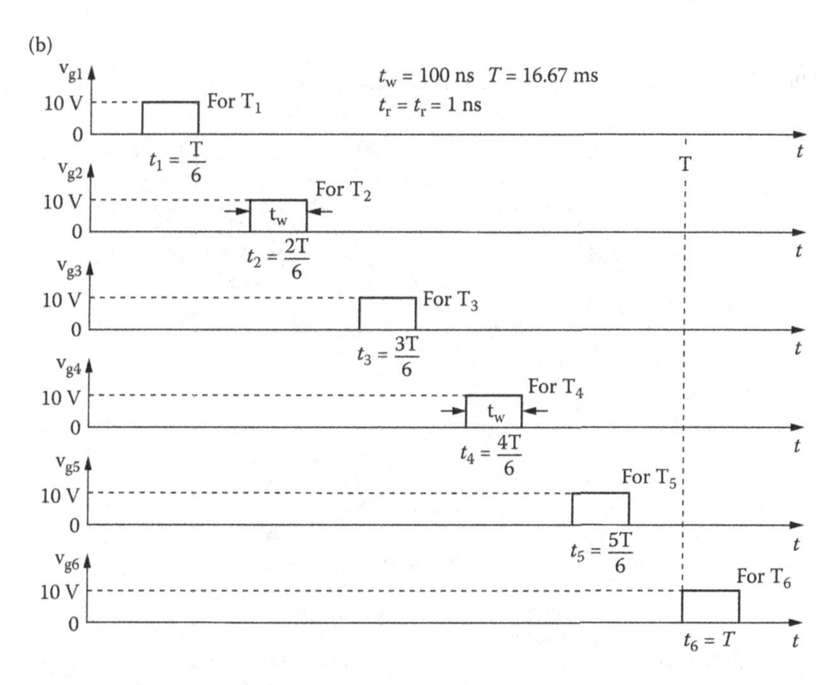

FIGURE 12.17 Three-phase full-wave delta-connected controller for PSpice simulation. (a) Circuit. (b) Gate signal voltages.

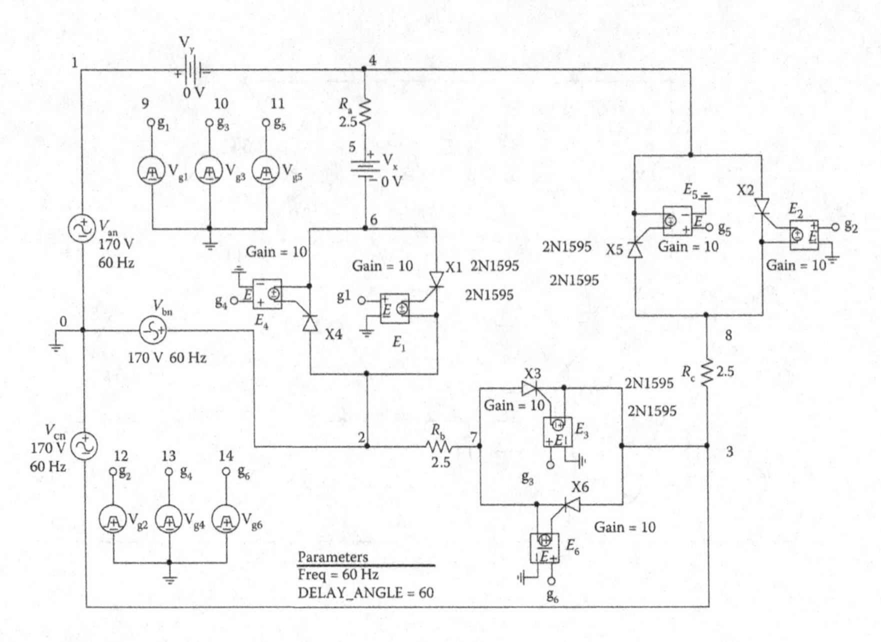

FIGURE 12.18 PSpice schematic for Example 12.4.

```
                 Vg4 13 6 PULSE (0 10 {{{Delay_Angle}+180}/
                 {360*{Freq}}} 1ns 1ns 100us {1/{Freq}})
                 Vg5 11 4 PULSE (0 10 {{{Delay_Angle}+240}/
                 {360*{Freq}}} 1ns 1ns 100us {1/{Freq}})
                 Vg6 14 7 PULSE (0 10 {{{Delay_Angle} + 300}/
                 {360*{Freq}}} 1ns 1ns 100us {1/{Freq}})
                 Ra 4 5 2.5
                 VX 5 6 DC 0V ; Load current ammeter
                 Rb 2 7 2.5
                 Rc 3 8 2.5
                 VY 1 4 DC 0V ; Line current ammeter
CIRCUIT  ■■      * Subcircuit calls for thyristor model:
                 XT1 6 2 9 2 ASCR ; Thyristor T1
                 XT3 7 3 10 3 ASCR ; Thyristor T3
                 XT5 8 4 11 4 ASCR ; Thyristor T5
                 XT2 4 8 12 8 ASCR ; Thyristor T2
                 XT4 2 6 13 6 ASCR ; Thyristor T4
                 XT6 3 7 14 7 ASCR ; Thyristor T6
                 * Subcircuit ASCR, which is missing, must be
                 * inserted.
ANALYSIS ■■■     .TRAN 0.1US 33.33MS ; Transient analysis
                 .PROBE ; Graphics post-processor
                 .OPTIONS ABSTOL=1.00N RELTOL=0.01 VNTOL=0.01
                 ITL5=10000;
                 *                                   Convergence
                 .FOUR 60HZ I(VX)
            .END
```

Note the following:

a. The PSpice plots of the instantaneous output phase voltage V(4,5) and the line current I(VY) are shown in Figure 12.19. The output voltage and

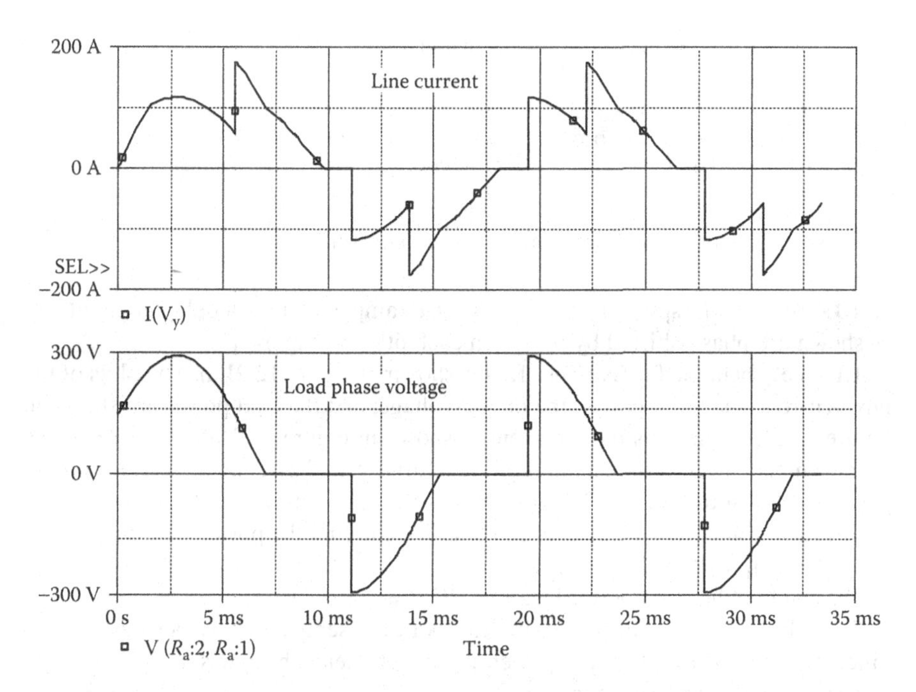

FIGURE 12.19 Plots for Example 12.4 for a delay angle of $\alpha = 60°$.

input current are discontinuous. Note that the output voltage waveform is discontinuous and has a high harmonic content.

b. The Fourier coefficients of the load phase current are as follows.

Fourier Components of Transient Response I(VX)DC

Component = −8.217138E−06

Harmonic Number	Frequency (Hz)	Fourier Component	Normalized Component	Phase (Deg)	Normalized Phase (Deg)
1	6000E+01	6.929E+01	1.000E+00	−2.624E+00	0.000E+00
2	1.200E+02	1.711E−05	2.469E−07	1.321E+01	1.583E+01
3	1.800E+02	3.726E+01	5.377EY01	1.796E+02	1.822E+02
4	2.400E+02	1.747E−05	2.521E−07	1.294E+02	1.321E+02
5	3.000E+02	1.243E+01	1.795E−01	5.899E+01	6.161E+01
6	3.600E+02				
7	4.200E+02	1.241E+01	1.791E−01	−6.115E+01	−5.852E+01
8	4.800E+02	2.332E−05	3.366E−07	2.827E+01	3.089E+01
9	5.400E+02	7.466E+00	1.077E−01	1.783E+02	1.809E+02

Total harmonic distortion = 6.042025E+01%

From the Fourier components, we get

DC component = −8.21 μA
Peak fundamental component at 60 Hz = 69.29 A
Fundamental phase delay $\phi_1 = -2.62°$
Total harmonic distortion = −60.42%

The input power factor can be calculated from

$$PF_i = \frac{1}{\sqrt{1+THD^2}}\cos\phi_1 = \frac{1}{\sqrt{1+0.6042^2}}\cos(-2.62) = 0.855 \text{ (lagging)}$$

Note: The three thyristors form a delta connection.

LTspice: The LTspice gating signals for Example 12.4 for a delay angle of 90° as shown are phase-shifted by 60° from each other in Figure 12.20.

LTspice schematic for Example 12.4 is shown in Figure 12.21(a). The plots of the input current, the input power, the output voltage, and the load power are shown in Figure 12.21(b). The rms input current as shown in Figure 12.21(c) is 121.83 A, the average input power as shown in Figure 12.21(d) is 41.117 kW, and the average output power as shown in Figure 12.21(e) is 40.409 kW. The rms value of the output voltage as shown in Figure 12.21(f) is 183.5 V. We can find the power efficiency of the converter as $\eta = \frac{P_{out}}{P_{in}} = \frac{40409}{41117} = 98.3\%$

We can find the input power factor as $PF_i = \frac{P_{in}}{3V_s I_s} = \frac{41117}{3\times120\times121.83} = 93.75$

Note: For a low output power, the losses in the semiconductor devices are significant compared to the output power and the efficiency becomes low. There is no power loss in the load inductor.

Example 12.5:

Finding the Performance of a Three-Phase Three-Thyristor Delta-Connected Controller

A three-phase three-thyristor controller is supplied from a three-phase wye-connected supply as shown in Figure 12.22(a). The input phase voltage has a peak of 169.7 V, 60 Hz. The load resistance per phase is $R = 2.5\Omega$. The delay angle is

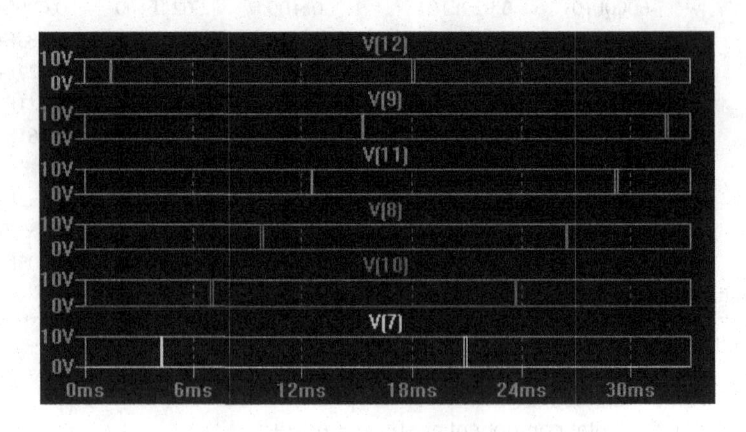

FIGURE 12.20 LTspice gating signals for Example 12.4.

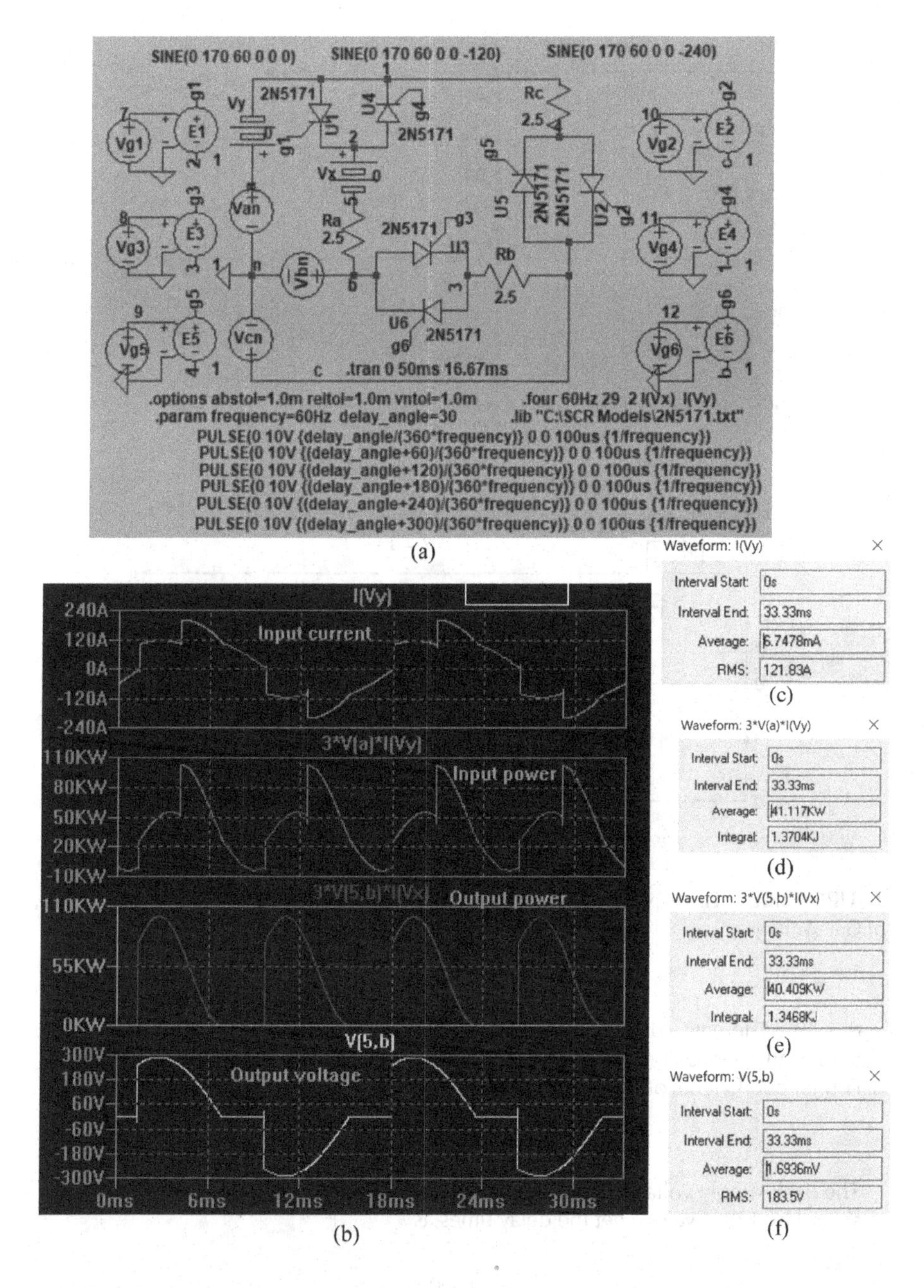

FIGURE 12.21 LTspice schematic for Example 12.4. (a) LTspice schematic. (b) Output waveforms. (c) Input current. (d) Input power. (e) Output power. (f) Load voltage.

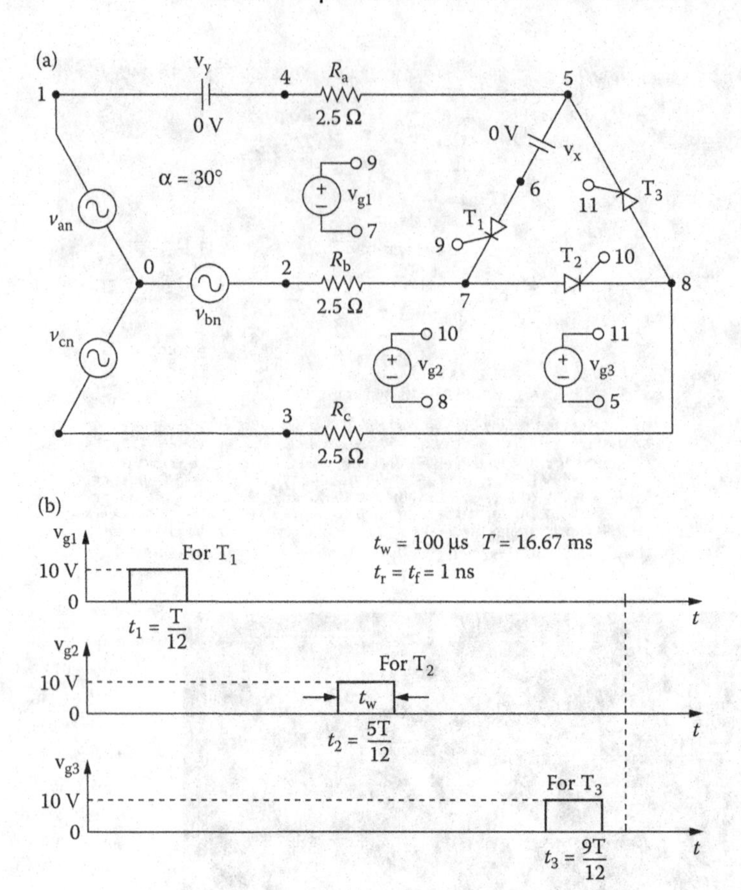

FIGURE 12.22 Three-phase three-thyristor controller for PSpice simulation. (a) Circuit. (b) Gate voltages.

α = 30°. The gate voltages are shown in Figure 12.22(b). Use PSpice to (a) plot the instantaneous line current i_a and thyristor T_1 current i_{T1} and (b) calculate the Fourier coefficients of the line current i_a.

SOLUTION

The peak supply voltage per phase V_m = 169.7 V. for α = 90°, since each phase is shifted by 120°, we can get the delay times as

$$\text{time delay } t_1 = \frac{30}{360} \times \frac{1000}{60\text{Hz}} \times 1000 = 1388.9\,\mu\text{s}$$

$$\text{time delay } t_2 = \frac{150}{360} \times \frac{1000}{60\text{Hz}} \times 1000 = 6944.5\,\mu\text{s}$$

$$\text{time delay } t_3 = \frac{270}{360} \times \frac{1000}{60\text{Hz}} \times 1000 = 12{,}500\,\mu\text{s}$$

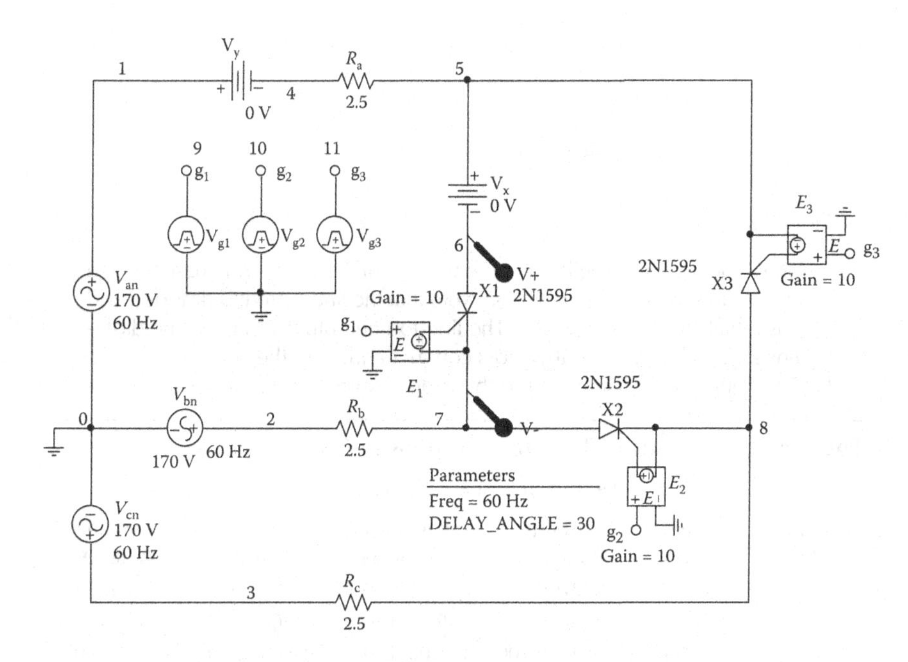

FIGURE 12.23 PSpice schematic for Example 12.5.

The PSpice schematic with SCRs is shown in Figure 12.23. The model name of the SCR, 2N1595, is changed to ASCR, whose subcircuit definition is listed in Section 12.2. By varying the delay cycle a, the output voltage can be changed. The supply frequency {FREQ} and the duty cycle {DELAY_ANGLE} are defined as variables.

Referring to Figure 12.13(a), the listing of the circuit file is as follows:

Three-Phase Three-Thyristor AC Controller

```
SOURCE   ■    Van 1 0 SIN (0 169.7V 60HZ)
              Vbn 2 0 SIN (0 169.7V 60HZ 0 0 -120DEG)
              Vcn 3 0 SIN (0 169.7V 60HZ 0 0 -240DEG)
              .PARAM DELAY_ANGLE=30 FREQ=60Hz
              Vg1 9 7 PULSE (0 10 {{Delay_Angle}/{360*{Freq}}}
              1ns 1ns 100us {1/{Freq}})
              Vg2 10 8 PULSE (0 10 {{{Delay_Angle} + 120}/
              {360*{Freq}}} 1ns 1ns 100us {1/{Freq}})
              Vg3 11 5 PULSE (0 10 {{{Delay_Angle} + 240}/
              {360*{Freq}}} 1ns 1ns 100us {1/{Freq}})
CIRCUIT  ■■   Ra 4 5 2.5
              VX 5 6 DC 0V ; Load current ammeter
              Rb 2 7 2.5
              Rc 3 8 2.5
              VY 1 4 DC 0V ; Line current ammeter
              * Subcircuit calls for thyristor model:
              XT1 6 7 9 7 ASCR ; Thyristor T1
              XT2 7 8 10 8 ASCR ; Thyristor T3
              XT3 8 5 11 5 ASCR ; Thyristor T5
              * Subcircuit ASCR, which is missing, must be
              * inserted.
```

```
ANALYSIS ■■■ .TRAN 1US 33.33MS 00.1MS ; Transient analysis
             .PROBE ; Graphics post-processor
             .OPTIONS ABSTOL = 1.00N RELTOL = 0.01 VNTOL = 0.01
             ITL5 = 10000
             .FOUR  60HZ I(VX)
        .END
```

Note the following:

a. The PSpice plots of the instantaneous line current I(VY) and phase voltage V(6,7) are shown in Figure 12.24. Note that the line current is distorted and has a high harmonic content. The line-to-line output voltage is the difference $v_{an}-v_{bn}$, while two thyristors conduct and carry the current.
b. The Fourier coefficients of the thyristor T_1 current are as follows.

Fourier Components of Transient Response I(VX)

DC Component = 1.614633E+01

Harmonic Number	Frequency (Hz)	Fourier Component	Normalized Component	Phase (Deg)	Normalized Phase (Deg)
1	6.00E+01	2.614E+01	1.000E+00	8.992E+00	0.000E+00
2	1.200E+02	1.393E+01	5.328E–01	5.685E+01	-6.584E+01
3	1.800E+02	8.074E+00	3.089E–01	-9.054E+01	-9.953E+01
4	2.400E+02	5.129E+00	1.962E–01	–1.365E+02	–1.455E+02
5	3.000E+02	3.097E+00	1.185E–01	–1.210E+02	–1.300E+02
6	3.600E+02	5.598E+00	2.142E–01	–1.713E+02	–1.803E+02
7	4.200E+02	3.119E+00	1.193E–01	1.192E+02	1.102E+02
8	4.800E+02	2.366E+00	9.053E–02	1.559E+02	1.469E+02
9	5.400E+02	4.030E+00	1.542E–01	8.903E+01	8.004E+01

Total harmonic distortion = 7.237835E+01%

The Fourier coefficients of the input line current, which can be found in the PSpice output file, are as follows.

Fourier Components of Transient Response I(V_VY)

DC Component = –3.824458E–03

Harmonic Number	Frequency (Hz)	Fourier Component	Normalized Component	Phase (Deg)	Normalized Phase (Deg)
1	6.000E+01	4.523E+01	1.000E+00	–2.102E+01	0.000E+00
2	1.200E+02	2.404E+01	5.315E–01	–2.658E+01	1.546E+01
3	1.800E+02	5.530E–03	1.223E–04	–1.033E+02	–4.021E+01
4	2.400E+02	8.844E+00	1.956E–01	–1.666E+02	–8.249E+01
5	3.000E+02	5.415E+00	1.197E–01	–8.998E+01	1.511E+01
6	3.600E+02	3.271E–03	7.232E–05	–6.083E+01	6.527E+01
7	4.200E+02	5.327E+00	1.178E–01	9.000E+01	2.371E+02
8	4.800E+02	4.194E+00	9.272E–02	–1.741E–02	–5.938E+00
9	5.400E+02	6.090E–03	1.346E–04	–4.595E+01	1.432E+02
10	6.000E+02	3.287E+00	7.267E–02	–6.085E–00	2.041E+02

Total harmonic distortion = 6.023730E+01%

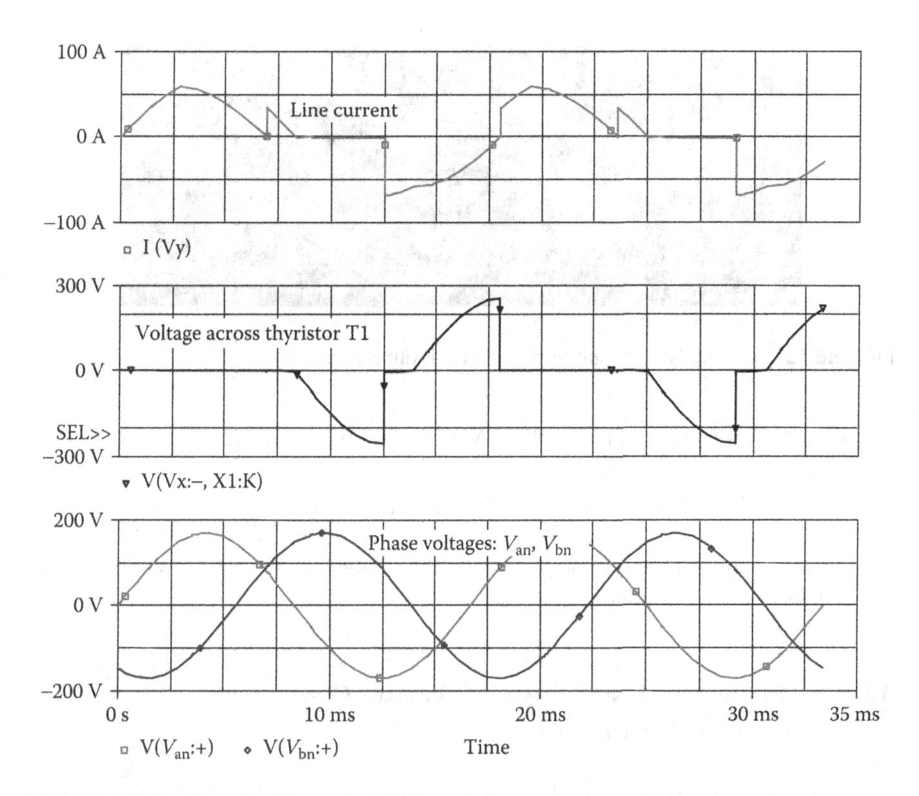

FIGURE 12.24 Plots for Example 12.5 for a delay angle of $\alpha = 30°$.

From the Fourier components, we get

DC component = −3.82 mA
Peak Fundamental component at 60 Hz = 45.23 A
Fundamental phase delay $\phi_1 = -21.02°$
Total harmonic distortion = 60.23%
The input power factor can be calculated from

$$PF_i = \frac{1}{\sqrt{1+THD^2}}\cos\phi_1 = \frac{1}{\sqrt{1+0.6023^2}}\cos(21.02) = 0.805\ \text{(lagging)}$$

LTspice: The gating signals for a delay angle of 30° as shown are phase-shifted by 120° from each other are shown in Figure 12.25.

LTspice schematic for Example 12.5 is shown in Figure 12.26(a). The plots of the input current, the input power, the output voltage, and the load power are shown in Figure 12.26(b). The rms input current as shown in Figure 12.26(c) is 37.57 A, the average input power as shown in Figure 12.26(d) is 10798 W, and the average output power as shown in Figure 12.26(e) is 10584 W. The rms value of the output voltage

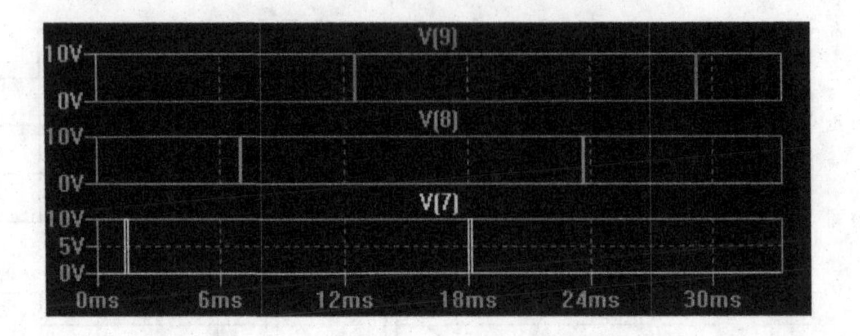

FIGURE 12.25 The LTspice gating signals for Example 12.5.

as shown in Figure 12.26(f) is 93.06 V. We can find the power efficiency of the converter as $\eta = \frac{P_{out}}{P_{in}} = \frac{10584}{10708} = 98.8\%$

We can find the input power factor as $PF_i = \frac{P_{in}}{3V_sI_s} = \frac{10798}{3 \times 120 \times 37.57} = 0.792$

Note: For a low output power, the losses in the semiconductor devices are significant compared to the output power and the efficiency becomes low. There is no power loss in the load inductor.

12.4.3 Examples of Single-Phase AC Voltage Controllers with an Output Filter

Example 12.6:

Finding the Performance of a Single-Phase AC Voltage Controller with an Output Filter

A capacitor of 780 μF is connected across the output of the single-phase full-wave controller of Figure 12.2(a). This is shown in Figure 12.27(a). The input voltage has a peak of 169.7 V, 60 Hz. The load inductance L is 6.5 mH and the load resistance $R = 2.5\ \Omega$. The delay angles are equal: $\alpha_1 = \alpha_2 = 60°$. The gate voltages are shown in Figure 12.27(b). Use PSpice to (a) plot the instantaneous output voltage v_o and the load current i_o and (b) calculate the Fourier coefficients of the input current i_a and the input power factor PF.

Note: R_s and C_s are used as a snubber circuit to protect the devices from high dv/dt.

SOLUTION

The peak supply voltage V_m = 169.7 V, for $\alpha_1 = \alpha_2 = 60°$, the time delay t_1 = (60/360) × (1000/60 Hz) × 1000 = 2777.7 μs. A series snubber with $C_s = 0.1$ μF and $R_s = 750\ \Omega$ is connected across the thyristor to cope with the transient voltage due to the inductive load. The PSpice schematic with SCRs is shown in Figure 12.28. The model name of the SCR, 2N1595, is changed to ASCR, whose subcircuit definition is listed in Section 12.2. By varying the delay cycle a, the output voltage can be varied. The supply frequency {FREQ} and the duty cycle {DELAY_ANGLE} are defined as variables.

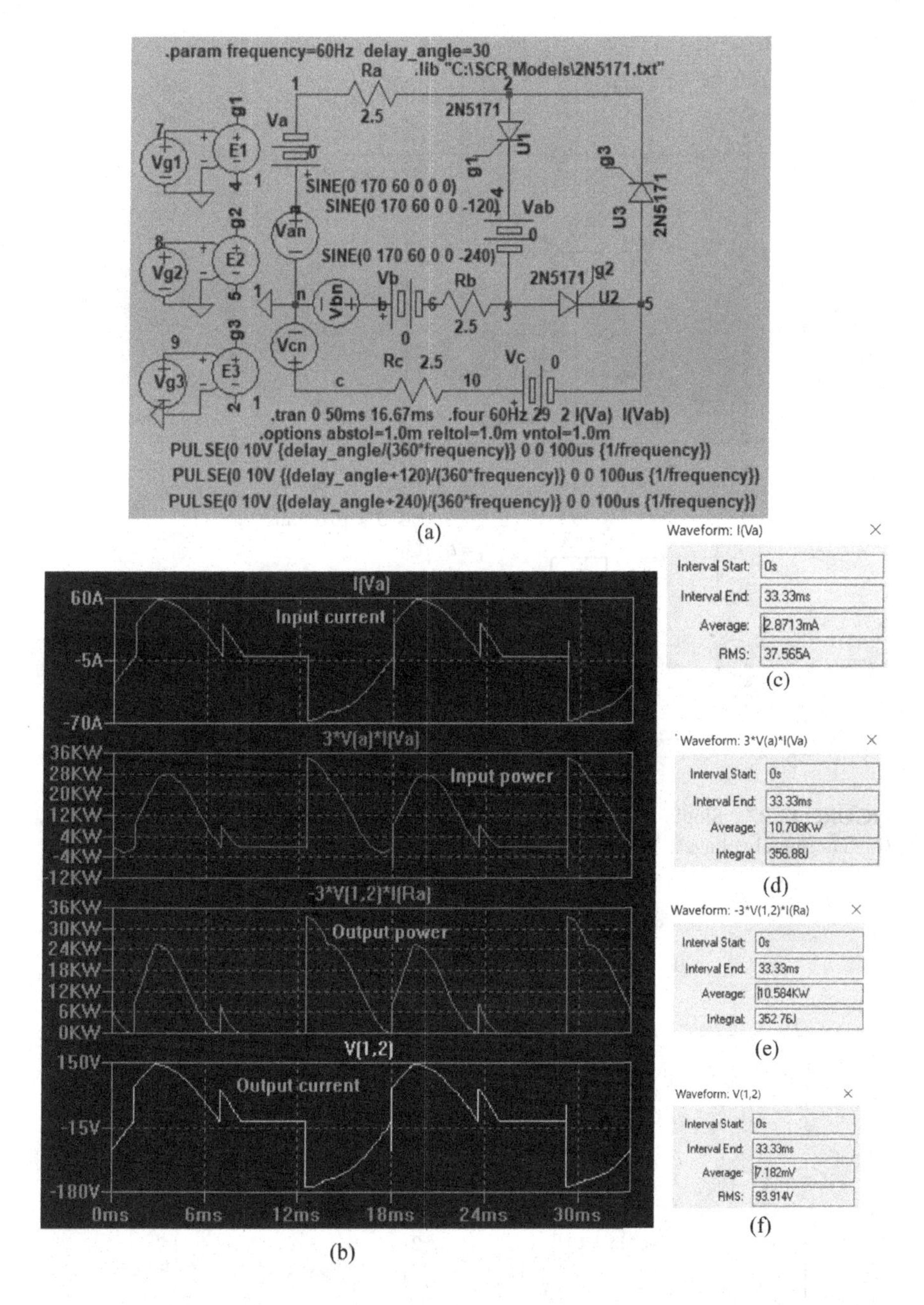

FIGURE 12.26 LTspice schematic for Example 12.5. (a) LTspice schematic. (b) Output waveforms. (c) Input current. (d) Input power. (e) Output power. (f) Load voltage.

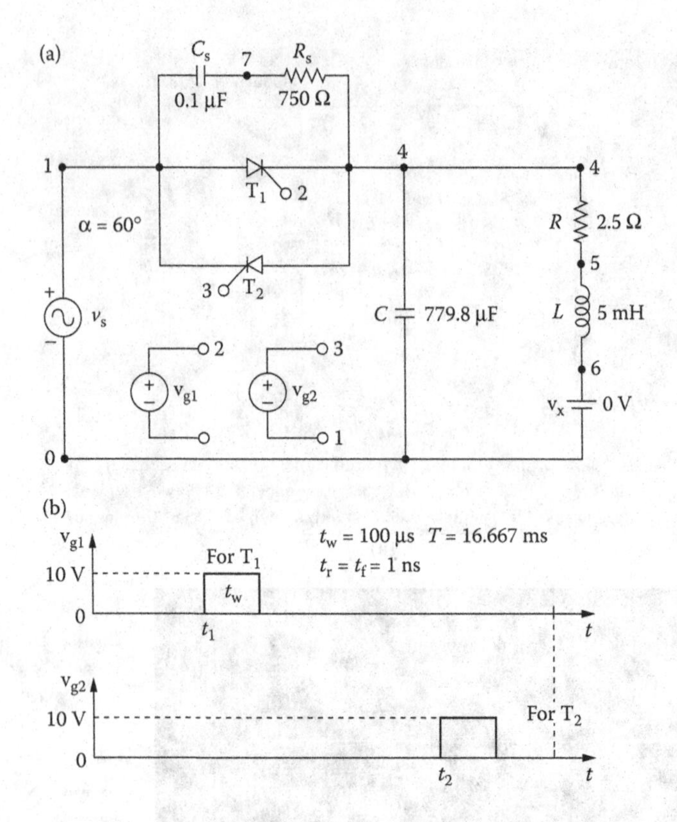

FIGURE 12.27 Single-phase full-wave AC controller for PSpice simulation. (a) Circuit. (b) Gate signal voltages.

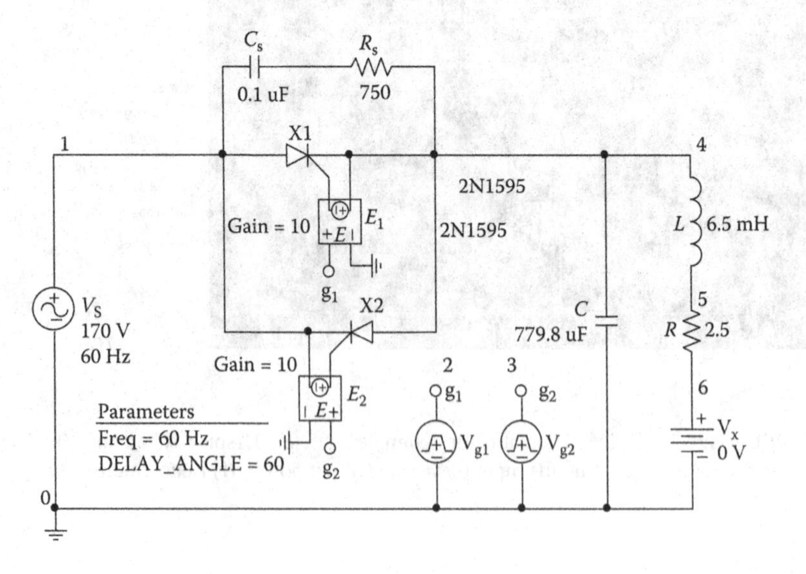

FIGURE 12.28 PSpice schematic for Example 12.6.

Referring to Figure 12.21(a), the listing of the circuit file is as follows:

Single-Phase AC Voltage Controller with Output Filter

```
SOURCE   ■    VS 1 0 SIN (0 169.7V 60HZ)
              .PARAM DELAY_ANGLE=60  FREQ=60Hz
              Vg1 2 4 PULSE (0 10 {{Delay_Angle}/
              {360*{Freq}}} 1ns 1ns 100us {1/{Freq}})
              Vg2 3 1 PULSE (0 10 {{{Delay_Angle}+180}/
              {360*{Freq}}} 1ns 1ns 100us {1/{Freq}})
CIRCUIT  ■■   R 4 5 2.5
              L 5 6 6.5MH
              VX 6 0 DC 0V ; Voltage source to measure the
              load current
              C 4 0 780UF ; Output filter capacitance
              CS 1 7 0.1UF
              RS 7 4 750
              * Subcircuit calls for thyristor model:
              XT1 1 4 2 4 ASCR ; Thyristor T1
              XT2 4 1 3 1 ASCR ; Thyristor T2
              * Subcircuit ASCR, which is missing, must be
              * inserted.
ANALYSIS ■■■  .TRAN 1US 33.33MS ; Transient analysis
              .PROBE ;Graphics post-processor
              .OPTIONS ABSTOL=1.00N RELTOL=1.0M VNTOL=1.0M
              ITL5=10000
              .FOUR 60HZ I(VX) ;Fourier analysis
          .END
```

Note the following:

a. The PSpice plots of instantaneous output voltage V(4) and load current I(VX) are shown in Figure 12.29. Note that the load current is more sinusoidal and smoother as compared to the waveform in Figure 12.4.
b. The Fourier series of the input current is as follows:

Fourier Components of Transient Response I(V_VY)

DC Component = −1.681664E−02

Harmonic Number	Frequency (Hz)	Fourier Component	Normalized Component	Phase (Deg)	Normalized Phase (Deg)
1	6.000E+01	3.314E+01	1.000E+00	−6.980E+01	0.000E+00
2	1.200E+02	1.606E−02	4.845E−04	1.053E+02	1.751E+02
3	1.800E+02	4.381E+00	1.322E−01	1.816E+01	8.795E+01
4	2.400E+02	4.887E−03	1.475E−04	4.289E+01	1.127E+02
5	3.000E+02	1.404E+00	4.238E−02	−1.722E+02	−1.024E+02
6	3.600E+02	2.921E−03	8.815E−05	4.669E+01	1.165E+02
7	4.200E+02	7.014E−01	2.116E−02	5.241E+00	7.504E+01
8	4.800E+02	1.879E−03	5.670E−05	2.298E+01	9.278E+01
9	5.400E+02	4.193E−01	1.265E−02	−1.764E+02	−1.066E+02

Total harmonic distortion = 1.410080E+01%

Total harmonic distortion of input current $THD = 14.1\% = 0.141$

Displacement angle $\phi_1 = -69.88°$

Displacement factor $DF = \cos\phi_1 = \cos(-69.88) = 0.344$ (lagging)

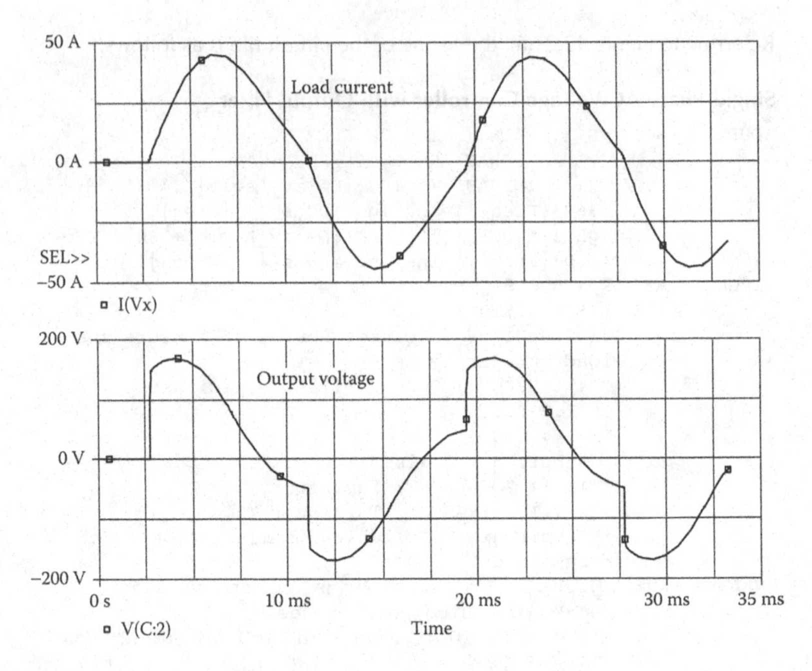

FIGURE 12.29 Plots for Example 12.6 for delay angle $\alpha = 60°$.

From Equation 7.3, the input power factor is

$$PF = \frac{1}{(1+THD^2)^{1/2}}\cos\phi_1 = \frac{1}{(1+0.141^2)^{1/2}} \times 0.344 = 0.341\,(\text{lagging})$$

Note: The *THD* is reduced, but the displacement power factor is also reduced, causing a reduction in the input power factor.

LTspice: The gating signals for a delay angle of 90° as shown in Figure 12.30 are phase-shifted by 180°

LTspice schematic for Example 12.6 is shown in Figure 12.31(a). The plots of the input current, the input power, the output current, and the load power are shown in Figure 12.31(b). The rms input current as shown in Figure 12.315(c) is 168.9 A, the average input power as shown in Figure 12.31(d) is 2404W, and the average output power as shown in Figure 12.31(e) is 1383 W. The rms value of the load current as

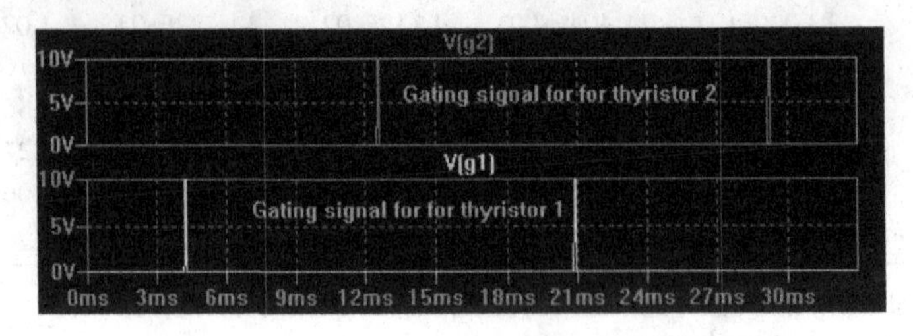

FIGURE 12.30 LTspice gating signals for Example 12.6

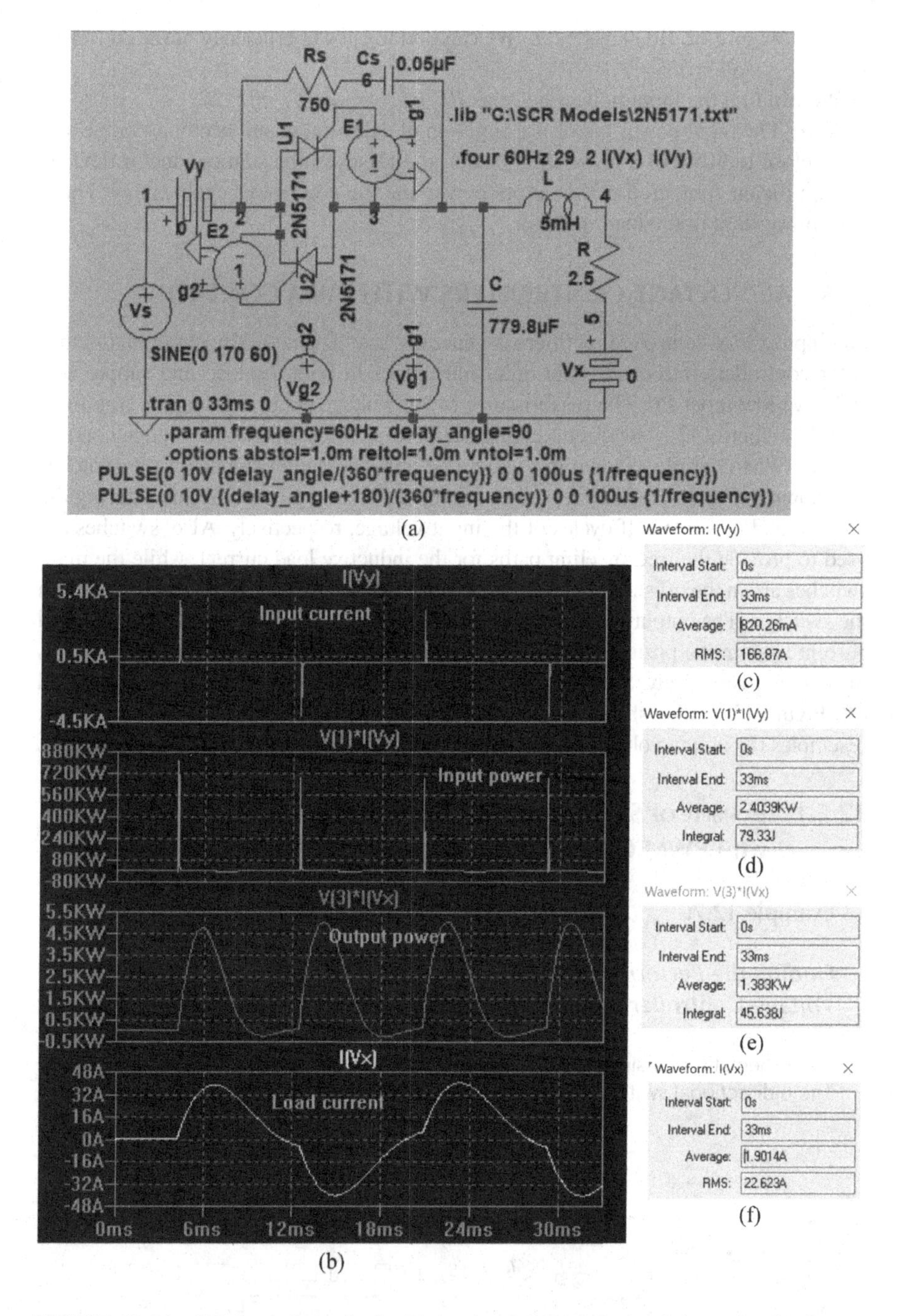

FIGURE 12.31 LTspice schematic for Example 12.6. (a) LTspice Schematic. (b) Output waveforms. (c) Input current. (d) Input power. (e) Output power. (f) Load current.

shown in Figure 12.31(f) is 22.62 A. We can find the power efficiency of the converter as $\eta = \frac{P_{out}}{P_{in}} = \frac{1383}{2404} = 58\%$

We can find the input power factor as $PF_i = \frac{P_{in}}{V_s I_s} = \frac{2404}{120 \times 166.9} = 0.12$

Note: The input power factor is a function of delay angle and becomes worse at a value close to 90°. For a low output power, the losses in the semiconductor devices are significant compared to the output power and the efficiency becomes low. There is no power loss in the load inductor.

12.5 AC VOLTAGE CONTROLLERS WITH PWM CONTROL

The input *PF* of controlled rectifiers is generally low. The naturally commutated thyristor controllers introduce lower-order harmonics in both the load and supply side and have low-input *PF*. The performance of AC voltage controllers can be improved by PWM control [2]. The circuit configuration of a single-phase AC voltage controller for PWM control is shown in Figure 12.32(a). The gating signals of the switches are shown in Figure 12.32(b). Switches are turned on and off several times during the positive and negative half-cycles of the input voltage, respectively. Also, switches are used to provide the freewheeling paths for the inductive load current, while the main switches are in the off state. The diodes prevent reverse voltages from appearing across the switches. The output voltage v_o is shown in Figure 12.33. With an *RL* load, the load current rises in the positive or in the negative direction when either switch S_1 or S_2 is turned on, respectively. Similarly, the load current falls when either S_1 or S_2 is turned off. Figure 12.33 also shows the load current i_o for a resistive load. The load current resembles the output voltage and there is no need for the freewheeling switches.

12.5.1 Example of Single-Phase AC Voltage Controller with PWM Control

Example 12.7:

Finding the Performance of a Single-Phase AC Voltage Controller with PWM Control

The schematic of a single-phase full-wave controller is shown in Figure 12.34(a). The bidirectional switch is implemented with an IGBT and four diodes. The input

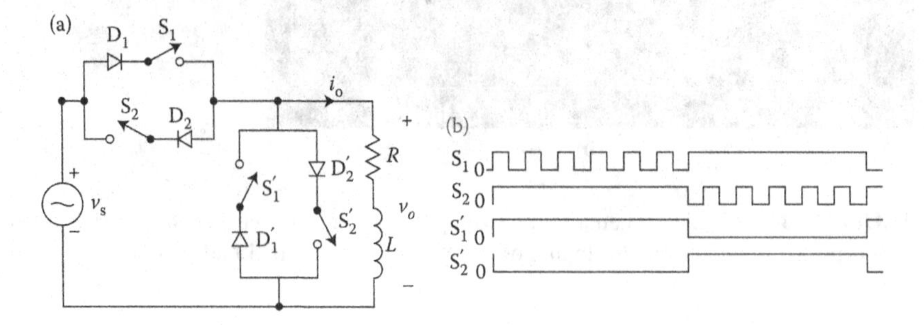

FIGURE 12.32 AC voltage controller with PWM control. (a) Circuit. (b) Gating signals.

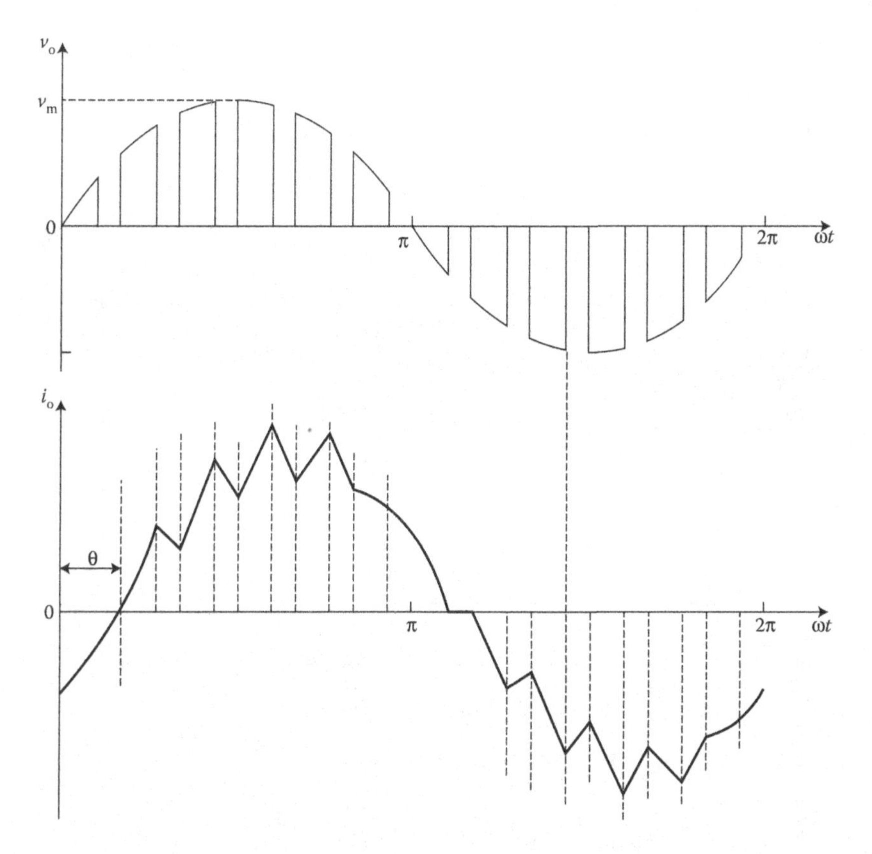

FIGURE 12.33 Output voltage and current of an AC voltage controller with PWM control and with an RL load.

voltage has a peak of 169.7 V, 60 Hz. The load resistance is $R = 100\ \Omega$. There are five pulses per half-cycle and the duty cycle is $k = 8\%$. The gate signal voltage is generated within the hierarchy block (HB1). Hierarchy generates the gate signal by comparing a triangular wave with a DC modulation signal of $k = 0.8$. Use PSpice to (a) plot the instantaneous output voltage v_o and (b) calculate the Fourier coefficients of the input current i_a and the input power factor *PF*.

SOLUTION

The peak supply voltage $V_m = 169.7$ V at $f = 60$ Hz. The duty cycle $k = 0.8$, and the number of pulses per half-cycle is $p = 5$. The switching frequency, which is the frequency of the reference triangular signal, is $f_{ref} = 2pf\text{–}2 \times 5 \times 60 = 600$ Hz.

The PSpice schematic is shown in Figure 12.34.

The plot of the output voltage is shown in Figure 12.35. There are five pulses per half-cycle as expected. The output voltage is symmetrical and is in phase with the input supply voltage. The on-and-off switching of the output voltage is not sharp due to the limitations of PSpice in processing pulses with fast rise and fall time. The PSpice introduces delay time in the comparator in Figure 12.34(b).

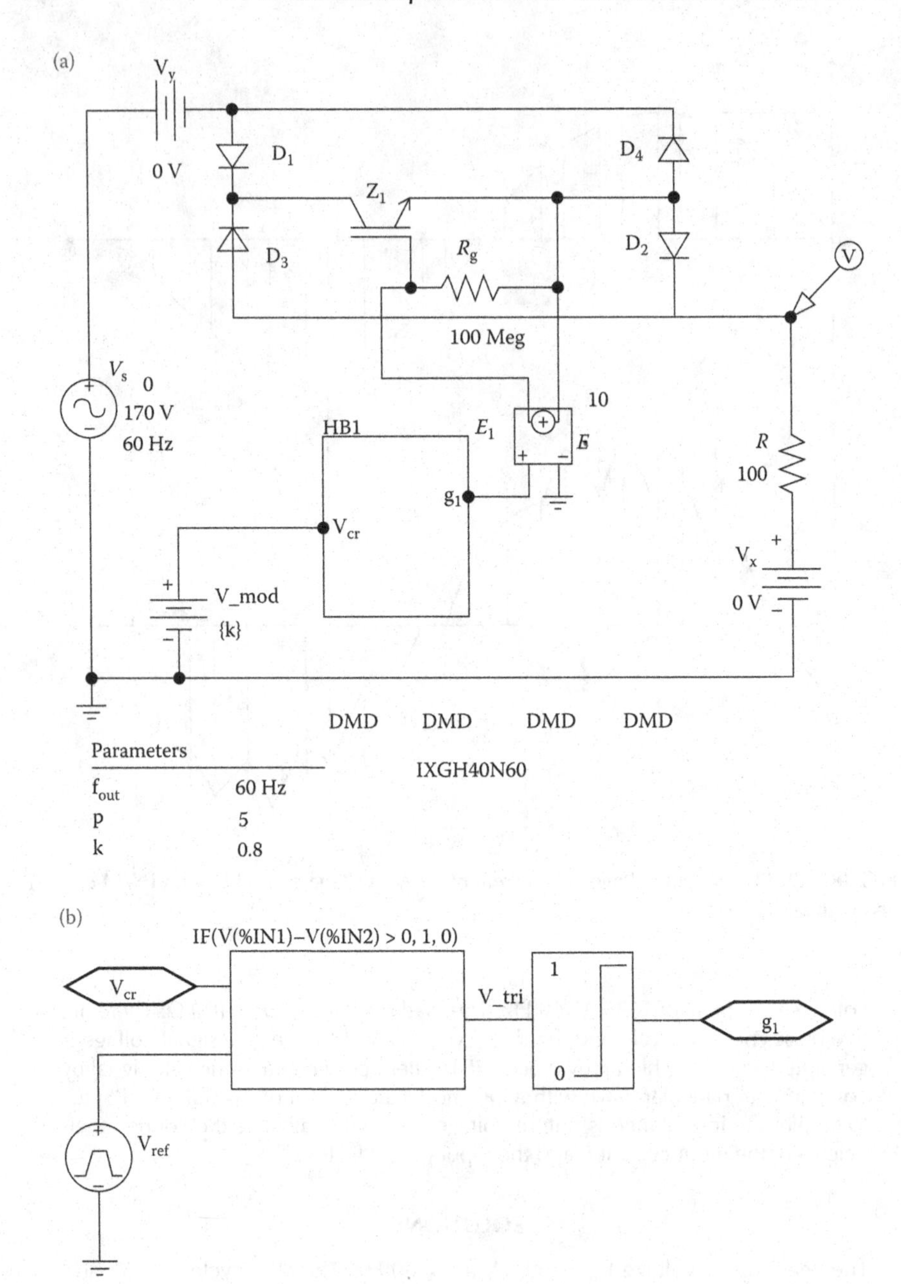

FIGURE 12.34 PSpice schematic of an AC voltage controller with PWM control. (a) PSpice schematic. (b) Gate control signal generation.

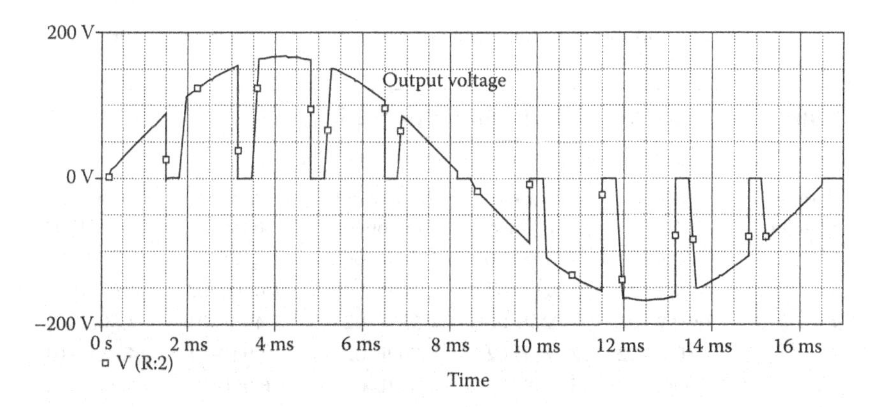

FIGURE 12.35 Output voltage for Example 12.7.

The listing of the circuit file for the inverter is as follows:

A Single-Phase AC Voltage Controller with PWM Control

```
SOURCE ■        V_Vx           1   0     0V
                V_Vs           2   0     SIN(0 170V 60Hz 0 0 0)
                V_Vy           2   3     0V
                E_E1           5   6     11        0         10
                V_HB1_Vref     8   0
                +PULSE 1 0 0 {1/(2*{2*{p}*{fout}})} {1/
                (2*{2*{p}*{fout}})} Ins {1/{1*{2*{p}+{fout}}}}
                E_HB1_ABM21    10     0   VALUE {IF(V(9)-V(8)>0, 1, 0)}
                E_HB1_L1M1T1 11       0   VALUE {LIMIT(V(10),0,1)}
                V_V_mod           9   0   {k}
CIRCUIT ■■      D_D1    3       4         DMD
                R_Rg    5       6         100Meg
                Z_Z1    4       5         6         IXGH40N60
                D D4    6       3         DMD
                D_D3    7       4         DMD
                R_R     1       7         100
                D_D2    6       7         DMD
ANALYSIS ■■■    .PARAM  fout=60Hz         p=5       K=0.8
                .tran   0ns               16.7ms
                .four   60Hz              10        1(V_Vy)
                .OPTIONS  ABSTOL=1nA CHGTOL=0.01nC RFLTOL= 0.01
                VNTOL=1mV
                .MODEL  IXGH40N60   NIGBT (VT= 4.1822
                KF=0.36047   CGS=31.942n1
                + COXD 53.188nF VTD=2.657 TAU=287.56ns
                KP=50.034
                + AREA=37.5u AGD=18.75u)
                .MODEL DMD D (IS=2.22E-15 BV=1.2E+03 CJO=1.0nF)
                .OP
                .probe
                .END
```

The Fourier components of the input current, which can be found in the PSpice output file, are as follows:

Fourier Components of Transient Response I(V_VY)

DC Component = –1.681664E–02

Harmonic Number	Frequency (Hz)	Fourier Component	Normalized Component	Phase (Deg)	Normalized Phase (Deg)
1	6.000E+01	1.277E+00	1.000E+00	–2.240E–02	0.000E+00
2	1.200E+02	3.970E–02	3.108E–02	–1.680E+02	–1.679E+02
3	1.800E+02	2.022E–02	1.583E–02	7.495E+01	7.502E+01
4	2.400E+02	2.183E–02	1.709E–02	3.994E+01	4.003E+01
5	3.000E+02	5.789E–03	4.532E–03	–6.91E+01	–6.899E+01
6	3.600E+02	8.915E–03	6.979E–03	1.749E+02	1.751E+02
7	4.200E+02	4.124E–02	3.228E–02	1.342E+02	1.343E+02
8	4.800E+02	3.004E–02	2.352E–02	2.942E+01	2.960E+01
9	5.400E+02	3.575E–01	2.799E–01	–1.623E+00	–1.422E+00
10	6.000E+02	4.809E–03	3.765E–03	–1.597E+02	–1.595E+02

Total harmonic distortion = 2.855440E+01%

From the Fourier components, we get

DC component = –1.46 mA
Peak fundamental component at 60 Hz = 1.27 A
Fundamental phase delay $\phi_1 = -0.022°$
Total harmonic distortion = 28.55%

The input power factor can be calculated from

$$PF_i = \frac{1}{\sqrt{1+THD^2}}\cos\phi_1 = \frac{1}{\sqrt{1+0.2855^2}}\cos(-0.0022) = 0.962 \text{ (lagging)}$$

Notes:

1. The power factor of a WM control is high as compared to phase-controlled AC voltage controllers.
2. The simulation runs smoothly with a resistive load. If we have an inductive load, we need to provide a freewheeling path for the inductor current to flow when the main switch is turned off as shown in Figure 12.36(a). The gate signal g2 is the logic inverse of gate signal g1 as shown in Figure 12.36(b).
3. If the gate signals are not carefully selected, there are possibilities of short-circuited paths, for example, through D_1, D_2, D_7, and D_8. The simulation will run into convergence problems and fail giving no final output. PSpice often gives convergence problems due to the fast rising and falling of voltages and currents in a circuit. One could try to adjust the OPTIONS parameters to make the circuit converge but it is not always an easy task.

LTspice: The gating signals are generated by comparing a triangular wave with the magnitude of a sinewave as shown Figure 12.37.

LTspice schematic for Example 12.7 is shown in Figure 12.38(a). The plots of the input current, the input power, the output current, and the load power are shown in

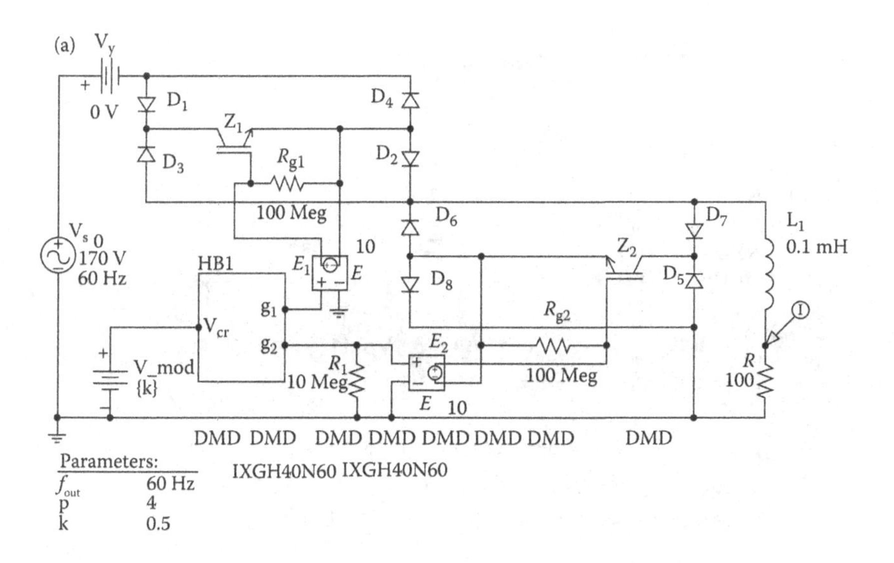

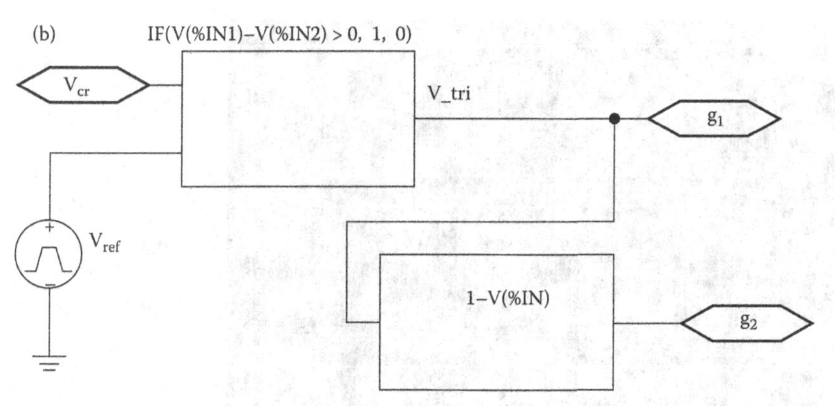

FIGURE 12.36 PSpice schematic of an AC voltage controller with RL load and PWM control. (a) PSpice schematic. (b) Gate control signal generation.

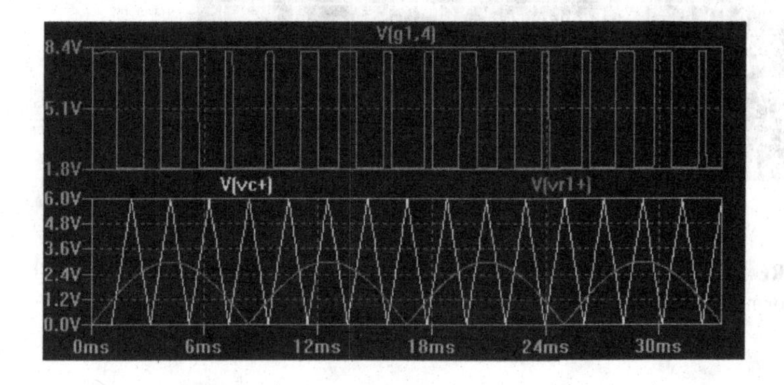

FIGURE 12.37 LTspcie gating signals for Example 12.7.

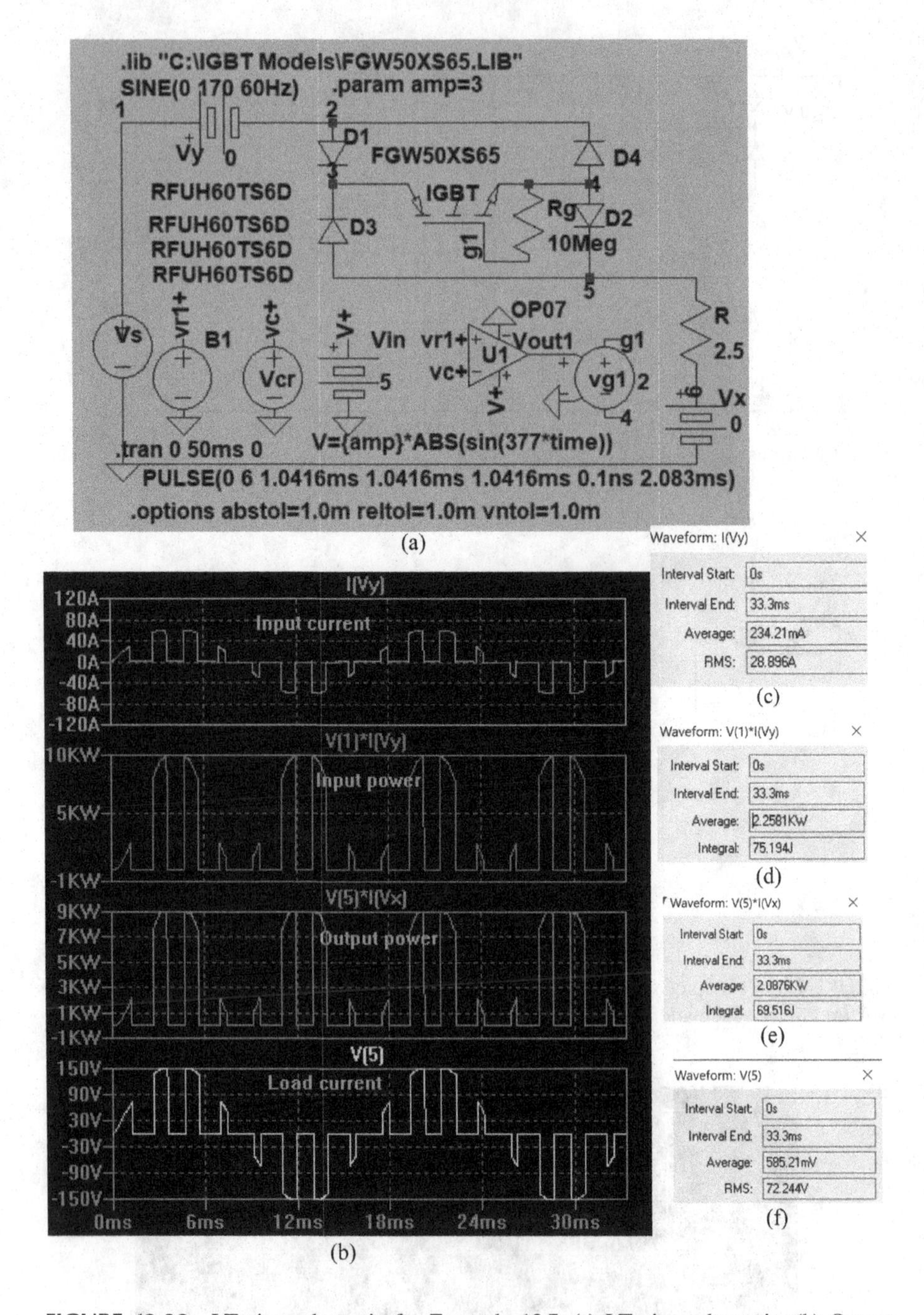

FIGURE 12.38 LTspice schematic for Example 12.7. (a) LTspice schematic. (b) Output waveforms. (c) Input current. (d) Input power. (e) Output power. (f) Load voltage.

Figure 12.38(b). The rms input current as shown in Figure 12.38(c) is 28.9 A, the average input power as shown in Figure 12.38(d) is 2259 W, and the average output power as shown in Figure 12.38(e) is 2088 W. The rms value of the load voltage shown in Figure 12.38(f) is 72.24 V. We can find the power efficiency of the converter as $\eta = \frac{P_{out}}{P_{in}} = \frac{2088}{2259} = 92.4\%$

We can find the input power factor as $PF_i = \frac{P_{in}}{V_s I_s} = \frac{2259}{120 \times 28.9} = 0.65$

Note: The input power factor is a function of delay angle and becomes worse at a value close to 90°. For a low output power, the losses in the semiconductor devices are significant compared to the output power and the efficiency becomes low. There is no power loss in the load inductor.

12.6 CYCLOCONVERTERS

The AC voltage controllers provide a variable output voltage but the frequency of the output voltage is fixed and in addition the harmonic content is high, especially at a low output voltage range. A variable output voltage at a variable frequency can be obtained from two-stage conversions: fixed AC to variable DC (e.g., controlled rectifiers) and variable DC to variable AC at variable frequency (e.g., inverters). However, cycloconverters can eliminate the need for one or more intermediate converters. A cycloconverter is a direct-frequency changer that converts AC power at one frequency to AC power at another frequency by AC–AC conversion, without an intermediate conversion link.

The majority of cycloconverters are naturally commutated and the maximum output frequency is limited to a value that is only a fraction of the source frequency.

A single-phase/single-phase cycloconverter as shown in Figure 12.39(a) consists of two single-phase controlled converters which are operated as bridge rectifiers. However, their delay angles are such that the output voltage of one converter is equal and opposite to that of the other converter. If converter P is operating alone, the average output voltage is positive and if converter N is operating, the output voltage is negative. Figure 12.39(b) shows the simplified equivalent circuit of the dual converter. Figure 12.40 shows the waveforms for the output voltage and gating signals of positive and negative converters, with the positive converter on for time $T_0/2$ and the negative converter operating for time $T_0/2$. The frequency of the output voltage is $f_c = 1/T_0$.

If α_p is the delay angle of the positive converter, the delay angle of the negative converter is $\alpha_n = \pi - \alpha_p$. The average output voltage of the positive converter is equal and opposite to that of the negative converter.

$$V_{dc2} = -V_{dc1}$$

12.6.1 Example of Single-Phase Cycloconverter

Example 12.8:

Finding the Performance of a Single-Phase Cycloconverter

A single-phase cycloconverter is shown in Figure 12.41 which consists of two single-phase full-bride converters. The input voltage has a peak of 169.7 V, 60 Hz. The load

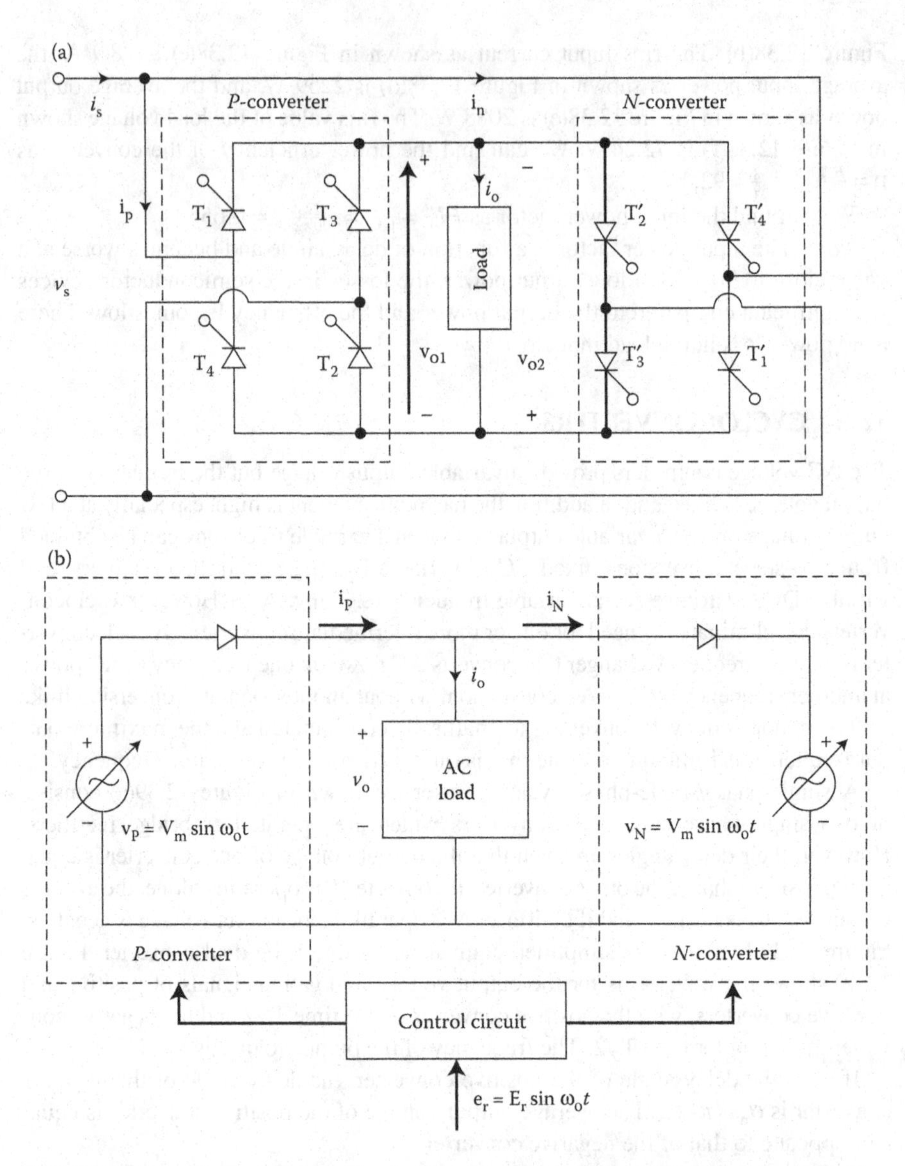

FIGURE 12.39 Single-phase/single-phase cycloconverter. Circuit. (b) Equivalent circuit.

inductances are $L_1 = L_2 = 2.5$ mH and the load resistance is $R = 10\ \Omega$. The delay angles are equal: $\alpha_1 = \alpha_2 = 60°$. The output frequency is $f_{out} = 20$ Hz. Use PSpice to (a) plot the instantaneous input current i_s and the load current i_o and (b) calculate the Fourier coefficients of the input current i_s, the load current i_o, and the input power factor *PF*.

SOLUTION

The peak supply voltage $V_m = 169.7$ V, for $\alpha_1 = \alpha_2 = 60°$, the time delay $t_1 = (60/360) \times (1000/60 \text{ Hz}) \times 1000 = 2777.7\ \mu s$. The generation of gate control signals is shown in Figure 12.41(b). Note that Figure 12.42(b) could be included

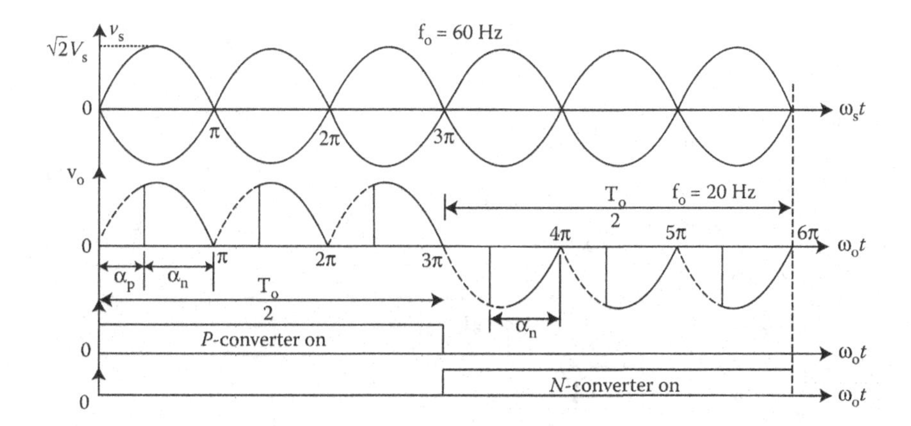

FIGURE 12.40 Output voltage of a single-phase/single-phase cycloconverter with resistive load.

within a hierarchy block. The *P*-converter, which produces the positive output voltage, operates during the positive half-cycle of the output voltage $f_{out}/2$. The *N*-converter, which produces the negative output voltage, operates during the negative half-cycle $f_{out}/2$. This is shown in Figure 12.34(a). The gate signals are shown in Figure 12.42(b).

The plots of the output current and the output voltage are shown in Figure 12.43. There are three pulses per half-cycle of the AC output current (as expected). The output current is symmetrical and is in phase with the input supply voltage. The frequency of the input current is 60 Hz while the frequency of the AC output current is 60/3 = 20 Hz. It is important to note that the output frequency cannot be greater than the supply frequency of 60 Hz.

The listing of the circuit file for the inverter is as follows:

A Single-Phase Cycloconverter

```
SOURCE  ■   E_E2      1     0      g1     0   10
            E_E3      2     3      g3     0   10
            E_E1      4     3      g1     0   10
            E_E4      5     gs+    g3     0   10
            E_E3p     11    6      g3p    0   10
            E_E1p     10    6      g1p    0   10
            E_E4p     8     gs+    g3p    0   10
            E_E2p     9     0      g1p    0   10
            V_Vy      12    gs+    0V
            V_Vx      14    6      0V
            V_Vp      15    0      PULSE 0 1 0 1ns 1ns {0.5/{fout}}
            {1/{fout}}
            E_MULTIg1 0 VALUE {V(16)*V(15)}
            V_Vg3     17    0
            +PULSE 0 1 {{{Delay_Angle}+180}/{360*{Freq}}} 1ns
            1ns 200us {1/{Freq}}          E_MULT6 g1 p 0 VALUE
            {V(18)*V(16)} E_MULT7      g3p 0 VALUE {V(17)*V(18)}
```

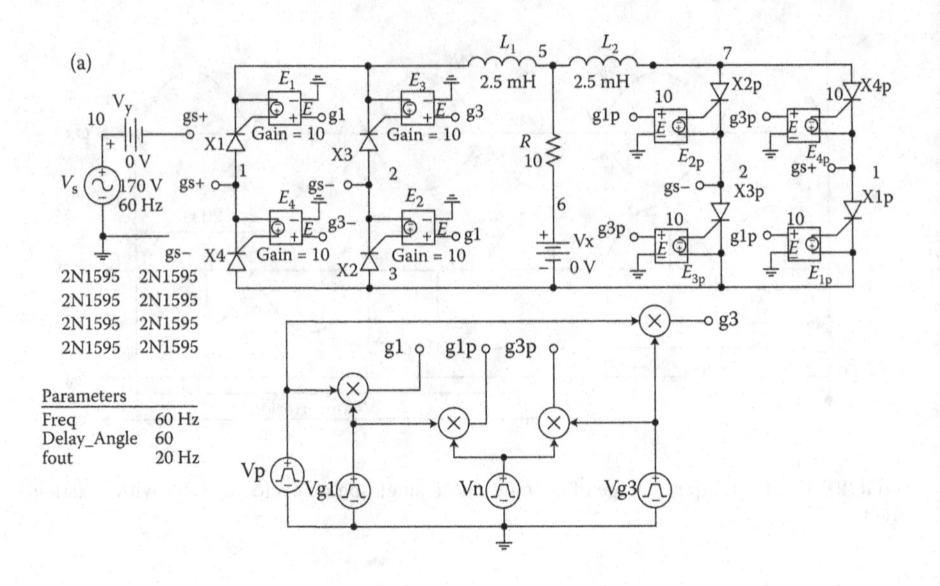

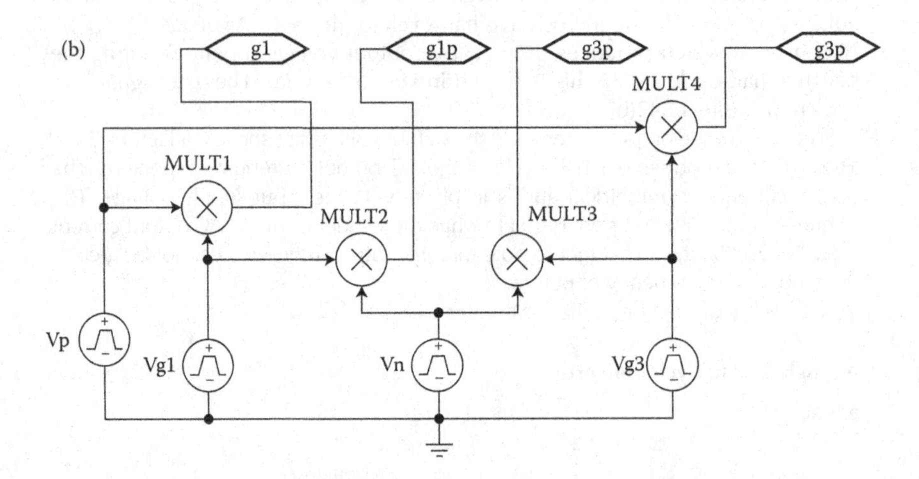

FIGURE 12.41 PSpice schematic of a single-phase cycloconverter. (a) PSpice schematic. (b) Gate control signal generation.

```
E_MULT2         g3 0 VALUE {V(17)*V(15)}
V_Vgl     16    0
+PULSE 0 1 {{Delay Angle}/{360*{Freq}}} 1ns 1ns
200us {1/{Freq}}
V_Vn      18    0
+PULSE 0 1 {0.5/{fout}} 1ns 1ns {0.5/{fout}} {1/
{fout}}
X_X3p     0     11    6        SCRMOD
V_Vs      12    0     AC       0          SIN 0170V 60Hz 0
0 0
```

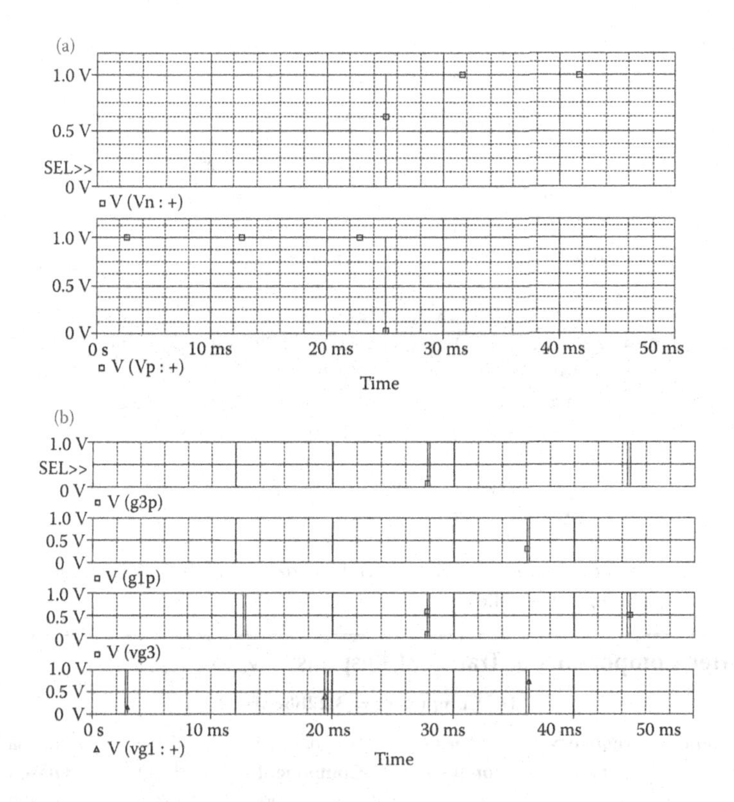

FIGURE 12.42 Gate control signals for single-phase cycloconverter. (a) On and off time of the *P*-converter and *N*-converter. (b) Gate control signals.

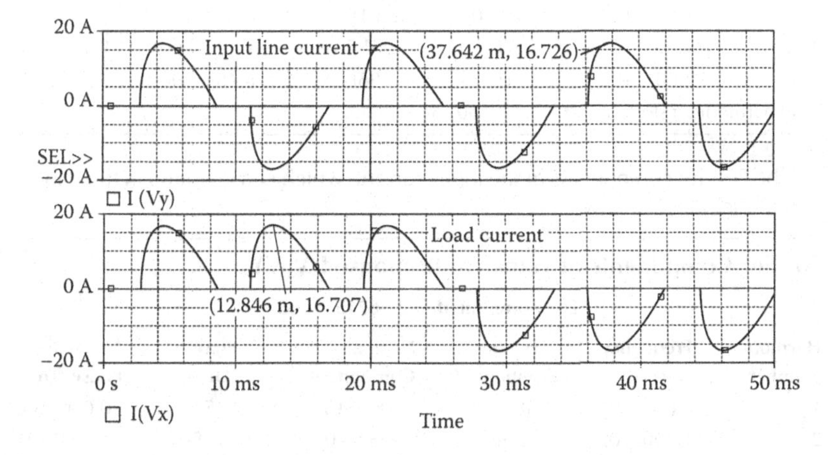

FIGURE 12.43 Output voltage for Example 12.8.

```
CIRCUIT ■■  X_X1    gs+  4     3       SCRMOD
            X_X3    0    2     3       SCRMOD
            X_X4    6    5     gs+     SCRMOD
            X_X2    6    1     0       SCRMOD
            X_X4p   7    8     gs+     SCRMOD
            X_X2p   7    9     0       SCRMOD
            X_X1p   gs+  10    6       SCRMOD
            L_L1    3    13    2.5mH
            L_L2    13   7     2.5mH
            R_R     14   13    10
ANALYSIS ■■■ .PARAM      Freq = 60Hz       Delay_Angle = 60 fout = 20Hz
            .tran        1us   50ms
            .four        60    10          I(V_Vy)
            .OPTIONS ABSTOL = 1u RELTOL = 0.01 TNOM = 1m
             VNTOL = 0.01
            .MODEL VSWITCH SMOD (RON=0.0125 ROFF=10.0E+06
             VON=0.5 VOFF=0
```

The Fourier components of the output current, which can be found in the PSpice output file, are as follows.

Fourier Components of Transient Response I(v_Vx)

DC Component = 3.228885E–03

Harmonic Number	Frequency (Hz)	Fourier Component	Normalized Component	Phase (Deg)	Normalized Phase (Deg)
1	2.000E+01	1.040E+01	1.000E+00	–7.226E+00	0.000E+00
2	4.000E+01	6.449E–03	6.204E–04	8.895E+01	1.034E+02
3	6.000E+01	4.626E+00	4.451E–01	–2.138E+01	2.996E–01
4	8.000E+01	6.448E–03	6.203E–04	8.788E+01	1.168E+02
5	1.000E+02	7.246E+00	6.971E–01	–3.441E+01	1.721E+00
6	1.200E+02	6.443E–03	6.198E–04	8.674E+01	1.301E+02
7	1.400E+02	4.847E+00	4.663E–01	1.355E+02	1.861E+02
8	1.600E+02	6.446E–03	6.201E–04	8.572E+01	1.435E–02
9	1.800E–02	1.305E+00	1.255E–01	1.340E+02	1.990E+02
10	2.000E+02	6.441E–03	6.196E–04	8.465E+01	1.569E+02

Total harmonic distortion = 9.577122E+01%

The Fourier components of the input current, which can be found in the PSpice output file, are as follows.

Fourier Components of Transient Response I(V_VY)

DC Component = 2.667157E–06

Harmonic Number	Frequency (Hz)	Fourier Component	Normalized Component	Phase (Deg)	Normalized Phase (Deg)
1	6.000E+01	1.389E+01	1.000E+00	–2.145E+01	0.000E+00
2	1.200E+02	5.529E–06	3.982E–07	–1.518E+01	2.773E+01

(Continued)

Fourier Components of Transient Response I(V_VY) ***(Continued)***

DC Component = 2.667157E−06

Harmonic Number	Frequency (Hz)	Fourier Component	Normalized Component	Phase (Deg)	Normalized Phase (Deg)
3	1.800E+02	3.902E+00	2.810E−01	1.342E+02	1.986E+02
4	2.400E+02	5.002E−06	3.602E−07	1.765E+02	2.623E+02
5	3.000E+02	2.127E+00	1.532E−01	3.482E+01	1.421E+02
6	3.600E+02	6.293E−06	4.532E−07	9.379E+01	2.225E+02
7	4.200E+02	9.101E−01	6.554E−02	-9.275E+01	5.743E+01
8	4.800E+02	4.811E−06	3.465E−07	-4.610E+01	1.255E+02
9	5.400E+02	7.892E−01	5.683E−02	1.278E+02	3.209E+02
10	6.000E+02	4.718E−06	3.398E−07	−1.789E+02	3.569E+01

Total harmonic distortion = 3.316340E+01%

From the Fourier components, we get

DC component = 2.66 mA
Peak fundamental component at 60 Hz = 13.89 A
Fundamental phase delay $\phi_1 = -21.45°$
Total harmonic distortion = 33.16%

The input power factor can be calculated from

$$PF_i = \frac{1}{\sqrt{1+THD^2}}\cos\phi_1 = \frac{1}{\sqrt{1+0.3316^2}}\cos(-21.45) = 0.883 \text{ (lagging)}$$

Notes:

1. The input power factor will depend on the delay angle.
2. The *P*-converter supplies the load current during the positive half-cycle of the output voltage or load current at the output frequency while the *N*-converter supplies the load current during the negative half-cycle of the output voltage of current. However, the shape of the input current is the same as that of a single-phase converter.

LTspice: The gating signals are generated by multiplying short pulses of 100 μs duration and the frequency of these pulses is twice the supply frequency of 60Hz. of short duration. These pulses are multiplied by a square wave at the desired output frequency, e.g., 30 Hz with a 50% duty cycle of 50% on-time and 50% off-time. The number of desired gating pulses of the positive converter depends on the output frequency of the cycloconverter. As shown in Figure 12.44 for a frequency of 30Hz, there are 3 pulses per half-cycle of the output period. The pulses for the negative converter are generated by multiplying the gating pulses by an inverted square wave at the output frequency with a 50% duty cycle.

LTspice schematic for Example 12.8 is shown in Figure 12.45(a). The plots of the input current, the output current, and load power are shown in Figure 12.45(b). The rms input current as shown in Figure 12.45(c) is 1510 A, and the rms load current as

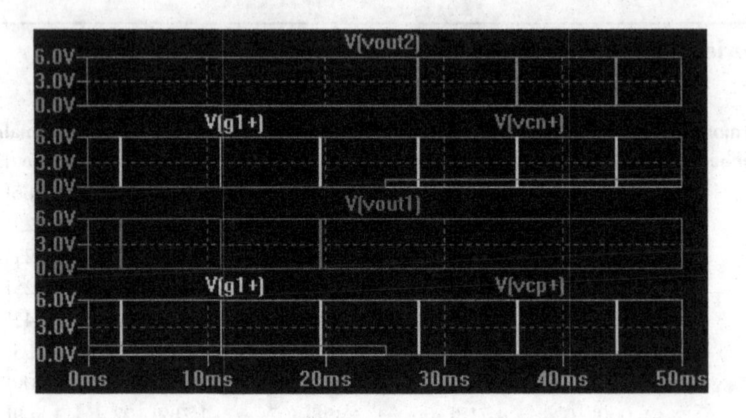

FIGURE 12.44 Ltspice gating signals for Example 12.8.

shown in Figure 12.45(d) is 4.98 A. The average output power as shown in Figure 12.45(e) is 248.4 W.

Note: The input current is significantly high due to the circulating current between the positive and the negative converters. If we run the simulation with a small circulating inductor, e.g., $L_1 = L_2 = 6.5\mu H$, we should see higher input currents. Readers are encouraged to see the effects of the circulation current limiting inductors.

12.7 LABORATORY EXPERIMENTS

The following two experiments are suggested to demonstrate the operation and characteristics of thyristor AC controllers:

Single-phase AC voltage controller
Three-phase AC voltage controller

Warning: If you are building and testing the experimental circuits, you must isolate the gate signals v_{g1}, v_{g2}, v_{g3}, and so on from the power circuit through pulse transformers or opto-coupler. You must also connect the snubber circuits to protect the power semiconductor devices from excessive dv/dt. You may also need to mount the power devices into heat sinks to avoid thermal runaway.

12.7.1 Experiment AC.1

Single-Phase AC Voltage Controller	
Objective	To study the operation and characteristics of a single-phase AC voltage controller under various load conditions.
Applications	The single-phase AC voltage controller is used to control power flow in industrial and induction heating, pumps and fans, light dimmers, food blenders, and so on.
Textbook	See Ref. 2, Sections 11.4 and 11.5.

(Continued)

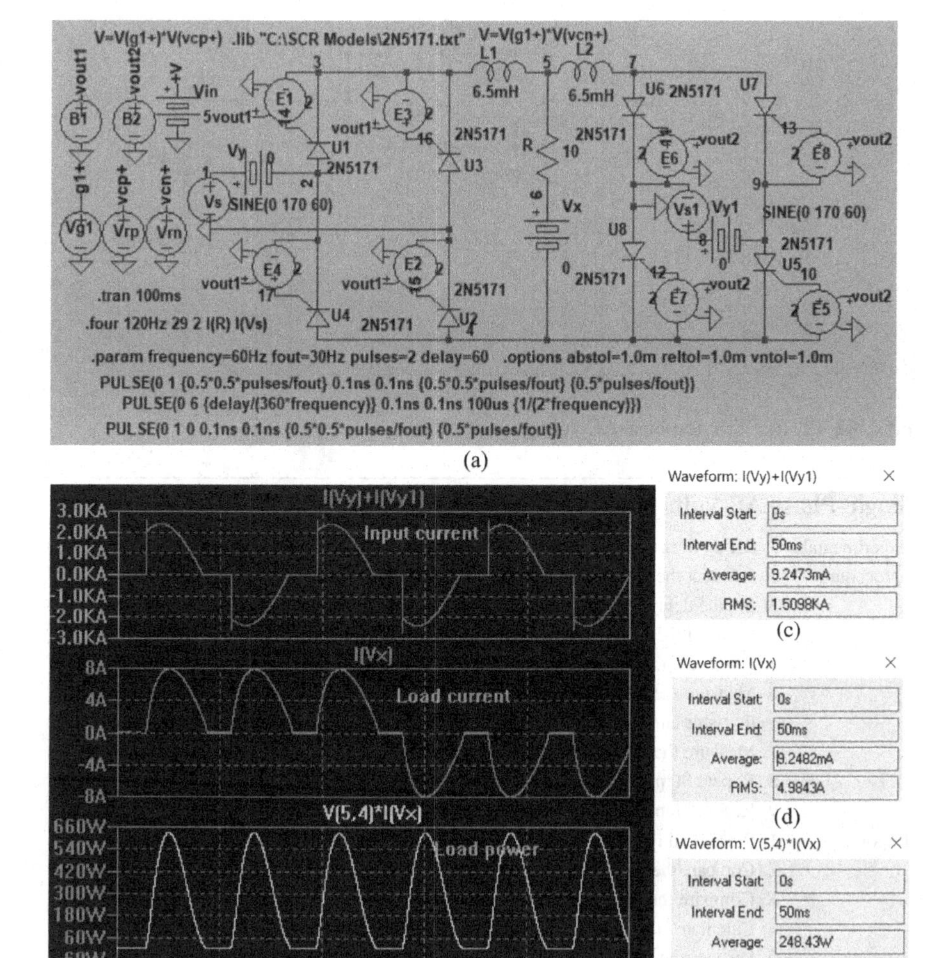

FIGURE 12.45 LTspice schematic for Example 12.8. (a) LTspice schematic. (b) Output waveforms. (c) Input current. (d) Load current. (e) Load power.

Single-Phase AC Voltage Controller *(Continued)*

Apparatus

1. Two phase-controlled thyristors with ratings of at least 50 A and 400 V, mounted on heat sinks.
2. A firing pulse generator with isolating signals for gating thyristors.
3. An *RL* load.
4. One dual-beam oscilloscope with floating or isolating probes.
5. AC voltmeters and ammeters and one noninductive shunt.

Warning

Before making any circuit connection, switch off the AC power. Do not switch on the power unless the circuit is checked and approved by your laboratory instructor. Do not touch the thyristor heat sinks, which are connected to live terminals.

(Continued)

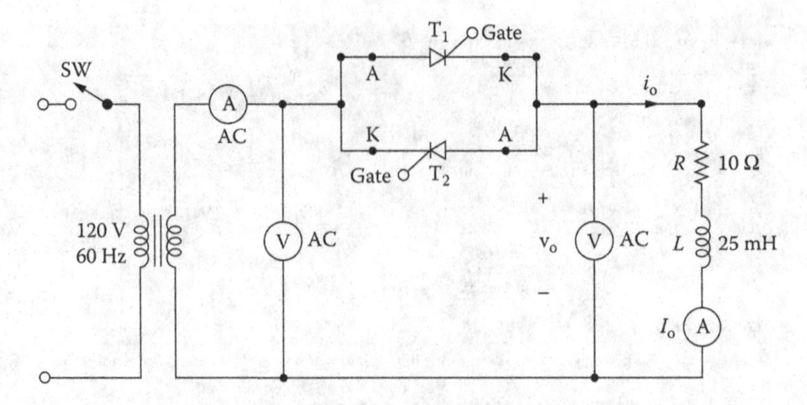

FIGURE 12.46 Single-phase AC voltage controller.

Single-Phase AC Voltage Controller *(Continued)*

Experimental procedure

1. Set up the circuit as shown in Figure 12.46. Use a load resistance R only.
2. Connect the measuring instruments as required.
3. Set the delay angle to $\alpha = \pi/3$.
4. Connect the firing pulses to appropriate thyristors.
5. Observe and record the waveforms of the load voltage v_o and the load current i_o.
6. Measure the RMS load voltage $V_{o(RMS)}$), the RMS load current $I_{o(RMS)}$), the average thyristor current $I_{r(RMS)}$, the RMS input voltage $V_{o(RMS)}$, and the load power P_L.
7. Measure the conduction angle of thyristor T_1.
8. Repeat Steps 2–7 with a load inductance L only.
9. Repeat Steps 2–7 with both load resistance R and load inductance L.

Report

1. Present all recorded waveforms and discuss all significant points.
2. Compare the waveforms generated by SPICE with the experimental results, and comment.
3. Compare the experimental results with the predicted results.
4. Calculate and plot the RMS output voltage V_o against the delay angle α.
5. Discuss the advantages and disadvantages of this type of controller.

12.7.2 Experiment AC.2

Three-Phase AC Voltage Controller

Objective — To study the operation and characteristics of a three-phase AC voltage controller under various load conditions.

Applications — The three-phase controller is used to control power flow in industrial and induction heating, lighting, speed control of induction motor-driven pumps and fans, and so on.

Textbook — See Ref. 2, Section 11.6.

Apparatus

1. Six phase-controlled thyristors with ratings of at least 50 A and 400 V, mounted on heat sinks.
2. A firing pulse generator with isolating signals for gating thyristors.
3. RL loads.
4. One dual-beam oscilloscope with floating or isolating probes.
5. AC voltmeters and ammeters and one noninductive shunt.

(Continued)

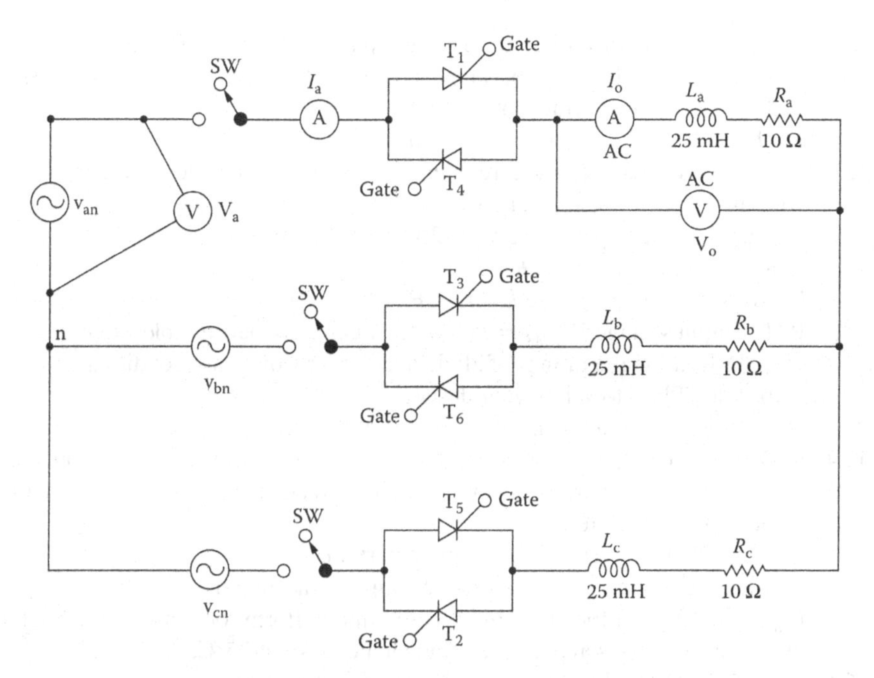

FIGURE 12.47 Three-phase AC voltage controller.

Three-Phase AC Voltage Controller *(Continued)*

Warning	See Experiment AC.1.
Experimental procedure	Repeat the steps of Experiment AC.1 for the circuit of Figure 12.47.
Report	See Experiment AC.1.

SUMMARY

The statements for an AC thyristor are as follows.

```
* Subcircuit call for switched AC thyristor model:
XT1     NA      NC         +NC        -NC        ASCR
     anode   cathode    +control   -control    model
*                       voltage    voltage     name
```

DESIGN PROBLEMS

12.1 It is required to design the single-phase AC voltage controller of Figure 12.46 with the following specifications:
AC supply voltage $V_s = 120$ V (RMS), 60 Hz
Load resistance $R = 5\ \Omega$
Load inductance $L = 15$ mH
RMS output voltage $V_{o(RMS)} = 75\%$ of the maximum permissible value
  a. Determine the ratings of all devices under worst-case conditions.
  b. Use SPICE to verify your design.
  c. Provide a component cost estimate of the circuit.

12.2 a. Design an output C filter for the single-phase full–wave AC voltage controller of Problem 12.1. The harmonic content of the load current should be less than 10% of the average load current.
b. Use SPICE to verify your design in part (a).

12.3 It is required to design the three-phase AC voltage controller of Figure 12.47 with the following specifications:
AC supply voltage per phase $V_s = 120$ V (RMS), 60 Hz
Load resistance per phase $R = 5\ \Omega$
Load inductance per phase $L = 15$ mH
RMS output voltage $V_{o(RMS)} = 75\%$ of the maximum permissible value
a. Determine the ratings of all devices under worst-case conditions.
b. Use SPICE to verify your design.
c. Provide a component cost estimate of the circuit.

12.4 a. Design an output C filter for the three-phase AC voltage controller of Problem 12.3. The harmonic content of the load current should be <10% of the value without the filter.
b. Use SPICE to verify your design in part (a).

12.5 Use PSpice to find the worst-case minimum and maximum output voltages ($V_{o(min)}$ and $V_{o(max)}$) for Problem 12.1. Assume uniform tolerances of ±15% for all passive elements and an operating temperature of 25°C.

12.6 Use PSpice to find the worst-case minimum and maximum output voltages ($V_{o(min)}$ and $V_{o(max)}$) for Problem 12.2. Assume uniform tolerances of ±15% for all passive elements and an operating temperature of 25°C.

12.7 Use PSpice to find the worst-case minimum and maximum output voltages ($V_{o(min)}$ and $V_{o(max)}$) for Problem 12.3. Assume uniform tolerances of ±15% for all passive elements and an operating temperature of 25°C.

12.8 Use PSpice to find the worst-case minimum and maximum output voltages ($V_{o(min)}$ and $V_{o(max)}$) for Problem 12.4. Assume uniform tolerances of ±15% for all passive elements and an operating temperature of 25°C.

SUGGESTED READING

1. L.J. Giacoletto, Simple SCR and TRAIC PSPICE computer models, *IEEE Transactions on Industrial Electronics*, *36*(3), 1989, 451–455.
2. M.H. Rashid, *Power Electronics: Circuits, Devices, and Applications*, Second Edition, Englewood Cliffs, NJ: Prentice-Hall, 2014, Chapter 11.
3. M.H. Rashid, *Power Electronics Handbook*, Burlington, MA: Butterworth-Heinemann, 2010, Chapter 18.
4. A.A. Zekry and G.T. Sayah, SPICE model of thyristors with amplifying gate and emitter-shorts, *IET Power Electronics*, 7(3), March 2014, 724–735, https://doi.org/10.1049/iet-pel.2013.0158

13 Control Applications

After completing this chapter, students should be able to do the following:

- Model operational amplifier (op-amp) circuits for control applications and specify their model parameters.
- Design analog behavioral models (ABMs) for signal conditioning applications.
- Perform analog simulations and analog behavioral modeling of control systems.
- Perform transient analysis of control systems.
- Perform worst-case analysis of AC voltage controllers for parametric variations of model parameters and tolerances.

13.1 INTRODUCTION

In practical applications, power converters are normally operated under closed-loop control, which requires comparing the desired output with the actual output. The control implementation requires summing, differentiating, and integrating signals to obtain the desired control strategy. SPICE can be used in the following modeling:

- Op-amp circuits
- Control systems
- Signal conditioning

13.2 OP-AMP CIRCUITS

An op-amp may be modeled as a linear amplifier to simplify the design and analysis of op-amp circuits. The linear models give reasonable results, especially for determining the approximate design values of op-amp circuits. The simulation of the actual behavior of op-amps is required in many applications to obtain accurate responses of electronic circuits. An op-amp can be simulated from the circuit arrangement of the particular type of op-amp. The μ741 type of op-amp consists of 24 transistors and is beyond the capability of the student (or demo) version of PSpice. However, a macromodel, which is a simplified version of the op-amp and requires only two transistors, is quite accurate for many applications and can be simulated as a subcircuit or library file. Some op-amp manufacturers supply macromodels of their op-amps [1]. In the absence of a complex op-amp model, the characteristics of op-amp circuits may be represented approximately by one of the following models:

- DC linear models
- AC linear models
- Nonlinear macromodels

DOI: 10.1201/9781003284451-13

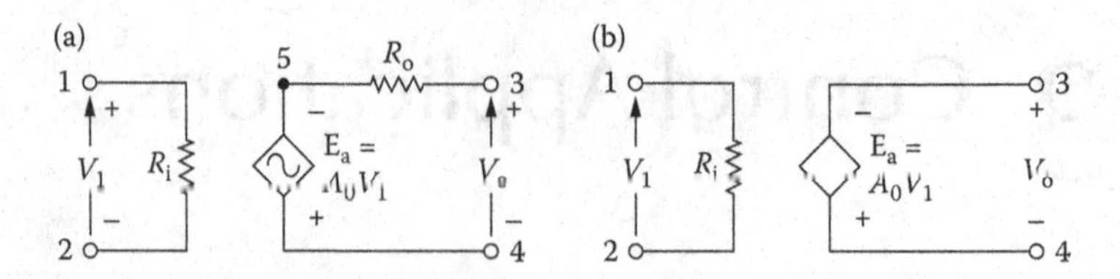

FIGURE 13.1 DC linear op-amp models. (a) DC model. (b) Simple DC model.

13.2.1 DC Linear Models

An op-amp may be modeled as a voltage-controlled voltage source, as shown in Figure 13.1(a). The input resistance is high, typically 2 MΩ, and the output resistance is very low, typically 75 Ω. For an ideal op-amp, the model of Figure 13.1(a) can be reduced to Figure 13.1(b). These models do not take into account the saturation effect and slew rate that exist in practical op-amps. The voltage gain is also assumed to be independent of the frequency, but in practical op-amps the gain falls with the frequency. These simple models are normally suitable for DC or low-frequency applications.

13.2.2 AC Linear Models

The frequency response of an op-amp can be approximated by a single break frequency, as shown in Figure 13.2(a). This characteristic can be modeled by the circuit

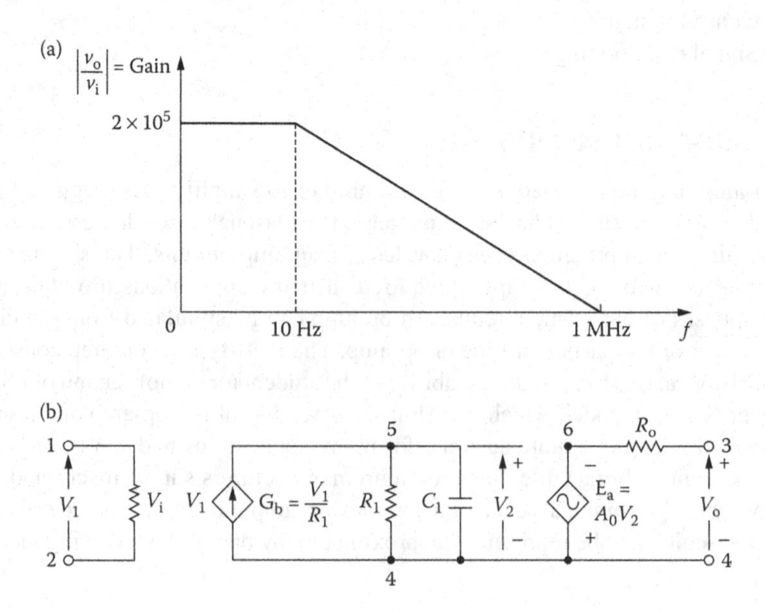

FIGURE 13.2 AC linear model with single break frequency. (a) Frequency response. (b) Circuit model.

of Figure 13.2(b), a high-frequency op-amp model. If an op-amp has more than one break frequency, it can be represented by using as many capacitors as there are breaks. R_{in} is the input resistance and R_o the output resistance.

The dependent sources in the op-amp model of Figure 13.2(b) have a common node. Without this, the PSpice will give an error message because there is no DC path from the nodes of the dependent current source. The common node could be in either the input stage or the output stage. This model does not take into account the saturation effect and is suitable only if the op-amp operates within the linear region.

The output voltage can be expressed as

$$V_o = A_o V_2 = \frac{A_o V_i}{1 + R_1 C_1 s}$$

Substituting $s = j2\pi f$ yields

$$V_o = \frac{A_o V_i}{1 + j2\pi f R_1 C_1} = \frac{A_o V_i}{1 + j(f/f_b)}$$

where $f_b = 1/2(2\pi R_1 C_1)$ is called the *break frequency* (in hertz) and A_o is the *large-signal* (or *DC*) *gain* of the op-amp. Thus, the open-loop voltage gain is

$$A(f) = \frac{V_o}{V_i} = -\frac{A_o}{1 + j(f/f_b)}$$

for µ741 op-amps, $f_b = 10$ Hz, $A_o = 2 \times 10^5$, $R_i = 2$ MΩ, and $R_o = 75$ Ω. Letting $R_1 = 10$ kΩ, $C_1 = 1/(2\mu \times 10 \times 10 \times 10^3) = 1.15619$ µF

Note: We could choose a value of C_1 and then find the value of R_1.

13.2.3 Nonlinear Macromodels

The circuit arrangement of the op-amp macromodel is shown in Figure 13.3 [2,3]. The macromodel can be used as a subcircuit with a .SUBCKT command. However, if an op-amp is used in various circuits, it is convenient to have the macromodel as a library file (i.e., EVAL.LIB), and it is not necessary to type the statements of the macromodel in every circuit in which the macromodel is employed. The library file EVAL.LIB that comes with the student version of PSpice has macromodels for op-amps, comparators, diodes, MOSFETs, BJTs, IGBTs, and SCRs. The professional version of PSpice supports library files for many devices. Check the name of the current library file by listing the files of the PSpice programs.

The macromodel of the µ741 op-amp is simulated at room temperature. The library file EVAL.LIB contains the op-amp macromodel as a subcircuit definition UA741 with a set of .MODEL statements. This op-amp model contains nominal, not worst-case devices, and does not consider the effects of temperature.

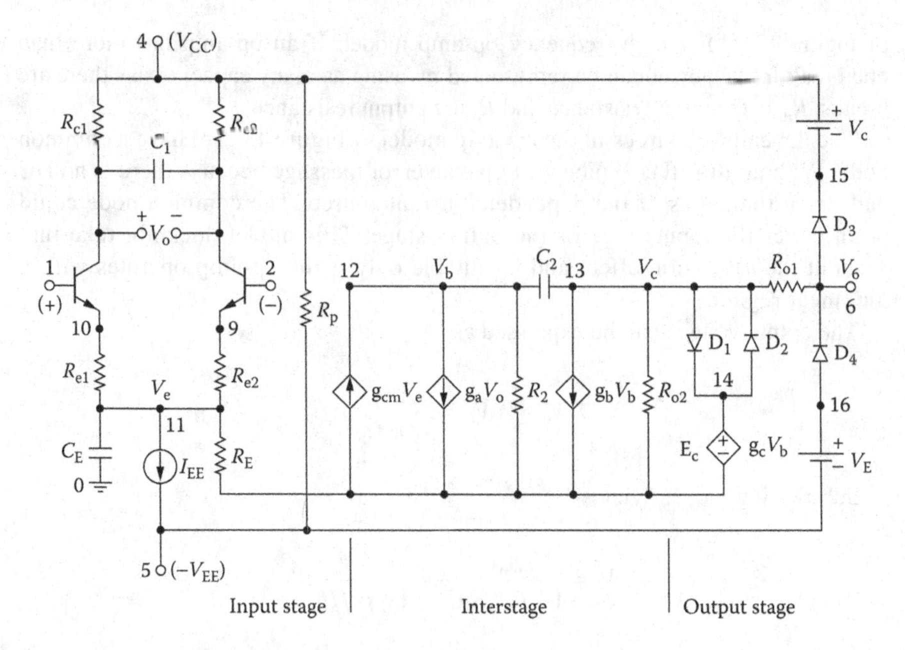

FIGURE 13.3 Circuit diagram of the op-amp macromodel.

The listing of the subcircuit UA741 in the library file EVAL.LIB is as follows:

```
* Subcircuit for µ741 op-amp
* connections:        noninverting input
*                     :      inverting input
*                     :      :
*                     :      :      positive power supply
*                     :      :      :      negative power supply
*                     :      :      :      :      output
*                     :      :      :      :      :
.SUBCKT        UA741  1      2      4      5      6
*                     Vi+    Vi-    Vp+    Vp-    Vout
Q1      7      1      10     UA71QA
Q2      8      2      9      UA741QB
RC1     4      7      5.305165D+03
RC2     4      8      5.305165D+03
C1      7      8      5.459553D-12
RE1    10     11      2.151297D+03
RE2     9     11      2.151297D+03
IEE    11      5      1.666000D-05
CE     11      0      3.000000D-12
RE     11      0      1.200480D+07
GCM     0     12     11      0      5.960753D-09
GA     12      0      8      7      1.884955D-04
R2     12      0      1.000000D+05
```

```
C2      12     13       3.000000D-11
GB      13      0      12      0       2.357851D+02
RO2     13      0       4.500000D+01
D1      13     14        UA741DA
D2      14     13        UA741DA
EC      14      0       6      0       1.0
RO1     13      6       3.000000D+01
D3       6     15        UA741DB
VC       4     15       2.803238D+00
D4      16      6        UA741DB
VE      16      5       2.803238D+00
RP       4      5      18.16D+03
* Models for diodes and transistors:
.MODEL   UA741DA      D(IS = 9.762287D-11)
.MODEL   UA741DB      D(IS = 8.000000D-16)
.MODEL   UA741QA      NPN (IS = 8.000000D-16 BF = 9.166667D+01)
.MODEL   UA741QB      NPN (IS = 8.309478D-16 BF = 1.178571D+02)
.ENDS    UA741                    ;End of subcircuit definition
```

13.2.4 Examples of Op-Amp Circuits

Example 13.1:

Finding the Performance of an Op-Amp Inverting Amplifier

An inverting amplifier is shown in Figure 13.4. Use PSpice to plot the DC transfer characteristic if the input is varied from –1 to +1 V with a 0.2-V increment. (a) Use the op-amp of Figure 13.1(a) as a subcircuit; its parameters are $A_o = 2 \times 10^5$, R_i = 2 MΩ, and R_o = 75Ω (b) Use the op-amp of Figure 13.2 as a subcircuit; its parameters are R_i = 2 Ω, R_o = 75Ω, C_1 = 1.5619 μF, and R_1 = 10 kΩ. (c) Use the macromodel of Figure 13.3 for the UA741.

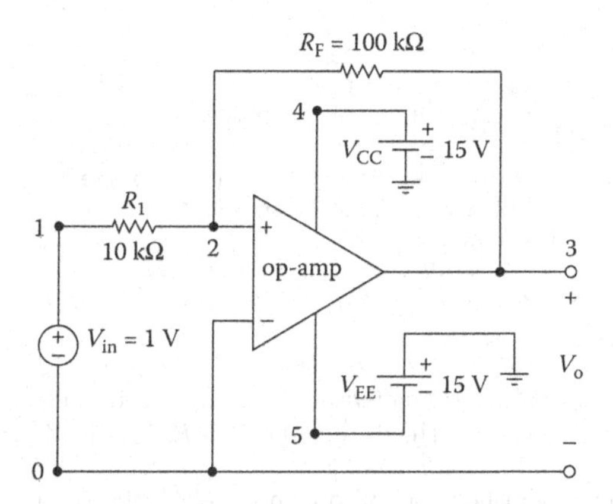

FIGURE 13.4 Inverting amplifier.

SOLUTION

The listing of the subcircuit OPAMP-DC for Figure 13.1(a) is as follows:

```
*Subcircuit definition for OPAMP-DC:
.SUBCKT      OPAMP-DC         1        2        3        4
*            model name       Vi-      Vi+      Vo+      Vo-
RIN     1      2        2MEG
RO      5      3       75
EA      5      4       2 1 2E+5; Voltage-controlled voltage source
*End of subcircuit definition:
.ENDS     OPAMP-DC            ; End of subcircuit definition
```

The listing of the subcircuit OPAMP-AC for Figure 13.2 is as follows:

```
*Subcircuit definition for OPAMP-AC:
.SUBCKT      OPAMP-AC         1        2        3        4
*            model name       Vi-      Vi+      Vo+      Vo-
RI      1      2        2MEG
GB      4      5      1      2      0.1M      ;Voltage-controlled
                                               current source
R1      5      4      10K
C1      5      4      1.5619UF
EA      4      6      5      4      2E+5      ;Voltage-controlled
                                               voltage source
RO      6      3      75
.ENDS       OPAMP-AC                          ;End of subcircuit
                                               definition
```

a. The PSpice schematic for a DC op-amp model is shown in Figure 13.5, and the DC op-amp model is shown in Figure 13.5(b)

The listing of the circuit file with the DC op-amp model is as follows:

Inverting Amplifier with DC Op-Amp Model

```
SOURCE   ■    VIN   1     0    DC 1V
CIRCUIT  ■■   R1    1     2    10K
              RF    2     3    100K
              * Calling subcircuit OPAMP-DC:
              XA1   2     0     3     0     OPAMP-DC
              *     Vi-   Vi+   Vo+   Vo-   model name
              * Subcircuit definition OPAMP-DC must be
                inserted.
ANALYSIS ■■■  .DC VIN  -1V 1V 0.1V ; DC sweep
              .PROBE               ; Graphics post-processor
         .END
```

The plot of the transfer characteristic using the DC op-amp model of Figure 13.5 is shown in Figure 13.6. The expected gain is $R_F/R_1 = 100/10 = 10$.

b. The PSpice schematic for an AC op-amp model is shown in Figure 13.7 (a), and the AC op-amp model is shown in Figure 13.7(b).

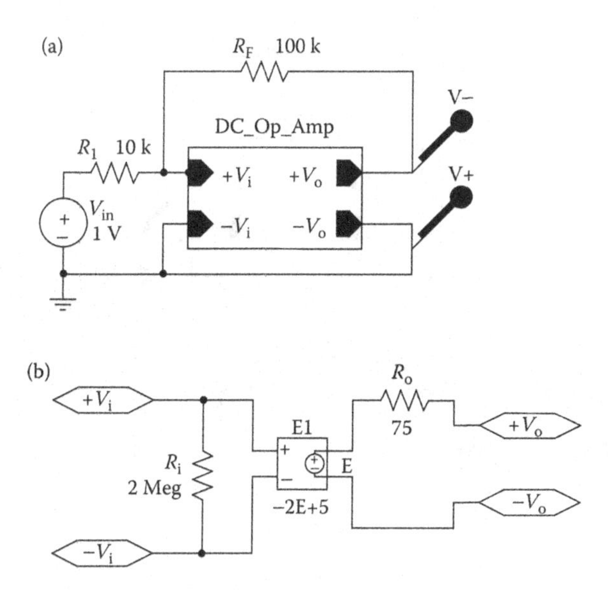

FIGURE 13.5 PSpice schematic for Example 13.1 with a DC op-amp model. (a) Schematic. (b) DC op-amp model.

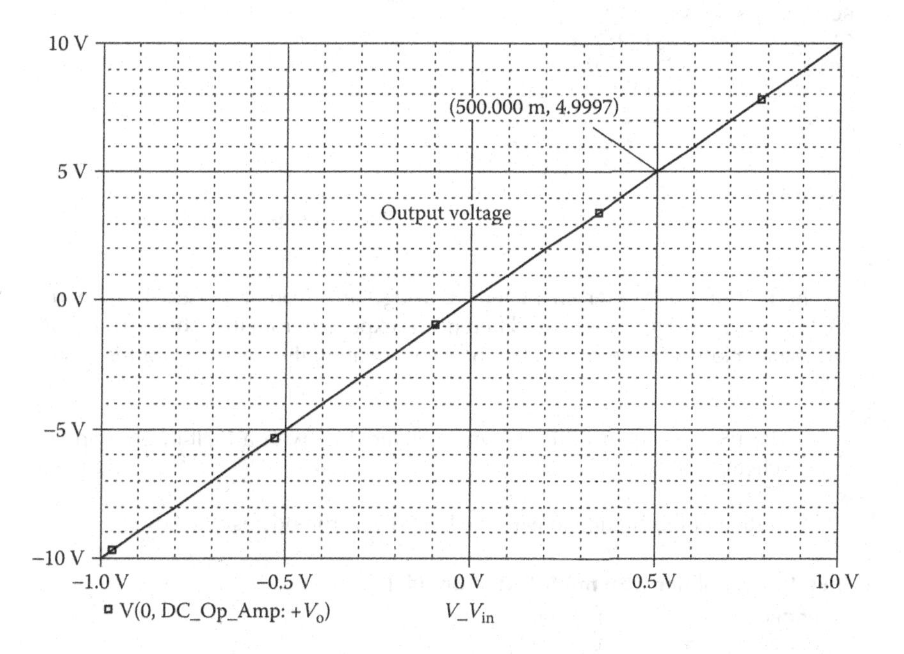

FIGURE 13.6 Plot for Example 13.1 with DC op-amp model.

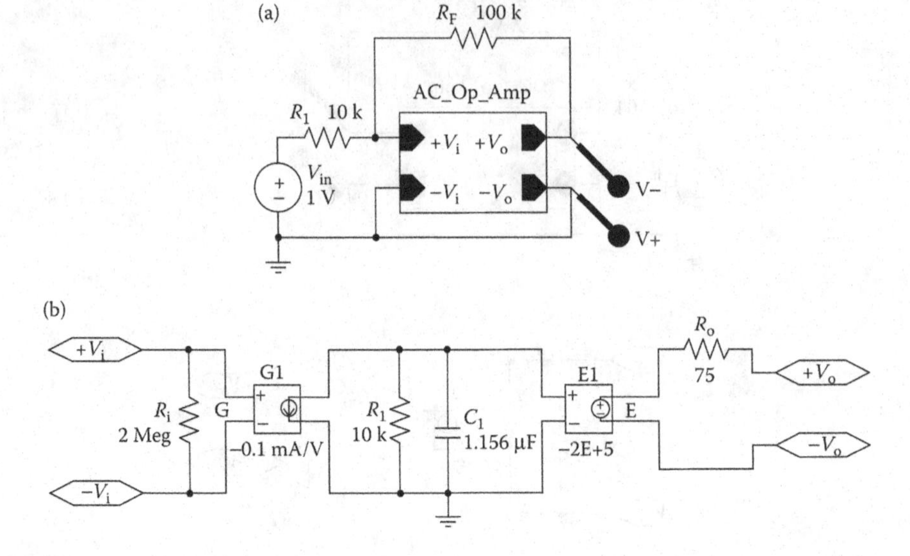

FIGURE 13.7 PSpice schematic for Example 13.1 with an AC op-amp model. (a) Schematic. (b) AC op-amp model.

The listing of the circuit file with the AC op-amp model is as follows:

Inverting Amplifier with AC Op-Amp Model

```
SOURCE   ■    VIN 1 0 DC 1V
CIRCUIT  ■■   R1 1 2 10K
              RF 2 3 100K
              *Calling subcircuit OPAMP-DC:
              XA1  2       0       3        0       OPAMP-DC
              *    Vi-     Vi+     Vo+      Vo-     model name
              *    Subcircuit definition OPAMP-DC must be inserted
ANALYSIS ■■■  .DC VIN   -1V 1V 0.1V   ;DC sweep
              .PROBE                  ;Graphics post-processor
         .END
```

The plot of the transfer characteristic using the AC op-amp model of Figure 13.2 is shown in Figure 13.8. No difference is expected from that obtained with a DC signal. However, the output will be dependent on the frequency of the input signal.

c. The PSpice schematic is shown in Figure 13.9 with a built-in op-amp schematic model.

The listing of the circuit file with the UA741 macromodel is as follows:

Inverting Amplifier with mA741 Macromodel

```
SOURCE   ■    VIN 1 0 DC 1V
CIRCUIT  ■■   R1   1 2 10K
              RF   2 3 100K
```

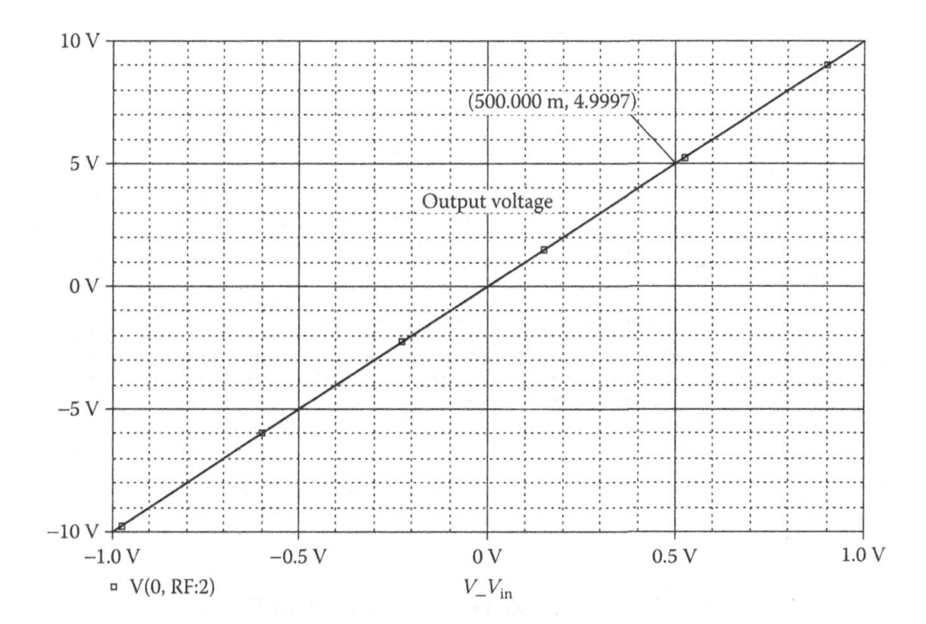

FIGURE 13.8 Plot for Example 13.1 with AC op-amp model.

```
                 VCC 4 0 DC 15V
                 VEE 0 5 DC 15V
                 * Calling subcircuit op-amp UA741:
                 XA1  2     0    4    5     3       UA741
                 *    Vi-   Vi+  Vp+  Vp-   Vout    model name
                 * Subcircuit definition UA741-DC must be inserted.
ANALYSIS ■■■     .DC VIN    -1V 1V 0.1V ; DC sweep
                 .PROBE                 ; Graphics post-processor
             .END
```

The plot of the transfer characteristic using the UA741 macromodel of Figure 13.3 is shown in Figure 13.10. A macromodel affects the characteristics because it has saturation limits. The voltage gain is $\Delta V_o/V_i = 5/0.5 = 10$.

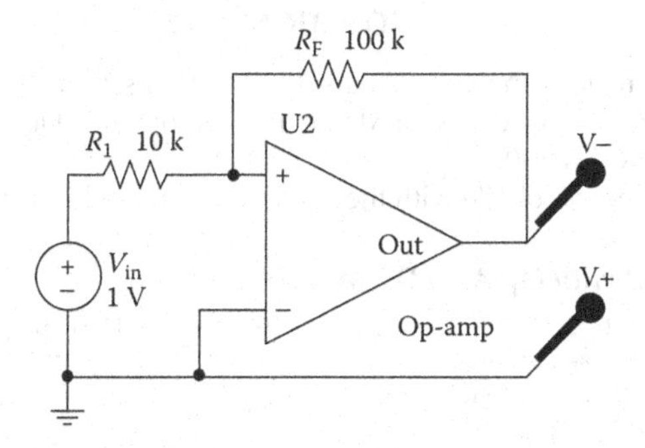

FIGURE 13.9 Pspice schematic for Example 13.1 with a schematic op-amp model.

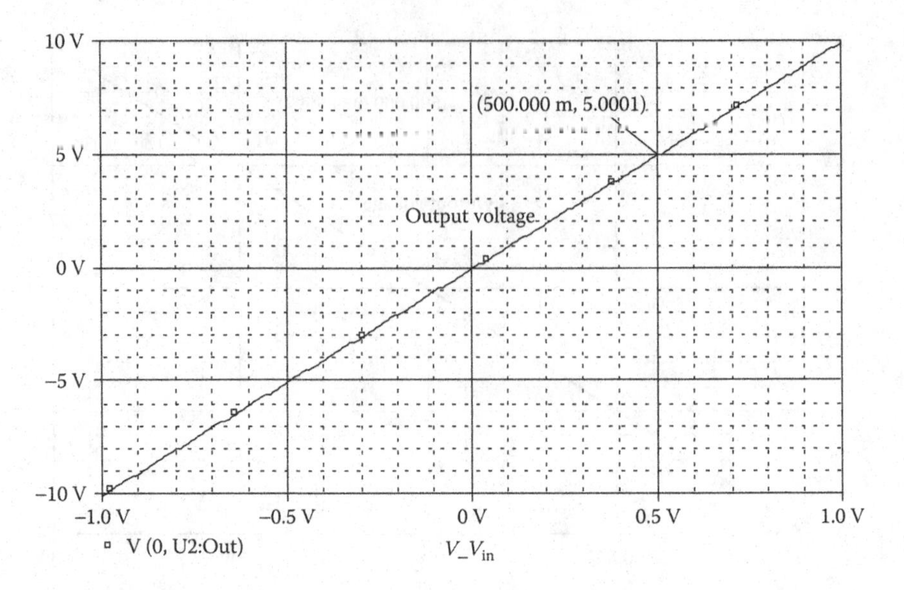

FIGURE 13.10 Plot for Example 13.1 with the UA741 macromodel.

LTspice: LTspice schematic of an op-amp inverting amplifier [10] for Example 13.1 is shown in Figure 13.11(a). The plots of the output voltage for $R_1 = 10k$, $15k$, $20k$ are shown in Figure 13.11(b). As expected, the voltage gain $A_f = -(R_F/R_1)$ decreases with the increase in R_1, and the output voltage is limited to the saturation voltage limit, less than the DC supply voltage of 10 V.

Example 13.2:

Finding the Performance of an Op-Amp Integrator

An op-amp integrator circuit is shown in Figure 13.12. The input voltage is shown in Figure 13.12(b). Use PSpice to plot the transient response of the output voltage for a duration of 0–4 ms in steps of 50 μs.

SOLUTION

The PSpice schematic is shown in Figure 13.13 with a schematic op-amp model, $V_{POS} = 15$ V and $V_{NEG} = -15$ V. The ABM integrator is shown in Figure 13.13(c) which has a gain of $1/(R_1C_1) = 400$.

The listing of the circuit file with the op-amp DC model is as follows:

Integrator Circuit with Op-Amp DC Model

```
SOURCE   ■    VIN 1 0 PULSE (-1V 1V 1MS 1NS 1NS 1MS 2MS) ;
              Pulse waveform
CIRCUIT  ■■   R1 1 2 2.5K
              RF 2 3 1MEG
              C1 2 3 0.1UF IC = 0V ; Set initial condition
              * Calling subcircuit OPAMP-DC:
              XA1 2 0 3 0 OPAMP-DC
```

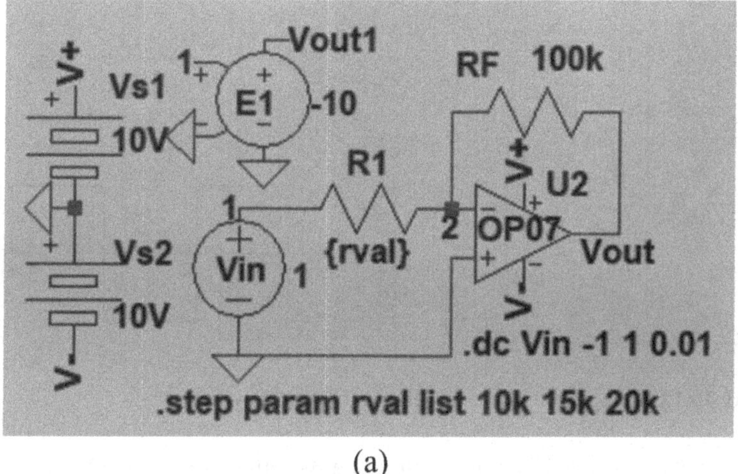

(a)

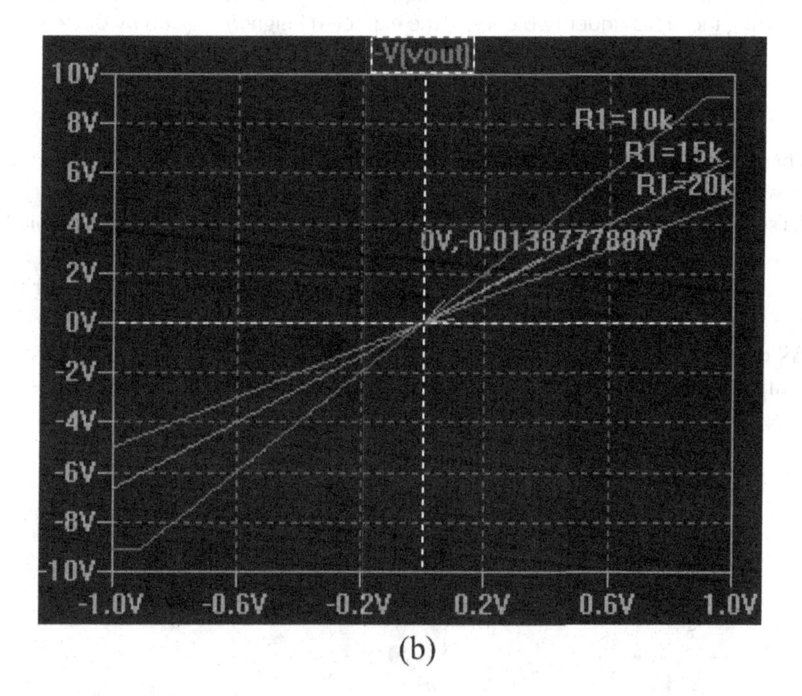

(b)

FIGURE 13.11 LTspice schematic for Example 13.1. (a) LTspice Schematic. (b) Plot of Transfer functions.

```
               * Vi-  Vi+  Vo+  Vo-  model name
               * Subcircuit definition OPAMP-DC must be inserted
ANALYSIS ■■■   .TRAN 10US 4MS UIC ; Use initial condition in
               transient analysis
               .PLOT TRAN V(3) V(1)  ; Prints on the output file
               .PROBE                ; Graphics post-processor
         .END
```

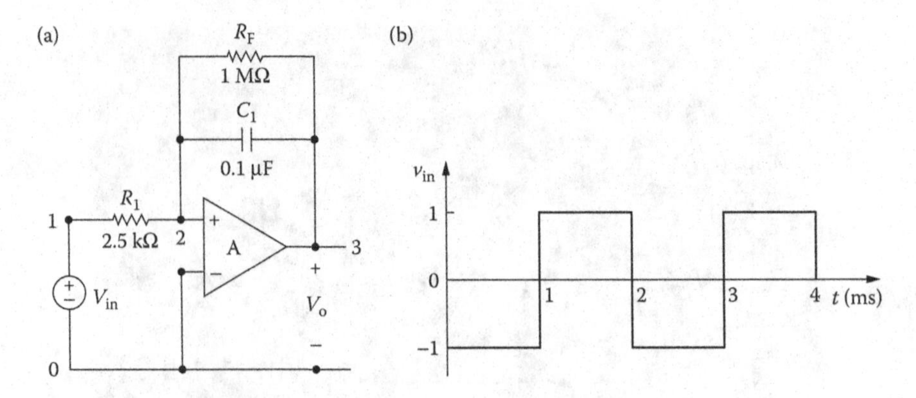

FIGURE 13.12 Integrator circuit. (a) Circuit. (b) Input waveform.

The plot of the transient response using the op-amp DC model is shown in Figure 13.15. The output voltage is triangular in response to a square-wave input. Equating the areas under two curves, the expected height h is given by 0.5 × 1 ms or $h = (1\ V + 1\ V) \times 1$ ms, or $h = 4$ V, which is verified by SPICE simulation.

Note: Students are encouraged to simulate the circuit in Figure 13.13(b).

The plot of the Transient response for Example 13.2 with the op-amp DC model is shown in Figure 13.14.

LTspice: LTspice schematic of an op-amp inverting integrator [10] for Example 13.2 is shown in Figure 13.15(a). The plots of the output voltage for $R_F = 1M\Omega, R_1 = 2.5k\Omega$ and $C_1 = 0.1\mu F$ are shown in Figure 13.15(b). It can be shown that the transfer function of the integrator is given by $A_f(s) = \frac{-R_F/R_1}{1+sR_FC_1}$

As expected, a square pulsed wave produces an inverted triangular wave, and the output voltage is limited to the saturation voltage limit, less than the DC supply voltage of 10 V.

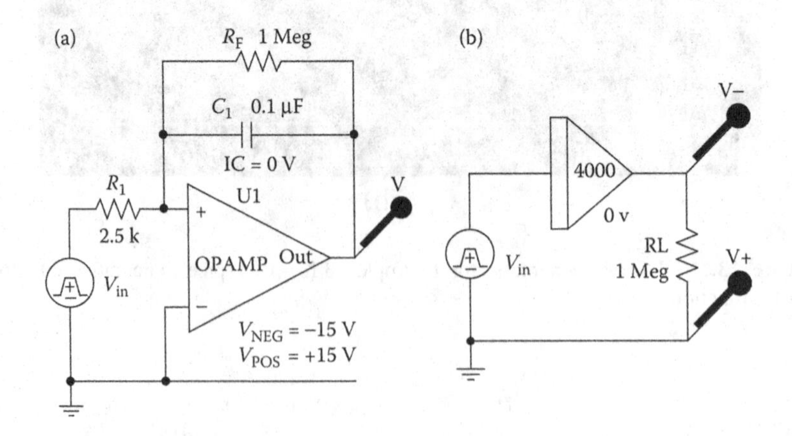

FIGURE 13.13 PSpice schematic for Example 13.2 with a schematic op-amp model. (a) Op-amp integrator. (b) ABM integrator.

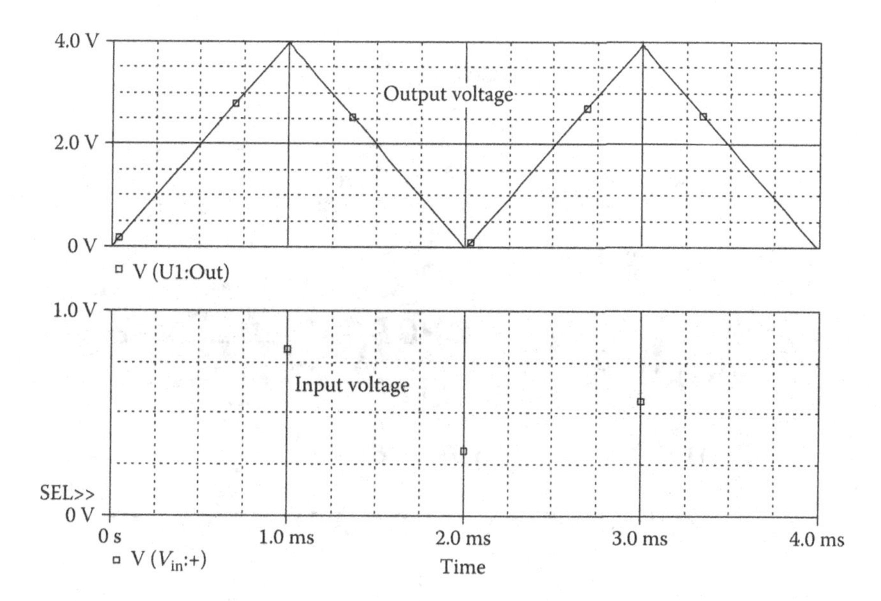

FIGURE 13.14 Transient response for Example 13.2 with the op-amp DC model.

Example 13.3:

Finding the Performance of a Differentiator

A practical differentiator circuit is shown in Figure 13.16(a). The input voltage is shown in Figure 13.16(b). Use PSpice to plot the transient response of output voltage for a duration of 0-4 ms in steps of 50 μs.

SOLUTION

The PSpice schematic is shown in Figure 13.17 with a schematic op-amp model, V_{POS} = 15 V and V_{NEO} = −15 V. The ABM differentiator is shown in Figure 13.14(b) which has a gain of (R_1C_1) = 40 μs.

The listing of the circuit file is as follows:

Differentiator Circuit

```
SOURCE   ■    VIN 1 0 PULSE (0 1V 0 1MS 1MS 1NS 2MS) ; Pulse waveform
CIRCUIT  ■■   R1 1 2 100
              C1 2 3 0.4UF
              RF 3 4 10K
              * Calling subcircuit OPAMP-DC:
              XA1    3     0      4      0     OPAMP-DC
              *     Vi-    Vi+    Vo+    Vo-   model name
              * Subcircuit definition OPAMP-DC must be
              * inserted.
ANALYSIS ■■■ .TRAN 10US 4MS ; Transient analysis
              .PLOT TRAN V(4) V(1) ; Prints on the output file
              .PROBE                ; Graphics post-processor
        .END
```

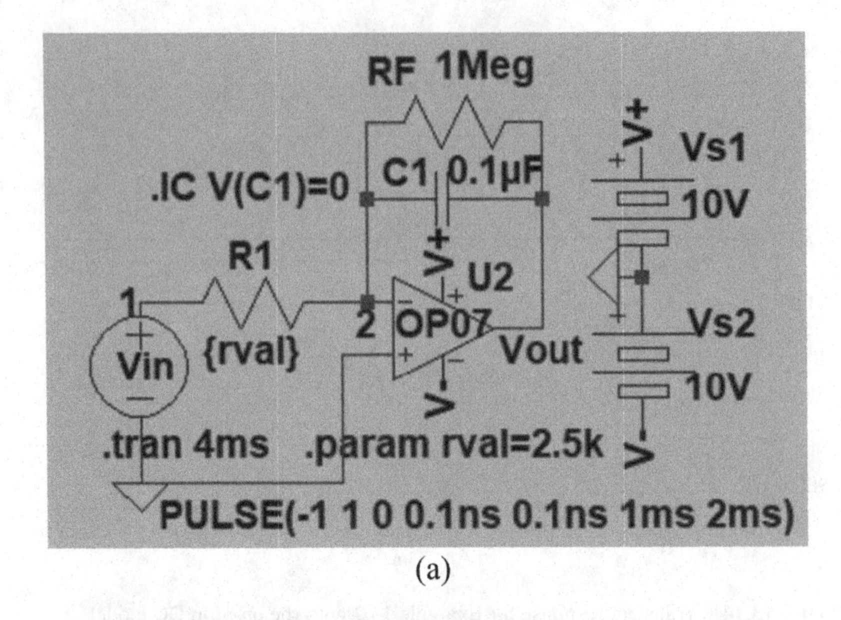

(a)

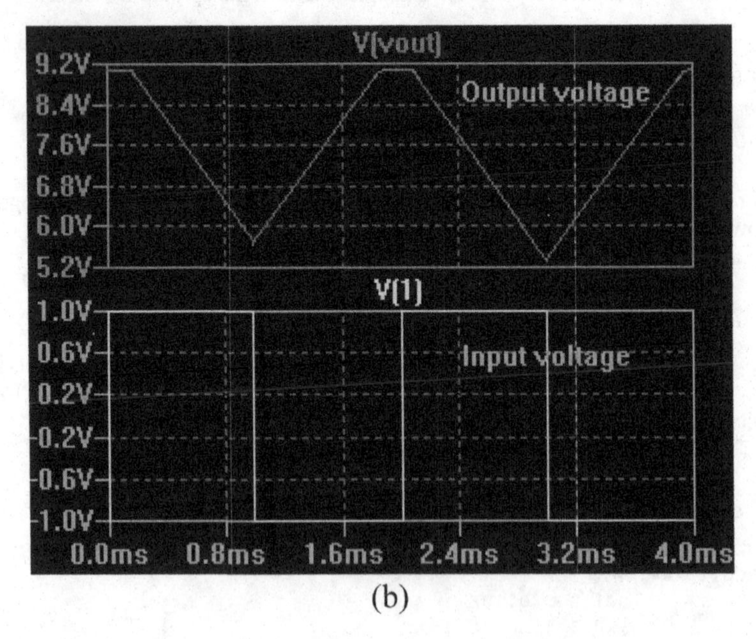

(b)

FIGURE 13.15 LTspice schematic for Example 13.2. (a) LTspice schematic. (b) Plot of Transfer functions.

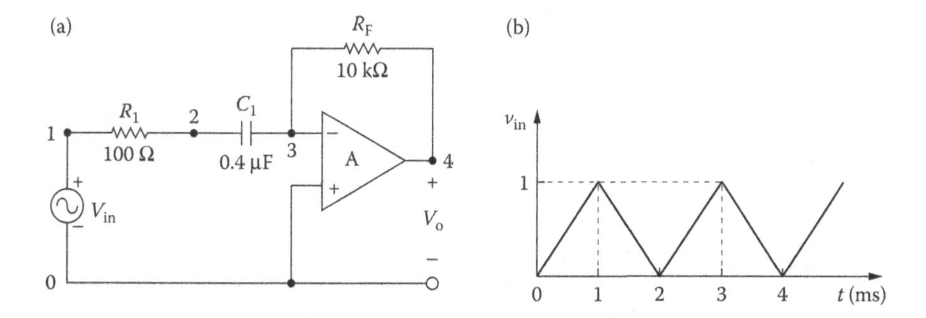

FIGURE 13.16 Differentiator circuit. (a) Circuit. (b) Input waveform.

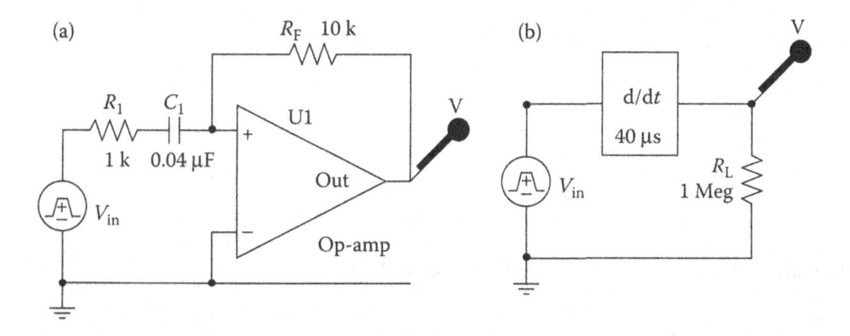

FIGURE 13.17 PSpice schematic for Example 13.3 with a schematic op-amp model. (a) Op-amp differentiator. (b) ABM differentiator.

The plot of the transient response is shown in Figure 13.18. The output voltage is a square wave in response to a triangular input. The time constant of the circuit limits the sharp rise and fall of the output voltage.

LTspice: LTspice schematic of an op-amp inverting differentiator [10] for Example 13.3 is shown in Figure 13.19(a). The plots of the output voltage for $R_1 = 2.5k\Omega$ are shown in Figure 13.19(b). The plots of the output voltage for $R_F = 10k\Omega, R_1 = 1k\Omega$ and $C_1 = 0.4\mu F$ are shown in Figure 13.19(b). It can be shown that the transfer function of the differentiator is given by $A_f(s) = \frac{-sR_FC_1}{1+sR_1C_1}$

As expected, a square pulsed wave produces an inverted triangular wave and the output voltage is limited to the saturation voltage limit, less than the DC supply voltage of 10 V.

13.3 CONTROL SYSTEMS

13.3.1 Examples of Control Circuits

The simulation of control systems requires integrators, multipliers, summing amplifiers, and function generators. These features can easily be simulated by PSpice. Additional PSpice features, such as Polynomial, Table, Frequency, Laplace, Parameter, Value, and Step, make PSpice a versatile tool for simulating complex control systems.

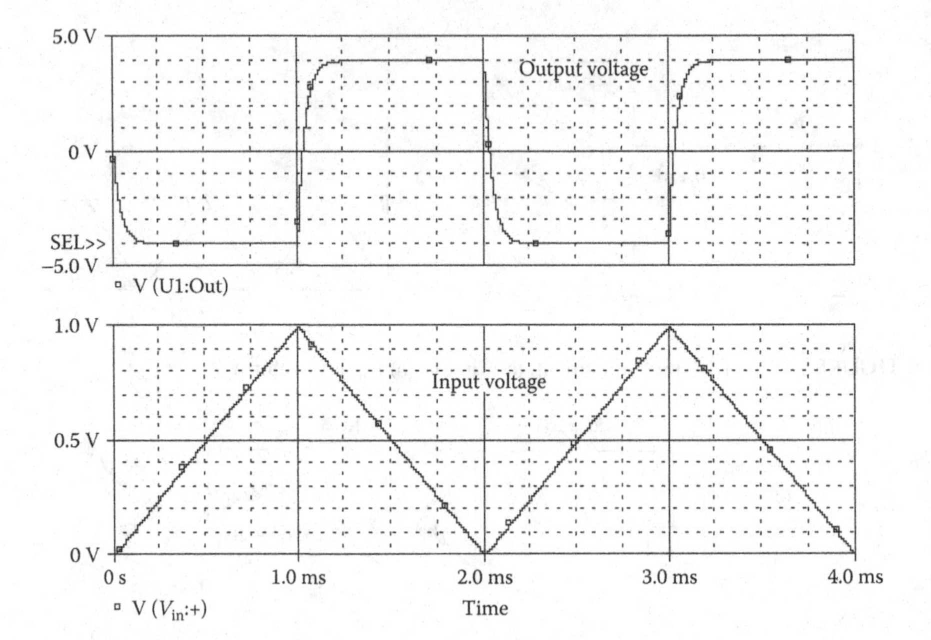

FIGURE 13.18 Transient response for Example 13.3.

Example 13.4:

Finding the Step Response of Unity Feedback Control System

A unity feedback control system is shown in Figure 13.20. The reference input v_r is a step voltage of 1 V. Use PSpice to plot the transient response of the output voltage v_o for a duration of 0–10 s in steps of 10 ms. The gain K is to be varied from 0 0.5 to 2 with an increment of 0.5. Assume that all initial conditions are zero.

SOLUTION

The relations among V_r, V_o, and V_e, in Laplace's s domain, are as follows:

$$V_e(s) = V_r(s) - V_o(s)$$

$$\frac{V_0(s)}{V_e(s)} = \frac{K}{s(1+s)(1+0.2s)}$$

which gives

$$\left[s(1+s)(1+0.2s)\right]V_o(s) = KV_e(s)$$

$$\left[s + 1.2s^2 + 0.2s^3\right]V_o(s) = KV_e(s)$$

which can be written in the time domain as

$$0.2\frac{d^3V_o}{dt^3} + 1.2\frac{d^2V_o}{dt^2} + \frac{dv_o}{dt} = Kv_e = K(v_r - v_o) \tag{13.1}$$

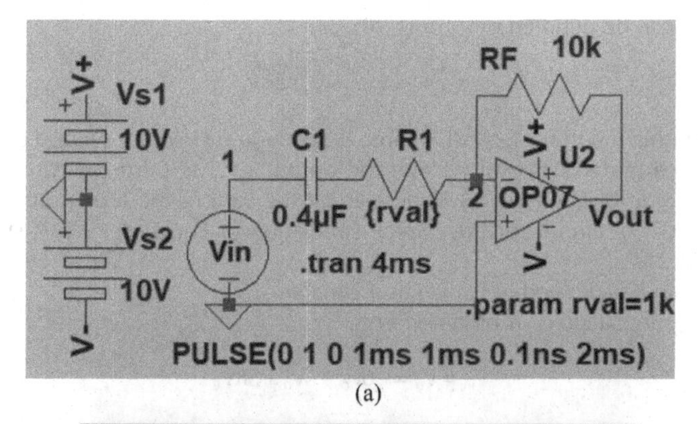

(a)

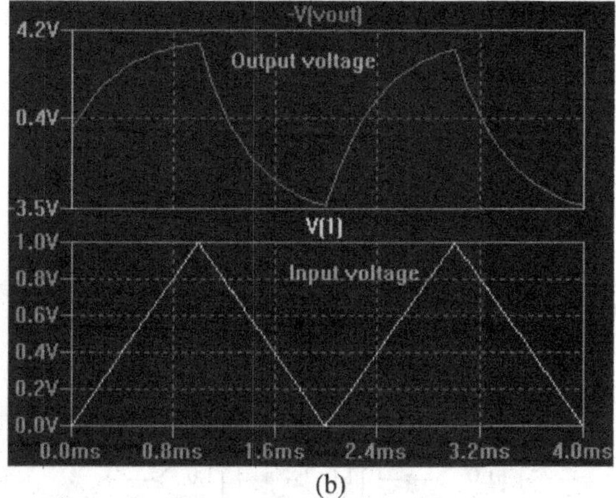

(b)

FIGURE 13.19 LTspice schematic for Example 13.3. (a) LTspice Schematic. (b) Plot of Transfer functions.

Dividing both sides by 0.2 gives

$$\frac{d^3 v_o}{dt^3} + 6\frac{d^2 v_o}{dt^2} + 5\frac{dv_o}{dt} = 5Kv_e$$

where $v_e = V_r - V_o$, or

$$\frac{d^3 v_o}{dt^3} = -6\frac{d^2 v_o}{dt^2} - 5\frac{dv_o}{dt} + 5Kv_e \tag{13.2}$$

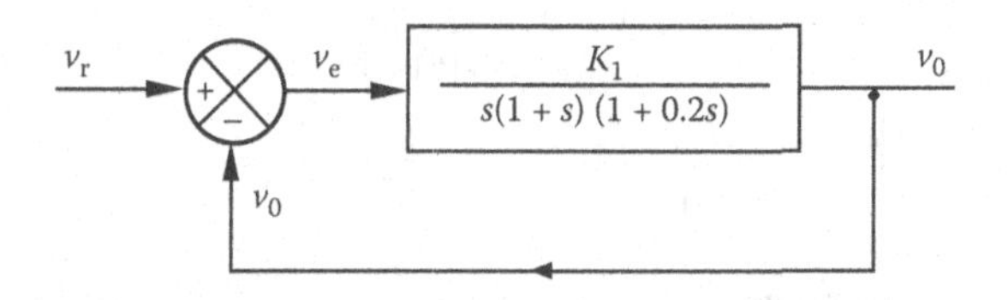

FIGURE 13.20 Unity feedback control system.

which can be denoted by

$$\dddot{v}_o = -6\ddot{v}_o - 5\dot{v}_o + 5Kv_e \tag{13.3}$$

Integrating the third derivative three times should yield the second derivative of v_o, the first derivative of v_o, and the output v_o. The third derivative on the left-hand side of Equation 13.3 must equal the sum of the terms on the right-hand side. The circuit for the PSpice simulation of Equation 13.2 is shown in Figure 13.21.

The gain K is defined in PSpice as a variable.

The input voltage v_{in} is obtained from

$$v_{in} = \dddot{v}_o = -6\ddot{v}_o - 5\dot{v}_o + 5Kv_e$$

It should be noted that the output signal of an integrator is inverted. This should be taken into account in summing signals v_{in} and v_e.

We can use the ABMs to represent the transfer function and the comparator as shown in Figure 13.21(b).

The listing of the circuit file is as follows:

Unity Feedback Control System with a Step Input

```
SOURCE   ■   Vr 8 0 PWL (0 0V 1NS 1V 10MS 1V) ; Reference voltage
             Rg 8 0 10MEG
CIRCUIT  ■■ .PARAM VAL=10K ; Parameter VAL
```

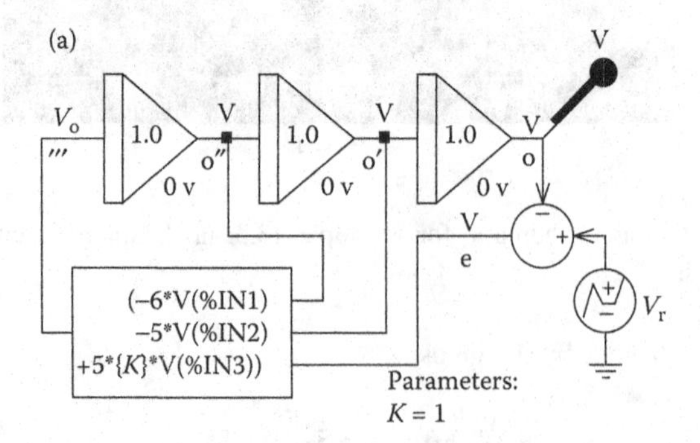

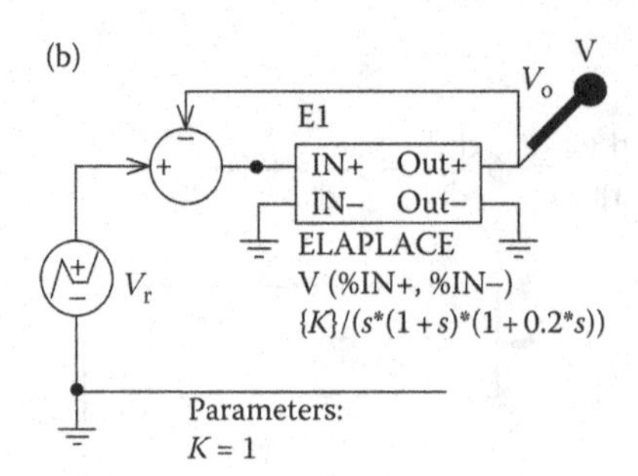

FIGURE 13.21 Unity feedback control system for PSpice schematic of Example 13.4. (a) Analog representation. (b) ABM representation.

```
            .STEP PARAM VAL 10K 40K 10K ; Step change of
            parameter VAL
            R1  1  2 1K
            C1  2  3 0.001 IC = 0V ; Set initial condition
            R2  3  4 1K
            C2  4  5 0.001 IC = 0V ; Set initial condition
            R3  5  6 1K
            C3  6  7 0.001 IC = 0V ; Set initial condition
            R   9 10 20K
            RF 10 11 {VAL}
            * Calling subcircuit OPAMP-DC:
            XA1  2 0  3 0 OPAMP-DC
            XA2  4 0  5 0 OPAMP-DC
            XA3  6 0  7 0 OPAMP-DC
            XA4 10 0 11 0 OPAMP-DC
            E1   9 0 POLY(2) 8 0 7 0  0 1 1
            E2   1 0 POLY(3) 3 0 5 0 11 0 0 6.0  -5.0  -5.0
           * Subcircuit definition OPAMP-DC must be inserted.
ANALYSIS ■■■ .TRAN 0.01S 10S UIC ; Use initial condition in
            transient analysis
            .PLOT TRAN V(7)      ; Prints on the output file
            .OPTIONS ABSTOL = 1.00N RELTOL = 0.01
            VNTOL = 0.1 ITL5 = 0
            .PROBE               ; Graphics post-processor
        .END
```

The plot of the transient response for the feedback control system is shown in Figure 13.22. A higher value of K gives more overshoot and the system tends to be unstable. The transient should settle to the input signal level.

LTspice: LTspice schematic of an op-amp inverting differentiator for Example 13.4 is shown in Figure 13.23(a). Three integrators are cascaded to obtain the output voltage

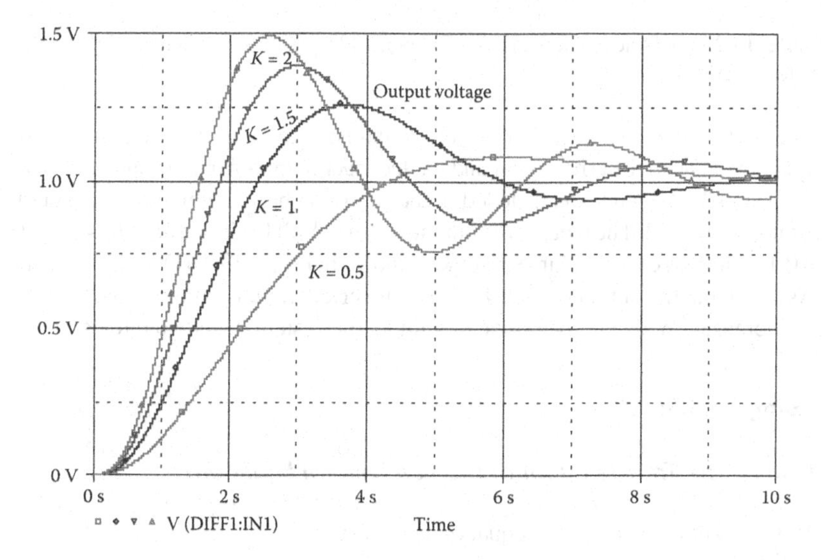

FIGURE 13.22 Transient response for Example 13.4.

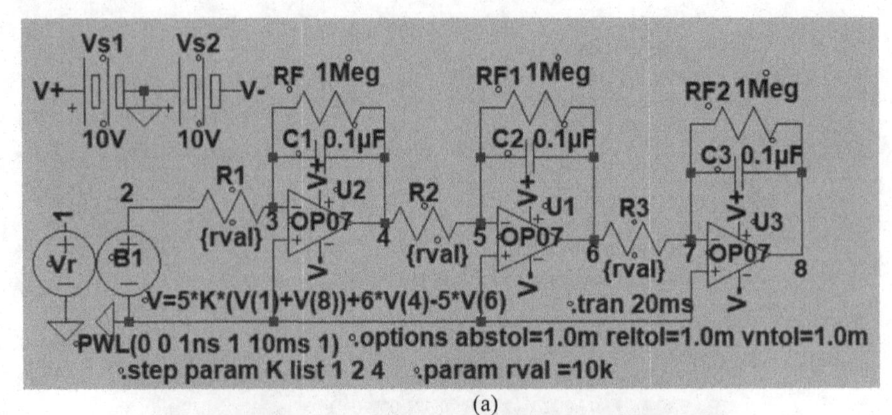

(a)

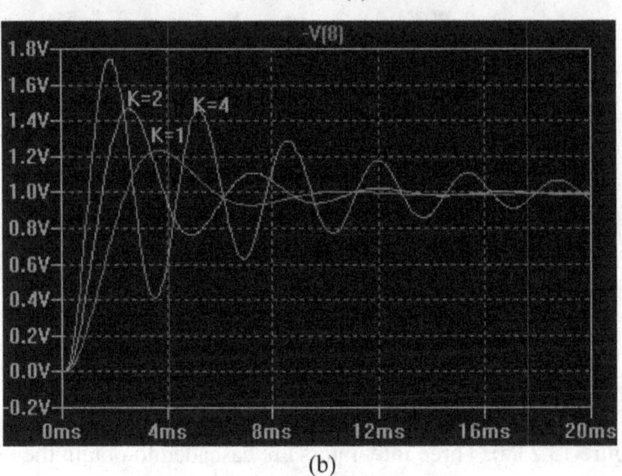

(b)

FIGURE 13.23 LTspice schematic for Example 13.4. (a) Ltspice Schematic. (b) Plot of Transfer functions.

at node 8 V(8). Vr is a step input signal block and B1 is a summing block that compares the output voltage to the error signal and the feedback signals from the output of integrators 1, 2, and 3. The gain K is included in the summing back, nothing that output of the integrator is inverted. The plots of the output voltage –V(8) for $R_F = 1M\Omega$, $R_1 = 10k\Omega$ and $C_1 = 0.1\mu F$ for three identical integrators are shown in Figure 13.23(b) for K = 1, 2, and 4.

As expected, the output of node 8, V(8) is inverted. As the gain increases, the overshoot increases. With a higher value of gain K, the system would go into oscillations.

Example 13.5:

Finding the Transient Plot of the Van Der Pol Equation

The well-known Van der Pol equation is represented by

$$\ddot{y} - \mu\left(1 - y^2\right)\dot{y} + y = 0$$

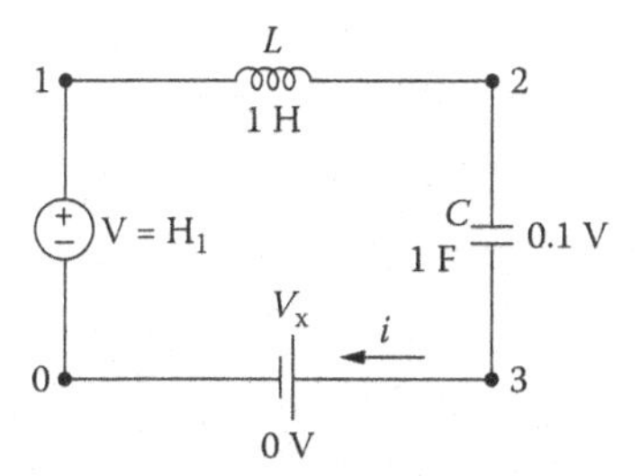

FIGURE 13.24 LC circuit for representing the Van der Pol equation.

Use PSpice to plot the transient response of the output signal for a duration of 0–20 s in steps of 10 ms, and the phase plane (dy/dt against y). Assume a constant $\mu = 2$ and an initial disturbance of 0.1 unit.

SOLUTION

The Van der Pol equation can be written as

$$\frac{d^3y}{dt^2} = \mu\left(1 - y^2\right)\frac{dy}{dt} - y \tag{13.4}$$

This is identical to that for the LC circuit of Figure 13.24. Let us assume that $L = 1$ H and $C = 1$ F. The equation describing the LC circuit of Figure 13.24 is as follows

$$v = L\frac{di}{dt} + \frac{1}{C}\int i\,dt = \frac{di}{dt} + \int i\,dt$$

Differentiating both sides yields

$$\frac{dv}{dt} = \frac{d^2i}{dt^2} + i$$

Solving for d^2i/dt^2, we get

$$\frac{d^2i}{dt^2} = \frac{dv}{dt} - i \tag{13.5}$$

Equation 13.4 will be identical to Equation 13.5 if the current i represents y and dv/dt equals

$$\frac{dv}{dt} = \mu\left(1 - y^2\right)\frac{dy}{dt}\left(1 - i^2\right)\frac{di}{dt} \tag{13.6}$$

which can be integrated to give

$$v = \int \mu\left(1 + i^2\right)di = \mu\left(i - \frac{i^3}{3}\right) \tag{13.7}$$

Equation 13.7 is a polynomial of the form

$$v = P_o + P_1 i + P_2 i^2 + P_3 i^3$$

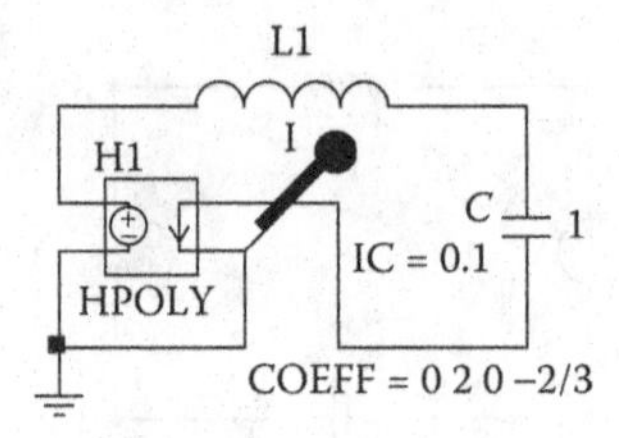

FIGURE 13.25 PSpice schematic for Example 13.5.

where $P_0 = P_2 = 0$, $P_1 = 2$, and $P_3 = -\mu/3$.

The voltage across the inductor $v_L = di/dt$ represents dy/dt.
The PSpice schematic is shown in Figure 13.25
The listing of the circuit file is as follows:

Van der Pol Equation

```
CIRCUIT  ■  H1 1 0 POLY(1) VX 0 2 0 −2/3
            L  1 2 1
            C  2 3 1 IC = 0.1 ; Set initial condition
            VX 3 0 DC 0V ; Senses the circuit current
ANALYSIS ■■ .TRAN 0.01S 20S UIC ; Use initial condition in
            transient analysis
            .PLOT TRAN I(VX) V(1) ; Prints on the output file
            .OPTIONS ABSTOL = 1.00N  RELTOL = 0.01
            VNTOL = 0.1 ITL5 = 0
            .PROBE                 ; Graphics post-processor
        .END
```

The plots of the transient response and the phase plane for the Van der Pol equation are shown in Figures 13.26 and 13.27, respectively. The system is

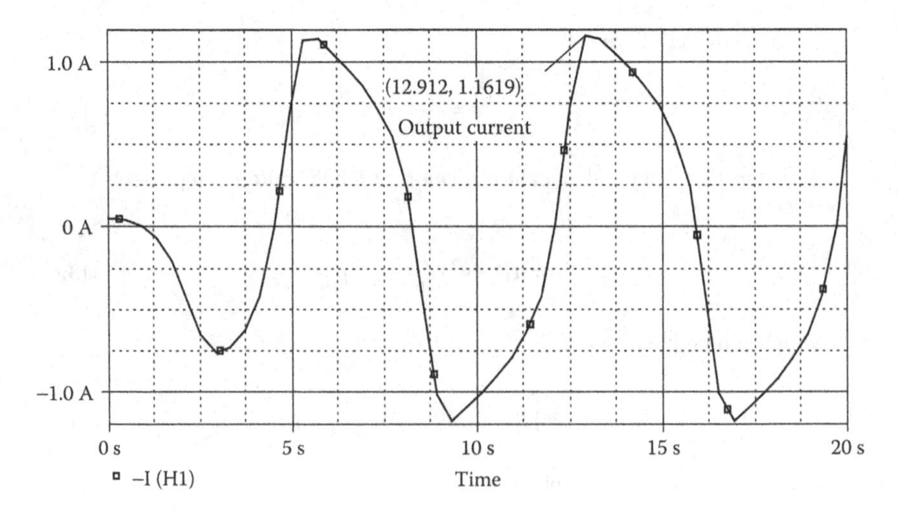

FIGURE 13.26 Transient response of the Van der Pol equation in Example 13.5.

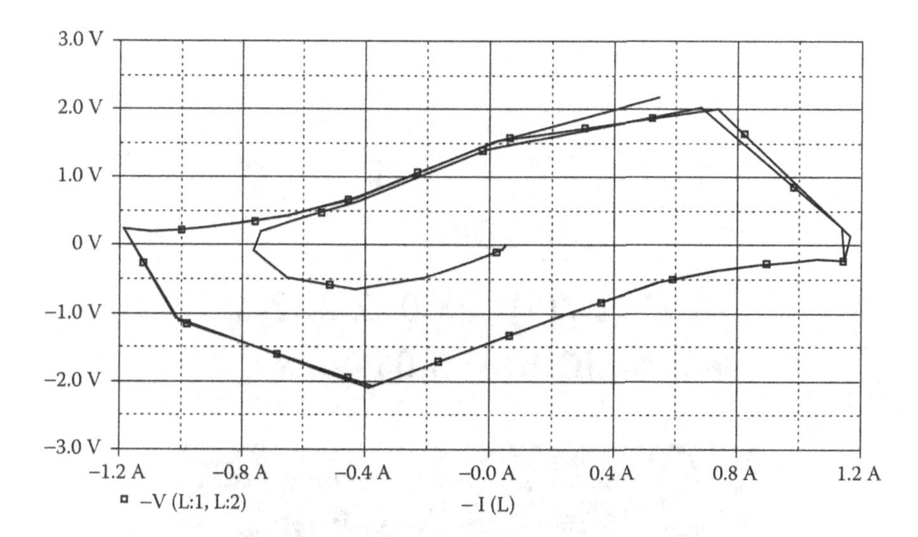

FIGURE 13.27 Phase plane plot of the Van der Pol equation in Example 13.5.

unstable and oscillates between limit cycles. A smoother curve can be obtained by reducing the printing time by 0.01 s.

LTspice: LTspice schematic of the Van der Pol equation for Example 13.5 is shown in Figure 13.28(a). B1 is a first-order polynomial of a current-dependent I(Vx) voltage source and acts as an input to the LC circuit. The initial capacitor voltage of V(3) = 0.01 is included to initial the oscillation. The transient plot of the current is shown in Figure 13.28(b). The plot of the capacitor voltage versus the current exhibiting the characteristic of the Van der Pol equation is shown in Figure 13.28(c). Changing the value of L or C should change the number of loops of the Van der Pol equation.

13.4 SIGNAL CONDITIONING CIRCUITS

An op-amp can be used for wave shaping to yield a desired control characteristic. The student version of PSpice allows only one op-amp macromodel; it also increases the computation time. Simple *RC* circuits can be used to perform the functions of an op-amp, thereby reducing the computation time. We illustrate the applications of *RC* circuits in performing the following operations:

Integration
Averaging
RMS
RMS
Hysteresis
Analog behavioral models (ABMs)

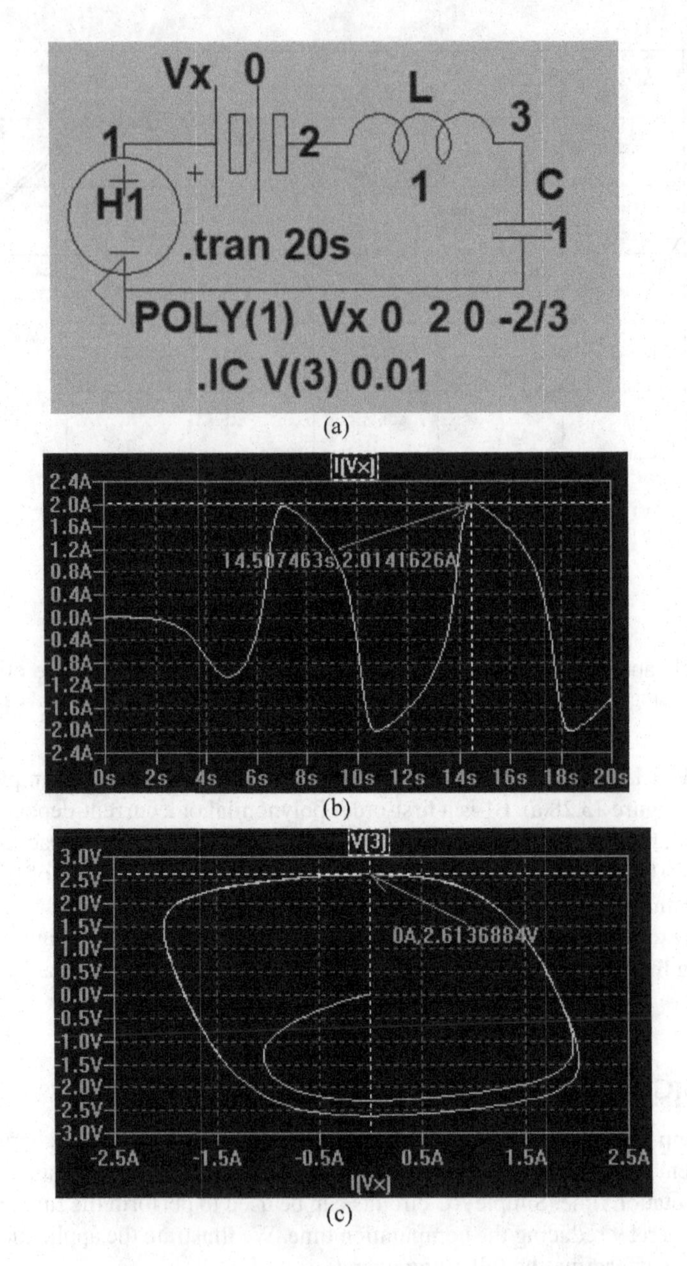

FIGURE 13.28 LTspice schematic for Example 13.5. (a) LTspice schematic. (b) Transient response of the current. (c) Voltage versus current Van der Pol equation.

13.4.1 Examples of Signal Conditioning Circuits

Example 13.6:

Finding the Transient Response of an Integrating Circuit

The input signal v_{in} = sin(120πt + 90°) is to be integrated and then fed to a load resistance R_L = 1 GΩ. This is shown in Figure 13.29(a). Use PSpice to plot the transient response of the input and output voltages for a duration of 0–33.33 ms in steps of 10 μs.

SOLUTION

The integration is accomplished by the *RC* circuit of Figure 13.29(b). The input frequency is *f* = 60 Hz.

The PSpice schematic is shown in Figure 13.30(a) with a capacitor integrator. Figure 13.30(b) shows an implementation with an ABM integrator. Note that a large resistor of 10 GΩ is connected as a load.

The listing of the circuit file is as follows:

Integrating Circuit

```
SOURCE   ■   VIN 1 0 SIN (0 1V 60HZ 0 0 90DEG)
CIRCUIT  ■■  .PARAM FREQ = 60HZ ; Input frequency in hertz
             .PARAM TWO-PI = 6.2832
             RL 2 0 1G
             * Calling subcircuit INTG:
             X1    1    2    0     INTG
             *   Vi-  Vo+  Vo-  model name
             * Subcircuit definition for INTG:
             .SUBCKT      INTG      1    3    2 PARAMS: FREQ = 60HZ
             *          model name Vi+  Vo+  Vo-
             RI    1 0 10G
             GINTG 2  4  1 0 1
```

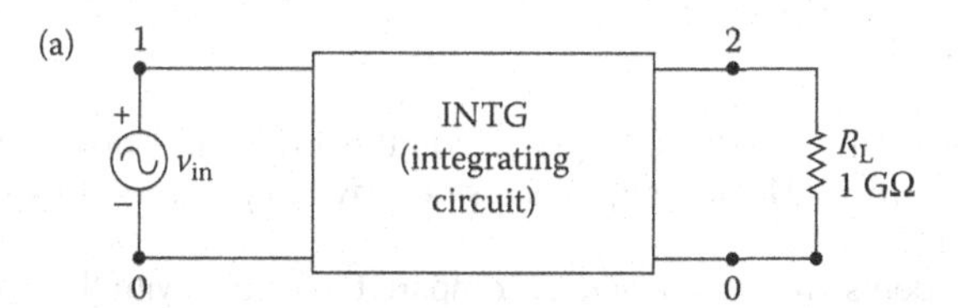

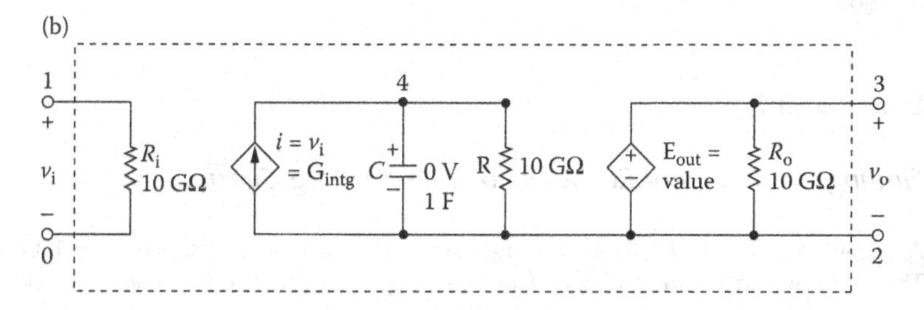

FIGURE 13.29 Integrating circuit. (a) Circuit. (b) Subcircuit.

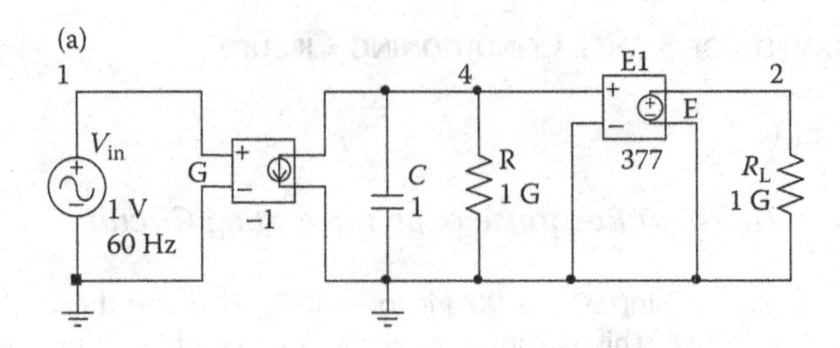

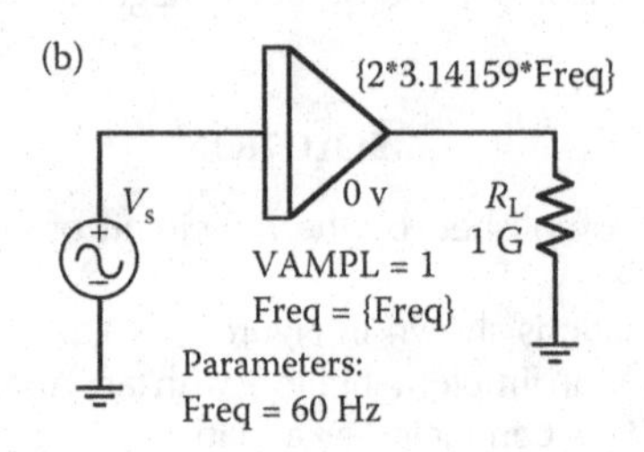

FIGURE 13.30 PSpice schematic for Example 13.6. (a) Capacity integrator. (b) ABM integrator.

```
                  C      4  2   1 IC=0V
                  R      4  2   10G
                  EOUT   3  2   VALUE = {V(4, 2) *TWO-PI*FREQ}
                  RO     3  2   10G
                  .ENDS INTG                  ; End of subcircuit definition
ANALYSIS ■■■      .TRAN 10US 33.33MS UIC ; Use initial condition
                  in transient analysis
                  .PLOT TRAN V(2) V(1)    ; Prints on the output file
                  .OPTIONS ABSTOL = 1.00N RELTOL = 0.01
                  VNTOL = 0.1 ITL5 = 0
                  .PROBE                      ; Graphics post-processor
              .END
```

The plot of the transient response for the input and output voltages are shown in Figure 13.31 The output is a sine wave in response to a cosine signal as expected.

Note: Students are encouraged to compare the results with those of the circuit in Figure 13.30(b).

Example 13.7:

Finding the Transient Response of an Averaging Circuit

An input signal is to be averaged and then fed to a load resistance R_L = 1 GΩ. This is shown in Figure 13.32(a). The input signal is a triangular wave, as shown in Figure 13.32(b). Use PSpice to plot the transient response of input and output voltages for a duration of 0–33.33 ms in steps of 10 μs.

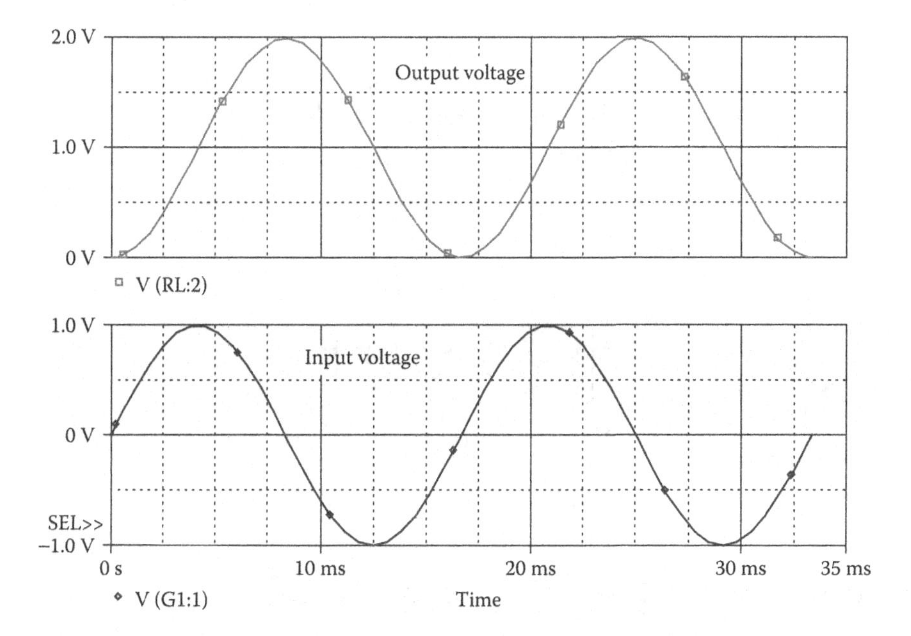

FIGURE 13.31 Transient response of the integrator circuit of Example 13.6.

SOLUTION

The averaging is accomplished by the *RC* circuit of Figure 13.32(c). This is similar to Example 13.6, except that the value is divided by time. The input frequency is $f = 1$ kHz.

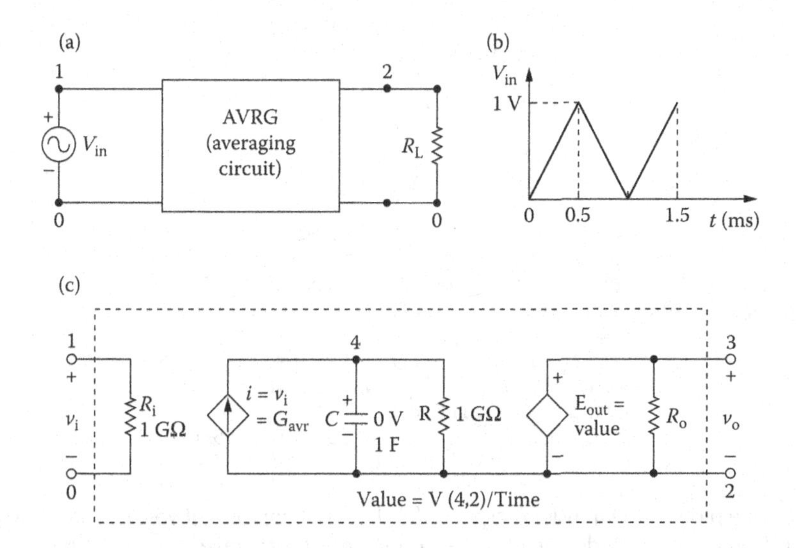

FIGURE 13.32 Averaging circuit. (a) Circuit. (b) Input signal. (c) Subcircuit.

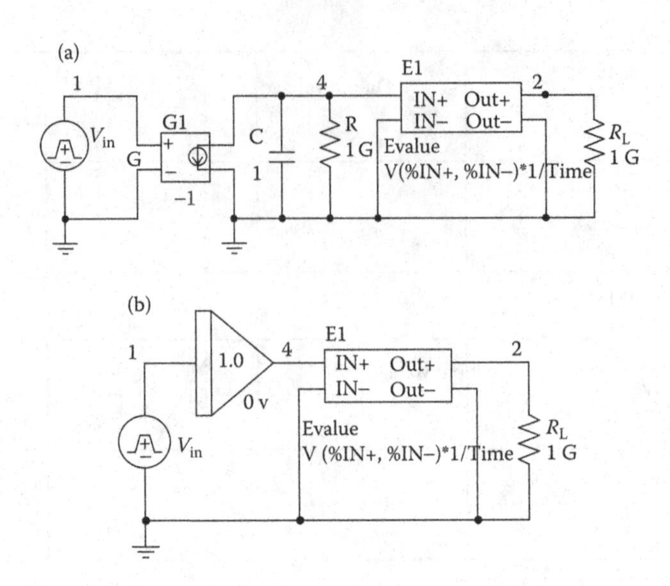

FIGURE 13.33 PSpice schematic for Example 13.7. (a) Averaging circuit. (b) ABM integrator.

The PSpice schematic is shown in Figure 13.33(a) with an averaging circuit. Figure 13.33(b) shows an implementation with an ABM integrator.

The listing of the circuit file is as follows:

Averaging Circuit

```
SOURCE   ■   VIN 1 0 PULSE (0 1V 0 0.5MS 0.5MS 1NS 1MS)
CIRCUIT  ■■  RL 2 0 1G
             * Calling subcircuit AVRG:
             X1     1    2    0        INTG
             *      Vi-  Vo+  Vo-  model name
             * Subcircuit definition for AVRG:
             .SUBCKT      AVRG        1    3    2
             *          model name  Vi+  Vo+  Vo-
             RI    1  0   1G
             GAVR 2   4   1   0 1
             C     4  2   1   IC=0V
             R     4  2   1G
             EOUT 3 2 VALUE={V(4,2)/TIME}
             RO    3 2 1G
             .ENDS AVRG ; End of subcircuit definition
ANALYSIS ■■■ .TRAN 10US 4MS UIC ; Use initial condition in
             transient analysis
             .PLOT TRAN V(2) V(1) ; Prints on the output file
             .OPTIONS ABSTOL = 1.00N RELTOL = 0.01
             VNTOL = 0.1 ITL5 = 0
             .PROBE                   ; Graphics post-processor
         .END
```

The plots of the transient response for the input and output voltages are shown in Figure 13.34. Under steady-state conditions, the average value of a triangular wave remains constant at $0.5bh/T = 0.5 \times 1\ \text{ms} \times 1\ \text{V}/1\ \text{ms} = 0.5\ \text{V}$.

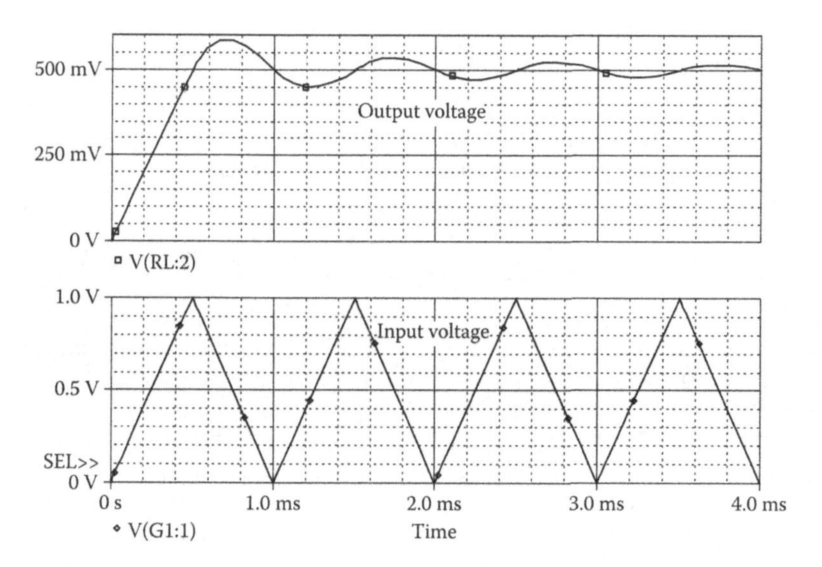

FIGURE 13.34 Transient response for the averaging circuit of Example 13.7.

Example 13.8:

Finding the Transient Response of an RMS Circuit

The RMS value of an input signal $v_{in} = \sin(120\pi t)$ is to be fed to a load resistance R_L = 1 GΩ. This is shown in Figure 13.35(a). Use PSpice to plot the transient response of the input and output voltages for a duration of 0–33.33 ms in steps of 10 μs.

SOLUTION

The RMS value is determined by the *RC* circuit of Figure 13.35(b). The input frequency is f = 60 Hz.

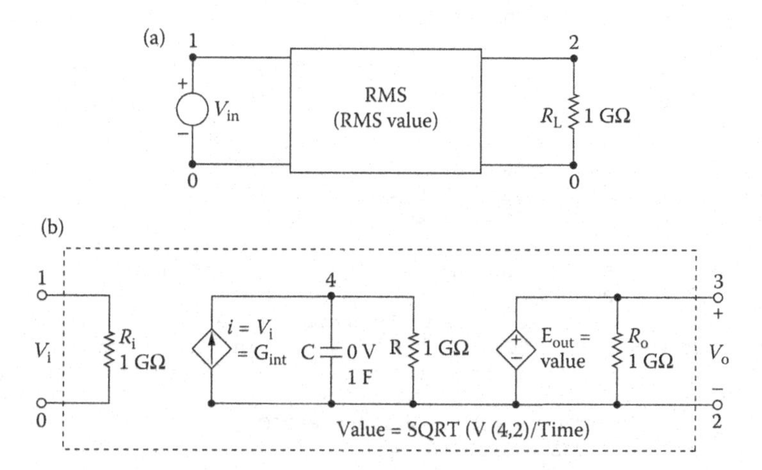

FIGURE 13.35 RMS value circuit. (a) Circuit. (b) Subcircuit.

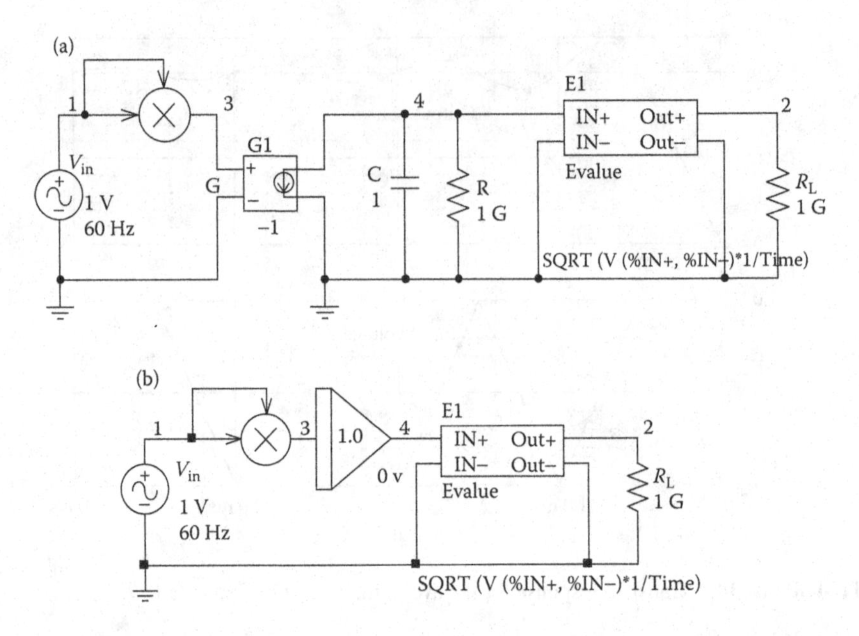

FIGURE 13.36 PSpice schematic for Example 13.8. (a) RMS circuit. (b) ABM integrator.

The PSpice schematic is shown in Figure 13.36(a) for an RMS circuit. Figure 13.36(b) shows an implementation with an ABM integrator.

Note: The input signal v_{in} is squared with the multiplier and then the signal is integrated. The ABM Evalue performs the square root of the signal to produce the output signal. A large resistor of 1 GΩ is connected as a load.

The listing of the circuit file is as follows:

RMS Circuit

```
SOURCE   ■    VIN 1 0 SIN (0 1V 60HZ)
CIRCUIT  ■■   RL 2 0  1G
              * Calling subcircuit RMS:
              X1   1    2    0         RMS
              *    Vi+  Vo+  Vo-   model name
              * Subcircuit definition for RMS:
              .SUBCKT        RMS        1    3    2
              *          model name    Vi+  Vo+  Vo-
              RI   1  0  1G
              GAVR 2  4  VALUE = {V(1)*V(1)}
              C    4  2  1 IC=0V
              R    4  2  1G
              EOUT 3  2  VALUE = {SQRT (V(4,2)/TIME)}
              RO   3  2  1G
              .ENDS RMS ; End of subcircuit definition
ANALYSIS ■■■  .TRAN 10US 33.33MS UIC ; Use initial condition
              in transient analysis
              .PLOT TRAN V(2) V(1)    ; Prints on the output file
              .OPTIONS ABSTOL = 1.00N RELTOL = 0.01
              VNTOL = 0.1 ITL5 = 0
              .PROBE ; Graphics post-processor
          .END
```

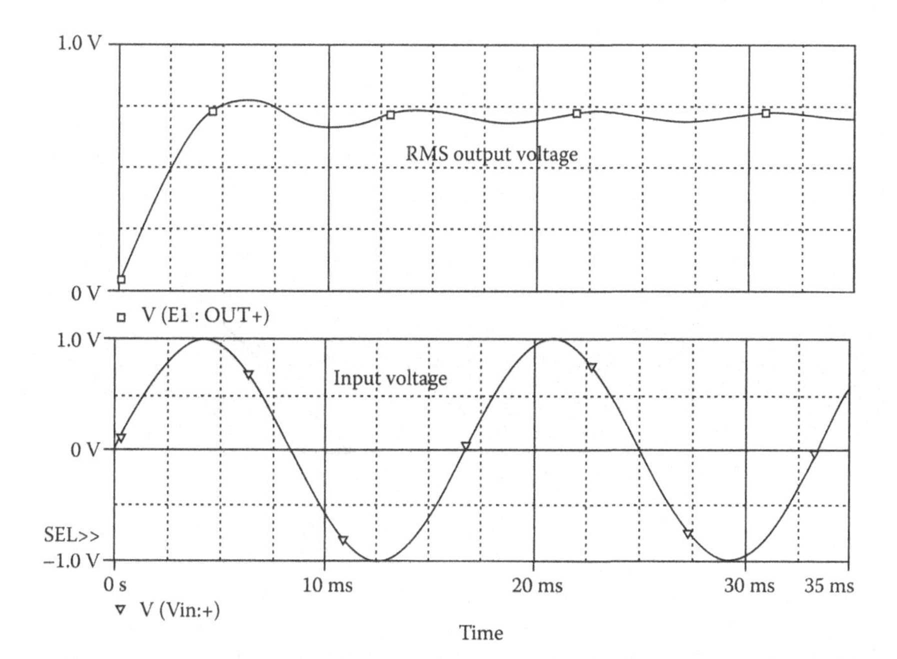

FIGURE 13.37 Transient response for the RMS circuit of Example 13.8.

The plots of the transient response for the input and output voltages are shown in Figure 13.37. The RMS value of a sine wave with 1 V peak is $V/\sqrt{2} = 0.707\,V$.

Example 13.9:

Finding the Transient Response of an Integrating Circuit with a Hysteresis Loop

The input signal to the integrating circuit of Figure 13.38(a) is $v_{in} = 2\sin(120\pi f)$. The plot of current i against V_{in}/H is shown in Figure 13.38(b). Use PSpice to calculate the transient response for a duration of 0–33.33 ms in steps of 10 μs, and to plot the output voltage V_o against the input voltage V_{in}.

SOLUTION

The input frequency is $f = 60$ Hz. The PSpice schematic is shown in Figure 13.39 for an integrating circuit with a hysteresis loop. Note that the hysteresis loop is simulated with a TABLE. We could use the ABM integrator instead of the RC integrator in Figure 13.39.

The listing of the circuit file is as follows:

Hysteresis Loop

```
SOURCE   ■    VIN 1 0 SIN (0 2V 60HZ)
CIRCUIT  ■■   .PARAM  H = 2 ; Hysteresis
              RL   3 0 1G
              RI   1 0 1G
```

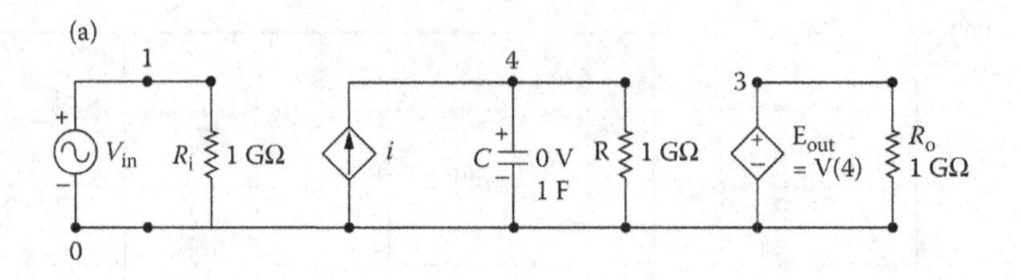

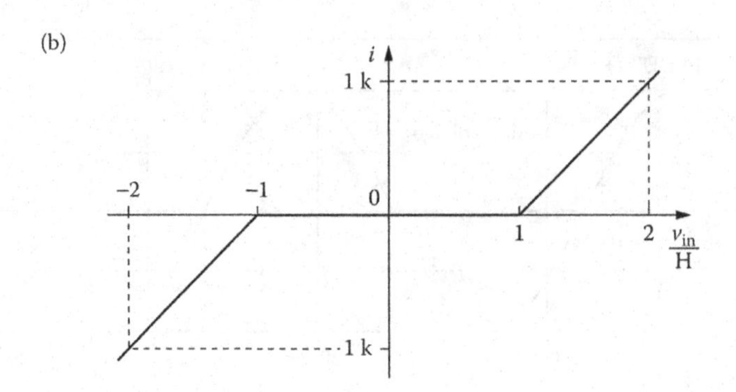

FIGURE 13.38 Hysteresis circuit.

```
            GTAB 0 4 TABLE      {V(1)/(H/2)}   =  ;
            Table for ratio Vin/(H/2)
            +  (-2 -1K)  (-1, 0)  (1, 0)  (2 1K)
            C     4 0 1 IC = 0V
            R     4 0 1G
            EOUT 3 0 4 0 1
ANALYSIS ■■■ .TRAN 0.1MS 16.67MS UIC ; Use initial condition
            in transient analysis
            .PLOT TRAN V(3) V(1)      ; Prints on the output file
            .OPTIONS ABSTOL = 1.00N RELTOL = 0.01 VNTOL = 0.1
             ITL5 = 0
            .PROBE                    ; Graphics post-processor
        .END
```

The plot of the output voltage V(3) against the input voltage V(1) is shown in Figure 13.40. The output remains constant during the hysteresis band.

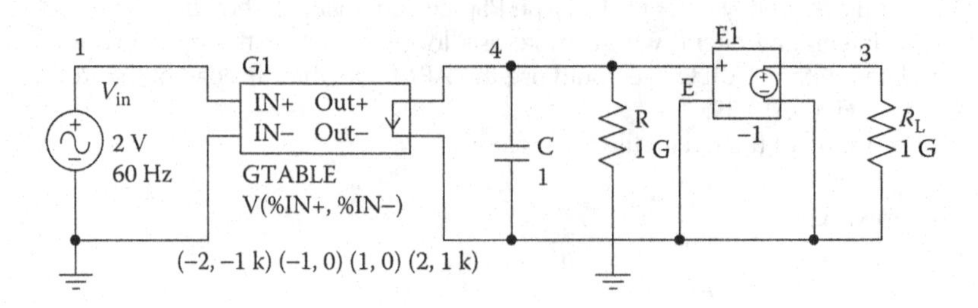

FIGURE 13.39 PSpice schematic for Example 13.9.

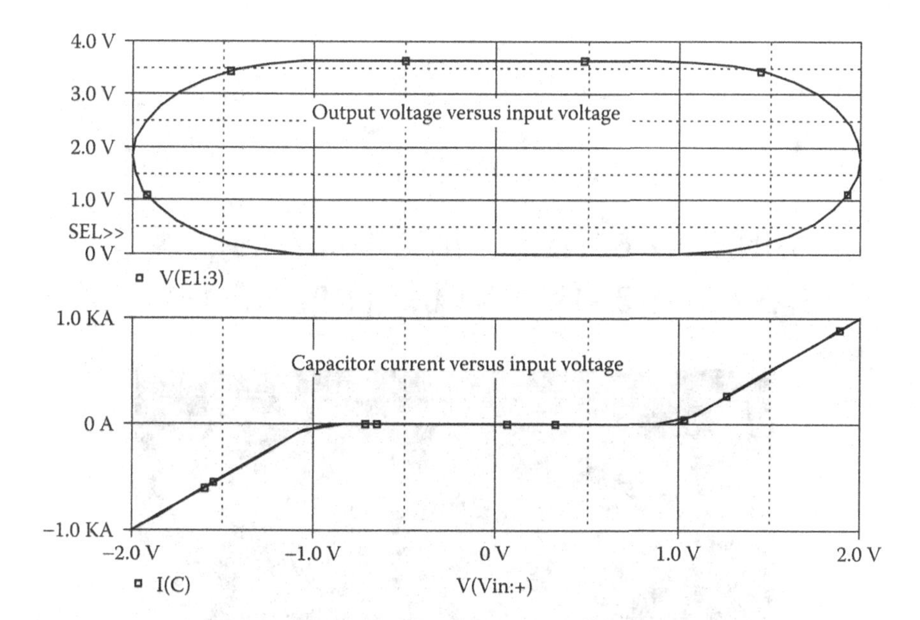

FIGURE 13.40 Output voltage against the input voltage for Example 13.9.

LTspice: LTspice schematic for Example 13.9 is shown in Figure 13.41(a). G1 is a voltage-controlled voltage source represented as a Table to create the Hysteresis band listing data points with time. The plot of the output voltage versus input voltage is shown in Figure 13.41(b). The plot of the capacitor current versus the input voltage exhibiting the hysteresis band is shown in Figure 13.41(c). Changing the value of C should change the output voltage level in Figure 13.41(b).

13.5 CLOSED-LOOP CURRENT CONTROL

Closed-loop control is used in many power electronics circuits to control the shape of a particular current (e.g., inductor motor with current control, and current source inverter). In the following example, we illustrate the simulation of a current-controlled rectifier circuit such that the output current of the rectifier is half a sine wave, thereby giving a sine wave at the input side of the rectifier.

13.5.1 Examples of Closed-Loop Control

Example 13.10:

Finding the Closed-Loop Response of the Output Voltage for the Diode Rectifier-Boost Converter

A diode rectifier followed by a boost converter is shown in Figure 13.42. The input voltage is $v_s = 170\ \sin(120\pi t)$. The circuit is operated closed loop, so that

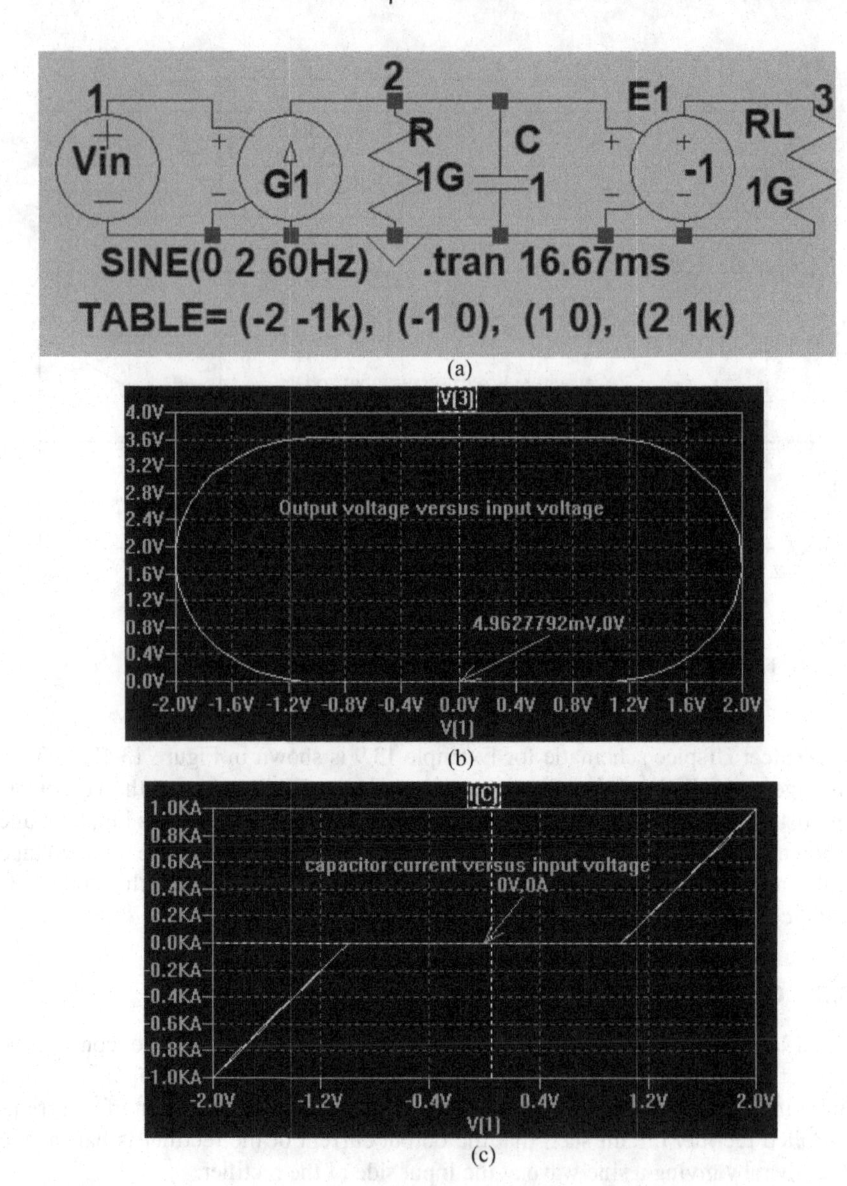

FIGURE 13.41 LTspice schematic for Example 13.9. (a) LTspice Schematic. (b) Output voltage versus input voltage. (c) Capacitor current versus input voltage.

the output voltage is $V_o = 220 \pm 0.2$ V, and the input current is sinusoidal with an error of ±0.2 A.

Use PSpice to plot the transient response of output voltage v_o and the input current i_s.

The model parameters for the IGBT are as follows.

```
TAU=287.56E9 KP=50.034 AREA=37.500E-6 AGD=18.750E6 VT=4.1822
KF=.36047 CGS=31.942E9 COXD=53.188E9 VTD=2.6570
```

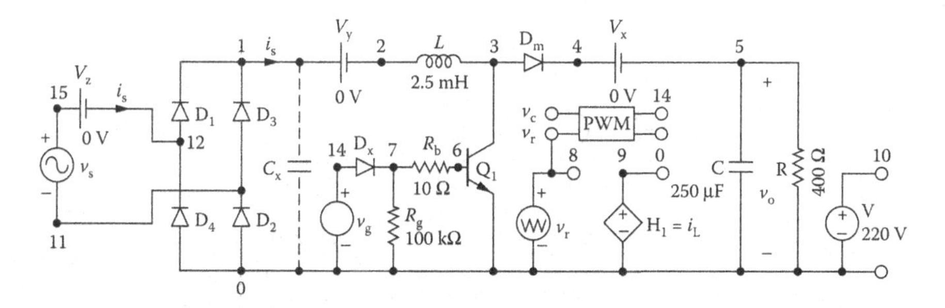

FIGURE 13.42 Diode rectifier with input current control.

The model parameters for the BJT are as follows:

```
IS=2.33E-27 and BF=13
```

and those for the diode are as follows:

```
IS=2.2E-15, BV=1200V, CJO=1PF, and TT=0.
```

SOLUTION

The peak input voltage is $V_m = 170$ V. Assuming a diode drop of 1 V, the peak-rectified voltage becomes 170 − 2 = 168 V. The input frequency is $f = 60$ Hz. The block diagram for closed-loop control is shown in Figure 13.43(a). The error of the output voltage after passing through a controller generates the reference current for the current controller, whose output is the carrier signal for the PWM generator. Equating rectified output power to the load power, we get

$$V_{in(DC)} I_{in(DC)} = \frac{V_0^2}{R} \tag{13.8}$$

Because for a rectified sine wave $V_{(average)} = 2V_{(peak)}/\pi$, Equation 13.8 becomes

$$\frac{2V_m}{\pi}\frac{2I_m}{\pi} = \frac{V_0^2}{R}$$

which gives the multiplication constant δ as

$$\delta = \frac{I_m}{V_m} = \frac{\pi^2 V_0^2}{4RV_m^2} \tag{13.9}$$

$$= \frac{\pi^2 \times 220^2}{4 \times 400 \times 168^2} = 0.01$$

A value of δ higher than 0.01 would increase the sensitivity and give a better transient response. But the PWM generator will be operated in the overmodulation region if it is too high. Let us assume δ = 0.05.

We use a proportional controller with an error band of ±1. The circuit for the PSpice simulation of the current controller is shown in Figure 13.43(b). The

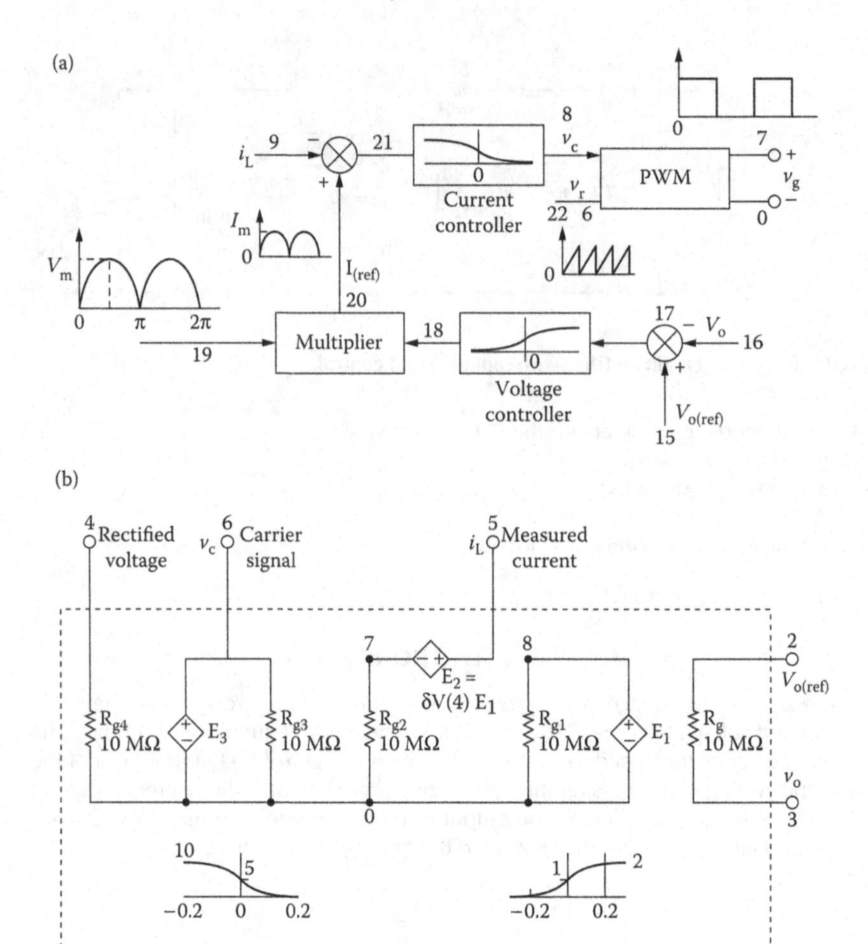

FIGURE 13.43 Subcircuit CONTR current controller. (a) The block diagram for closed-loop control. (b) Circuit for PSpice simulation of current controller.

multiplication is implemented with VALUE. The voltage and current controllers are implemented by TABLE.

The listing of the subcircuit CONTR for Figure 13.43(b) is as follows:

```
* Subcircuit definition for CONTR current controller:
.SUBCKT CONTR   2       3        4            5              6
*       model desired output rec. input input  current  carrier
*       name  voltage voltage  voltage  signal (voltage) signal
E1 8 0 TABLE{V(2,3)} =              ; Voltage controller
+ (-0.2, 2) (0, 1) (0.2, 0)
E2 5 7 VALUE = {0.05*V(4)*V(8)} ; Reference current
E3 6 0 TABLE{V(7)} = ; Current controller
+ (-0.2, 0) (0, 5) (0.2, 10)
.ENDS CONTR
```

The PWM subcircuit is implemented with an ABM, ABM2 with two inputs. Note that the output signal switches between 0 and 10 if [V(2) – V(1)] > 0. The listing of the subcircuit PWM is as follows:

```
* Sub circuit definition PWM:
.SUBCKT PWM        1          2        3
*       model  reference  carrier  output
*       name    voltage    signal  voltage
E_ABM21  3        0     VALUE {IF(V(2) -V(1)>0, 10, 0)}
.ENDS PWM
```

Using 20 pulses per half cycle of the input voltage, the switching frequency of the PWM generator is $f_s = 40 \times 60 = 2.4$ kHz, and the switching period is $T_s = 416.67$ μs.

The PSpice schematic is shown in Figure 13.44 for a unity power factor diode with a boost converter.

The listing of the circuit file for Figure 13.44 is as follows:

Diode Rectifier with PWM Current Control

```
.PARAM Vo_ref=220V  Freq=2.0kHz
Vr 8 0 PULSE (0 10 0 {1/({2*{Freq}}) - 1ns} 1ns 1ns {1/
{2*{Freq}}} ) ; Reference Signal
Vo_ref 16 0 DC {Vo_ref}
SOURCE   ■   VS 4 11 SIN (0V 170V 60HZ)
CIRCUIT  ■■  VZ 4 12 DC 0V ; Measures supply current
             D1 12 1 DMOD ; Rectifier diodes
             D2 0 11 DMOD
             D3 11 1 DMOD
             D4 0 12 DMOD
             VY 1  2 DC 0V ; Voltage source to measure input
             current
             H1 9  0 VY 0.034
             Rg 6  0 100K
             RB 7  6 250  ; Transistor base resistance
             L  2  3 2.5MH
             DM 3  4 DMOD ; Freewheeling diode
             VX 4  5 DC 0V ; Voltage source to measure
             inductor current
             R  5  0 100 ; Load resistance
             C  5  0 250UF IC = 0V ; Load filter capacitor
             .MODEL DMOD D(IS = 2.2E - 15 BV = 1200V CJO = 1PF
             TT = 0)       ; Diode model parameters
             .MODEL DM D(IS = 2.2E - 15 BV = 1200V CJO = 0 TT = 0) ;
             Diode model parameters
             Z1 3 6 0 IXGH40N60   ; IGBT switch
             * .MODEL IXGH40N60D NPN (IS=2.33E-27 BF=13)
             .MODEL IXGH40N60 NIGBT (TAU=287.56E-9
             KP=50.034 + AREA=37.500E-6 + AGD=18.750E-6
             VT=4.1822 KF=.36047 + CGS=31.942E-9 COXD=53.188E-9
             VTD=2.6570)
             * Subcircuit call for PWM control:
             XPW 8 13 10 PWM ;Control voltage for transistor Z1
             * Subcircuit call for CONTR current controller:
```

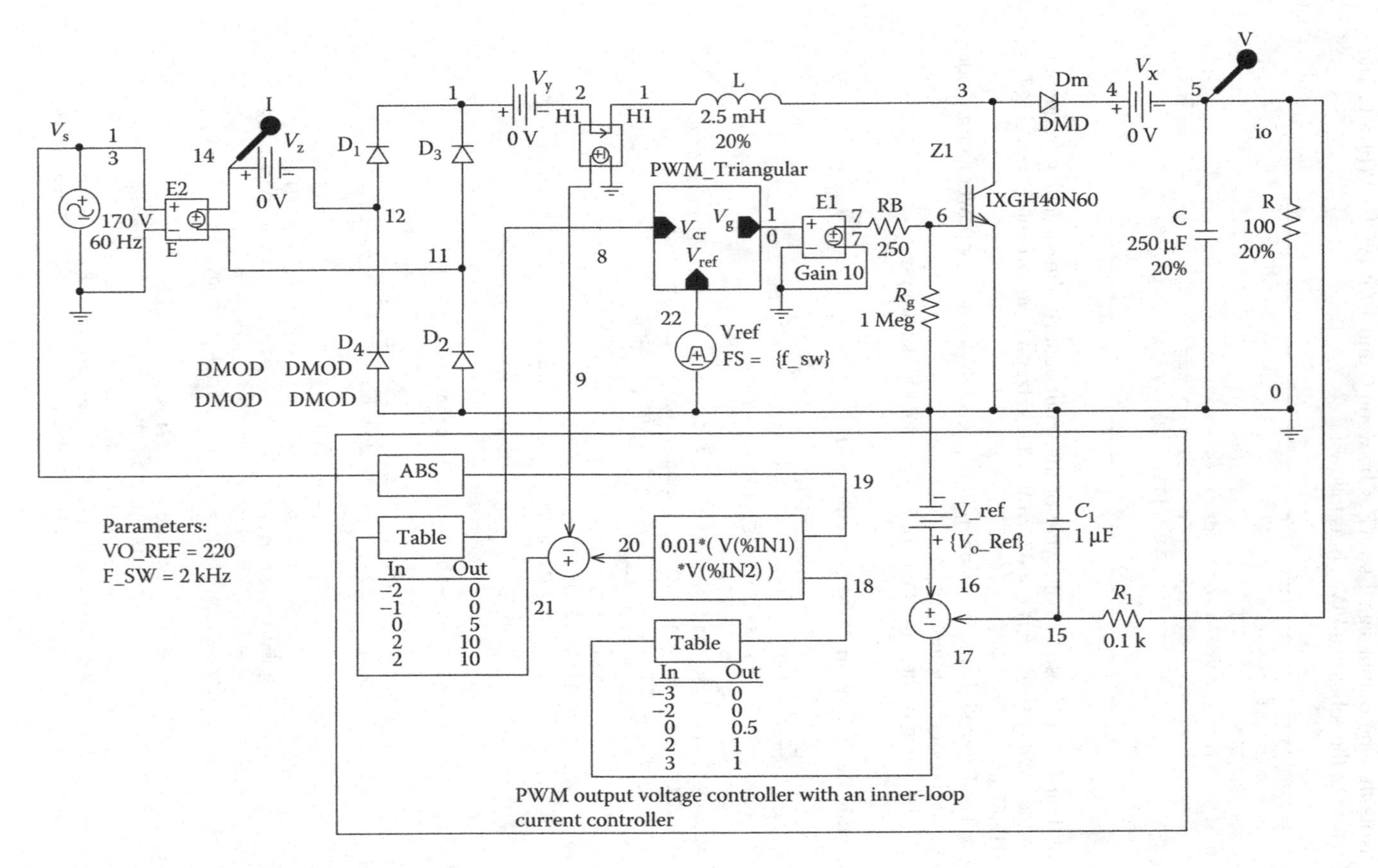

FIGURE 13.44 PSpice schematic for Example 13.10.

```
            XCONT 16 15 17 9 13 CONTR
            * Subcircuit definition CONTR must be inserted.
            * Subcircuit definition PWM must be inserted.
ANALYSIS ■■■ .TRAN 1US 35MS          ; Transient analysis
            .PROBE                   ; Graphics post-processor
            .FOUR 60HZ I(VZ)         ; Fourier analysis
            .OPTIONS ABSTOL = 1.00U RELTOL = 0.01
            VNTOL = 0.1 ITL5 = 0   ; Convergence
      .END
```

The plots of the output voltage V(5) and input current I(VZ) are shown in Figure 13.45(a). The expanded waveforms are shown in Figure 13.45(b). The current controller forces the input current to be in phase with the input supply voltage and follows a sinusoidal reference current. This improves the input power factor. A small filter can remove the high-frequency components of the input current. The upper bound of the output voltage is limited by the proportional controller. It should be noted that the load voltage has not yet reached the steady-state condition. The lower limit depends on the time constant of the load circuit. It requires careful design to determine the values of L and C, the controller parameters, and the switching frequency. A proportional-integral (PI) controller, coupled with a large value for the load filter capacitor, should give a faster response to the output voltage.

The Fourier components of the input current are as follows:

DC Component = 3.606477E–02

Harmonic Number	Frequency (Hz)	Fourier Component	Normalized Component	Phase (Deg)	Normalized Phase (Deg)
1	6.000E+01	8.195E+00	1.000E+00	2.291E+00	0.000E+00
2	1.200E+02	6.478E–02	7.905E–03	5.133E+01	5.591E+01
3	1.800E+02	2.039E+00	2.488E01	1.678E+02	1.747E+02
4	2.400E+02	5.545E–02	6.766E–03	1.601E+02	1.509E+02
5	3.000E+02	7.729E–01	9.432E–02	3.090E+01	1.944E+01
6	3.600E+02	4.388E–02	5.354E–03	3.395E+01	2.020E+01
7	4.200E+02	2.740E–01	3.344E–02	1.189E+02	1.350E+02
8	4.800E+02	6.214E02	7.582E03	1.638E+02	1.455E+02
9	5.800E+02	8.997E–02	1.094E02	1.693E+02	1.899E+02
10	6.000E+02	8.997E–02	1.098E–02	2.126E+01	4.417E+01

Total harmonic distortion = 2.689696E+01 percent

For $THD = 26.89\%$ and $\phi_1 = -2.291°$, Equation 7.3 gives the input power factor of the rectifier as

$$PF_i = \frac{1}{\sqrt{1+\left(\frac{\%THD}{100}\right)^2}}\cos\phi_1 = \frac{1}{\sqrt{1+\left(\frac{26.89}{100}\right)^2}}\cos(-2.291°) = 0.965\ \text{(lagging)}$$

LTspice: LTspice schematic for Example 13.10 is shown in Figure 13.46(a) [9,11]. The gating signals are generated by comparing a triangular carrier signal with a reference sine wave through an op-amp comparator with 10 pulses per half-cycle as shown in Figure 13.46(b). The absolute value of a sinewave is generated as special function B1 at the supply output input frequency of 60 Hz.

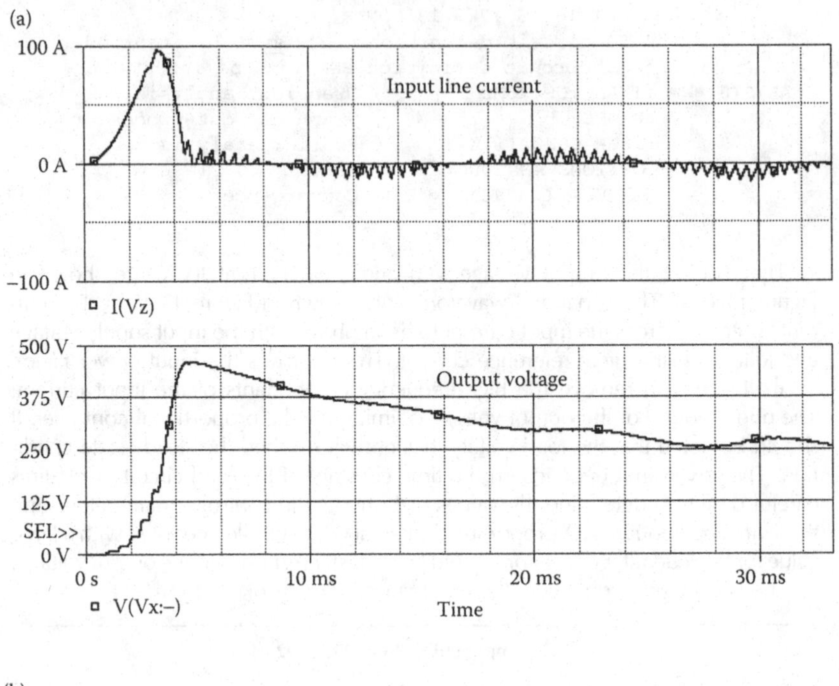

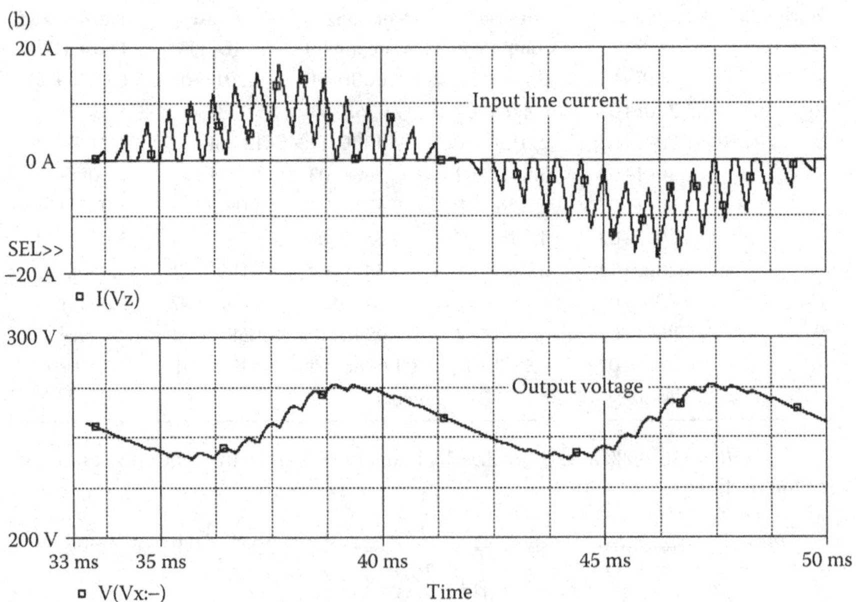

FIGURE 13.45 Output voltage and the input line current for Example 13.10. (a) Input line current and the output voltage. (a) Expanded timescale of the input line current and the output voltage.

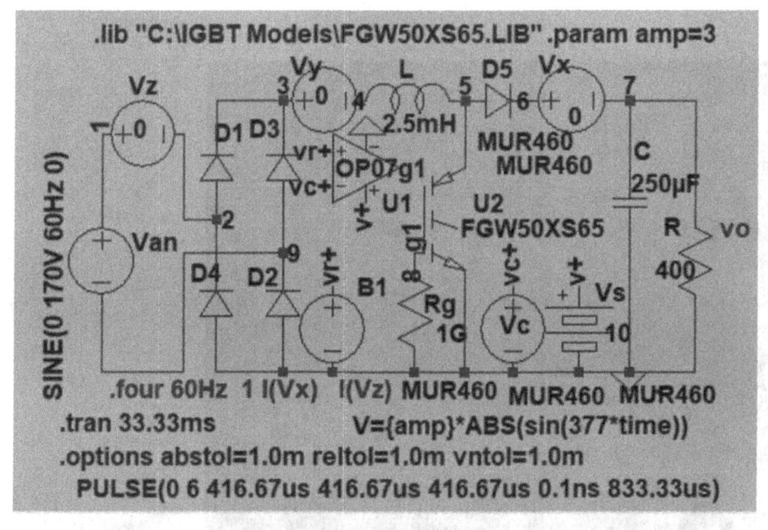

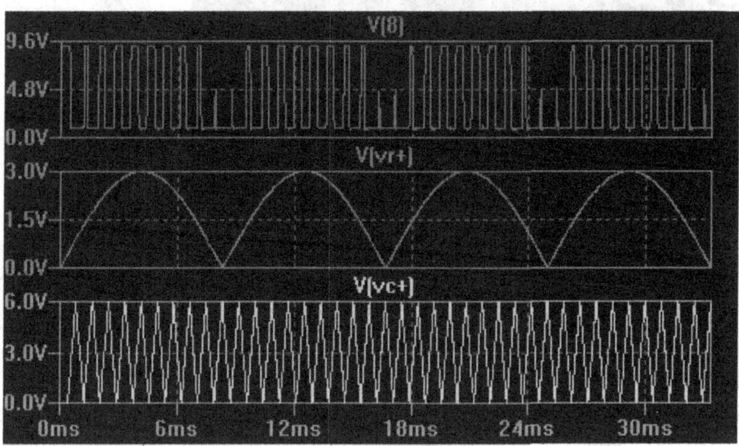

FIGURE 13.46 LTspice schematic for Example 13.10. (a) Schematic. (b) Generation of gating signals.

The plots of the input current, the input power, the output voltage, and the load power are shown in Figure 13.47(a). The rms input current as shown in Figure 13.47(b) is 18.895 A, the average input power as shown in Figure 13.47(c) is 1436.6, and the average output power as shown in Figure 12.47(d) is 1335.7. The rms value of the output voltage is 440.76 V. We can find the power efficiency of the converter as $\eta = \frac{P_{out}}{P_{in}} = \frac{1335.7}{1436.6} = 93\%$

We can find the input power factor as $PF_i = \frac{P_{in}}{V_s I_s} = \frac{1436.6}{120 \times 18.895} = 0.63$

Note: For a low output power, the losses in the semiconductor devices are significant compared to the output power and the efficiency becomes low. There is no power loss in the load inductor.

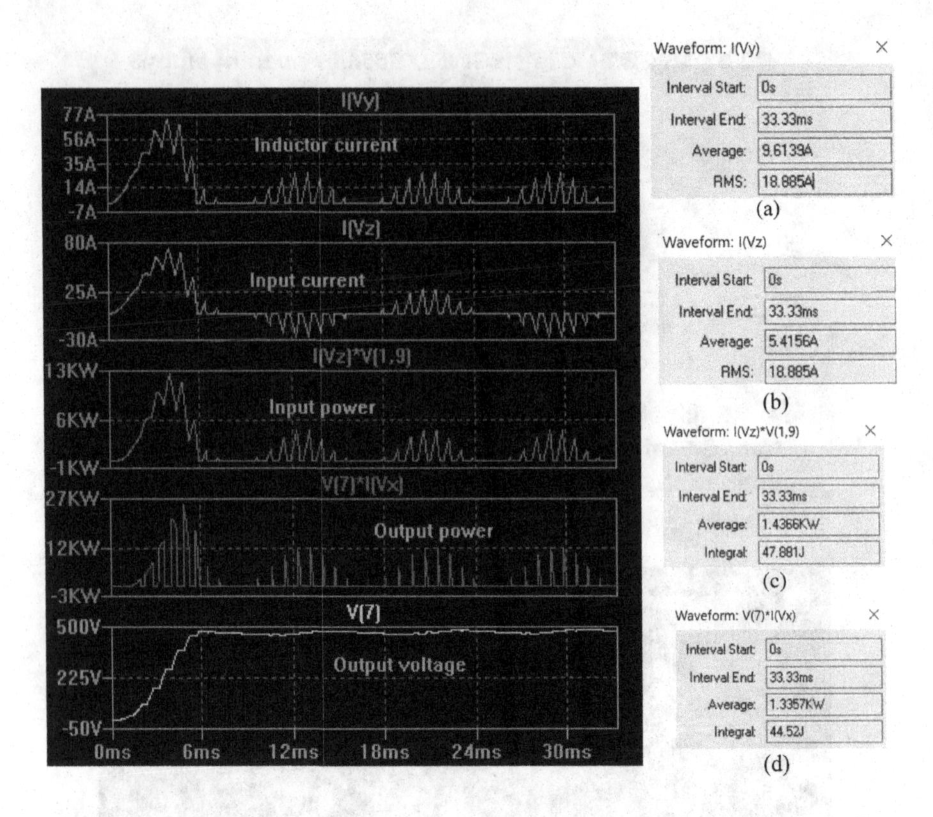

FIGURE 13.47 Output waveforms for (a) Inductor current. (b) Input current. (c) Input power. (d) Output power.

PROBLEMS

13.1 A full-wave precision rectifier is shown in Figure P13.1. The input voltage is $v_i = 0.1 \sin(2000\pi t)$. Plot the transient response of the output voltage for a duration of 0–1 ms in steps of 10 μs. The op-amp μA741 can be modeled as a macromodel as shown in Figure 13.3. The supply voltages are $V_{CC} = 12$ V and $V_{EE} = 12$ V.

13.2 Plot the DC transfer characteristics for Figure P13.2. The input voltage is varied from −10 to +10 V in steps of 0.1 V. The zener voltages are $V_{Z1} = V_{Z2} = 6.3$ V. The op-amp, which is modeled by the circuit in Figure 13.1a, has $R_i = 2$ MΩ, $R_o = 75$ Ω, $C_1 = 1.5619$ μF, and $R_1 = 10$ kΩ.

13.3 A feedback control system is shown in Figure P13.3. The reference input v_r is the step voltage of 1 V. Use PSpice to plot the transient response of the output voltage for a duration of 0–10 s in steps of 10 ms. The gain K is to be varied from 0.5 to 2 with an increment of 0.5. Assume that all initial conditions are zero and that $K_1 = 0.5$.

13.4 The constant μ of the Van der Pol equation is 4. Use PSpice to plot (a) the transient response of the output signal for a duration of 0–20 s in steps of 10 ms, and (b) the phase plane (dy/dt against y). Assume an initial disturbance of 1.

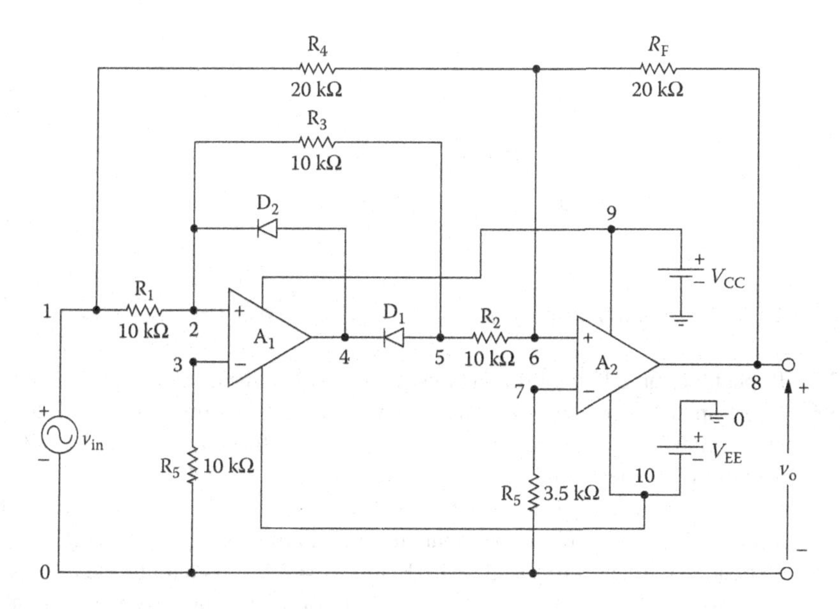

FIGURE P13.1 Full-wave precision rectifier.

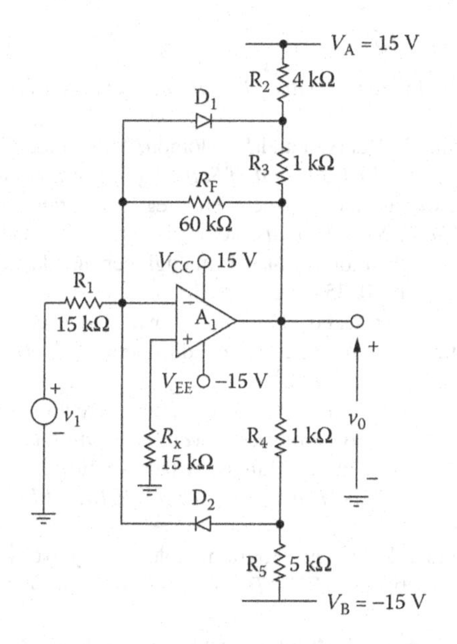

FIGURE P13.2 Op-amp limiting circuit.

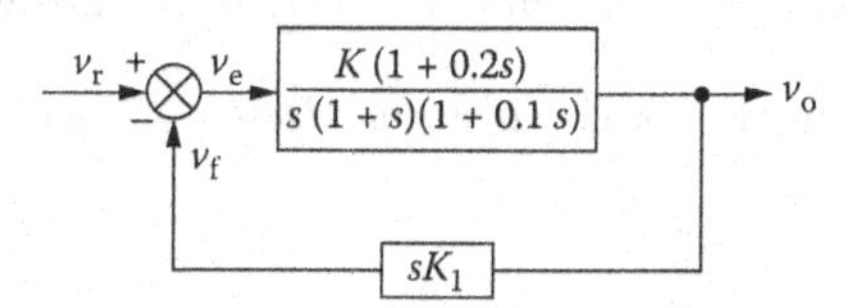

FIGURE P13.3 Feedback control system.

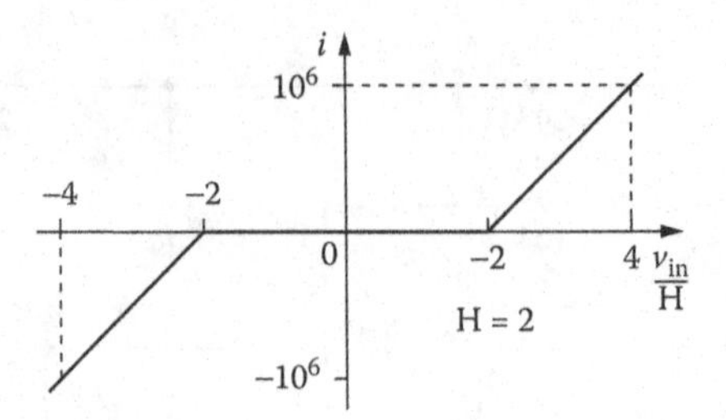

FIGURE P13.4 Backlash input signal.

13.5 Repeat Example 13.9 for the i against V_{in}/H characteristic shown in Figure P13.4

13.6 Use PSpice to find the worst-case minimum and maximum output voltages ($V_{o(min)}$ and $V_{o(max)}$) for Example 13.10. Assume uniform tolerances of ±15% for all passive elements and an operating temperature of 25°C.

13.7 Repeat Example 13.10 for $L = 5$ mH. What are the effects of increasing the value of L on the *THD* of the input current and the input power factor?

13.8 Repeat Example 13.10 for $C_1 = 1$ μF and $R_1 = 1$ kΩ. What are the effects of increasing the time constant of the *RC* filter of the voltage feedback circuit?

SUGGESTED READING

1. Texas Instruments, *Linear Circuits: Operational Amplifier Macromodels*, Dallas, TX, 1990.
2. G. Boyle, B. Cohn, D. Pederson and J. Solomon, Macromodeling of integrated circuit operational amplifiers, *IEEE Journal of Solid-State Circuits*, SC-*9*(*6*), 1974, 353–364.
3. I. Getreu, A. Hadiwidjaja and J. Brinch, An integrated-circuit comparator macromodel, *IEEE Journal of Solid-State Circuits*, SC-*11*(*6*), 1976, 826–833.
4. S. Progozy, Novel applications of SPICE in engineering education, *IEEE Transactions on Education*, *32*(*1*), 1990, 35–38.
5. M. Kazerani, P.D. Ziogas and G. Ioos, A novel active current wave shaping technique for solid-state input power factor conditioners, *IEEE Transactions on Industrial Electronics*, IE*38*(*1*), 1991, 72–78.
6. A.R. Prasad, P.D. Ziogas, An active power factor correction technique for three phase diode rectifiers, *IEEE Transactions on Power Electronics*, *6*(*1*), 1991, 83–92.
7. A.R. Prasad, P.D. Ziogas and S. Manias, A passive current wave shaping method for three phase diode rectifiers, *Proceedings of the IEEE APEC-91 Conference Record*, 1991, pp.319–330.
8. M.S. Dawande and G.k. Dubey, Programmable input power factor correction method for switch-mode rectifiers, *IEEE Transactions on Power Electronics*, *II*(*4*), 1996, 585–591.
9. M. H. Rashid, *Power Electronics - circuits, devices and applications*, Pearson Publishing, Fourth Edition, 2014, Chapter 5.
10. M.H. Rashid, *Microelectronic Circuits - Analysis and Design*, Cengage Publishing, 2017, Chapter 2.
11. M.H. Rashid, *Control Systems: Analysis and Design*, Cengage India Publishing, 2022, Chapter 14.

14 Characteristics of Electrical Motors

After completing this chapter, students should be able to do the following:

- Model the characteristics of DC motors and specify their model parameters.
- Model the characteristics of induction motors and specify their model parameters.
- Perform transient analysis of electric motors.
- Perform worst-case analysis of electric motors for parametric variations of model parameters and tolerances.

14.1 INTRODUCTION

The PSpice simulation of power converters can be combined with the equivalent circuit of electrical machines to obtain their control characteristics. The machines can be represented in SPICE by a linear or nonlinear magnetic circuit or in a function or table form. We use linear circuit models to obtain:

DC motor characteristics
Induction motor characteristics

14.2 DC MOTOR CHARACTERISTICS

The motor back emf is given by

$$E_g = K\omega I_f$$

and the developed motor torque

$$T_d = KI_a I_f$$

can be represented by polynomial sources. The torque T_d is related to the load torque T_L and motor speed ω by

$$T_d = J\frac{d\omega}{dt} + B\omega + T_L$$

$$T_d - T_L - B\omega = J\frac{d\omega}{dt}$$

DOI: 10.1201/9781003284451-14

Thus, integrating the net torque will give the motor speed ω, which after further integration gives the shaft position θ. The behavioral models in PSpice allow performing many functions such as addition, subtraction, multiplication, integration, and differentiation.

14.2.1 Examples of DC Motor Controlled by DC–DC Converter

Example 14.1:

Finding the Performance of a Separately Excited Motor Controlled by a DC–DC Converter

The armature of a separately excited DC motor is controlled by a DC–DC converter operating at a frequency of f_c = 1 kHz and a duty cycle of k = 0.8. The DC–DC supply voltage to the armature is V_s = 220 V. The field current is also controlled by a DC–DC converter operating at a frequency of f_s = 1 kHz and a duty cycle of δ = 0.5. The DC supply voltage to the field is V_f = 280 V. The motor parameters are as follows:

Armature resistance, R_m = 0.1 Ω
Armature inductance, L_m = 10 mH
Field resistance, R_f = 10 Ω
Field inductance, L_f = 20 mH
Back emf constant, K = 0.91
Viscous torque constant, B = 0.3
Motor inertia, J = 1
Load torque, T_L = 50, 100, and 150 N · m

Use PSpice to plot the transient response of the armature and field current, the torque developed, and the motor speed for a duration of 0–30 ms in steps of 10 μs.

SOLUTION

The armature and field circuits for PSpice simulation are shown in Figure 14.1(a) and (b), respectively. The net torque, which is obtained by subtracting the viscous (T_B) and load torque (T_L) from the torque developed (T_d), is integrated to obtain the motor speed as shown in Figure 14.1(c). The motor speed is integrated to obtain the shaft position as shown in Figure 14.1(d).

The PSpice schematic is shown in Figure 14.2, which comprises three separate blocks: the motor field, motor armature, and motor load consisting of inertia J and viscous torque constant B. It has one PWM generator for the armature control and one PWM generator for the motor field control. The listing of the circuit file is as follows:

DC Separately Excited Motor with Variable Load Torques

```
SOURCE   ■   VS 1 0 DC 220V      ; Armature supply
             PARAM Duty_a=0.5    ; duty cycle of the armature
             circuit
             .PARAM Duty_f=0.8 ; duty cycle of the field
             circuit
             .PARAM Freq=1kHz    ; switching frequency
```

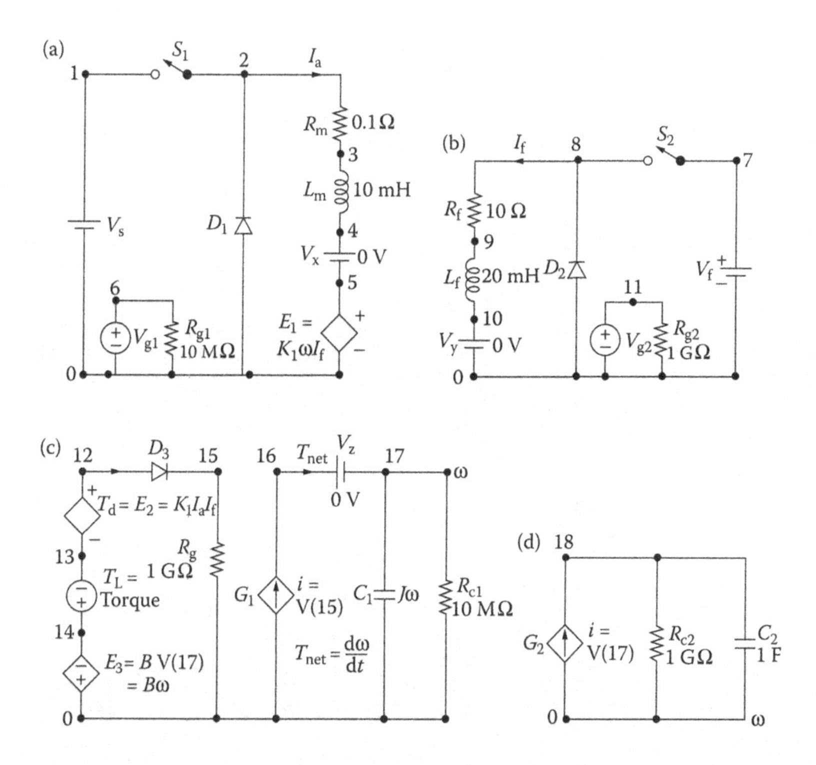

FIGURE 14.1 DC–DC converter-controlled DC motor for PSpice simulation. (a) Armature circuit. (b) Field circuit. (c) Integration network. (d) Integrating the speed ω.

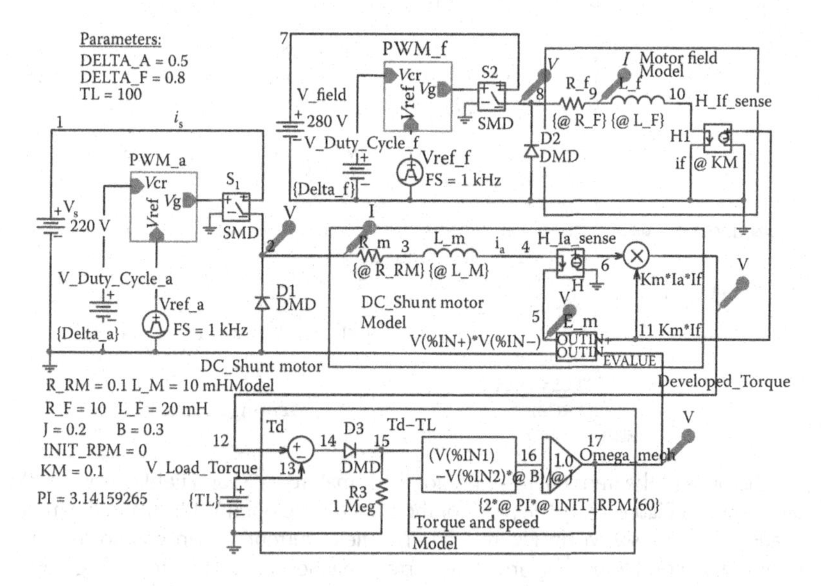

FIGURE 14.2 The PSpice schematic for Example 14.1.

```
                  .PARAM Km=0.1
                  Vg1  6 0 PULSE (0 20V 0 1ns 1ns {{Duty_a}/
                  {Freq}-2ns} {1/{Freq}}
                  Vg2 11 0 PULSE (0 20V 0 1ns 1ns {{Duty_f}/
                  {Freq}-2ns} {1/{Freq}}
                  Rg1  6 0 10MEG
                  VF   7 0 DC 280V   ; Field supply
                  Rg2 11 0 10MEG
CIRCUIT      ■■   .PARAM VISCOUS=0.3 ; Viscous constant
                  .PARAM J=1 ; Motor inertia
                  .PARAM TL=100 ; Load torque
                  .STEP PARAM TL 50 150 50 ; Load torque varied
                  S1   1  2 6 0 SMOD ; Voltage-controlled switch
                  .MODEL SMOD VSWITCH (RON=0.01 ROFF=10E+6 VON=10V
                  VOFF=5V)
                  D1   0  2 DMOD
                  .MODELDMODD (IS=2.2E-15 BV=1200V CJO=0
                  TT=0) ; Diode model parameters
                  RM   2  3 0.1
                  LM   3  4 10MH
                  VX   4  5 DC  0V      ; Senses the armature
                  current
                  E1   5  0 VALUE={{km} *V(17) *I(VY)}
                  RF   8  9 10
                  LF   9 10 20MH
                  VY  10  0 DC  0V      ; Senses the field current
                  S2   7  8 11  0  SMOD ; Voltage-controlled
                  switch
                  D2   0  8 DMOD
                  E2  12 13 VALUE={{km} *I(VX) *I(VY)} ; Torque
                  developed
                  VL  14 13 {TL}        ; Load torque
                  E3   0 14 VALUE={VISCOUS*V (17)} ; Viscous torque
                  D3  12 15 DMOD
                  Rg  15  0 1G
                  G1   0 16 15  0 1     ; Net torque
                  VZ  16 17 DC  0V      ; Measures the net torque
                  C1  17  0 {J} IC=0V   ; Load inertia
                  Rc1 17  0 1G
                  G2   0 18 17 0 1 ; Velocity to position
                  C2  18  0 1 IC=0V
                  RC2 18  0 1G
ANALYSIS     ■■■  .TRAN 10US 30MS UIC   ; Transient analysis
                  with initial condition
                  .PLOT TRAN V(3) V(1)  ; Prints on the
                  output file
                  .OPTIONS ABSTOL=1.00N RELTOL=0.01
                  VNTOL=0.1
                  ITL5=50000
                  .PROBE                ; Graphics post-processor
              .END
```

The plots of the transient response for the armature I(VX) and field currents I(VY) are shown in Figure 14.3. The plots for the torque developed V(12,13) and the motor speed V(17) are shown in Figure 14.4. The field current shown has not reached steady-state conditions. The armature current reaches a peak before settling down. The torque is pulsating because of pulsating armature and field currents.

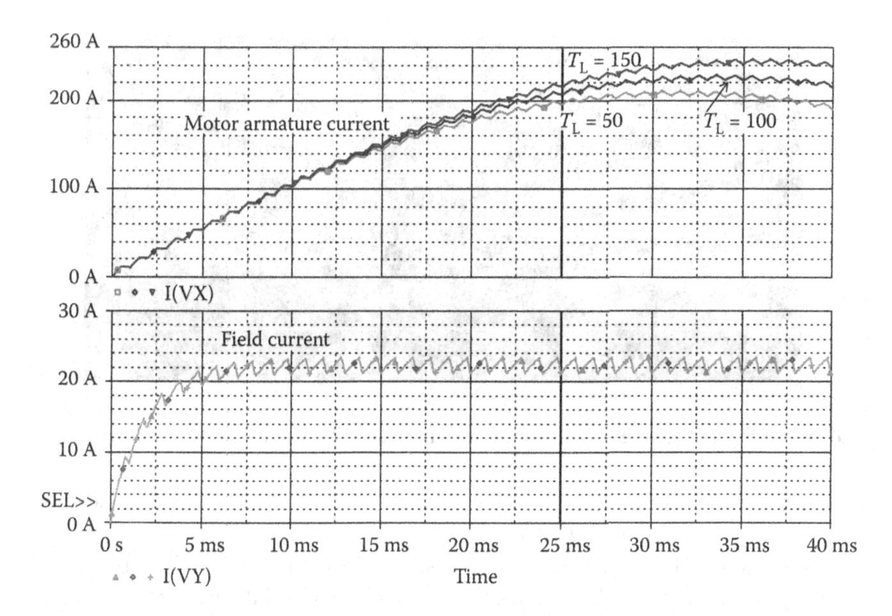

FIGURE 14.3 Plots of armature and field currents for Example 14.1.

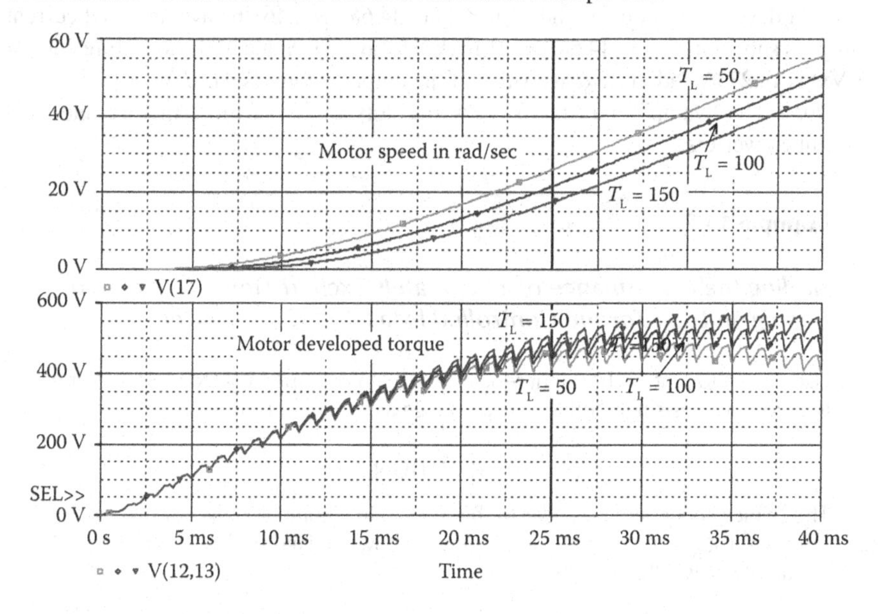

FIGURE 14.4 Motor speed and motor developed torque for Example 14.1.

Note: There are two separate DC–DC converters—one for the armature voltage control and one for the field voltage control.

LTspice: Figure 14.5 shows the LTspice gating signals for the armature and the field converter are generated by comparing a triangular carrier signal with a DC

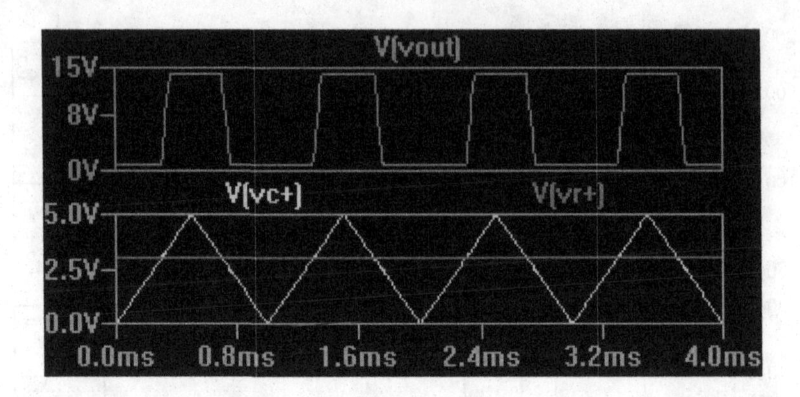

FIGURE 14.5 The gating signals for the armature and the field converter for Example 14.1.

reference signal "vamp" for the armature converter and "famp" for the field converter for the armature gating signal.

LTspice schematic for Example 14.1 is shown in Figure 14.6(a). The plots of the armature load current, the load voltage, the field current, and the net torque developed by the motor are shown in Figure 14.6(b) for vamp = 3, famp = 3, and speed = 50 rad/s. The net developed torque as shown in Figure 14.6(c) is 4.46, the average filed current I(Vy) as shown in Figure 14.6(d) is 12.16 A. The average value of the armature current I(Vx) is 18.273A and the average value of the output voltage is 94.47 V.

Note: The "vamp" and "famp" vary the duty cycles of the armature and filed circuit converters.

Example 14.2:

Finding the Performance of a Separately Excited Motor with a Step Change in Load Torque Controlled by a DC–DC Converter

Use PSpice to plot the transient response of the motor speed of Example 14.1 if the load torque T_L is subjected to a step change as shown in Figure 14.7.

SOLUTION

The PSpice schematic is similar to that shown in Figure 14.2, except that the load torque is a step pulse from 50 to 250 ms as shown in Figure 14.8. The listing of the circuit file is as follows:

DC Separately Excited Motor with Step Load Torque Change

```
SOURCE    ■   VS 1 0 DC 220V      ; Armature supply
              .PARAM Duty_a=0.5   ; duty cycle of the
              armature circuit
              .PARAM Duty_f=0.8 ; duty cycle of the field
              circuit
              .PARAM Freq=1kHz    ; switching frequency
              .PARAM Km=0.1
```

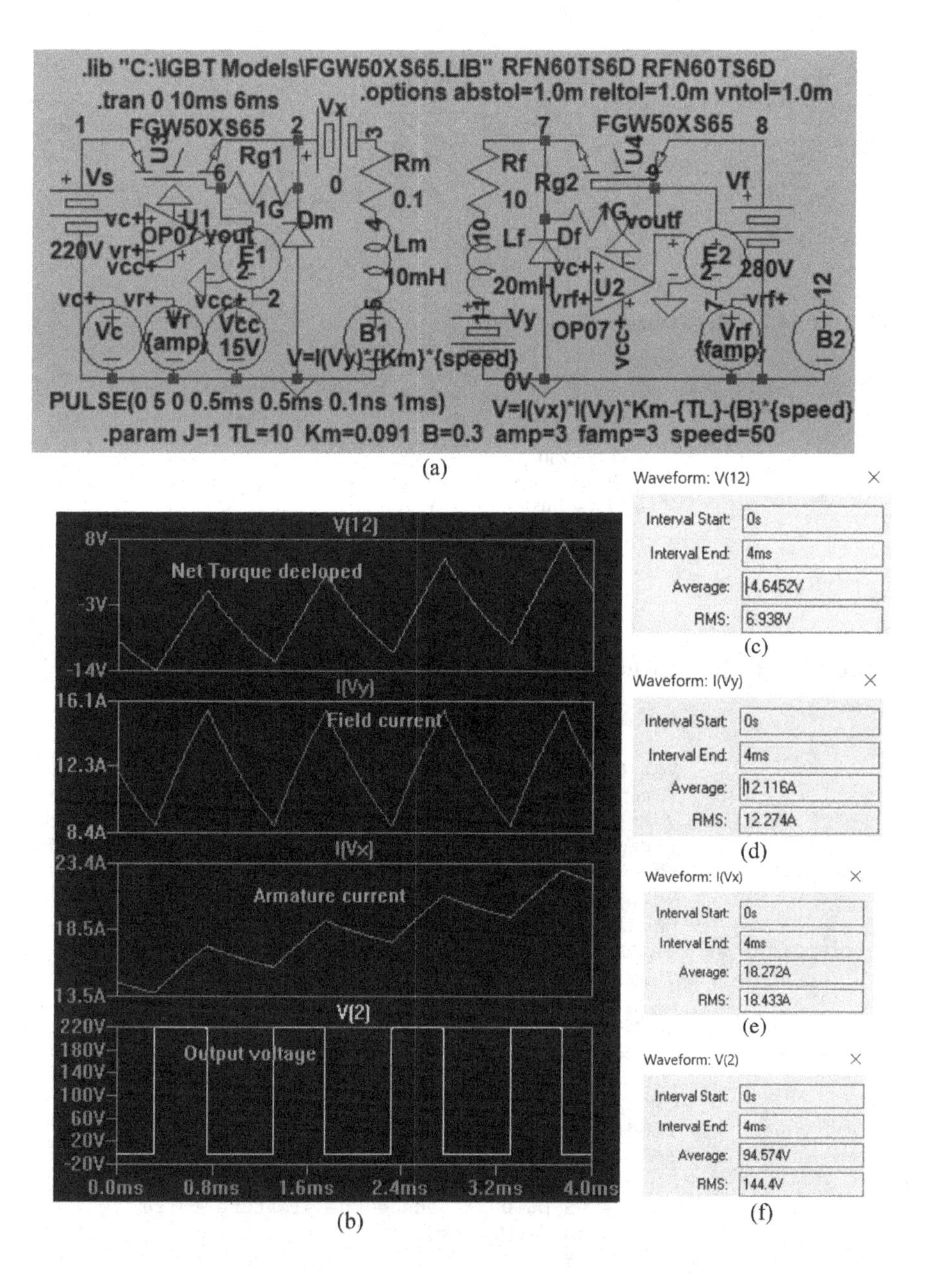

FIGURE 14.6 LTspice schematic for Example 14.1. (a) LTspice schematic. (b) Output waveforms. (c) Net torque. (d) Field current. (e) Armature current. (f) Output voltage.

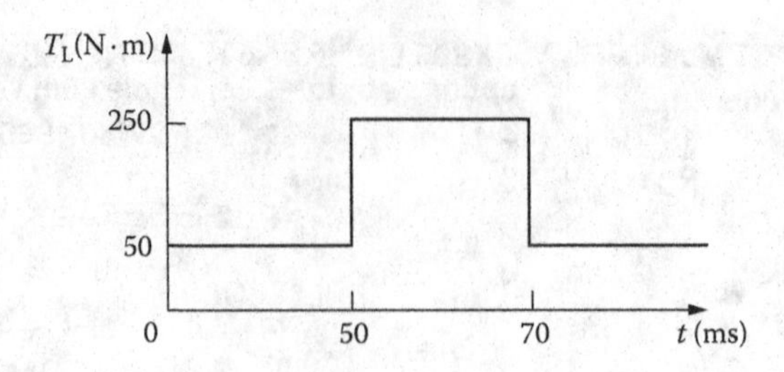

FIGURE 14.7 Step change of load torque.

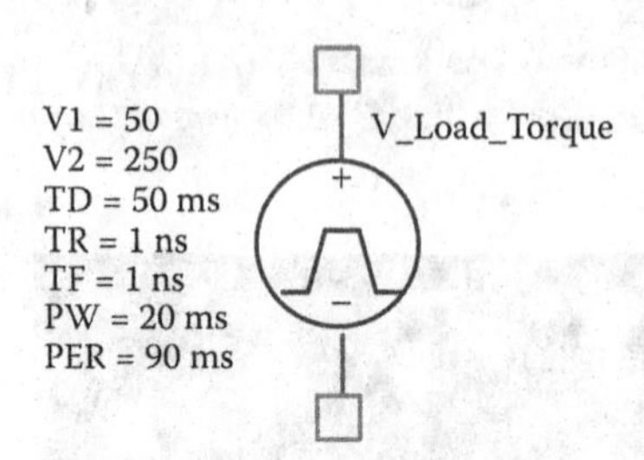

FIGURE 14.8 Step load torque for Example 14.2.

```
Vg1 6 0 PULSE (0 20V 0 1ns 1ns { {Duty_a}/
{Freq}-2ns} {1/{Freq}}
Vg2 11 0 PULSE (0 20V 0 1ns 1ns { {Duty_f}/
{Freq}-2ns} {1/{Freq}}
Rg1 6 0 10MEG
VF   7 0 DC 280V   ; Field supply
Rg2 11 0 10MEG
```

CIRCUIT ■■

```
.PARAM VISCOUS=0.3  J=0.2 TL_min=150
TL_max=250
S1   1 2 6 0 SMOD ; Voltage-controlled swit
.MODEL SMOD VSWITCH (RON=0.01ROFF=10E+6VON=1
VOFF=5V ; Switch model
D1   0 2 DMOD
.MODEL DMOD D(IS=2.2E-15BV=1200VCJO=0TT=0)
Diode model parameters
RM   2  3 0.1
LM   3  4 35MH
VX   4  5 DC 0V ; Senses the armature curre
E1   5  0 VALUE={{km} *V(15) * I(VY)}
RF   8  9 10
LF   9 10 20MH
VY  10  0 DC 0V ; Senses the field current
S2   7  8 11 0 SMOD ; Voltage-controlled
switch
D2   0  8 DMOD
E2     12 13 VALUE={{km} *I(VX) *I(VY)} ;
Torque developed
VL  14 13 PULSE ({TL_min} {TL_max} 40MS 1NS
1NS 20MS 90MS) ; Load torque
```

```
        E3    0 14 VALUE={VISCOUS*V (17)} ; Viscous
        torque
        D3   12 15 DMOD
        Rg   15   0 1G
        G1    0 16 15 0 1 ; Net torque
        VZ   16 17 DC 0V ; Measures the net torque
        C1   17   0 1 IC=0V
        Rc1 17   0 1G
        G2    0 18 17 0 1 ; Velocity to position
        C2   18   0 1 IC=0V
        Rc2 18   0 1G
ANALYSIS ■■■ .TRAN 10US 90MS UIC    ; Transient analysis
        with initial condition
        .OPTIONS ABSTOL=1.00N RELTOL=0.01 VNTOL=0.1
        ITL5=0
        .PROBE                  ; Graphics post-processor
    .END
```

The plots of the transient response of the load torque V(14,13) and the motor speed V(17) are shown in Figure 14.9. As expected, an increase in load torque slows down the speed.

LTspice: Figure 14.10 shows the LTspice gating signals for the armature and the field converter are generated by comparing a triangular carrier signal with a dc reference signal "vamp" for the armature converter and "famp" for the field converter for the armature gating signal.

LTspice schematic for Example 14.2 is shown in Figure 14.11(a). The plots of the net developed torque, the field current, the armature current, the gating signal of the field converter, and the generation of the gating signals of the field converter are shown in Figure 14.11(b) for vamp = 2, famp = 1 and 4, and speed = 50 rad/s.

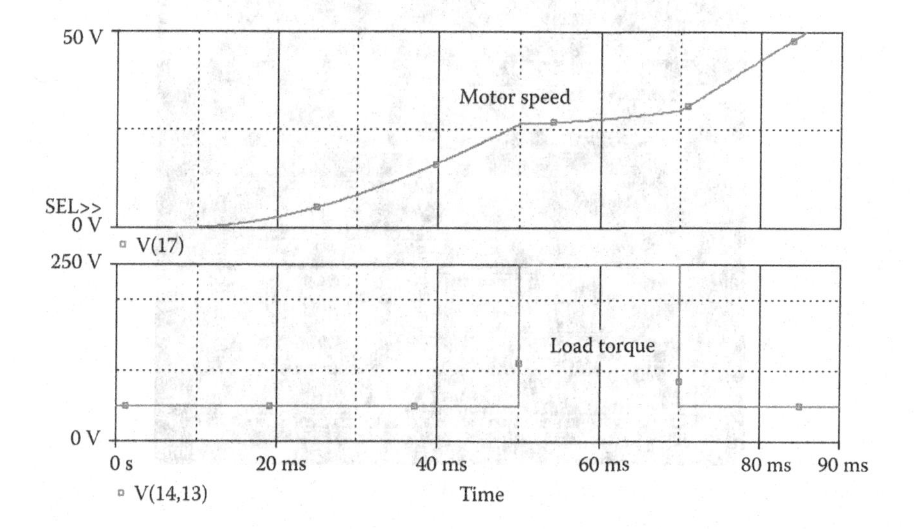

FIGURE 14.9 Plots of the load torque and the motor speed for Example 14.2.

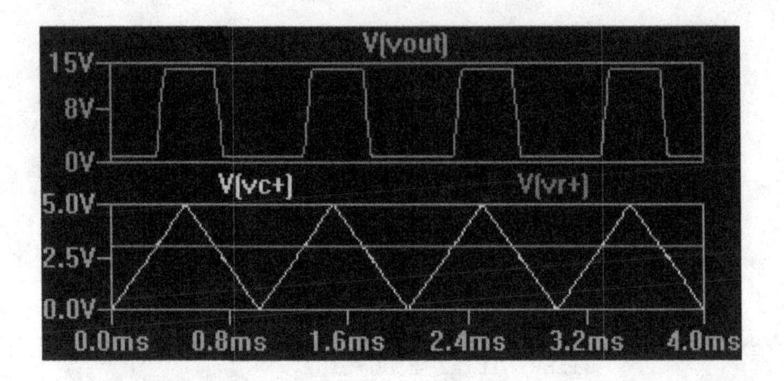

FIGURE 14.10 The gating signals for the armature and the field converter for Example 14.1.

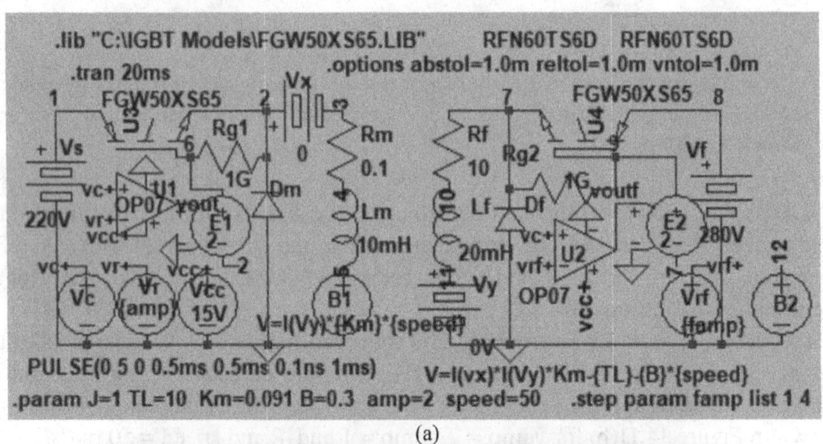

(a)

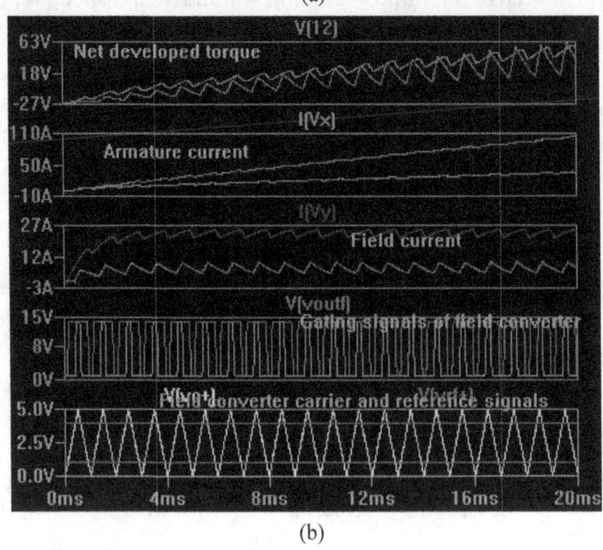

(b)

FIGURE 14.11 LTspice schematic for Example 14.2. (a) LTspice schematic. (b) Output waveforms.

Note: As the field current decreases, the back emf of the armature decreases and the armature current increases maintaining the motor torque. It is a common practice in motor control to reduce the field current at a higher speed to develop the needed power so that torque x speed remains constant.

14.3 INDUCTION MOTOR CHARACTERISTICS

PSpice can simulate the equivalent circuit of an induction motor and generator to determine their control characteristics. Parameters such as supply frequency, rotor resistance, and rotor slip can be varied to find the effects of control variables on the performance of motor drives. An inverter that controls the motor voltage, current, and frequency can be added to the motor circuit to simulate an inverter-fed induction motor drive.

14.3.1 EXAMPLES OF INDUCTION MOTOR CHARACTERISTICS

Example 14.3:

Finding the Torque-Speed Characteristic of an Induction Motor for Varying Slip

The equivalent circuit of an induction motor is shown in Figure 14.12. Use PSpice to plot the torque developed against frequency for slip $s = 0.1$, 0.25, 0.5, and 0.75. The supply frequency is to be varied from 0.1 to 100 Hz. The motor parameters are as follows.

Stator resistance, $R_s = 0.42\ \Omega$
Stator inductance, $L_s = 2.18$ mH
Rotor resistance, $R_r = 0.42\ \Omega$ (referred to stator)
Rotor inductance, $L_r = 2.18$ mH (referred to stator)
Magnetizing inductance, $L_m = 58.36$ mH (referred to stator)

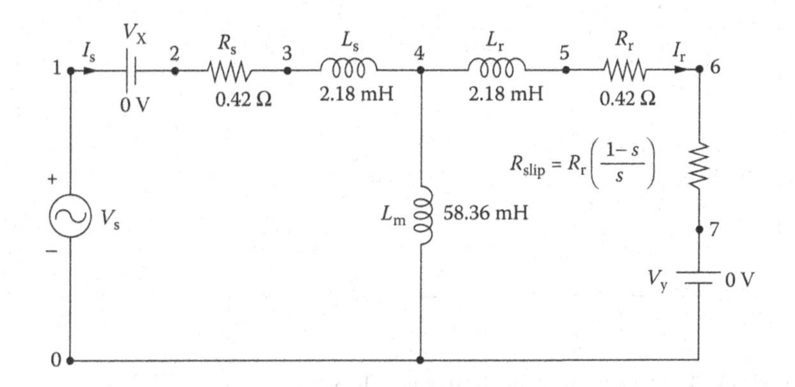

FIGURE 14.12 Equivalent circuit for induction motor.

SOLUTION

The effective resistance due to slip is

$$R_{slip} = R_r \frac{1-s}{s}$$

By varying the slip, R_{slip}, the developed torque T_L can be varied. The PSpice schematic is shown in Figure 14.13. The slip is varied by using the PSpice parametric command .PARAM.

The listing of the circuit file is as follows:

Torque-Speed Characteristic of Induction Motor

```
SOURCE    ■    VS 1 0 AC  170V ; input voltage of 170V
CIRCUIT   ■■   .PARAM SLIP=0.05   ; Slip
               .PARAM RRES=0.42   ; Rotor resistance
               .PARAM Freq=1kHz      ; switching frequency
               .PARAMRSLIP={RRES* (1-slip)/slip}
               .STEP PARAM SLIP LIST 0.1 0.25 0.5 0.75 ;
               Slip values
               VX 1 2 DC 0V          ; Senses the stator
               current
               RS 2 3 {RRES}
               LS 3 4 2.18MH
               LM 4 0 58.36MH
               LR 4 5 2.18MH
               RR 5 6 {RRES}
               RX 6 7 {RSLIP}
               VY 7 0 DC 0V          ; Senses the rotor
               current
ANALYSIS  ■■■  .AC DEC 100 0.1HZ 100HZ      ; Ac analysis
               .OPTIONS ABSTOL=1.00N  RETOL=0.1
               VNTOL=0.1 ITL5=0
               .PROBE                      ; Graphics
               post-processor
               .END
```

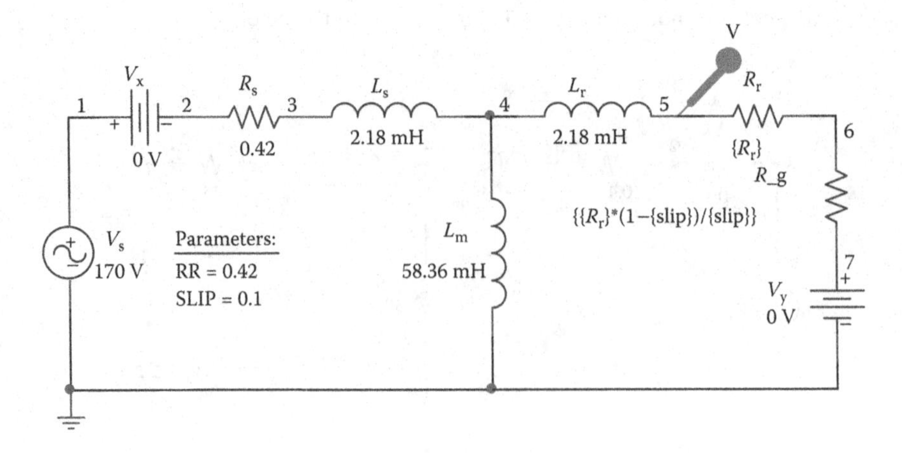

FIGURE 14.13 PSpice schematic for Example 14.3.

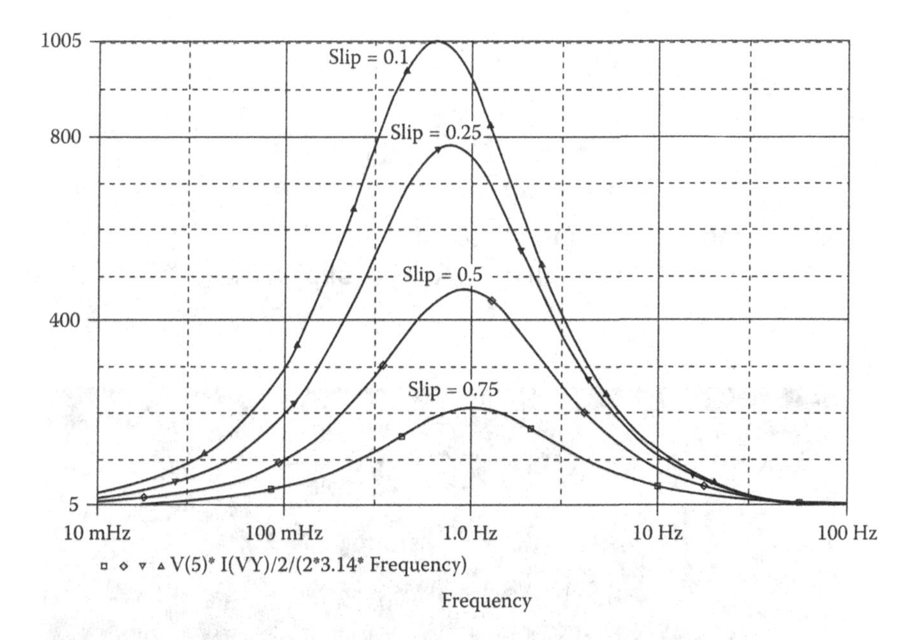

FIGURE 14.14 Plots of the developed torque versus frequency for Example 14.3.

The plots of the torque developed $T_d = V(5) \times I(Vy)/2/(2 \times 3.14 \times \text{frequency})$ versus frequency for various slips are shown in Figure 14.14. The torque developed increases as the slip is reduced. For a fixed slip, there is a region of constant torque.

The developed torque of the induction motor can be expressed by [3]

$$T_d = \frac{3R_r V_S^2}{s\omega_s \left[\left(R_s + \frac{R_r}{s} \right)^2 + (X_s + X_r)^2 \right]}$$

where $\omega_s = 2\pi f_s$, $X_s = \omega_s L_s$, and $X_r = \omega_s L_r$.

LTspice: LTspice schematic for Example 14.3 is shown in Figure 14.15(a). The plots of the gap power against the frequency are shown in Figure 14.15(b) for various slips. The gap power decreases with the increase in slips. The peak occurs at a particular frequency that depends on the motor parameters. For induction motor control, the gap power can be varied by varying the slip frequency.

Example 14.4:

Finding the Torque–Speed Characteristic of an Induction Motor for Varying Rotor Resistance

Repeat Example 14.3 for rotor resistance $R_r = 0.1$, 0.2, 0.3, and 0.42 Ω. The slip is kept fixed at $s = 0.1$.

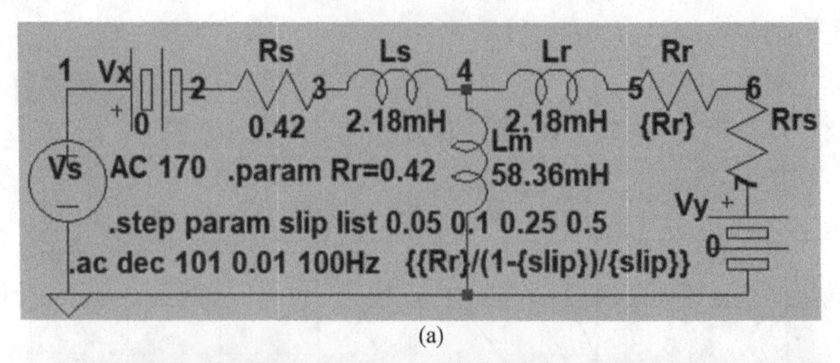

(a)

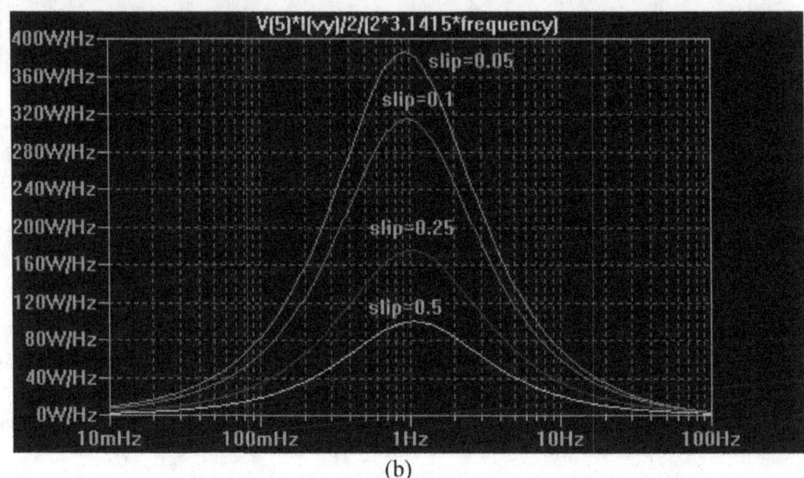

(b)

FIGURE 14.15 LTspice schematic for Example 14.3. (a) LTspice schematic. (b) Plots of gap power against frequency.

SOLUTION

The PSpice schematic is similar to that shown in Figure 14.13. The rotor resistance is varied by using the PSpice parametric command .PARAM.

The listing of the circuit file is as follows:

Torque–Speed Characteristic with Variable Rotor Resistance

```
SOURCE    ■    VS 1 0 AC 170V              ; Input voltage of 170V
CIRCUIT   ■■   .PARAM SLIP=0.1             ; Slip
               .PARAM RRES=0.42            ; Rotor resistance
               .PARAM RSLIP={RRES* (1-slip)/slip}
               .STEP PARAM RRES LIST 0.1 0.2 0.3 0.42 ; List
               values
               VX 1 2 DC 0V                 ; Senses the stator
               current
```

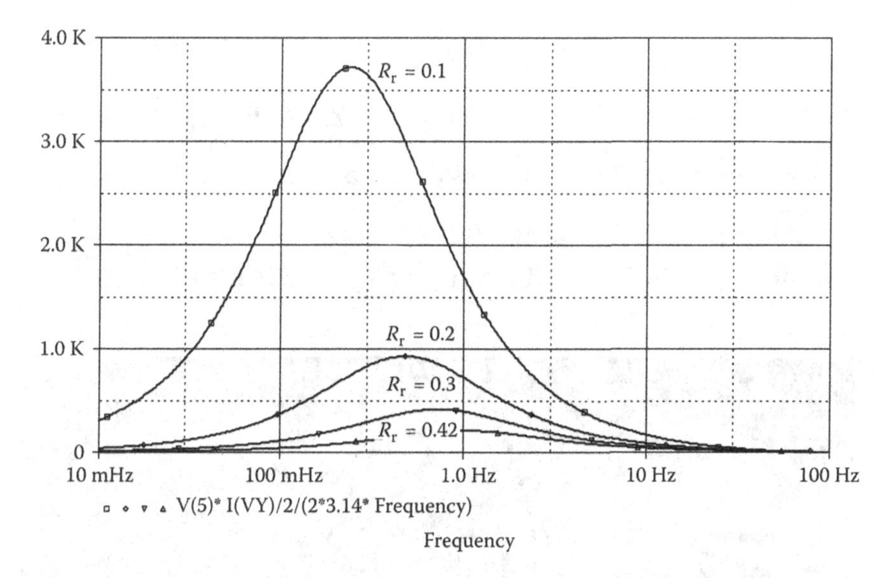

FIGURE 14.16 Plots of the developed torque versus frequency for Example 14.4.

```
              RS 2 3 {RRES}
              LS 3 4 2.18MH
              LM 4 0 58.36MH
              LR 4 5 2.18MH
              RR 5 6 {RRES}
              RX 6 7 {RSLIP}
              VY 7 0 DC 0V                ; Senses the rotor
              current
ANALYSIS ■■■ .AC DEC 100 0.1HZ 100HZ     ; AC analysis
              .OPTIONS  ABSTOL=1.00N RELTOL=0.01
              VNTOL=0.1 ITL5=0
              .PROBE                      ; Graphics
              post-processor
          .END
```

The plots of the torque developed $T_d = V(5) \times I(Vy)/2/(2 \times 3.14 \times \text{frequency})$ versus frequency for various slips are shown in Figure 14.16. The torque increases as the rotor resistance is reduced.

LTspice: LTspice schematic for Example 14.4 is shown in Figure 14.17(a). The plots of the gap power against the frequency are shown in Figure 14.17(b) for various rotor resistances. The gap power decreases with the increase in rotor resistances. The peak occurs at a particular frequency that depends on the motor parameters. For induction motor control, the gap power can be varied by varying the rotor resistance.

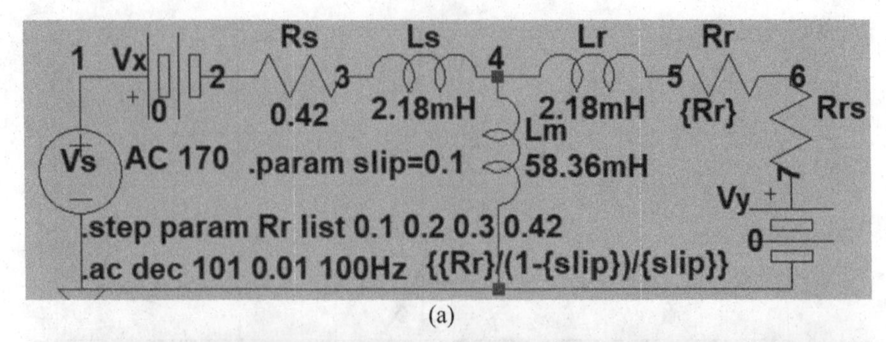

(a)

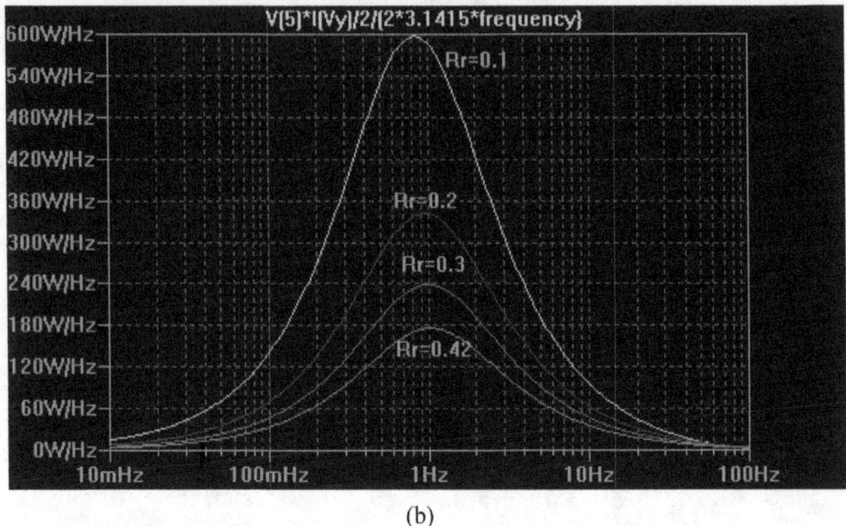

(b)

FIGURE 14.17 LTspice schematic for Example 15.4. (a) LTspice schematic. (b) Plots of gap power against frequency.

PROBLEMS

14.1 The DC–DC converter of Figure 14.1(a) is operating at a frequency of $f_c = 2$ kHz and a duty cycle of $k = 0.75$. The DC supply voltage to the armature is $V_s = 220$ V. The field current is also controlled by a DC–DC converter operating at a frequency of $f_s = 2$ kHz and a duty cycle of $\delta = 0.75$. The DC supply voltage to the field is $V_f = 220$ V. The motor parameters are as follows:
Armature resistance, $R_m = 1\ \Omega$
Armature inductance, $L_m = 5$ mH
Field resistance, $R_f = 10\ \Omega$
Field inductance, $L_f = 10$ mH
Back emf constant, $K = 0.91$
Viscous torque constant, $B = 0.4$
Motor inertia, $J = 0.8$
Load torque, $T_L = 10, 100, 200$ N°m

Use PSpice to plot the transient response of (a) the armature and field currents, (b) the torque developed, and (c) the motor speed for a duration of 0–4 ms in steps of 10 μs.

14.2 Use PSpice to plot the transient response of motor speed in Problem 14.1 if the load torque is subjected to a step change as shown in Figure P14.1.

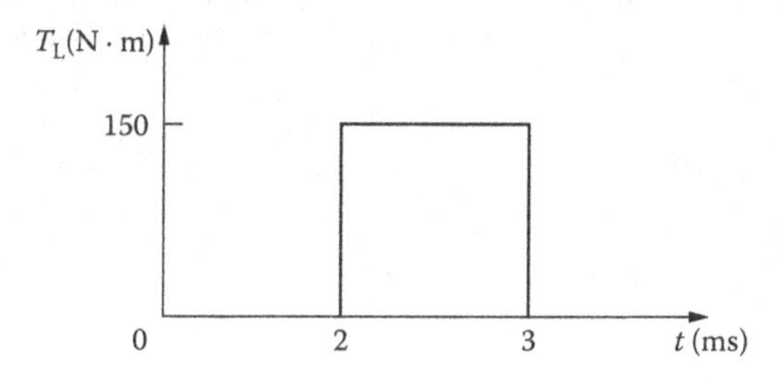

FIGURE P14.1 Step torque change.

14.3 The parameters of the induction motor equivalent circuit shown in Figure 14.12 are as follows:
Stator resistance, $R_s = 1.01\ \Omega$
Stator inductance, $L_s = 3.4$ mH
Rotor resistance, $R_r = 0.69\ \Omega$ (referred to stator)
Rotor inductance, $L_r = 5.15$ mH (referred to stator)
Magnetizing inductance, $L_m = 115.4$ mH (referred to stator)
Use PSpice to plot the torque developed versus frequency for slip 5 = 0.1, 0.2, 0.4, and 0.6. The supply frequency is to be varied from 0.1 to 200 Hz.

14.4 Repeat Problem 14.3 for rotor resistances of $R_r = 0.1, 0.2, 0.5$, and 0. 69 Ω. The slip is kept fixed at $s = 0.15$.

14.5 Use PSpice to find the worst-case minimum and maximum motor speed ($\tau_{m(min)}$ and $\tau_{m(max)}$) and motor torque ($T_{d(min)}$ and $T_{d(max)}$) for Example 14.1. Assume uniform tolerances of ±15% for all passive elements and an operating temperature of 25°C.

14.6 Use PSpice to find the worst-case minimum and maximum motor-developed torque ($T_{d(min)}$ and $T_{d(max)}$) for Example 14.3. Assume uniform tolerances of ±15% for all passive elements and an operating temperature of 25°C.

SUGGESTED READING

1. J.F., Lindsay and M.H. Rashid, *Electromechanics and Electrical Machinery*, Englewood Cliffs, NJ: Prentice-Hall, 1986.
2. Y.C. Liang and V.J. Gosbel, DC machine models for SPICE2 simulation, *IEEE Transactions on Power Electronics*, 1990. 16–20.
3. M.H., Rashid, *Power Electronics: Circuits, Devices, and Applications*, Third Edition, Englewood Cliffs, NJ: Prentice-Hall, 2014.
4. A.E. Emannuel, Electromechanical transients simulations by means of PSpice, *IEEE Transactions on Power Systems*, *PES*(1), 1991, 72–78.

5. M. Giesselmann and N. Mohan, *Advanced Simulation of Motor Drives and Power Electronics Using PSpice*, Tutorial #3, 1998 IAS Annual Meeting, St. Louis, MO, *PES*(1), 1998, 1–64.
6. J. Rivas and J.M. Zamarro, C. Pereria, Simple approximation for magnetization curves and hysteresis loops, *IEEE Transactions on Magnetics*, MAG-*17(4)*, 1981, 1498–1502.
7. K.H. Carpenter, *Using Spice to Solve Coupled Magnetic/Electric Circuit Problems*, Tutorial #3, 26th North American Power Symposium, 1994, pp.147–150.
8. M.H. Rashid, *Power Electronics Handbook*, Burlington, MA: Butterworth-Heinemann, 5/e, 2023.

15 Simulation Errors, Convergence Problems, and Other Difficulties

The learning objectives of this chapter are to develop the following abilities:

- Familiarizing oneself with common types of simulation errors in SPICE and how to overcome them.
- Handling convergence problems that are common in PSpice, especially in circuits with rapidly switching voltages or currents or both.
- Using the options setup and their values.

15.1 INTRODUCTION

An input file may not run for various reasons, and it is necessary to know what to do when the program does not work. To run a program successfully requires knowledge of what would not work and why, and how to fix the problem. There could be many reasons why a program does not work, and in this chapter we cover the commonly encountered problems and their solutions. The problems could be due to one or more of the following:

Large circuits
Running multiple circuits
Large outputs
Long transient runs
Convergence
Analysis accuracy
Negative component values
Power-switching circuits
Floating nodes
Nodes with fewer than two connections
Voltage source and inductor loops

15.2 LARGE CIRCUITS

The entire description of an input file must fit into RAM during the analysis. However, the analysis results are not stored in RAM. All results (including intermediate results for the .PRINT and .PLOT statements) go to the output file or one of the temporary files. Therefore, whether the run would fit into RAM depends on how big the input file is.

DOI: 10.1201/9781003284451-15

The size of an input file can be found by using the ACCT option in the .OPTIONS statement and looking at the MEMUSE number printed at the end of runs that terminate successfully. MEMUSE denotes the peak memory use of the circuit. If the circuit file does not fit into RAM, the possible remedies are as follows:

1. Break up the file into pieces, and run them separately.
2. Reduce the amount of memory taken up by other resident software (e.g., DOS, utilities). The total memory available can be checked by the DOS command CHKDSK.
3. Buy more memory (up to 640 kilobytes, the most PCDOS recognizes).

15.3 RUNNING MULTIPLE CIRCUITS

A set of circuits may be run as a single job by putting all of them into one input file. Each circuit begins with a title statement and ends with a .END command, as usual. PSPICE1.EXE will read all the circuits in the input file and process them in sequence. The output file will contain the outputs from each circuit in the same order as they appear in the input file. This technique is most suitable for running a set of large circuits overnight, especially with SPICE or the professional version of PSpice. However, Probe can be used in this situation, because only the results of the last circuit will be available for graphical output by Probe.

15.4 LARGE OUTPUTS

A large output file will be generated if an input file is run with several circuits, or for several temperatures, or with sensitivity analysis. This will not be a problem with a hard disk. For a PC with floppy disks, the diskette may not be able to accommodate the output file. The best solution is to

1. Direct the output to a printer instead of a file, or
2. Direct the output to an empty diskette instead of the one containing PSPICE1. EXE by assigning the PSpice programs to drive A: and the input and output files to drive B:. The command to run a circuit file would be as follows:

```
A: PSPICEB:EX2-1. CIRB: EX2-1.OUT
```

It is recommended that SPICE be run on a hard disk.
Note: Direct the output to the hard drive, if possible.

15.5 LONG TRANSIENT RUNS

Long transient analysis runs can be avoided by choosing the appropriate limit options. The limits that affect transient analysis are as follows:

1. Number of print steps in a run, LIMPTS
2. Number of total iterations in a run, ITL5
3. Number of data points that Probe can handle

The number of print steps in a run is limited to the value of the LIMPTS option. It has a default value of 0 (meaning no limit) but can be set to a positive value as high as 32,000 (e.g., .OPTIONS LIMPTS = 6000). The number of print steps is simply the final analysis time divided by the print interval time (plus one). The size of the output file that is generated by PSpice can be limited if errors occur, by using the LIMPTS option.

The total number of iterations in a run is limited to the value of the ITL5 option. It has a default value of 5000 but can be set as high as 2×10^9 (e.g., .OPTIONS ITL5 = 8000). The limit can be turned off by setting ITL5 = 0. This is the same as setting ITL5 to infinity and is often more convenient than setting it to a positive number. It is advisable to set ITL5 = 0.

Probe limits the data points to 16,000. This limit can be overcome by using the third parameter on the .TRAN statement to suppress part of the output at the beginning of the run. for a transient analysis from 0 to 10 ms in steps of 100 μs and printing output from 8 to 10 ms, the command would be as follows:

```
.TRAN 10US 10MS 8MS
```

Note: The limit options from Tables 6.1 and 6.2 can be typed into the circuit file as an .OPTIONS command. Alternatively, the limit options can be set from the Change Options menu, in which case PSpice would write these options into the circuit file automatically.

15.6 CONVERGENCE

PSpice uses iterative algorithms. These algorithms start with a set of node voltages, and each iteration calculates a new set, which is expected to be closer to a solution of Kirchhoff's voltage and current laws. That is, an initial guess is used and the successive iterations are expected to converge to the solution. Convergence problems may occur in

- DC sweep
- Bias-point calculation
- Transient analysis

15.6.1 DC Sweep

If the iterations do not converge to a solution, the analysis fails. The DC sweep skips the remaining points in the sweep. The most common cause of failure of the DC sweep analysis is an attempt to analyze a circuit with regenerative feedback, such as a Schmitt trigger. The DC sweep is not appropriate for calculating the hysteresis of such circuits, because it is required to jump discontinuously from one solution to another at the crossover point.

To obtain the hysteresis characteristics, it is advisable to use transient analysis with a PWL voltage source to generate a very slowly rising ramp. There is no CPU-time penalty for this because PSpice will adjust the internal time step to be large away from the crossover point and small close to it. A very slow ramp ensures that the switching time of the circuit will not affect hysteresis levels. This is similar to changing the input voltage slowly until the circuit switches. With a PWL source in transient analysis, the hysteresis characteristics due to upward and downward switching can be calculated.

Example 15.1:

Finding The Hysteresis Characteristic of an Emitter-Coupled Circuit

An emitter-coupled Schmitt trigger circuit is shown in Figure 15.1(a). Plot the hysteresis characteristics of the circuit from the results of transient analysis. The input voltage, which is varied slowly from 1.5 to 3.5 V and from 3.5 to 1.5 V, is shown in Figure 15.1(b). The model parameters of the transistors are IS = 2.33E–27, BF = 13, CJE = 1PE, CJC = 607.3PF, and TF = 26.5NS. Print the job statistical summary of the circuit.

SOLUTION

The input voltage is varied very slowly from 1.5 to 3.5 V and from 3.5 to 1.5 V as shown in Figure 15.1(b).

The PSpice schematic is shown in Figure 15.2(a). The ACCT option is selected from the Options menu as shown in Figure 15.2(b).

The listing of the circuit file is as follows:

Emitter-Coupled Trigger Circuit

```
SOURCE     ■   VDD 5 0 DC 5V     ; DC supply voltage of 5V
               VIN 1 0 PWL (0  1.5V  2 3.5V  4  1.5V) ; PWL waveform
CIRCUIT   ■■   R1 5 2 4.9K
               R2 5 3 3.6K
               RE 4 0 1K
               Q1 2 1 4 4 2N6546    ; Transistor Q1
               Q2 3 2 4 4 2N6546    ; Transistor Q2
```

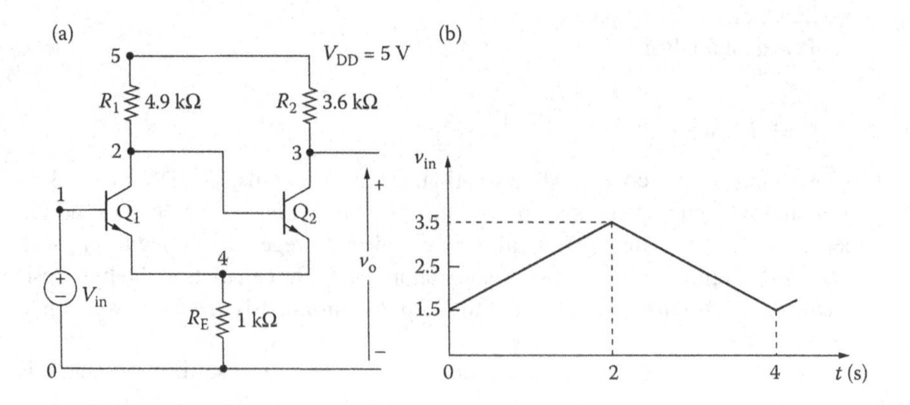

FIGURE 15.1 Schmitt trigger circuit. (a) Circuit. (b) Input voltage.

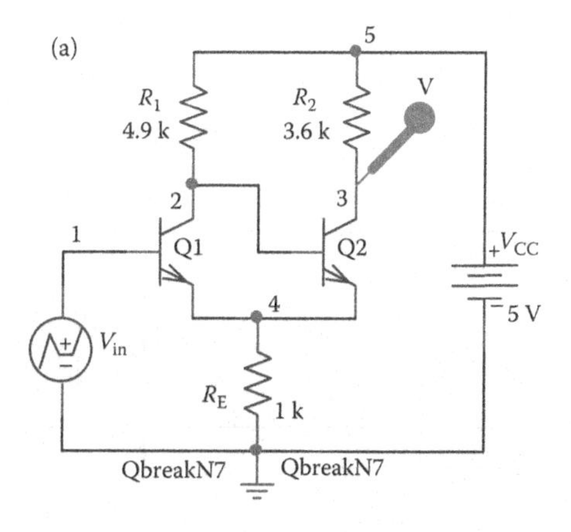

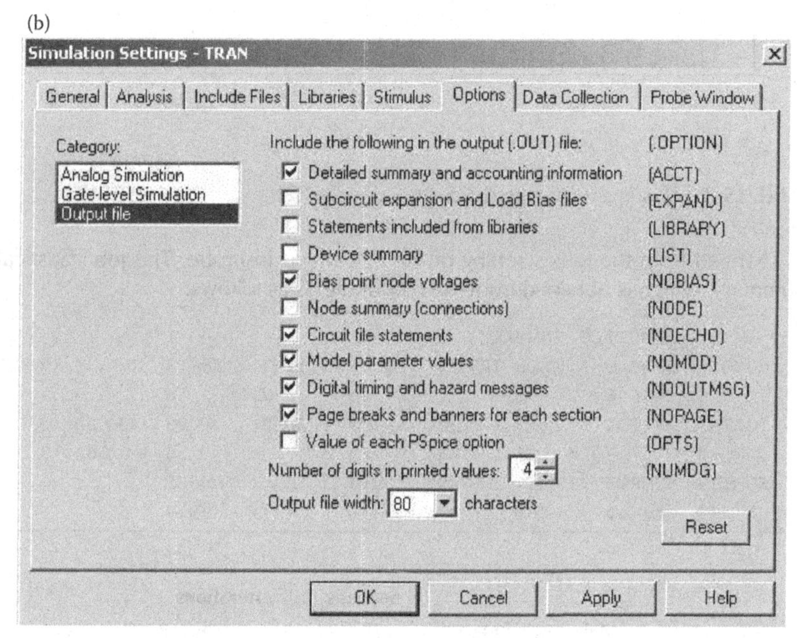

FIGURE 15.2 PSpice schematic for Example 15.1. (a) Schematic. (b) Setup for enabling ACCT.

```
               .MODEL 2N6546 NPN (IS=2.33E-27 BF=13
               CJE=1PF CJC=607.3PF TF=26.5NS)
               .OPTIONS ACCT        ; Printing the accounts
                                      summary
ANALYSIS ■■■   .TRAN 0.01 4         ; Transient analysis form 0
               to 4 s in steps of 0.01 s
               .PROBE               ; Graphics post-processor
          .END
```

The hysteresis characteristic for Example 15.1 is shown in Figure 15.3. The default x-axis for transient analysis is time. The x-axis setting as shown in Figure 15.3

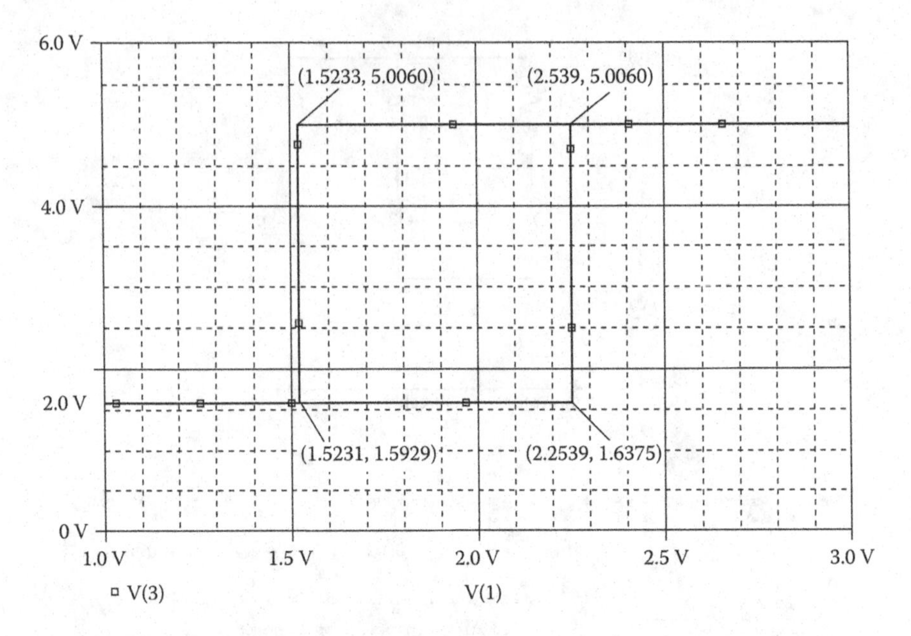

FIGURE 15.3 Hysteresis characteristics for Example 15.1.

is changed from the x-axis setting of the Plot menu in Probe. The job statistical summary, which is obtained from the output file, is as follows:

```
*** JOB STATISTICS SUMMARY
   NUNODS NCNODS NOMNOD NUMEL DIODES    BJTS   JFETS   MFETS GASFETS
        6      6      6     7      0       2       0       0       0
 NDIGITAL   NSTOP  NTTAR NTTBR  NTTOV  IFILL    IOPS PERSPA
        0       8     23    23     54      0      36 64.063
   NUMTTP NUMRTP NUMNIT DIGTP DIGEVT DIGEVL MEMUSE
      210     40    896     0      0      0   9914
```

	Seconds	Iterations
Matrix solution	0.77	5
Matrix load	4.42	
Reading	0.50	
Setup	0.11	
DC sweep	0.00	0
Bias point	0.99	92
AC and noise	0.00	0
Transient analysis	9.56	896
Output	0.00	
Total job time	10.54	

LTspice: LTspice schematic for Example 15.1 is shown in Figure 15.4(a). The input voltage V(1) as shown in Figure 15.4(b) increases from 1.5 V to 2.5 V and falls

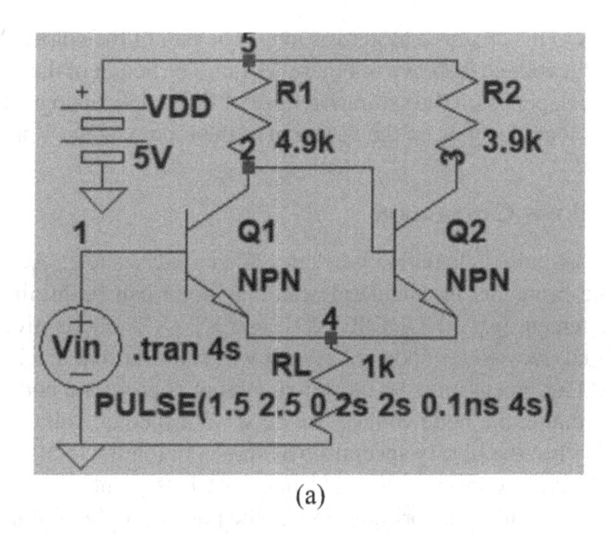

(a)

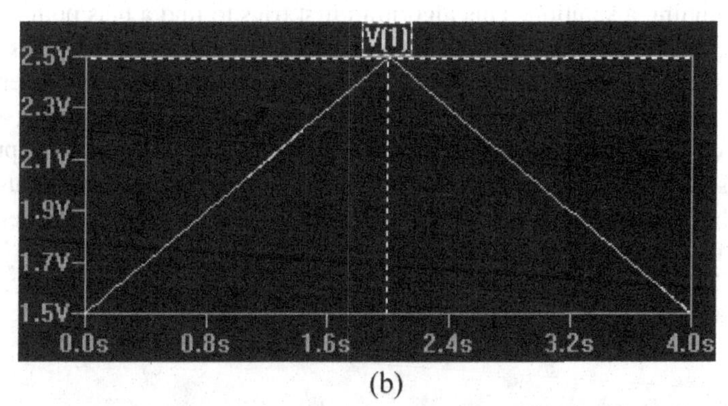

(b)

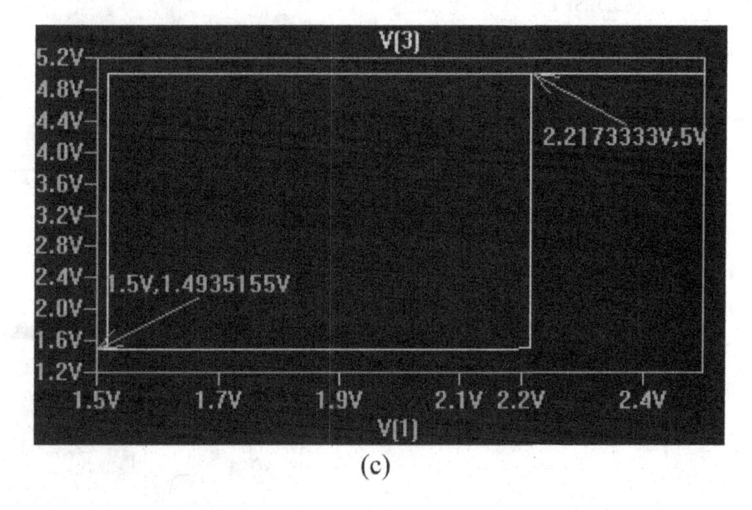

(c)

FIGURE 15.4 LTspice schematic for Example 15.1. (a) LTspice schematic. (b) Input voltage. (c) Hysteresis loop.

from 2.5 to 1.5 V. This creates a hysteresis loop. The plot of the voltage V(3) at node 3 against the input voltage is shown in Figure 15.4(c). The height of the loop as shown in Figure 15.4(c) depends on the saturation limit of the supply voltage and the width of the hysteresis loop depends on the variation of the input voltage Vin.

15.6.2 Bias-Point Calculation

Failure of the bias-point calculation precludes other analyses (e.g., AC analysis and sensitivity). The problems in calculating the bias point can be minimized by the .NODESET statement [e.g., .NODESET V(1) = 0 V]. If PSpice is given "hints" in the form of initial guesses for node voltages, it will start out that much closer to the solution. A little judgment must be used in assigning appropriate node voltages. In the PSpice schematic, the node voltage can be set at a specific voltage as shown in Figure 15.5(a) within the library special.slb as shown in Figure 15.5(b).

It is rare to have a convergence problem in the bias-point calculation. This is because PSpice contains an algorithm to scale the power supplies automatically if it has trouble finding a solution. This algorithm first tries to find a bias point with the power supplies at full scale. If there is no convergence, the power supplies are cut back to πth strength, and the program tries again. If there is still no convergence, the supplies are cut by another factor of 4 to 1/16th strength, and so on. At power supplies of 0 V, the circuit definitely has a solution with all nodes at 0 V, and the program will find a solution for some supply value scaled far enough back. It then uses that solution to help it work its way back up to a solution with the power supplies at full strength. If this algorithm is in effect, a message such as

```
Power supplies cut back to 25%
```

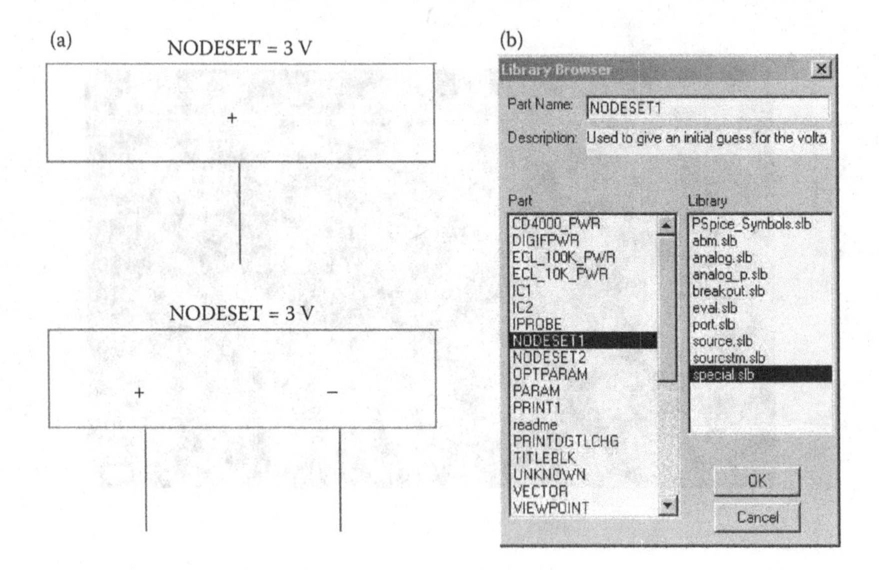

FIGURE 15.5 Setup for node voltages. (a) Node voltages. (b) Setup for node set.

(or some other percentage) appears on the screen while the program calculates the bias point.

15.6.3 Transient Analysis

In the case of failure due to convergence, the transient analysis skips the remaining time. The few remedies that are available for transient analysis are as follows:

1. To change the relative accuracy RELTOL from 0.001 to 0.01.
2. To set the iteration limits at any point during transient analysis using the option ITL4. Setting ITL4 = 50 (by the statement `OPTIONS ITL4=50`) will allow 50 iterations at each point. As a result of more iteration points, a longer simulation time will be required. It is not recommended for circuits that do not have a convergence problem in transient analysis.

15.7 ANALYSIS ACCURACY

The accuracy of PSpice's results is controlled by the parameters RELTOL, VNTOL, ABSTOL, and CHGTOL in the .OPTIONS statement. The most important of these is RELTOL, which controls the relative accuracy of all the voltages and currents that are calculated. The default value of RELTOL is 0.001 (0.1%).

VNTOL, ABSTOL, and CHGTOL set the best accuracy for the voltages, currents, and capacitor charges/inductor fluxes, respectively. If a voltage changes its sign and approaches zero, RELTOL will force PSpice to calculate more accurate values of that voltage because 0.1% of its value becomes a tighter and tighter tolerance. This would prevent PSpice from ever letting the voltage cross zero. To prevent this problem, VNTOL can limit the accuracy of all voltages to a finite value, and the default value is 1 μV. Similarly, ABSTOL and CHGTOL can limit the currents and charges (or fluxes), respectively.

The default values for the error tolerances in PSpice are the same as in Berkeley SPICE2. However, they differ from that of the commercial HSpice program, as shown in Table 15.1. RELTOL = .001 (0.1%) is more accurate than is necessary for many applications. The speed can be increased by setting RELTOL = 0.01 (1%), and this would increase the average speed-up by a factor of 1.5. In most power electronics circuits, the default values can be changed without affecting the results significantly.

TABLE 15.1
Tolerances

	PSpice	SPICE2	HSpice
RELTOL	0.001	0.001	0.01
VNTOL	1 μV	1 μV	50 μV
ABSTOL	1 μA	1 μA	1 nA

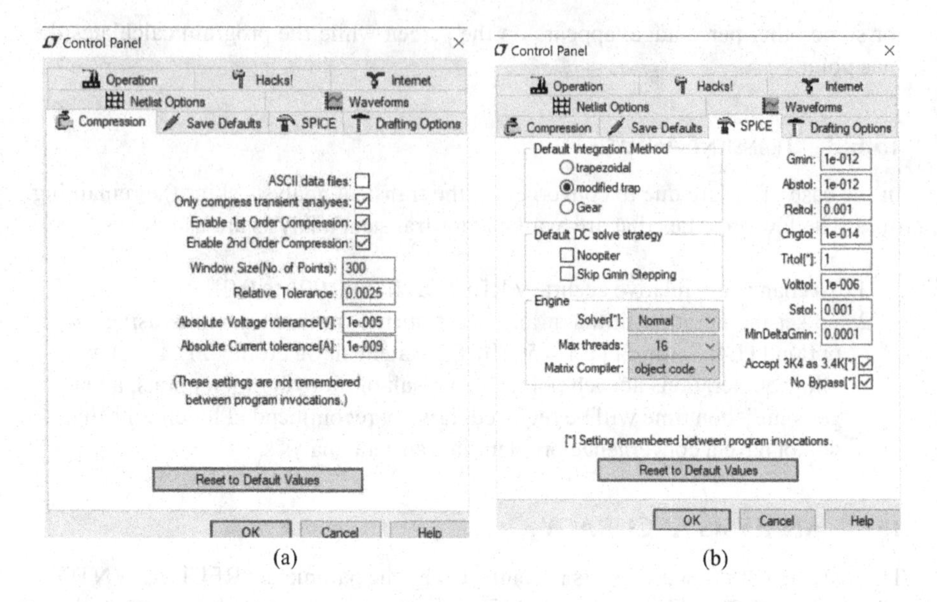

(a) (b)

FIGURE 15.6 LTspice Simulation Control panel. (a) Control panel. (b) SPICE menu.

Note: The limit options from Table 6.1 can be typed into the circuit file as an .OPTIONS command. Alternatively, the limit options can be set from the Change Options menu, in which case PSpice would write these options into the circuit file automatically.

LTspice simulation control panel is shown in Figure 15.6(a) and the SPICE menu is shown in Figure 15.6(b). The default values are too small and could cause simulation failure. An option statement with the control values should resolve the convergence problem as follows.

```
.options abstol=1.0m reltol=1.0m vntol=1.0m
```

It is advisable to complete the simulation with the highest tolerance values and then lower the tolerances.

15.8 NEGATIVE COMPONENT VALUES

PSpice allows negative values for resistors, capacitors, and inductors. It should calculate a bias point or DC sweep for such a circuit. The .AC and .NOISE analyses can handle negative components. In the case of resistors, their noise contribution comes from their absolute values, and the components are not allowed to generate negative noise. However, negative components, especially negative capacitors and inductors, may cause instabilities in time, and the transient analysis may fail for a circuit with negative components.

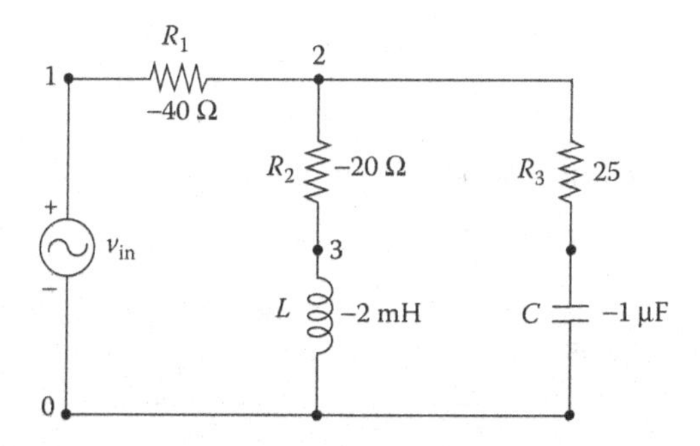

FIGURE 15.7 Circuit with negative components.

Example 15.2:

Transfer Function Analysis of a Circuit with a Negative Resistor

A circuit with negative components is shown in Figure 15.7. The input voltage is V_{in} = 120 V (peak), 60 Hz. Use PSpice to calculate the currents I(R1), I(R2), and I(R3) and the voltage V(2).

SOLUTION

The PSpice schematic with a negative resistance is shown in Figure 15.8.

The listing of the circuit file is as follows:

Circuit with Negative Components

```
SOURCE   ■ VIN 1 0 AC 120V     ; AC input voltage of 120 V
CIRCUIT  ■■ R1 1 2 -40         ; Negative resistances
            R2 2 3 -20
            R3 2 4  25
            L  3 0 -2MH        ; Negative inductance
            C  4 0 -1UF        ; Negative capacitance
```

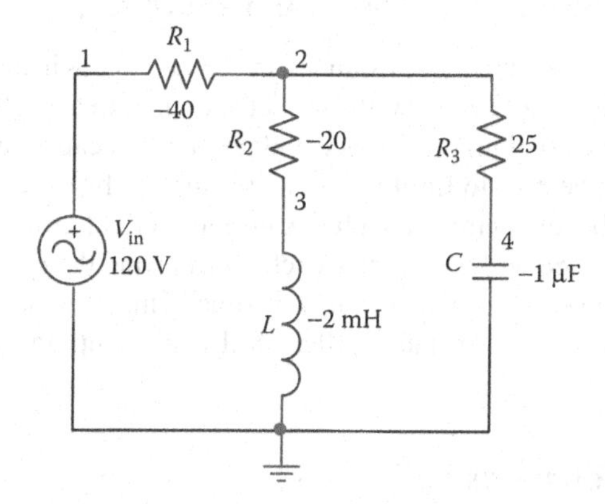

FIGURE 15.8 PSpice schematic for Example 15.2.

```
ANALYSIS ■■■ .AC LIN 1 60HZ 60HZ      ; AC analysis
           .PRINT AC 1M(R1) IM(R2) IM(R3) VM(2)
        .END
```

The results of the PSpice simulation are as follows:

```
****          AC ANALYSIS             TEMPERATURE = 27.000 DEG C
FREQ               IM(R1)     IP(R1)  VM(2)          VP(2)
6.000E+01     2.000E+00 1.794E+02 4.003E+01      1.151E+00
         JOBCONCLUDED
         TOTAL JOB TIME 1.26
```

LTspice: LTspice schematic for Example 15.2 is shown in Figure 15.9(a). The results of the AC analysis displaying the voltages, currents, and phase angles at a frequency of 60 Hz are listed in Figure 15.9(b). Although circuit elements are assigned negative values, there are no physical resistances or inductors. However, it simply means that the v-i slope of the vi-i relationship has a negative slope. That is, if the voltage increases, the current falls and the power loss across a resistor or energy stored in an inductor is positive.

15.9 POWER-SWITCHING CIRCUITS

The SPICE program was developed to simulate integrated circuits containing many small, fast transistors. Because of the integrated circuit emphasis, the default values of the overall parameters are not optimal for simulating power circuits. Convergence problems can be minimized by paying special attention to

Model parameters of diodes and transistors
Error tolerances
Snubbing resistor
Quasi–steady-state conditions

15.9.1 Model Parameters of Diodes and Transistors

The default values of all parasitic resistances and capacitances in .MODEL statements are zero. If the parameters RS and CJO are not specified in a .MODEL statement for a device, it will have no ohmic resistance and no junction capacitance. With RS = 0, the circuit may not be able to limit the forward current through the device, and the current can easily become large enough to cause numerical problems. With CJO = 0 (and TT = 0), the device will have zero switching time and the transient analysis may find itself trying to make a transition in zero time. This will cause PSpice to make the internal time step smaller and smaller until it gives up and reports a transient convergence problem.

15.9.2 Error Tolerances

The main error tolerance is RELTOL (default = 0.001 = .1%). It is not affected by power circuits. VNTOL is used for setting the most accurate voltage (default = 1 μV),

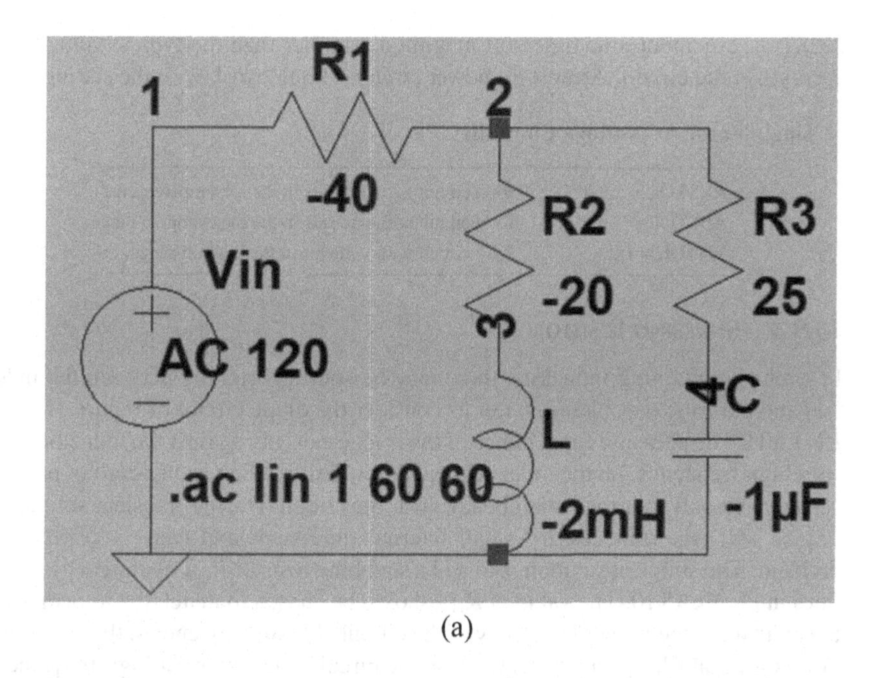

(a)

```
--- AC Analysis ---

frequency:    60           Hz
V(1):         mag:         120 phase:            0°
V(2):         mag:     40.0304 phase:      1.15108°
V(3):         mag:     1.50811 phase:       89.068°
V(4):         mag:     40.0286 phase:      1.69106°
I(C):         mag:   0.0150904 phase:     -88.3089°
I(L):         mag:      2.0002 phase:      178.992°
I(R3):        mag:   0.0150904 phase:     -88.3089°
I(R2):        mag:      2.0002 phase:      178.992°
I(R1):        mag:     1.99954 phase:      179.424°
I(Vin):       mag:     1.99954 phase:    -0.576077°
```

(b)

FIGURE 15.9 LTspice schematic for Example 15.2. (a) LTspice schematic. (b) Results of AC analysis.

and ABSTOL is used for setting the most accurate current (default = 1 pA). The dynamic range of PSpice is about 12 orders of magnitude. In a circuit with currents in the kiloampere range, ABSTOL = 1 pA will exceed this range and may cause a convergence problem. For power circuits, it is often necessary to adjust ABSTOL higher than its default value of 1 pA. The recommended settings for VNTOL and

ABSTOL are about nine orders of magnitude smaller than the typical voltages and currents in the circuit. Almost all power circuits should work with the settings.

Single-Phase AC Voltage Controller

ABSTOL = 1 μA	for a circuit with currents in the kiloampere range
ABSTOL = 1 mA	for a circuit with currents in the megaampere range
VNTOL = 1 μV	for a circuit with voltages in the kilovolt range

15.9.3 Snubbing Resistor

In circuits containing inductors, there may be spurious ringing between the inductors and parasitic capacitances. Let us consider the diode circuit of Figure 15.7 with $L = 1$ mH. The parasitic capacitance of the bridge can ring against the inductor with a very high frequency, on the order of megahertz. This ringing is the result of parasitic capacitance only, not the actual behavior of the circuit. During transient simulation, PSpice will take unnecessary small internal time steps and cause a convergence problem. The simplest solution is to add a snubbing resistor R_{snub} as shown by dashed lines in Figure 15.10. The value of R_{snub} should be chosen to match the impedance of the inductor at the corner frequency of the circuit. At low frequencies, the impedance of L is low, and R_{snub} has little effect on the circuit's behavior. At high frequencies, the impedance of L is high, and R_{snub} prevents it from supporting the ringing. The action of R_{snub} is similar to the physical mechanisms, primarily eddy current losses, which limit the frequency response of an inductor.

If components in series with an inductor switch off while current is still flowing in the inductor, di/dt can be high, causing large spikes and convergence problems. A snubbing resistor can keep such spikes to a large but tractable size and thereby eliminate such convergence problems.

15.9.4 Quasi-Steady-State Condition

Running transient analysis on power-switching circuits can lead to long run times. PSpice must keep the internal time step short compared to the switching period, but

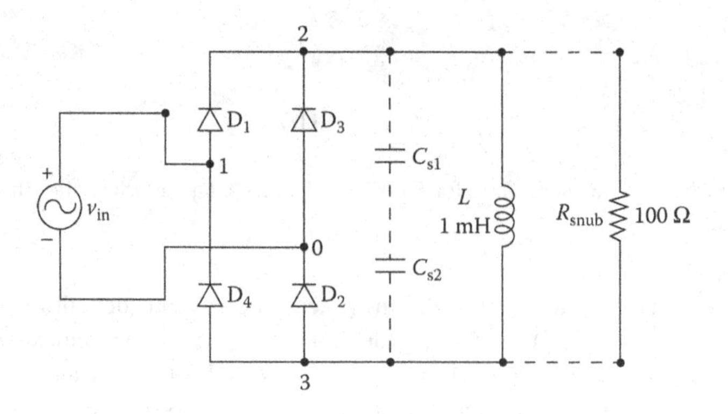

FIGURE 15.10 Diode circuit with snubbing resistor.

the circuit's response generally extends over many switching cycles. This problem can be solved by transforming the switching circuit into an equivalent circuit without switching. The equivalent circuit represents a kind of quasi-steady-state of the actual circuit and can accurately model its response as long as the inputs do not change too fast. This is illustrated in Example 15.3.

Example 15.3:

Transient Response of a Single-Phase Full-Bridge Inverter with Voltage-Controlled Switches

A single-phase bridge-resonant inverter is shown in Figure 15.11. The transistors and diodes can be considered as switches whose on-state resistance is 10 mΩ and whose on-state voltage is 0.2 V. Plot the transient response of the capacitor voltage and the current through the load from 0 to 2 ms in steps of 10 μs. The output frequency of the inverter is $f_o = 4$ kHz.

SOLUTION

When transistors Q_1 and Q_2 are turned on, the voltage applied to the load will be V_s, and the resonant oscillation will continue for the entire resonant period, first through Q_1 and Q_2 and then through diodes D_1 and D_2. When transistors Q_3 and Q_4 are turned on, the load voltage will be $-V_s$, and the oscillation will continue for another entire period, first through Q_3 and Q_4 and then through diodes D_3 and D_4. The resonant period of the circuit is calculated approximately as

$$\omega_r = \left(\frac{1}{LC} - \frac{R^2}{4L^2}\right)^{1/2}$$

For $L = L_1 = 50$ μH, $C = C_1 = 6$ μF, and $R = R_1 + R_{1(sat)} + R_{2(sat)} = 0.5 + 0.2 + 0.\ 2 = 0.54$ Ω, $\omega_r = 57572.2$ rad/s, and $f_r = \omega_r/2\pi = 9162.9$ Hz. The resonant period is $T_r = 1/f = 1/9162.9 = 109.1$ μs. The period of the output voltage is $T_o = 1/f_o = 1/4000 = 250$ μs.

The switching action of the inverter can be represented by two voltage-controlled switches as shown in Figure 15.12(a). The switches are controlled by

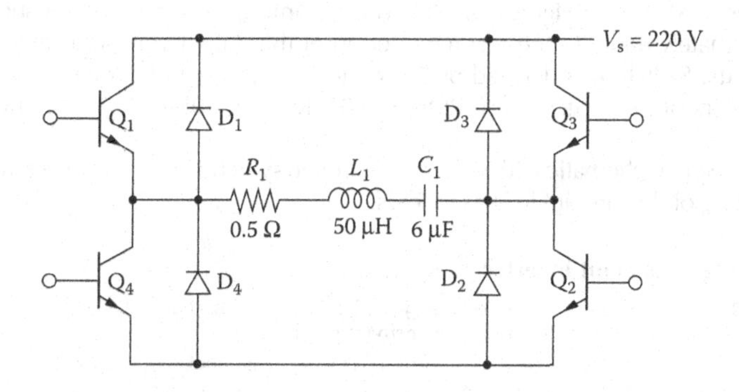

FIGURE 15.11 Single-phase bridge-resonant inverter.

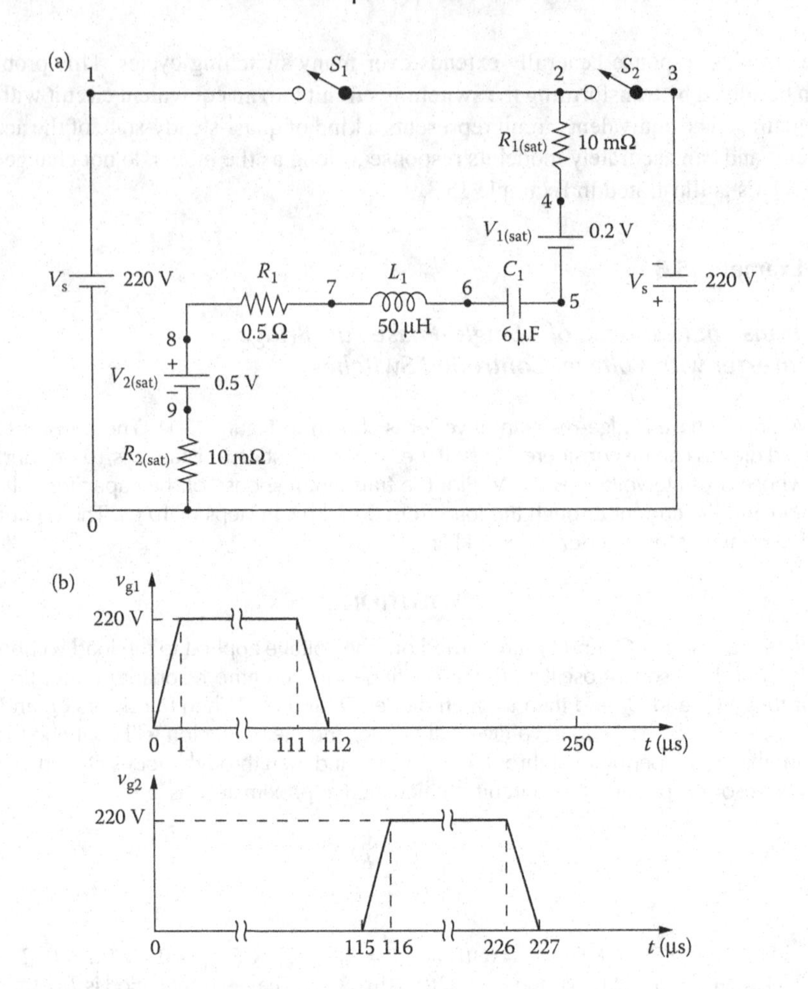

FIGURE 15.12 Equivalent circuit for Figure 15.8. (a) Equivalent circuit. (b) Controlling voltages.

voltages as shown in Figure 15.12(b). The on-time of switches, which should be approximately equal to the resonant period of the output voltage, is assumed to be 112 μs. Switch S_2 is delayed by 115 μs to take account of overlap. The model parameters of the switches are RON = 0.01, ROFF = 10E+6, VON = 0.001, and VOFF = 0.0.

The PSpice schematic with voltage-controlled switches is shown in Figure 15.13. The listing of the circuit file is as follows:

Full-Bridge Resonant Inverter

```
SOURCE  ■   * The controlling voltage for switch S1:
            V1 1 0 PULSE(0220V01US1US110US250US)
            * The controlling voltage for switch S2 with a
            delay time of 115 μs:
            V2 3 0 PULSE(0-220V115US1US1US110US250US)
```

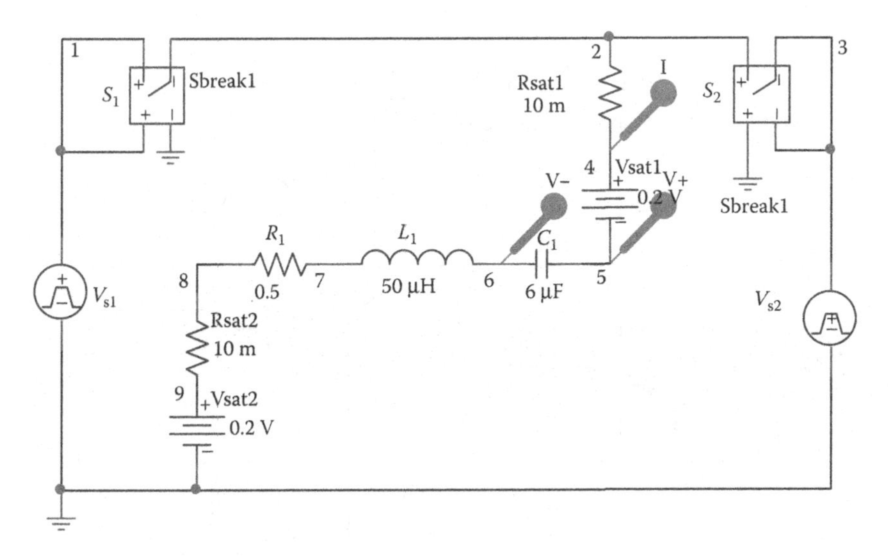

FIGURE 15.13 PSpice schematic for Example 15.3.

```
CIRCUIT ■■  S1 1 2 1 0 SMOD      ; Voltage-controlled switches
            S2 2 3 0 3 SMOD
            * Switch model parameters for SMOD:
            .MODEL SMOD VSWITCH (RON=0.01 ROFF=10E+10E+6
            VON=0.001 VOFF=0.0)
            RSAT1 2 4 10M
            VSAT1 4 5 DC  0.2V
            RSAT2 9 0 10M
            VSAT2 8 9 DC  0.2V
            * Assuming an initial capacitor voltage of
            -250V to reduce settling time:
            C1 5 6 6UF  IC=-250V
            L1 6 7 50UH
            R1 7 8 0.5
ANALYSIS ■■■ *Transient analysis with UIC condition:
            .TRAN 1US 500US UIC  ; Transient analysis with
            UIC condition
            .PROBE               ; Graphics post-processor
        .END
```

The transient response for Example 15.3 is shown in Figure 15.14.

LTspice: The gating signals for the switching devices for Example 14.3 have a duty cycle of 50% (see Figure 15.15). The signals for switches U3 and U4 are inverted to that for U1 and U2.

LTspice schematic for Example 14.3 is shown in Figure 15.16(a). The plots of the load current and the capacitor voltage are shown in Figure 15.16(b). The rms value of the load current as shown in Figure 15.16(c) is 48.56 A and that of the capacitor voltage as shown in Figure 15.16(d) is 248.27 V. Although the voltage and the current are AC waveforms, the waveforms have DC components as shown in Figure 15.16(c) and (b).

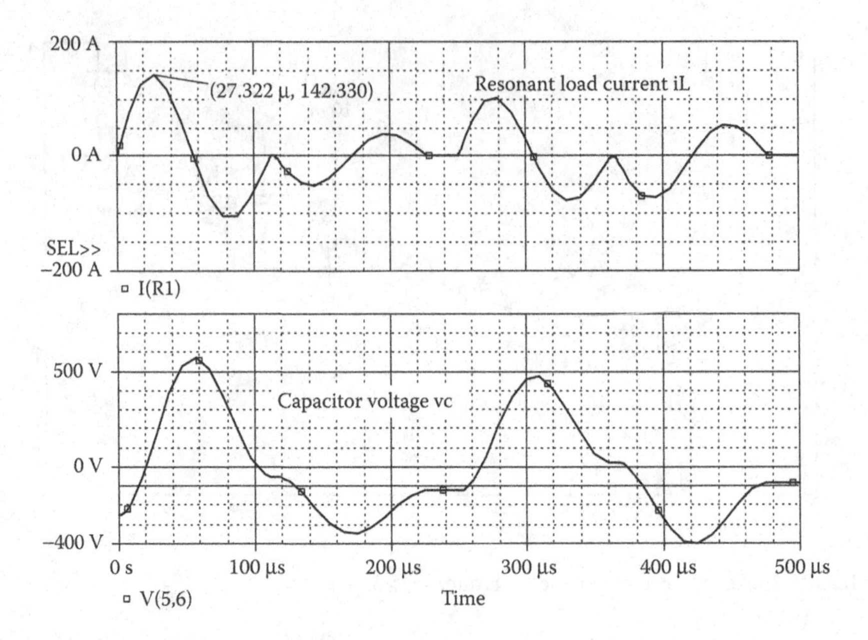

FIGURE 15.14 Transient responses for Example 15.3.

15.10 FLOATING NODES

PSpice requires that there be no floating nodes. If there are any, PSpice will indicate a read-in error on the screen, and the output file will contain a message similar to

```
ERROR: Node 15 is floating
```

This means that there is no DC path from node 15 to the ground. A DC path is a path through resistors, inductors, diodes, and transistors. This is a very common problem, and it can occur in many circuits, as shown in Figure 15.17.

Node 4 in Figure 15.17(a) is floating and does not have a DC path. This problem can be avoided by connecting node 4 to node 0 as shown by dashed lines (or by

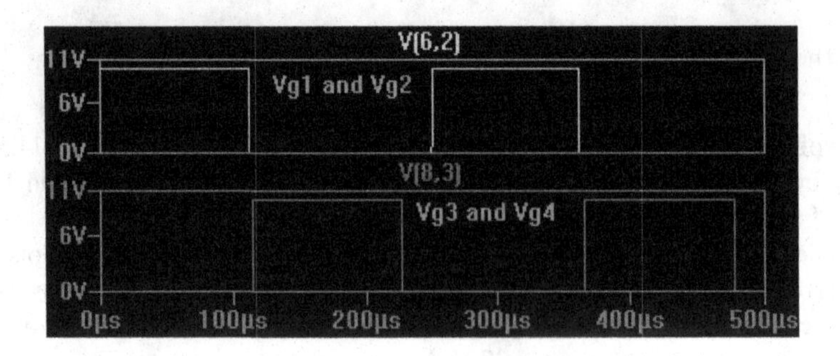

FIGURE 15.15 LTspice gating signals for the switching devices.

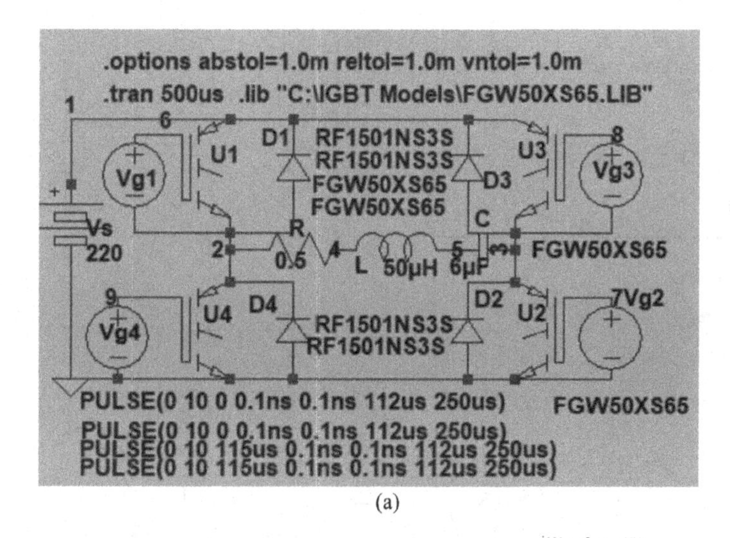

(a)

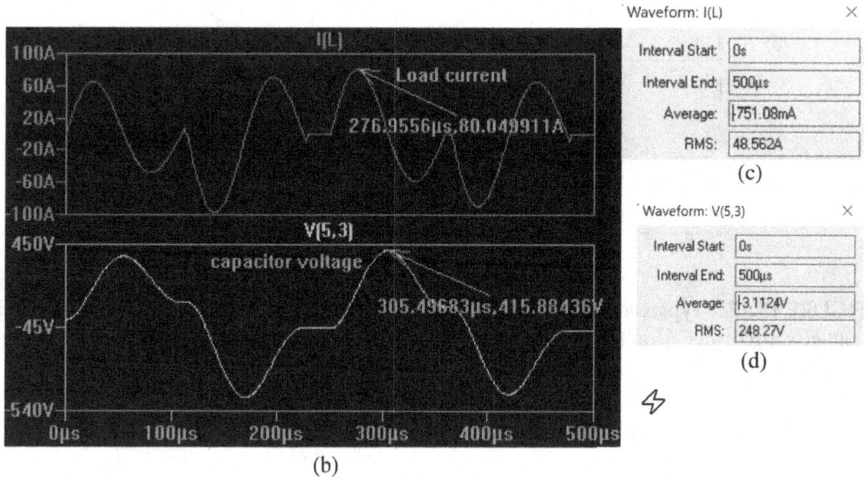

(b)

(c)

(d)

FIGURE 15.16 LTspice schematic for Example 15.3. (a) LTspice schematic. (b) Output waveforms. (c) Load current. (d) Capacitor voltage.

connecting node 3 to node 2). A similar situation can occur in voltage-controlled and current-controlled sources as shown in Figure 15.17(b) and Figure 15.17(c). The model of op-amps as shown in Figure 15.17(d) has many floating nodes, which should be connected to provide DC paths to the ground. For example, nodes 0, 3, and 5 could be connected together or, alternatively, nodes 1, 2, and 4 may be joined together.

The two sides of a capacitor have no DC path between them. If there are many capacitors in a circuit as shown in Figure 15.18, nodes 3 and 5 do not have DC paths. DC paths can be provided by connecting a very large resistance R_3 (say, 100 MΩ) across capacitor C_3, as shown by dashed lines.

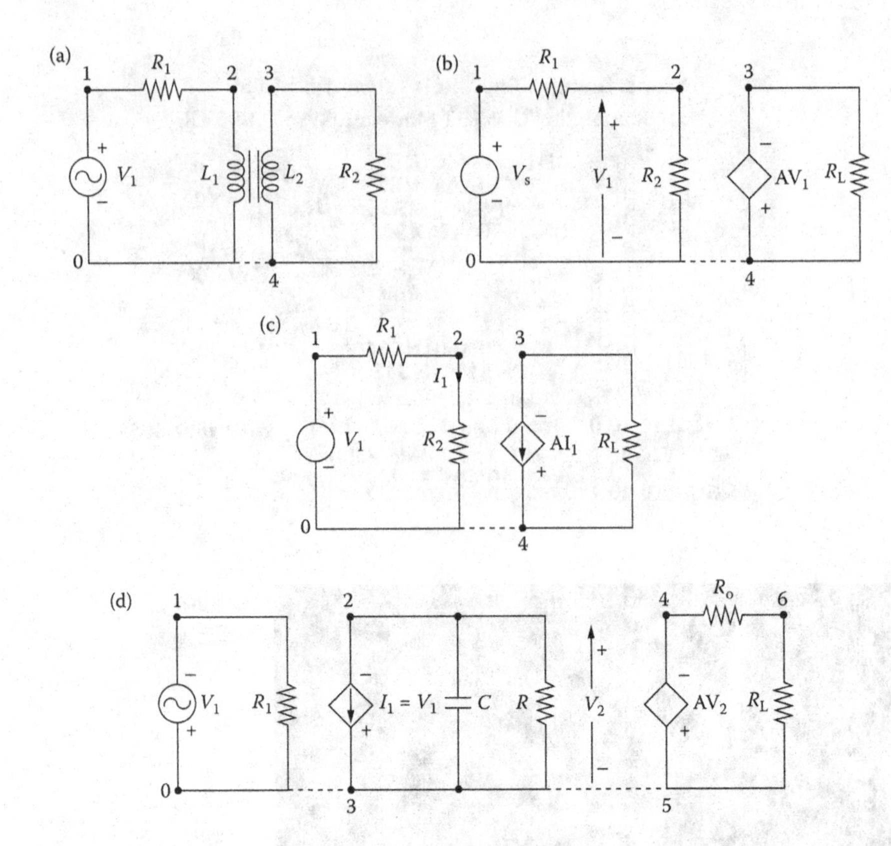

FIGURE 15.17 Typical circuits with floating nodes. (a) Transformer. (b) Voltage-controlled voltage source. (c) Current-controlled current source. (d) Op-amp model.

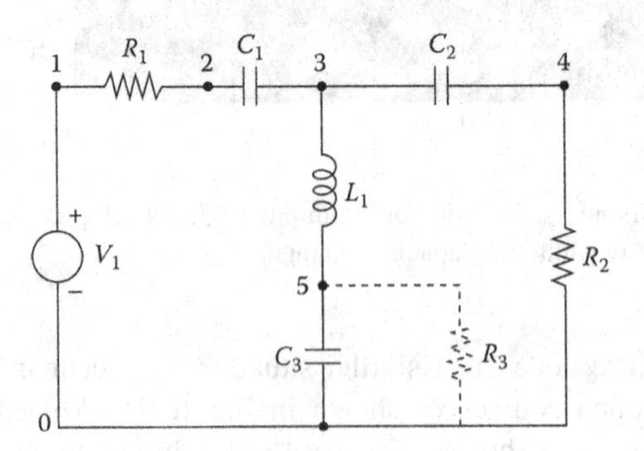

FIGURE 15.18 Typical circuit without DC path.

Example 15.4:

Providing DC Paths to a Passive Filter

A passive filter is shown in Figure 15.19. The output is taken from node 9. Plot the magnitude and phase of the output voltage separately against the frequency. The

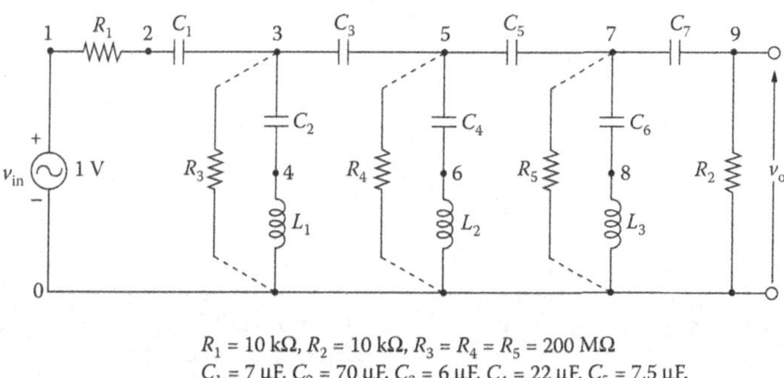

$R_1 = 10\ \text{k}\Omega$, $R_2 = 10\ \text{k}\Omega$, $R_3 = R_4 = R_5 = 200\ \text{M}\Omega$
$C_1 = 7\ \mu\text{F}$, $C_2 = 70\ \mu\text{F}$, $C_3 = 6\ \mu\text{F}$, $C_4 = 22\ \mu\text{F}$, $C_5 = 7.5\ \mu\text{F}$,
$C_6 = 12\ \mu\text{F}$, $C_7 = 10.5\ \mu\text{F}$
$L_1 = 1.5\ \text{mH}$, $L_2 = 1.75\ \text{mH}$, $L_3 = 2.5\ \text{mH}$

FIGURE 15.19 Passive filter.

frequency should be varied from 100 Hz to 10 kHz in steps of one decade and 101 points per decade.

SOLUTION

The nodes between C_1 and C_3, C_3 and C_5, and C_5 and C_7 do not have DC paths to the ground. Therefore, the circuit cannot be analyzed without connecting resistors R_3, R_4, and R_5 as shown in Figure 15.19 by dashed lines. If the values of these resistances are very high, say 200 MΩ, their influence on the AC analysis would be negligible.

The PSpice schematic of a passive filter with DC paths is shown in Figure 15.20. The listing of the circuit file is as follows:

Passive Filter

```
SOURCE   ■    VIN 1 0 AC 1V        ; Input voltage is 1 V peak
CIRCUIT  ■■   R1 1 2 10K
              R2 9 0 10K
              * Resistances R3, R4, and R5 are connected to
              *provide DC paths:
              R3 3 0 200MEG
              R4 5 0 200MEG
              R5 7 0 200MEG
```

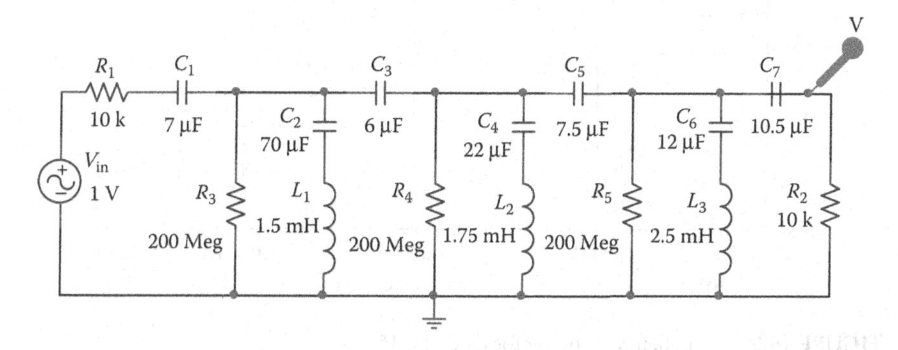

FIGURE 15.20 PSpice schematic for Example 15.4.

```
                C1 2 3 7NF
                C2 3 4 70NF
                C3 3 5 6NF
                C4 5 6 22NF
                C5 5 7 7.5NF
                C6 7 8 12NF
                C7 7 9 10.5NF
                L1 4 0 1.5MH
                L2 6 0 1.75MH
                L3 8 0 2.5MH
ANALYSIS ■■■    * AC analysis for 100 Hz to 10 kHz with a decade
                * increment and 101 points per decade:
                .AC DEC 101 100 10KHZ
                * Plot the results of AC analysis for the magnitude
                * of voltage at node 9.
                .PLOT AC VM(9) VP(9) ; Plots on output file
                .PLOT AG VP(9) file ; Plots on output
                .PROBE ; Graphics post-processor
            .END
```

The frequency response for Example 15.4 is shown in Figure 15.21.

LTspice: LTspice schematic for Example 15.4 is shown in Figure 15.22(a). The plots of the magnitude of the output voltage at node 9, V(9), and its phase angle are shown in Figure 15.22(b). Resistors R3, R4, and R5 have high values and are connected to prove a DC path to the circuit. The default plot for an ac analysis is a Bode

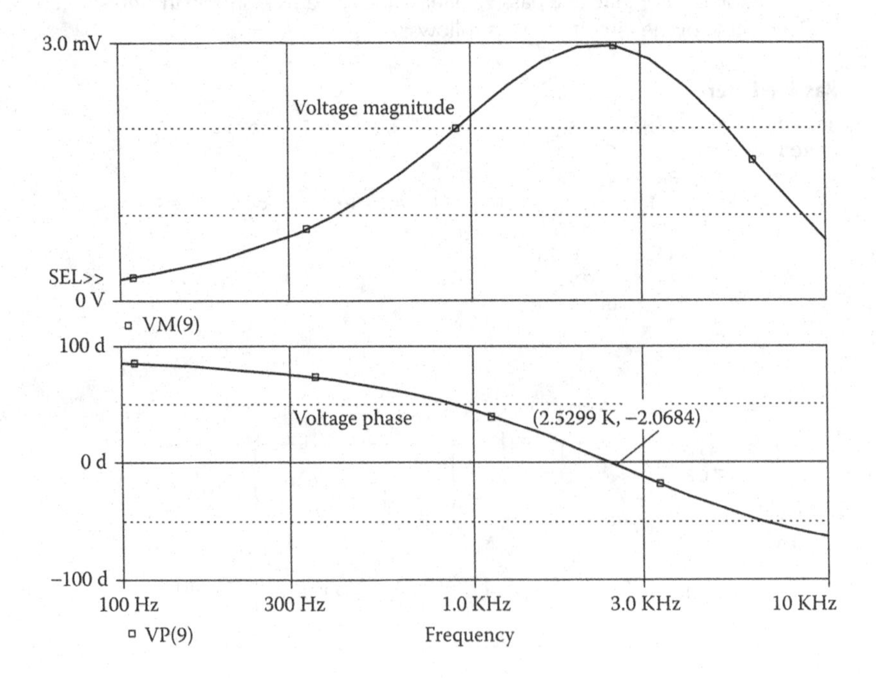

FIGURE 15.21 Frequency response for Example 15.4.

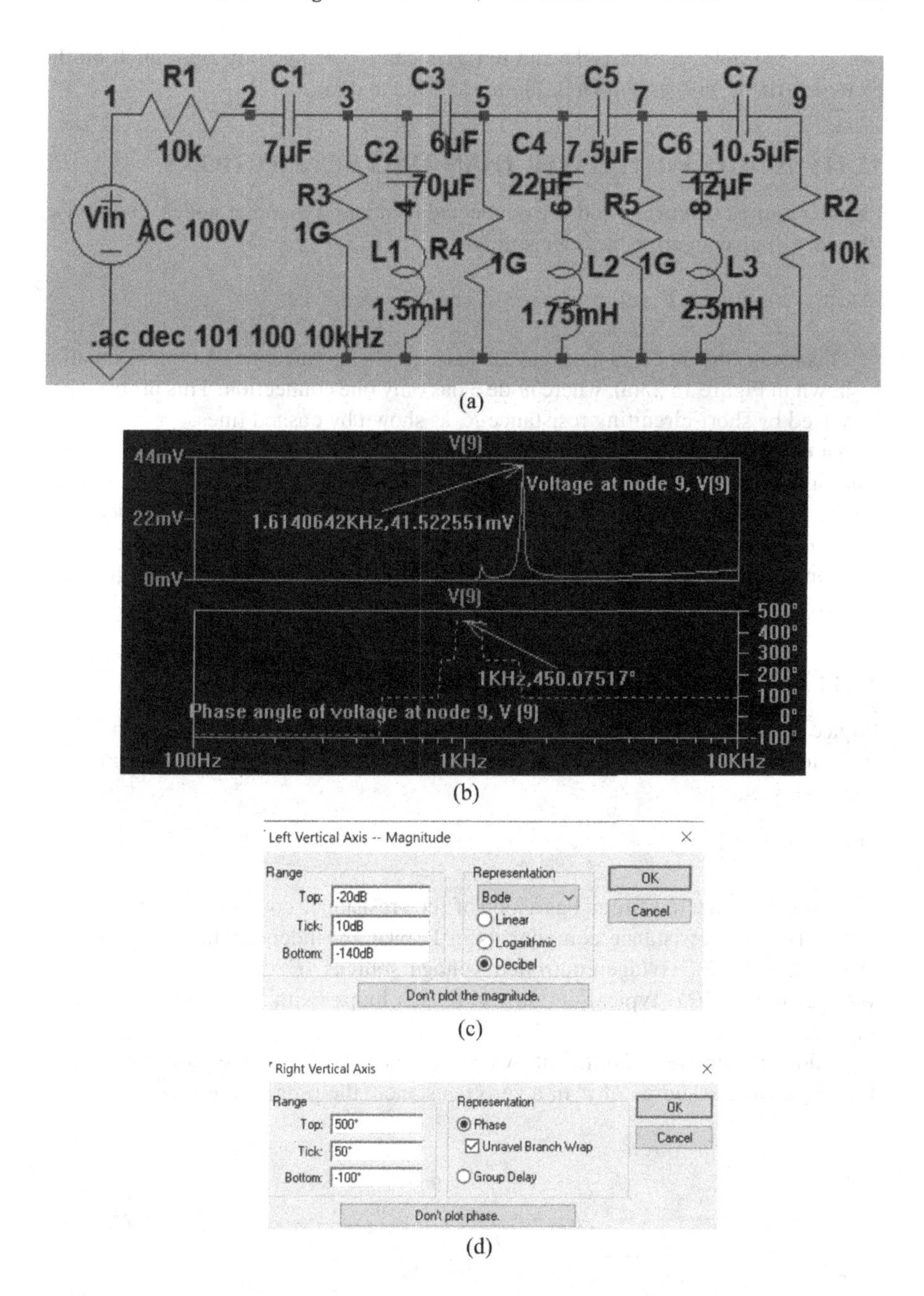

FIGURE 15.22 LTspice schematic for Example 15.4. (a) LTspice schematic. (b) Output waveforms. (c) Left axis menu. (d) Right axis menu.

plot. The variables on the right and left axis can be changed by right click on the plotted variable V(9) as shown in Figure 15.22(c) and (d).

15.11 NODES WITH FEWER THAN TWO CONNECTIONS

PSpice requires that every node be connected to at least two other nodes. Otherwise, PSpice will give an error message similar to

```
ERROR: Fewer than two connections at node 10
```

This means that node 10 must have at least another connection. A typical situation is shown in Figure 15.23(a), where node 3 has only one connection. This problem can be solved by short-circuiting resistance R_3 as shown by dashed lines.

An error message may be indicated in the output file for a circuit with voltage-controlled sources as shown in Figure 15.23(b). The input to the voltage-controlled source will not be considered to have connections during the check by PSpice. This is because the input draws no current, and it has infinite impedance. A very high resistance (say, 10 GΩ) may be connected from the input to the ground as shown by dashed lines.

15.12 VOLTAGE SOURCE AND INDUCTOR LOOPS

PSpice requires that there be no loops with zero resistance. Otherwise, PSpice will indicate a read-in error on the screen, and the output file will contain a message similar to

```
ERROR: Voltage loop involving V5
```

This means that the circuit has a loop of zero-resistance components, one of which is V5. The zero-resistance components in PSpice are independent voltage sources (*V*), inductors (*L*), voltage-controlled voltage sources (*E*), and current-controlled voltage sources (*H*). Typical circuits with such loops (with *V* and *L*) are shown in Figure 15.24.

It does not matter whether the values of the voltage sources are zero or not. Having a voltage source of *E* in a zero resistance, the program will need to divide

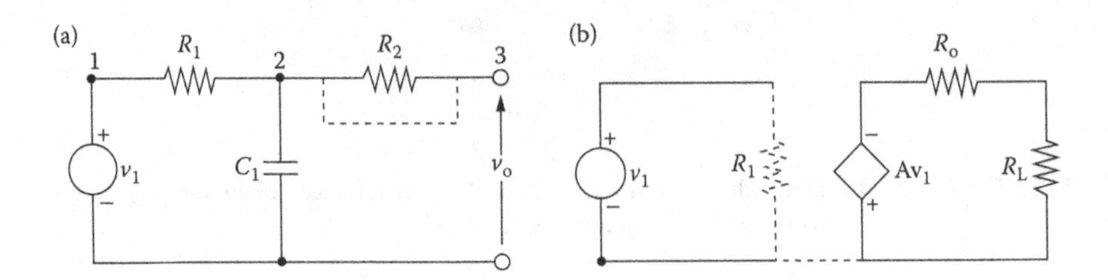

FIGURE 15.23 Typical circuits with less than two connections at a node. (a) Node with one connection. (b) Voltage-controlled source.

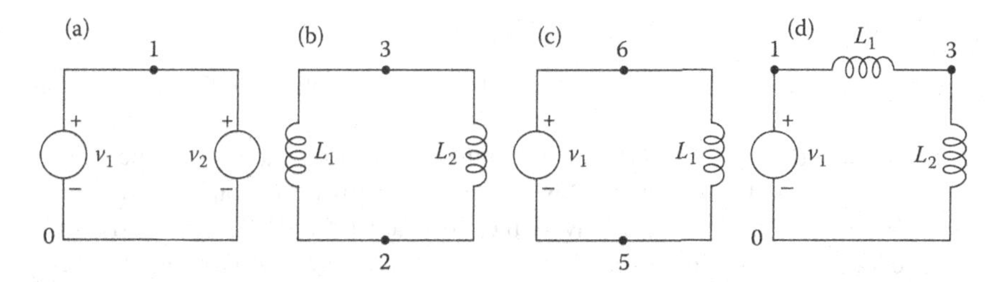

FIGURE 15.24 Typical circuits with zero-resistance loops. (a) Voltage loop. (b) Inductor loop. (c) Voltage with inductor. (d) Voltage with inductors.

E by 0. But because $E = 0$ V, the program will need to divide 0 by 0, which is also impossible. It is therefore the presence of a zero-resistance loop that is the problem, not the values of the voltage sources. A simple solution is to add a series resistance to at least one component in the loop. The resistor's value should be small enough so that it does not disturb the operation of the circuit. However, the resistor's value should not be less than 1 μΩ.

15.13 RUNNING PSpice FILES ON SPICE

PSpice will give essentially the same results as Berkeley SPICE 2G (referred to as SPICE). There could be small differences, especially for values crossing zero due to the corrections made for convergence problems. The semiconductor device models are the same as in SPICE.

There are a number of PSpice features that are not available in SPICE. These are as follows:

1. Extended syntax for output variables (e.g., in .PRINT and .PLOT). SPICE allows only voltages of the form V(x) or $V(x, y)$ and currents through voltage sources. Group delay is not available.
2. Extra devices:
   - Gallium arsenide model.
   - Nonlinear magnetic (transformer) model.
   - Voltage- and current-controlled switch models.
3. Optional models for resistors, capacitors, and inductors. Temperature coefficients for capacitors and inductors and exponential temperature coefficients for resistors are not available.
4. The model parameters RG, RDS, L, W, and WD are not available in the MOSFET's .MODEL statement in SPICE.
5. Extensions to the DC sweep. SPICE restricts the sweep variable to be the value of an independent current or voltage source. SPICE does not allow sweeping of model parameters or temperature.
6. The .LIB and .INCLUDE statements.
7. SPICE requires the input (.CIR) file to be uppercase.

15.14 RUNNING SPICE FILES ON PSpice

PSpice can run any circuit that Berkeley SPICE 2G can run with the following exceptions:

1. Circuits that use .DISTO (small-signal distortion) analysis, which has errors in Berkeley SPICE. Also, the special distortion output variables (HD2, DIM3, etc.) are not available. Instead of the .DISTO analysis, we recommend running a transient analysis and looking at the output spectrum with the Fourier transform mode of the Probe. This technique shows the distortion (spectral) products for both small-signal and large-signal distortion.
2. The IN = option of the .WIDTH statement is not available. PSpice always reads the entire input file regardless of how long the input lines are.
3. Temperature coefficients for resistors must be put into a .MODEL statement instead of the resistor statement. Similarly, the voltage coefficients for capacitors and the current coefficients for inductors are used in the .MODEL statements.

15.15 USING EARLIER VERSION OF SCHEMATICS

PSpice Schematics is being continuously improved with new versions. However, all versions have the same platform, and the schematic files have the extension .slb. Thus, a new version can run schematic files drawn in earlier versions.

OrCAD, Inc. owns Microsim PSpice. The OrCAD platform is slightly different from the PSpice Schematics platform. The file extension is .OPJ. The PSpice schematic files can be imported to OrCAD Capture as shown in Figure 15.25(a). It

(a)

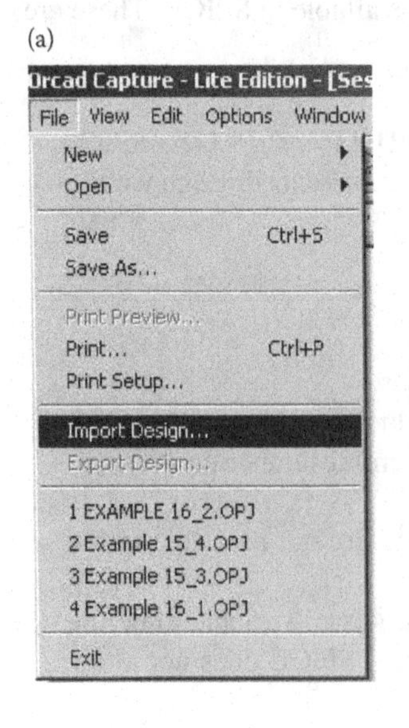

(b)

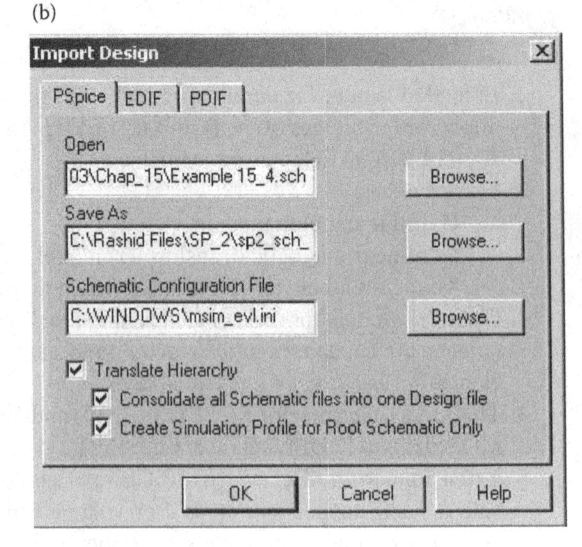

FIGURE 15.25 Importing PSpice schematic files to OrCAD Capture. (a) OrCAD Capture. (b) Importing PSpice schematic to OrCAD Capture.

will require identifying the location of the schematic configuration file: C:\WINNT\msim_evl.ini as shown in Figure 15.25(b). OrCAD's PSpice AD Lite Edition can run all PSpice files with extension .CIR.

PROBLEMS

15.1 For the inverter circuit in Figure P15.1, plot the hysteresis characteristics. The input voltage is varied slowly from –5 to +5 V and from +5 to –5 V. The model parameters of the PMOS are VTO = –2.5, KP = 4.5E–3, CBD = 5PF, CBS = 2PF, CGSO = 1PF, CGDO = 1PF, and CGBO = 1PF.

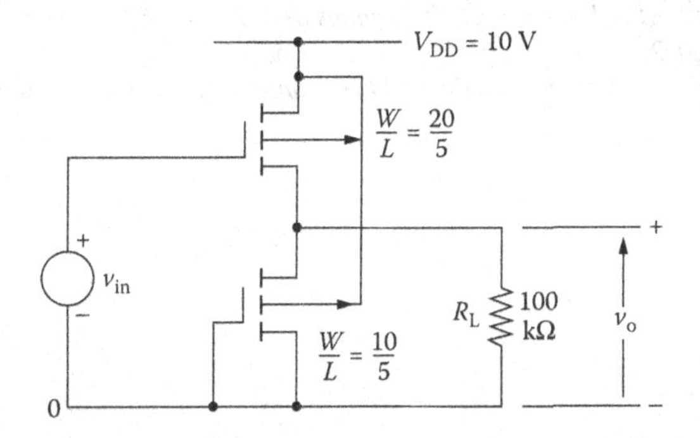

FIGURE P15.1 Inverter circuit.

15.2 For the circuit in Figure P15.2, plot the hysteresis characteristics from the results of the transient analysis. The input voltage is varied slowly from −4 to +4 V and from +4 to −4 V. The op-amp can be modeled as a macromodel, as shown in Figure 13.3. The description of the macromodel is listed in the library file EVAL.LIB. The supply voltages are V_{CC} = 12 V and V_{EE} = −12 V.

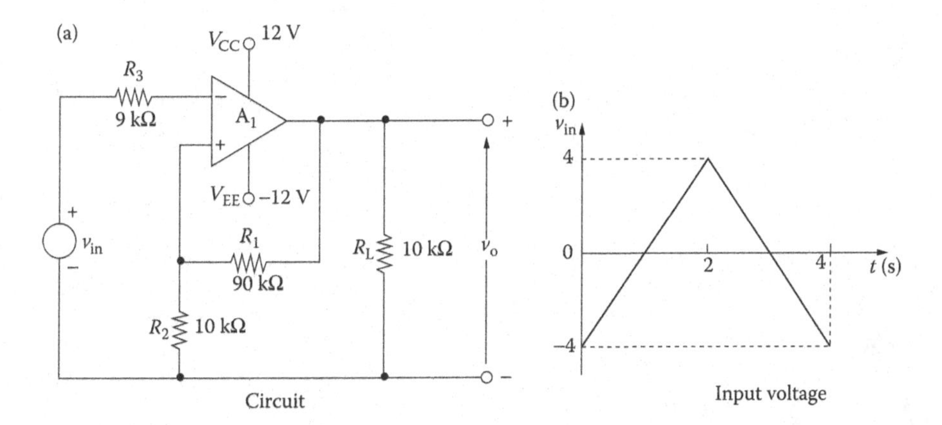

FIGURE P15.2 Op-amp hysteresis circuit.

SUGGESTED READING

1. MicroSim Corporation, *PSpice Manual*, Irvine, CA, 1992.
2. W. Blume, Computer circuit simulation, *Byte*, *11*(7), 1986, 165.
3. *Cadence Design Systems, PSpice 9.1 Student Version*, SanJose, CA, 2001. http://www.cadencepcb.com/products/downloads/PSpicestudent/default.asp.
4. *Cadence Design Systems*, Orcad 9.2 Demo, SanJose, CA, 2001. http://www.cadencepcb.com/products/downloads/orcaddemo/default.asp.
5. Cadence Design Systems, *PSpice Design Community*, SanJose, CA, 2001. http://www.PSpice.com.
6. M.H. Rashid, *Introduction to PSpice using OrCAD for circuits and electronics*, Upper Saddle River, NJ: Pearson/Prentice Hall, 3/e, 2004.
7. M.H Rashid, *SPICE for Power Electronics and Electric Power*, Boca Raton, FL: CRC Pres, 2/e, 2012.
8. M.H. Rashid, *SPICE for Circuits and Electronics*, 4/e, Cengage India Publishing, 2019.

Index

Note: *Italic* and **Bold** page numbers refer to *figures* and **tables**, respectively.

E

F

G

H

I

M

N

O

Q

R

U

V

W

Z

Printed in the United States
by Baker & Taylor Publisher Services

Printed in the United States
by Baker & Taylor Publisher Services